Wavelets
and
Filter Banks

Gilbert Strang
Massachusetts Institute of Technology

Truong Nguyen
University of Wisconsin

Wellesley-Cambridge Press
Box 812060
Wellesley MA 02181

Wavelets and Filter Banks

Printed in the United States of America

ISBN 0-9614088-7-1

QA 403.3.S87

Other texts from Wellesley-Cambridge Press

Introduction to Applied Mathematics, Gilbert Strang, ISBN 0-9614088-0-4.

An Analysis of the Finite Element Method, Gilbert Strang and George Fix, ISBN 0-9614088-8-X.

Calculus, Gilbert Strang, ISBN 0-9614088-2-0.

Introduction to Linear Algebra, Gilbert Strang, ISBN 0-9614088-5-5.

Linear Algebra, Geodesy, and GPS, Gilbert Strang and Kai Borre, ISBN 0-9614088-6-3.

Wellesley-Cambridge Press
Box 812060
Wellesley MA 02181 USA
(781) 431-8488 FAX (617) 253-4358
gs@math.mit.edu
http://www-math.mit.edu/~gs

MATLAB® is a registered trademark of The MathWorks, Inc. info@mathworks.com
24 Prime Park Way Natick MA 01760 (508) 647-7000 http://www.mathworks.com

The MATLAB Wavelet Toolbox was created by Michel Misiti, Yves Misiti, Georges Oppenheim, and Jean-Michel Poggi.

A Solutions Manual for instructors is available by email from the publisher.

Homepage for the book:
http://saigon.ece.wisc.edu/~waveweb/Tutorials/book.html

Book design and typesetting using LaTeX2ε by
Martin Stock, Winchester, Mass. mstock@mit.edu

The cover design is by Gail Corbett and Tracy Baldwin.

Contents

Chapter 10 **Design Methods**

Chapter 11 **Applications**

Preface

This book is about filter banks and wavelets. Those are new ways to see and represent a signal. They are alternatives to the Fourier transform, using short wavelets instead of long waves. We will explain the advantages (and disadvantages!) of the new methods. The final decision will depend on the application itself, the actual signal, and its bandwidth.

To design and understand wavelets we still use Fourier techniques — the connections between time and frequency. This idea remains at the center of signal processing. Wavelets are "alternatives" rather than replacements. The classical transforms will survive very well. But other ideas have come quickly forward, to be understood and applied.

In a word, the new transforms are much more *local*. An event stays connected to the time when it occurs. Instead of transforming a pure "time description" into a pure "frequency description", the new methods find a good compromise — a *time-frequency description*. This is like a musical score, with specified frequencies at specified times. Remember that the Heisenberg uncertainty principle stands in the way of perfection! We lose accuracy in time when we gain accuracy in frequency. The musician cannot and would not change frequencies instantly. But our eyes and ears succeed to give location as well as frequency, and the new wavelet transforms have the same purpose.

The extreme case is an instantaneous impulse, with all frequencies in equal amounts. Its Fourier transform has constant magnitude over the whole spectrum. By contrast, a wavelet transform will involve only a small fraction of the wavelets — those that overlap the impulse. Figure 0.1 shows a sum of two extreme cases — an impulse and a pure wave $2\cos\omega n$. The Fourier transform spreads the impulse while it concentrates the wave (at frequencies ω and $-\omega$ of $e^{i\omega n} + e^{-i\omega n}$). The wavelet transform in the third figure is large at the *time* of the impulse and also large at the *frequency* of the wave.

Purpose of the Book

There are already good books on this subject. The ideas and applications are beautiful, and the word has spread. The bibliography lists many of those books, which have special strengths. Our purpose is different. We believe that a *textbook* is needed. Our text explains filter banks and wavelets from the beginning — in several ways and at least two languages. The examples and exercises come from our courses at M.I.T. and Wisconsin, which brought students from all over engineering and science.

Whether you are working individually or in class, we hope this book clarifies the central ideas. Implicit in that goal is the recognition that we cannot describe every filter and prove every theorem. The book concentrates on the underpinning of the subject, which is now stable. There is a special "*glossary*" to organize and define the terms that are constantly used, some from signal processing and others from mathematics. The central idea is a *perfect reconstruction filter bank*, with properties and purposes selected from this list:

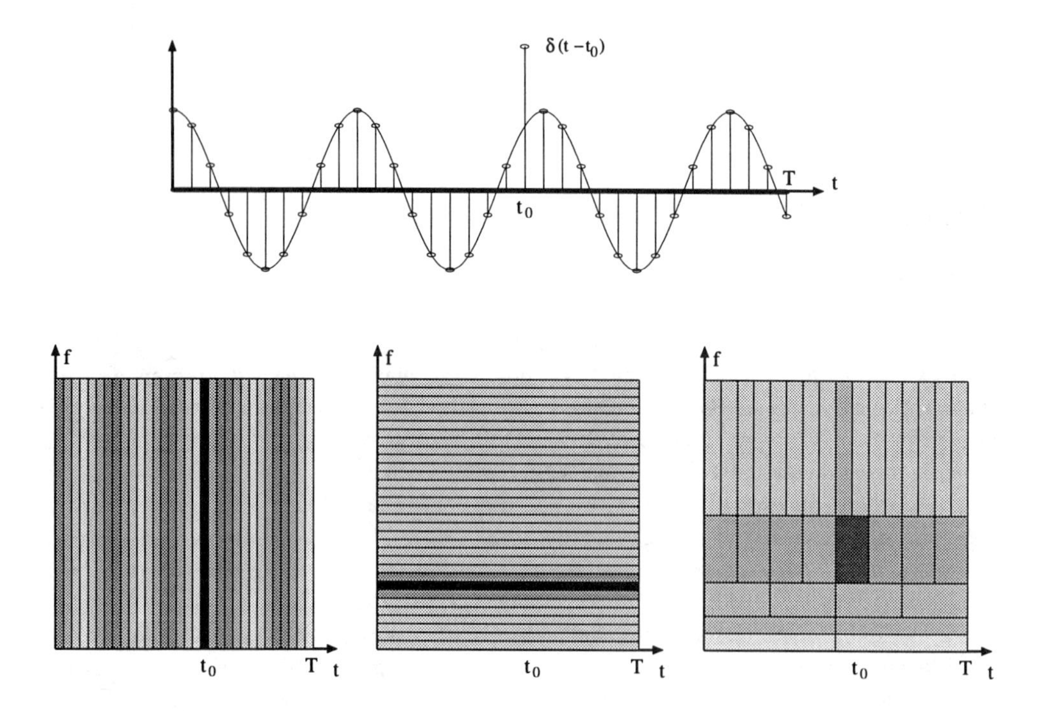

Figure 0.1: Impulse plus sinusoid, in time and frequency and time-frequency(time-scale).

Properties: Orthogonality, symmetry, short length, good attenuation.

Purposes: Audio/video compression, echo cancellation, radar, image analysis, communications, medical imaging, ...

Design and analysis techniques: Time domain, frequency domain, z-transform.

If the reader is willing, we would like to develop specific examples in this preface. Traditionally, the opening pages thank those who helped to create the book. Our debt to friends will soon be very gratefully acknowledged. But first, we go directly to transforms.

Transforms

Start with the basic idea and its purpose. The transform of a signal (a vector) is a *new representation* of that signal. The components $\boldsymbol{x}(0), \boldsymbol{x}(1), \boldsymbol{x}(2), \boldsymbol{x}(3)$ of a four-dimensional signal are replaced by four other numbers. Those numbers $\boldsymbol{y}(0), \boldsymbol{y}(1), \boldsymbol{y}(2), \boldsymbol{y}(3)$ are combinations of the $\boldsymbol{x}$'s. Our transforms are linear, so these are *linear combinations*—for example sums and differences:

$$\begin{aligned} \boldsymbol{y}(0) &= \boldsymbol{x}(0)+\boldsymbol{x}(1) \qquad & \boldsymbol{y}(2) &= \boldsymbol{x}(2)+\boldsymbol{x}(3) \\ \boldsymbol{y}(1) &= \boldsymbol{x}(0)-\boldsymbol{x}(1) \qquad & \boldsymbol{y}(3) &= \boldsymbol{x}(2)-\boldsymbol{x}(3). \end{aligned}$$

What is the purpose of the $\boldsymbol{y}$'s? And can we get back to the $\boldsymbol{x}$'s? The second question is easier and we answer it first.

This transform can be inverted. If you add the equations for $\boldsymbol{y}(0)$ and $\boldsymbol{y}(1)$, you get $2\boldsymbol{x}(0)$. Subtracting those equations yields $2\boldsymbol{x}(1)$. The ***inverse transform*** uses addition and subtraction (like the original!), and then division by 2:

$$\begin{aligned} \boldsymbol{x}(0) &= 0.5\,(\boldsymbol{y}(0)+\boldsymbol{y}(1)) \qquad & \boldsymbol{x}(2) &= 0.5\,(\boldsymbol{y}(2)+\boldsymbol{y}(3)) \\ \boldsymbol{x}(1) &= 0.5\,(\boldsymbol{y}(0)-\boldsymbol{y}(1)) \qquad & \boldsymbol{x}(3) &= 0.5\,(\boldsymbol{y}(2)-\boldsymbol{y}(3))\,. \end{aligned}$$

The $\boldsymbol{y}$'s allow perfect reconstruction of the $\boldsymbol{x}$'s, by the inverse transform. Looking ahead for a brief moment, we divide transforms into three groups:

(a)	**Lossless (orthogonal) transforms**	(orthogonal and unitary matrices)
(b)	**Invertible (biorthogonal) transforms**	(invertible matrices)
(c)	**Lossy transforms**	(not invertible)

A lossless unitary transform is like a rotation. The transformed signal has the same length as the original. This is true of the Fourier transform and all its real versions (DCT = discrete cosine transform, DST = discrete sine transform, HT = Hartley transform). The same signal is measured along new perpendicular axes.

For biorthogonal transforms, lengths and angles may change. The new axes are not necessarily perpendicular, but no information is lost. Perfect reconstruction is still available. We just invert. *Orthogonal wavelets give orthogonal matrices and unitary transforms, biorthogonal wavelets give invertible matrices and perfect reconstruction.* These transforms don't remove any information (or any noise), they just move it around—aiming to separate out the noise and decorrelate the signal. The irreversible step is to destroy small components, as we do below in "compression." Then invertibility is lost.

Matrices will appear very early, and we make no apology. A matrix displays the details of the transform. (The key texts in signal processing are amazingly empty of matrices. With the importance of systems like MATLAB, this must change.) Our book maintains a time-domain description by matrices, in parallel with the frequency-domain description by functions of ω. When the inputs $\boldsymbol{x}(n)$ and outputs $\boldsymbol{y}(n)$ are seen as vectors, the transform from $\boldsymbol{x}$ to $\boldsymbol{y}$ is executed by a matrix:

$$\begin{bmatrix} y(0) \\ y(1) \\ y(2) \\ y(3) \end{bmatrix} = \begin{bmatrix} 1 & 1 & 0 & 0 \\ 1 & -1 & 0 & 0 \\ 0 & 0 & 1 & 1 \\ 0 & 0 & 1 & -1 \end{bmatrix} \begin{bmatrix} x(0) \\ x(1) \\ x(2) \\ x(3) \end{bmatrix}.$$

The sum $\boldsymbol{x}(0)+\boldsymbol{x}(1)$ comes from the first row. The difference $\boldsymbol{x}(0)-\boldsymbol{x}(1)$ comes from the second row. Readers may recognize this as the 2-point DFT, and now comes its inverse. The matrix that recovers the $\boldsymbol{x}$'s from the $\boldsymbol{y}$'s is changed only by the factor $\frac{1}{2}$:

$$\begin{bmatrix} x(0) \\ x(1) \\ x(2) \\ x(3) \end{bmatrix} = \frac{1}{2} \begin{bmatrix} 1 & 1 & 0 & 0 \\ 1 & -1 & 0 & 0 \\ 0 & 0 & 1 & 1 \\ 0 & 0 & 1 & -1 \end{bmatrix} \begin{bmatrix} y(0) \\ y(1) \\ y(2) \\ y(3) \end{bmatrix}.$$

If you multiply these matrices, the result is the 4×4 identity matrix.

Small note When the first matrix is divided by $\sqrt{2}$, *it becomes an orthogonal matrix.* The rows of length $\sqrt{1+1} = \sqrt{2}$ will become unit vectors. The inner products between rows remain at zero. The transform becomes unitary (lossless) when the perpendicular axes—the four rows—are correctly normalized. ***The inverse is the transpose.***

The length squared of the transform is now

$$\left(\frac{x(0)+x(1)}{\sqrt{2}}\right)^2 + \left(\frac{x(0)-x(1)}{\sqrt{2}}\right)^2 + \left(\frac{x(2)+x(3)}{\sqrt{2}}\right)^2 + \left(\frac{x(2)-x(3)}{\sqrt{2}}\right)^2.$$

This simplifies to $\boldsymbol{x}(0)^2+\boldsymbol{x}(1)^2+\boldsymbol{x}(2)^2+\boldsymbol{x}(3)^2$. Thus $\|y\|^2 = \|x\|^2$ and the transform is unitary. For real matrices we also use the word orthogonal.

Purpose of the Transform

One important purpose is to see patterns in the $\boldsymbol{y}$'s that are not so clear in the $\boldsymbol{x}$'s. This means that the computer should see the pattern. What it sees best is *large versus small.* The computer will notice nothing special about the four numbers

$$\boldsymbol{x}(0) = 1.2 \quad \boldsymbol{x}(1) = 1.0 \quad \boldsymbol{x}(2) = -1.0 \quad \boldsymbol{x}(3) = -1.2.$$

Your eye notices various things in Figure 0.2 (I hope). Maybe the small movement and then the jump. Maybe the antisymmetry. To see this signal in the $\boldsymbol{y}$ representation, compute sums and differences:

$$\boldsymbol{y}(0) = 2.2 \quad \boldsymbol{y}(1) = 0.2 \quad \boldsymbol{y}(2) = -2.2 \quad \boldsymbol{y}(3) = 0.2.$$

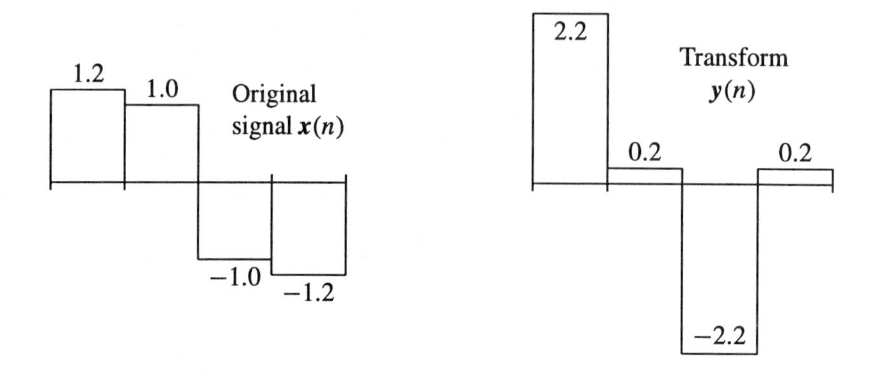

Figure 0.2: The $\boldsymbol{x}$'s transform to the $\boldsymbol{y}$'s by sums and differences.

First point: $\boldsymbol{y}(1)$ and $\boldsymbol{y}(3)$ are much smaller than $\boldsymbol{y}(0)$ and $\boldsymbol{y}(2)$. The differences are an order of magnitude smaller than the sums. Our transform has *almost lined up the signal* in the $\boldsymbol{y}(0)$–$\boldsymbol{y}(2)$ plane. Those two components of $\boldsymbol{y}$ do most of the work of four. It is true that we need all four $\boldsymbol{y}$'s to reconstruct perfectly all four $\boldsymbol{x}$'s. But if we change the small numbers $\boldsymbol{y}(1)$ and $\boldsymbol{y}(3)$—just cancel them!—the compressed signal $\boldsymbol{y}_c$ is

$$\boldsymbol{y}_c(0) = 2.2 \quad \boldsymbol{y}_c(1) = 0 \quad \boldsymbol{y}_c(2) = -2.2 \quad \boldsymbol{y}_c(3) = 0.$$

Those numbers 0.2 were below our threshold. In the compressed $\boldsymbol{y}_c$ they are gone. Figure 0.3 shows the signal $\boldsymbol{x}_c$ reconstructed from $\boldsymbol{y}_c$—the small difference between $\boldsymbol{x}(0)$ and $\boldsymbol{x}(1)$ is lost.

For comparison we compute the 4-point DFT, expecting and seeing the imaginary number $i = j = \sqrt{-1}$:

$$\begin{bmatrix} \widehat{\boldsymbol{x}}(0) \\ \widehat{\boldsymbol{x}}(1) \\ \widehat{\boldsymbol{x}}(2) \\ \widehat{\boldsymbol{x}}(3) \end{bmatrix} = \begin{bmatrix} 1 & 1 & 1 & 1 \\ 1 & i & i^2 & i^3 \\ 1 & i^2 & i^4 & i^6 \\ 1 & i^3 & i^6 & i^9 \end{bmatrix} \begin{bmatrix} 1.2 \\ 1.0 \\ -1.0 \\ -1.2 \end{bmatrix} = \begin{bmatrix} 0 \\ 2.2(1+i) \\ 0.4 \\ 2.2(1-i) \end{bmatrix}.$$

That zero in $\widehat{\boldsymbol{x}}$ reflects the antisymmetry. To see it in the sum-difference transform, we must go to the next level.

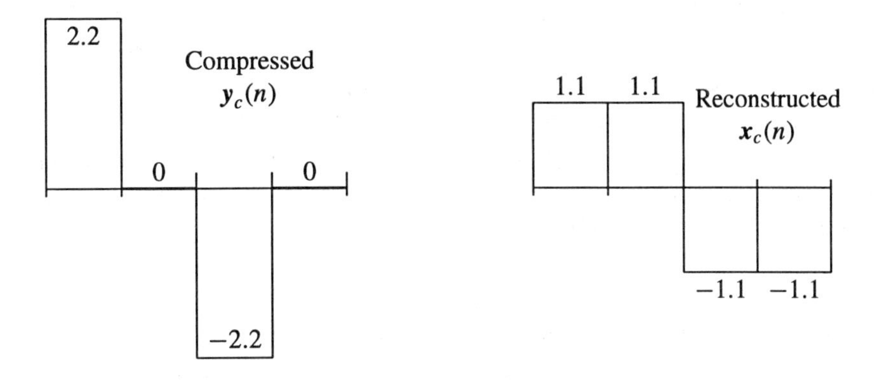

Figure 0.3: y_c and x_c are close to y and x, when small becomes zero.

Multilevel Transforms by Recursion

A key idea for wavelets is the concept of "*scale*." Sums and differences of neighbors are at the finest scale. We move to a larger picture by taking sums and differences again. This is recursion — the same transform at a new scale. It leads to a ***multiresolution*** of the original signal $x(0), x(1), x(2), x(3)$. Averages and details will appear at different scales.

The wavelet formulation keeps the differences $y(1)$ and $y(3)$ at the finest level, and *iterates only on* $y(0)$ *and* $y(2)$. Iteration means sum and difference of the transform:

$$z(0) = y(0) + y(2) = 0 \quad \text{and} \quad z(2) = y(0) - y(2) = 4.4.$$

At this level there is extra compression. *One component* $z(2)$ now does most of the work of the original four. The wavelet transform $z(0), y(1), z(2), y(3)$ is in Figure 0.4.

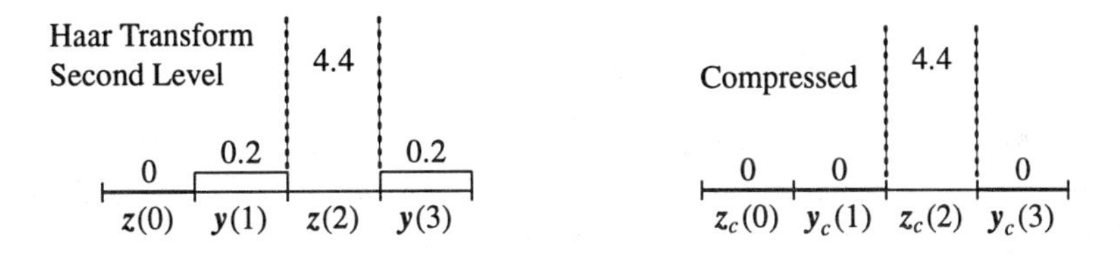

Figure 0.4: $z(0)$ and $z(2)$ are the sum and difference of $y(0)$ and $y(2)$. Compress to $z(2)$.

The key point is the multilevel construction, which is clear in a flow-graph:

x → sums $y(0), y(2)$ → sum $z(0)$
→ difference $z(2)$
x → differences $y(1), y(3)$

This two-step wavelet transform is executed by a product of two matrices:

$$\begin{bmatrix} 1 & & 1 & \\ & 1 & & \\ 1 & & -1 & \\ & & & 1 \end{bmatrix} \begin{bmatrix} 1 & 1 & 0 & 0 \\ 1 & -1 & 0 & 0 \\ 0 & 0 & 1 & 1 \\ 0 & 0 & 1 & -1 \end{bmatrix} = \begin{bmatrix} 1 & 1 & 1 & 1 \\ 1 & -1 & 0 & 0 \\ 1 & 1 & -1 & -1 \\ 0 & 0 & 1 & -1 \end{bmatrix}.$$

That is invertible! To draw the flow-graph of the inverse, just reverse the arrows. To invert the matrix on the right, multiply the inverses of the matrices on the left (in reverse order of course).

The wavelet transform becomes unitary as before, when we divide sums and differences by $\sqrt{2}$. You will see this number appearing throughout the book, to compensate for scale changes.

Perhaps one more transform could be mentioned. It is the Walsh-Hadamard transform, which iterates *also on the differences* $\boldsymbol{y}(1)$ *and* $\boldsymbol{y}(3)$. Instead of the "logarithmic tree" for wavelets, we have a complete binary tree for Walsh-Hadamard:

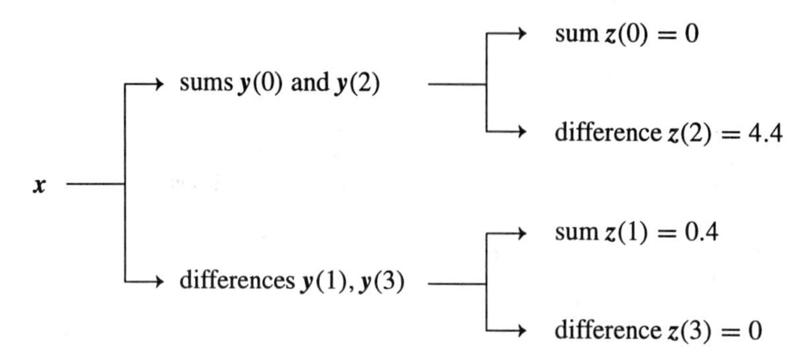

This also counts as a *wavelet packet* (those include every binary tree). You would not want to miss the "Hadamard matrix" for this transform, which is exceptional. All entries are 1 and -1 and all rows are orthogonal:

$$\begin{bmatrix} 1 & & 1 & \\ & 1 & & 1 \\ 1 & & -1 & \\ & 1 & & -1 \end{bmatrix} \begin{bmatrix} 1 & 1 & & \\ 1 & -1 & & \\ & & 1 & 1 \\ & & 1 & -1 \end{bmatrix} = \begin{bmatrix} 1 & 1 & 1 & 1 \\ 1 & -1 & 1 & -1 \\ 1 & 1 & -1 & -1 \\ 1 & -1 & -1 & 1 \end{bmatrix}.$$

When divided twice by $\sqrt{2}$ this matrix is orthogonal and also symmetric. The original $\boldsymbol{x}$'s can be reconstructed from the $\boldsymbol{y}$'s or the $\boldsymbol{w}$'s or the wavelet outputs $\boldsymbol{y}(1), \boldsymbol{y}(3), \boldsymbol{z}(0), \boldsymbol{z}(2)$. From compressed outputs we get approximate (but good) reconstruction.

Don't forget the disadvantage compared to wavelets! All sixteen entries are nonzero. The complete tree takes more computation than the wavelet tree. For signals of length n (a power of two), the Hadamard matrix has n^2 nonzeros and the wavelet matrix has only $n + n\log_2 n$. Those are the costs *without recursion*, jumping from the original $\boldsymbol{x}$'s directly to the final $\boldsymbol{z}$'s.

When the computations are recursive, as they always should be, we have a product of very sparse matrices. The flow-graph gives $2n\log_2 n$ for the complete Walsh-Hadamard transform (*exactly matching the Fast Fourier Transform*). This is good but wavelets are slightly better. The transform to $\boldsymbol{z}(0), \boldsymbol{y}(1), \boldsymbol{z}(2), \boldsymbol{y}(3)$ computes the sum and difference of $n-1$ pairs. This needs only $2n-2$ calls to memory.

The wavelet transform achieves the Holy Grail of complexity theory (or simplicity theory). *The transform is an $O(n)$ computation.* But does it separate the true signal from noise? Does it allow compression of 4:1 or 8:1 or 20:1? This sum-difference transform is analyzed further in Chapter 1 — it is such a good example — but we admit here that it is too simple. Better filters are needed, and they lead to better wavelets (still with $O(n)$ operations). That is the subject of our book.

Applications

The choice of waves or wavelets, Fourier analysis or filter analysis, depends on the signal. It must. Signals coming from different applications have different characteristics. It is helpful to see a broad picture:

Audio: Use many subband filters or a windowed (short time) Fourier transform.

Speech: The time variation is irregular and requires nonuniform intervals.

Images: Finite length signals need special treatment to reduce blocking. Symmetric extension goes with symmetric filters.

Video: Use motion prediction (optical flow of images) or space-time filtering.

The best choice for medical imaging is not clear. The legal questions arising from lossy compression are not clear either. Identification from synthetic aperture radar (SAR images) is an enormous problem. So is de-noising, which is at the heart of signal representation. Ultimately we are trying to choose a good basis.

The problem is to represent typical signals *with a small number of conveniently computable functions*. The traditional bases (Fourier, Bessel, Legendre, ...) come from differential equations. Wavelets do not come from differential equations! One reason is, those equations don't include dilation. *A dilation equation* $\phi(t) = \sum 2\boldsymbol{h}(k)\phi(2t-k)$ *involves two time scales*. Its solution $\phi(t)$ is nonzero only on a finite interval. Then $\phi(2t)$ is nonzero on half of that interval. The basis is localized. It is quickly (and recursively) computed from the numbers $\boldsymbol{h}(k)$. Those are the coefficients in the corresponding lowpass filter.

This relation of filters to functions is the heart of the book.

Acknowledgments

Many friends have created this theory. A lot of them also helped us to write about it. The bibliography points to their original work (we wish it were possible to be complete — that is better attempted electronically). Our personal thanks must begin with Vasily Strela and Chris Heil. They read most of the words and discussed all the ideas. Their unselfish help is deeply appreciated.

Other friends will see, at specific places in the book, exactly where their ideas and suggestions made a difference. We are enormously grateful to Kevin Amaratunga, Ross Barmish, Christopher Brislawn, Charles Chui, Albert Cohen, Adriannus Djohan, Dave Donoho, Jeff Geronimo, Doug Hardin, Peter Heller, Tom Hopper, Angela Kunoth, Seng Luan Lee, Michael Lightstone, Eric Majani, Stéphane Mallat, Henrique Malvar, Ricardo de Queiroz, Jianhong Shen, Wim Sweldens, Pankaj Topiwala, Steffen Trautmann, Chi Wah Kok, Victor Wickerhauser, and Zhifeng Zhang. There is no way to thank you enough for such generous help.

We want to recognize separately the outstanding leadership of four authors, P. P. Vaidyanathan and Martin Vetterli in signal processing and Ingrid Daubechies and Yves Meyer in mathematics. The two subjects are thoroughly mixed, thanks to these four! Their work has been a model in every way, personal (to us) as well as scientific (for all).

Our writing of the book was shared in a natural way — words mostly from the first author, filter designs and applications from the second author, ideas from both. Vasily Strela and Robert Becker helped to get Versions 1.0 to 9.9 into TeX, and Martin Stock made it more beautiful. The whole project grew out of our courses and workshops (which will continue). Those were tremendous sources of inspiration, thanks to the participants.

This subject is not just theory. The ideas have to be implemented, and MATLAB is the outstanding way to do that. It is a pleasure to have this book closely linked to the *Wavelet Toolbox* offered by MathWorks. Exercises that use this Toolbox are in the book; the reader can *see* the wavelets and the multiresolution they produce. A more extensive Wavelet Manual will come from the same sources: Wellesley-Cambridge Press and MathWorks.

For correspondence about this whole subject, and for comments and corrections of any kind, the authors thank the readers. Especially we thank Jill and Thuy Duong for their patience and support. It is finished at last! We hope you enjoy this book.

Gilbert Strang and Truong Nguyen

M.I.T. and Wisconsin, November 1995

gs@math.mit.edu and nguyen@ece.wisc.edu

http://saigon.ece.wisc.edu/~waveweb/Tutorials/book.html

Guide to the Book

This book has a two-part subject. One part is discrete, the other is continuous. In discrete time we develop the idea and applications of ***filter banks***. In continuous time we have ***scaling functions*** $\phi(t)$ and ***wavelets*** $w(t)$. By a natural limiting process, iteration of the lowpass filter leads to the scaling function. One highpass filter then produces a wavelet. Our goal is to make this connection clear. We find the conditions on the discrete coefficients that lead to good filter banks and good wavelets.

Historically and mathematically, the filters come first. Perfect reconstruction filter banks were developed in the early 1980's. The excitement around wavelets started later (and grew quickly). This excitement was not universal — designers of filter banks naturally asked what was new. Part of the answer is precisely in that process of ***iteration***. For a filter to behave well in practice, when it is combined with subsampling and repeated five times, it must have an extra property — not built into earlier designs. This property expresses itself in the frequency domain by a sufficient number of "zeros at π". Then the frequency band can be successfully separated into five octaves.

The underlying problem is to choose a good basis. We want to represent a signal well, by a small number of basic signals. These can be sinusoids and they can be wavelets. On a discrete grid, $\omega = \pi$ is the highest frequency at which a signal can oscillate. Those oscillations $\boldsymbol{x}(n) = e^{i\pi n} = (-1)^n$ are stopped by the lowpass filter with a "zero at π". The highpass filter lets fast oscillations through, and the synthesis filters can reconstruct the exact input. But ***compression*** may come between analysis and synthesis. Frequencies that are barely represented will be intentionally lost. That mostly means high frequencies but the filter bank is impartial — it keeps the basis functions that are important to the specific signal. We want to show when, and why, filter banks and wavelets are effective in reconstruction and signal representation and compression.

Filter Banks

Some readers will begin this book with Chapter 1. Others will jump forward to a topic of particular interest. This brief guide is for both, especially to tell the first readers what is coming and the second group where to look. We are pointing to places where preparation and explanation come together, to design and study new structures.

For filter banks, that place is Section 4.1. There we identify the two conditions for perfect reconstruction (in the absence of lossy compression). One condition removes distortion, the other condition removes aliasing. The anti-distortion condition applies to the products $\boldsymbol{F}_0\boldsymbol{H}_0$ and $\boldsymbol{F}_1\boldsymbol{H}_1$ along the channels of the filter bank. Then the anti-aliasing condition controls how those products can be separated into the four filters.

The design of a perfect reconstruction filter bank is a choice of $\boldsymbol{F}_0\boldsymbol{H}_0$ and then a factorization. To understand the conditions on distortion and aliasing, we apply the techniques of multirate filtering. Those techniques are explained in Chapters 1–3, with examples throughout. Of course filters are to be defined! But we can go forward even now, to illustrate a filter bank that gives perfect reconstruction.

The analysis bank is on the left. It has a lowpass filter $\boldsymbol{H}_0$, and a highpass filter $\boldsymbol{H}_1$, and decimation by $(\downarrow 2)$—which removes the odd-numbered components after filtering. The analysis bank yields two "half-length" outputs. Then the synthesis bank on the right begins with the upsampling operation $(\uparrow 2)$—which inserts zeros in those odd components:

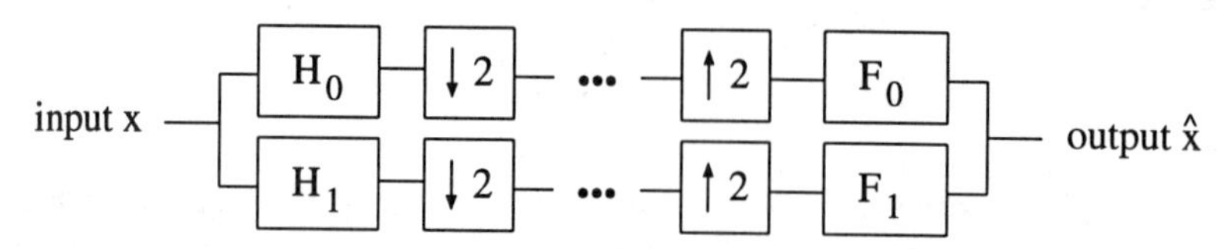

Two-channel filter bank: Separate the input into frequency bands (filter and downsample). Then reassemble (upsample and filter).

The gap in the center indicates where the subband signals are compressed or enhanced. The applications of this structure are extremely widespread. We believe that any reader interested in signal processing (and image processing) will find that filter bank analysis is extremely useful.

The filters $\boldsymbol{H}_0, \boldsymbol{H}_1, \boldsymbol{F}_0$, and $\boldsymbol{F}_1$ are linear and time-invariant. The operators $(\downarrow 2)$ and $(\uparrow 2)$ are ***not*** time-invariant. These multirate operations are responsible for *aliasing* and for *imaging*—they create undesirable and extraneous signals that the filters must cancel. To understand how that happens, and to design good filters, we use the tools developed in Chapters 1–3: especially transformation to the frequency domain and z-domain. We try to explain the analysis of multirate filtering, with $(\downarrow 2)$ and $(\uparrow 2)$, clearly and memorably.

The theory and the design of filter banks and wavelets will dominate Chapters 4–6. This is the heart of the book. The structure of an ***orthogonal*** bank is very special, and the next figure shows how the filters are related. For length 4 all filters use the four coefficients a, b, c, d that Daubechies derived:

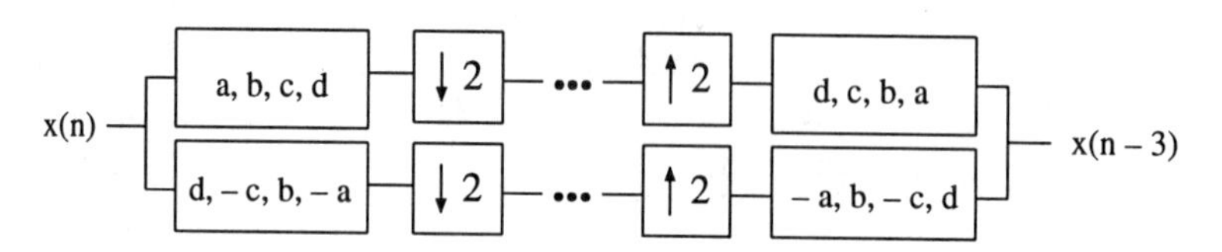

The form of an orthogonal filter bank with four coefficients.

How did she choose a, b, c, d? Part of the answer will have to wait, but here is the essential idea. The product along the top channel gives a particular "halfband filter":

$$(a,\ b,\ c,\ d) * (d,\ c,\ b,\ a) = (-1,\ 0,\ 9,\ 16,\ 9,\ 0,\ -1)/16.$$

This convolution is a multiplication of two polynomials, when $a,\ b,\ c,\ d$ are the coefficients:

$$(a + bz^{-1} + cz^{-2} + dz^{-3})\,(d + cz^{-1} + bz^{-2} + az^{-3}) =$$
$$(-1 + 9z^{-2} + 16z^{-3} + 9z^{-4} - z^{-6})/16.$$

The four coefficients are pleasant to calculate. The serious job in Chapter 4 is to explain what is special about that 6th-degree polynomial in which z^{-1} and z^{-5} are missing.

A filter bank also gives perfect reconstruction if it is ***biorthogonal***. This design is less restricted. The product $\boldsymbol{F}_0\boldsymbol{H}_0$ must skip the same odd powers of z^{-1}, but $\boldsymbol{F}_0$ does not have to be the transpose (the flip) of $\boldsymbol{H}_0$. Here are specific numbers for the filter coefficients — not the only choice and maybe not the best. They show how the filters $\boldsymbol{F}_0$ and $\boldsymbol{F}_1$ on the synthesis side are related to the analysis filters $\boldsymbol{H}_1$ and $\boldsymbol{H}_0$ (by alternating signs):

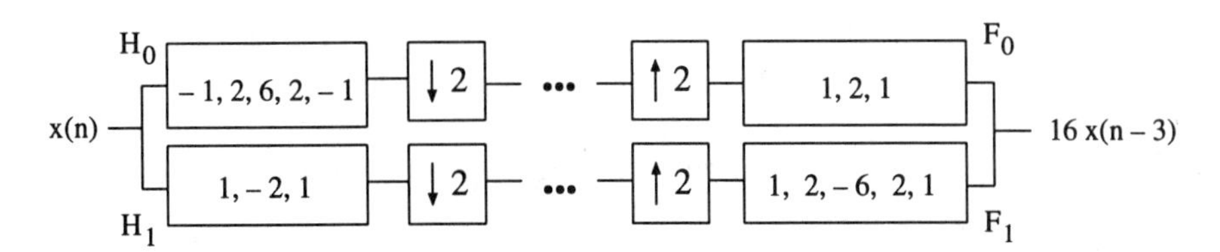

A biorthogonal filter bank: Perfect reconstruction with 3 delays.

For filters, we stop here. This time $F_0(z) = 1 + 2z^{-1} + z^{-2}$. Multiplied by $H_0(z)$ it gives the same important 6th-degree polynomial as before. To understand why the zero coefficients are necessary in that polynomial, and why $-\frac{1}{16}$ and $\frac{9}{16}$ are desirable, I am afraid that you have to read the book!

Our discussion went this far (further than we intended) so as to make a basic and encouraging point: *The construction of new filter banks need not be complicated.* This subject is accessible to new ideas and experiments.

Wavelets

Wavelets are localized waves. Instead of oscillating forever, they drop to zero. They come from the ***iteration*** of filters (with rescaling). The link between discrete-time filters and continuous-time wavelets is in the limit of a logarithmic filter tree:

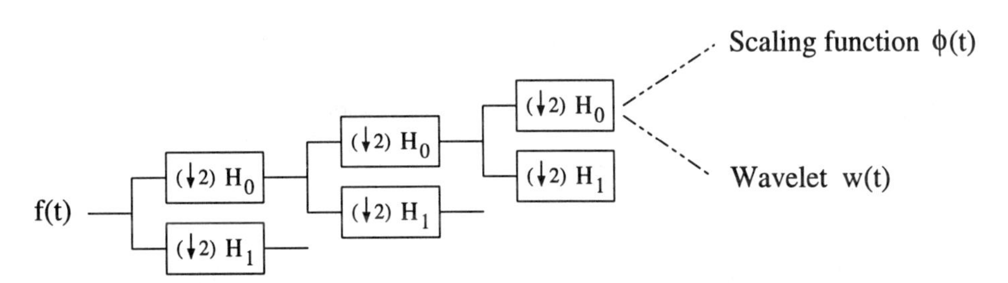

Scaling function and wavelets from iteration of the lowpass filter.

Scaling functions and wavelets have remarkable properties. They inherit orthogonality, or biorthogonality, from the filter bank. Because of the repeated rescaling that produces them, wavelets decompose a signal into details at all scales. The wavelet $w(t)$ and its shifts $w(t-k)$ are at unit scale. The wavelets $w(2^j t)$ and $w(2^j t - k)$ are at scale 2^{-j}. The biorthogonal functions $\widetilde{\phi}(t)$ and $\widetilde{w}(t)$ come from iterating the synthesis bank.

Wavelets produce a natural ***multiresolution*** of every image, including the all-important edges. Where the low frequency part of the Fourier transform is often a blur, the output from the lowpass channel is a useful compression.

Sections 5.5 and 6.2 study the particular wavelets created by Ingrid Daubechies. They are orthogonal, with the advantages and limitations that this property brings. Sections 4.1 and 6.5 study biorthogonal alternatives, which come from different factorizations of the same polynomial (as above). This polynomial corresponds to a "maxflat halfband filter", and we hope you will like the connections.

More than that, we hope you enjoy the whole book. This subject is a beautiful combination of mathematical analysis and signal processing applications. The analysis and the applications are based on designs that give perfect reconstruction. To explain both sides of this subject, we need words from mathematics and words from digital signal processing. The Glossary at the end is a dictionary of their meanings. Above all we need ***ideas*** from both sides, and from a tremendous range of application areas. It is to the understanding of filters and wavelets, and the growth of successful applications, that this book is dedicated.

Summary of the Theory

There are four conditions that play a central part in this book. Because of their importance we highlight them here. They apply directly to the coefficients in the filter banks — and the consequences are felt (after iteration!) in the scaling functions and wavelets. Here are the four conditions — some might say in decreasing order of importance:

PR Condition *Perfect reconstruction.*
The synthesis bank inverts the analysis bank, with ℓ delays.
Biorthogonal banks with no aliasing and no distortion.

Condition O *Orthogonality.*
The analysis bank is inverted by its transpose.
The wavelets are orthogonal to all their dilates and translates.

Condition A_p *Accuracy of order p for approximation by scaling functions $\phi(t-k)$.*
p vanishing moments in the wavelets.
pth order decay of wavelet coefficients for smooth $f(t)$.

Condition E *Eigenvalue condition on the cascade algorithm.*
Determines convergence to $\phi(t)$ and smoothness of wavelets.
Equivalent to stability of the wavelet basis.

One step further and this Guide is ended. The four fundamental conditions will be stated explicitly for a two-channel filter bank, with the sections in which they appear. We continue to use the polynomials $H_0(z)$, $H_1(z)$, $F_0(z)$, and $F_1(z)$, whose coefficients come directly from the filters. By convention, these are polynomials in z^{-1}, and the lowpass analysis filter is represented by $H_0(z) = \boldsymbol{h}(0) + \boldsymbol{h}(1)z^{-1} + \cdots + \boldsymbol{h}(N)z^{-N}$. Here are the conditions that give filters and wavelets with good properties:

1. Perfect Reconstruction (***PR condition in Section 4.1***)

$$F_0(z)H_0(z) + F_1(z)H_1(z) = 2z^{-\ell} \quad \text{and} \quad F_0(z)H_0(-z) + F_1(z)H_1(-z) = 0.$$

The second equation gives the anti-aliasing choices $F_0(z) = H_1(-z)$ and $F_1(z) = -H_0(-z)$.

2. Orthogonality (***Condition O in Section 5.3***)

The filter coefficients are reversed by $F_0(z) = z^{-N} H_0(z^{-1})$ and $F_1(z) = z^{-N} H_1(z^{-1})$. Then perfect reconstruction depends on the "double-shift orthogonality" of the lowpass coefficients $\boldsymbol{h}(k)$:

$$\sum \boldsymbol{h}(k)\, \boldsymbol{h}(k+2n) = \delta(n).$$

In terms of the polynomials this is $H_0(z)H_0(z^{-1}) + H_0(-z)H_0(-z^{-1}) = 2$.

3. Accuracy of order p (***Condition A_p in Sections 5.5 and 7.1***)

The lowpass filter has a zero of order p at $z = -1$:

$$H_0(z) = \left(\frac{1+z^{-1}}{2}\right)^p Q(z).$$

4. Convergence and Stability (***Condition E in Section 7.2***)

The transition matrix $\boldsymbol{T}$ has $\lambda = 1$ as simple eigenvalue and all other $|\lambda(\boldsymbol{T})| < 1$.

Final note: The sixth degree polynomial in the examples above has *four zeros* at $z = -1$:

$$-1 + 9z^{-2} + 16z^{-3} + 9z^{-4} - z^{-6} = (1 + z^{-1})^4\,(-1 + 4z^{-1} - z^{-2}).$$

These zeros give flat responses near $\omega = \pi$ and also $\omega = 0$. The absence of z^{-1} and z^{-5} is the key to perfect reconstruction. Polynomials of higher degree, also with zeros at $z = -1$ and also with only one odd power, factor into $F_0(z)H_0(z)$ to give the best filters for iteration. In the limit of the iterations, these filters give good wavelets.

Chapter 1

Introduction

1.1 Overview and Notation

We begin with an overview of ***filters***, ***filter banks***, and ***wavelets***. We want to indicate, first in rough outline and then in detail, the connections between these three topics. Our immediate purpose is to open up the problem and the language — starting with the filter coefficients $\boldsymbol{h}(n)$. The choice of those coefficients is the crucial decision. Their properties govern all that follows.

Each step is a natural development from the one before:

(1) A ***filter*** is a linear time-invariant operator. It acts on input vectors $\boldsymbol{x}$. The output vector $\boldsymbol{y}$ is the convolution of $\boldsymbol{x}$ with a fixed vector $\boldsymbol{h}$. The vector $\boldsymbol{h}$ contains the filter coefficients $\boldsymbol{h}(0), \boldsymbol{h}(1), \boldsymbol{h}(2), \ldots$. Our filters are digital, not analog, so the coefficients $\boldsymbol{h}(n)$ come at discrete times $t = nT$. The sampling period T is assumed to be 1 here. The inputs $\boldsymbol{x}(n)$ and outputs $\boldsymbol{y}(n)$ come at all times $t = 0, \pm 1, \pm 2, \ldots$:

$$y(n) \quad = \quad \sum_k \boldsymbol{h}(k)\boldsymbol{x}(n-k) \quad = \quad \text{convolution } \boldsymbol{h} * \boldsymbol{x} \text{ in the time domain.}$$

One input $\boldsymbol{x} = (\ldots, 0, 1, 0, \ldots)$ has special importance — a unit impulse at time zero. The input has $\boldsymbol{x}(n-k) = 0$ except when $n = k$. The sum in the convolution has only one term, and that term is $\boldsymbol{h}(n)$. This output $\boldsymbol{y}(n) = \boldsymbol{h}(n)$ is the response at time n to the unit impulse $\boldsymbol{x}(0) = 1$. It is the *impulse response* $\boldsymbol{h}(0),\ \boldsymbol{h}(1),\ ,\ldots,\ \boldsymbol{h}(N)$.

In a moment the same filter will be described in the frequency domain. Convolution with the vector $\boldsymbol{h}$ will become *multiplication* by a function H. It is the simplicity of multiplication that makes this subject a success. The action of a filter in time and frequency is the foundation on which signal processing is built.

(2) A ***filter bank*** is a set of filters. The analysis bank often has two filters, lowpass and highpass. They separate the input signal into frequency bands. Those subsignals can be compressed much more efficiently than the original signal. Then they can be transmitted or stored. We are describing "subband coding" and its applications. At any time the signals can be recombined (by the *synthesis bank*).

It is not necessary to preserve the full outputs from the analysis filters. Normally they are *downsampled*. ***We keep only the even components of the lowpass and highpass filter outputs.***

If there are M filters, then keeping every Mth component of each output gives a total of the same length as the input. Critical sampling is the key to subband coding.

This book explains how two or more filters, with downsampling, can jointly achieve properties that are impossible for a single filter. We are particularly interested in "perfect reconstruction FIR filter banks". In this case the reconstructed output $\widehat{\boldsymbol{x}}(n)$ from the synthesis bank is identical to the original input $\boldsymbol{x}$ to the analysis bank (with only a time delay). In matrix language, a banded matrix (for the analysis bank) has a banded inverse (the synthesis bank).

In the frequency domain, each filter leads to a multiplication. But downsampling is *not a time-invariant operation.* If we delay all components of $\boldsymbol{y}$ by one time unit, the output from downsampling is totally different. The new samples $\boldsymbol{y}(-1), \boldsymbol{y}(1), \boldsymbol{y}(3)$ are entirely separate and independent from the original samples $\boldsymbol{y}(0), \boldsymbol{y}(2), \boldsymbol{y}(4)$. Those two subsampled signals are two "phases" of $\boldsymbol{y}$, not connected. Therefore downsampling alters the multiplication picture in the frequency domain. In fact it introduces *aliasing*.

Chapter 4 will show how the simplicity of multiplication can be rescued by looking at each phase separately. Each phase of $\boldsymbol{y}$ comes from filtering the phases of $\boldsymbol{x}$ (using phases of $\boldsymbol{h}$). These separate pieces are multiplications in the frequency domain. The whole operation together, filtering followed by downsampling, becomes a matrix multiplication — by the *polyphase matrix.*

This is the foundation of filter bank theory (still to be explained in detail!). The analysis polyphase matrix $\boldsymbol{H}_p$ will reveal the correct synthesis bank for perfect reconstruction. That synthesis filter bank uses $\boldsymbol{H}_p^{-1}$.

(3) ***Wavelets*** are basis functions $w_{jk}(t)$ in continuous time. A basis is a set of linearly independent functions that can be used to produce all admissible functions $f(t)$:

$$f(t) = \text{ combination of basis functions } = \sum_{j,k} b_{jk}\, w_{jk}(t). \tag{1.1}$$

The special feature of the wavelet basis is that all functions $w_{jk}(t)$ are constructed from a single mother wavelet $w(t)$. This wavelet is a small wave (a pulse). Normally it starts at time $t = 0$ and ends at time $t = N$.

The shifted wavelets w_{0k} start at time $t = k$ and end at time $t = k+N$. The rescaled wavelets w_{j0} start at time $t = 0$ and end at time $t = N/2^j$. Their graphs are compressed by the factor 2^j, where the graphs of w_{0k} are translated (shifted to the right) by k:

$$\textit{compressed:} \quad w_{j0} = w\left(2^j t\right) \qquad\qquad \textit{shifted:} \quad w_{0k}(t) = w(t-k).$$

A typical wavelet w_{jk} is compressed j times and shifted k times. Its formula is

$$w_{jk}(t) = w\left(2^j t - k\right).$$

The remarkable property that is achieved by many wavelets is *orthogonality*. The wavelets are orthogonal when their "inner products" are zero:

$$\int_{-\infty}^{\infty} w_{jk}(t)\, w_{JK}(t)\, dt = \text{ inner product of } w_{jk} \text{ and } w_{JK} = 0. \tag{1.2}$$

In this case the wavelets form an *orthogonal basis* for the space of admissible functions. This basis corresponds to a set of axes that meet at 90° angles — as most good axes do. Orthogonality leads to a simple formula for each coefficient b_{JK} in the expansion for $f(t)$. Multiply the expansion displayed in equation (1.1) by $w_{JK}(t)$ and integrate:

$$\int_{-\infty}^{\infty} f(t)\, w_{JK}(t)\, dt = b_{JK} \int_{-\infty}^{\infty} (w_{JK}(t))^2\, dt. \tag{1.3}$$

All other terms in the sum disappear because of orthogonality. Equation (1.2) eliminates all integrals of w_{jk} times w_{JK}, except the one term that has $j = J$ and $k = K$. That term produces $(w_{JK}(t))^2$. Then b_{JK} is the ratio of the two integrals in equation (1.3).

As we describe the connection between filter banks and wavelets, you will see that it is the "*highpass filter*" that leads to $w(t)$. The "*lowpass filter*" leads to a scaling function $\phi(t)$. In most constructions the lowpass filter comes first—***the scaling function is obtained before the wavelet.*** In fact the scaling function (in continuous time) comes from infinite repetition $\boldsymbol{L\,L\,\ldots\,L}$ of the lowpass filter, with rescaling at each iteration. The wavelet follows from $\phi(t)$ by just *one* application of the highpass filter.

Multiresolution

At a given resolution of a signal or an image, the scaling functions $\phi\left(2^j t - k\right)$ are a basis for the set of signals. The level is set by j, and the time steps at that level are 2^{-j}. The new details at level j are represented by the wavelets $w\left(2^j t - k\right)$. Then the smooth signal plus the details, the ϕ's plus the w's, combine into a ***multiresolution*** of the signal at the finer level $j+1$. Averages come from the scaling functions, details come from the wavelets:

$$\begin{array}{ccc} \text{signal at level } j \text{ (local averages)} \searrow & & \\ + & & \text{signal at level } j+1 \\ \text{details at level } j \text{ (local differences)} \nearrow & & \end{array}$$

That is multiresolution for one signal. When we apply it to all signals, we have multiresolution for *spaces* of functions:

$$\begin{array}{ccc} V_j = \text{scaling space at level } j \searrow & & \\ \oplus & & V_{j+1} = \text{scaling space at level } j+1 \\ W_j = \text{wavelet space at level } j \nearrow & & \end{array}$$

This idea of multiresolution is absolutely basic to wavelet analysis. Again, we are only introducing it. We are sending a coarse signal to the reader, not the details. You only have the input at level 1.

Thus the signal is divided into different ***scales*** of resolution, rather than different frequencies. The "time-scale plane" takes the place for wavelets that the "time-frequency plane" takes for filters. Multiresolution divides the frequencies into *octave bands*, from ω to 2ω, instead of uniform bands from ω to $\omega + \Delta\omega$. The compression of a graph, when $f(t)$ is replaced by $f(2t)$, means expansion of its Fourier transform from $F(\omega)$ to $\frac{1}{2}F\left(\frac{\omega}{2}\right)$. Frequencies shift upward by an octave, when time is rescaled by two. You will see how the time-frequency plane is partitioned naturally into *rectangles of constant area* (Figure 1.1).

This matching of long time with low frequency and short time with high frequency occurs in a natural way for wavelets. It is one of the attractions of a wavelet decomposition.

To the reader: We have reprinted in Appendix A an article on wavelets published in the American Scientist of May 1994. This article introduces wavelet notation through its correspondence with *musical notation*. In music, each note specifies a frequency and a position in time.

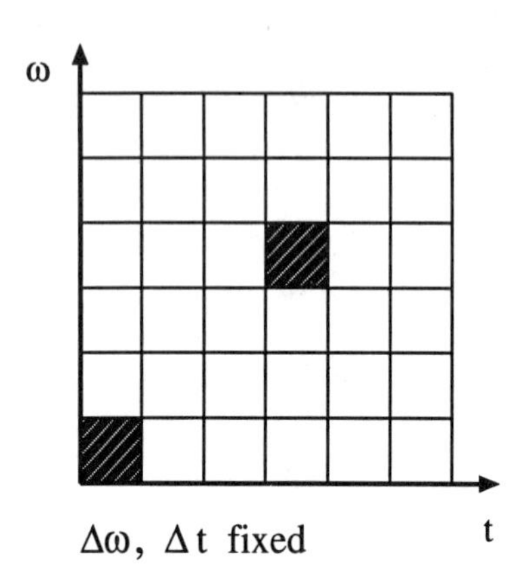

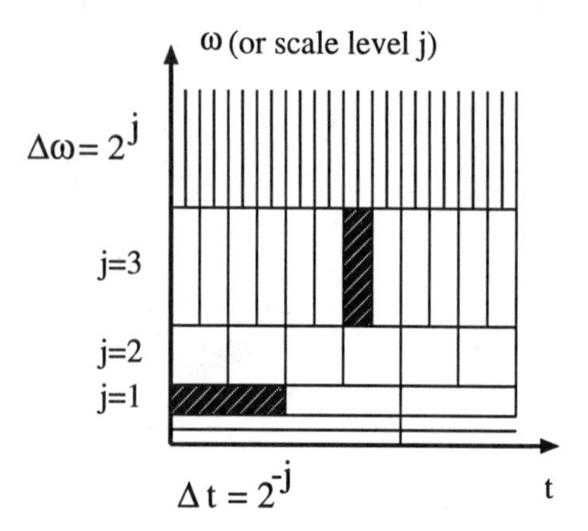

Figure 1.1: Time-frequency squares for Fourier decompositions become rectangles for wavelets. Short time intervals are natural for high frequencies.

Its vertical placement gives frequency, its horizontal placement indicates time. A musical score is almost a wavelet decomposition — except that it has fractional jumps in frequency. There are notes between middle C and high C, while wavelets jump by octaves. The shortest note I have seen is a 32nd note, corresponding to level $j = 5$ (because $2^{-5} = \frac{1}{32}$). Wavelets often stop there too, in practice. But in principle the scale level j goes to infinity.

That article on wavelets was written for a nontechnical audience, but it aims to explain the essential ideas. In the wavelet decomposition, all instruments play the same tune! They have different amplitudes and they play at different speeds and different times. The basses contribute the coarsest signals $b_{0k}w(t-k)$, starting at integer times $t = k$ and overlapping. The cellos play an identical tune but twice as fast. They contribute $b_{1k}w(2t - k)$, starting at half-integer times $t = k/2$ and again overlapping. The violas and violins add details at levels $j = 2$ and $j = 3$. Those details are wavelets $w(4t)$ and $w(8t)$ and their translates. It is the *orthogonality* of all these tunes, and especially the *localization* of each tune into a short time interval, that makes it possible to decompose the symphony efficiently.

Similarly it is orthogonality (or biorthogonality) and localization that make wavelet decompositions attractive for other signals.

Frequency Domain and Notation

To see a filter as a multiplication, we must take Fourier transforms. This will be the *discrete-time Fourier transform*, since the vectors $\boldsymbol{x}(n)$ and $\boldsymbol{h}(n)$ and $\boldsymbol{y}(n)$ are discrete. The time index n goes from $-\infty$ to ∞. (A vector with zero components at all negative times is called *causal*.) The transform of $\boldsymbol{x}$ has two reasonable notations. They both stand for the same transform, which we denote by X:

$$X(e^{j\omega}) = \sum_{-\infty}^{\infty} \boldsymbol{x}(n)\, e^{-jn\omega} \quad \text{(signal processing notation)}$$

$$X(\omega) = \sum_{-\infty}^{\infty} \boldsymbol{x}(n)\, e^{-in\omega} \quad \text{(reduced notation).}$$

You see two differences. On the right side, one uses j and the other uses i. Both represent $\sqrt{-1}$. On the left side, the standard signal processing notation uses $e^{j\omega}$ while the reduced notation writes only ω. (Each has advantages and we are prepared to print the book both ways!)

The standard notation allows a direction conversion of the Fourier transform to the *z-transform*. The transform is still X but the variable becomes z:

$$X(z) = \sum_{-\infty}^{\infty} \boldsymbol{x}(n)\, z^{-n}.$$

We simply replace $e^{j\omega}$ by z, extending the formal definition of X from $e^{j\omega}$ on the unit circle to z in the whole complex plane. (Remember: $e^{j\omega}$ has magnitude 1.) The Fourier transform will dominate the first part of the book, but the z-transform appears more frequently in the end.

The reduced notation has the advantage of outstanding simplicity. There will be many, many occasions to write $X(\omega)$ and $X(\omega+\pi)$ or to write $X(e^{j\omega})$ and $X(e^{j(\omega+\pi)})$. The first occasion is the most important right now, and we want to express the action of a filter both ways:

Convolution by h in time becomes multiplication by H in frequency:

$$\begin{aligned} Y(e^{j\omega}) &= H(e^{j\omega})\, X(e^{j\omega}) && \text{in signal processing notation} \\ Y(\omega) &= H(\omega)\, X(\omega) && \text{in reduced notation.} \end{aligned}$$

This is the transform of $\boldsymbol{y}(n) = \sum \boldsymbol{h}(k)\boldsymbol{x}(n-k)$. Exercise 10 will ask you to verify this fundamental fact. It is the "convolution rule". In the z-domain it becomes $Y(z) = H(z)\, X(z)$. The only inputs to the proof are the definition of convolution and of the transform.

Decision on notation. Simplicity often wins. We keep the freedom to write $X(\omega)$ rather than $X(e^{j\omega})$. In reduced notation, the *frequency response* is $H(\omega)$:

$$H(\omega) = \sum \boldsymbol{h}(n)\, e^{-in\omega}.$$

This is the response at frequency ω to a unit input at that frequency. When the input at *each* frequency is $X(\omega) = 1$, the output at each frequency is $H(\omega)$. Those inputs are coming from an impulse (all $X(\omega)$ are equal). ***Then the frequency response is the transform of the impulse response:***

When the input is a unit impulse $\boldsymbol{x}(0) = 1$, the output is $\boldsymbol{y}(n) = \boldsymbol{h}(n)$.
When the input is a unit impulse $X(\omega) \equiv 1$, the output is $Y(\omega) = H(\omega)$.

Please note the difference between the vector $\boldsymbol{x} = (\ldots, 0, 0, 1, 0, 0, \ldots)$ in discrete time and the function $X(\omega) \equiv 1$ in continuous frequency. If the impulse is delayed to time $t = 1$, the input is $(\ldots, 0, 0, 0, 1, 0, \ldots)$. The output is equally delayed to $\boldsymbol{y}(n) = \boldsymbol{h}(n-1)$. This is ***time-invariance:*** *A delay of the input just produces a delay of the output.*

In the frequency domain, the delayed impulse is $X(\omega) = e^{-i\omega}$. The corresponding output is $Y(\omega) = e^{-i\omega} H(\omega)$. In $e^{j\omega}$ notation this is $Y\left(e^{j\omega}\right) = e^{-j\omega} H\left(e^{j\omega}\right)$. It is the transform of the delayed impuse response $\boldsymbol{h}(n-1)$.

Convolution by Hand

A good way to compute $y = h * x$ is to arrange it as an ordinary multiplication — but don't carry digits from one column to the next:

$$\begin{array}{rrrrr} & & x(2) & x(1) & x(0) \\ & & h(2) & h(1) & h(0) \\ \hline & & (2) & (1) & (0) \\ & (3) & (2) & (1) & \\ (4) & (3) & (2) & & \\ \hline y(4) & y(3) & y(2) & y(1) & y(0) \end{array} \qquad \begin{array}{rrrrrl} & & 3 & 2 & 4 & = x \\ & & 1 & 5 & 2 & = h \\ \hline & & 6 & 4 & 8 & \\ & 15 & 10 & 20 & & \\ 3 & 2 & 4 & & & \\ \hline 3 & 17 & 20 & 24 & 8 & = y \end{array}$$

The coefficients $x(n) = 4,\ 2,\ 3$ add to $X(0) = 9$. The sum of $h(n) = 2,\ 5,\ 1$ is $H(0) = 8$. Then notice that the sum for $y = h * x$ is $H(0)X(0) = 72$.

Another check is at $\omega = \pi$. The alternating sum $4 - 2 + 3$ gives $X(\pi) = 5$. Similarly $H(\pi) = -2$. Then necessarily $Y(\pi) = (5)(-2)$. This agrees with the alternating sum $3 - 17 + 20 - 24 + 8 = -10$.

Problem Set 1.1

1. Suppose the only nonzero components of x and h are $x(0) = 1, x(1) = 3$, and $h(0) = \frac{1}{2}$, $h(1) = \frac{1}{2}$. Compute the outputs $y(n)$. Verify in the frequency domain that $Y(\omega) = H(\omega)X(\omega)$.

2. What are the components $h(n)$ for the filter to become a simple advance? For any input vector $x(n)$, the output is $y(n) = x(n+1)$. Find $H(\omega)$ for this filter and verify that $Y(\omega) = H(\omega)X(\omega)$.

3. If the input filter vector x and the vector h are both causal, explain why the output y is also causal (meaning that $y(n) = 0$ for negative n). If h is causal and if $x(n) = 0$ for all $n < 8$, what can you conclude about $y(n)$?

4. If the output vector y is causal *whenever* the input x is causal, explain why the vector h must be causal.

5. If $\delta = (\ldots, 0, 0, 1, 0, 0, \ldots)$ is the unit impulse at time zero, show that convolution with any vector v leaves that vector unchanged. Translate this statement $v * \delta = v$ into the frequency domain.

6. What are the seven components of $h * h$, if $h(0) = h(1) = h(2) = h(3) = 1$ (all other $h(n) = 0$)? Use the long multiplication format in the text.

7. The long multiplication format corresponds to thinking of h and x as $h(0) + 10h(1) + 100h(2) + \cdots$ and $x(0) + 10x(1) + 100x(2) + \cdots$. The product begins with $h(0)x(0) + 10(\_\_) + 100(\_\_)$.

   This is the convolution rule $Y(z) = H(z)X(z)$ with $z = 10$.

8. Verify the convolution rule $Y = HX$ in the important special case when $h(1) = 1$ and all other $h(n) = 0$. Thus, $h = (\ldots, 0, 0, 0, 1, 0, \ldots)$.

   1. What is $y = h * x = h * (\ldots, x(-1), x(0), x(1), \ldots)$?
   2. What is $H(e^{j\omega})$, also written $H(\omega)$?
   3. In this special case the filter H represents a ____.
   4. Verify that $\sum y(n)\, e^{-jn\omega}$ agrees with $H(e^{j\omega})\, X(e^{j\omega})$.

9. Repeating the previous exercise k times shows that $Y = HX$ is still correct when H is a k-step delay: $\boldsymbol{h}(k) = 1$ and all other $\boldsymbol{h}(n) = 0$. The frequency response for this delay is $H =$____.

   An arbitrary vector $\boldsymbol{h}$ is a linear combination of delays (and also advances!). By linearity, $Y = HX$ is true in general.

10. **(Important)** A direct approach to the convolution rule $Y = HX$. What is the coefficient of z^{-n} in $\left(\sum \boldsymbol{h}(k)\, z^{-k}\right)\left(\sum \boldsymbol{x}(\ell)\, z^{-\ell}\right)$? Show that your answer agrees with $\sum \boldsymbol{h}(k)\, \boldsymbol{x}(n-k)$.

    The exponent $-n$ appears in H times X when $k + l$ equals ____.

11. The *autocorrelation* $\boldsymbol{p} = \boldsymbol{h} * \boldsymbol{h}^T$ is the convolution of $\boldsymbol{h}$ with its time reversal (or transpose): $\boldsymbol{h}^T(n) = \boldsymbol{h}(-n)$. Express the kth component $\boldsymbol{p}(k)$ as a sum of terms.

1.2 Lowpass Filter = Moving Average

We go forward with this introduction by studying the simplest lowpass filter. Its output at time $t = n$ is the average of the input $\boldsymbol{x}(n)$ at that time and the input $\boldsymbol{x}(n-1)$ at the previous time:

$$\boldsymbol{y}(n) = \tfrac{1}{2}\boldsymbol{x}(n) + \tfrac{1}{2}\boldsymbol{x}(n-1). \tag{1.4}$$

The filter coefficients are $\boldsymbol{h}(0) = \frac{1}{2}$ and $\boldsymbol{h}(1) = \frac{1}{2}$. Equation (1.4) fits the standard form $\sum \boldsymbol{h}(k)\boldsymbol{x}(n-k)$, with only two terms $k = 0$ and $k = 1$ in the sum. This is a convolution $\boldsymbol{y} = \boldsymbol{h} * \boldsymbol{x}$. It is a *moving average*, because the output averages the current component $\boldsymbol{x}(n)$ with the previous one. Old components drop away as the average moves forward with the time.

Suppose the input is the unit impulse $\boldsymbol{x} = (\ldots, 0, 0, 1, 0, 0, \ldots)$. Then there are only two nonzero components in the output. The input vector has $\boldsymbol{x}(0) = 1$; all other input components are zero. The output vector has $\boldsymbol{y}(0) = \frac{1}{2}$, from equation (1.4) with $n = 0$. It also has $\boldsymbol{y}(1) = \frac{1}{2}$, from equation (1.4) with $n = 1$. All other outputs, or moving averages, are zero. Thus the impulse response is the vector $\boldsymbol{y} = \left(\ldots, 0, 0, \frac{1}{2}, \frac{1}{2}, 0, \ldots\right)$. Its components agree with the filter coefficients $\boldsymbol{h}(n)$ as they should.

We want to see this filter as a linear time-invariant operator. It is a combination of two special operators, the "identity" which yields output = input and the "delay" whose output is the input one time earlier:

$$\text{averaging filter} = \tfrac{1}{2}(\text{identity}) + \tfrac{1}{2}(\text{delay}).$$

Every linear operator acting on the signal vector $\boldsymbol{x}$ can be represented by a matrix. Since the vectors are infinite, so are the matrices. Infinitely many components in $\boldsymbol{x}$ and $\boldsymbol{y}$ mean infinitely many entries in the filter matrix $\boldsymbol{H}$. The matrix has a special structure which you see immediately in $\boldsymbol{y} = \boldsymbol{H}\boldsymbol{x}$:

$$\begin{bmatrix} \cdot \\ \boldsymbol{y}(-1) \\ \boldsymbol{y}(0) \\ \boldsymbol{y}(1) \\ \cdot \end{bmatrix} = \begin{bmatrix} \cdot & & & & \\ \frac{1}{2} & \frac{1}{2} & & & \\ & \frac{1}{2} & \frac{1}{2} & & \\ & & \frac{1}{2} & \frac{1}{2} & \\ & & & \cdot & \cdot \end{bmatrix} \begin{bmatrix} \cdot \\ \boldsymbol{x}(-1) \\ \boldsymbol{x}(0) \\ \boldsymbol{x}(1) \\ \cdot \end{bmatrix}$$

The numbers $\frac{1}{2}$ on the main diagonal come from $\frac{1}{2}$ (identity). The numbers $\frac{1}{2}$ on the subdiagonal come from $\frac{1}{2}$ (delay). Substitute the unit impulse for $\boldsymbol{x}$, with $\boldsymbol{x}(0) = 1$ as its only nonzero component. Matrix multiplication produces the impulse response $\boldsymbol{y}$. This vector has components $\frac{1}{2}$

and $\frac{1}{2}$. The response y is the filter vector h, in the middle column of the matrix, when x is the unit impulse.

This is a first occasion to see a filter as a ***constant-diagonal matrix***. The coefficient $h(0)$ appears constantly down the main diagonal. It represents $h(0)$ times the identity matrix and it yields $h(0)\,x(n)$. The coefficient $h(1)$ appears down the first subdiagonal, to represent $h(1)$ times a delay and to yield $h(1)\,x(n-1)$. If there is a coefficient $h(2)$ down the next diagonal, it multiplies a two-step delay and yields $h(2)\,x(n-2)$. The total output $y(n)$ is the sum of these special outputs:

$$\begin{aligned} y(n) &= h(0)\,x(n) + h(1)\,x(n-1) + h(2)\,x(n-2) + \cdots \\ &= \sum_k h(k)\,x(n-k). \end{aligned}$$

The reader notices that $h(-1)$ is not being allowed. We are dealing with ***causal filters***; the output cannot come earlier than the input. This makes $h(n) = 0$ for negative n, and it makes the filter matrix *lower triangular.*

Our example has only a finite number (two) of nonzero filter coefficients $h(n)$. The filter has a ***finite impulse response***. It is an "FIR filter". For large n, the coefficients $h(n)$ that give distant responses to the unit impulse are all zero.

To repeat: A causal FIR filter has $h(n) = 0$ for all negative n and for large positive n. The matrix is banded and lower triangular. Only a finite number of coefficients $h(0), h(1), \ldots, h(N)$ can be nonzero. The filter has $N+1$ "taps". The matrix has bandwidth N; all other diagonals contain zeros. We concentrate almost exclusively on causal FIR filters.

Frequency Response

To find the frequency response to the filter, we change the input vector. Instead of an impulse, which combines all frequencies, the vector x will have pure frequency ω. Its components are

$$x(n) = e^{in\omega}. \tag{1.5}$$

This vector extends over all time, $-\infty < n < \infty$. Its special feature is that *the output vector y is a multiple (depending on ω) of the input vector x.* A linear time-invariant operator, which is represented by a constant-diagonal matrix, has a pure frequency response to a pure frequency input. For our moving average this response is

$$\begin{aligned} y(n) &= \tfrac{1}{2}x(n) + \tfrac{1}{2}x(n-1) \\ &= \tfrac{1}{2}e^{in\omega} + \tfrac{1}{2}e^{i(n-1)\omega} \\ &= \left(\tfrac{1}{2} + \tfrac{1}{2}e^{-i\omega}\right)e^{in\omega}. \end{aligned} \tag{1.6}$$

You see in parentheses the ***frequency response function*** $H(\omega) = \frac{1}{2} + \frac{1}{2}e^{-i\omega}$. Notice that $H = \frac{1}{2} + \frac{1}{2} = 1$ when $\omega = 0$. At zero frequency (direct current) the signal $x = (\ldots, 1, 1, 1, \ldots)$ comes out unchanged as $y = (\ldots, 1, 1, 1, \ldots)$. The low frequencies, near $\omega = 0$, have $H(\omega)$ near 1. Thus the name "lowpass filter".

There will be a similar $H(\omega)$ for other filters. When the filter coefficients are $h(n)$, the response to $x(n) = e^{in\omega}$ is $y(n) = h(0)e^{in\omega} + h(1)e^{i(n-1)\omega} + \cdots$. The input is multiplied by the frequency response $H(\omega)$:

$$\begin{aligned} H(\omega) &= h(0) + h(1)\,e^{-i\omega} + h(2)\,e^{-2i\omega} + \cdots \\ &= \sum h(n)\,e^{-in\omega}. \end{aligned} \tag{1.7}$$

This function is always periodic: $H(\omega+2\pi) = H(\omega)$. When we add 2π to the frequency ω, we add $2\pi n$ to the angle $n\omega$. The cosine, sine and complex exponential $e^{-in\omega} = \cos n\omega - i \sin n\omega$ are not changed.

Note that the response function $H(\omega)$ involves complex numbers. We strongly prefer $e^{i\omega}$ or $e^{j\omega}$ to its rectangular form $\cos\omega + i\sin\omega$. Separate formulas for the real and imaginary parts are much more complicated than a single formula for $H(\omega)$. But a single graph cannot so easily represent this complex function. One way to do it is to plot the magnitude $|H(\omega)|$ separately from the phase angle $\phi(\omega)$, recalling that

$$H(\omega) = |H(\omega)|\, e^{i\phi(\omega)}.$$

Our example has $H(\omega) = \frac{1}{2} + \frac{1}{2}e^{-i\omega}$. We factor out $e^{-i\omega/2}$ to leave a symmetric quantity $\frac{1}{2}\left(e^{i\omega/2} + e^{-i\omega/2}\right)$. This quantity is a perfect cosine:

$$H(\omega) = \left(\cos\frac{\omega}{2}\right) e^{-i\omega/2}. \tag{1.8}$$

This displays the magnitude and also the phase:

$$|H(\omega)| = \cos\frac{\omega}{2} \quad \text{and} \quad \phi(\omega) = -\frac{\omega}{2}. \tag{1.9}$$

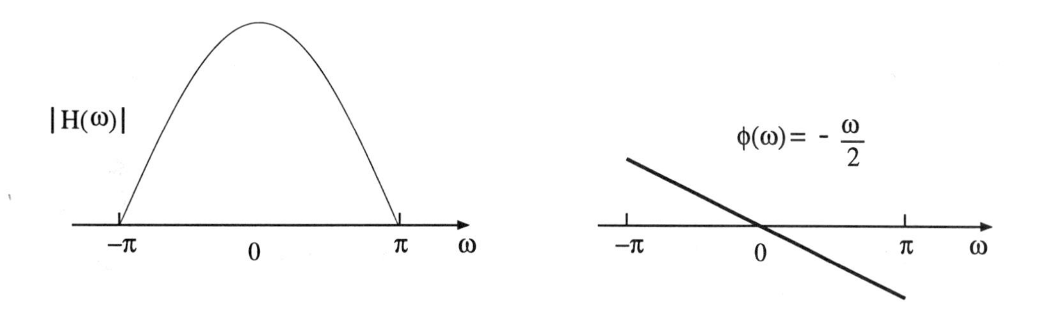

Figure 1.2: The magnitude $|H(\omega)| = \cos\left(\frac{\omega}{2}\right)$ and the phase of $H(\omega)$.

Figure 1.2 shows the plot of the magnitude $|H(\omega)|$ against the frequency ω. The cosine of $\frac{\omega}{2}$ drops to zero at $\omega = \pi$. This high frequency is wiped out, when the filter takes a moving average. In the time domain, $\omega = \pi$ corresponds to the input vector $\boldsymbol{x}(n) = e^{i\pi n}$ with components $(-1)^n$. This input vector is

$$\boldsymbol{x} = (\ldots, 1, -1, 1, -1, 1, \ldots).$$

The moving average of these components is constantly zero! This confirms $H(\pi) = 0$ as the correct response to the frequency $\omega = \pi$.

This is a lowpass filter. The lowest frequency $\omega = 0$, which is the DC term (direct current), is exactly preserved because $\cos 0 = 1$. The input vector $(\ldots, 1, 1, 1, \ldots)$ is equal to its moving average.

The phase $\phi(\omega)$ is the angle from the horizontal when the complex number $H(\omega)$ is plotted in the complex plane. For this particular response, those points $H(\omega) = \frac{1}{2} + \frac{1}{2}e^{-i\omega}$ lie along a circle. The constant term $\frac{1}{2}$ is the center. Then $\frac{1}{2}e^{-i\omega}$ produces a circle of radius $\frac{1}{2}$ around that center. Figure 1.3 shows this "*Nyquist diagram*".

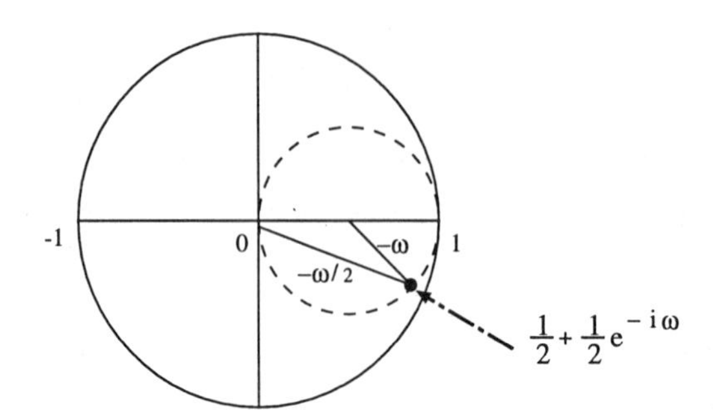

Figure 1.3: Nyquist diagram of the points $H(\omega)$ in the complex plane.

The graph of $\phi(\omega) = -\frac{\omega}{2}$ is a straight line. This is an example of "*linear phase*", an important property that some special filters possess. It reflects the fact that the filter coefficients $\frac{1}{2}$ and $\frac{1}{2}$ are symmetric. Reverse the order of coefficients and nothing changes.

Note that if the filter coefficients were symmetric *around zero*, so that $\boldsymbol{h}(-1) = \boldsymbol{h}(1)$ and $\boldsymbol{h}(-2) = \boldsymbol{h}(2)$, the frequency response would be *real*:

$$\begin{aligned} H(\omega) &= \boldsymbol{h}(0) + \boldsymbol{h}(1)\left(e^{i\omega} + e^{-i\omega}\right) + \cdots && \text{(symmetric coefficients)} \\ &= \boldsymbol{h}(0) + \boldsymbol{h}(1)\,(2\cos\,\omega) + \cdots && \text{(real response function)} \end{aligned}$$

In this case the phase angle is $\phi = 0$. The response has "zero phase". Similarly, coefficients that are *anti*symmetric around zero produce a pure imaginary $H(\omega)$. The phase is $\frac{\pi}{2}$ or $-\frac{\pi}{2}$. If $\boldsymbol{h}(-n) = -\boldsymbol{h}(n)$ then $\boldsymbol{h}(0) = 0$:

$$\begin{aligned} H(\omega) &= \boldsymbol{h}(1)\,(-e^{i\omega} + e^{-i\omega}) + \boldsymbol{h}(2)\left(-e^{2i\omega} + e^{-2i\omega}\right) + \cdots && \text{(antisymmetry)} \\ &= -2i\;[\boldsymbol{h}(1)\sin\omega + \boldsymbol{h}(2)\sin 2\omega + \cdots] && \text{(imaginary } H(\omega)\text{)} \end{aligned}$$

Zero phase is ruled out for a causal filter. The coefficients can be symmetric or antisymmetric, but not around $n = 0$. Causal filters have *linear* phase when their coefficients are symmetric or antisymmetric around the central coefficient:

$$\textbf{Linear phase} \quad \begin{aligned} \boldsymbol{h}(k) &= \boldsymbol{h}(N-k) && \text{(symmetry)} \\ \boldsymbol{h}(k) &= -\boldsymbol{h}(N-k) && \text{(antisymmetry)} \end{aligned}$$

Moving the center of the symmetry from 0 to $N/2$ produces a factor $e^{-iN\omega/2}$ in $H(\omega)$. This means a ***linear*** term $-N\omega/2$ in the phase. The magnitude $|H(\omega)|$ is an even (symmetric) function because $H(-\omega)$ is the complex conjugate of $H(\omega)$. The graph of $|H(\omega)|$ for $0 \le \omega \le \pi$ displays complete information.

The exercises ask you to compute $|H(\omega)|^2 = \frac{1}{2}(1 + \cos\,\omega)$. Then a trigonometric identity produces our special form $\cos\frac{\omega}{2}$:

$$|H(\omega)|^2 = \tfrac{1}{2}\,(1 + \cos\,\omega) = \cos^2\frac{\omega}{2} \quad \text{so that} \quad |H(\omega)| = \left|\cos\frac{\omega}{2}\right|.$$

For other filters, $|H(\omega)|^2$ is a cosine series but its square root $|H(\omega)|$ has no simple formula.

Problem Set 1.2

1. The magnitude squared is $H(\omega)$ times its complex conjugate $\overline{H(\omega)}$. Show that $|H(\omega)|^2 = \frac{1}{2}(1 + \cos\omega)$ for the moving average filter.

2. Obtain the same result from $|H(\omega)|^2 =$ (real part)^2+ (imaginary part)^2, with $H(\omega) = \frac{1}{2} + \frac{1}{2}(\cos\omega - i\sin\omega)$.

3. Why is the formula $|H(\omega)| = \cos\frac{\omega}{2}$ wrong beyond the frequency $\omega = \pi$? Draw the graph of $|H(\omega)|$ from -2π to 2π.

4. In the complex plane, draw $\tan\phi(\omega)$ as the ratio of the imaginary part $\operatorname{Im} H(\omega)$ to the real part $\operatorname{Re} H(\omega)$. Simplify this ratio to $-\tan\frac{\omega}{2}$.

Solution. The phase $\phi(\omega)$ is the angle from the horizontal, so the tangent of $\phi(\omega)$ is the ratio of imaginary part to real part:

$$\tan\phi(\omega) = \frac{\operatorname{Im} H(\omega)}{\operatorname{Re} H(\omega)} = \frac{-\frac{1}{2}\sin\omega}{\frac{1}{2} + \frac{1}{2}\cos\omega}.$$

Again trigonometric identities make this unusually simple:

$$\tan\phi(\omega) = \frac{-\sin\frac{\omega}{2}\cos\frac{\omega}{2}}{\cos^2\frac{\omega}{2}} = -\tan\frac{\omega}{2} \quad \text{and} \quad \phi(\omega) = -\frac{\omega}{2}.$$

5. Find the frequency response $H(\omega)$ and its magnitude and phase, for the 3-point moving average filter with $h(0) = h(1) = h(2) = \frac{1}{3}$. Does it have zero phase, constant phase, or linear phase?

6. What infinite matrix $\boldsymbol{H}$ represents the filter with $\boldsymbol{h}(0) = \boldsymbol{h}(1) = \boldsymbol{h}(2) = \frac{1}{3}$? Find two input vectors $\boldsymbol{x}(n)$ for which the output is $\boldsymbol{H}\boldsymbol{x} = \boldsymbol{0}$.

Solution. $\boldsymbol{x} = (\ldots, 2, -1, -1, 2, -1, -1, \ldots)$ and $\boldsymbol{x}' =$ delay of $\boldsymbol{x}$.

7. What is the phase $\phi(\omega)$ of a symmetric filter with $\boldsymbol{h}(k) = \boldsymbol{h}(8 - k)$?

8. What constant-diagonal matrix represents the anticausal filter $\boldsymbol{H}'$ with coefficients $\boldsymbol{h}'(0) = \frac{1}{2}$ and $\boldsymbol{h}'(-1) = \frac{1}{2}$? What matrix represents the symmetric filter $\boldsymbol{F} = \boldsymbol{H}'\boldsymbol{H}$ and what is the frequency response of $\boldsymbol{F}$?

9. Find causal filters whose magnitude responses are $|H(\omega)| = |\cos\omega|$ and $|H(\omega)| = (\cos\frac{\omega}{2})^2$. Are they unique?

10. Iterate the averaging filter four times to get $\boldsymbol{K} = \boldsymbol{H}^4$. What is $K(\omega)$ and what is the impulse response $\boldsymbol{k}(n)$?

11. Why is an antisymmetric filter, with $\boldsymbol{h}(k) = -\boldsymbol{h}(N - k)$, never lowpass?

12. Consider an averaging filter with four coefficients $\boldsymbol{h}(0) = \boldsymbol{h}(1) = \boldsymbol{h}(2) = \boldsymbol{h}(3) = \frac{1}{4}$ and the input $\boldsymbol{x}(n) = (-1)^n$. What is the output $\boldsymbol{y}(n) = \boldsymbol{x}(n) * \boldsymbol{h}(n)$? Without computing $H(\omega)$, explain why $\boldsymbol{h}(n)$ is lowpass.

13. Let $H(z)$ be a lowpass filter with $N + 1$ coefficients and let $G(z) = z^{-N} H(z^{-1})$. Find $\boldsymbol{g}(n)$ in terms of $\boldsymbol{h}(n)$. What is the phase of $G(\omega)$ if H is symmetric? Is $G(z)$ lowpass or highpass?

1.3 Highpass Filter = Moving Difference

A lowpass filter takes "averages". It smoothes out the bumps in the signal. A bump is a high-frequency component, which the lowpass filter reduces or removes. The response is small or zero near the highest discrete-time frequency $\omega = \pi$.

A *highpass filter takes "differences"*. It picks out the bumps in the signal. The smooth parts are low-frequency components, which the highpass filter reduces or removes. Now the frequency response is small or zero for frequencies near $\omega = 0$ (which is direct current and has no bumps).

The lowpass filter that outputs the moving average $\frac{1}{2}(\boldsymbol{x}(n) + \boldsymbol{x}(n-1))$ has a twin (or mirror) filter. This is the highpass filter that computes *moving differences*:

$$\textbf{\textit{Highpass:}} \quad y(n) = \tfrac{1}{2}x(n) - \tfrac{1}{2}x(n-1). \tag{1.10}$$

The new filter coefficients are $\boldsymbol{h}(0) = \frac{1}{2}$ and $\boldsymbol{h}(1) = -\frac{1}{2}$. Equation (1.10) is a convolution $\boldsymbol{y} = \boldsymbol{h} * \boldsymbol{x}$, and the vector $\boldsymbol{h}$ for this highpass filter is

$$\boldsymbol{h} = \left(\ldots, 0, 0, \tfrac{1}{2}, -\tfrac{1}{2}, 0, \ldots\right). \tag{1.11}$$

This $\boldsymbol{h}$ is exactly the response to the impulse $\boldsymbol{x} = (\ldots, 0, 0, 1, 0, 0, \ldots)$. At time zero, the difference $\frac{1}{2}(\boldsymbol{x}(0) - \boldsymbol{x}(-1))$ is $\frac{1}{2}(1-0)$. The next difference $\frac{1}{2}(\boldsymbol{x}(1) - \boldsymbol{x}(0))$ is $\frac{1}{2}(0-1)$. Those numbers $\frac{1}{2}$ and $-\frac{1}{2}$ are the coefficients in $\boldsymbol{h}$.

This unit impulse at time zero we will denote by $\boldsymbol{\delta}$. In the language of convolution, we are saying that $\boldsymbol{\delta} = (\ldots, 0, 0, 1, 0, 0, \ldots)$ always yields

$$\boldsymbol{h} * \boldsymbol{\delta} = \boldsymbol{h}. \tag{1.12}$$

In the language of matrices, multiplying any matrix times the special column vector $\boldsymbol{\delta}$ always picks out the *zeroth column of the matrix*. The matrix for our highpass filter does have $\frac{1}{2}$ and $-\frac{1}{2}$ in its zeroth column:

$$\begin{bmatrix} \cdot \\ y(-1) \\ y(0) \\ y(1) \\ \cdot \end{bmatrix} = \begin{bmatrix} \cdot & \cdot & & & \\ -\frac{1}{2} & \frac{1}{2} & & & \\ & -\frac{1}{2} & \frac{1}{2} & & \\ & & -\frac{1}{2} & \frac{1}{2} & \\ & & & \cdot & \cdot \end{bmatrix} \begin{bmatrix} \cdot \\ x(-1) \\ x(0) \\ x(1) \\ \cdot \end{bmatrix}. \tag{1.13}$$

This is again a constant-diagonal matrix because the filter is again time-invariant. (We define and discuss time-invariance in the next chapter.) The main diagonal entries produce $\frac{1}{2}$(identity). The subdiagonal entries give $-\frac{1}{2}$(delay). The filter as a whole is

$$\text{highpass filter} = \tfrac{1}{2}\,(\text{identity}) - \tfrac{1}{2}\,(\text{delay}) = \tfrac{1}{2}\mathbf{I} - \tfrac{1}{2}\mathbf{S}. \tag{1.14}$$

This is another causal FIR filter with two taps, but its frequency response is completely different from the lowpass function $H_0(\omega) = \frac{1}{2}\left(1 + e^{-i\omega}\right)$.

Frequency Response

At frequency ω, the input vector is $\boldsymbol{x}(n) = e^{in\omega}$. The highpass output is

$$\begin{aligned} y(n) &= \tfrac{1}{2}\, e^{in\omega} - \tfrac{1}{2}\, e^{i(n-1)\omega} \\ &= \left(\tfrac{1}{2} - \tfrac{1}{2}\, e^{-i\omega}\right) e^{in\omega} \\ &= H_1(\omega)\, e^{in\omega}. \end{aligned} \tag{1.15}$$

This quantity $H_1(\omega) = \frac{1}{2} - \frac{1}{2}\, e^{-i\omega}$ is the highpass response. We want to graph it and compare it with $H_0(\omega)$ — the lowpass response. *We are introducing the subscripts* 0 *and* 1 *for lowpass and highpass.*

As before, we take out a factor $e^{-i\omega/2}$. This leaves a sine not a cosine:

$$H_1(\omega) = \tfrac{1}{2}\left(e^{i\omega/2} - e^{-i\omega/2}\right) e^{-i\omega/2} = \sin\left(\frac{\omega}{2}\right) i e^{-i\omega/2}. \tag{1.16}$$

The magnitude is $|H_1(\omega)| = \left|\sin \frac{\omega}{2}\right|$. Since the sine function can be negative, we must take its absolute value.

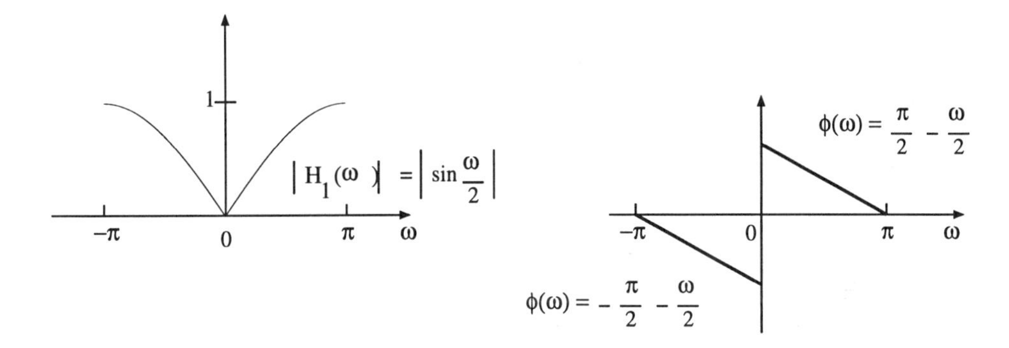

Figure 1.4: The magnitude and phase of the highpass filter $\frac{1}{2} - \frac{1}{2}\, e^{-i\omega}$.

The graph of $|H_1(\omega)|$ is in Figure 1.4a. It shows zero response to direct current, because $\sin 0 = 0$. It shows unit response at the highest frequency $\omega = \pi$, because $\sin \frac{\pi}{2} = 1$. In the time domain, these numbers come from taking differences of the lowest and highest frequency inputs:

$$\boldsymbol{x}_{\text{low}} = (\ldots, 1, 1, 1, 1, 1, \ldots) \quad \text{and} \quad \boldsymbol{x}_{\text{high}} = (\ldots, 1, -1, 1, -1, 1, \ldots).$$

The response to $\boldsymbol{x}_{\text{low}}$ is $0\,\boldsymbol{x}_{\text{low}}$. The response to $\boldsymbol{x}_{\text{high}}$ is $1\,\boldsymbol{x}_{\text{high}}$. The first vector has no bumps and the second vector is all bumps.

The phase factor in $H_1(\omega)$ is a little tricky. When the magnitude $\left|\sin \frac{\omega}{2}\right|$ is removed, we are left with a sign change at $\omega = 0$:

$$e^{i\phi(\omega)} = \begin{cases} -i\, e^{-i\omega/2} & \text{for} \quad -\pi < \omega < 0 \\ +i\, e^{-i\omega/2} & \text{for} \quad 0 < \omega < \pi. \end{cases}$$

Changing i to $e^{i\pi/2}$, we see a *discontinuity in the phase.* Figure 1.4b shows how $\phi(\omega)$ jumps from $-\frac{\pi}{2}$ to $+\frac{\pi}{2}$ at $\omega = 0$. At other points the graph is linear, so we turn a blind eye to this discontinuity and say that the filter is still *linear phase.* It is the zero at $\omega = 0$ that causes this discontinuity in phase.

Invertibility and Noninvertibility

In any reasonable sense, the averaging and differencing filters are *not invertible.* The constant signal $\boldsymbol{x} = (\dots, 1, 1, 1, 1, 1, \dots)$ is wiped out by $\boldsymbol{H}_1$. There cannot exist a filter $\boldsymbol{H}_1^{-1}$ that recovers this $\boldsymbol{x}$ from the zero vector. A linear operator cannot raise the dead—it only recovers $\mathbf{0}$ from $\mathbf{0}$. *If a filter $\boldsymbol{H}$ has an inverse,* the only vector in its nullspace is that zero vector:

$$\text{If } \boldsymbol{H} \text{ is invertible and } \boldsymbol{H}\boldsymbol{x} = \mathbf{0}, \text{ then } \boldsymbol{H}^{-1}\boldsymbol{H}\boldsymbol{x} = \mathbf{0} \text{ and thus } \boldsymbol{x} = \mathbf{0}.$$

The frequency response of an invertible filter must have $H(\omega) \neq 0$ at all frequencies. Our filters are not invertible because $H_0(\pi) = 0$ and $H_1(0) = 0$.

The lowpass filter wipes out the alternating signal $\boldsymbol{x} = (\dots, 1, -1, 1, -1, \dots)$. We cannot recover this $\boldsymbol{x}$ from zero output.

Note. When the inverse filter exists, it has frequency response $1/H(\omega)$. Multiplication will recover the input, because $H(\omega)/H(\omega) = 1$. We are safe as long as $H(\omega)$ is nonzero.

Suppose we attempt this inversion for the moving average. It is doomed to failure because $H_0(\pi) = 0$, but we can compute the filter coefficients that come from $1/H_0(\omega)$:

$$\frac{1}{\frac{1}{2} + \frac{1}{2}e^{-i\omega}} = 2\left(1 - e^{-i\omega} + e^{-2i\omega} - e^{-3i\omega} + \cdots\right). \tag{1.17}$$

Those coefficients $2, -2, 2, -2, \dots$ go down the diagonals of the inverse filter matrix:

$$\boldsymbol{H}^{-1}\boldsymbol{H} = \begin{bmatrix} \cdot & & & & \\ -2 & 2 & & & \\ 2 & -2 & 2 & & \\ -2 & 2 & -2 & 2 & \\ \cdot & \cdot & \cdot & \cdot & \cdot \end{bmatrix} \begin{bmatrix} \cdot & & & & \\ \frac{1}{2} & \frac{1}{2} & & & \\ & \frac{1}{2} & \frac{1}{2} & & \\ & & \frac{1}{2} & \frac{1}{2} & \\ & & & \frac{1}{2} & \frac{1}{2} \end{bmatrix} = \boldsymbol{I}.$$

We seem to have found an inverse, but it is not a legal filter. *This $\boldsymbol{H}^{-1}$ is not stable.* The bounded input $\boldsymbol{y} = (\dots, 1, -1, 1, -1, 1, \dots)$ would produce an unbounded output $\boldsymbol{H}^{-1}\boldsymbol{y}$. The series expansion in (1.17) breaks down completely at $\omega = \pi$, where $1 + e^{-i\pi} = 0$:

$$\frac{1}{0} = \frac{1}{\frac{1}{2} + \frac{1}{2}e^{-i\pi}} = 2(1 + 1 + 1 + 1 + \dots).$$

But also note: If the input signal contains no frequencies near $\omega = \pi$, then we *could* reconstruct $\boldsymbol{x}$ from $\boldsymbol{y}$. The output $\boldsymbol{y}$ would also have no high frequencies. Dividing by $H(\omega)$ would not involve dividing by zero.

The lowpass filter with coefficients $\boldsymbol{h}(0) = \frac{2}{3}$ and $\boldsymbol{h}(1) = \frac{1}{3}$ is safely invertible. Its frequency response $\frac{2}{3} + \frac{1}{3}e^{-i\omega}$ is never zero. The response at $\omega = 0$ is $\frac{2}{3} + \frac{1}{3} = 1$. The response at $\omega = \pi$ is $\frac{2}{3} - \frac{1}{3} = \frac{1}{3}$. The convolution with $\boldsymbol{h} = (\dots, 0, \frac{2}{3}, \frac{1}{3}, 0, \dots)$ can be undone by *deconvolution.* To find the deconvolution coefficients, which enter the inverse filter, divide by $H(\omega)$:

$$\frac{1}{\frac{2}{3} + \frac{1}{3}e^{-i\omega}} = \frac{3}{2}\left(1 - \frac{1}{2}e^{-i\omega} + \frac{1}{4}e^{-2i\omega} - \frac{1}{8}e^{-3i\omega} + \dots\right). \tag{1.18}$$

Those coefficients $\frac{3}{2}, -\frac{3}{4}, \frac{3}{8}, -\frac{3}{16}, \dots$ go on the diagonals of the inverse filter matrix. It is an IIR filter, with infinitely many coefficients.

The main point is analysis in the frequency domain. Dividing by $\frac{2}{3}+\frac{1}{3}e^{-i\omega}$ inverts the filter (and the constant-diagonal infinite matrix) for deconvolution.

Final remark. The inverse filter is very rarely FIR, because $1/H(\omega)$ is not a polynomial. To invert a finite length filter with a finite length filter, we need to go to a *filter bank.* That comes next.

Problem Set 1.3

1. Can a symmetric filter, with $\boldsymbol{h}(k)=\boldsymbol{h}(N-k)$, be a highpass filter?

2. Draw the graph for $|\omega| \le \pi$ of "ideal" lowpass and highpass filters $H_0(\omega)$ and $H_1(\omega)$ with $H_0(\omega)+H_1(\omega)\equiv 1$. Why don't we use these filters exclusively?

3. Which of the following filters are invertible? Find the inverse filters:

$$\begin{array}{ll}(a) & \boldsymbol{h}(0)=\frac{2}{3} \text{ and } \boldsymbol{h}(1)=-\frac{1}{3}\\ (b) & \boldsymbol{h}(0)=2 \text{ and } \boldsymbol{h}(2)=1\\ (c) & \boldsymbol{h}(n)=\frac{1}{n!}\ \ (n=0,1,2,\ldots)\end{array}$$

4. Invent a highpass filter K with three or four taps (coefficients) that is better than the moving difference H_1: the goal is

$$|K(\omega)| < |H_1(\omega)| \text{ for } 0<|\omega|<\frac{\pi}{2}$$

and

$$|H_1(\omega)| < |K(\omega)| < 1 \text{ for } \frac{\pi}{2}<|\omega|<\pi.$$

5. Find all input signals $\boldsymbol{x}(n)$ that are cut in half by the moving difference $\boldsymbol{H}_1$, so that $\boldsymbol{H}_1\boldsymbol{x}(n)=\frac{1}{2}\boldsymbol{x}(n)$. Answer in the frequency domain.

6. ***Important*** If $H_0(\omega)$ is the response of a lowpass filter, what is the response $H_1(\omega)$ of a corresponding highpass filter? If $\boldsymbol{h}(0),\ldots,\boldsymbol{h}(N)$ are the coefficients of H_0, what are the coefficients of your H_1?

7. Let $H(z)$ be symmetric lowpass with $2K+1$ coefficients. What is the phase of $G(z)=z^{-K}-H(z)$? Sketch the amplitude response if $H(\omega)$ is $\frac{1}{3}$-band (near 1 for $|\omega|\le\frac{\pi}{3}$). Is G lowpass or highpass?

8. A simple highpass filter design is $G(z)=H(-z)$. What is the relation between $\boldsymbol{g}(n)$ and $\boldsymbol{h}(n)$? With $H(\omega)$ as in Problem 7, sketch $G(\omega)$.

9. If $G(z)=H(-z^{-1})$ find $\boldsymbol{g}(n)$ in terms of $\boldsymbol{h}(n)$. If $H(z)$ is highpass show that $G(z)$ is lowpass.

1.4 Filter Bank = Lowpass and Highpass

Separately, the lowpass and highpass filters are not invertible. $\boldsymbol{H}_0$ removes the highest frequency $\omega=\pi$, and $\boldsymbol{H}_1$ removes the lowest frequency $\omega=0$. Together, these filters do something very desirable. *They separate the signal into frequency bands.* The filtered output $\boldsymbol{H}_0\boldsymbol{x}$ is weighted towards low frequencies, and $\boldsymbol{H}_1\boldsymbol{x}$ is in some way the complement. The cutoff is not sharp, because these filters are so crude. But they make up the beginning of a *filter bank.*

One difficulty: the signal length has doubled. If the input $\boldsymbol{x}$ is nonzero over a time T, so are the outputs from both filters. If the first nonzero input is $\boldsymbol{x}(0)$ and the last is $\boldsymbol{x}(T)$, then $\boldsymbol{H}_0\boldsymbol{x}$ and $\boldsymbol{H}_1\boldsymbol{x}$ end at time $T+1$ (because of the final average and difference). One extra component is no problem, but doubling the storage by keeping two full-length outputs is unacceptable.

The solution is to downsample (or decimate).

Downsampling

We can keep *half* of $\boldsymbol{H}_0\boldsymbol{x}$ and $\boldsymbol{H}_1\boldsymbol{x}$, and still recover $\boldsymbol{x}$. This is an essential part of the filter bank, to *downsample* the outputs from the separate filters. We shall save only the *even-numbered components* of the two outputs. The odd-numbered components are removed:

$$(\downarrow 2)\boldsymbol{y} = (\ldots, \boldsymbol{y}(-4), \boldsymbol{y}(-2), \boldsymbol{y}(0), \boldsymbol{y}(2), \boldsymbol{y}(4), \ldots). \tag{1.19}$$

The symbol $\downarrow 2$ indicates downsampling or decimation. This is a linear operation (of course). But it is not time-invariant and it is not invertible. We study it closely in Chapter 3, and already this first filter bank indicates how it works in practice.

One note about normalization. To compensate for losing half the components in $(\downarrow 2)$, we multiply the surviving $\boldsymbol{y}(2n)$ by $\sqrt{2}$. This normalizing factor (explained below) is usually included with the filter bank, so that

$$\begin{array}{lll} \textit{lowpass:} & H_0(\omega) \text{ changes to } & C(\omega) = \sqrt{2}\, H_0(\omega) \\ \textit{highpass:} & H_1(\omega) \text{ changes to } & D(\omega) = \sqrt{2}\, H_1(\omega). \end{array}$$

In the averaging filter, the coefficients increase to $\frac{\sqrt{2}}{2}$. We keep the notation $\boldsymbol{h}(0), \boldsymbol{h}(1)$ for the original coefficients, and we introduce $\boldsymbol{c}(0), \boldsymbol{c}(1)$ for the new (renormalized) coefficients of $\boldsymbol{C}$:

$$\boldsymbol{c}(0) = \boldsymbol{c}(1) = \frac{\sqrt{2}}{2} \quad \text{(which is } \frac{1}{\sqrt{2}}\text{)}.$$

Similarly the highpass coefficients $\frac{1}{2}$ and $-\frac{1}{2}$ are multiplied by $\sqrt{2}$, in $\boldsymbol{D}$:

$$\boldsymbol{d}(0) = \frac{\sqrt{2}}{2} = \frac{1}{\sqrt{2}} \quad \text{and} \quad \boldsymbol{d}(1) = -\frac{\sqrt{2}}{2} = -\frac{1}{\sqrt{2}}.$$

Systematically in this book, we will write $\boldsymbol{C}$ and $\boldsymbol{D}$ for $\sqrt{2}\boldsymbol{H}_0$ and $\sqrt{2}\boldsymbol{H}_1$. The response at frequency $\omega = 0$ is $C = \sqrt{2}$ rather than $H_0 = 1$.

Decimation Filters in the Time Domain

Downsampling follows the filter $\boldsymbol{C}$, in operating on $\boldsymbol{x}$. These two steps, filtering and decimation, can be done with *one matrix* $\boldsymbol{L}$. Decimation removes the odd-numbered components. To obtain $(\downarrow 2)\,\boldsymbol{Cx}$ in one step, *remove the odd-numbered rows of the filter matrix* $\boldsymbol{C}$. The combination of filtering by $\boldsymbol{C}$ and decimation by $(\downarrow 2)$ is represented by a *rectangular matrix* $\boldsymbol{L}$ that no longer has constant diagonals. It has 1×2 blocks:

$$\boldsymbol{L} = (\downarrow 2)\,\boldsymbol{C} = \begin{bmatrix} \frac{1}{\sqrt{2}} & \frac{1}{\sqrt{2}} & & & \\ & & \frac{1}{\sqrt{2}} & \frac{1}{\sqrt{2}} & \\ & & & & \ddots \end{bmatrix}.$$

The entries are $\boldsymbol{c}(0)$ and $\boldsymbol{c}(1)$ but half the rows of $\boldsymbol{C}$ have disappeared. Similarly the decimated highpass filter is represented by a rectangular matrix $\boldsymbol{B} = (\downarrow 2)\boldsymbol{D}$. Removing half the rows of $\boldsymbol{D}$ leaves the matrix $\boldsymbol{B}$ with a ***double-shift***:

$$\boldsymbol{B} = (\downarrow 2)\,\boldsymbol{D} = \begin{bmatrix} -\frac{1}{\sqrt{2}} & \frac{1}{\sqrt{2}} & & & \\ & & -\frac{1}{\sqrt{2}} & \frac{1}{\sqrt{2}} & \\ & & & & \ddots \end{bmatrix}.$$

When the lowpass $\boldsymbol{L}$ and the highpass $\boldsymbol{B}$ go into one matrix, you will see why the normalization by $\sqrt{2}$ is desirable. The rectangular $\boldsymbol{L}$ and $\boldsymbol{B}$ fit into a square matrix:

$$\begin{bmatrix} (\downarrow 2)\boldsymbol{C} \\ (\downarrow 2)\boldsymbol{D} \end{bmatrix} = \begin{bmatrix} \boldsymbol{L} \\ \boldsymbol{B} \end{bmatrix} = \frac{1}{\sqrt{2}} \begin{bmatrix} 1 & 1 & & & \\ & & 1 & 1 & \\ & & & & \cdot\ \cdot \\ -1 & 1 & & & \\ & & -1 & 1 & \\ & & & & \cdot\ \cdot \end{bmatrix}.$$

This matrix represents the whole analysis bank. It executes the lowpass channel and the highpass channel (both decimated). All rows are unit vectors (because of the division by $\sqrt{2}$). Those row vectors are mutually orthogonal. At the same time, the columns are also *orthogonal unit vectors*.

The combined square matrix is invertible. ***The inverse is the transpose***:

$$\begin{bmatrix} \boldsymbol{L} \\ \boldsymbol{B} \end{bmatrix}^{-1} = \begin{bmatrix} \boldsymbol{L}^T & \boldsymbol{B}^T \end{bmatrix} = \frac{1}{\sqrt{2}} \begin{bmatrix} 1 & & & -1 & & \\ 1 & & & 1 & & \\ & 1 & & & -1 & \\ & 1 & & & 1 & \\ & & \vdots & & & \vdots \end{bmatrix}.$$

Multiplying in either order yields the identity matrix. *The second matrix* $\begin{bmatrix} \boldsymbol{L}^T & \boldsymbol{B}^T \end{bmatrix}$ *represents the synthesis bank.* This is an ***orthogonal filter bank***, because *inverse = transpose.* We pause to summarize what you have seen in this example.

The channels $\boldsymbol{L} = (\downarrow 2)\boldsymbol{C}$ *and* $\boldsymbol{B} = (\downarrow 2)\boldsymbol{D}$ *of an orthogonal filter bank are represented in the time domain by a combined orthogonal matrix:*

$$\begin{bmatrix} \boldsymbol{L}^T & \boldsymbol{B}^T \end{bmatrix} \begin{bmatrix} \boldsymbol{L} \\ \boldsymbol{B} \end{bmatrix} = \boldsymbol{L}^T\boldsymbol{L} + \boldsymbol{B}^T\boldsymbol{B} = \boldsymbol{I}.$$

The synthesis bank is the transpose of the analysis bank. When one follows the other we have perfect reconstruction. For causality we add a delay.

That summary is a foretaste of later chapters. We will construct longer and better filters, but the underlying problem will be close to this one. There is an analysis bank and a synthesis bank. When they are transposes as well as inverses, the whole filter bank is called ***orthogonal***. When they are inverses but not necessarily transposes, the filter bank is ***biorthogonal***.

In the biorthogonal case, the coefficients in the synthesis bank are different from $\boldsymbol{c}(n)$ and $\boldsymbol{d}(n)$ in the analysis bank. Our ***Haar filter bank*** has orthogonal filters. This chapter pursues it further, all the way to Haar wavelets.

Block Form of a Filter Bank

The best way to represent a two-channel filter bank is by a block diagram (Figure 1.5). The input is a vector $\boldsymbol{x}$. The blocks are linear operators and the output is two half-length vectors: $\boldsymbol{Lx} = (\downarrow 2)\,\boldsymbol{Cx}$ and $\boldsymbol{Bx} = (\downarrow 2)\,\boldsymbol{Dx}$.

For this special filter bank, we want to display all vectors in detail. Later we could supply only the essential information: the coefficients $\boldsymbol{c}(n)$ and $\boldsymbol{d}(n)$. There will be several ways to display those coefficients — the filter bank form, the modulation form, and the polyphase form. We

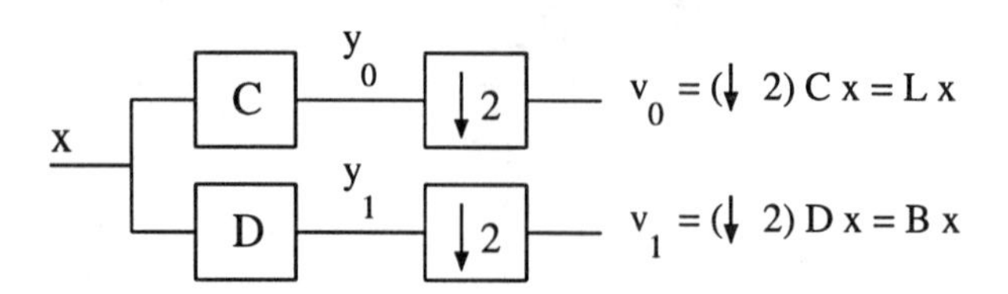

Figure 1.5: Schematic of the analysis half of a two-channel filter bank.

always need the coefficients! Here we use the direct filter bank form, and write the components of each output vector:

$$\boldsymbol{Lx} = \frac{1}{\sqrt{2}}\begin{bmatrix} & \cdot & \\ x(0) & + & x(-1) \\ x(2) & + & x(1) \\ x(4) & + & x(3) \end{bmatrix} \qquad \boldsymbol{Bx} = \frac{1}{\sqrt{2}}\begin{bmatrix} & \cdot & \\ x(0) & - & x(-1) \\ x(2) & - & x(1) \\ x(4) & - & x(3) \end{bmatrix}.$$

The odd-numbered components of $\boldsymbol{Cx}$ involved $x(1)+x(0)$ and $x(3)+x(2)$. Those are gone in $\boldsymbol{Lx}$. Similarly the odd-numbered components of $\boldsymbol{Dx}$ are gone in $\boldsymbol{Bx}$. We are left with two half-length signals. Nevertheless we have enough information to recover the full-length input $\boldsymbol{x}$ — and you see how.

To recover the zeroth component, add $x(0)+x(-1)$ to the difference $x(0)-x(-1)$. That gives $2x(0)$. Since our sums and differences are already divided by $\sqrt{2}$ in the analysis bank, we need another $\sqrt{2}$ in synthesis:

$$x(0) = \frac{1}{\sqrt{2}}\left(\frac{x(0)+x(-1)}{\sqrt{2}} + \frac{x(0)-x(-1)}{\sqrt{2}}\right). \tag{1.20}$$

For $x(-1)$, use the same components of $\boldsymbol{Lx}$ and $\boldsymbol{Bx}$, and *subtract*:

$$x(-1) \quad = \quad \frac{1}{\sqrt{2}}\left(\frac{x(0)+x(-1)}{\sqrt{2}} - \frac{x(0)-x(-1)}{\sqrt{2}}\right).$$

Addition and subtraction are simple for this example, but the synthesis steps must be organized in a way that extends to other examples. At the end of the reconstruction, we want to reach these two vectors $\boldsymbol{w}_0$ and $\boldsymbol{w}_1$:

$$\boldsymbol{w}_0 = \frac{1}{2}\begin{bmatrix} & \cdot & \\ x(0) & + & x(-1) \\ x(0) & + & x(-1) \\ x(2) & + & x(1) \\ x(2) & + & x(1) \\ & \cdot & \end{bmatrix} \text{ and } \boldsymbol{w}_1 = \frac{1}{2}\begin{bmatrix} & \cdot & \\ -x(0) & + & x(-1) \\ x(0) & - & x(-1) \\ -x(2) & + & x(1) \\ x(2) & - & x(1) \\ & \cdot & \end{bmatrix}. \tag{1.21}$$

Notice how the signs in $\boldsymbol{w}_1$ are adjusted so that $\boldsymbol{w}_0+\boldsymbol{w}_1$ recovers the input vector. Actually we are getting $x(n-1)$ instead of $x(n)$. *The total effect of the whole filter bank is a delay.* The input is $x(n)$ and the output is the delayed $x(n-1)$. The sum $\boldsymbol{w}_0+\boldsymbol{w}_1$ is almost, but not quite, the original $\boldsymbol{x}$.

A delay is built in because all filters are *causal*. We analyzed $\boldsymbol{x}$ into low and high frequencies. Now we synthesize to reach $\boldsymbol{w}_0$ and $\boldsymbol{w}_1$ and recover $\boldsymbol{x}$.

The Synthesis Bank

A well-organized synthesis bank is the inverse of the analysis bank. The analysis bank had two steps, *filtering* and *downsampling*. The synthesis bank also has two steps, ***upsampling and filtering***. Notice how the order is reversed — as it always is for inverses.

The first step is to bring back full-length vectors. The downsampling operation $(\downarrow 2)$ is not invertible, but *upsampling* is as close as we can come. The odd-numbered components are *returned as zeros by upsampling*. Applied to a half-length vector $\boldsymbol{v}$, upsampling inserts zeros:

$$\textbf{\textit{Upsampling}} \qquad (\uparrow 2)\begin{bmatrix} \cdot \\ v(0) \\ v(1) \\ v(2) \\ \cdot \end{bmatrix} = \begin{bmatrix} \cdot \\ v(0) \\ 0 \\ v(1) \\ 0 \\ v(2) \\ \cdot \end{bmatrix}. \tag{1.22}$$

Upsampling is denoted by $(\uparrow 2)$. To understand it, look at the result of downsampling to get $\boldsymbol{v} = (\downarrow 2)\boldsymbol{y}$ and upsampling to get $\boldsymbol{u} = (\uparrow 2)(\downarrow 2)\boldsymbol{y}$:

$$\boldsymbol{y} = \begin{bmatrix} \cdot \\ y(0) \\ y(1) \\ y(2) \\ y(3) \\ y(4) \\ \cdot \end{bmatrix} \qquad (\downarrow 2)\boldsymbol{y} = \begin{bmatrix} \cdot \\ y(0) \\ y(2) \\ y(4) \\ \cdot \end{bmatrix} \qquad (\uparrow 2)(\downarrow 2)\boldsymbol{y} = \begin{bmatrix} \cdot \\ y(0) \\ 0 \\ y(2) \\ 0 \\ y(4) \\ \cdot \end{bmatrix}. \tag{1.23}$$

The odd-numbered components of $\boldsymbol{y}$ are replaced by zeros. We will see that $(\uparrow 2)$ is the *transpose* of $(\downarrow 2)$. Fortunately, transposes come in reverse order exactly as inverses do. So synthesis can be the *transpose* of analysis — apart from our ever-present delay.

Small note: Also $(\uparrow 2)$ is a *right-inverse* of $(\downarrow 2)$. If we put in zeros and remove them, we recover $\boldsymbol{y}$. Thus $(\downarrow 2)(\uparrow 2) = \boldsymbol{I}$. The order $(\uparrow 2)(\downarrow 2)$ that we actually use is displayed above, and it inserts zeros. Properly speaking, $(\uparrow 2)$ is the "pseudoinverse" of $(\downarrow 2)$ — which has no inverse.

The vectors reached by upsampling have zeros in their odd components:

$$\mathbf{u}_0 = \frac{1}{\sqrt{2}}\begin{bmatrix} x(0) + x(-1) \\ 0 \\ x(2) + x(1) \\ 0 \\ x(4) + x(3) \\ \cdot \end{bmatrix} \quad \text{and} \quad \mathbf{u}_1 = \frac{1}{\sqrt{2}}\begin{bmatrix} x(0) - x(-1) \\ 0 \\ x(2) - x(1) \\ 0 \\ x(4) - x(3) \\ \cdot \end{bmatrix}. \tag{1.24}$$

The second step in the synthesis bank, after upsampling, is *filtering*. The two vectors $\mathbf{u}_0$ and $\mathbf{u}_1$ are the inputs to the two filters. The vectors $\boldsymbol{w}_0$ and $\boldsymbol{w}_1$ are the desired outputs. Schematically, the structure of the synthesis bank is in Figure 1.6.

Normally we would construct the synthesis filters $\boldsymbol{F}$ and $\boldsymbol{G}$ based on the analysis filters $\boldsymbol{C}$ and $\boldsymbol{D}$. That will be our procedure in the rest of the book. For this example, when we know the

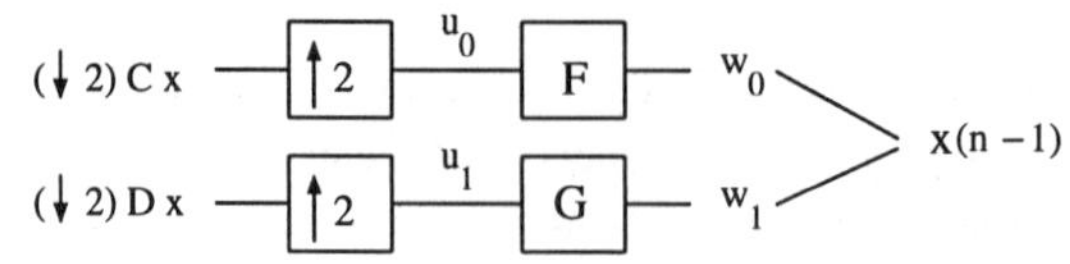

$$(\downarrow 2)\,Cx \to \boxed{\uparrow 2} \xrightarrow{u_0} \boxed{F} \to w_0, \qquad (\downarrow 2)\,Dx \to \boxed{\uparrow 2} \xrightarrow{u_1} \boxed{G} \to w_1, \qquad w_0 + w_1 = x(n-1)$$

Figure 1.6: The synthesis half of a filter bank: upsample, filter, and add.

desired outputs, we proceed more directly. The filter $\boldsymbol{F}$ that produces $\boldsymbol{w}_0$ from $\boldsymbol{u}_0$ is an addition filter:

$$\boldsymbol{F} \text{ filters } \frac{1}{\sqrt{2}}\begin{bmatrix} \boldsymbol{x}(0)+\boldsymbol{x}(-1) \\ 0 \\ \boldsymbol{x}(2)+\boldsymbol{x}(1) \\ 0 \\ \cdot \end{bmatrix} \text{ to give } \frac{1}{2}\begin{bmatrix} \boldsymbol{x}(0)+\boldsymbol{x}(-1) \\ \boldsymbol{x}(0)+\boldsymbol{x}(-1) \\ \boldsymbol{x}(2)+\boldsymbol{x}(1) \\ \boldsymbol{x}(2)+\boldsymbol{x}(1) \\ \cdot \end{bmatrix} = \boldsymbol{w}_0.$$

This is the output we want. It comes from a time-invariant causal filter $\boldsymbol{F}$. There is no separate treatment of even and odd components! When the input to $\boldsymbol{F}$ has components $\boldsymbol{u}(n)$, the output has components

$$\boldsymbol{F}\boldsymbol{u}(n) = \frac{1}{\sqrt{2}}(\boldsymbol{u}(n) + \boldsymbol{u}(n-1)). \tag{1.25}$$

The filter coefficients are $\boldsymbol{G}(0) = \boldsymbol{f}(1) = \frac{1}{\sqrt{2}}$.

The second synthesis filter $\boldsymbol{G}$ is a subtraction filter. Its coefficients are $\frac{-1}{\sqrt{2}}$ and $\frac{1}{\sqrt{2}}$. For an arbitrary input vector $\boldsymbol{u}$, the output has components $\frac{1}{\sqrt{2}}(-\boldsymbol{u}(n) + \boldsymbol{u}(n-1))$. Notice especially how $\boldsymbol{G}$ acts on $\boldsymbol{u}_1 = (\uparrow 2)(\downarrow 2)\,\boldsymbol{D}\boldsymbol{x}$:

$$\boldsymbol{G} \text{ filters } \frac{1}{\sqrt{2}}\begin{bmatrix} \boldsymbol{x}(0)-\boldsymbol{x}(-1) \\ 0 \\ \boldsymbol{x}(2)-\boldsymbol{x}(1) \\ 0 \\ \cdot \end{bmatrix} \text{ to give } \frac{1}{2}\begin{bmatrix} -\boldsymbol{x}(0) & + & \boldsymbol{x}(-1) \\ \boldsymbol{x}(0) & - & \boldsymbol{x}(-1) \\ -\boldsymbol{x}(2) & + & \boldsymbol{x}(1) \\ \boldsymbol{x}(2) & - & \boldsymbol{x}(1) \\ & \cdot & \end{bmatrix} = \boldsymbol{w}_1.$$

The filter gives the right result, again without treating even and odd components differently. We caution that the highpass coefficients are in the order $-\frac{1}{\sqrt{2}}, \frac{1}{\sqrt{2}}$. When you transpose $\boldsymbol{D}$, the order of coefficients is reversed. Then a delay makes $\boldsymbol{G}$ a causal filter (and the same for $\boldsymbol{F}$).

The only other caution concerns the factors $\frac{1}{\sqrt{2}}$. They are present in all four filters. A less perfect symmetry would have $\frac{1}{2}$ in one bank and 1 in the other bank. This is exactly like the two-point discrete Fourier transform, where we often allow $\frac{1}{2}$ and 1 in the matrix and its inverse. The orthogonal matrix has $\frac{1}{\sqrt{2}}$ in both:

$$\begin{bmatrix} 1 & 1 \\ 1 & -1 \end{bmatrix}^{-1} = \frac{1}{2}\begin{bmatrix} 1 & 1 \\ 1 & -1 \end{bmatrix} \qquad \text{inverse matrices with 1 and } \tfrac{1}{2}$$

$$\begin{bmatrix} 1/\sqrt{2} & 1/\sqrt{2} \\ 1/\sqrt{2} & -1/\sqrt{2} \end{bmatrix}^{-1} = \begin{bmatrix} 1/\sqrt{2} & 1/\sqrt{2} \\ 1/\sqrt{2} & -1/\sqrt{2} \end{bmatrix} \qquad \text{orthogonal matrix with } 1/\sqrt{2}.$$

An orthogonal (or unitary) matrix has orthogonal rows and orthogonal columns normalized to be unit vectors. *The inverse is the transpose* (or conjugate transpose). It is an accident for this example that the matrix is real and symmetric. You can't see that it was transposed and conjugated.

Important. The connection between this Haar filter bank and the 2-point DFT is no accident. The simplest M-band filter bank comes from an M-point DFT. It is called a *uniform DFT filter bank.* We are seeing the case $M = 2$, written as an ordinary filter bank (and made causal by a delay).

We summarize this section with a schematic of the whole filter bank (Figure 1.7).

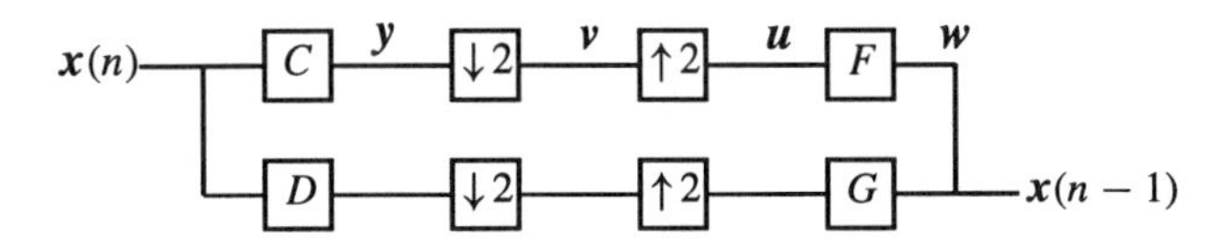

Figure 1.7: The analysis bank followed by the synthesis bank.

This allows us to indicate the symbols $\boldsymbol{y}$, $\boldsymbol{u}$, $\boldsymbol{v}$, $\boldsymbol{w}$ for the outputs at the four stages. For a two-channel bank, we mark those vectors in the lowpass and highpass channels by subscripts 0 and 1. When there are M channels, we need subscripts $0, 1, \ldots, M-1$.

From the filter bank, the next step is to wavelets.

Problem Set 1.4

1. Write down the matrix $(\downarrow 2)$ that executes downsampling: $(\downarrow 2)\,\boldsymbol{x}(n) = \boldsymbol{x}(2n)$.
2. Write down the transpose matrix $(\uparrow 2) = (\downarrow 2)^T$. Multiply the matrices $(\uparrow 2)\ (\downarrow 2)$ and $(\downarrow 2)\ (\uparrow 2)$. Describe the output from $(\uparrow 2)\,\boldsymbol{y}(n)$.
3. Describe the output from $(\downarrow 2)^2\,\boldsymbol{x}(n)$ and $(\uparrow 2)^2\,\boldsymbol{y}(n)$.
4. Send the signal with $\boldsymbol{x}(0) = \boldsymbol{x}(1) = \boldsymbol{x}(2) = 1$ through the whole filter bank, and give the output at every step.
5. Put each row $[\ -1 \quad 1\ ]/\sqrt{2}$ of $\boldsymbol{B} = (\downarrow 2)\,\boldsymbol{D}$ after the row $[\ 1 \quad 1\ ]/\sqrt{2}$ of $\boldsymbol{L} = (\downarrow 2)\,\boldsymbol{C}$. In this order we see a *block transform*. What is the inverse transform?
6. Show how to delay an anticausal (= upper triangular) filter matrix so it becomes causal. If the coefficients $\boldsymbol{h}(0), \ldots, \boldsymbol{h}(N)$ are on the diagonals of the anticausal matrix, what are the diagonals after the delay? Give the two frequency responses, anticausal and causal.
7. The 4-channel bank analogous to Haar is based on the 4-point DFT matrix $\boldsymbol{F}_4$. Find $\boldsymbol{F}_4$ and the four analysis filters and the outputs after downsampling by $(\downarrow 4)$.
8. Suppose a matrix has the property that $\boldsymbol{Q}^T\boldsymbol{Q} = \boldsymbol{I}$. Show that the columns of $\boldsymbol{Q}$ are mutually orthogonal unit vectors. Does it follow that $\boldsymbol{Q}\boldsymbol{Q}^T = \boldsymbol{I}$?
9. In a transmultiplexer, the synthesis bank comes *before* the analysis bank. Compute $\boldsymbol{L}\boldsymbol{L}^T$ and $\boldsymbol{L}\boldsymbol{B}^T$ and $\boldsymbol{B}\boldsymbol{B}^T$ to verify that the Haar transmultiplexer still gives perfect reconstruction:
$$\begin{bmatrix} \boldsymbol{L} \\ \boldsymbol{B} \end{bmatrix} \begin{bmatrix} \boldsymbol{L}^T & \boldsymbol{B}^T \end{bmatrix} = \begin{bmatrix} \boldsymbol{L}\boldsymbol{L}^T & \boldsymbol{L}\boldsymbol{B}^T \\ \boldsymbol{B}\boldsymbol{L}^T & \boldsymbol{B}\boldsymbol{B}^T \end{bmatrix} = \begin{bmatrix} \boldsymbol{I} & \boldsymbol{0} \\ \boldsymbol{0} & \boldsymbol{I} \end{bmatrix}.$$
10. Compute the subband outputs $\boldsymbol{v}_0(n)$ and $\boldsymbol{v}_1(n)$ for the average-difference filter bank with input $\boldsymbol{x}(n) = 0,\ 1,\ -1,\ 2,\ 5,\ 1,\ 7,\ 0$. Reconstruct the signal by feeding $\boldsymbol{v}_k(n)$ into the synthesis bank in Figure 1.6. Verify that the output is $\boldsymbol{x}(n-1)$.
11. If $H_0(z) = 1$ and $H_1(z) = z^{-1}$ (no filtering) write the entries of $\begin{bmatrix} \boldsymbol{L} \\ \boldsymbol{B} \end{bmatrix} \begin{bmatrix} \boldsymbol{L}^T & \boldsymbol{B}^T \end{bmatrix} = \boldsymbol{I}$.

1.5 Scaling Function and Wavelets

Corresponding to the lowpass filter, with $\boldsymbol{h}(0) = \frac{1}{2}$ and $\boldsymbol{h}(1) = \frac{1}{2}$, there is a continuous-time scaling function $\phi(t)$. Corresponding to the highpass filter, with coefficients $\frac{1}{2}$ and $-\frac{1}{2}$, there is a wavelet $w(t)$. We now describe the *dilation equation* that produces $\phi(t)$ and the *wavelet equation* for $w(t)$.

You will see how the filter coefficients (the $\boldsymbol{c}$'s and $\boldsymbol{d}$'s) enter these equations. Two time scales also appear. This joint appearance of t and $2t$ is the novelty of the dilation equation. It is also the source of difficulty! We have a "two-scale difference equation". Later we develop the background of these equations and a general method for solving them. Here we quickly recognize the box function as $\phi(t)$. Then we construct the wavelet $w(t)$ and use it.

The dilation equation for the scaling function $\phi(t)$ is

$$\phi(t) = \sqrt{2}\sum_{k=0}^{N} \boldsymbol{c}(k)\,\phi(2t-k). \tag{1.26}$$

In terms of the original lowpass coefficients $\boldsymbol{h}(k)$, the extra factor is 2:

$$\phi(t) = 2\sum_{k=0}^{N} \boldsymbol{h}(k)\,\phi(2t-k). \tag{1.27}$$

This involves a function $\phi(t)$ in continuous time, and a set of coefficients $\boldsymbol{c}(k)$ or $\boldsymbol{h}(k)$ from discrete time. The presence of t and $2t$ is the key. Without the 2 we would have an ordinary constant-coefficient equation (look for exponential solutions $e^{\lambda t}$). With two time scales, there are major changes:

1. There may or may not be a solution $\phi(t)$.
2. The solution is zero outside the interval $0 \le t < N$.
3. The solution seldom has an elementary formula.
4. The solution is not likely to be a smooth function.

Formally, we can find an expression for the Fourier transform of $\phi(t)$. It is an infinite product (Section 6.4). The inverse Fourier transform yields $\phi(t)$ as a "distribution"—not necessarily continuous, possibly involving delta functions. (Those are impulses. Their integrals are jumps. More cautious people call them steps.) In our Haar example, the solution $\phi(t)$ lies just outside the class of continuous functions—it has a jump.

For the coefficients $2\boldsymbol{h}(0) = 1$ and $2\boldsymbol{h}(1) = 1$, the dilation equation is

$$\phi(t) = \phi(2t) + \phi(2t-1). \tag{1.28}$$

The graph of $\phi(t)$ is *compressed by* 2, to give the graph of $\phi(2t)$. When that is *shifted to the right by* $\frac{1}{2}$, it becomes the graph of $\phi(2t-1)$. We ask the two compressed graphs to combine into the original graph. Figure 1.8 shows that this occurs when $\phi(t)$ is the *box function*:

$$\phi(t) = \begin{cases} 1 & \text{for } 0 \le t < 1 \\ 0 & \text{otherwise.} \end{cases}$$

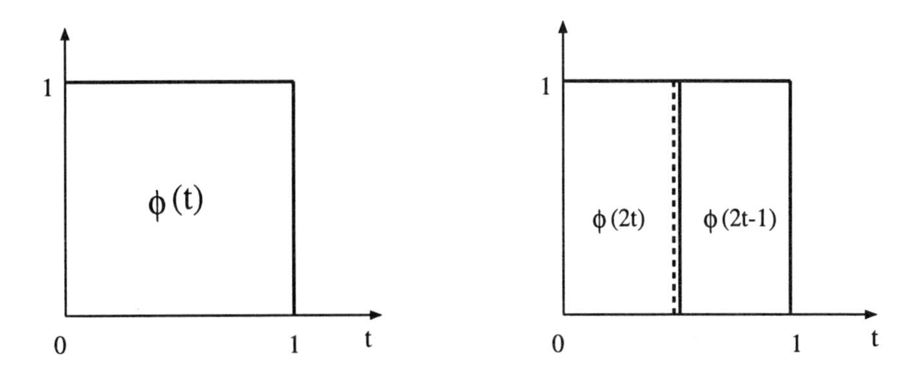

Figure 1.8: The box function with dilation and translation.

The graphs of $\phi(2t)$ and $\phi(2t-1)$ are half-size boxes. Their sum is the full-size box $\phi(t)$.

As planned, we wrote down the solution rather than deriving it. The dilation equation is linear, so any multiple of the box function is also a solution. It is convenient to normalize so that *the integral of $\phi(t)$ from $-\infty$ to ∞ equals one.* Note that a solution $\phi(t)$ covering unit area is only possible when the coefficients in the dilation equation add up to 2.

Theorem 1.1 *If* $\int_{-\infty}^{\infty} \phi(t)\,dt = 1$ *then* $2h(0) + 2h(1) + \cdots + 2h(N) = 2.$

Proof: The graph of $\phi(2t)$ and every $\phi(2t-k)$ is compressed to area $\frac{1}{2}$:

$$2\int_{-\infty}^{\infty} \phi(2t-k)\,dt = \int_{-\infty}^{\infty} \phi(u)\,du = 1 \qquad (\text{set } u = 2t-k). \tag{1.29}$$

So integrating both sides of the dilation equation $\phi(t) = 2\sum h(k)\phi(2t-k)$ gives $1 = h(0) + h(1) + \cdots + h(N)$. This is our lowpass filter convention.

Important. For the filter, then the filter bank, and finally the dilation equation, *the normalization is different.* This is clear from the actual numbers in our lowpass filter. The sum of coefficients is 1 or $\sqrt{2}$ or 2:

$$\begin{aligned} h:&\quad \tfrac{1}{2} \text{ and } \tfrac{1}{2} \text{ in the single filter, adding to } 1 \\ c:&\quad \tfrac{1}{\sqrt{2}} \text{ and } \tfrac{1}{\sqrt{2}} \text{ in the filter bank, adding to } \sqrt{2} \\ 2h:&\quad 1 \text{ and } 1 \text{ in the dilation equation, adding to } 2. \end{aligned}$$

The single lowpass filter has sum $H(0) = 1$. That preserves the zero frequency DC term: output $Y(0) =$ input $X(0)$. The filter bank has a factor $\sqrt{2}$ to account for the downsampling step. There are only half as many components and half as many terms in the energy, when $\sum (y(n))^2$ is replaced by $\sum (y(2n))^2$. To compensate we must multiply $y(2n)$ by $\sqrt{2}$. That gives the normalization $C(0) = \sqrt{2}$ in place of $H(0) = 1$.

For the dilation equation, we have just seen why rescaling the time requires renormalizing the coefficients by 2 —to preserve area.

There are certainly filters in which $H(0)$ is not exactly 1. If $H(\omega)$ stays very near 1 over an interval around $\omega = 0$, this still deserves the name "lowpass filter". *Such filters do not lead to wavelets!* The requirements for wavelets are very strict, at $\omega = 0$ and $\omega = \pi$. Only a subset of special filters can pass the test, which starts with $H(0) = 1$ and $H(\pi) = 0$.

The box function is like the averaging filter — it smoothes the input. Convolution with the box function gives a moving average in continuous time, just as the filter coefficients $\boldsymbol{h} = \frac{1}{2}, \frac{1}{2}$ did in discrete time:

$$\boldsymbol{h} * (\ldots, \boldsymbol{x}(0), \boldsymbol{x}(1), \ldots) = \left(\ldots, \frac{\boldsymbol{x}(0)+\boldsymbol{x}(-1)}{2}, \frac{\boldsymbol{x}(1)+\boldsymbol{x}(0)}{2}, \ldots\right)$$

$$\phi(t) * x(t) = \int_{t-1}^{t} x(s)\, ds = \text{average over moving interval.} \tag{1.30}$$

There is a similar convolution with $w(t)$. Instead of picking up the smooth low frequency part of the function, the wavelet will lead to the high-frequency details. The coefficients for the Haar wavelet are 1 and -1.

The Wavelet Equation

The equation for the wavelet involves the highpass coefficients $\boldsymbol{d}(k)$. It is a direct equation that gives $w(t)$ immediately and explicitly from $\phi(t)$:

$$w(t) = \sqrt{2} \sum \boldsymbol{d}(k)\, \phi(2t - k). \tag{1.31}$$

In terms of the original coefficients $\boldsymbol{h}_1(k)$, the factor $\sqrt{2}$ becomes 2:

$$\textit{Wavelet equation} \qquad w(t) = 2 \sum \boldsymbol{h}_1(k)\, \phi(2t - k). \tag{1.32}$$

In our example, $\phi(t)$ is a box function and its dilations $\phi(2t-k)$ are half-boxes. Then the wavelet is a *difference of half-boxes*:

$$w(t) = \phi(2t) - \phi(2t - 1). \tag{1.33}$$

Explicitly, $w(t) = 1$ for $0 \le t < \frac{1}{2}$ and $w(t) = -1$ for $\frac{1}{2} \le t < 1$. This is the ***Haar wavelet***. Its graph is in Figure 1.9 along with the graphs of $w(2t)$ and $w(2t - 1)$. Those are wavelets at scale $2t$; their graphs are compressed and shifted. They join the original $w(t)$, and all its other dilations and translations, in the *wavelet basis*.

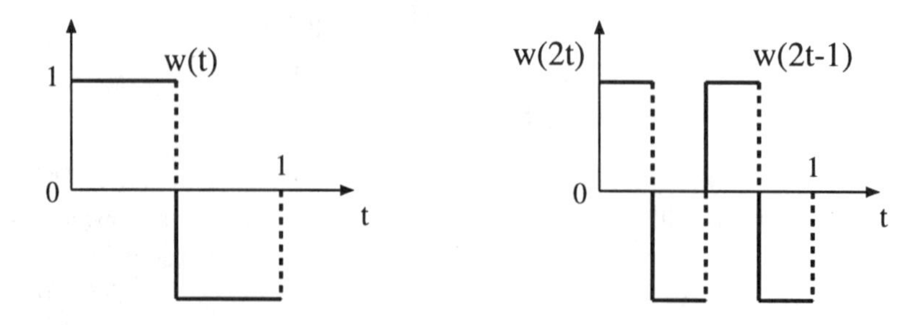

Figure 1.9: The Haar wavelet $w(t)$; the rescaled wavelets $w(2t)$ and $w(2t - 1)$.

It goes without saying that Alfred Haar did not call his function a "***wavelet***". He was writing in 1910 about this particular function; all other wavelets came later. That name emerged from the literature on geophysics, by a route through France. The word *onde* led to *ondelette*. In translation, the word wave led to wavelet. ***A wavelet is a small wave***. In Haar's case it happened to be a square wave.

Comment. The square wave $w(t)$ has compact support. It comes from an FIR filter, with finite length. We use the word "support" for the closed interval in continuous time, here [0, 1], outside which $w(t)$ is zero. When we close the set where it is nonzero, to include the jump location $t = \frac{1}{2}$ as well as the endpoints, we have found the support of the function $w(t)$. The words "compact support" mean that this closed set is bounded. The wavelet is zero outside a bounded interval: ***compact support corresponds to FIR***.

Wavelets need not have compact support! They can come from IIR filters instead of FIR filters. Historically, since the connection of wavelets to filters was not immediately recognized, the wavelets after Haar were constructed in other ways (with considerable difficulty). They oscillated above and below zero along the whole line $\infty < t < \infty$, decaying as $|t| \to \infty$. This still qualifies as a wavelet. It is a localized pulse that decreases to zero and *has integral zero*. Ingrid Daubechies showed in 1988 that compact support was possible for other wavelets than Haar's.

We will tell that story more completely in Chapter 6. From Haar onwards, one property that most designers hoped for was **orthogonality**. This means: $w(t)$ is orthogonal to all its dilations and translations. The wavelet basis, containing all these functions $w(2^j t - k)$, is an *orthogonal basis*. For the first few Haar wavelets that is easy to verify:

$$\begin{aligned} \text{inner product} &= \textstyle\int_{-\infty}^{\infty} w(t)w(2t)\,dt &&= 0 \\ \text{inner product} &= \textstyle\int_{-\infty}^{\infty} w(t)w(2t-1)\,dt &&= 0 \\ \text{inner product} &= \textstyle\int_{-\infty}^{\infty} w(2t)w(2t-1)\,dt &&= 0. \end{aligned}$$

In the first integral, $w(t) = 1$ in Figure 1.9 where $w(2t)$ is positive and then negative. The integral is zero. Similarly $w(t) = -1$ on the second half-interval where $w(2t-1)$ is plus and minus. The second integral is therefore zero. The third integral vanishes for a different reason — the functions $w(2t)$ and $w(2t-1)$ do not overlap. One is zero where the other is nonzero. So the product $w(2t)\,w(2t-1)$ is zero everywhere.

The pattern continues for all translations by k and dilations by 2^j. Haar wavelets at the same scaling level (same j) do not overlap. They are orthogonal in the strictest way. When Haar wavelets at different levels j and J do overlap, the coarse one is constant where the fine one goes up and down. All integrals are zero, giving an orthogonal basis $w_{jk}(t) = w\left(2^j t - k\right)$:

$$\text{inner product} = \int_{-\infty}^{\infty} w\left(2^j t - k\right) w\left(2^J t - K\right) dt = 0. \tag{1.34}$$

A perfect basis is not only orthogonal but orthonormal. The functions have length 1. Like a unit vector, the inner product $< w(t), w(t) >$ is normalized to 1:

$$\text{length squared} = \int_{-\infty}^{\infty} (w(t))^2\,dt = 1.$$

This is true for the Haar wavelet. To make it true for the dilations of that wavelet, we multiply by $2^{j/2}$ — otherwise the compressed graphs cover less area. The same factor will apply to other wavelets, and we record it now:

Theorem 1.2 *The rescaled Haar wavelets* $w_{jk}(t) = 2^{j/2}w\left(2^j t - k\right)$ *form an orthonormal basis:*

$$\int_{-\infty}^{\infty} w_{jk}(t)\, w_{JK}(t)\, dt = \delta(j-J)\, \delta(k-K). \tag{1.35}$$

This Kronecker delta symbol equals zero except when $j = J$ and $k = K$. In that case the integral of $(w_{JK}(t))^2$ equals one. We need two indices j and k because there are two operations (dilation and translation) in constructing the basis.

Important note. For these Haar wavelets, orthogonality was verified by direct integration. For future wavelets, this integration is not desirable and not possible. We will not have elementary formulas for $\phi(t)$ and $w(t)$. Instead, we will know the coefficients $c(k)$ in the dilation equation and $d(k)$ in the wavelet equation. ***All information about*** $\phi(t)$ ***and*** $w(t)$ ***— their support interval, their orthogonality, their smoothness, and their vanishing moments — will be determined by and from the c's and d's.***

Second note. We have not said ***which space of functions*** has the wavelets $w_{jk}(t)$ as a basis. Actually there are many choices. The space starts with *finite* combinations $g(t) = \sum b_{jk} w_{jk}(t)$. Those functions are piecewise constant on binary intervals — length $1/2^j$ and endpoints $m/2^j$. Other functions $f(t)$ are in the space *if they are limits of these piecewise constant* g's:

$$\|f(t) - g_n(t)\| \to 0 \quad \text{for some sequence } g_n(t).$$

The choice of function space is decided by the choice of the norm $\|f - g_n\|$. A function $f(t)$ might be a limit of piecewise constants in the maximum norm but not in an integral norm, or vice versa. The most frequent choice is the L^2 norm, where the superscript 2 signals that we integrate the *square*:

$$\|f(t) - g(t)\| = \left(\int_{-\infty}^{\infty} |f(t) - g(t)|^2\, dt\right)^{1/2}.$$

This choice of the L^2 norm is popular for four major reasons:

1. The norm is directly connected to the inner product:

By definition	$\|f(t)\|^2$	$=$	$< f(t),\ f(t) >$
By the Schwarz inequality	$\|f(t)\|\,\|g(t)\|$	$\geq$	$\lvert < f(t),\ g(t) > \rvert$.

2. Minimizing the L^2 norm leads to linear equations. This is familiar from ordinary least squares problems. The word "squares" introduces the L^2 norm. When the functions $g(t)$ are restricted to a subspace, the closest one to $f(t)$ is $g(t) =$ *projection of* $f(t)$ *onto the subspace*. Projection leads to right angles and linear equations.

3. The Fourier transform *preserves the* L^2 *norm and the inner product* $< f, g >$. This is the Parseval identity:

$$\|f\|^2 = \int_{-\infty}^{\infty} |f(t)|^2\, dt = \int_{-\infty}^{\infty} |\widehat{f}(\omega)|^2\, d\omega = \|\widehat{f}\|^2 \tag{1.36}$$

$$< f, g > = \int_{-\infty}^{\infty} f(t)\,\overline{g(t)}\, dt = \int_{-\infty}^{\infty} \widehat{f}(\omega)\,\overline{\widehat{g}(\omega)}\, d\omega = < \widehat{f}, \widehat{g} >. \tag{1.37}$$

The reader will know that irrelevant but necessary factors of 2π should enter these equations. They depend on how we define the Fourier transforms. Since $\widehat{f}(\omega)$ and $\widehat{g}(\omega)$ can be complex, we introduced absolute values in (1.36) and complex conjugates in (1.37). Of course (1.37) becomes (1.36) if $g(t) = f(t)$.

4. For an orthonormal basis (like the Haar wavelet basis), the L^2 norm $\left\|\sum b_{jk}w_{jk}(t)\right\|^2$ equals the sum of squares of the coefficients b_{jk}:

$$\begin{aligned}\int \left(\sum b_{jk}\, w_{jk}(t)\right)^2 dt &= \sum\sum b_{jk}\, b_{JK} \int w_{jk}(t)\, w_{JK}(t)\, dt \\ &= \sum \left(b_{jk}\right)^2 .\end{aligned}$$

For all these reasons, and more, the L^2 norm is our choice. But we must mention that wavelets provide an unusually convenient basis for other norms and other function spaces. This makes them popular in functional analysis and harmonic analysis, which deal with the properties of functions. If we change the exponent from 2 to p, the space L^p contains all functions for which $\left(\int |f(t)|^p\right)^{1/p}$ is finite. This norm $\|f\|_p$ is not quite equal to $C\left(\sum |b_{jk}|^p\right)^{1/p}$, as it was for $p = 2$. But if the b_{jk} are wavelet coefficients, rather than Fourier coefficients, this sum lies between fixed bounds $A_p\,\|f\|_p$ and $B_p\,\|f\|_p$. Thus the absolute values $|b_{jk}|$ still indicate which functions have finite norm and belong to L^p.

In shorthand: The wavelets are an ***unconditional basis*** for L^p (no condition on the signs of the b_{jk}). The complex exponentials are conditional; phase information is needed on the b_{jk}.

We recognize that these last comments are "pure mathematics". The reader is invited to start learning that language too — if desired and not already achieved. It is rewarding and not difficult. The *wavelet transform* that connects $f(t)$ to its wavelet coefficients b_{jk} is absolutely central — to theory and also to applications.

We turn to the practical problem: *How to compute the coefficients b_{jk} quickly?* This has a very good answer for wavelet transforms. There is a recursive Fast Wavelet Transform comparable in speed and stability to the FFT for Fourier transforms.

Problem Set 1.5

1. If $w(t)$ has unit norm, so that $\int (w(t))^2 dt = 1$, show that the function $w_{jk}(t) = 2^{j/2}w(2^j t - k)$ also has unit norm.

2. What combination of the half-boxes $\phi(2t)$ and $\phi(2t-1)$ is closest in L^2 (least squares) to $f(t) = t^2$ for $0 \le t \le 1$?

3. The box function $\phi_s(t)$, shifted to the interval [3, 4], is what combination of the functions $\phi_s(2t-k)$?

4. Show that the convolution of the box function with a continuous-time signal is the continuous average in (1.30). The convolution formula is

$$\phi(t) * x(t) = \int_{-\infty}^{\infty} \phi(t-s)\, x(s)\, ds.$$

5. Given a combination $a_0\phi(2t) + a_1\phi(2t-1)$, express it as $A_0\phi(t) + B_0 w(t)$. Then invert to find a_0 and a_1 from A_0 and B_0.

6. What scaling function satisfies the three-scale equation $\phi(t) = \phi(3t) + \phi(3t-1) + \phi(3t-2)$? Find ***two wavelets*** that are combinations of $\phi(3t-k)$, orthogonal to $\phi(t)$ and to each other.

1.6 Wavelet Transforms by Multiresolution

The wavelet transform operates in continuous time (on functions) and in discrete time (on vectors). The input is $f(t)$ or $\boldsymbol{x}(n)$. The output is the set of coefficients b_{jk}, which express the input in the wavelet basis. For functions and infinite signals, this basis is necessarily infinite. For finite length vectors with L components, there will be L basis vectors and L coefficients. The discrete wavelet transform, from L components of the signal to L wavelet coefficients, is expressed by an L by L matrix.

To begin, we derive formulas for the coefficients b_{jk}. In continuous time they involve integrals of $f(t)$ times $w_{jk}(t)$. In discrete time we are solving a linear system. The inverse transform involves the inverse matrix.

Then we show how the b_{jk} can be found recursively. Levels j and $j-1$ are connected. This reorganizes the matrix multiplication. The discrete wavelet transform (DWT) becomes the *fast* wavelet transform (FWT). The central idea in this fast recursion is *multiresolution*.

First come the two directions, synthesis and analysis, for an orthonormal basis:

$$\begin{aligned} &\textbf{Synthesis}\text{ of a function:} \quad f(t) = \sum_{j,k} b_{jk}\, w_{jk}(t) \\ &\textbf{Analysis}\text{ of a function:} \quad b_{JK} = \int_{-\infty}^{\infty} f(t)\, w_{JK}(t)\, dt. \end{aligned} \tag{1.38}$$

In the matrix case, the wavelets are ordinary vectors. They go into the columns of the ***wavelet matrix S***. To maintain the parallel with the continuous case, we use a double index jk for the column number. A single index would go from 1 to L, but the double index is more natural. Each wavelet vector has a position in time given by k and a position in frequency (better to say, in *scale*) given by j. The columns of the L by L matrix $\boldsymbol{S}$ are the discrete wavelets:

$$\textbf{Synthesis}\text{ in discrete time: } \boldsymbol{x} = \boldsymbol{S}\boldsymbol{b}.$$

The rows of the L by L matrix $\boldsymbol{A}$ contain the "analyzing" wavelets:

$$\textbf{Analysis}\text{ in discrete time: } \boldsymbol{b} = \boldsymbol{A}\boldsymbol{x}.$$

For Haar and all orthonormal wavelets, the columns of $\boldsymbol{S}$ are the same as the rows of $\boldsymbol{A}$. Analysis and synthesis are related by $\boldsymbol{A} = \boldsymbol{S}^T$. In general they are related by $\boldsymbol{A} = \boldsymbol{S}^{-1}$.

The synthesis equation $\boldsymbol{x} = \boldsymbol{S}\boldsymbol{b}$ multiplies each coefficient b_{jk} by the basis vector in column jk of $\boldsymbol{S}$, and adds. This is just matrix multiplication: $\boldsymbol{S}\boldsymbol{b}$ is a combination of the columns of $\boldsymbol{S}$.

The analysis equation $\boldsymbol{A}\boldsymbol{x} = \boldsymbol{A}\boldsymbol{S}\boldsymbol{b} = \boldsymbol{b}$ is completely parallel to the continuous formula $\int w_{jk}(t) f(t)\, dt = b_{jk}$. The left side is the inner product of $\boldsymbol{x}$ with each analyzing vector. The wavelet basis consists of *unit vectors*, so all formulas are simple. The basis is orthonormal. In the continuous case, $\int \left(w_{jk}(t)\right)^2 dt = 1$. In the discrete case $\boldsymbol{S}^T\boldsymbol{S} = \boldsymbol{I}$. Then $\boldsymbol{b} = \boldsymbol{S}^T\boldsymbol{x}$ and no division by length squared is required.

Analysis is the ***inverse*** of synthesis. Why did we multiply by the *transpose matrix*, when we should have introduced the *inverse matrix*? For an orthonormal basis, the reason is fundamental: The inverse is the same as the transpose!

$$\text{Orthonormal columns: } \boldsymbol{S}^T\boldsymbol{S} = \boldsymbol{I} \text{ means } \boldsymbol{S}^{-1} = \boldsymbol{S}^T.$$

When the basis is only orthogonal, and not normalized to unit vectors, $\boldsymbol{S}^T\boldsymbol{S}$ is diagonal. By using the word *basis*, we ensured that all these matrices are square. In the rectangular case $\boldsymbol{S}$ would not have an inverse. There are too many columns to be independent; instead of a basis we have a *frame* (Section 2.6). Our formulas would give the "pseudoinverse" S^+ instead of S^{-1}.

Without orthogonality, the rows of $A = S^{-1}$ are *biorthogonal* to the columns of S:

$$(\text{row } i \text{ of } A) \cdot (\text{column } j \text{ of } S) = \delta(i - j).$$

Each row of S^{-1} is orthogonal to $L - 1$ columns of S. This is ***biorthogonality***. The columns of S are the *synthesis basis*, and the rows of $A = S^{-1}$ are the *analysis basis*.

Tree-Structured Filter Bank

We move from transforms toward *fast* transforms. The wavelet basis and the Fourier basis have special properties, beyond orthogonality. The scales j and $j - 1$ are closely related, just as the frequencies ω and $\omega/2$ are closely related. By taking advantage of these relations, the multiplications by S and S^T can be reorganized. We will explain the special properties, and then derive fast transforms (starting in this section with Haar wavelets).

In the Fourier case, the matrices become F and F^T — or actually $\overline{F}^T$, since the Fourier vectors are complex exponentials. These are the L-point DFT matrix and its inverse. The Fast Fourier Transform is summarized in Section 8.1 (on periodic problems). You will see that the FWT is asymptotically faster than the FFT, requiring only $O(L)$ steps instead of $O(L \ln L)$. (The Walsh basis, which is not local, brings back $L \ln L$. The components are ± 1, with no zeros.) All these transforms are so important and successful that we do not overemphasize the comparison. Our goal is to understand wavelets and Fourier both.

The recursive nature of wavelets is clearest when we construct a tree of filter banks (Figure 1.10). The highpass filter D computes differences of the input. The downsampling step symbolized by $(\downarrow 2)$ keeps the even-numbered differences $(x(2k) - x(2k - 1))/\sqrt{2}$. These are final outputs because they are not transformed again. They are at the end of their branch, in this "logarithmic tree". These outputs b_{jk} are at the fine-mesh level. The factor $r = 1/\sqrt{2}$ is included to produce unit vectors in the matrix S, below.

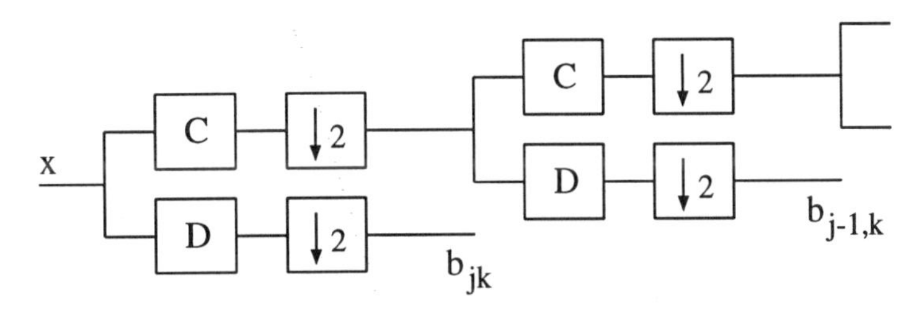

Figure 1.10: The logarithmic tree that leads to wavelets.

The lowpass filter C computes averages. Again the $(\downarrow 2)$ step keeps the even samples. These averages are not final outputs, because they will be filtered again by D and C. The averages and differences of all levels follow the fundamental recursion:

$$\begin{aligned} &\textit{Averages (lowpass filter)} && a_{j-1,k} = \tfrac{1}{\sqrt{2}}\left(a_{j,2k} + a_{j,2k+1}\right) \\ &\textit{Differences (highpass filter)} && b_{j-1,k} = \tfrac{1}{\sqrt{2}}\left(a_{j,2k} - a_{j,2k+1}\right). \end{aligned} \tag{1.39}$$

These filters are anticausal. ***There is a time-reversal here***. That requirement is built in to convolutions and inner products. We will discuss it again below, for functions in continous time. The essential thing is to see *filters with downsampling* in equation (1.39). Each step takes us from a finer level j to a coarser level $j-1$, with half as many outputs.

You can see this logarithmic tree as a ***pyramid*** of averages and differences. The averages are sent up the pyramid, to be averaged again (and also differenced). Whenever a difference is computed, it is final. The sum of $\frac{1}{2}+\frac{1}{4}+\frac{1}{8}+\ldots$ gives 1, representing 100% of the output in the limit.

Figure 1.11 shows a finite pyramid. The input vector $\boldsymbol{x}$ at the bottom has length $L=2^J$. It is at level J, where we find 2^{J-1} differences and averages. The averages are the inputs at the next level $J-1$. Eventually we reach level 0 with an overall average and an overall difference (second half average $-$ first half average). Keep this overall average as the final component—you might say the zeroth component—of the wavelet transform. Starting at level $J=3$, where the input $\boldsymbol{x}$ has $L=2^J=8$ components, the count of wavelet coefficients is

$$4 \text{ differences} + 2 \text{ differences} + 1 \text{ difference} + \text{overall average} = 8.$$

The seven differences are wavelet coefficients b_{jk}. The overall average can be denoted for convenience by a_{00}. In the model case of infinite length, the iteration of the lowpass filter can go on forever.

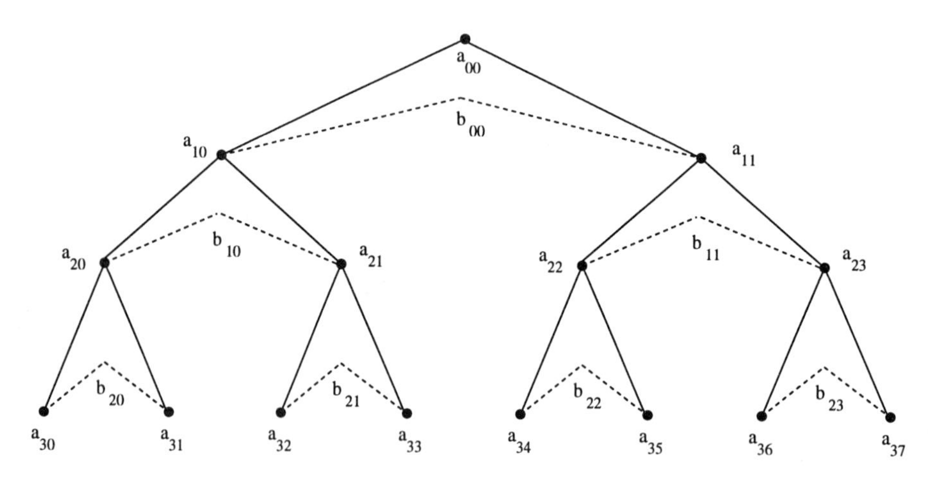

Figure 1.11: Averages a_{jk} go up the pyramid. Differences b_{jk} stop.

You must see the finite case in terms of matrices. With length $L=4$, there are two fine differences, one coarse difference, and the overall average. The columns of S are the basis vectors for the Haar wavelets:

$$\begin{bmatrix} 1 & 1 & 1 & 0 \\ 1 & 1 & -1 & 0 \\ 1 & -1 & 0 & 1 \\ 1 & -1 & 0 & -1 \end{bmatrix} \text{ is scaled by } r=\frac{1}{\sqrt{2}} \text{ to } S=\begin{bmatrix} r^2 & r^2 & r & 0 \\ r^2 & r^2 & -r & 0 \\ r^2 & -r^2 & 0 & r \\ r^2 & -r^2 & 0 & -r \end{bmatrix}. \quad (1.40)$$

The scaling gives unit vectors in the columns and rows, because $2r^2=1$ and $4r^4=1$. The inverse matrix appears in analysis, as we create the tree of averages and differences:

$$A = S^{-1} = S^T = \begin{bmatrix} r^2 & r^2 & r^2 & r^2 \\ r^2 & r^2 & -r^2 & -r^2 \\ r & -r & 0 & 0 \\ 0 & 0 & r & -r \end{bmatrix}. \tag{1.41}$$

The first row gives the overall average a_{00}. The second row gives b_{00}. The third and fourth rows give the finer differences, coming earlier in the tree of filters. The transform $\boldsymbol{b} = A\boldsymbol{x}$ contains these four numbers.

Fast Wavelet Transform (FWT)

The tree of filters is the fast wavelet transform!! The FWT expresses the analysis matrix A as a product of simple average-difference matrices, coming from the downsampled filters in the tree:

$$A = \begin{bmatrix} r & r & & \\ r & -r & & \\ & & 1 & \\ & & & 1 \end{bmatrix} \begin{bmatrix} r & r & & \\ & & r & r \\ r & -r & & \\ & & r & -r \end{bmatrix} \tag{1.42}$$

The matrix on the right is first in the tree. The coefficients r, r are in the lowpass filter, and downsampling leaves a matrix L in the top two rows. The rows of L have a *double shift.* The coefficients $r, -r$ come from the highpass (or bandpass) filter, downsampled to leave B. ***There is again a time-reversal from*** $-r, r$. The pattern that you see in the 4 by 4 Haar matrix is the product in (1.42):

$$\textit{Analysis tree:} \quad A = \left[\begin{array}{c|c} \begin{smallmatrix} L \\ B \end{smallmatrix} & \\ \hline & I \end{array}\right] \begin{bmatrix} L \\ B \end{bmatrix}.$$

This pattern applies to transforms of all lengths $L = 2^J$. ***It will apply to all wavelets!*** The fast transform expresses A (and later S) using *matrices with many zeros*. It is the matrix form of the *pyramid algorithm.*

For length $L = 2^J$, there will be J levels in the tree and J matrices in A. The matrix on the right, from the start of the tree, has two nonzeros in each row. Filters with T coefficients will produce T nonzeros in each row. Then the finest factor has TL nonzero entries.

The next factor has $TL/2$ coefficients in the top half, processing a shorter input. (The identity matrix in the lower right costs nothing. It leaves the differences alone.) The third stage of the tree has $TL/4$ coefficients. The total for the fast wavelet transform (= factored form of W^{-1}) comes from J factors:

$$TL\left(1 + \frac{1}{2} + \frac{1}{4} + \cdots + \frac{1}{2^{J-1}}\right) < 2TL. \tag{1.43}$$

Theorem 1.3 *The fast wavelet transform computes the L coefficients $\boldsymbol{b} = A\boldsymbol{x}$ in less than $2TL$ multiplications.*

The same count applies to the synthesis step $\boldsymbol{x} = S\boldsymbol{b}$. The factors of S give the inverse of equation (1.42). The synthesis tree has the inverse matrices in opposite order:

$$S = \begin{bmatrix} r & & r & \\ r & & -r & \\ & r & & r \\ & r & & -r \end{bmatrix} \begin{bmatrix} r & r & & \\ r & -r & & \\ & & 1 & \\ & & & 1 \end{bmatrix}. \tag{1.44}$$

The matrix on the right inverts the last analysis filter (still with time-reversal). It is the first step in the synthesis tree. The product of J matrices is the whole synthesis bank, which reconstructs $\boldsymbol{x}$ in Figure 1.12.

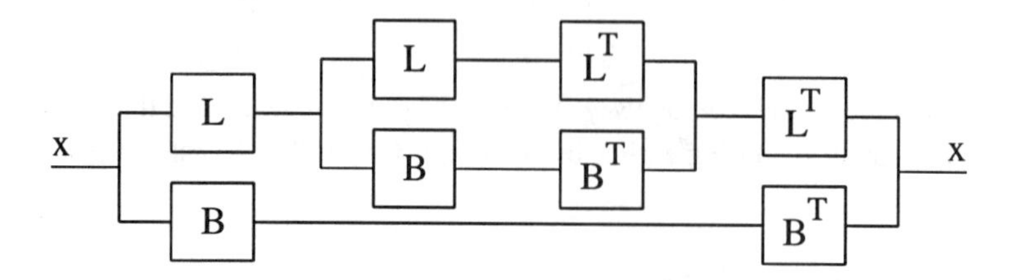

Figure 1.12: Analysis bank followed by synthesis bank: $\boldsymbol{SAx} = \boldsymbol{x}$. The matrices are factored by the tree, producing the FWT.

Haar Wavelets and Recursion

We move now to continuous time. The input $f(t)$ is a function instead of a vector. The output is the set of coefficients b_{jk} that multiply the wavelet basis functions $w_{jk}(t)$. These coefficients are inner products of $f(t)$ with $w_{jk}(t)$. The basis functions are normalized to unit length by the factor $2^{j/2}$:

$$b_{jk} = < f, w_{jk} > = \int_{-\infty}^{\infty} f(t)\, 2^{j/2}\, w\left(2^j t - k\right) dt. \tag{1.45}$$

For Haar, the wavelets are piecewise constant. The original wavelet $w(t) = w_{00}(t)$ is $+1$ on the interval $[0, \frac{1}{2})$ and -1 on the interval $[\frac{1}{2}, 1)$. The basis function w_{jk} is $2^{j/2}$ on a subinterval of length $\frac{1}{2}2^{-j}$ and $-2^{j/2}$ on the next subinterval. There are four wavelets at level $j = 2$, when we start on the unit interval $[0, 1)$. There are 2^j wavelets at level j.

To compute all the inner product integrals at levels $j = 0$, $j = 1$, and $j = 2$, we can integrate $f(t)$ over all eight subintervals of length $\frac{1}{8}$. This gives eight numbers. *How do those numbers produce b_{2k} and b_{1k} and b_{00} and a_{00}?*

The answer is beautiful. Those eight numbers are exactly the level 3 averages $a_{30}, a_{31}, a_{32}, \ldots, a_{37}$. We act on them exactly as in discrete time: filter and downsample! *The numbers at level $j-1$ come directly from the numbers at level j.* This is because the functions at level $j-1$ come from the functions at level j. The box function is a sum of half-boxes and the wavelet is a difference of half-boxes:

$$\phi(t) = \phi(2t) + \phi(2t-1) \text{ gives } \phi_{j-1,k}(t) = \frac{1}{\sqrt{2}}\left[\phi_{j,2k}(t) + \phi_{j,2k+1}(t)\right] \tag{1.46}$$

$$w(t) = \phi(2t) - \phi(2t-1) \text{ gives } w_{j-1,k}(t) = \frac{1}{\sqrt{2}}\left[\phi_{j,2k}(t) - \phi_{j,2k+1}(t)\right]. \tag{1.47}$$

Please look at these equations. They are the dilation equation and wavelet equation. The scaling function $\phi_{jk}(t) = 2^{j/2}\phi(2^j t - k)$ is constant on the interval of length 2^{-j} starting at $t = k2^{-j}$. When we take its inner product with $f(t)$, we are integrating $f(t)$ over this subinterval. Multiply the two equations by $f(t)$ and integrate:

$$\begin{array}{ll} \textit{Scaling coefficient:} & a_{j-1,k} = \frac{1}{\sqrt{2}}(a_{j,2k} + a_{j,2k+1}) \\ \textit{Wavelet coefficient:} & b_{j-1,k} = \frac{1}{\sqrt{2}}(a_{j,2k} - a_{j,2k+1}). \end{array} \tag{1.48}$$

The coefficients follow the pyramid algorithm. Notice again the time-reversal in which $2k+1$ appears. Our coefficients $a_{jk} = < f, \phi_{jk}(t) >$ and $b_{jk} = < f, w_{jk}(t) >$ are inner products, and a filter (a convolution $\sum \boldsymbol{h}(k)\, \boldsymbol{x}(n-k)$) has this reversal.

The main point is the pyramid. This beautiful connection between wavelets and filter banks was discovered by Stéphane Mallat. The pyramid algorithm is also called the Mallat algorithm (don't pronounce the 't'). It is a tree of butterflies in the Preface, it is a tree of filters in Figure 1.11, and a third form of the tree shows the trunk of averages a_{jk} and the branches of differences b_{jk}:

a_{jk} ⟶ $a_{j-1,k}$ • • • ⟶ a_{1k} ⟶ a_{0k}

↘ $b_{j-1,k}$ ↘ b_{1k} ↘ b_{0k}

Wavelet coefficients at level $j+1$ are differences of scaling coefficients at level j.

This pyramid is also an equality of functions. A function at fine resolution j is equal to a combination of "average plus detail" at coarse resolution $j-1$:

$$\sum_k a_{jk}\, \phi_{jk}(t) = \sum_k a_{j-1,k}\, \phi_{j-1,k}(t) + \sum_k b_{j-1,k}\, w_{j-1,k}(t). \tag{1.49}$$

Multiresolution in Continuous Time

The equality of functions in (1.49) is also an equality of *function spaces*. On the left side is a combination of ϕ's at level j. Let V_j denote the space of all such combinations. On the right side is a combination of ϕ's at level $j-1$. This is a function in the scaling space V_{j-1}. Also on the right is a combination of w's at level $j-1$. This is a function in the wavelet space W_{j-1}. The key statement of multiresolution is

$$V_j = V_{j-1} \oplus W_{j-1} \tag{1.50}$$

The symbol $+$ for vector spaces means that every function in V_j is a sum of functions in V_{j-1} and W_{j-1}, as in equation (1.49). The symbol $\oplus$ for "direct sum" means that those smaller spaces meet only in the zero function. This is guaranteed when ***the two subspaces are orthogonal***. In that case the direct sum $\oplus$ becomes an "orthogonal sum". For emphasis we restate the definition of the subspaces:

$$V_j = \text{all combinations } \sum_k a_{jk}\phi_{jk}(t) \text{ of scaling functions at level } j$$
$$W_j = \text{all combinations } \sum_k b_{jk}w_{jk}(t) \text{ of wavelets at level } j.$$

V_j is spanned by translates of $\phi\left(2^j t\right)$ and W_j is spanned by translates of $w\left(2^j t\right)$. The time scale is 2^{-j}. At each level, all inner products are zero. We have two orthogonal bases for V_j, either the ϕ's at level j or the ϕ's and w's at level $j-1$.

Notice how this multiresolution grows to three levels or more:

$$V_3 = V_2 \oplus W_2 = V_1 \oplus W_1 \oplus W_2 = V_0 \oplus W_0 \oplus W_1 \oplus W_2. \tag{1.51}$$

On the left side are all piecewise constant functions on intervals of length $\frac{1}{8}$. On the right side is the same space of functions, differently expressed. The functions in V_0 are constant on $[0, 1)$. The functions in W_0, W_1, W_2 are combinations of wavelets. The function $f(t)$ in V_3 has a piece $f_j(t)$ in each wavelet subspace W_j (plus V_0):

$$f(t) = \sum_k a_{0k}\phi_{0k}(t) + \sum_k b_{0k}w_{0k}(t) + \sum_k b_{1k}w_{1k}(t) + \sum_k b_{2k}w_{2k}(t). \tag{1.52}$$

Note to the reader. The parallels between a filter tree in discrete time and multiresolution in continuous time are almost perfect. The filter bank separates lower and lower frequencies, as we iterate. Multiresolution uses longer and longer wavelets, as we climb the pyramid. We are speaking of the analysis half, where inputs are separated by *scale*.

Discrete time	***Continuous time***
filter bank tree	multiresolution
downsampling $\omega \to \frac{\omega}{2}$	rescaling $t \to 2t$
lowpass filter	averaging with $\phi(t)$
highpass filter	detailing with $w(t)$
orthogonal matrices	orthogonal bases
analysis bank output	wavelet coefficients
synthesis bank output	sum of wavelet series
product of filter matrices	fast wavelet transform

The reader will understand that in writing about Haar wavelets, we are writing about all wavelets. The pattern is fundamental, the pieces in the pattern can change. The actual $\boldsymbol{c}(k)$ and $\boldsymbol{d}(k)$ and $\phi(t)$ and $w(t)$ are at our disposal. The next chapters move to *filter design*, where we make choices. Those choices determine the wavelet design. Some filters and wavelets are better than others. We will not allow ourselves to forget the pattern that makes all of them succeed.

Caution. The Haar wavelets are orthogonal. Thus they are biorthogonal *to themselves*. We are not seeing a clear difference between analysis and synthesis filters, $\boldsymbol{C}$ and $\boldsymbol{D}$ versus $\boldsymbol{F}$ and $\boldsymbol{G}$. For the same reason we are not seeing the dual functions $\widetilde{\phi}(t)$ and $\widetilde{w}(t)$. A clue to this shadow world (or tilde world) is in the time-reversals, which involve the *next* sample $2k+1$ instead of the previous sample $2k-1$. This suggests filters $\boldsymbol{C}^T$ and $\boldsymbol{D}^T$ rather than $\boldsymbol{C}$ and $\boldsymbol{D}$.

The biorthogonal case comes in Section 6.5. It has ***two*** multiresolutions, $\widetilde{V}_{j+1} = \widetilde{V}_j \oplus \widetilde{W}_j$ in parallel with $V_{j+1} = V_j \oplus W_j$. The pyramid and the fast wavelet transform go one way in analysis (with tilde). The inverse transform goes the other way in synthesis (without tilde).

Problem Set 1.6

1. (a) Show that the exact sum in equation (1.43) is $2T(N-1)$.

(b) How many matches are needed to decide the winner in a knockout tournament with N players? The average is the winner that goes to the next round (next filter). The difference is the loser that stops.

2. Write out the factorization of Haar's A for $N = 8$ (following 1.42).

3. For $N = 8$, write out the factorization of S corresponding to (1.44).

4. Draw the synthesis tree, the reverse of Mallat's analysis tree. A similar pyramid algorithm was proposed by Burt and Adelson before wavelets were named.

5. Split the function $f(t)$ into a scaling function plus a wavelet.

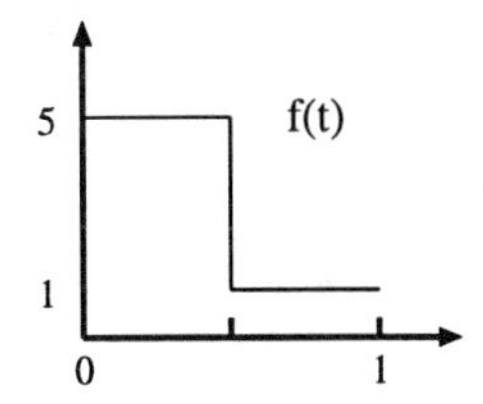

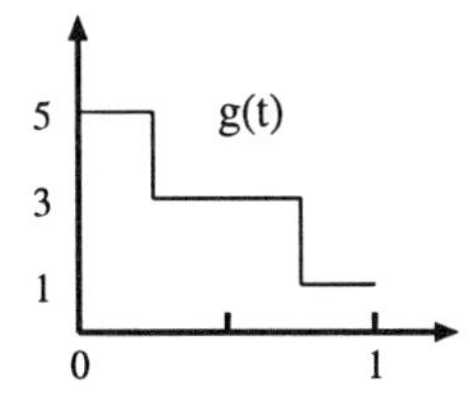

6. Split the function $g(t)$ into its pieces in V_0, W_0, and W_1 (box plus up-down coarse wavelet plus two fine wavelets).

7. The function $g(t)$ is in V_2 (its scale is $\frac{1}{4}$). The pyramid splits it first into a function in V_1 (two half-size boxes) plus a function in W_1 (two half-size wavelets). Find those pieces. Then split the piece in V_1 into a full-size box plus a full-size wavelet.

8. These three pieces are in V_0 and W_0 and W_1. Synthesize $f_1(t)$ in V_1 from the first two pieces. Then add the details in the third piece to synthesize $f_2(t)$ in V_2.

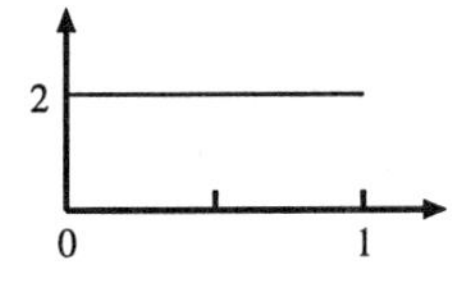

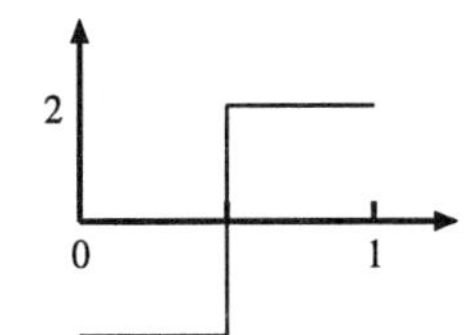

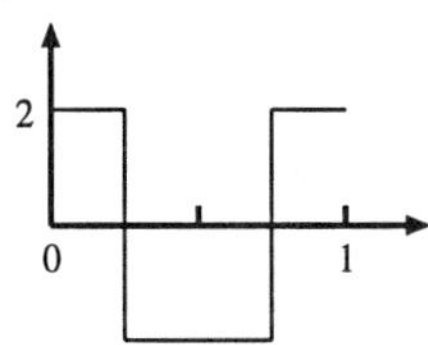

9. Suppose $H(t - \frac{1}{3})$ is the unit step function with jump at $t = \frac{1}{3}$. Its inner products a_{jk} with the boxes $\phi_{jk}(t)$ will be nonzero for about two-thirds of the 2^j boxes on $[0, 1]$. How many inner products b_{jk} with the wavelets $w_{jk}(t)$ will be nonzero?

This example shows the compression of step functions by Haar wavelets.

Chapter 2

Filters

This chapter comes between the Haar example of Chapter 1 and the full development of filter banks (leading to wavelets). We decided to collect other definitions that belong to this circle of ideas. Here is an indication of our plan:

Basic filters: Ideal filters and then FIR filter design
Basic tools: Fourier methods and functional analysis
Bases and frames: A matrix $\boldsymbol{T}$ has a two-sided or only a one-sided inverse
Integral transforms: windows in time-frequency, wavelets in time-scale.

Signal processing is an enormous subject. The input signal can arrive in many forms: continuous time, discrete time, finite time. It can be processed in many ways. Our greatest interest is a signal $\boldsymbol{x}(n)$ in discrete time that is processed by a linear time-invariant operator. *If the input is shifted in time then the output is equally shifted.* These operators are ***filters*** — the fundamental actors in signal processing.

Filters can be expressed in three domains: n, ω, z. In each domain the filter is a multiplication:

(n) : Multiplication by a ***Toeplitz matrix*** with $\boldsymbol{h}(n)$ on the nth diagonal.
(ω) : Multiplication by the ***frequency response*** $H(e^{j\omega}) = \sum \boldsymbol{h}(n)e^{-j\omega n}$.
(z) : Multiplication by the ***transfer function*** $H(z) = \sum \boldsymbol{h}(n)z^{-n}$.

We want to explain these three forms and the connections between them. *If we emphasize the matrix form more than usual, it is because that form is less well known.* We believe that the matrix formulation must become familiar, as the teaching and practice of signal processing rely increasingly on computer systems like MATLAB.

Our goal is to understand filters, through impulse responses and frequency responses and transfer functions. If you know signal processing as in the Oppenheim–Schafer text, go past the first sections. If you are learning the whole subject from scratch, these sections can help. Do not hesitate to use this chapter for reference, as you reach Chapter 4 and the heart of the book.

We begin with signals.

2.1 Signals, Samples, and Time-invariance

A discrete-time signal is a sequence of numbers. The sequence could be finite or infinite. Most signals in this book are *doubly infinite*; the index n goes from $-\infty$ to $+\infty$. Time has no start and no finish. The signals look like

$$\boldsymbol{x} = (\ldots, x_{-1}, x_0, x_1, x_2, \ldots) \qquad \text{or} \qquad \boldsymbol{x} = \begin{bmatrix} \cdots \\ \boldsymbol{x}(-1) \\ \boldsymbol{x}(0) \\ \boldsymbol{x}(1) \\ \boldsymbol{x}(2) \\ \cdots \end{bmatrix}.$$

Those components are real or complex numbers (usually real). One particular signal is of tremendous value. It is the *unit impulse* $\boldsymbol{x} = \boldsymbol{\delta}$:

$$\boldsymbol{\delta} = (\ldots, 0, 0, 1, 0, 0, \ldots) \quad \text{has components} \quad \delta(n) = \begin{cases} 0, & n \neq 0 \\ 1, & n = 0. \end{cases} \tag{2.1}$$

The continuous-time analogue is the "delta function" $\delta(t)$ — also called a Dirac impulse. In one case n is an integer, in the other t is a real number. The standard notations are $n \in \mathbf{Z}$ and $t \in \mathbf{R}$.

Together with the special vector $\boldsymbol{\delta}$ goes the delayed impulse $S\boldsymbol{\delta}$, where the unit component appears one sample later at $n = 1$. The whole vector is shifted by one time step. The symbol $\boldsymbol{S}$ stands for the *shift* or *delay* that has this effect on the vector $\boldsymbol{\delta}$:

$$S\boldsymbol{\delta} = (\ldots, 0, 0, 0, 1, 0, \ldots) \quad \text{has components} \quad \delta(n-1) = \begin{cases} 0, & n \neq 1 \\ 1, & n = 1. \end{cases}$$

It is worth emphasizing that a shift *to the right* (a delay) produces the *minus sign* in the expression $n - 1$. It is the same in continuous time. The graph of $f(t)$, when it is shifted one unit to the right, is the graph of $f(t-1)$. The delayed function at $t = 1$ equals the original function at $t = 0$.

When the components are shifted to the left, the impulse comes sooner (at $n = -1$). This shift is an *advance* instead of a delay. The symbol is S^{-1}. The operator "S inverse" has an effect opposite to S:

$$S^{-1}\boldsymbol{\delta} = (\ldots, 0, 1, 0, 0, 0, \ldots) \quad \text{has components} \quad \delta(n+1) = \begin{cases} 0, & n \neq -1 \\ 1, & n = -1. \end{cases}$$

Of course $S^{-1}S\boldsymbol{\delta} = \boldsymbol{\delta}$. The impulse is delayed by S and then advanced by S^{-1}. Also $SS^{-1}\boldsymbol{\delta} = \boldsymbol{\delta}$. The operator S^{-1} is a "two-sided inverse" of S.

The vector $\boldsymbol{\delta}$ could be defined, and Dirac's delta function should be defined, by what happens for inner products. The inner product (dot product) with any vector $\boldsymbol{x}(n)$ or any continuous function $x(t)$ picks out $\boldsymbol{x}(0)$:

$$\boldsymbol{x}^T\boldsymbol{\delta} = \sum_{-\infty}^{\infty} \boldsymbol{x}(n)\delta(n) = \boldsymbol{x}(0) \quad \text{and} \quad \langle x(t), \delta(t)\rangle = \int_{-\infty}^{\infty} x(t)\delta(t)\,dt = \boldsymbol{x}(0).$$

The number $\boldsymbol{x}(0)$ is a sample of the function $x(t)$. Many discrete signals come from sampling continuous signals. This *analog to digital* (A/D) conversion is a central part of communications technology. We sketch two continous-time signals $x(t)$, a step and an exponential, and their discrete-time samples $\boldsymbol{x}(n)$.

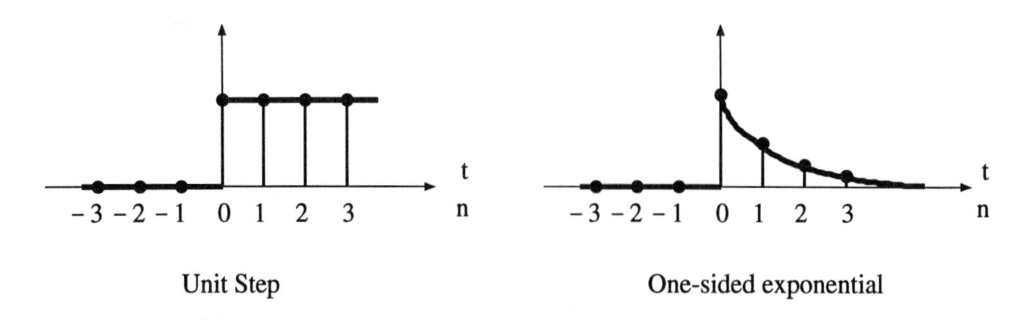

Unit Step One-sided exponential

The samples of $\cos \omega t$ are of special importance. The continuous signal has frequency ω (often normalized as $f = \frac{\omega}{2\pi}$). Everything depends on the sampling period T and the sampling rate $f_s = \frac{1}{T}$. The sampling may be fast enough to catch the oscillations in $\cos \omega t$, or it may be too slow. We may catch the oscillations in $\cos \omega t$, or miss some. The borderline is the ***Nyquist rate***. The sampling rate is exactly the Nyquist rate when $f_s = 2f$ and $\frac{1}{T} = \frac{\omega}{\pi}$ and $\omega T = \pi$. This is the rate (two samples per oscillation) in the first figure. Those samples have the fastest oscillation that a discrete vector can achieve: $\boldsymbol{x}(n) = (-1)^n$.

To repeat: The Nyquist rate gives the highest possible frequency $\omega T = \pi$.

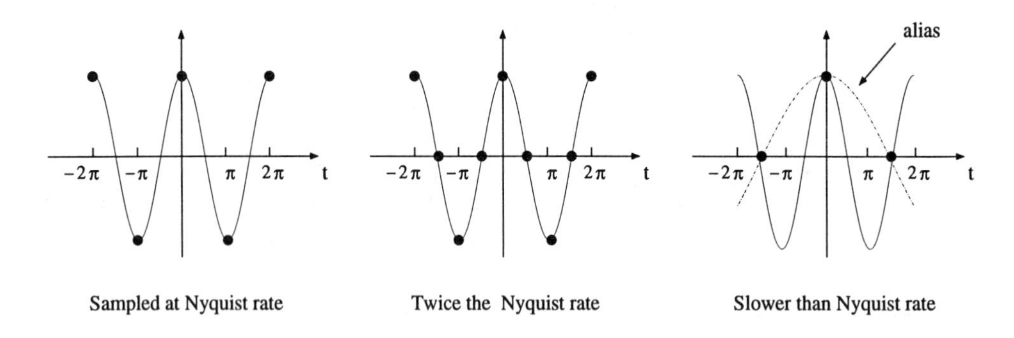

Sampled at Nyquist rate Twice the Nyquist rate Slower than Nyquist rate

The second sampling rate is *faster than Nyquist.* So the digital frequency ωT is less than π. In this figure, the sampling frequency f_s is twice the Nyquist rate, four samples (*four bullets*) per oscillation. The sampled signal is $\boldsymbol{x}(n) = \cos \omega n T = \cos \frac{n\pi}{2}$. Therefore the samples are $1, 0, -1, 0, 1, 0, \ldots$.

The third sampling rate is *slower than Nyquist.* The sample after $\cos 0$ is $\cos \frac{3\pi}{2}$. The sampling rate is $\frac{2}{3}$ of the Nyquist rate ($\frac{1}{T} = \frac{2}{3}\frac{\omega}{\pi}$). At this slow rate, the samples are the same $1, 0, -1, 0, 1, 0, \ldots$ as in the second figure! We cannot tell whether the true frequency of the continuous signal is ω (as drawn with solid line) or a slower frequency $\frac{\omega}{3}$ (as drawn by dotted line). ***This is aliasing***. The slow frequency $\frac{\omega}{3}$ is an alias for the true frequency ω, because the discrete samples at this rate will be exactly the same.

An extreme case of slow sampling is when $\omega T = 2\pi$. All the samples are $\cos \omega n T = \cos 2\pi n = 1$. Every sample is at the top of the wave. The digital frequency 2π looks identical to frequency zero (which is the alias of 2π).

When the continuous signal is a combination of many frequencies ω (or $f = \frac{\omega}{2\pi}$), the largest one ω_{max} sets the Nyquist rate $2f_{max}$. The corresponding Nyquist period T has $\omega_{max} T = \pi$. As long as the sampling period is smaller than this T, the sampling rate is faster than the Nyquist rate. Then there is no aliasing (by a lower frequency than the true frequency). *The continuous signal $x(t)$ can be recovered from its samples $\boldsymbol{x}(n)$.* The Shannon sampling formula, to achieve that recovery, is in Section 2.2.

Impulses and Delays in Three Domains

Impulses are the building blocks for all signals. Delays are the building blocks for all filters. We will present signals and filters in three ways — with time variable n, and frequency variable ω, and complex variable z.

$$\begin{array}{llcl} \text{Signal in the time domain} & \boldsymbol{x}(n) & = & (\ldots, \boldsymbol{x}(-1), \boldsymbol{x}(0), \boldsymbol{x}(1), \ldots) \\ \text{Signal in the frequency domain} & X(e^{j\omega}) & = & \sum \boldsymbol{x}(n) e^{-j\omega n} \ \ \text{(standard)} \\ & X(\omega) & = & \sum \boldsymbol{x}(n) e^{-i\omega n} \ \ \text{(reduced)} \\ \text{Signal in the } z\text{-domain} & X(z) & = & \sum \boldsymbol{x}(n) z^{-n}. \end{array}$$

The standard notation and reduced notation were compared in Section 1.1. We use both! Sometimes the standard notation is clearer; it allows direct replacement of $e^{j\omega}$ by z. Sometimes the reduced notation is simpler, as in $Y(\omega) = H(\omega)X(\omega)$.

The impulse $\boldsymbol{\delta} = (\ldots, 0, 0, 1, 0, 0, \ldots)$ becomes the constant function "1" in the frequency domain and z-domain. The only nonzero component $\boldsymbol{\delta}(0) = 1$ is in the constant term. The delayed impulse $\boldsymbol{y} = (\ldots, 0, 0, 0, 1, 0, \ldots)$ looks more interesting. In the other domains this $\boldsymbol{y}(n) = \boldsymbol{\delta}(n-1)$ is $Y(e^{j\omega}) = e^{-j\omega}$ and $Y(z) = z^{-1}$.

Note that $\boldsymbol{y}(1) = 1$ multiplies the *negative power* of z, by the signal processing convention. If the impulse is advanced instead of delayed, the nonzero occurs at $n = -1$. The transform is z instead of z^{-1}. This signal is no longer causal. The advance operator $\boldsymbol{S}^{-1}$ is not a causal filter.

Now we define filters in general and study delays in particular.

A ***digital filter*** is a combination $\boldsymbol{H} = \sum \boldsymbol{h}(n)\boldsymbol{S}^n$ of delays $\boldsymbol{S}$ and advances $\boldsymbol{S}^{-1}$.

The filter is completely determined by its coefficients $\boldsymbol{h}(n)$. When this sequence is finite, we have an "FIR filter". When $\boldsymbol{h}(n) = 0$ for negative n, we have a "causal filter". Our greatest interest is in causal FIR filters like

$$\boldsymbol{H} = \frac{1+\sqrt{3}}{8}\boldsymbol{I} + \frac{3+\sqrt{3}}{8}\boldsymbol{S} + \frac{3-\sqrt{3}}{8}\boldsymbol{S}^2 + \frac{1-\sqrt{3}}{8}\boldsymbol{S}^3 \quad \text{(Daubechies } D_4 \text{ filter)}.$$

Suppose this filter acts on the impulse $\boldsymbol{\delta}$. The output is a combination of $\boldsymbol{\delta}$ and its delays $\boldsymbol{S\delta}$ and $\boldsymbol{S}^2\boldsymbol{\delta}$ and $\boldsymbol{S}^3\boldsymbol{\delta}$:

$$\boldsymbol{H\delta} = (\ldots, \frac{1+\sqrt{3}}{8}, \frac{3+\sqrt{3}}{8}, \frac{3-\sqrt{3}}{8}, \frac{1-\sqrt{3}}{8}, 0, \ldots)$$

This is the "impulse response." It is equal to the vector $\boldsymbol{h}$ of filter coefficients:

The ***impulse response*** (causal and FIR) is $\boldsymbol{h} = (\boldsymbol{h}(0), \boldsymbol{h}(1), \ldots, \boldsymbol{h}(N))$.

You see why $\boldsymbol{H} = \sum \boldsymbol{h}(n)\boldsymbol{S}^n$ acting on $\boldsymbol{\delta}$ produces this output $\boldsymbol{h}$. Each term $\boldsymbol{h}(n)\boldsymbol{S}^n$ produces one response $\boldsymbol{h}(n)$ at time n.

As displayed, the filter is FIR and causal with $N+1$ "taps". The filter length is even when N is odd! Some authors end at $\boldsymbol{h}(N-1)$, so the length is N — but then the power $z^{-(N-1)}$ enters into a large number of formulas. We prefer to have sums from 0 to N, and scaling functions and wavelets on the interval $0 \le t < N$, and z^{-N} in all those formulas. The Daubechies filter has length 4 because $N = 3$.

The delay $\boldsymbol{S}$ takes $\boldsymbol{x} = (\ldots, \boldsymbol{x}(0), \boldsymbol{x}(1), \ldots)$ into $\boldsymbol{y} = (\ldots, \boldsymbol{x}(-1), \boldsymbol{x}(0), \ldots)$. Every linear operator like $\boldsymbol{S}$ is represented by a matrix. Since $\boldsymbol{x}$ and $\boldsymbol{y} = \boldsymbol{Sx}$ have infinitely many components, the shift matrix $\boldsymbol{S}$ has infinitely many rows and columns. This causal operator becomes a ***lower triangular*** matrix:

$$\boldsymbol{Sx} = \begin{bmatrix} \cdot & \cdot & \cdot & \cdot & \cdot & \cdot \\ \cdot & 0 & 0 & 0 & 0 & \cdot \\ \cdot & 1 & 0 & 0 & 0 & \cdot \\ \cdot & 0 & 1 & 0 & 0 & \cdot \\ \cdot & 0 & 0 & 1 & 0 & \cdot \\ \cdot & \cdot & \cdot & \cdot & \cdot & \cdot \end{bmatrix} \begin{bmatrix} \cdot \\ \boldsymbol{x}(-1) \\ \boldsymbol{x}(0) \\ \boldsymbol{x}(1) \\ \boldsymbol{x}(2) \\ \cdot \end{bmatrix} = \begin{bmatrix} \cdot \\ \boldsymbol{x}(-2) \\ \boldsymbol{x}(-1) \\ \boldsymbol{x}(0) \\ \boldsymbol{x}(1) \\ \cdot \end{bmatrix}.$$

The only nonzero coefficient for this filter is $\boldsymbol{h}(1) = 1$. This coefficient goes along diagonal ***one***. In general $\boldsymbol{h}(n)$ goes on diagonal n.

What does the delay do in the z-domain? The input $X(z)$ and output $Y(z)$ are

$$X(z) = \cdots + \boldsymbol{x}(0) + \boldsymbol{x}(1)z^{-1} + \cdots \quad \text{and} \quad Y(z) = \cdots + \boldsymbol{x}(0)z^{-1} + \boldsymbol{x}(1)z^{-2} + \cdots$$

The delay has multiplied $X(z)$ by z^{-1} to produce $Y(z)$. This transfer function z^{-1} is exactly the z-transform of the vector $(\ldots, 0, 1, 0, 0, \ldots)$ of filter coefficients. The pattern is always $Y(z) = H(z)X(z)$. Here is the special result for this particular filter, a delay $\boldsymbol{H} = \boldsymbol{S}$:

$$X(e^{j\omega}) \text{ is multiplied by } e^{-j\omega} \text{ and } X(z) \text{ is multiplied by } z^{-1}.$$

Time-invariant Filters

Our filters $\boldsymbol{H}$ are *linear*. This means in particular that "zero in produces zero out". If $\boldsymbol{x} = 0$ then necessarily $\boldsymbol{y} = 0$. The output from $2\boldsymbol{x}$ is $2\boldsymbol{Hx}$. The output from $\boldsymbol{x} + \boldsymbol{z}$ is $\boldsymbol{Hx} + \boldsymbol{Hz}$. This has an important consequence: $\boldsymbol{H}$ *is represented by a matrix.*

Our filters are also *time-invariant* (meaning shift-invariant). This leads to a special constant-diagonal property of the matrix:

$\boldsymbol{H}(\boldsymbol{Sx}) = \boldsymbol{S}(\boldsymbol{Hx})$: A shift of the input produces a shift of the output.
Each column of $\boldsymbol{H}$ is a delay of the previous column.
Each diagonal of $\boldsymbol{H}$ is constant and the nth diagonal contains $\boldsymbol{h}(n)$.

Those are different statements of time-invariance. They imply that $\boldsymbol{H}$ is a combination of shift operators: Every filter has the form $\boldsymbol{H} = \sum \boldsymbol{h}(n)\boldsymbol{S}^n$. The filter $\boldsymbol{H} = \boldsymbol{I} + 4\boldsymbol{S} + 3\boldsymbol{S}^2$ has coefficients $\boldsymbol{h} = (1, 4, 3)$. These are the entries down every column of the Toeplitz matrix.

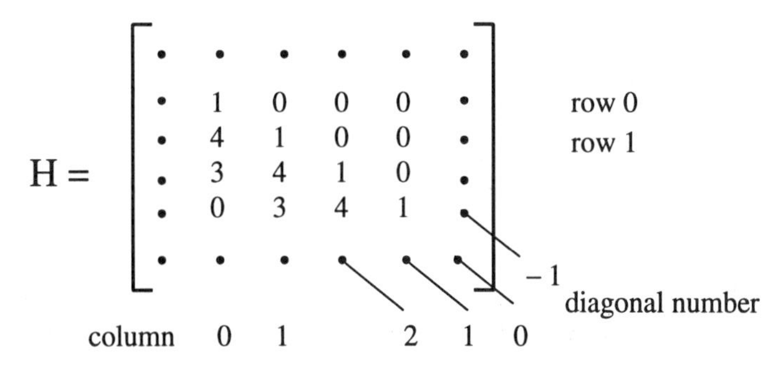

Toeplitz matrix = constant-diagonal matrix with entries $\mathbf{H}_{ij} = \mathbf{h}(i-j)$.

The numbers 1, 4, 3 also appear in every row—***but the order is reversed***. This is a causal FIR filter. It is time-invariant, because $\boldsymbol{HS}$ or $\boldsymbol{SH}$ (they are the same!) is $\boldsymbol{S} + 4\boldsymbol{S}^2 + 3\boldsymbol{S}^3$. All columns and all diagonals shift down by one, from the delay.

The difference between the row number i and the column number j is the diagonal number $k = i - j$. The entries of $\boldsymbol{H}$ depend only on k. This is a constant-diagonal matrix and a convolution matrix and a ***Toeplitz matrix***.

Matrix Multiplication and Vector Convolution

The jth column of $\boldsymbol{H}$ is $\boldsymbol{S}^j\boldsymbol{h}$. We can compute $\boldsymbol{Hx}$ as a combination $\sum \boldsymbol{x}(j)(\boldsymbol{S}^j\boldsymbol{h})$ of those columns. We can also compute $\sum \boldsymbol{h}(k)(\boldsymbol{S}^k\boldsymbol{x})$. Best to see the nth component $\boldsymbol{Hx}(n)$:

$$\sum_j \boldsymbol{x}(j)\boldsymbol{h}(n-j) \quad = \quad \sum_k \boldsymbol{h}(k)\boldsymbol{x}(n-k).$$

↑ nth component of delayed $\boldsymbol{h}$ ↑ nth component of delayed $\boldsymbol{x}$

The equality comes by changing j to $n-k$. The sums are over all integers, so the change is allowed. The finite sum to $j = N$ would not equal the sum to $k = N$, unless $\boldsymbol{h}$ has period N. (Then the N by N matrix $\boldsymbol{H}$ would be a ***circulant matrix***. This "wraparound" is the easiest way to deal with finite length signals, but not generally the best way.)

In both formulas for $\boldsymbol{Hx}(n)$, the indices add to n. The zeroth component is row zero of $\boldsymbol{H}$ times $\boldsymbol{x}$:

$$\boldsymbol{y}(0) = \boldsymbol{Hx}(0) = \boldsymbol{h}(N)\boldsymbol{x}(-N) + \cdots + \boldsymbol{h}(1)\boldsymbol{x}(-1) + \boldsymbol{h}(0)\boldsymbol{x}(0). \tag{2.2}$$

Each pair of indices on the right adds to zero—which is the index on the left. The numbers $\boldsymbol{h}(N), \ldots, \boldsymbol{h}(0)$ are like a *moving window* that multiplies $\boldsymbol{x}$. The nth component of the output has indices adding to n:

$$\boldsymbol{Hx}(n) = \boldsymbol{h}(N)\boldsymbol{x}(n-N) + \cdots + \boldsymbol{h}(1)\boldsymbol{x}(n-1) + \boldsymbol{h}(0)\boldsymbol{x}(n). \tag{2.3}$$

This is the pattern that produces ***convolution***:

The output vector is $\boldsymbol{Hx} = \boldsymbol{h} * \boldsymbol{x} =$ *convolution of* $\boldsymbol{h}$ *with* $\boldsymbol{x}$.

Convolution Rule Indices automatically add when they are the exponents in a polynomial. Multiply $H(z)X(z)$ and their coefficients undergo a convolution:

$$H(z)X(z) = (\boldsymbol{h}(0) + \boldsymbol{h}(1)z^{-1} + \cdots + \boldsymbol{h}(N)z^{-N})(\cdots \boldsymbol{x}(-1)z + \boldsymbol{x}(0) + \boldsymbol{x}(1)z^{-1} + \cdots).$$

The coefficient of z^0 is $\boldsymbol{h}(0)\boldsymbol{x}(0) + \boldsymbol{h}(1)\boldsymbol{x}(-1) + \cdots + \boldsymbol{h}(N)\boldsymbol{x}(-N)$. This is $\boldsymbol{Hx}(0)$.

The coefficient of z^{-n} in the product $H(z)X(z)$ is the nth component of $\boldsymbol{Hx}$. Multiplying polynomials means collecting terms with the same exponent. This is convolution.

Example 2.1. The filter matrix has $\boldsymbol{h}(0) = 1, \boldsymbol{h}(1) = 4$, and $\boldsymbol{h}(2) = 3$. Suppose the input has $\boldsymbol{x}(0) = 1$ and $\boldsymbol{x}(1) = 1$. Then we multiply polynomials or we take convolution of vectors:

$$\begin{array}{rrrr} & 3 & 4 & 1 \\ & & 1 & 1 \\ \hline & 3 & 4 & 1 \\ 3 & 4 & 1 & \\ \hline 3 & 7 & 5 & 1 \end{array} \qquad \begin{array}{rcl} (1 + 4z^{-1} + 3z^{-2})(1 + z^{-1}) & = & 1 + 5z^{-1} + 7z^{-2} + 3z^{-3} \\ (1, 4, 3) \ast (1, 1) & = & (1, 5, 7, 3). \end{array}$$

At $z = 1$ this is 8 times 2 equals 16. At $z = -1$ we check 0 times 0 equals 0.

Example 2.2. Multiplying two filter matrices (Toeplitz matrices) is also a convolution. The product $\boldsymbol{FH}$ is another time-invariant filter, and its coefficients are in $\boldsymbol{f} \ast \boldsymbol{h}$. This is just like multiplying polynomials, with the shift $\boldsymbol{S}$ in place of the complex variable z^{-1}:

$$\boldsymbol{FH} = (\boldsymbol{I} + \boldsymbol{S})(\boldsymbol{I} + 4\boldsymbol{S} + 3\boldsymbol{S}^2) = \boldsymbol{I} + 5\boldsymbol{S} + 7\boldsymbol{S}^2 + 3\boldsymbol{S}^3.$$

Note again that the order is not important: $\boldsymbol{FH} = \boldsymbol{HF}$. This will change when there are sampling operators $(\downarrow 2)$ and $(\uparrow 2)$ between the filters.

Example 2.3. The convolution of $(1, a, a^2, \ldots)$ with $(1, b, b^2, \ldots)$ is

$$(1 + bz^{-1} + b^2z^{-2} + \cdots)(1 + az^{-1} + a^2z^{-2} + \cdots) = 1 + (a + b)z^{-1} + (a^2 + ab + b^2)z^{-2} + \cdots$$

The power z^{-2} in the product comes from z^0 times z^{-2}, and from z^{-1} times z^{-1}, and from z^{-2} times z^0. The sum of exponents is -2, to give z^{-2} in the answer. This is the pattern for the indices k and $n - k$. Their sum is always n, to give $\boldsymbol{y}(n)$ in the convolution.

The Inverse of a Time-invariant H

The filter $\boldsymbol{H}$ is ***invertible*** if and only if

$$\begin{array}{ll} H(\omega) \neq 0 & \text{for all frequencies } \omega \\ H(z) \neq 0 & \text{for all } |z| = |e^{j\omega}| = 1. \end{array} \tag{2.4}$$

Then $\boldsymbol{H}^{-1}$ is also a constant-diagonal matrix. Its frequency response is $1/H(\omega)$.

Invertibility is the first of many properties that become infinitely simpler by transforming convolution to $H(\omega)X(\omega)$. The inverse of multiplication is division! We recover $X(\omega)$ from $Y(\omega)/H(\omega)$. The requirement is $H(\omega) \neq 0$.

To emphasize: If we know a frequency ω_0 for which $H(\omega_0) = 0$, then we know an input $\boldsymbol{x}$ for which $\boldsymbol{Hx} = \boldsymbol{0}$. That input has the pure frequency ω_0. It is the vector with components $\boldsymbol{x}(n) = e^{-j\omega_0 n}$. The pure frequency is selectively killed by $H(\omega_0) = 0$. Then $H(\omega_0)X(\omega_0) = 0$ and $\boldsymbol{H}^{-1}$ fails.

A moving average with equal weights $\boldsymbol{h}(0) = \boldsymbol{h}(1) = \frac{1}{2}$ is not invertible. The frequency response $H(\omega) = \frac{1}{2}(1 + e^{-j\omega})$ is zero at $\omega = \pi$. The vector with components $\boldsymbol{x}(n) = e^{-j\pi n} = (-1)^n$ is exactly the vector that has $\boldsymbol{Hx} = \boldsymbol{0}$. By changing to two unequal weights *the system becomes invertible.*

Example 2.4. Suppose $\boldsymbol{h}(0) = 1$ and $\boldsymbol{h}(1) = -\beta$. The frequency response is $H(\omega) = 1 - \beta e^{-i\omega}$. If we select β smaller than one, then $1 \neq \beta e^{-i\omega}$. Thus $H(\omega) \neq 0$. The matrix $\boldsymbol{H}$ has 1 on the main diagonal and $-\beta$ on the diagonal below. To invert in the frequency domain, divide by $H(\omega)$. To invert in the time domain, practice with a 4 by 4 matrix:

$$\begin{bmatrix} 1 & & & \\ -\beta & 1 & & \\ & -\beta & 1 & \\ & & -\beta & 1 \end{bmatrix}^{-1} = \begin{bmatrix} 1 & & & \\ \beta & 1 & & \\ \beta^2 & \beta & 1 & \\ \beta^3 & \beta^2 & \beta & 1 \end{bmatrix}. \qquad (2.5)$$

This suggests the correct diagonals $1, \beta, \beta^2, \ldots$ for the infinite matrix $\boldsymbol{H}^{-1}$. If $\boldsymbol{H}$ is $\boldsymbol{I} - \beta\boldsymbol{S}$, its inverse $\boldsymbol{I} + \beta\boldsymbol{S} + \beta^2\boldsymbol{S}^2 + \cdots$ has the frequency response $1/H(\omega)$:

$$\frac{1}{1 - \beta e^{-i\omega}} = 1 + \beta e^{-i\omega} + (\beta e^{-i\omega})^2 + (\beta e^{-i\omega})^3 + \cdots$$

The most important of all infinite series (the *geometric series*) gives us this inverse: $\frac{1}{1-\beta} = 1 + \beta + \beta^2 + \cdots$. The sum is restricted to $|\beta| < 1$. Otherwise the series diverges.

Example 2.5. What if β is larger than 1? For a finite matrix we don't notice the difference. The 4 by 4 inverse above is still correct. But the infinite series has to be written *in powers of* $1/\beta$. *The inverse matrix changes from causal to anticausal.* Look first at the frequency response $1/H(\omega)$:

$$\frac{1}{1 - \beta e^{-i\omega}} = \frac{e^{i\omega}/\beta}{(e^{i\omega}/\beta) - 1} = -\frac{e^{i\omega}}{\beta} - \frac{e^{2i\omega}}{\beta^2} - \frac{e^{3i\omega}}{\beta^3} - \cdots. \qquad (2.6)$$

This involves positive powers. We have *advances instead of delays* in the inverse.

The difference between $|\beta| < 1$ and $|\beta| > 1$ is the difference between a zero *inside* the unit circle and a zero *outside* that circle. $H(z) = 1 - \beta z^{-1}$ has its only zero at $z = \beta$. Here is the general rule for inverses of causal FIR systems:

No inverse when a zero is *on* the unit circle : $\boldsymbol{I} + \boldsymbol{S}$ has no inverse.

Causal inverse when all zeros are *inside* the unit circle: $1 - \beta z^{-1}$ has $|\beta| < 1$.

Anticausal inverse when all zeros are *outside* $|z| = 1$: $1 - \beta z^{-1}$ has $|\beta| > 1$.

A zero on the unit circle gives a particular frequency ω_0 at which $H = 0$. Then $1/H$ breaks down and the system has no inverse.

Problem Set 2.1

1. What matrix represents the inverse shift $y = S^{-1}x$? In the z-domain, the input is $X(z) = \cdots + x(0) + x(1)z^{-1} + \cdots$ and the output is $Y(z) =$ ___. The advance S^{-1} multiplies the z-transform by ___.

2. Express the filter $H = S + S^{-1}$ with coefficients $h(-1) = 1$ and $h(1) = 1$ in all three domains. Write the matrix H with two nonzero diagonals. Write the transfer function $H(z)$ with two terms. Write the frequency response $H(e^{j\omega})$ or $H(\omega)$, and check that this is the transform of the impulse response $h = H\delta$.

3. Show that H=S+S^{-1} is not invertible in three ways. Find a nonzero input x such that $Hx = 0$. Find a frequency ω that has response $H(e^{j\omega}) = 0$. Find a number with $|z| = 1$ such that $H(z) = 0$.

4. What are the matrix H and coefficient vector h for the 3-term moving average $Hx(n) = \frac{1}{3}(x(n) + x(n-1) + x(n-2))$? This is not invertible. Find two vectors x for which $Hx = 0$. Find two numbers with $|z| = 1$ such that $H(z) = 0$. Find two frequencies such that $H(\omega) = 0$.

5. Express this 3-term moving average in the form $H = \sum h(n)S^n$. What is N? Find the output y when the input has $x(0) = x(1) = 1$. In the z-domain what are $X(z)$ and $Y(z)$, and how are they related?

6. For matrices show that $SS^{-1} = I$. What is the corresponding statement in the z-domain, about the transfer functions of S and S^{-1}?

7. Multiply the matrix S by itself. The product $H = S^2$ corresponds to what coefficient vector h and what transfer function $H(z)$?

8. Every filter $\sum h(n)S^n$ commutes with a delay: $\sum h(n)S^{n+1}$ is HS and also SH. Why does every filter commute with every other filter?

9. If the continuous-time signal is $x(t) = \cos t$, what is the period T that gives sampling exactly at the Nyquist rate? What samples $x(nT)$ do you get at this rate? What samples do you get from $x(t) = \sin t$?

10. If the sampling period is $T = 1$ and the continuous signal is $x(t) = e^{2\pi i t/5}$, describe the discrete signal $x(n)$. Is it periodic? Find two other frequencies ω such that $x(t) = e^{i\omega t}$ would give the same samples.

11. If the signal $x(t)$ has bandwidth 3 Khz, then the sampling rate must be at least ___ to avoid aliasing.

    Note that the sampling period is generally normalized to $T = 1$. Then the largest digital frequency is $\omega = \pi$. Our graphs of $H(\omega)$ do not extend beyond π.

12. Why is the downsampling operator $(\downarrow 2)\, x(n) = x(2n)$ not time-invariant? Give an example with $(\downarrow 2)\, Sx \neq S(\downarrow 2)x$.

13. When are these filters invertible? Which has a causal inverse? Which has an FIR inverse? Which is allpass with $|H(e^{j\omega})| = 1$?

$$\begin{aligned} H_1(z) &= (1-\alpha z^{-1})(1-\beta z^{-1}) \\ H_2(z) &= 1 + \beta z^{-1} + \beta^2 z^{-2} + \beta^3 z^{-3} + \cdots \\ H_3(z) &= (z-\beta)/(1-\beta z^{-1}) \\ H_4(z) &= 1 - \beta z^{-1} + z^{-2} \end{aligned}$$

14. Determine the range of α and β for which the LTI system with impulse response $h(n) = \begin{cases} \alpha^n; & n \geq 0 \\ \beta^n; & n < 0 \end{cases}$ is stable. Find the output $y(n)$ when $x(n) = (-1)^n$.

15. Determine the impulse response for the cascade of the two LTI systems having impulse responses $h_1(n) = \left(\frac{1}{2}\right)^n u(n)$ and $h_2(n) = \left(\frac{1}{4}\right)^n u(n)$ using the convolution formula $h(n) = \sum h_1(k)h_2(n-k)$. Here $u(n)$ is the unit-step sequence.

16. Let H_1 be a system that throws away odd-indexed samples: $y(n) = x(2n)$. Is H_1 linear and time-invariant? H_1 is the downsampling block and its operation is discussed in Chapter 3.

17. Suppose H_2 inserts an extra zero between samples of the input: $y(n) = \begin{cases} x(n/2), & \text{even } n \\ 0, & \text{odd } n \end{cases}$.

Is H_2 a linear time-invariant system? H_2 is the upsampling block and its operation is discussed in Chapter 3.

18. Which cascade of downsampling and upsampling is time-invariant and what is its impulse response?

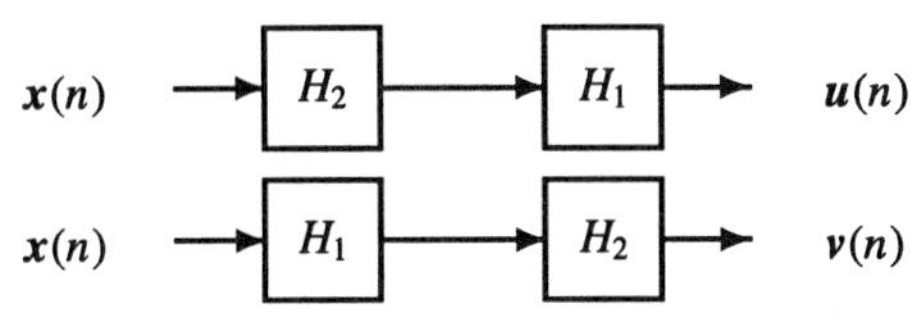

19. Give two examples of LTI systems, two examples of linear time-varying systems, and two examples of nonlinear systems.

2.2 Ideal Filters, Shannon Sampling, Sinc Wavelets

The word *filter* suggests that $\boldsymbol{H}$ selects a band of frequencies. It rejects another band. For ω in the ***passband***, the frequency response is near to $H(\omega) = 1$. For ω in the ***stopband***, the response is near to $H(\omega) = 0$. Any realizable non-ideal filter has a ***transition band*** in between, where $H(\omega)$ changes from pass to stop (from near 1 to near 0).

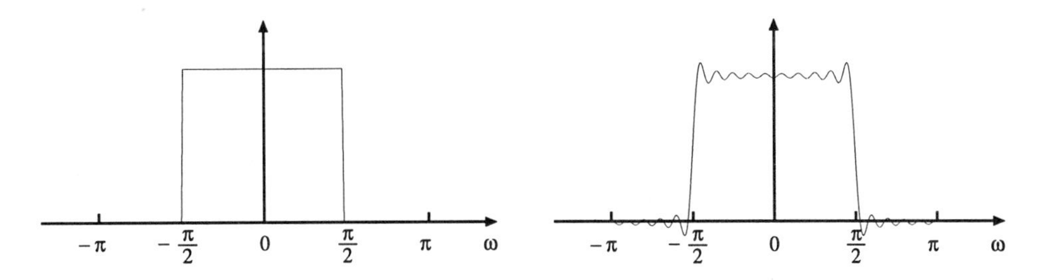

Figure 2.1: Ideal lowpass filter and best mean-square approximation with 20 terms.

We begin with an ideal filter, which has no transition band. Its responses are exactly $H(\omega) = 1$ and $H(\omega) = 0$. This is often called a *brick wall filter*, because of the step function in its graph. The response from an ideal lowpass filter is shown in Figure 2.1. This is a halfband filter, with sharp cutoff at $\omega = \frac{\pi}{2}$. The response $H(\omega) = \sum \boldsymbol{h}(k)e^{-ik\omega}$ is 2π-periodic, and the ideal response is zero in the high frequency band from $\frac{\pi}{2}$ to π:

$$\textbf{\textit{Ideal lowpass}} \quad H(\omega) = \sum_{-\infty}^{\infty} \boldsymbol{h}(k)e^{-ik\omega} = \begin{cases} 1, & 0 \le |\omega| < \frac{\pi}{2} \\ 0, & \frac{\pi}{2} \le |\omega| < \pi. \end{cases} \tag{2.7}$$

What filter coefficients $\boldsymbol{h}(k)$ produce this response? *Multiply the equation for $H(\omega)$ by $e^{in\omega}$ and integrate from $-\pi$ to π:*

$$\int_{-\pi}^{\pi} H(\omega)e^{in\omega}\,d\omega = \int_{-\pi}^{\pi} \Big(\sum_{k=-\infty}^{\infty} h(k)e^{-ik\omega} \Big) e^{in\omega}\,d\omega. \tag{2.8}$$

On the right side, there is an integral of $e^{-ik\omega}e^{in\omega}$ for each k. The great property of complex exponentials is that this integral is zero except when $k = n$:

$$\int_{-\pi}^{\pi} e^{-ik\omega}e^{in\omega}\,d\omega = \left[\frac{e^{i(n-k)\omega}}{i(n-k)}\right]_{-\pi}^{\pi} = 0 \quad \text{if} \quad k \neq n. \tag{2.9}$$

The function in brackets is periodic. It has the same value at $-\pi$ and π. After substituting those limits, the definite integral is zero. ***The complex exponentials are orthogonal.***

Now equation (2.8) has only one term on the right, from $k = n$. Integrating this constant from $-\pi$ to π gives the result $2\pi \boldsymbol{h}(n)$. This equals the left side:

$$\int_{-\pi}^{\pi} H(\omega)e^{in\omega}\,d\omega = 2\pi \boldsymbol{h}(n). \tag{2.10}$$

The brick wall filter has $H(\omega) = 1$ on the part $|\omega| < \frac{\pi}{2}$:

$$2\pi \boldsymbol{h}(n) = \int_{-\frac{\pi}{2}}^{\frac{\pi}{2}} e^{in\omega}\,d\omega = \left[\frac{e^{in\omega}}{in}\right]_{-\frac{\pi}{2}}^{\frac{\pi}{2}} = \frac{2}{n}\sin\frac{\pi n}{2}. \tag{2.11}$$

The coefficients in the ideal lowpass filter are samples of a sinc function:

$$\boldsymbol{h}(n) = \frac{\sin\frac{\pi n}{2}}{\pi n} = \begin{cases} \frac{1}{2}, & n = 0 \\ \pm\frac{1}{\pi n}, & n \text{ odd} \\ 0, & n \text{ even}, n \neq 0. \end{cases} \tag{2.12}$$

The halfband cutoff has produced a halfband filter! The coefficient $\boldsymbol{h}(0) = \frac{1}{2}$ is the "DC term" = average value of $H(\omega)$. *All other even-numbered coefficients are* $\boldsymbol{h}(n) = 0$. When $H(\omega)$ is antisymmetric around the halfband frequency $\omega = \frac{\pi}{2}$, the filter is always halfband.

For odd n, the numbers $\boldsymbol{h}(n)$ alternate sign and decay slowly:

$$\boldsymbol{h}(1) = \boldsymbol{h}(-1) = -\frac{1}{\pi}, \quad \boldsymbol{h}(3) = \boldsymbol{h}(-3) = \frac{1}{3\pi}, \quad \boldsymbol{h}(5) = \boldsymbol{h}(-5) = -\frac{1}{5\pi}, \ldots$$

The series that adds up to the brick wall (= square wave = ideal lowpass response) is

$$H(\omega) = \frac{1}{2} - \frac{e^{i\omega} + e^{-i\omega}}{\pi} + \frac{e^{i3\omega} + e^{-i3\omega}}{3\pi} - \frac{e^{i5\omega} + e^{-i5\omega}}{5\pi} + \cdots. \tag{2.13}$$

At $\omega = \frac{\pi}{2}$, the only nonzero term is halfway down the brick wall: $H\left(\frac{\pi}{2}\right) = \frac{1}{2}$. Most important is the behavior *close to this jump at* $\omega = \frac{\pi}{2}$, as shown in Figure 2.1 and below. Suppose we chop off the series after N terms:

> The ripple at $\omega = \frac{\pi}{2}$ gets narrower as $N \to \infty$ but its height approaches a constant (about 0.09). This is the ***Gibbs phenomenon***.

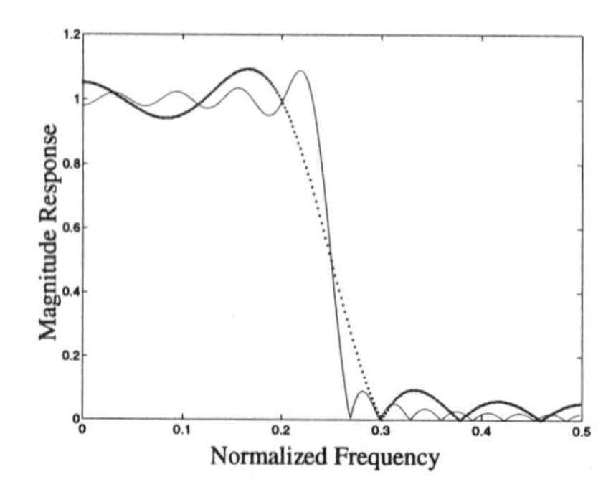

This Gibbs phenomenon can be a disaster numerically. The ripple represents error. It is expensive to take a large number of terms and impossible to take all terms. A finite N gives the best approximation in the *mean square sense* — but the tall ripple remains. This "sidelobe" shows up as an echo in audio filtering and as a ghost in image processing. Practical design turns toward ***equiripple filters***, which have many ripples of equal height. This design minimizes the maximum ripple height instead of the total ripple energy.

Note however that equiripple filters do not behave well in iteration. *They do not lead to good wavelets.*

Minimax filter design is implemented by the Parks-McClellan algorithm, which computes best approximations to ideal filters. Those filters have a passband, a stopband, and a "don't care" transition band. The ripple heights (maximum errors) decay exponentially with the filter length N. If the acceptable error is specified, there is a formula for N (see *equiripple* in the Glossary). An alternative is *eigenfilter design*. For many problems this allows a simple mean square calculation, but without the big sidelobe from the Gibbs phenomenon.

Historical note. It is surprising to read the original paper by Gibbs. He completely missed the Gibbs phenomenon. His correction published later was even shorter — about three important lines. This correction must have the highest signal to noise ratio in the history of science.

Fourier's Series by J. Willard Gibbs [*Nature*, vol. LIX, p. 200, December 29, 1898.]

... Let us write $f_n(x)$ for the sum of the first n terms of the series

$$\sin x - \tfrac{1}{2}\sin 2x + \tfrac{1}{2}\sin 3x - \tfrac{1}{4}\sin 4x + \text{etc.}$$

As n increases without limit, the curve defined by $y = 2f_n(x)$ approaches a limiting form, which may be thus described. Let a point move from the origin in a straight line at an angle of 45° with the axis of X to the point (π, π), thence vertically in a straight line to the point $(\pi, -\pi)$, thence obliquely in a straight line to the point $(3\pi, \pi)$. The broken line thus described (continued indefinitely forwards and backwards) is the limiting form of the curve as the number of terms increases indefinitely....

Correction [in *Nature*, vol. LIX, p. 606, April 27, 1899.]

I should like to correct a careless error which I made in describing the limiting form of the family of curves represented in the equation

$$y = 2\left(\sin x - \tfrac{1}{2}\sin 2x \cdots \pm \tfrac{1}{n}\sin nx\right) \tag{2.14}$$

as a zigzag line consisting of alternate inclined and vertical portions. The inclined portions were correctly given, but the vertical portions, which are bisected by the axis of X, extend beyond the points where they meet the inclined portions, their total lengths being expressed by four times the definite integral $\int_0^\pi \frac{\sin u}{u} du$.... But this limiting form of the graphs of the functions expressed by the sum is different from the graph of the function expressed by the limit of that sum.

I think this distinction important, for (with exception of what relates to my unfortunate blunder described above) whatever differences of opinion have been expressed on this subject seem due, for the most part, to the fact that some writers have had in mind the *limit of the graphs*, and others the *graph of the limit* of the sum.

The Gibbs phenomenon means that convergence to a brick wall is *not uniform*, as N increases. The coefficients $\boldsymbol{h}(n)$ approach zero but the sum of absolute values $1 + \frac{1}{3} + \frac{1}{5} + \cdots$

is infinite. These are multiplied by $\pm \sin n\omega$, so the actual series for $H(\omega)$ does not blow up. But the slow $1/n$ decay prevents uniform convergence and allows the large sidelobe. This ripple always appears in the Fourier series near a jump discontinuity.

Ideal Filter with Downsampling

In the time domain, $\boldsymbol{h}(n)$ is on the nth diagonal of the filter matrix $\boldsymbol{H}$. Writing a and b in place of $\boldsymbol{h}(1) = \boldsymbol{h}(-1)$ and $\boldsymbol{h}(3) = \boldsymbol{h}(-3)$, three rows are

$$\boldsymbol{H} = \begin{matrix} \cdots & 0 & b & 0 & a & 0.5 & a & 0 & b & 0 & \cdots & & \\ & \cdots & 0 & b & 0 & a & 0.5 & a & 0 & b & 0 & \cdots & \\ & & \cdots & 0 & b & 0 & a & 0.5 & a & 0 & b & 0 & \cdots \end{matrix}$$

Those rows are not orthogonal! The dot product of the first two rows is a. Only an allpass filter has orthonormal rows (and columns). Then $|H(\omega)| = 1$ for all ω.

What is fundamental for this book is that *the even-numbered rows of $\boldsymbol{H}$ are orthogonal.* When the downsampling operator $(\downarrow 2)$ removes half of the rows, this leaves a ***double shift*** in the remaining rows — the rows of $(\downarrow 2)\boldsymbol{H}$:

$$(\downarrow 2)\boldsymbol{H} = \begin{matrix} \cdots & 0 & b & 0 & a & 0.5 & a & 0 & b & 0 & \cdots & & & \\ & & \cdots & 0 & b & 0 & a & 0.5 & a & 0 & b & 0 & \cdots & \\ & & & & \cdots & 0 & b & 0 & a & 0.5 & a & 0 & b & 0 \; \cdots \end{matrix}$$

Orthogonality is not so clear in this time domain. Moving into the frequency domain, the double shift is a multiplication by $e^{-i2\omega}$ and row 0 of $\boldsymbol{H}$ is orthogonal to row 2:

$$(\text{row } 0) \cdot (\text{row } 2) \quad = \quad \int_{-\pi}^{\pi} H(\omega)\overline{e^{-i2\omega} H(\omega)}\, d\omega \quad = \quad \int_{-\pi/2}^{\pi/2} e^{i2\omega}\, d\omega \quad = \quad 0.$$

Similarly row 0 is orthogonal to row 4, because the integral of $e^{i4\omega}$ is zero. This integral is over the half-period where $H(\omega) = 1$. The integral of $e^{i\omega}$ is not zero over this half-period, and row 0 of $\boldsymbol{H}$ is not orthogonal to row 1.

The Ideal Highpass Filter

Haar's lowpass filter has coefficients $\frac{1}{2}$ and $\frac{1}{2}$, where the highpass filter has $\frac{1}{2}$ and $-\frac{1}{2}$. Those are clearly orthogonal. Now the ideal lowpass filter has infinitely many coefficients. We want to construct a highpass filter $\boldsymbol{H}_1$, so that the rows of $(\downarrow 2)\boldsymbol{H}_1$ will be orthogonal to the rows of $(\downarrow 2)\boldsymbol{H}$.

It is easy to make $H_1(\omega)$ orthogonal to the ideal lowpass $H(\omega)$. We set $H_1 = 1$ in the intervals where $H = 0$:

$$\text{The ideal } H_1(\omega) \text{ can be } \begin{cases} 0 & \text{when} \quad 0 \le |\omega| < \frac{\pi}{2} \\ 1 & \text{when} \quad \frac{\pi}{2} \le |\omega| < \pi. \end{cases}$$

Shifted by π, the ideal $H(\omega)$ produces $H_1(\omega)$. In the time domain, that shift by π reverses the signs of the *odd-numbered* components (because $\boldsymbol{h}(k)$ changes to $\boldsymbol{h}(k)e^{ik\pi}$):

$$\boldsymbol{h}_1(0) = \frac{1}{2}, \quad \boldsymbol{h}_1(1) = \boldsymbol{h}_1(-1) = +\frac{1}{\pi}, \quad \boldsymbol{h}_1(3) = \boldsymbol{h}_1(-3) = -\frac{1}{3\pi}, \; \ldots$$

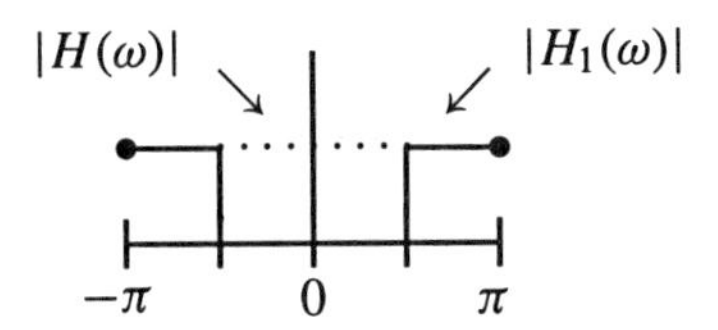

Figure 2.2: Ideal magnitude frequency responses: low $|H(\omega)|$ and high $|H_1(\omega)|$.

The graphs of $|H(\omega)|$ and $|H_1(\omega)|$ are in Figure 2.2. We explain below why absolute values suddenly appeared. Whatever the phase, orthogonality is sure because of no overlap.

We note that $|H(\omega)| + |H_1(\omega)| = 1$. This is not the identity that really matters. It is the sum of *squares* that applies not only in this case but in all orthogonal filter banks:

$$|H(\omega)|^2 + |H(\omega+\pi)|^2 \equiv 1. \tag{2.15}$$

For the Haar example this is the identity $\cos^2 \frac{\omega}{2} + \sin^2 \frac{\omega}{2} = 1$. The ideal case is deceptive because $1 = 1^2$ and $0 = 0^2$. The orthogonality requirement (2.15) will be established in Section 5.2.

The Alternating Flip (with Odd Shift)

There is an unusual point about the step from $\boldsymbol{H}$ to $\boldsymbol{H}_1$. In the time domain this usually comes from three operations on the coefficients $\boldsymbol{h}(n)$: *reverse the order, alternate the signs, and shift by 1* (or any odd N). This takes the lowpass coefficients $\boldsymbol{h}(n)$ into an orthogonal highpass sequence:

$$\textbf{\textit{Alternating flip}} \quad \boldsymbol{h}_1(n) = (-1)^n \boldsymbol{h}(N-n). \tag{2.16}$$

For a finite sequence $\boldsymbol{h}(0), \boldsymbol{h}(1), \ldots, \boldsymbol{h}(N)$ — assuming N is odd! — you immediately see the flip in the next figure and the orthogonality between rows:

low	$\boldsymbol{h}(0)$	$\boldsymbol{h}(1)$	$\cdots$	$\boldsymbol{h}(N-1)$	$\boldsymbol{h}(N)$
high	$\boldsymbol{h}(N)$	$-\boldsymbol{h}(N-1)$	$\cdots$	$\boldsymbol{h}(1)$	$-\boldsymbol{h}(0)$

Important: There is also orthogonality of double shifts, as in

shift low by 2	$\boldsymbol{h}(2)$	$\boldsymbol{h}(3)$	$\cdots$	$\boldsymbol{h}(N-1)$	$\boldsymbol{h}(N)$
high	$\boldsymbol{h}(N)$	$-\boldsymbol{h}(N-1)$	$\cdots$	$\boldsymbol{h}(3)$	$-\boldsymbol{h}(2)$

$\boldsymbol{h}(2)\boldsymbol{h}(N)$ cancels $-\boldsymbol{h}(N)\boldsymbol{h}(2)$. This happens for all double shifts. It does not usually happen for single shifts! A single shift of Haar to $0, \frac{1}{2}, \frac{1}{2}$ is not orthogonal to the highpass $\frac{1}{2}, -\frac{1}{2}, 0$. Double shifts are all we care about, because it is double-shifted rows in $(\downarrow 2)\boldsymbol{H}$ that are orthogonal — and now the highpass rows in $(\downarrow 2)\boldsymbol{H}_1$ complete the filter bank.

The alternating flip is the key to orthogonality. In the frequency domain it has three steps: *Multiply by $e^{i\omega}$ to shift, take complex conjugates to flip, shift by π to alternate signs*:

$$H_1(\omega) = e^{-i\omega}\overline{H(\omega+\pi)}. \tag{2.17}$$

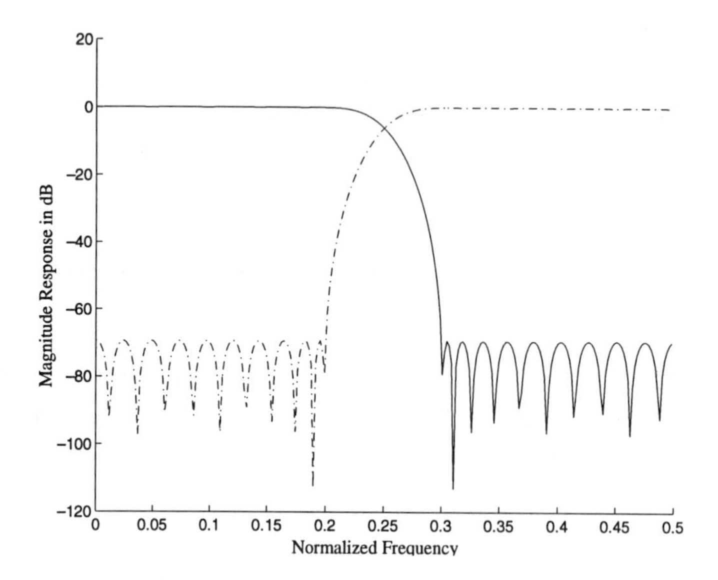

Theorem 2.1 *The alternating flip makes the rows of* $(\downarrow 2)\boldsymbol{H}$ *orthogonal to the rows of* $(\downarrow 2)\boldsymbol{H}_1$.

This was verified by eye, in the low and high rows above. You can verify it again for the ideal filters with infinitely many $\boldsymbol{h}$'s:

row 0 of $\boldsymbol{H}$	$\cdots$	b	0	a	0.5	a	0	b	0	$\cdots$
row 0 of $\boldsymbol{H}_1$	$\cdots$	0	$-b$	0	$-a$	0.5	$-a$	0	$-b$	$\cdots$

In the frequency domain we will see orthogonality in another form:

$$H(\omega)\overline{H_1(\omega)} + H(\omega+\pi)\overline{H_1(\omega+\pi)} = H(\omega)e^{i\omega}H(\omega+\pi) + H(\omega+\pi)e^{i(\omega+\pi)}H(\omega) \equiv 0. \tag{2.18}$$

All great, but for the ideal filters there is a very strange point. There was no odd shift! From the brick wall $H(\omega)$ on $[-\frac{\pi}{2}, \frac{\pi}{2}]$, we shifted by π to build the highpass brick wall. The wall is real, so conjugation has no effect. Still there should have been a phase shift from $e^{i\omega}$ and there wasn't.

Apparently row 0 of $\boldsymbol{H}$ is orthogonal to $[-b \ \ 0 \ \ -a \ \ .5 \ \ -a \ \ 0 \ \ -b \ \ 0 \ \cdots]$ *without the shift*. This unusual point occurs because $H_1(\omega)$ does not overlap $H(\omega)$. We can give $H_1(\omega)$ any phase we desire. It was natural to make it real. It is more consistent to include the phase shift $e^{i\omega}$.

This oddity (actually it is a lack of oddity) will reappear below for ideal wavelets. The sinc wavelets should have a shift in time, from the phase factor $e^{i\omega}$ — but generally they are taken from the unshifted $\boldsymbol{H}_1$. Before turning to scaling functions and wavelets, we include a short discussion of the Sampling Theorem — to recover a band-limited function $x(t)$ from its samples. This famous theorem appears everywhere, so we focus on a particular aspect involving $(\downarrow 2)$.

Shannon (Down-)Sampling Theorem

Our main theme for filter banks is perfect reconstruction of all signals. With two filters this will be achieved in spite of downsampling. The Sampling Theorem restricts the input to a subspace of band-limited signals. Then *one* downsampled output is enough to recover the input.

The signal lies in the lower halfband $|\omega| < \frac{\pi}{2}$; no higher frequencies are allowed. In an ideal filter bank, nothing comes out of the highpass channel. Full information must be in $(\downarrow 2)\boldsymbol{Hx}$. For a half-range of *input frequencies*, we only need a half-range of *output samples*.

Suppose the output $(\downarrow 2)\boldsymbol{x}$ *is an impulse* $\boldsymbol{\delta} = (\ldots, 0, 1, 0, \ldots)$. *What was* $\boldsymbol{x}$*?* A first suggestion is $\boldsymbol{x} = \boldsymbol{\delta}$, since $(\downarrow 2)\boldsymbol{\delta}$ is equal to $\boldsymbol{\delta}$. But this is wrong—because $\boldsymbol{\delta}$ is not band-limited. The impulse has all frequencies in equal amounts.

The correct input to yield $(\downarrow 2)\boldsymbol{x} = \boldsymbol{\delta}$ has *a halfband of frequencies in equal amounts*. The graph of $X(\omega)$ is a square wave:

$$X(\omega) = \begin{cases} 2 & \text{for} \quad 0 \le |\omega| < \frac{\pi}{2} \\ 0 & \text{for} \quad \frac{\pi}{2} \le |\omega| < \pi. \end{cases} \tag{2.19}$$

Downsampling doubles every frequency, so $(\downarrow 2)\boldsymbol{x}$ has a full band of frequencies in equal amounts. It equals $\boldsymbol{\delta}$ as required.

What signal $\boldsymbol{x}$ has the transform $X(\omega) =$ square wave? The inverse transform is

$$\boldsymbol{x}(n) = \frac{1}{2\pi}\int_{-\pi}^{\pi} X(\omega)e^{in\omega}\,d\omega = \frac{1}{\pi}\int_{-\frac{\pi}{2}}^{\frac{\pi}{2}} e^{in\omega}\,d\omega = \frac{2}{n\pi}\sin\frac{n\pi}{2}. \tag{2.20}$$

The input before downsampling has been recovered as the **sinc vector**:

$$\boldsymbol{x}_{\text{sinc}}(n) = \frac{\sin\frac{n\pi}{2}}{\frac{n\pi}{2}} \quad \text{with the convention} \quad \boldsymbol{x}_{\text{sinc}}(0) = 1. \tag{2.21}$$

Downsampling gives $\boldsymbol{\delta}$ *because if* n *is even then* $\sin\frac{n\pi}{2} = 0$.

The recovery problem is now solved when $(\downarrow 2)\boldsymbol{x}$ is $\boldsymbol{\delta}$. The band-limited input was $\boldsymbol{x}_{\text{sinc}}$. But every output is a combination of impulses at different times:

$$(\downarrow 2)\boldsymbol{x} = (\ldots, \boldsymbol{x}(0), \boldsymbol{x}(2), \ldots) = \cdots + \boldsymbol{x}(0)\boldsymbol{\delta} + \boldsymbol{x}(2)S\boldsymbol{\delta} + \cdots$$

Delaying the input by 2 delays $(\downarrow 2)\boldsymbol{x}$ by 1. This leads us to the correct input:

Downsampling Theorem The halfband signal that produces $(\downarrow 2)\boldsymbol{x} = (\ldots, \boldsymbol{x}(0), \boldsymbol{x}(2), \ldots)$ is

$$\boldsymbol{x}(n) = \cdots + \boldsymbol{x}(0)\boldsymbol{x}_{\text{sinc}}(n) + \boldsymbol{x}(2)\boldsymbol{x}_{\text{sinc}}(n-2) + \cdots = \sum_{-\infty}^{\infty} \boldsymbol{x}(2k)\frac{\sin\left((n-2k)\frac{\pi}{2}\right)}{(n-2k)\frac{\pi}{2}}.$$

For even n all terms are zero, except $2k = n$ which yields $\boldsymbol{x}(n)$. The input signal $\boldsymbol{x}$ is halfband because $\boldsymbol{x}_{\text{sinc}}$ and its shifts are halfband.

By changing $2k$ to k we would get the ordinary Shannon Sampling Theorem.

Sinc Wavelets (Shannon Wavelets)

In the Haar example, the lowpass $\boldsymbol{H}$ is the averaging filter (coefficients $\frac{1}{2}$ and $\frac{1}{2}$). By iteration we reached a continuous-time box function. That function satisfies the dilation equation with coefficients 1 and 1. The box is the "scaling function," and there is a corresponding up-down Haar wavelet.

Now $\boldsymbol{H}$ is the ideal lowpass filter. *Its scaling function is the sinc function* $\phi(t) = \frac{\sin \pi t}{\pi t}$. Since the ideal filter is IIR, the steps in our discussion will go a little more steeply. After this page, our main theme is FIR filter banks.

Section 6.4 has an infinite product formula for the Fourier transform of $\phi(t)$:

$$\widehat{\phi}(\omega) = \prod_{j=1}^{\infty} H\left(\frac{\omega}{2^j}\right). \tag{2.22}$$

In the ideal case, every factor $H(\omega/2^j)$ is one for $|\omega| < \pi$. The jth factor is zero for $2^{j-1}\pi \le |\omega| < 2^j\pi$. The infinite product gives a box function for $\widehat{\phi}(\omega)$, stretching from $-\pi$ to π. (Note that $H(\omega)$ is 2π-periodic, but the infinite product $\widehat{\phi}(\omega)$ is not.) The inverse transform $\phi(t)$ of this box is a sinc function:

The ideal scaling function is $\phi(t) = \frac{1}{2\pi}\int_{-\pi}^{\pi} e^{i\omega t}\, d\omega = \frac{\sin \pi t}{\pi t}$.

The wavelet $w(t) = \sum 2h_1(k)\phi(2t-k)$ comes from one application of the highpass filter (with downsampling) to $\phi(t)$. In the frequency domain this is

$$\widehat{w}(\omega) = H_1\left(\frac{\omega}{2}\right)\widehat{\phi}\left(\frac{\omega}{2}\right) = \begin{cases} 1 & \text{for } \pi \le |\omega| < 2\pi \\ 0 & \text{otherwise.} \end{cases} \tag{2.23}$$

The ideal filter bank cuts the frequency band in half. The upper half of the band goes through the highpass filter (discrete time). In continuous time it is a combination of wavelets. The lower half of the frequency band goes through the lowpass filter (discrete time). In continuous time this half is a combination of scaling functions ϕ—ready to be split again into wavelets and scaling functions at the next finer scale. We have an *octave decomposition* = *logarithmic decomposition* = "*constant-Q decomposition*" of the line of frequencies:

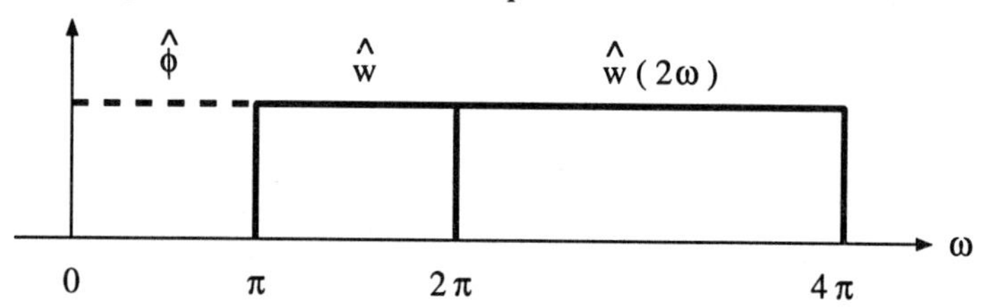

Note 1 Meyer smoothed out this ideal picture to produce band-limited wavelets with fast decay (IIR of course). The key is to keep $\sum |\widehat{w}_{\text{Meyer}}(2^n\pi\omega)|^2 \equiv 1$. This can be done smoothly with overlap of nearest neighbors only [D]. All band-limited wavelets have two bumps ($\pm\omega$) like the Shannon and Meyer wavelets.

Problem Set 2.2

1. Show that the inverse transform of $\widehat{w}$ in (2.23) is the ***sinc wavelet*** $w(t) = 2\text{sinc}(2t) - \text{sinc}(t)$.
2. Find the shifted sinc wavelet by inverse transform when the factor $e^{i\omega}$ is included in $H_1(\omega)$. This is the odd shift that we normally need for orthogonality of w to ϕ.
3. What are the coefficients $\boldsymbol{h}(n)$ for the ideal quarterband filter, with $H(\omega) = 1$ on $[-\frac{\pi}{4}, \frac{\pi}{4}]$? What scaling function $\widehat{\phi}(\omega)$ comes from the infinite product formula? Is $\phi(t)$ orthogonal to its translates?
4. ***(Important)*** Show that $H(\omega)$ has halfband symmetry (odd around its value at $\omega = \frac{\pi}{2}$) when $\boldsymbol{h}(n)$ is a **halfband filter**. This means $(\downarrow 2)\boldsymbol{h} = \boldsymbol{\delta}$.
5. Let $\boldsymbol{h}_{LP}(n)$ denote an ideal lowpass filter with cutoff frequency ω_c.

(a) Compute $H_{LP}(e^{j\omega})$ and normalized $\boldsymbol{h}_{LP}(n)$ such that $H_{LP}(e^{j0}) = 1$.

(b) The ideal highpass filter $\boldsymbol{h}_{LP}(n)$ can be designed by

$$\boldsymbol{h}_{HP}(n) = \begin{cases} 1 - \boldsymbol{h}_{LP}(n); & n = 0, \\ -\boldsymbol{h}_{LP}(n); & \text{otherwise.} \end{cases}$$

Find $H_{HP}(e^{j\omega})$.

(c) Design a highpass filter with cutoff frequency at $\pi - \omega_c$, using $\boldsymbol{h}_{LP}(n)$.

6. Let $H(z) = (1-2z^{-1}+3z^{-2}-3z^{-3}+2z^{-4}-z^{-5})$. Compute $|H(e^{j\omega})|$ and the phase response $\phi(\omega)$ and the group delay $\phi'(\omega)$.

7. Let $H(z)$ be an FIR lowpass filter of length $(N+1)$. Define $G_1(z) = z^{-N}\,H(z^{-1})$, $G_2(z) = H(-z)$ and $G_3(z) = z^{-N}H(-z^{-1})$.

(a) What are $\boldsymbol{g}_1(n), \boldsymbol{g}_2(n)$ and $\boldsymbol{g}_3(n)$ in terms of $\boldsymbol{h}(n)$?

(b) If $H(z)$ is an even-length symmetric filter, what is the symmetry or antisymmetry of $G_1(z)$, $G_2(z)$ and $G_3(z)$?

(c) If z_0 is a zero of $H(z)$, what are the corresponding zeros of $G_1(z)$, $G_2(z)$ and $G_3(z)$?

(e) What are the relations of $|H(e^{j\omega})|$, $|G_1(e^{j\omega})|$, $|G_2(e^{j\omega})|$, and $|G_3(e^{j\omega})|$?

8. Show that $G(z) = H(z)H(z^{-1})$ is a symmetric filter. What type of filter is $G(z)$ (lowpass, bandpass, highpass) if $H(z)$ is highpass? What is the (constant) phase of $G(z)$?

9. We have stated that $\delta_p = \delta_s$ in a halfband filter. Prove this.

2.3 Lowpass and Highpass Filter Design

The previous section dealt with ideal filters (necessarily IIR). This section deals with real FIR filters — often symmetric or antisymmetric (thus linear phase). We indicate the goals of filter design and we briefly discuss design methods.

For ideal brick walls, the transition from $H(\omega) = 1$ to $H(\omega) = 0$ happens instantly. For FIR filters this is not possible. It is important to see an actual magnitude response graph (Figure 2.3). In normal scale, we can observe details in the passband but not in the stopband. In logarithmic scale (dB scale, plotting $20\log_{10}|H|$), it is the other way around. The stopband details are visible in dB scale but the passband details are lost.

Before moving to good lowpass filters, we review two very short and rather poor filters. This gives us a chance to emphasize the four types of linear phase filters — odd and even length, symmetric and antisymmetric.

Example 2.6. The impulse response is $\boldsymbol{h} = \left(\frac{1}{2}, \frac{1}{2}\right)$. The frequency response is

$$H(\omega) = \tfrac{1}{2}\left(1 + e^{-i\omega}\right) = \left(\frac{e^{i\omega/2} + e^{-i\omega/2}}{2}\right)e^{-i\omega/2} = \left(\cos\frac{\omega}{2}\right)e^{-i\omega/2}. \tag{2.24}$$

In that last form you see the magnitude $|H(\omega)| = \cos\frac{\omega}{2}$ and the phase $\phi(\omega) = -\frac{\omega}{2}$. The magnitude is $\cos 0 = 1$ at zero frequency — this is a lowpass filter. The magnitude drops to zero at $\omega = \pi$. The *phase* $\phi(\omega)$ is the *angle* $-\omega/2$ in the polar form $re^{i\phi}$. *This phase function* $-\frac{\omega}{2}$ *is linear in* ω. The noninteger 1/2 reflects the fact that the coefficients in $\boldsymbol{h}$ are symmetric about the "1/2" position.

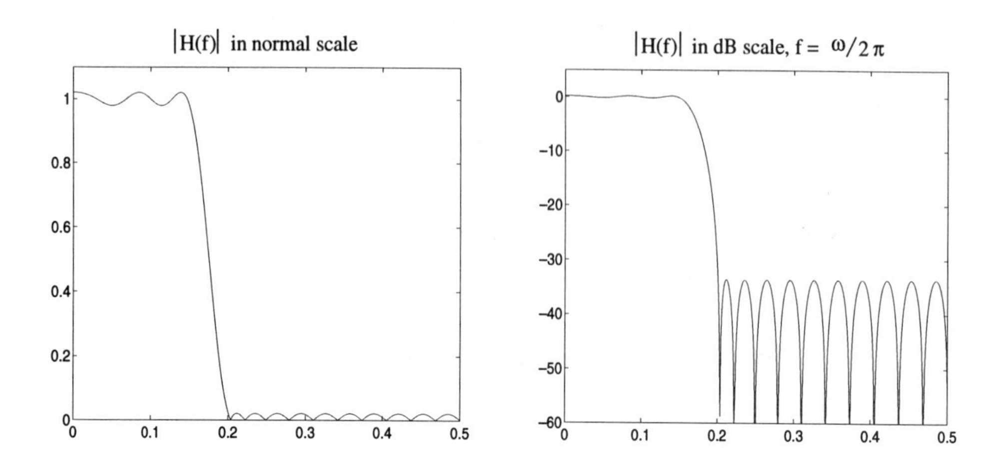

Figure 2.3: Magnitude response of a lowpass filter in normal and dB scale.

We will compute the phase $\phi(\omega)$ for other filters before analyzing its significance. Here we only mention: linear phase is a desirable property.

Example 2.7. By cascading the previous example we square its matrix $\boldsymbol{H}$ and we square its frequency response. The new filter coefficients are in

$$\begin{bmatrix} 0.5 & & \\ 0.5 & 0.5 & \\ 0 & 0.5 & 0.5 \\ \ddots & \ddots & \ddots \end{bmatrix}^2 = \begin{bmatrix} 0.25 & & \\ 0.50 & 0.25 & \\ 0.25 & 0.50 & 0.25 \\ \ddots & \ddots & \ddots \end{bmatrix} = \boldsymbol{H}_{\text{new}}.$$

The new impulse response is $\left(\frac{1}{4}, \frac{1}{2}, \frac{1}{4}\right)$. It is the convolution of $\left(\frac{1}{2}, \frac{1}{2}\right)$ with itself. Cascading filters means convolution of impulse responses (time domain). In the frequency domain we multiply responses $H(\omega)$:

$$H_{\text{new}}(\omega) = [H_{\text{old}}(\omega)]^2 = \tfrac{1}{4}(1 + e^{i\omega})^2 = \left(\cos^2 \frac{\omega}{2}\right)e^{-i\omega}. \tag{2.25}$$

This squares the magnitude and doubles the phase. The new phase $-\omega$ is still linear. It corresponds to a time shift of 1. The impulse response is a unit delay of a symmetric $\boldsymbol{h}$. $\boldsymbol{H}_{\text{new}}$ is a shift (= delay) times a symmetric matrix.

Every FIR matrix can be made causal, by sufficiently many delays. A symmetric matrix (not causal) corresponds to phase = zero because $H(\omega)$ is *real*:

$$H_{\text{sym}}(\omega) = \tfrac{1}{4}e^{i\omega} + \tfrac{1}{2} + \tfrac{1}{4}e^{-i\omega} = \tfrac{1}{2}(1 + \cos\omega). \tag{2.26}$$

Was this cascade desirable? Neither $\boldsymbol{H}$ or $\boldsymbol{H}_{\text{new}}$ is very impressive. The frequency responses are far from ideal. $H_{\text{new}}(\omega)$ has better attenuation in the stopband, because the cosine is squared. But it is also smaller in the passband—farther away from the ideal $H \equiv 1$.

By alternating signs in the lowpass coefficients they become *highpass*: $\boldsymbol{h}_1 = (0.5, -0.5)$ and $\boldsymbol{h}_1 * \boldsymbol{h}_1 = (0.25, -0.50, 0.25)$. These are still linear phase. The alternation of signs is a phase shift (a *modulation*) by π. The first is antisymmetric but the second is still symmetric.

It is useful to tabulate the four types of linear phase filters with real coefficients. N can be odd or even. The coefficients can satisfy

$$\boldsymbol{h}(n) = \boldsymbol{h}(N-n) \text{ for symmetric,} \quad \boldsymbol{h}(n) = -\boldsymbol{h}(N-n) \text{ for antisymmetric.}$$

The linear phase $\phi(\omega) = -\omega N/2$ and the (real) amplitude response $H_R(\omega)$ are seen in $H\left(e^{j\omega}\right) = ce^{-j\omega N/2}H_R(\omega)$. The table shows that with odd N, symmetry guarantees a zero at $\omega = \pi$ and antisymmetry guarantees a zero at $\omega = 0$. The responses $H_R(\omega)$ in the table have factors $\cos\frac{\omega}{2}$ and $\sin\frac{\omega}{2}$. Remember that the filter length (number of taps) is $N+1$. Here $c = 1$ for symmetric and $c = j$ for antisymmetric filters.

Type 1	Type 2	Type 3	Type 4
even $N = 2K$ symmetric	odd $N = 2K+1$ symmetric	even $N = 2K$ antisymmetric	odd $N = 2K+1$ antisymmetric
$H_R = \sum_0^K b_n \cos n\omega$	$\cos\frac{\omega}{2}\sum_0^K b_n \cos n\omega$ zero at $\omega = \pi$	$\sin\omega\sum_0^{K-1} b_n \cos n\omega$ zeros at $\omega = 0, \pi$	$\sin\frac{\omega}{2}\sum_0^K b_n \cos n\omega$ zero at $\omega = 0$

Now consider the ideal lowpass filter with cutoff frequency ω_c. This is a band-limited filter in the frequency domain, therefore its support in the time domain is infinite. It must be IIR. Its impulse response is $\boldsymbol{h}_I(n) = \pi\frac{\sin(\omega_c n)}{\omega_c n}$. Since the time-support is infinite, one needs to approximate it by a finite impulse response. The sequence $\boldsymbol{h}(n)$ becomes time-limited, therefore not band-limited. The magnitude response $|H\left(e^{j\omega}\right)|$ typically has errors δ_p and δ_s in the passband and stopband (Figure 2.4). Those bands have cutoff frequencies ω_p and ω_s.

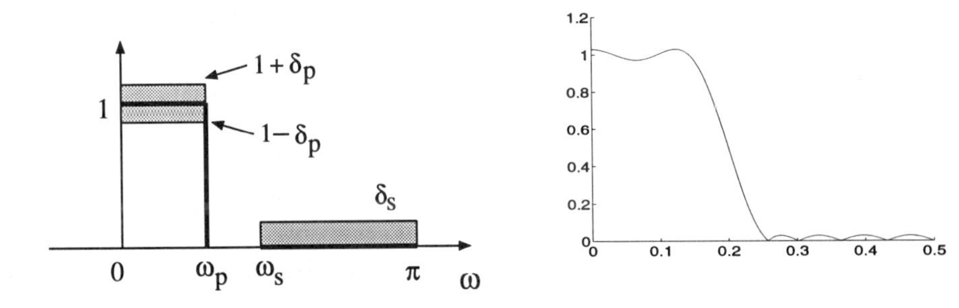

Figure 2.4: Idealized and typical magnitude responses of a lowpass filter.

Given ω_p, ω_s, and N, the errors δ_p and δ_s cannot both be small. Let $W = b/a$ be the relative error weighting. Increasing W in the design algorithm will decrease δ_s and increase δ_p. Figure 2.4 shows a magnitude response plot for a lowpass filter with length 19 and $\omega_p = 0.2\pi$ and $\omega_s = 0.3\pi$. The error weighting W is large (and thus yields a small stopband error).

In the sections below, several methods for the design of FIR digital filters are reviewed. These are based on windowing, minimax criteria, and weighted least squares.

Design by Windowing

The simplest way to truncate the ideal response $\boldsymbol{h}_I$ is by a rectangular window:

$$\boldsymbol{h}(n) = \boldsymbol{h}_I(n)\boldsymbol{w}(n) \quad \text{where} \quad \boldsymbol{w}(n) = \begin{cases} 1; & |n| \leq N/2 \\ 0; & \text{otherwise.} \end{cases}$$

This $H\left(e^{j\omega}\right)$ is the best least squares approximation to $H_I\left(e^{j\omega}\right)$. But chopping off the impulse response manifests as passband and stopband ripples in the frequency response. As the window size increases, the ripples get closer to the cutoff frequency ω_c, but these error heights do not decrease. This is the notorious Gibbs phenomenon. (Section 2.2 showed the magnitude response for halfband filters, when $\omega_c = \frac{\pi}{2}$. The solid line has $N = 30$, the dotted line has $N = 10$.) To design FIR filters with better error characteristics, we can smooth out the window $\boldsymbol{w}(n)$:

Hamming window	$w(n) = \alpha + (1-\alpha)\cos\left(\frac{2\pi n}{N}\right)$	$\lvert n\rvert \leq \frac{N-1}{2}$
Hanning window	$w(n) = \frac{1}{2} + \frac{1}{2}\cos\left(\frac{2\pi n}{N}\right)$	$\lvert n\rvert \leq \frac{N-1}{2}$
Kaiser window	$w(n) = \frac{1}{2} I_0\left[\beta\sqrt{1-\left(\frac{2n}{N}\right)^2}\right] / I_0(\beta)$	$\lvert n\rvert \leq \frac{N}{2}$

The parameter β in the Kaiser window controls the attenuation of the lowpass filter. $I_0(x)$ is the modified Bessel function and practical designs use about 20 terms of

$$I_0(x) = 1 + \sum_{k=1}^{\infty}\left[\frac{(0.5x)^2k}{k!}\right]^2.$$

Minimax Criteria (Equiripple Filter)

The filter with the smallest maximum error in passband and stopband is an *equiripple filter*. The equal heights of the ripples (and the number of ripples) assure that the error cannot be reduced—some ripple sizes will go up if others go down. A polynomial of degree N cannot have alternating signs at all ripples. The Remez algorithm to equalize the ripples was adapted to filter design by Parks and McClellan.

Figure 2.5 shows the magnitude response plot of an equiripple lowpass filter. Given a frequency specification in terms of cutoff frequencies (ω_p, ω_s), filter length $(N+1)$ and relative errors (δ_p, δ_s), the equiripple filter has the smallest maximum error in the frequency interval $0 \leq \omega \leq \pi$. The design algorithm for equiripple filters is the Remez exchange (McClellan-Parks) algorithm.

The order (N) of an equiripple filter is estimated by

$$N \approx \frac{-20\log_{10}\sqrt{\delta_1\delta_2} - 13}{14.6\Delta f} \tag{2.27}$$

where $\Delta f = (\omega_s - \omega_p)/2\pi$ is the transition band.

Equiripple designs are optimal in important respects—*but not optimal for iteration.* The reason is that they have at most one zero at $\omega = \pi$. The sampling operators $(\downarrow 2)$ will mix up the frequency bands that an equiripple filter so carefully separates! *The Daubechies filters go to the other extreme—no ripples at all and maximum flatness at* $\omega = \pi$. Then iteration of

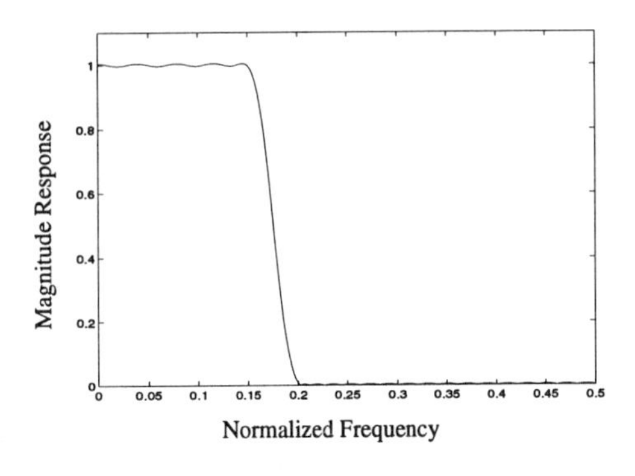

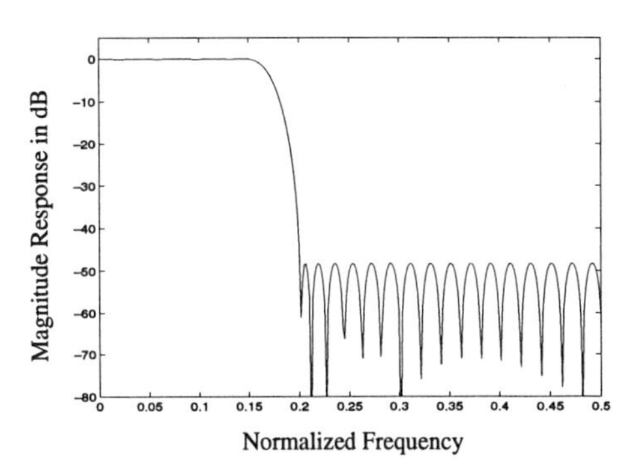

Figure 2.5: Magnitude response of an equiripple filter: normal and dB scales.

$(\downarrow 2)\boldsymbol{H}$ is very stable. But Section 5.5 will show that the transition band $\Delta\omega$ widens from N^{-1} for equiripple to $N^{-\frac{1}{2}}$ for the maxflat Daubechies filters.

A compromise is certainly possible, constraining the minimax design to have a fixed number p of zeros at $\omega = \pi$.

Weighted Least Squares (Eigenfilters)

The eigenfilter approach chooses the filter $\boldsymbol{H}$ to minimize a (weighted) integral error

$$E = \int \left|D(\omega) - H(e^{j\omega})\right|^2 \text{ (weight) } d\omega \tag{2.28}$$

The integral is over the passband and stopband, not the transition band. $D(\omega)$ is the desired frequency response, possibly one and zero in the two bands. The weighting function is optional. The goal is to express the error as a quadratic form $E = \boldsymbol{h}^T\boldsymbol{P}\boldsymbol{h}$. The unknown filter coefficients are in the vector $\boldsymbol{h}$.

The matrix $\boldsymbol{P}$ is symmetric positive definite because $E > 0$. If the normalization has the form $\boldsymbol{h}^T\boldsymbol{h} = 1$ then the minimization is an eigenvalue problem $\boldsymbol{P}\boldsymbol{h} = \lambda_{\text{min}}\boldsymbol{h}$ in linear algebra:

$$\text{The minimum of } E = \frac{\boldsymbol{h}^T\boldsymbol{P}\boldsymbol{h}}{\boldsymbol{h}^T\boldsymbol{h}} \text{ is } \lambda_{\min}(\boldsymbol{P}).$$

If the normalization is changed to $\boldsymbol{h}^T\boldsymbol{Q}\boldsymbol{h} = 1$, the eigenvalue problem $\boldsymbol{P}\boldsymbol{h} = \lambda\boldsymbol{Q}\boldsymbol{h}$ involves both matrices. When there are several quadratic constraints $\boldsymbol{h}^T\boldsymbol{Q}_k\boldsymbol{h} = 1$, we go beyond an eigenvalue problem — to the Quadratic Constrained Least Squares algorithm. Section 5.4 will apply this QCLS method to filter design. The applications of eigenfilters (one quadratic constraint) are very extensive, and we give two examples: lowpass filters and halfband filters.

Lowpass eigenfilter design: A symmetric filter $\boldsymbol{h}(n) = \boldsymbol{h}(2L - 1 - n)$ of length $2L$ has response

$$H(e^{j\omega}) = \sum_{n=0}^{2L-1} \boldsymbol{h}(n)e^{-j\omega n} = e^{-j\omega(L-\frac{1}{2})}\sum_{0}^{L-1} \boldsymbol{h}(n)\boldsymbol{c}(\omega).$$

The last sum is $H_{\text{real}}(\omega) = \boldsymbol{h}^T \boldsymbol{c}(\omega)$ with $\boldsymbol{h} = [\boldsymbol{h}(0) \ \cdots \ \boldsymbol{h}(L-1)]$. The vector $\boldsymbol{c}(\omega)$ has components $2\cos(L-\frac{1}{2})\omega, \ldots, 2\cos\frac{\omega}{2}$. In the stopband, from ω_s to π, where $D = 0$ is the desired response, the error is

$$E_{\text{stop}} = \int H_{\text{real}}^2(\omega) d\omega = \boldsymbol{h}^T \int_{\omega_s}^{\pi} \boldsymbol{c}(\omega)\boldsymbol{c}(\omega)^T d\omega\, \boldsymbol{h} \ = \ \boldsymbol{h}^T \boldsymbol{P}_{\text{stop}}\, \boldsymbol{h}.$$

The entries of $\boldsymbol{P}_{\text{stop}}$ are known integrals of cosines. In the passband from 0 to ω_p, we can normalize the desired constant response to be $D = \boldsymbol{h}^T \boldsymbol{c}(0)$. Then the passband error involves the difference between that desired response and the attained response $\boldsymbol{h}^T \boldsymbol{c}(\omega)$:

$$E_{\text{pass}} = \boldsymbol{h}^T \int_0^{\omega_p} (\boldsymbol{c}(0) - \boldsymbol{c}(\omega))(\boldsymbol{c}(0) - \boldsymbol{c}(\omega))^T d\omega\, \boldsymbol{h} \ = \ \boldsymbol{h}^T \boldsymbol{P}_{\text{pass}}\, \boldsymbol{h}.$$

The entries of $\boldsymbol{P}_{\text{pass}}$ are known integrals of $4[1 - \cos(n + \frac{1}{2})\omega][1 - \cos(m + \frac{1}{2})\omega]$.

We can weight the errors by $E = \alpha E_{\text{stop}} + (1-\alpha) E_{\text{pass}}$. The matrix whose lowest eigenvector is the best $\boldsymbol{h}$ will be $\boldsymbol{P} = \alpha \boldsymbol{P}_{\text{stop}} + (1 - \alpha)\boldsymbol{P}_{\text{pass}}$. Figure 2.6 shows the magnitude response of a lowpass filter of length 55 and cutoff frequencies $\omega_p = 0.2\pi$ and $\omega_s = 0.3\pi$. Here, the weight α is 0.5.

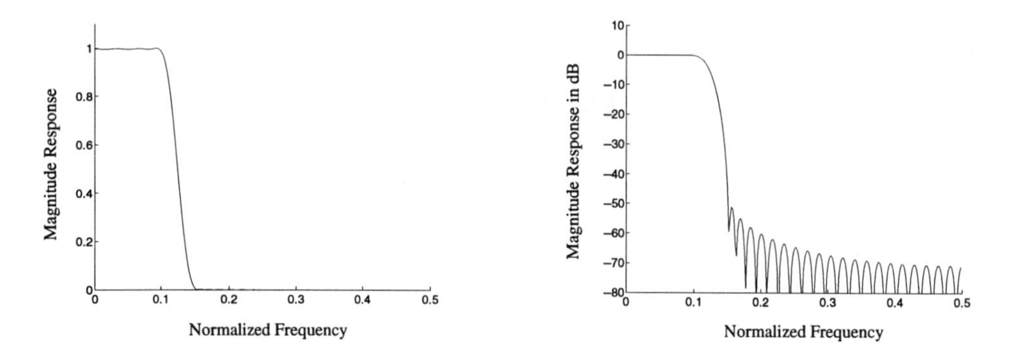

Figure 2.6: Magnitude response plot of a lowpass eigenfilter in normal and dB scale.

Halfband and Mth-band filter design: In many digital systems, a change in the sampling rate is essential for efficiency in real time. The design of suitable filters for the rate changing operations is important. When the subsampling is by a factor of $M = 2$, we are led to *halfband filters*. When we keep every Mth sample, we need *Mth–band filters*. The center coefficient is $\boldsymbol{h}(0) = 1$ and all coefficients at multiples of M are zero:

$$\textbf{\textit{Mth-band:}} \quad \boldsymbol{h}(nM) = \delta(n) \text{ or in other words } (\downarrow M)\boldsymbol{h} = \delta. \tag{2.29}$$

The filter banks have two channels or M channels. The design begins with a halfband filter or an Mth-band filter. This is the product filter of Section 4.1, which is factored into analysis times synthesis. We shift the filters to make them causal, for implementation. But for design we keep them centered, *and simply zero out the coefficients* $\boldsymbol{h}(nM)$ for $n \neq 0$. This zeros out the corresponding rows and columns of the error matrix $\boldsymbol{P}$ in weighted least squares. The optimum $\boldsymbol{h}$ is the lowest eigenvector of the reduced matrix $\boldsymbol{P}_{\text{red}}$, and this $\boldsymbol{h}$ is Mth–band.

***Design procedure*:** Given N, M, ω_p, ω_s for an Mth-band filter, find $\boldsymbol{P}$ using the eigenfilter formulation. Find $\boldsymbol{P}_{\text{red}}$ by deleting the rows and columns of $\boldsymbol{P}$ that correspond to zero coefficients in $\boldsymbol{h}$. Find the eigenvector $\boldsymbol{h}_{\text{red}}$ corresponding to the minimum eigenvalue of $\boldsymbol{P}_{\text{red}}$. The optimal impulse response $\boldsymbol{h}(n)$ is obtained from $\boldsymbol{h}_{\text{red}}(n)$ by inserting the zero coefficients.

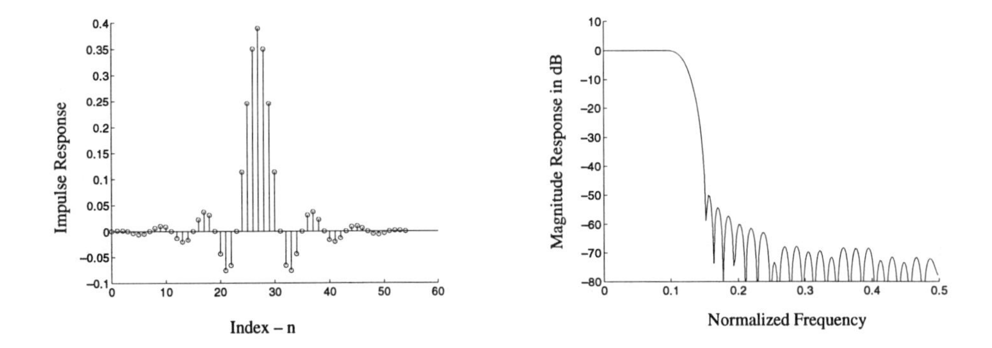

Figure 2.7: Impulse response and magnitude response of a 4th-band FIR filter.

Figure 2.7 shows the magnitude response of an 4th-band filter with $\omega_p = 0.2\pi$ and $\omega_s = 0.3\pi$. Note that $\omega_p + \omega_s = 0.5\pi = \frac{2\pi}{M}$, as required for a symmetric 4th-band filter. The length is 55 (odd length is also required).

Maximally Flat Filter *(Maxflat filter)*: The design of a maxflat filter begins with a maxflat polynomial $\widetilde{H}(y)$. This polynomial of degree $2p-1$ is determined by p conditions at $y=0$ and at $y=1$:

$$\widetilde{H}^{(k)}(0) = \delta(k) \text{ and } \widetilde{H}^{(k)}(1) = 0 \text{ for } 0 \le k < p.$$

The pth order zero at $y=1$ means that $\widetilde{H}(y) = (1-y)^p Q(y)$. Then the conditions at $y=0$ determine $Q(y)$. The details are in Section 5.5, leading to the Daubechies filter by factoring this halfband filter. Here we highlight a remarkable result: *$Q(y)$ consists of the first p terms of the series for* $(1-y)^{-p}$. With $p=4$, for example, $\widetilde{H}(y) = (1-y)^4(1+4y+10y^2+20y^3) \approx (1-y)^4(1-y)^{-4}$.

Those coefficients 1, 4, 10, 20 come from the binomial series for $(1-y)^{-4}$. They also appear in $(1+y+y^2+y^3)^4$. They are binomial numbers. The product $\widetilde{H}(y)$ then has three zero derivatives at $y=0$. Its graph is in the figure below.

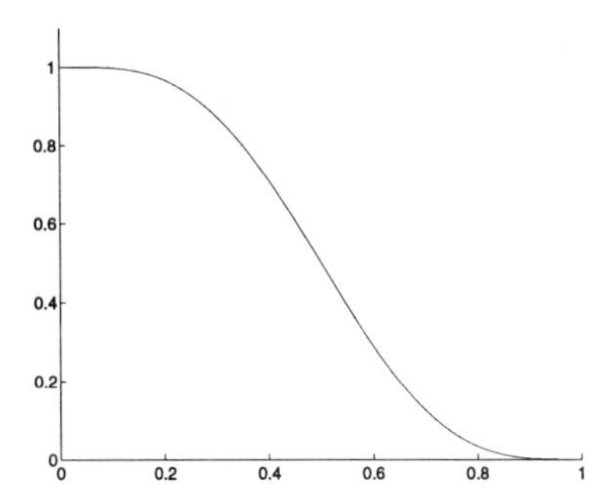

The relations among the variables y and ω and z are

$$\begin{aligned} y &= \frac{1-\cos\omega}{2} &= \tfrac{1}{4}(2-z-z^{-1}) &= -z\Big(\frac{1-z^{-1}}{2}\Big)^2. \\ 1-y &= \frac{1+\cos\omega}{2} &= \tfrac{1}{4}(2+z+z^{-1}) &= z\Big(\frac{1+z^{-1}}{2}\Big)^2. \end{aligned} \tag{2.30}$$

The zeros at $y = 1$ become zeros at $\omega = \pi$ and at $z = -1$. The flatness at $y = 0$ becomes flatness at $\omega = 0$ and at $z = 1$. We give the frequency response for the centered halfband Daubechies filter:

$$H(e^{j\omega}) = \Big(\frac{1+\cos\omega}{2}\Big)^p \sum_{k=0}^{p-1} \binom{p-1+k}{k} \Big(\frac{1-\cos\omega}{2}\Big)^k. \tag{2.31}$$

It is a binomial exercise to show that $H = \frac{1}{2}$ at $\omega = \frac{\pi}{2}$. To find $H(z)$, substitute using equations (2.30). Then shift by z^{-p} to make the filter causal.

Note Another form of this polynomial (which is so important in wavelet theory) is the "Bernstein form"

$$H(e^{j\omega}) = \sum_{k=p}^{2p-1} \binom{2p-1}{k} \Big(\frac{1+\cos\omega}{2}\Big)^k \Big(\frac{1-\cos\omega}{2}\Big)^{2p-1-k}. \tag{2.32}$$

As p increases, $H(e^{j\omega})$ approaches the ideal lowpass response (one for $\cos\omega > 0$ and zero for $\cos\omega < 0$). The $p^{-1/2}$ width of the transition band is established in Section 5.5, together with the approximation of the zeros of $H(z)$.

The ideal is infinitely flat. Of course it needs infinitely many coefficients.

Problem Set 2.3

1. Truncate the ideal lowpass filter after three nonzero coefficients. What is this windowed filter $\boldsymbol{h}(n)$? Sketch the graph of $H(\omega)$.
2. Truncate the ideal lowpass filter after 4 terms and 8 terms. Draw the frequency responses $H(\omega)$ to see the Gibbs phenomenon.
3. Compare the graphs of the ideal brick wall filter truncated after 20 terms (rectangular window) and the Kaiser window $\boldsymbol{w}(n)$. Choose a suitable Kaiser parameter β. What are the maximum errors in the passband?
4. Use MATLAB to design equiripple halfband filters of length 8 and 20. Compute the height of the ripples.
5. What is the frequency response for the maxflat Daubechies filter (2.31) with $p = 2$? Graph $H(\omega)$ by hand or by computer. What are its symmetries?
6. Graph the maxflat Daubechies filter response (2.31) for $p = 8$. What are the differences from the truncated ideal filter and the equiripple filter?
7. Construct a 4-tap lowpass filter that you approve of. What properties have you achieved?
8. Derive the Bernstein form (2.32). Hint: Show that the sum starting at $k = 0$ would be 1. Deduce that the given sum equals $1 + O(y^p)$ as Daubechies required.
9. Formulate the eigenfilter design for a highpass filter with cutoff frequency ω_c.
10. Compute $\boldsymbol{h}(n)$ for the halfband Daubechies filter with $p = 5$. Verify that $H(e^{j\omega})$ has four zero derivatives at $\omega = 0$ and $\omega = \pi$.

2.4 Fourier Analysis

This section might be called "Notes on Fourier Analysis." The subject is enormous — too large for our book! We pick out key points that are needed for signal processing and for wavelets.

The Fourier transform of a signal $\boldsymbol{x}(n)$ is a function $X(\omega)$. The signal is in the time domain, $X(\omega)$ is in the frequency domain. The time variable n is discrete, the frequency variable ω is continuous. $X(\omega)$ is 2π-periodic because each exponential $e^{in\omega}$ is 2π-periodic:

$$X(\omega) = \sum_{n=-\infty}^{\infty} \boldsymbol{x}(n)e^{-in\omega}. \tag{2.33}$$

Fourier analysis studies the connections between $\boldsymbol{x}(n)$ and $X(\omega)$ — how the properties of the signal are reflected in its transform. The ***inverse Fourier transform*** recovers $\boldsymbol{x}(n)$:

$$\boldsymbol{x}(n) = \frac{1}{2\pi}\int_{-\pi}^{\pi} X(\omega)e^{in\omega}d\omega. \tag{2.34}$$

This formula synthesizes $\boldsymbol{x}$ by combining the complex exponentials. To find $\boldsymbol{x}(N)$, *multiply equation* (2.33) *by* $e^{iN\omega}$ *and integrate from* $-\pi$ *to* π. The integral of $e^{-in\omega}$ times $e^{iN\omega}$ is zero except when $n = N$. That integral is 2π, leading to (2.34).

Fourier analysis usually starts with $f(t)$ and computes its coefficients. Signal processing starts with the coefficients $\boldsymbol{x}(n)$ and transforms to $X(\omega)$.

Note about orthogonality. Real vectors are orthogonal (perpendicular) when $\boldsymbol{x} \cdot \boldsymbol{y} = 0$. Real functions are orthogonal when $\int X(\omega)Y(\omega)d\omega = 0$. If the vectors or the functions are complex, there is a small but important change. We take complex conjugates of one vector (say $\boldsymbol{x}$, in the physics convention) and of one function $X(\omega)$. *Orthogonality in the complex case means*

$$\overline{\boldsymbol{x}} \cdot \boldsymbol{y} = \sum \overline{\boldsymbol{x}(n)}\boldsymbol{y}(n) = 0 \quad \text{and} \quad \langle X, Y\rangle = \int_{-\pi}^{\pi} \overline{X(\omega)}Y(\omega)\,d\omega = 0. \tag{2.35}$$

The integration $\int e^{-in\omega}e^{iN\omega}\,d\omega = 0$ does not say that $e^{-in\omega}$ is orthogonal to $e^{iN\omega}$. It says that $e^{in\omega}$ is orthogonal to $e^{iN\omega}$ (for $N \neq n$). It is this orthogonality that allows the inverse transform to have the clean and simple formula (2.34).

The discrete analogue of an orthonormal transform is a square matrix with orthonormal columns. This is an "orthogonal" matrix if real, a "unitary" matrix if complex. For such a matrix U, the inverse is again clean and simple. It equals the conjugate transpose $\overline{U}^T$. This applies to orthonormal filter banks, when the rows of $(\downarrow 2)\boldsymbol{H}_0$ and $(\downarrow 2)\boldsymbol{H}_1$ are orthonormal. Their transposes are the columns of $\boldsymbol{F}_0(\uparrow 2)$ and $\boldsymbol{F}_1(\uparrow 2)$ — also orthonormal.

Is there a connection between discrete and continuous orthogonality? If two signals are orthogonal (in discrete time), are their transforms orthogonal (in continuous frequency)? The question is important and the answer is ***yes***.

This answer follows from the orthogonality of the exponentials $e^{in\omega}$ — on which the whole theory depends. For any $\boldsymbol{x}$ and $\boldsymbol{y}$, the inner products in the time and frequency domains are equal up to a factor of 2π:

$$2\pi \sum_{n=-\infty}^{\infty} \overline{\boldsymbol{x}(n)}\boldsymbol{y}(n) = \int_{-\pi}^{\pi} \overline{X(\omega)}Y(\omega)\,d\omega. \tag{2.36}$$

For proof, substitute $\sum \overline{x(N)} e^{iN\omega}$ for $\overline{X(\omega)}$ on the right. Also substitute $\sum y(n)e^{-in\omega}$ for $Y(\omega)$. Integrate from $-\pi$ to π. By orthogonality, the only terms with nonzero integrals are those with $N = n$. The integral is 2π, multiplied by the number $\overline{x(n)}y(n)$. The left side of (2.36) is the sum of these nonzero terms.

Special case when $x = y$. Now the two vectors are the same. Their transforms are the same. We are integrating $\overline{X(\omega)}X(\omega)$, which is $|X(\omega)|^2$, because $a + ib$ times its conjugate $a - ib$ is $a^2 + b^2$. For the same reason $\overline{x(n)}x(n) = |x(n)|^2$. With $x = y$, the energy in the signal (times 2π) equals the energy in the transform:

$$2\pi \sum_{-\infty}^{\infty} |x(n)|^2 = \int_{-\pi}^{\pi} |X(\omega)|^2 \, d\omega. \tag{2.37}$$

Example 2.8. Suppose x is $(1, \beta, \beta^2, \ldots)$. Its energy is $1 + \beta^2 + \beta^4 + \cdots = 1/(1 - \beta^2)$. Its transform is a one-sided sum, in this case a geometric series:

$$X(\omega) = \sum_{0}^{\infty} \beta^n e^{-in\omega} = 1 + \beta e^{-i\omega} + \left(\beta e^{-i\omega}\right)^2 + \cdots = \frac{1}{1 - \beta e^{-i\omega}}.$$

Consider the inverse transform from $X(\omega)$ back to $x(n)$:

$$x(n) = \frac{1}{2\pi} \int_{-\pi}^{\pi} X(\omega) e^{in\omega} d\omega = \frac{1}{2\pi} \int_{-\pi}^{\pi} \frac{e^{in\omega} d\omega}{1 - \beta e^{-i\omega}}.$$

How do we see that this integral gives the correct signal $(1, \beta, \beta^2, \ldots)$? The direct way is to write the integrand as $e^{in\omega}\left(1 + \beta e^{-i\omega} + \beta^2 e^{-2i\omega} + \cdots\right)$. Integration picks out the correct power β^n. (If $n < 0$ then the integral gives zero.) The indirect way is to substitute z for $e^{i\omega}$, and integrate $z^n/(z - \beta)$ around the unit circle:

$$\frac{1}{2\pi} \int_{-\pi}^{\pi} \frac{e^{in\omega} d\omega}{1 - \beta e^{-i\omega}} = \frac{1}{2\pi i} \int_{|z|=1} \frac{z^n (dz/z)}{1 - (\beta/z)}.$$

There is a pole at $z = \beta$. The *residue* at this pole is β^n. This is the answer we want. Again the case $n < 0$ is separate (with two poles) and gives zero.

The actual calculation of such an inversion integral is generally difficult or impossible. We seldom need to do it. For a ratio of polynomials it can be done in an emergency by the residue method of complex integration.

Example 2.9. An allpass filter has $|H(\omega)| = 1$ so that $|Y(\omega)| = |X(\omega)|$. The filter conserves energy. The integral of $|Y(\omega)|^2$ equals the integral of $|X(\omega)|^2$, because these functions are the same:

$$\textit{Allpass:} \quad \int_{-\pi}^{\pi} |Y(\omega)|^2 d\omega = \int_{-\pi}^{\pi} |H(\omega)X(\omega)|^2 d\omega = \int_{-\pi}^{\pi} |X(\omega)|^2 d\omega.$$

Therefore ***the output energy equals the input energy:***

$$\sum_{-\infty}^{\infty} |y(n)|^2 = \sum_{-\infty}^{\infty} |x(n)|^2 \quad \text{or} \quad \|y\|^2 = \|x\|^2. \tag{2.38}$$

Please do not think that each $|y(n)| = |x(n)|$. It is the *frequency* response that has $|H(\omega)| = 1$. The energy in each frequency band is conserved by an allpass filter, not the energy in each time band.

Convergence of the Fourier Series

In defining $X(\omega)$, we have been assuming that this series converges. Otherwise what meaning do we give to $X(\omega)$? We are touching here on a central problem of mathematical analysis (with a literature that goes back for centuries). Touching this problem is as much as we can do—by identifying three types of convergence, and the signals that produce each type.

As with all infinite series, a few terms have nothing to do with convergence. By changing a finite number of inputs $\boldsymbol{x}(n)$ we certainly change the transform—but we do not alter convergence or divergence. It is the behavior of $\boldsymbol{x}(n)$ *for large n* that is crucial. Here are three types of convergence:

1. Uniform convergence with $\sum |\boldsymbol{x}(n)| < \infty$
2. Strong convergence (in L^2) with $\sum |\boldsymbol{x}(n)|^2 < \infty$
3. Weak convergence allowing polynomial growth in $\boldsymbol{x}(n)$.

1. Uniform convergence Suppose the magnitudes $|\boldsymbol{x}(n)|$ have a finite sum. These are also the magnitudes of $\boldsymbol{x}(n)e^{-in\omega}$, because $|e^{-in\omega}| = 1$. Those terms have different phases, which may produce cancellation when we add them. When $\sum |\boldsymbol{x}(n)|$ converges, *we don't need that help.* The series of magnitudes converges "absolutely," and the series $\sum \boldsymbol{x}(n)e^{-in\omega}$ converges "uniformly." Then the sum $X(\omega)$ is a *continuous function*—with no jumps.

In the example with $\boldsymbol{x} = (1, \beta, \beta^2, \ldots)$ we imposed $|\beta| < 1$. That produced uniform convergence. The transform $X(\omega) = 1/(1 - \beta e^{-i\omega})$ is a continuous function. But not all continuous functions have $\sum |\boldsymbol{x}(n)| < \infty$.

In the brick wall filter, the odd coefficients have magnitude $|\boldsymbol{h}(n)| = \frac{1}{\pi n}$. The sum of magnitudes does *not* converge. The terms $\frac{1}{\pi n}$ do not go to zero quickly enough. The sum $H(\omega) = \sum \boldsymbol{h}(n)e^{-in\omega}$ does not converge uniformly. And, in fact, $H(\omega)$ is a step function (or square wave) with a jump.

2. Convergence in energy (L^2 convergence) Suppose the *squared* magnitudes $|\boldsymbol{x}(n)|^2$ have a finite sum. By squaring, small terms become much smaller. Convergence is easier to achieve. If the sum of $|\boldsymbol{x}(n)|$ is finite, the sum of squares is certainly finite. Then the Fourier series converges in L^2. This "squared" test is passed by the Fourier series for a step function:

$$|\boldsymbol{h}(n)| = \frac{1}{\pi n} \quad \text{has} \quad \sum |\boldsymbol{h}(n)|^2 = \sum \frac{1}{\pi^2 n^2} = \text{convergent series.}$$

Comparing the sums of $1/n$ and $1/n^2$ is like comparing the integrals of $1/x$ and $1/x^2$. One integral is $\log x$, which becomes large as $x \to \infty$. The area under $1/x^2$ stays finite as $x \to \infty$.

When the squared magnitudes $|\boldsymbol{x}(n)|^2$ have a finite sum, the squared magnitude $|X(\omega)|^2$ has a finite integral. They are equal apart from 2π. Then $X(\omega)$ is a function in the Hilbert space denoted by L^2, just as $\boldsymbol{x}(n)$ is a vector in the Hilbert space denoted by ℓ^2. These spaces contain all functions and vectors with finite energy: *the square of the L^2 norm or the ℓ^2 norm is the energy.*

Functions in Hilbert space may have jumps. Those jumps have no energy (no contribution to the integral). The *derivative* of a jump is a Dirac delta function. Its coefficients are all $\boldsymbol{x}(n) = \frac{1}{2\pi}$ and its energy is infinite, so this function is outside the space L^2.

But the delta function is included in the third type of convergence.

3. Weak convergence (to a distribution) Distributions $F(\omega)$ are defined by their inner products with smooth functions $G(\omega)$. *For $\int F(\omega)G(\omega)\,d\omega$, use integration by parts.* We integrate $F(\omega)$, which needs it, and we differentiate $G(\omega)$, which can take it. The indefinite integral of the delta function $F(\omega) = \delta(\omega)$ is the unit step $H(\omega)$. The definite integral of $\delta(\omega)G(\omega)$ is the number $G(0)$:

$$\int_{-\pi}^{\pi} \delta(\omega)G(\omega)\,d\omega = [H(\omega)G(\omega)]_{-\pi}^{\pi} - \int_{-\pi}^{\pi} H(\omega)G'(\omega)\,d\omega = G(\pi) - \int_{0}^{\pi} G'(\omega)\,d\omega = G(0).$$

This defines $\delta(\omega)$. It is a distribution, a derivative of a true function. Still it has Fourier coefficients that are easy to find, with $G(\omega) = e^{in\omega}$:

$$\boldsymbol{x}(n) = \frac{1}{2\pi}\int_{-\pi}^{\pi} \delta(\omega)e^{in\omega}\,d\omega = \frac{1}{2\pi}\cdot 1.$$

All frequencies are present in the same amount. The Fourier series is

$$\delta(\omega) = \frac{1}{2\pi}\sum_{n=-\infty}^{\infty} e^{in\omega}. \tag{2.39}$$

On the left is a distribution. On the right is a divergent series. The terms don't even approach zero. But in a weak sense those terms cancel each other to produce zero, away from the spike at $\omega = 0$ where they reinforce.

Weak convergence is based on the same idea of testing inner products with smooth $G(\omega)$. The series *converges weakly*, say at $\omega = 0$:

$$G(0) = \int_{-\pi}^{\pi} \delta(\omega)G(\omega)\,d\omega = \frac{1}{2\pi}\sum_{n}\left(\int_{-\pi}^{\pi} e^{in\omega}G(\omega)\,d\omega\right)e^{in0}. \tag{2.40}$$

You see the Fourier coefficients of $G(\omega)$ on the right side, adding to $G(0)$.

Numerically, this weak convergence is not so great. Figure 2.8 shows the sum of 41 terms. In a pointwise sense, and in area, those side lobes will not shrink to zero! In the L_2 sense, the energy is growing with every term. But in a weak sense, this sum is approaching $\delta(\omega)$. As the number of terms increases, the oscillations become faster (not smaller). Multiplied by a smooth $G(\omega)$, the main central lobe picks out $G(0)$ and the integral over the oscillations approaches zero.

Question: What is the weak limit of pure oscillations $e^{in\omega}$ as $n \to \infty$?

Answer: The limit is the zero function. The inner products $\int e^{in\omega}G(\omega)\,d\omega$ approach zero. Oscillations converge *weakly* to their average value.

Question: Does the series $\delta'(\omega) = \frac{i}{2\pi}\sum n\, e^{in\omega}$ for the derivative of a Dirac function also converge weakly? The Fourier coefficients are growing with n.

Answer: Yes. Integrate again by parts. The integral of $\delta'(\omega)G(\omega)$ is $-G'(0)$.

The ***Gibbs phenomenon*** is just the integral of $\sum e^{in\omega}$. The integral of $\delta(\omega)$ is a step function. The integral of $e^{in\omega}$ introduces $\frac{1}{n}$, so there is L^2 convergence to the step (but not uniform convergence). With integration, *the area under the side lobes become crucial.* The area shows as a height in the Gibbs figure (the integral of the delta function is a step). This undesirable oscillation of Fourier series at a jump (an *edge* in image processing) has brought forward localized bases like wavelets.

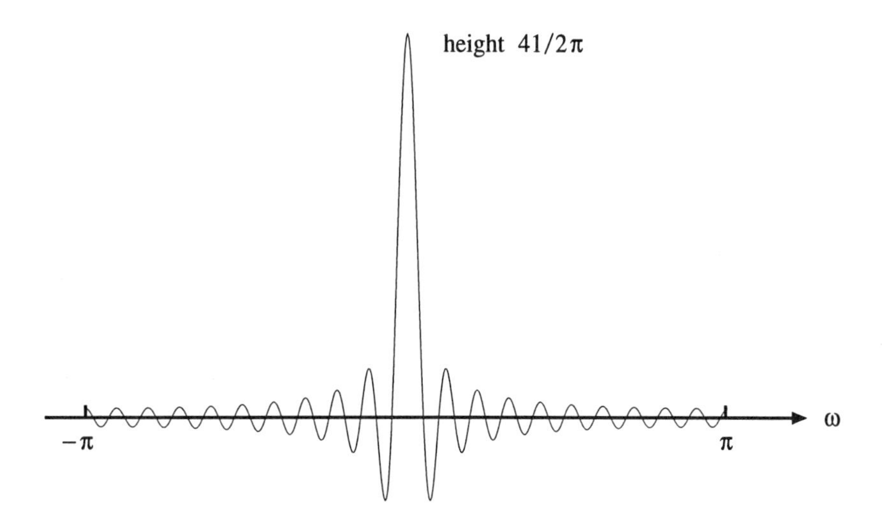

Figure 2.8: The Fourier series $\frac{1}{2\pi}\sum e^{-in\omega}$ converges weakly to $\delta(\omega)$. Its integral shows the Gibbs phenomenon (nonuniform convergence).

It is significant that the same Gibbs difficulty in shock calculations has been handled very differently. The finite difference schemes are made non-oscillatory by *nonlinear terms*. The nonlinearity is active where it is needed. It is inactive where the solution is smooth. Morel's book [Mo] gives a strong impetus to this nonlinear idea for image processing.

For iteration of the filter $\boldsymbol{h} = (\frac{1}{2}, 0, 0, \frac{1}{2})$, we will see weak convergence in Chapter 7. The cascade algorithm (= lowpass iteration) produces functions that oscillate between 1 and 0 on the interval $[0, 3]$. They have no limit in L^2. The weak limit of the oscillations is a stretched box $\phi(t)$ with constant value $\frac{1}{3}$.

Good filters give L^2 convergence (and usually uniform convergence) when iterated. The necessary and sufficient Condition E is in Section 7.2. But a "good filter" in the classical sense, with small errors in the passband and stopband, may fail in iteration!

The requirement for success in iteration is flatness of the response $H(\omega)$. *The number of zeros at $\omega = \pi$ is absolutely critical.* This is the new property that has become important. It makes the iteration process strong and not weak, regular and not oscillatory. It must be built in or iterations will oscillate — as happened with highly regarded filters.

Poisson's Summation Formula

The Fourier coefficients of the Dirac delta function on $[-\pi, \pi]$ are all $\frac{1}{2\pi}$. By periodicity, the Dirac delta becomes a *Dirac comb*. There is an impulse at every multiple of 2π. The Fourier series for this periodic train of delta functions is

$$\sum_{k=-\infty}^{\infty} \delta(\omega - 2\pi k) = \frac{1}{2\pi} \sum_{n=-\infty}^{\infty} e^{-in\omega}. \tag{2.41}$$

The left side is a sum of impulses. The right side is its Fourier series, *converging weakly* as above. The formula is often seen with t instead of ω, because Fourier analysis usually starts with continuous time. Then the frequencies are discrete. In signal processing it is the other way around.

Equation (2.41) is a short statement of the Poisson formula, but it involves weak convergence. The terms don't approach zero. To see ordinary convergence we take inner products with any smooth function $G(\omega)$. With integrals over the whole line, we ask $G(\omega)$ to decay as $|\omega| \to \infty$. The inner product with delta functions gives point values $G(2\pi k)$. The inner product on the right gives point values $\widehat{G}(n)$ of the Fourier transform. The formula says that the two sums are equal:

$$\textbf{\textit{Poisson's Summation Formula}} \qquad \frac{1}{2\pi} \sum_{k=-\infty}^{\infty} G(2\pi k) = \sum_{n=-\infty}^{\infty} \widehat{G}(n). \tag{2.42}$$

This is a remarkable formula. It relates the samples of G to the samples of its transform $\widehat{G}$. We can change the spacing of one set of samples to any $T > 0$, provided the other set of samples is spaced at $\frac{2\pi}{T}$.

Smoothness of $X(\omega)$ and Decay of $x(n)$

The regularity of a function is partly revealed in its Fourier coefficients. When $X(\omega)$ is continuous, its Fourier coefficients $\boldsymbol{x}(n)$ approach zero. But the coefficients can approach zero when $X(\omega)$ is *not* continuous; the step function is an example. There is a gap between $\boldsymbol{x}(n) \to 0$ and $\sum |\boldsymbol{x}(n)| < \infty$, which is at the heart of Fourier analysis:

$$\sum |\boldsymbol{x}(n)| < \infty \qquad \Longrightarrow \qquad \text{continuous } X(\omega) \qquad \Longrightarrow \qquad \boldsymbol{x}(n) \to 0.$$

One virtue of wavelets is that such gaps can be closed. Instead of necessary conditions for smoothness, and then sufficient conditions for smoothness, we can find *necessary and sufficient conditions*. These are conditions on the wavelet coefficients $\boldsymbol{x}(n)$, for the function $X(\omega)$ to have specified regularity. $X(\omega)$ lies in a specified function space when $\boldsymbol{x}(n)$ lies in a corresponding vector space [DeLu]. The function norm and vector norm are equivalent—as for $X(\omega)$ in L^2 and $\boldsymbol{x}(n)$ in ℓ^2 by Parseval's formula.

Fourier coefficients give partial information from the magnitudes $|\boldsymbol{x}(n)|$. Each extra order of smoothness in $X(\omega)$ is reflected in one extra order of decay in $|\boldsymbol{x}(n)|$.

Theorem 2.2 *Suppose $X(\omega)$ has s continuous derivatives. Then $n^s|\boldsymbol{x}(n)| \to 0$ as $|n| \to \infty$.*

When $s = 0$ this is the Riemann-Lebesgue Lemma. The Fourier coefficients approach zero (we really only need $\int |X(\omega)|d\omega$ to be finite). For integers $s = 1, 2, 3, \ldots$ we look at the Fourier coefficients of the derivative $X^{(s)}(\omega)$. Those coefficients are $(in)^s \boldsymbol{x}(n)$. They approach zero (again by Riemann-Lebesgue) when $X^{(s)}(\omega)$ is continuous. This is the theorem.

In compression of a signal, this decay of coefficients is crucial. Small coefficients are removed (partly or completely). For a Fourier basis, the smoothness of an image determines the decay of coefficients. For a wavelet basis, it is the *piecewise smoothness* that matters. Wavelets are well adapted to piecewise smooth functions (with edges). Wavelets are local where Fourier waves are global.

The theorem was stated for integers $s = 0, 1, 2, \ldots$ but fractions can be allowed. This is important for a satisfactory theory, because functions like $X(\omega) = |\omega|^{1/2}$ are more than continuous ($s = 0$) and less than differentiable ($s = 1$). The in-between smoothness is measured by the *Hölder exponent*. The function $X(\omega)$ belongs to the space C^s, $0 < s < 1$, if it satisfies the Hölder condition

$$|X(\omega_2) - X(\omega_1)| \leq \text{ constant } |\omega_2 - \omega_1|^s.$$

The square root function $|\omega|^{1/2}$ belongs to $C^{1/2}$. Similarly $|\omega|^s$ belongs to C^s. For $s > 1$ we would take $[s]$ derivatives first. The Hölder exponent of that derivative is the fractional part $s - [s]$. Then the Fourier coefficients decay to order s at least: $n^s|x(n)| \to 0$.

The Fourier basis automatically takes advantage of high regularity. *The wavelet basis takes full advantage of a high s, only if it is designed to do so.* The lowpass filter must have more than s zeros at π. Haar wavelet expansion does not converge faster as the function gets smoother. Therefore Haar compression is poor.

For biorthogonal wavelets, the analyzing filter $\boldsymbol{H}_0$ governs the decay of coefficients. The synthesizing filter $\boldsymbol{F}_0$ governs the smoothness of the output. *We would like both to have many zeros at* π*!* When forced to choose, whether to put zeros into $\boldsymbol{H}_0$ or $\boldsymbol{F}_0$, the preference often goes to $\boldsymbol{F}_0$—then the synthesis basis functions will be smooth.

We emphasize one more point. *In the* L^2 *norm,* where "mean square" replaces "maximum," *the Fourier coefficients exactly reflect the smoothness.* The coefficients are in ℓ^2 precisely when the function is in L^2. The energy is the same for both, by Parseval's identity. This equality extends to the sth derivative $X^{(s)}(\omega)$, whether s is an integer or not:

$$\int_{-\pi}^{\pi} \left|X^{(s)}(\omega)\right|^2 d\omega = 2\pi \sum_{-\infty}^{\infty} \left|n^s x(n)\right|^2. \tag{2.43}$$

When s is not an integer, we *define* this derivative $X^{(s)}(\omega)$ by its coefficients $(in)^s x(n)$. The equation above is just Parseval itself, for the derivative. Wavelet theory is greatly simplified by working in the Hilbert spaces of functions with s derivatives in L^2, rather than the Hölder spaces C^s of functions with s continuous derivatives.

The number p of zeros at π is crucial in both cases. For L^2 spaces we will find in Chapter 7 the exact smoothness of $\phi(t)$ and $w(t)$.

Heisenberg's Uncertainty Principle

The underlying property of wavelets is that they are pretty well localized in both time and frequency. The functions $e^{i\omega t}$ are perfectly localized at ω but they extend over all time. Wavelets are *not* at a single frequency, or even a finite range, but they are limited to finite time. As we rescale, the frequency goes up by 2^j and the time interval goes down by 2^j. This suggests that the *product* of frequency interval and time interval is a stable quantity. The Heisenberg Uncertainty Principle makes those definitions precise, and gives a *lower bound* for the product.

We emphasize that the wavelet can only be "pretty well localized." It cannot have finite support in both t and ω. (A famous theorem.) Instead of the support length we use the variances

$$\sigma^2 = \int_{-\infty}^{\infty} t^2|f(t)|^2\,dt \quad \text{and} \quad \widehat{\sigma}^2 = \frac{1}{2\pi}\int_{-\infty}^{\infty} \omega^2|\widehat{f}(\omega)|^2\,d\omega. \tag{2.44}$$

We are no longer working with periodic functions, so t and ω extend from $-\infty$ to ∞.

Theorem 2.3 (Heisenberg) *If* $\|f\| = 1$ *then the product* $\sigma\widehat{\sigma}$ *is at least* $\frac{1}{2}$.

The lower bound $\frac{1}{2}$ is attained by the Gaussian function $f(t) = e^{-t^2}$. Its transform $\widehat{f}(\omega)$ is also a Gaussian. These have infinite support! But they are as local as possible, measured by σ and $\widehat{\sigma}$. We can rescale time by c and frequency by $\frac{1}{c}$, we can shift to $t - t_0$, and we can modulate by $e^{i\omega_0 t}$. The variances σ^2 and $\widehat{\sigma}^2$ would be computed around t_0 and ω_0, and the bound $\frac{1}{2}$ is still reached.

This led Gabor to use Gaussians in constructing "time-frequency atoms." But numerically, finite support in time is better.

Proof of the Heisenberg Principle. The key is that $\sigma = \|tf(t)\|$ and $\widehat{\sigma} = \left\|f'(t)\right\|$. The principle applies to pairs of operators, like position $\boldsymbol{P}$ and momentum $\boldsymbol{Q}$, that have the property $\boldsymbol{QP}-\boldsymbol{PQ} = \boldsymbol{I}$. When this property holds, the proof comes directly from the Schwarz inequality:

$$1 = \|f\|^2 = \langle f, (\boldsymbol{QP} - \boldsymbol{PQ})f\rangle \le 2\ \|\boldsymbol{Q}f\|\ \|\boldsymbol{P}f\| = 2\widehat{\sigma}\sigma. \tag{2.45}$$

In our case $\boldsymbol{P}$ is multiplication by t and $\boldsymbol{Q}$ is differentiation:

$$(\boldsymbol{QP} - \boldsymbol{PQ})f(t) = \frac{d}{dt}\big(tf(t)\big) - t\frac{df}{dt} = \boldsymbol{I}f(t) \quad \text{as required.}$$

Problem 4 develops this proof directly from the definitions of σ and $\widehat{\sigma}$.

Problem Set 2.4

1. Add the terms $\frac{1}{2\pi}\sum_{-N}^{N} e^{in\omega}$. This is a partial sum of the Fourier series for $\delta(\omega)$. At what ω does it come down to zero, at the end of the main lobe in Figure 2.8? Where is the end of the first side lobe?

2. Describe a continuous function $X(\omega)$ whose coefficients have infinite sum $\sum |\boldsymbol{x}(n)| = \infty$.

3. What is Poisson's summation formula for the Gaussian $G(\omega) = e^{-\omega^2}$?

4. Heisenberg's Uncertainty Principle is $\sigma\widehat{\sigma} \ge \frac{1}{2}$. The Schwarz inequality gives

$$\left|\int tf(t)f'(t)\,dt\right|^2 \le \int |tf(t)|^2\,dt \int |f'(t)|^2\,dt.$$

Identify the right side as $\sigma^2\widehat{\sigma}^2$. Integrate by parts on the left side to get

$$\int_{-\infty}^{\infty} t\left[f(t)f'(t)\right]dt = t\frac{f(t)^2}{2}\Big]_{-\infty}^{\infty} - \int_{-\infty}^{\infty}\frac{f(t)^2}{2}\,dt.$$

If the integrated part is zero at $\pm\infty$, and if $\|f\| = 1$, deduce that $\sigma\widehat{\sigma} \ge \frac{1}{2}$. Equality holds when $tf(t)$ is proportional to $f'(t)$, which leads to Gaussians $f(t) = e^{-ct^2}$.

5. What power of $\frac{1}{n}$ gives the decay rate for the coefficients $\boldsymbol{x}(n)$ of

1. $X(\omega) = \omega =$ discontinuous function at $\omega = \pm\pi$.
2. $X(\omega) = |\omega| =$ continuous function with period 2π.
3. $X(\omega) = \omega^2 =$ continuous function with period 2π.
4. $X(\omega) =$ spline with jump in the third derivative $X'''(\omega)$.

6. Compute the Fourier coefficients and the energy for

(a) $X(\omega) = \dfrac{1}{1 - \frac{1}{2}e^{-i\omega}}$ (b) $X'(\omega)$

(c) Periodic box function $H(\omega) = \begin{cases} 1 & 0 \le \omega < \pi \\ 0 & -\pi \le \omega < 0 \end{cases}$

7. Show that $X(\omega) = (1-\beta e^{-i\omega})^{-1}$ has the same energy as $\boldsymbol{x} = (1, \beta, \beta^2, \ldots)$. Use a substitution in the integral of $|X(\omega)|^2$.

8. Show that the Heisenberg product $\sigma\widehat{\sigma}$ is not changed by *dilation*, *modulation*, and *translation*: $f(t)$ is transformed to $2^{j/2}f(2^jt)$ and $e^{i\omega t}f(t)$ and $f(t-s)$.

9. Determine the Fourier transform of the signal $\boldsymbol{x}(n) = \alpha^{|n|}$, $|\alpha| < 1$.

2.5 Bases and Frames

Above all, this book is about **the choice of a good basis**. In reality, that choice governs everything. Every transform is a change to a different basis. The contribution of filter banks and wavelets (and Fourier transforms and local cosines and wavelet packets) is *to offer new bases.* The whole subject has been reopened, by the demands of its applications.

What is a basis and what makes it good? A basis is a sequence of vectors $\boldsymbol{v}_1, \boldsymbol{v}_2, \ldots$ or functions $v_1(t), v_2(t), \ldots$ with the property of *unique representation*:

> Every vector $\boldsymbol{v}$ or function $v(t)$ in the space can be represented in one and only one way as $\boldsymbol{v} = \sum b_i \boldsymbol{v}_i$ or $v(t) = \sum b_i v_i(t)$.

There is exactly one representation of every vector and function. The zero vector and zero function can only be represented with all $b_i = 0$. The basis functions are *linearly independent.*

It is usual to require convergence in norm, $\|\boldsymbol{v} - \sum_1^N b_i \boldsymbol{v}_i\| \to 0$ as $N \to \infty$. For L^2 spaces this norm is the square root of the energy. Section 1.5 mentions four reasons for that choice.

There are two separate properties to be established for any proposed basis: (1) *linear independence* and (2) *completeness.* Adding extra vectors will destroy independence. Removing vectors from the basis will kill completeness. *Linear independence is automatic for orthonormal vectors.* When all angles are 90° and all lengths are unity, there is no chance of degeneracy. In infinite dimensions we meet the possibility that angles can approach zero without reaching zero. In that case the basis is unstable. The coefficients b_i are out of control. In the good case, the coefficients satisfy

$$A\,\|\boldsymbol{v}\|^2 \le \sum |b_i|^2 \le B\,\|\boldsymbol{v}\|^2 \quad \text{with} \quad A > 0. \tag{2.46}$$

This is the defining property of a *Riesz basis*, also known as a *stable basis* or an *unconditional basis.* A key assumption in wavelet theory is that the translates $\phi(t-n)$ of the scaling function are a Riesz basis (for the space V_0 inside L^2). We will find the test to be applied to $\widehat{\phi}(\omega)$, and the equivalent Condition E on the filter coefficients, to produce a Riesz basis of scaling functions.

Dual Bases and Dual Frames

In a few lines, we can give the main points about bases and frames. These ideas will be developed below — bases will come first. Here are the key points in a hurry:

> ***Dual bases*** come from columns $\boldsymbol{v}_n$ of $\boldsymbol{T}^{-1}$ and rows $\boldsymbol{r}_n$ of $\boldsymbol{T}$: $\boldsymbol{T}^{-1}\boldsymbol{T} = \boldsymbol{I}$
>
> ***Dual frames*** come from columns $\boldsymbol{v}_n^+$ of $\boldsymbol{T}^+$ and rows $\boldsymbol{r}_n$ of $\boldsymbol{T}$: $\boldsymbol{T}^+\boldsymbol{T} = \boldsymbol{I}$.

The difference is that *the frame vectors $\boldsymbol{v}_n^+$ need not be independent.* They can be redundant (Figure 2.9). We still require the coefficients $b_i = \langle \boldsymbol{r}_i, \boldsymbol{v}\rangle$ to satisfy (2.46), but other combinations of the $\boldsymbol{v}_n^+$ can reconstruct the same $\boldsymbol{v}$.

$\boldsymbol{T}^+$ is only a left-inverse for a frame. It becomes a two-sided inverse for a basis.

In N-dimensional space, frames contain $M > N$ vectors. The matrix $\boldsymbol{T}$ is $M \times N$, and its left-inverse $\boldsymbol{T}^+$ is $N \times M$. ***The equation $\boldsymbol{T}\boldsymbol{T}^+ = \boldsymbol{I}$ is not true.*** For finite dimensions the theory and examples are particularly clear — our eventual applications are to infinite dimensions.

Note 1 *Change of Basis* Suppose $\boldsymbol{T}$ is a bounded linear operator with a bounded inverse. If the sequence $\{\boldsymbol{v}_n\}$ is a Riesz basis, so is the sequence $\{\boldsymbol{T}^{-1}\boldsymbol{v}_n\}$. When we expand $\boldsymbol{T}\boldsymbol{v} = \sum c_n \boldsymbol{v}_n$ in the

Figure 2.9: Good basis, bad basis, good frame, bad frame, all in $\mathbf{R}^2$.

original basis and multiply by T^{-1}, this gives the expansion $v = \sum c_n(T^{-1}v_n)$ in the new basis: $A = B = 1$. The new Riesz constants A, B come from the original A, B and the norms of T and T^{-1}.

The classical transforms of mathematics have a stronger property than invertibility. ***T** is often a unitary operator*. The inverse of T becomes the conjugate transpose T^*. The norms are $\|T\| = \|T^{-1}\| = 1$. If the original basis is orthonormal so is the new basis: $A = B = 1$. The Fourier transform and wavelet transform are unitary operators — when we use orthonormal wavelets! Biorthogonal wavelets lead to non-unitary operators T and to Riesz bases — but not orthonormal. Fortunately $\|T\|$ and $\|T^{-1}\|$ are in practice surprisingly close to 1.

In finite dimensions, every basis is a Riesz basis. *The basis vectors for* $\mathbf{R}^N$ *are the columns of an invertible* $N \times N$ *matrix.* That matrix is T^{-1}. It is essential to distinguish the change of coordinates (which uses T) from the new basis (the columns of T^{-1}). The fundamental fact is $T^{-1}T = I$. We use it now, multiplying *columns times rows*:

$$v = T^{-1}Tv = \sum_{n=1}^{N}(\text{column } n \text{ of } T^{-1})(\text{row } n \text{ of } T)v$$

$$\text{The } n\text{th basis vector is } v_n = \text{column } n \text{ of } T^{-1}$$

$$\text{The } n\text{th coordinate is } b_n = (\text{row } n \text{ of } T)v = \langle r_n, v\rangle.$$

Those three lines give $v = \sum b_n v_n$.

Every $N \times N$ matrix is a bounded operator. When the inverse exists, it is also bounded. But infinite matrices can represent unbounded operators. The averaging filter with coefficients $\frac{1}{2}, \frac{1}{2}$ is bounded. The inverse filter with coefficients $2, -2, 2, -2, 2, -2, \ldots$ is unbounded. The Riesz requirement, that T and T^{-1} are both bounded, is needed for a stable change of basis. An orthonormal basis ($A = B = 1$) has condition number 1. Then the product $\|T\|\,\|T^{-1}\|$ is the **condition number** of the new basis.

Bases from Filter Banks

The important examples for us are the bases from filter banks (discrete time) and the bases from wavelets (continuous time). Chapter 1 gave an example of both. The discrete time basis vectors had only two nonzero components:

$$\ldots, \begin{bmatrix} \cdot \\ 1 \\ 1 \\ 0 \\ 0 \\ \cdot \end{bmatrix}, \begin{bmatrix} \cdot \\ 1 \\ -1 \\ 0 \\ 0 \\ \cdot \end{bmatrix}, \begin{bmatrix} \cdot \\ 0 \\ 0 \\ 1 \\ 1 \\ \cdot \end{bmatrix}, \begin{bmatrix} \cdot \\ 0 \\ 0 \\ 1 \\ -1 \\ \cdot \end{bmatrix}, \ldots$$

This is an orthogonal basis because the vectors are mutually perpendicular. It is an orthonormal basis when divided by $\sqrt{2}$.

The heart of the book is the construction of other (and better) filter banks. The filters will be longer and they will overlap. Orthogonality will not be automatic and it may not be true. The bases are *the impulse responses of the filters* — with a double shift as above. Here is an example with four nonzero components:

$$\ldots, \begin{bmatrix} \cdot \\ 1 \\ 3 \\ 3 \\ 1 \\ 0 \\ 0 \\ \cdot \end{bmatrix}, \begin{bmatrix} \cdot \\ 1 \\ 3 \\ -3 \\ -1 \\ 0 \\ 0 \\ \cdot \end{bmatrix}, \begin{bmatrix} \cdot \\ 0 \\ 0 \\ 1 \\ 3 \\ 3 \\ 1 \\ \cdot \end{bmatrix}, \begin{bmatrix} \cdot \\ 0 \\ 0 \\ 1 \\ 3 \\ -3 \\ -1 \\ \cdot \end{bmatrix}, \ldots$$

Those basis vectors are *not* orthogonal. We are alternating lowpass filters with highpass filters, but the first and third (lowpass and shifted lowpass) are not orthogonal. Remember that these vectors are the columns of $\boldsymbol{T}^{-1}$. In our other language ***they are the impulse responses from the synthesis filters***. The synthesis filters combine to reconstruct the signal, which is exactly what the basis vectors do.

In this nonorthogonal case, the inverse is not the transpose. The basis is not self-orthogonal. It is *biorthogonal* to a different basis — which we now discuss.

Biorthogonal Bases (Dual Bases)

The basis $\{\boldsymbol{r}_n\}$ is biorthogonal to the basis $\{\boldsymbol{v}_n\}$ if the inner products are

$$\langle \boldsymbol{r}_i, \boldsymbol{v}_j \rangle = \delta(i-j). \tag{2.47}$$

This is the same property that governs the rows of a matrix $\boldsymbol{T}$ and the columns of $\boldsymbol{T}^{-1}$. It comes directly from $\boldsymbol{T}\boldsymbol{T}^{-1} = \boldsymbol{I}$. So when the columns of $\boldsymbol{T}^{-1}$ are a basis (as above), the rows of $\boldsymbol{T}$ are the biorthogonal basis — the *dual basis*.

We transpose $\boldsymbol{T}$ to turn its rows into columns. We transpose $\boldsymbol{T}^{-1}\boldsymbol{T} = \boldsymbol{I}$ into $\boldsymbol{T}^*\boldsymbol{T}^{-*} = \boldsymbol{I}$. Then the same idea that gave the basis $\{\boldsymbol{v}_n\}$ from $\boldsymbol{T}^{-1}$ now gives the biorthogonal basis $\{\boldsymbol{r}_n\}$ from the columns of the transpose matrix $\boldsymbol{T}^*$:

$$\boldsymbol{v} = \boldsymbol{T}^*\boldsymbol{T}^{-*}\boldsymbol{v} = \sum_{n=1}^{N} (\text{column } n \text{ of } \boldsymbol{T}^*)(\text{row } n \text{ of } \boldsymbol{T}^{-*})\boldsymbol{v}$$

The nth dual basis vector is $\boldsymbol{r}_n = \text{column } n \text{ of } \boldsymbol{T}^*$

The nth dual coordinate is $d_n = (\text{row } n \text{ of } \boldsymbol{T}^{-*})\boldsymbol{v} = \langle \boldsymbol{v}_n, \boldsymbol{v} \rangle$.

When $\boldsymbol{T}^* = \boldsymbol{T}^{-1}$, the basis is self-dual and $\boldsymbol{v}_n = \boldsymbol{r}_n$. We have an orthonormal basis. In general the rows of $\boldsymbol{T}$ and columns of $\boldsymbol{T}^{-1}$ produce biorthogonal bases — provided always that $\boldsymbol{T}$ and $\boldsymbol{T}^{-1}$ are bounded. In this case we have one expansion of $\boldsymbol{v}$ coming from $\boldsymbol{T}^{-1}\boldsymbol{T} = \boldsymbol{I}$, and another expansion from $(\boldsymbol{T}^{-1}\boldsymbol{T})^* = \boldsymbol{I}$:

Theorem 2.4 *If $\boldsymbol{r}_n$ and $\boldsymbol{v}_n$ are biorthogonal bases then any $\boldsymbol{v}$ has two expansions:*

$$\boldsymbol{v} = \sum c_n \boldsymbol{v}_n = \sum \langle \boldsymbol{r}_n, \boldsymbol{v} \rangle \boldsymbol{v}_n \quad \text{and} \quad \boldsymbol{v} = \sum d_n \boldsymbol{r}_n = \sum \langle \boldsymbol{v}_n, \boldsymbol{v} \rangle \boldsymbol{r}_n. \tag{2.48}$$

In other words $\sum \boldsymbol{v}_n \boldsymbol{r}_n^T = \boldsymbol{I}$ and also $\sum \boldsymbol{r}_n \boldsymbol{v}_n^T = \boldsymbol{I}$.

Example 2.10. The four-tap filters in the example above are biorthogonal to these four-tap filters (after we divide by 16). Check the inner products:

$$\ldots, \begin{bmatrix} \cdot \\ -1 \\ 3 \\ 3 \\ -1 \\ 0 \\ 0 \\ \cdot \end{bmatrix}, \begin{bmatrix} \cdot \\ -1 \\ 3 \\ -3 \\ 1 \\ 0 \\ 0 \\ \cdot \end{bmatrix}, \begin{bmatrix} \cdot \\ 0 \\ 0 \\ -1 \\ 3 \\ 3 \\ -1 \\ \cdot \end{bmatrix}, \begin{bmatrix} \cdot \\ 0 \\ 0 \\ -1 \\ 3 \\ -3 \\ 1 \\ \cdot \end{bmatrix}, \ldots.$$

The construction seems magical. We want more like this. Chapters 4 and 5 show how to get more—and the filters need not have equal length. These are the synthesis filters (in T^{-1}) and the analysis filters (in the rows of T) of a *perfect reconstruction filter bank.* The filter bank is biorthogonal.

The norms of T and T^{-1}, and therefore the condition number $\|T\|\,\|T^{-1}\|$, are easily computed for these bases. The double shift in T means that the transform involves a 2×2 matrix function of ω. Maximizing its norm over ω yields $\|T\|$, and maximizing the norm of the 2×2 inverse yields $\|T^{-1}\|$.

A small note of caution. The scaling functions and wavelets will come from iterating the lowpass filter, with rescaling. Not every biorthogonal filter bank leads to biorthogonal bases. Sometimes the iteration diverges. There is a serious step from $L^2(\mathbf{Z})$ in discrete time to $L^2(\mathbf{R})$ in continuous time. The example above fails! Those particular filters with ± 1 and ± 3 are only good when they are not iterated too often.

Wavelet Packets and the Best Basis

In Chapter 1 only the lowpass filter was iterated. It was assumed that lower frequencies contained more important information than higher frequencies. For many signals this is not true. A *wavelet packet basis allows any dyadic tree structure* (Figure 2.10). At each point in the tree we have

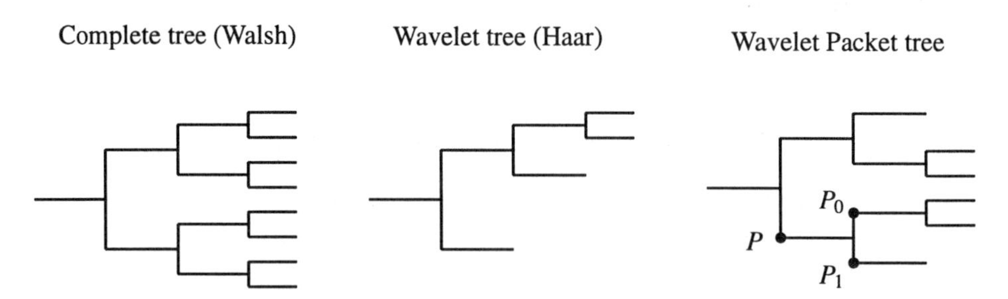

Figure 2.10: Each wavelet packet tree yields a basis, including Walsh and Haar.

the option to send the signal through the lowpass-highpass filter bank, ***or not***.

One possibility is the *logarithmic tree*, with lowpass iteration only. Another possibility is the *complete tree*, analogous to the Short Time Fourier Transform. Wavelet packets make up the entire family of bases. Each one is associated with a particular *quadtree*, because it comes from splitting into two (or not splitting) at each step. The decision to split or to merge should be aimed at achieving minimum distortion D—subject to cost and capacity constraints on the rate R.

The main point is that a library of wavelet packet bases is a practical possibility [W]. For a given signal and a given point P in the tree, we have the coefficients of the basis functions $w_P(t)$ for that point. If we choose to split, that set of functions is replaced by two half-size sets of functions—created by lowpass and highpass filtering:

$$\sum_{k=0}^{N} \boldsymbol{h}_0(k) w_P(2t-k) \quad \text{and} \quad \sum_{k=0}^{N} \boldsymbol{h}_1(k) w_P(2t-k).$$

Together with their time shifts, these are the new basis functions for the points P_0 and P_1 in the tree. At each of those new points we again have the option to split.

A point is at level j in the tree if it is reached after j splittings. The basis functions at the root (level $j = 0$) are the shifts $\phi(t-k)$ of the scaling function.

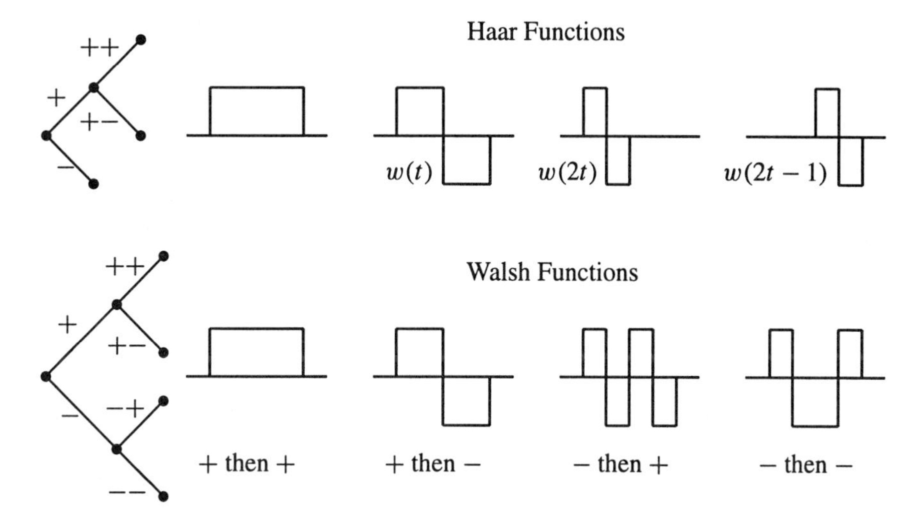

Figure 2.11: Haar iterates only the lowpass filter. Walsh iterates both filters.

Example 2.11. When $\phi(t)$ is the box function, the wavelet $w(t)$ is Haar's up-down square wave. The filter coefficients are $\boldsymbol{h}_0(k) = 1, 1$ for lowpass and $\boldsymbol{h}_1(k) = 1, -1$ for highpass, all divided by $\sqrt{2}$. We describe three special bases for the functions that are piecewise constant on intervals of length 2^{-J}. Within $[0, 1]$ this is a space of dimension 2^J:

1. ***Box basis:*** $\phi(2^J t - k)$ for $0 \le k < 2^J$.

2. ***Haar wavelet basis:*** $\phi(t-k)$ and $w(2^j t - k)$ for $j < J$ and $0 \le k < 2^j$.

3. ***Walsh basis:*** two functions $w_{J-1}(2t) \pm w_{J-1}(2t-1)$ from the basis for $J-1$.

A Walsh basis function takes values 1 *or* -1 *over the whole unit interval.* It comes from the complete tree with all branches. A wavelet basis function from the logarithmic tree is zero over most of $[0, 1]$. It is nonzero on an interval of length 2^{-j}, varying with j. The box basis functions are nonzero over intervals of fixed length 2^{-J}.

Figure 2.11 shows the four Haar and four Walsh functions (with + or − choices indicated). The *Walsh basis* chooses to split every time. The basis functions are like discrete cosines. *The wavelet basis* splits only along one branch. This gives dilation and translation. The *box basis* (scaling function basis) never splits. This gives translation only — a row of small boxes.

The frequency of the ordinary cosine is replaced by the number of *zero-crossings* (sign changes) of the piecewise constant Walsh function. A typical wavelet packet falls between Haar and Walsh. It makes the Walsh decisions, $+$ or $-$, up to certain points P on the tree. At those points it stops splitting and just rescales. The wavelet packet below could be the best basis for a signal with no low frequencies.

The wavelet packet bases are given by simple recursions (not by simple formulas). They inherit the properties of the filter bank — orthogonality or biorthogonality. The Riesz constants A, B and the condition number B/A are not necessarily in control as J increases. The condition number stays bounded for the logarithmic tree wavelet basis [CoDa]. The condition number is unbounded for the general wavelet packet tree.

We emphasize the recursive form of every wavelet packet decomposition. The coefficient sequence at a point P is treated in exactly the same way as the original sequence $\boldsymbol{x}(n)$ at the original root of the tree. The packet splits or not. If it splits then the two new branches end in two new decision points.

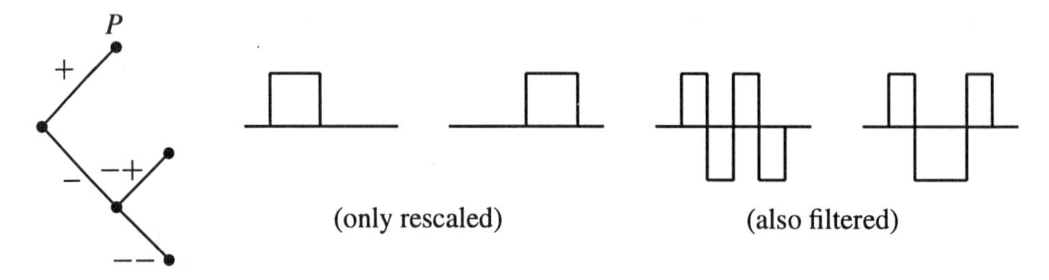

There is a corresponding Fourier packet. Its decision is whether to break the interval in two. Instead of a fixed length 8 for all DFT blocks, and 8×8 for an image, the splitting decision is made locally — depending on the signal. The "butterfly" in the FFT is like the lowpass-highpass bank for wavelets. But the FFT admits all frequencies 0, 1, 2, 3, ... where the wavelets are dyadic and octave-based.

Now we turn from bases to frames, and give up linear independence.

Frames and Frame Bounds

A frame $\{\boldsymbol{v}_n^+\}$ has one property of a Riesz basis $\{\boldsymbol{v}_n\}$ — every vector or function can be represented as $\sum c_n \boldsymbol{v}_n^+$ with control of $\sum |c_n|^2$. But the requirement of linear independence is dropped. A frame is associated with "oversampling" or "redundancy." There are too many vectors for a basis. We could even repeat the same basis vectors several times — this produces a frame but not an interesting one. More interesting is a set of functions like $\{e^{icnt}\}$, which are a basis for $L^2[0, \pi]$ when $c = 1$ and a tight frame for $c < 1$ (higher than Nyquist rate). For $c = \frac{1}{2}$ this frame is a union of two bases $\{e^{int}\}$ and $\{e^{int}e^{it/2}\}$.

The place to start is in finite dimensions. The M rows of a rectangular matrix give a frame for $\mathbf{R}^N$ — *provided the columns are independent*:

$\boldsymbol{T}$ is any $M \times N$ matrix with N independent columns.

In general $M > N$ and $\boldsymbol{T}^{-1}$ does not exist.

The left-inverse $\boldsymbol{T}^+ = (\boldsymbol{T}^*\boldsymbol{T})^{-1}\boldsymbol{T}^*$ does exist and $\boldsymbol{T}^+\boldsymbol{T} = \boldsymbol{I}$.

Linear algebra [S] says that T^*T always has the same rank as T. By assumption of independent columns, this rank is N. The $N \times N$ matrix T^*T with this rank is invertible. The identity $(T^*T)^{-1}\, T^*T = I$ shows that $T^+T = I$.

This matrix $T^+ = (T^*T)^{-1}\, T^*$ is a left-inverse and a "pseudo-inverse" of T. The columns of T^+ and the rows of T are two frames, *dual to each other*. The key is $T^+T = I$.

Example 2.12. The three rows r_1, r_2, r_3 of T constitute a frame for $\mathbf{R}^2$:

$$T = \begin{bmatrix} 2 & 0 \\ -1 & \sqrt{3} \\ -1 & -\sqrt{3} \end{bmatrix} \quad \text{has independent columns and} \quad T^*T = \begin{bmatrix} 6 & 0 \\ 0 & 6 \end{bmatrix}.$$

This example is actually a ***tight frame***, because T^*T is a multiple of I. The left-inverse is $(T^*T)^{-1}\, T^* = \frac{1}{6}T^*$. This matrix T^+ is 2×3. It cannot possibly be a right-inverse of T.

The frame vectors are not independent but they span the space. *We can recover v from the columns v_1^+, v_2^+, v_3^+ by using inner products $\langle r_1, v\rangle$ and $\langle r_2, v\rangle$ and $\langle r_3, v\rangle$.* The recovery is based on $T^+T = I$:

$$v = T^+Tv = \sum_{n=1}^{M} (\text{column } n \text{ of } T^+)(\text{row } n \text{ of } T)v$$

The nth analysis frame vector is $r_n = \text{row } n \text{ of } T$

The nth coordinate is $\langle r_n, v\rangle = (\text{row } n \text{ of } T)v$

The nth synthesis frame vector is $v_n^+ = \text{column } n \text{ of } T^+$.

In this example the column vector $v = [1\;\;1]'$ has coordinates $\langle r_n, v\rangle = 2$ and $-1+\sqrt{3}$ and $-1-\sqrt{3}$. To synthesize v from the three columns of T^+, multiply the columns by those coordinates:

$$T^+(Tv) = \frac{1}{6}\begin{bmatrix} 2 & -1 & -1 \\ 0 & \sqrt{3} & -\sqrt{3} \end{bmatrix} \begin{bmatrix} 2 \\ -1+\sqrt{3} \\ -1-\sqrt{3} \end{bmatrix} = \begin{bmatrix} 1 \\ 1 \end{bmatrix}.$$

The point of the pseudo-inverse T^+ (which is one of many left-inverses) is that the sum of squares of the coordinates is as small as possible. This is the key to frames: *control the sum of squares of coordinates.* Now we define a frame in infinite dimensions.

Definition. *An analysis frame is a set of vectors r_n such that*

$$A\|v\|^2 \;\le\; \sum |\langle r_n, v\rangle|^2 \;\le\; B\|v\|^2 \quad \textit{for all } v. \tag{2.49}$$

$A > 0$ and $B > 0$ are the "frame bounds." A tight frame has $A = B$.

The frame vectors r_n are not required to be independent. But they span the space! The only vector v orthogonal to every r_n is the zero vector, by (2.49).

The frame operator T transforms v into the sequence of numbers $\langle r_n, v\rangle$. In N dimensions these are M numbers, with $M > N$. In infinite dimensions we have infinitely many numbers, and the key is to recover v from these numbers. The purpose of the frame bounds A and B is to make T^*T and $(T^*T)^{-1}$ bounded operators:

$$\text{Note carefully that} \quad \langle v, T^*Tv\rangle = \langle Tv, Tv\rangle = \sum |\langle r_n, v\rangle|^2. \tag{2.50}$$

Inserting into (2.49) gives $A\|\boldsymbol{v}\|^2 \leq \langle \boldsymbol{v}, \boldsymbol{T}^*\boldsymbol{T}\boldsymbol{v}\rangle \leq B\|\boldsymbol{v}\|^2$. The operator $\boldsymbol{T}^*\boldsymbol{T}$ is bounded by B. Its inverse is bounded by $1/A$. When we choose the tightest A and B, which we might as well do, the norms of $\boldsymbol{T}^*\boldsymbol{T}$ and its inverse are exactly B and $1/A$. *The frame ratio B/A is the condition number of $\boldsymbol{T}^*\boldsymbol{T}$.*

The frame bounds ensure that $\boldsymbol{v}$ can be stably recovered from the components $\langle \boldsymbol{r}_n, \boldsymbol{v}\rangle$ of $\boldsymbol{T}\boldsymbol{v}$. The recovery operator — the synthesis operator — is the left-inverse $\boldsymbol{T}^+ = (\boldsymbol{T}^*\boldsymbol{T})^{-1}\boldsymbol{T}^*$:

$$\boldsymbol{v} = \boldsymbol{T}^+\boldsymbol{T}\boldsymbol{v} = \sum (\text{column } n \text{ of } \boldsymbol{T}^+)\, \langle \boldsymbol{r}_n, \boldsymbol{v}\rangle = \sum c_n \boldsymbol{v}_n^+. \tag{2.51}$$

The coordinates are $c_n = \langle \boldsymbol{r}_n, \boldsymbol{v}\rangle$. The synthesis frame vectors $\boldsymbol{v}_n^+$ are the columns of $\boldsymbol{T}^+$. When $\{\boldsymbol{r}_n\}$ is a basis, $\boldsymbol{T}^+$ is $\boldsymbol{T}^{-1}$ and the dual frames are dual bases.

Example 2.13. The three rows of $\boldsymbol{T}$ are an analysis frame (not tight) for $\mathbf{R}^2$:

$$\boldsymbol{T} = \begin{bmatrix} 1 & 1 \\ 1 & 0 \\ 1 & -1 \end{bmatrix} \quad \text{and} \quad \boldsymbol{T}^*\boldsymbol{T} = \begin{bmatrix} 3 & 0 \\ 0 & 2 \end{bmatrix}.$$

Any two rows of $\boldsymbol{T}$ are a basis. The three rows are a frame. The frame bounds are $A = 2$ and $B = 3$. Those are the extreme eigenvalues of $\boldsymbol{T}^*\boldsymbol{T}$ — easy to find in this example. There are infinitely many left-inverses of $\boldsymbol{T}$, and here are three:

$$\begin{bmatrix} 0 & 1 & 0 \\ 1 & -1 & 0 \end{bmatrix} \quad \text{and} \quad \begin{bmatrix} 0 & 1 & 0 \\ 0 & 1 & -1 \end{bmatrix} \quad \text{and} \quad \boldsymbol{T}^+ = \begin{bmatrix} \frac{1}{3} & \frac{1}{3} & \frac{1}{3} \\ \frac{1}{2} & 0 & -\frac{1}{2} \end{bmatrix}.$$

The last one is the best one! It is the pseudoinverse $(\boldsymbol{T}^*\boldsymbol{T})^{-1}\boldsymbol{T}^*$. You see $\frac{1}{3}$ and $\frac{1}{2}$ from inverting $\boldsymbol{T}^*\boldsymbol{T}$. The columns of $\boldsymbol{T}^+$ are the dual frame vectors $\boldsymbol{v}_1^+$ and $\boldsymbol{v}_2^+$ and $\boldsymbol{v}_3^+$.

Why is $\boldsymbol{T}^+$ best? Because it is the "*smallest*" left-inverse. We are reconstructing any vector $\boldsymbol{v}$ in $\mathbf{R}^2$ from the three components c_1, c_2, c_3 of $\boldsymbol{T}\boldsymbol{v}$. In exact arithmetic, the only solution to $\boldsymbol{T}\boldsymbol{v} = \boldsymbol{c}$ is $\boldsymbol{v}$ — and every left-inverse will find it. In actual arithmetic, and in actual measurements, there are errors in $\boldsymbol{c}$. We choose the least-squares solution to $\boldsymbol{T}\boldsymbol{v} = \boldsymbol{c}$. That comes from the normal equations $\boldsymbol{T}^*\boldsymbol{T}\boldsymbol{v} = \boldsymbol{T}^*\boldsymbol{c}$, whose solution is exactly $\boldsymbol{v} = \boldsymbol{T}^+\boldsymbol{c}$. The synthesis frame reconstructs the truest $\boldsymbol{v}$.

Frames are associated with *oversampling* and *redundancy*. We meet this in irregular sampling when it is difficult to sample at exactly the right rate. Better to sample too often than too seldom. Oversampling is usually better than undersampling. The interpolating functions $\operatorname{sinc}(t - t_n)$ are a basis for the band-limited space when the sampling rate is exactly right, and they are a frame when we oversample. They cannot reproduce all functions if we undersample.

For regular sampling at the times $t_n = nT$, the perfect rate is the Nyquist rate in Section 2.2. It requires two samples per period — the exact band of frequencies is $|\omega| \leq \pi/T$. In reality we often take more measurements than necessary, and reconstruct the signal by least squares.

The least-squares matrix is exactly $\boldsymbol{T}^*\boldsymbol{T}$! This is the matrix in the normal equations, when $\boldsymbol{T}$ has too many rows ($M > N$). The coefficient matrix $\boldsymbol{T}$ is $M \times N$ and not invertible. But $\boldsymbol{T}^*\boldsymbol{T}$ is invertible exactly when the rows of $\boldsymbol{T}$ are a frame. Instead of $\boldsymbol{T}^{-1}$ the least-squares solution uses $\boldsymbol{T}^+ = (\boldsymbol{T}^*\boldsymbol{T})^{-1}\boldsymbol{T}^*$.

Note that a pseudoinverse $\boldsymbol{T}^+$ is defined for all matrices, of arbitrary rank [S]. $\boldsymbol{T}^+$ is a left-inverse when the rank is N and the columns of $\boldsymbol{T}$ are independent and the rows of $\boldsymbol{T}$ are a frame. Those are equivalent statements about $\boldsymbol{T}$ in finite dimensions.

Mallat [Mt] identifies two approaches to the reconstruction of the signal $\boldsymbol{v}$. If $\boldsymbol{T}^+$ will be used often, we may compute it explicitly. Its columns are the synthesis frame vectors $\boldsymbol{v}_n^+$. For limited use we do not want the inverse of $\boldsymbol{T}^*\boldsymbol{T}$, only the solutions to specific equations with this coefficient matrix. Just as in linear equations $Ax = b$, we seldom want A^{-1} and we more often want $x = A^{-1}b$.

The iterative solution of the reconstruction problem for irregular sampling has been studied by Grochenig. This is an important application. His algorithms are much better than the simplest iterations.

We mention *Kadec's* $\frac{1}{4}$ *Theorem* for $L^2[0, 1]$: the exponentials $\left\{e^{2\pi i c_n t}\right\}$ are a basis if $|c_n - n|$ stays below a number $L < \frac{1}{4}$. This is nearly a regular sampling.

In our applications, the "rows" of $\boldsymbol{T}$ are usually functions $r_n(t)$. Then $\boldsymbol{T}$ maps functions $f(t)$ in L^2 to sequences $c_n = \langle r_n, f\rangle$ in ℓ^2. The matrix $\boldsymbol{T}\boldsymbol{T}^*$ contains the inner products of those rows with themselves. It is the "Gram matrix" whose (i, j) entry is the inner product $\langle r_i, r_j\rangle$. *The Riesz bounds A and B come from this matrix.* Its norm is B and the norm of its inverse is $\frac{1}{A}$.

Those numbers measure the linear independence of the functions $r_n(t)$. A case of special interest is when the rows $r_n(t)$ are translates $\phi(t - n)$ of a single scaling function. Theorem 6.6 will give a formula for A and B in this case, involving $\widehat{\phi}(\omega)$.

Summary: The bounds on $\boldsymbol{T}\boldsymbol{T}^*$ are the *Riesz constants* — measuring independence of the rows of $\boldsymbol{T}$. The bounds on $\boldsymbol{T}^*\boldsymbol{T}$ are the *frame bounds*. When $\boldsymbol{T}$ is invertible, these ideas are the same — we have a Riesz basis. When $\boldsymbol{T}$ is not invertible the Riesz lower bound is $A = 0$.

Question: Could a frame include the zero vector $\boldsymbol{r} = 0$?

Answer: Yes. The inner product $\langle 0, f\rangle = 0$ would not harm (or help) the frame bounds. A zero row of $\boldsymbol{T}$ has no effect at all on $\boldsymbol{T}^*\boldsymbol{T}$. And it destroys $\boldsymbol{T}\boldsymbol{T}^*$.

Question: Do a lowpass and highpass filter jointly produce a frame?

Answer: Yes, if their frequency responses have $|C(\omega)| + |D(\omega)| \geq A > 0$.

Question: When would the two filters jointly produce a tight frame?

Answer: When $|C(\omega)|^2 + |D(\omega)|^2 =$ constant. This happens in an orthonormal filter bank. The frame constant is 2! Subsampling removes the redundancy and reduces that constant to 1. The tight frame becomes an orthonormal basis.

Construction of Frames

In finite dimensions, any set of vectors that spans $\mathbf{R}^N$ is a frame. In infinite dimensions, the frame condition is not so simple. *The good constructions start with only one function.* We never want to compute with an infinite set of totally unrelated functions. The natural idea is to create an infinite family from that one function, in a systematic way.

Two systems are of special importance and they lead to ***windows*** and ***wavelets***:

1. ***Windows*** come from one window $g(t)$ by *modulation* and *translation*:

$$g_{mn}(t) = e^{im\Omega t} g(t - nT). \tag{2.52}$$

2. ***Wavelets*** come from one wavelet $w(t)$ by *dilation* and *translation*:

$$w_{jk}(t) = a^{j/2} w\left(a^j t - kT\right). \tag{2.53}$$

The indices m, n, j, and k extend over the set $\mathbf{Z}$ of all integers. We have a family $g_{mn}(t)$ of windows and a family $w_{jk}(t)$ of wavelets. The numbers Ω and T and a are fixed. *Those numbers decide whether the family is a frame.* Dividing T by 2 gives twice as many functions—increasing the likelihood of a frame. Dividing Ω by 2 works the same way; so does replacing a by $\sqrt{a}$. The crucial parameter is ΩT for the window family and $T \log a$ for the wavelet family. When these parameters are small, we have more functions and more likely a frame.

As T and Ω and $\log a$ approach zero, we begin to lose in efficiency. The ratio B/A eventually increases—the frame becomes excessively redundant. (Possibly B/A is a convex function of ΩT.) Most of this book is about $a = 2$ and $T = 1$ and special wavelets $w(t)$, leading not only to a frame but to a basis. Section 8.4 is about $\Omega = 1$ and $T = 1$ and special windows, again leading to a basis. Here we are only aiming for frames—much easier.

A point about normalization. The factor $a^{j/2}$ is included in the wavelets to keep the same norm. Without this factor the norms would go to zero as $j \to \infty$. The information to reconstruct $f(t)$ is still available in the inner products with the frame vectors, but the scaling is poor. This emphasizes that in infinite dimensions, the frame must span the space *stably*.

Window Frames: Examples and Theorems

Example 2.1. Suppose $g(t)$ is the unit box function on $[0, 1]$. For $T > 1$ the $g_{mn}(t)$ are not a frame. None of the boxes $g(t - nT)$ overlap the interval $[1, T]$. Any function $f(t)$ supported on this interval has zero inner product with all the windows. These spaced-out windows do not span L^2.

For $T = 1$ the translated boxes $g(t - n)$ fit tightly. Within each box we have exponentials $e^{im\Omega t}$. For $\Omega = 2\pi$ this is an orthonormal basis on the interval $[0, 1]$. For $\Omega < 2\pi$ it is a frame.

We see how a basis appears on the "edge" of a family of frames. As soon as Ω passes 2π, there are not enough exponentials $e^{im\Omega t}$ and the frame is lost. (For $\Omega = 4\pi$ we will completely miss $e^{i2\pi t}$.) In the next example, and often, no basis appears—we go directly from a frame for small ΩT to failure for larger ΩT.

Example 2.2. Suppose $g(t)$ is the Gaussian function $e^{-t^2/2}$ as in [Gabor]. It produces a frame if $\Omega T < 2\pi$. It does *not* produce a frame if $\Omega T = 2\pi$, although this was Gabor's favorite! At the edge of the frames, these functions $g_{mn}(t) = e^{2\pi imt}e^{-(t-n)^2/2}$ do span the space—but not stably. The inner products $\langle f, g_{mn}\rangle$ uniquely determine $f(t)$, but the operator $\boldsymbol{T}^*\boldsymbol{T}$ is not bounded below. The reconstruction formula from those inner products is not numerically stable.

For $\Omega T < 2\pi$ the tables in [D, p. 87] show how B/A depends on Ω and T.

We now state four general results about window frames, without proof. Three are positive, one is negative. The negative one is famous and comes first:

A. *Window frames are impossible with $\Omega T > 2\pi$. Smooth and decaying window frames are impossible with $\Omega T = 2\pi$.*

"Smooth and decaying" in this Balian-Low Theorem [BeHaWa] means that $g'(t)$ and $tg(t)$ are in $L^2(\boldsymbol{R})$. The window bases in Section 8.4 escape this restriction by a simple change in their construction (*cosines replace $e^{2\pi imt}$*). The impossibility for $\Omega T > 2\pi$ first came from Rieffel.

B. *The dual to a window frame comes from a dual window* $\widetilde{g}(t)$.

This makes it worthwhile to compute the dual window. One function $\widetilde{g}(t)$ gives the whole dual frame. In many other cases it is too much work to construct $(\boldsymbol{T}^*\boldsymbol{T})^{-1}\boldsymbol{T}^*$, whose columns contain the dual frame. Instead we solve a linear system for the synthesis coefficients in $\sum c_n \boldsymbol{r}_n$.

C. *Suppose* $\sum |g(t-nT)|^2$ *stays between positive constants for all* t. *Then there is a positive threshold* Ω_0 *such that* $\Omega < \Omega_0$ *yields a frame* [D, p. 82].

This assumes that $g(t)$ has compact support or $|t|^\alpha |g(t)| \to \infty$ for some $\alpha > 1$. ‘

D. The frame bounds satisfy $A \le 2\pi \|g\| / \Omega T \le B$.

Wavelet Frames: Examples and Theorems

Example 2.1. The *Mexican hat* $w(t) = (1-t^2)e^{-t^2/2}$ is the second derivative of the Gaussian. This wavelet is often used in analyzing vision. It has very rapid decay of both w and $\widehat{w}$. For $a = 2$ and small T, the tables in [D, p. 77] show that B/A is very near 1. As T increases we suddenly lose the frame.

Example 2.2. The *Morlet wavelet* is a modulated Gaussian (complex for real signals):

$$w(t) = e^{-i\omega_0 t} e^{-t^2/2} \quad \text{with} \quad \omega_0 = \pi\sqrt{\frac{2}{\ln 2}} = 5.336.$$

This shift in frequency almost gives $\int w(t)\,dt = 0$ as required for wavelets. The actual value is below 10^{-6}, and negligible. The *phase* plot of the wavelet coefficients $\langle f, w_{jk} \rangle$ becomes very useful [Mt] in locating singularities of $f(t)$.

Some applications of these two examples use N different wavelets, called *voices*. This usually produces a tighter frame; B/A comes near 1. A good way to spread the N voices over an octave is by *fractional dilation* of a single wavelet:

$$w_\ell(t) = w\left(2^{-\ell/N} t\right), \quad \ell = 0, \ldots, N-1. \tag{2.54}$$

The negative statement A and positive B in the window rules are reversed for wavelets.

A′. *There are wavelet frames* (even orthonormal bases) *for large values of* $T \log a$. *The restriction* $\Omega T < 2\pi$ *on smooth window frames does not apply to wavelets.*

B′. *The dual to a wavelet frame does not always come from one dual wavelet.*

C′. *If* $\sum \left|\widehat{w}\left(a^j \omega\right)\right|^2$ *stays between positive constants for all* ω, *and* $w(t)$ *is smooth, there is a positive threshold such that* $T < T_0$ *yields a frame* [D, p. 69].

D′. *The frame bounds* A *and* B *are related to the wavelet constant* C *by*

$$A \le \frac{C}{2T \log a} \le B \quad \textit{with} \quad C = 2\pi \int_{-\infty}^{\infty} |\widehat{w}(\omega)|^2 \, \frac{d\omega}{|\omega|}. \tag{2.55}$$

For a tight frame equality holds. For an orthonormal basis $A = B = 1$. This constant C will be crucial for the integral wavelet transform in the next section.

Problem Set 2.5

1. Find a formula for the Walsh function that comes from the choices $- + -$. With $J = 3$ this function $w_{-+-}(t)$ is equal to 1 or -1 on eight subintervals of $[0, 1]$.

2. Find a general formula for the Walsh functions at level $J = 3$. The numbers p, q, r give the three decisions $+1$ or -1.

3. Count how many wavelet packets do not go beyond the level $J = 3$.

4. In Example 2.13 with $\boldsymbol{T} = [1\ \ 1\ \ 1;\ \ 1\ \ 0\ \ -1]$ suppose the vector $\boldsymbol{v}$ is $[3\ \ 4]'$.

   1. What are its three inner products c_1, c_2, c_3 with the rows of $\boldsymbol{T}$?
   2. Verify that $2\|\boldsymbol{v}\|^2 \le c_1^2 + c_2^2 + c_3^2 \le 3\|\boldsymbol{v}\|^2$. The frame bounds are $A = 2$ and $B = 3$.
   3. Verify that $c_1\boldsymbol{v}_1^+ + c_2\boldsymbol{v}_2^+ + c_3\boldsymbol{v}_3^+$ reconstructs this $\boldsymbol{v}$.

5. Suppose the frame vectors are $\boldsymbol{r}_1 = [0\ \ 2]$ and $\boldsymbol{r}_2 = [1\ \ 1]$ and $\boldsymbol{r}_3 = [2\ \ 0]$. Compute $\boldsymbol{T}^*\boldsymbol{T}$ and its eigenvalues A and B (the frame bounds). Also compute $\boldsymbol{T}^+$ and the dual frame.

6. *The bounds for the dual frame* (in the columns of $\boldsymbol{T}^+$) *are* $\frac{1}{B}$ *and* $\frac{1}{A}$. Prove this by showing that $\boldsymbol{T}^+(\boldsymbol{T}^+)^*$ equals $(\boldsymbol{T}^*\boldsymbol{T})^{-1}$. The extreme eigenvalues are ________, and

$$\frac{1}{B}\|\boldsymbol{v}\|^2 \le \sum \left|\langle \boldsymbol{v}_n^+, v\rangle\right|^2 \le \frac{1}{A}\|\boldsymbol{v}\|^2.$$

7. A tight frame has $\boldsymbol{T}^*\boldsymbol{T} = A\boldsymbol{I}$. Explain why this is equivalent to $\sum |\langle \boldsymbol{r}_n, v\rangle|^2 = A\|\boldsymbol{v}\|^2$. The frame bounds A and B are equal. The analysis frame $\{\boldsymbol{r}_n\}$ and synthesis frame $\{\boldsymbol{v}_n^+\}$ are both tight.

8. The Mth roots of 1 are the complex numbers with coordinates $\left(\cos\frac{2\pi k}{M}, \sin\frac{2\pi k}{M}\right)$. Show that these M vectors in $\mathbf{R}^2$ form a tight frame.

9. Prove that $\left\{e^{icnt}\right\}$ is a tight frame for $c < 1$ with constant $A = \frac{1}{c}$.

   Hint: Change $\int_0^{2\pi} v(t)e^{-icnt}\,dt$ to $\int_0^{2\pi} g(s)e^{ins}\,ds$ where $g(s) = \frac{1}{c}v\left(\frac{s}{c}\right)$ up to $s = 2\pi c$ (then zero). The sum of squares of coefficients is $\|g\|^2$ because $\{e^{ins}\}$ is ________. Verify that $\|g\|^2 = \frac{1}{c}\|v\|^2$ so that $A = \frac{1}{c}$.

2.6 Time, Frequency, and Scale

This book emphasizes bases more than frames. The synthesis operators are inverses rather than pseudo-inverses. The analysis bank produces the "right" number of outputs—the data is critically sampled and not oversampled. We want bases that are convenient for computation (by fast transform) and well adapted to the signal (for high compression).

This section is different. First it looks at transforms in continuous time and continuous frequency and continuous scale. Reconstruction is by an integral instead of a sum. This allows a very wide choice of windows and wavelets, in the Short Time Fourier Transform and the Integral Wavelet Transform. The analysis and synthesis formulas are simple and general. But when we sample in time and frequency and scale—in order to select a discrete set as a basis or a frame—the conditions on the windows and wavelets become tighter.

At the end of the section we relax by using many more functions than necessary. Those are ***time-frequency atoms***—wavelets or windows or whatever. The search for the best representation from a big dictionary of atoms is a very active problem.

May we first contrast windows with wavelets? The competition between them is far from over. Both have the goal of localizing the basis functions. The windowing functions $g(t)$ achieve that by dropping quickly to zero—they can be Gabor windows with the Gaussian factor e^{-t^2}, or they can have finite length. The shifted window $g(t-s)$ is localized around time s. The expansion functions are oscillations inside the window:

$$\textbf{Windowed exponentials} \quad g_{\omega,s}(t) = g(t-s)e^{i\omega t}.$$

These are "time-frequency atoms." When sampled at discrete times $t = n$ and discrete frequencies $\omega = 2\pi k$, the localized exponentials are $g(t-n)e^{2\pi ikt}$. If $g(t)$ is a unit box, these functions give a basis—but not smooth. When $g(t)$ is a smooth window we don't have a basis (Balian-Low Theorem). The samples are over-complete (a frame) or they are incomplete—depending on $g(t)$ and the sampling rate. But Section 8.4 will show how smooth local cosines can give a very satisfactory basis when the frequencies are shifted to $k + \frac{1}{2}$.

Here we keep *all times s and all frequencies ω*. The familiar integral Fourier transform (no window) is a function of ω:

$$\widehat{f}(\omega) = \frac{1}{2\pi}\int_{-\infty}^{\infty} f(t)\, e^{-i\omega t}\, dt. \tag{2.56}$$

The windowed Fourier transform is a function of ω and also the position s:

$$\textbf{Windowed Transform} \quad F(\omega, s) = \frac{1}{2\pi}\int_{-\infty}^{\infty} f(t)\, g(t-s)e^{-i\omega t}\, dt. \tag{2.57}$$

This is the Fourier transform of the windowed functions $f(t)\,g(t-s)$ for all s. Without the windows, the reconstruction of $f(t)$ from $\widehat{f}(\omega)$ is famous:

$$f(t) = \int_{-\infty}^{\infty} \widehat{f}(\omega)e^{i\omega t}\, d\omega. \tag{2.58}$$

With the windows, we recover $f(t)\,g(t-s)$ by this integral over ω. Equation (2.58) for each s is

$$f(t)\,g(t-s) = \int_{-\infty}^{\infty} F(\omega, s)e^{i\omega t}\, d\omega. \tag{2.59}$$

Now multiply both sides by $g(t-s)$ (or $\overline{g(t-s)}$ if complex) and integrate over s:

$$f(t) = \frac{1}{\|g\|^2}\int_{-\infty}^{\infty}\int_{-\infty}^{\infty} F(\omega, s)\, g(t-s)e^{i\omega t}\, d\omega\, ds. \tag{2.60}$$

This is reconstruction from the time-frequency transform $F(\omega, s)$.

Compare this windowed STFT with wavelets, which begin with *one function* $w(t)$. The position variable s still comes from translation to $w(t-s)$. Now comes the difference. *Instead of the frequency variable we have a scale variable.* Instead of modulating the wavelets we rescale them. The "time-scale atoms" are translates and dilates of $w(t)$:

$$\textit{Wavelet functions} \quad w_{a,s}(t) = |a|^{-1/2} w\left(\frac{t-s}{a}\right).$$

The mother wavelet $w(t)$ is $w_{1,0}(t)$ at unit scale $a = 1$ and position $s = 0$. The factor $|a|^{-1/2}$ assures that the rescaled wavelets have equal energy $\|w_{a,s}\| = \|w\|$. We normalize so that all these functions have unit norm $\|w_{a,s}\| = 1$.

Notice how scaling the time by a or a^j automatically scales the translation steps by a^{-1} or a^{-j}:

$$w\left(a^j t - k\right) = w\left(a^j[t - ka^{-j}]\right).$$

The mesh length at level j is scaled down by a^{-j}. The "frequency" is scaled up by a^j. This hyperbolic scaling or dyadic scaling or octave scaling ($a = 2$) is a prime characteristic of wavelet analysis.

The integral wavelet transform is an inner product with wavelets, just as the windowed transform was an inner product with windowed exponentials:

$$\textit{Integral Wavelet Transform} \quad F_w(a, s) = |a|^{-1/2} \int_{-\infty}^{\infty} f(t) w\left(\tfrac{t-s}{a}\right) dt. \tag{2.61}$$

The transform F_w is defined on the *time-scale plane*. Again there are two variables, but the scale a has replaced the frequency ω. The subscript in F_w indicates this change.

The wavelet transform $F_w(a, s)$ is redundant, just as the windowed transform $F(\omega, s)$ was redundant. If we select a good wavelet, it will be enough to know F_w at a discrete set of scales and positions. That is the construction to implement digitally. Here we reconstruct $f(t)$ from its over-complete transform $F_w(a, s)$.

Important: The key to scaling is not a but $\log a$. The natural scale is *logarithmic*. The differential is not da but da/a. The frequency window da is proportional to a (this is often called *constant-Q*). Convolution with $\frac{1}{a} w\left(\frac{t}{a}\right)$ gives a unitary operator, when we integrate with respect to logarithmic measure da/a:

$$\int_0^{\infty} C_a C_a^* \frac{da}{a} = I \quad \text{for convolution } C_a \text{ with } \frac{1}{a} w\left(\frac{t}{a}\right).$$

This is *Calderón's identity*, rediscovered by Grossmann and Morlet. It is proved in Theorem 2.5 below, with a different normalization of the wavelet and wavelet transform (each by $|a|^{-1/2}$). It gives the reconstruction formula that inverts (2.61) and recovers $f(t)$:

Reconstruction from Wavelet Transform

$$f(t) = \frac{1}{C} \int_{-\infty}^{\infty} \int_{-\infty}^{\infty} F_w(a, s)|a|^{-1/2}\, w\left(\frac{t-s}{a}\right) \frac{da\, ds}{a^2}. \tag{2.62}$$

The constant for windows was $C = \|g\|^2$. The constant for wavelets is not $\|w\|^2$, which would be 2π times the integral of $|\widehat{w}|^2$. Instead the constant is $C = 2\pi \int |\widehat{w}|^2 d\omega/|\omega|$. Effectively, C is finite when the transform of the wavelet is zero at $\omega = 0$. This means that *the integral of the wavelet is zero*. Any smooth decaying function $w(t)$ is a mother wavelet for the integral transform provided $\int w(t)\, dt = 0$.

The Haar wavelet was not the box function. It was the up-down square wave with integral zero, one box minus another box. Other wavelets will be combinations of scaling functions $\phi(2t - k)$, *with coefficients that add to zero*. Those coefficients come from the highpass filter! They will be specially chosen, and $\phi(t)$ will be specially chosen, to give a discrete basis. In the continuous time-scale plane we only require that $\int w(t)\, dt = 0$.

The reconstruction formula (2.62) comes from the following theorem.

Theorem 2.5 *For any $f(t)$ and $g(t)$ in L^2, and C as above,*

$$C \int_{-\infty}^{\infty} f(t)\overline{g(t)}\, dt = \int_{-\infty}^{\infty} \int_{-\infty}^{\infty} F_w(a, s)\, \overline{G_w(a, s)} \frac{da\, ds}{|a|^2}. \tag{2.63}$$

With $f = g$ this is a "Parseval formula" in the time-scale plane. The energy in f equals the energy in its wavelet transform, when the time-scale area is measured correctly.

Ironically, the simplest proof of (2.63) uses the Fourier transform. For each a, the transform in (2.61) is a convolution of $f(t)$ with the scaled wavelet $|a|^{-1/2}w\left(\frac{-t}{a}\right)$. The Fourier transform of this convolution F_w is a multiplication $|a|^{1/2}\widehat{f}(\omega)\overline{\widehat{w}(a\omega)}$. Similarly, the transform of $\overline{G_w}$ is a multiplication $|a|^{1/2}\overline{\widehat{g}(\omega)}\widehat{w}(a\omega)$. Then for each a, the integral over s on the right side of (2.63) becomes

$$\int_{-\infty}^{\infty} F_w(a,s)\overline{G_w(a,s)}ds = 2\pi\int_{-\infty}^{\infty} |a|\widehat{f}(\omega)\,\overline{\widehat{g}(\omega)}\,|\widehat{w}(a\omega)|^2\,d\omega. \tag{2.64}$$

Integrate this with respect to $da/|a|^2$ and reverse the order of integration. The integral over a gives the predicted constant C, after changing variables to $a\omega$. The right side of (2.63) becomes

$$2\pi\int_{-\infty}^{\infty} \widehat{f}(\omega)\overline{\widehat{g}(\omega)}\int_{-\infty}^{\infty} |\widehat{w}(a\omega)|^2\,da\,d\omega = 2\pi C\int_{-\infty}^{\infty} \widehat{f}(\omega)\overline{\widehat{g}(\omega)}\,d\omega. \tag{2.65}$$

This equals the left side of (2.63) and completes the proof.

Suppose $g(t) = \delta(t)$ is the Dirac delta function (not in L^2). Then equation (2.63) formally becomes the reconstruction formula (2.62) at $t = 0$. The wavelet transform $G_w(a,s)$ of $\delta(t)$ is the wavelet $|a|^{-1/2}w\left(-\frac{s}{a}\right)$. This appears in (2.62) when $t = 0$. By shifting the delta function we reach the reconstruction formula at all t.

This use of the delta function is more legal and familiar than it seems. The ordinary Fourier reconstruction in (2.58) is reached the same way. The analogue of (2.63) is Parseval's equation

$$\int_{-\infty}^{\infty} f(t)\overline{g(t)}\,dt = 2\pi\int_{-\infty}^{\infty} \widehat{f}(\omega)\overline{\widehat{g}(\omega)}\,d\omega \quad \text{for all } f, g \text{ in } L^2. \tag{2.66}$$

With $g(t) = \delta(t)$ this is the reconstruction $f(0) = \int \widehat{f}(\omega)\,d\omega$ at $t = 0$. A shifted delta function gives the inversion formula at any other t. These integrals are correct *at each t* when f is smooth. They are correct *in the L^2 sense* for a general function f in L^2. Technically, we smooth f by restricting its transform to $|\omega| \le \Omega$ and then let $\Omega \to \infty$.

Note. For real wavelets $w(t)$, when $|\widehat{w}|^2$ is an even function, integrate only over $a > 0$ and compensate by taking only half of the constant C:

$$C \to \int_0^{\infty} |\widehat{w}(\omega)|^2\frac{d\omega}{\omega} = \tfrac{1}{2}C.$$

The Wigner-Ville Transform

The time-frequency analysis of a signal goes much further than windows and wavelets. By restricting to the choice of one function $g(t)$ or $w(t)$, we create good algorithms — which is our principal purpose. By allowing more general transforms, we can hope to get precise information about the "instantaneous frequency" and "instantaneous spectrum" of a signal. The windows and wavelets do a little smearing — not too much, and the time-frequency trace of the signal can be followed. But Wigner and Ville and many others wanted a sharp trace.

We shall devote just a very short space to this basic topic — the correlation of a signal with itself. Everywhere else we are correlating the signal with chosen windows or wavelets. As those move and rescale, we identify our signal from the correlations. The self-correlation gives an energy density that should automatically pick out the position and frequency and scale of the signal.

We define the *Wigner-Ville transform* of a finite energy function $f(t)$:

$$W(t,\omega) = \int_{-\infty}^{\infty} f\left(t+\frac{\tau}{2}\right)\overline{f\left(t-\frac{\tau}{2}\right)} e^{-i\omega\tau}\, d\tau. \tag{2.67}$$

Notice especially that W is quadratic in f. We expect $W(t,\omega)$ to be an "energy density"—like the *square* of the windowed transform and wavelet transform, but better. This density has several desirable properties:

$$\frac{1}{2\pi}\int W(t,\omega)\, d\omega = |f(t)|^2 \quad\text{and}\quad \int W(t,\omega)\, dt = \left|\widehat{f}(\omega)\right|^2. \tag{2.68}$$

$W(t-T, \omega-\Omega)$ is the transform of $e^{i\Omega t} f(t-T)$.

$W\left(\frac{t}{a}, a\omega\right)$ is the transform of $\frac{1}{\sqrt{a}} f\left(\frac{t}{a}\right)$.

$W(t,\omega)$ determines $f(t)$ up to a constant multiplier $|c| = 1$.

The transform goes from one variable to two variables. As with wavelets and windows, most functions of two variables are *not* transforms of any $f(t)$. The transform of $\widehat{f}$ flips the variables to reach $W(\omega, -t)$. And the Gaussian plays a special role. We get $W = e^{-t^2-\omega^2}$ from the unit Gaussian and we get modulations of $W = e^{-p^2t^2-q^2\omega^2 \pm pqt\omega}$ from all quadratic "chirps."

What good properties does this transform *not* have? First, it is not always positive. Second, the sum of two Gaussians (modulated and shifted) has a transform with *three terms*. Two terms are in the expected positions (T_1, Ω_1) and (T_2, Ω_2), as desired from the Gaussians. There is a rapidly oscillating *ghost* term in the wrong position $\frac{1}{2}(T_1 + T_2, \Omega_1 + \Omega_2)$.

"Analytic signals" have no negative frequencies: $\widehat{f}(\omega) = 0$ for $\omega < 0$. Restricted to these signals, the transform is a more perfect success. Then $W(t,\omega)$ earns the name "instantaneous spectrum" of $f(t)$. The instantaneous frequency—*the derivative of the phase* $\phi(t)$—equals the average $\int \omega W(t,\omega)\, d\omega / 2\pi\, |f(t)|^2$.

Summary: The integral transforms, by windows and wavelets, make minimal demands on the functions $g(t)$ and $w(t)$. We can analyze and synthesize (transform and inverse transform) with extremely general functions. Reality sets in when the transform becomes discrete—the shifts are multiples of T, the modulations are multiples of Ω, the scales are powers of a. The elegant books by Meyer give an overview of this whole picture.

The rest of this text deals with the discrete transform. We will have very strict conditions on the wavelets! The scales are powers of a fixed integer $a = M$, most often $a = M = 2$. Extra conditions are imposed on the wavelet. There are absolute requirements and optional properties. The requirements make it work, the extra conditions make it work well. The goal is to achieve these properties with a fast algorithm:

1. Two-scale or M-scale equation (with special coefficients)
2. Smoothness of the wavelet (optional but desirable)
3. Symmetry or antisymmetry (optional but very desirable)
4. Vanishing moments (one required, more desirable).

Atomic Decompositions

Whether they are wavelets or sinusoids or cosine packets or Gabor functions, we are approximating $f(t)$ by "atoms". A collection of atoms is a "dictionary". We need a reasonable algorithm to choose the nearly best M atoms for a given $f(t)$, out of a dictionary of P atoms. A single basis may be too inflexible, and *the algorithm must adapt to the signal.* Major effort is going into algorithms *not* based on one orthonormal basis.

We will only mention three ideas and their developers. The first is **matching pursuit** (S. Mallat and Z. Zhang). The M atoms are chosen one at a time. The choice at step $k+1$ is the atom that comes closest to the current difference $f_k(t) = f(t) - c_1\phi_1(t) - \cdots - c_k\phi_k(t)$. In practice matching pursuit takes (normalized) inner products $\langle f_k(t), \phi(t)\rangle$ and chooses a large one—an atom $\phi_{k+1}(t)$ that is highly correlated with $f_k(t)$. This is a *greedy and sub-optimal* selection process—greedy because each choice is ignorant of the later choices, sub-optimal because we don't insist on maximizing $\langle f_k(t), \phi(t)\rangle$. The complexity is $O(N \ln N)$ at each iteration, for a signal of length N. The pursuit ends at $k = M$. We can then compute the best combination of the M choices (this is *back-projection*). Experimentally the error approaches white noise as $M \to \infty$.

A second adaptive method is the **best basis** algorithm (R. Coifman and V. Wickerhauser). This begins with a dictionary of bases, often orthonormal. The algorithm chooses the best basis to represent $f(t)$. For wavelet packets from a family of binary trees, the method is particularly well adapted. Available software includes the Wavelet Packet Laboratory for Windows (AK Peters, Wellesley MA 02181-5910). A modification that is optimal in the rate-distortion sense, for compression, is described in [VK, p. 426].

With orthonormal bases at every step, rates and mean-square distortions of the two branches are additive. This allows a fast algorithm to choose the packet that is best adapted to the particular signal. The reader understands that an actual implementation leads to many options. The mean-square ℓ^2 norm may be replaced by other norms (losing additivity at each split but gaining in perceptual quality).

A third method is **basis pursuit** (D. Donoho). The dictionary is still overcomplete. The synthesis $f(t) = \sum c_i\phi_i(t)$ (modelled by $\boldsymbol{f} = \Phi\boldsymbol{c}$) is underdetermined. *Frame theory chooses* $\boldsymbol{c}$ *to have a small* l^2 *norm* (sum of c_i^2 is minimized, leading to linear equations and generalized inverses). *Basis pursuit chooses* $\boldsymbol{c}$ *to have a small* l^1 *norm* (sum of $|c_i|$ is minimized, leading to nonlinear equations and linear programming). This minimizer is generally much sparser—fewer c_i are nonzero.

The l^1 minimizer is also more expensive to compute. In place of the simplex method, interior point and log barrier algorithms associated with Karmarkar solve a sequence of weighted l^2 problems. This can give reasonable success for large dictionaries ($P = 10^4$ and $N = 10^3$). The method can distinguish two nearby bumps in $f(t)$, where matching pursuit will select one centered bump that has high correlation. Linear programming pursues the best basis—*which depends on the signal* $f(t)$. Donoho also studies "empirical atomic decomposition" for $P >> N$. This algorithm strongly controls the number M of atoms $\phi_k(t)$ in the approximation, by minimizing $||f(t) - \sum c_i\phi_i(t)||^2 + \lambda M$. The key is in the selection of $\lambda = \sigma\sqrt{2\log P}$. Software is on the web at `http://playfair.stanford.edu`.

The multiplier λ *becomes* $\sigma\sqrt{2\log N}$ *for de-noising with an orthonormal basis.* The noise is assumed Gaussian with standard deviation σ. The recommended solution is soft thresholding of the inner products $y_i = \langle \phi_i, f \rangle$ with threshold λ. Each scalar y is shifted toward zero but not beyond:

$$\textbf{\textit{Thresholding}} \quad c_{soft} = (y - \lambda)_+ \quad \text{for} \quad y > 0 \quad \text{and} \quad (y + \lambda)_- \quad \text{for} \quad y < 0.$$

Donoho notes that c_{soft} minimizes $\frac{1}{2}(y - c)^2 + \lambda |c|$. That $|c|$ is the l^1 norm again.

Finally we emphasize that the construction of wavelets has not ended. While writing the book we have learned of *directional* and *translation-invariant* and *steerable* wavelets. And the world of atoms is by no means restricted to wavelets!

Problem Set 2.6

1. Find the Wigner-Ville transform $W(t, \omega)$ when $f(t)$ is the Dirac function $\delta(t)$.

2. Show how $W(t, \omega)$ can also be expressed by an integral of $\widehat{f}$:

$$W(s, \xi) = \frac{1}{2\pi} \int \widehat{f}\left(\xi - \frac{\omega}{2}\right) \widehat{f}\left(\xi + \frac{\omega}{2}\right) e^{is\omega} \, d\omega.$$

Chapter 3

Downsampling and Upsampling

3.1 The Matrices for Downsampling and Upsampling

The previous chapter was about individual filters. The next chapter combines two or more filters into a ***filter bank***. The idea is to separate the incoming signal into frequency bands. Often we will use a bank of two filters to explain a point clearly. A lowpass filter C and a highpass filter D will split the signal into two parts. Those parts can be compressed and coded separately (and efficiently). Then they can be transmitted and the signal can be recovered.

When the synthesis bank recovers the input signal exactly (apart from a time delay), it is a *perfect reconstruction* filter bank. These PR banks interest us most.

One problem to overcome. We don't want two full length signals in place of one, if we can help it. It is not desirable to double the volume of data. The information has not doubled; the outputs Cx and Dx from the two filters must be redundant.

The solution to this problem is beautiful. *We keep only half of the outputs* Cx *and* Dx. More precisely, we ***downsample*** those vectors by removing every other component. This operation is called *decimation*. Downsampling is represented by the symbol $(\downarrow 2)$ (pronounced "down two"):

$$(\downarrow 2)\begin{bmatrix} \cdot \\ y(-1) \\ y(0) \\ y(1) \\ y(2) \\ \cdot \end{bmatrix} = \begin{bmatrix} \cdot \\ y(-2) \\ y(0) \\ y(2) \\ y(4) \\ \cdot \end{bmatrix}.$$

Make no mistake. This operation is not invertible. Most vectors y cannot be recovered from $(\downarrow 2)y$. The odd-numbered components are lost. Normally the even-numbered components of both $y = Cx$ and $y = Dx$ will be needed to recover all components of x.

We note that recovery of x from $(\downarrow 2)x$ is possible if the transform $X(\omega)$ is zero over a half-band of frequencies. Such a signal is "band-limited". It may be limited to the upper half-band or the lower half-band:

$$X(\omega) = 0 \text{ for } 0 \le |\omega| < \frac{\pi}{2} \quad \text{or} \quad X(\omega) = 0 \text{ for } \frac{\pi}{2} \le |\omega| < \pi.$$

It is amazing that for band-limited signals we can recover the odd-numbered components from the even-numbered components. This is achieved by the *Shannon Sampling Theorem* in Section 2.3.

Example 3.1. Downsampling a vector of alternating signs produces a vector of constant sign:

$$(\downarrow 2)\,(\ldots,1,-1,1,-1,\ldots) = (\ldots,1,1,\ldots)\,.$$

The -1's in the odd-numbered positions have disappeared. You will quickly think of other inputs that give the same output. One in particular is the input vector $(\ldots,1,1,1,1,\ldots)$ of all ones. Downsampling produces the same vector of ones. It is an eigenvector of the linear operator $(\downarrow 2)$, with eigenvalue $\lambda = 1$.

Another eigenvector is $\boldsymbol{\delta} = (\ldots,0,0,1,0,0,\ldots)$. This also has $\lambda = 1$, and $(\downarrow 2)\boldsymbol{\delta} = \boldsymbol{\delta}$.

It is valuable to compare the alternating and constant inputs. Vector $(\ldots,1,-1,1,-1,\ldots)$ is at the frequency $\omega = \pi$. That is the highest possible frequency in discrete time. No vector can oscillate faster. Its components are $\boldsymbol{x}(n) = e^{i\pi n}$, which is $(-1)^n$. By contrast, the input of all ones is at the lowest frequency $\omega = 0$. Its components are $\boldsymbol{x}(n) = e^{i0n} = 1$.

This is an example of ***aliasing***. Two different frequencies, after downsampling, cannot be distinguished. The "alias" of the high frequency $\omega = \pi$ is the low frequency $\omega = 0$.

Certainly we cannot recover the input (which input would it be?) from the common output $(\downarrow 2)\boldsymbol{x} = (\ldots,1,1,1,\ldots)$. If we know that all frequencies in the input satisfy $|\omega| < \frac{\pi}{2}$, then we can "invert" the downsampling operation. The only possible input would be the constant vector, the DC input with $\omega = 0$.

Example 3.2. When the impulse response $\boldsymbol{C\delta} = (\ldots,\boldsymbol{c}(0),\boldsymbol{c}(1),\boldsymbol{c}(2),\ldots)$ is downsampled, the odd-numbered responses are lost:

$$(\downarrow 2)\boldsymbol{C\delta} = (\ldots,\boldsymbol{c}(0),\boldsymbol{c}(2),\boldsymbol{c}(4),\ldots)\,.$$

This is the impulse response of a "decimation filter"—*filtering followed by downsampling*. The impulse $\boldsymbol{\delta}$ came at time zero, and downsampling leaves the even coefficients. The response is entirely different if the impulse is delayed—we compute that now. ***Downsampling is not time-invariant.*** The delayed impulse is $\boldsymbol{S\delta} = (\ldots,0,1,0,\ldots)$. The response from the time-invariant filter $\boldsymbol{C}$ is similarly delayed:

$$\boldsymbol{CS\delta} = (\ldots,0,\boldsymbol{c}(0),\boldsymbol{c}(1),\ldots)\,. \tag{3.1}$$

Because of the delay, downsampling now destroys the even-numbered components $\boldsymbol{c}(0), \boldsymbol{c}(2)$. We pick up the odd components:

$$(\downarrow 2)\boldsymbol{CS\delta} = (\ldots,0,\boldsymbol{c}(1),\boldsymbol{c}(3),\ldots)\,.$$

First conclusion: Complete information may still be there if we downsample *two* vectors. Here the vectors were $\boldsymbol{C\delta}$ and its shift $\boldsymbol{SC\delta} = \boldsymbol{CS\delta}$. Those outputs are not as "separated" as we want—the two pieces are probably far from orthogonal. Downsampling $\boldsymbol{C\delta}$ and $\boldsymbol{CS\delta}$ may be a waste of time, but we do see that the whole vector $(\boldsymbol{c}(0),\boldsymbol{c}(1),\boldsymbol{c}(2),\boldsymbol{c}(3),\ldots)$ is recoverable from the two samples. Upsampling will be the way to recover it.

A better filter bank will use two different filters $\boldsymbol{C}$ and $\boldsymbol{D}$. Part of the design problem is to ensure that $(\downarrow 2)\,\boldsymbol{Cx}$ and $(\downarrow 2)\,\boldsymbol{Dx}$ have all the information to recover $\boldsymbol{x}$. Even better if those vectors are orthogonal, which is not true for this easy choice $\boldsymbol{D} = $ shift of $\boldsymbol{C}$.

Definition 3.1 *The nth component of $\boldsymbol{v} = (\downarrow 2)\boldsymbol{x}$ is the $(2n)$th component of $\boldsymbol{x}$:*

$$\boldsymbol{v}(n) = \boldsymbol{x}(2n). \tag{3.2}$$

We will now identify the "downsampling matrix". *It is the identity matrix with odd-numbered rows removed.* You could say that the identity matrix has been downsampled! The matrix form of downsampling is

$$\begin{bmatrix} \cdot & 1 & & & & & \\ & 0 & 0 & 1 & & & \\ & & & 0 & 0 & 1 & \\ & & & & & \cdot & \cdot \end{bmatrix} \begin{bmatrix} \cdot \\ x(-2) \\ x(-1) \\ x(0) \\ x(1) \\ x(2) \\ \cdot \end{bmatrix} = \begin{bmatrix} x(-2) \\ x(0) \\ x(2) \\ \cdot \end{bmatrix}. \tag{3.3}$$

We are defining the linear operator $(\downarrow 2)$. Notice that the rows of the downsampling matrix are orthonormal. They are unit vectors and their dot products are all zero. The matrix multiplication $(\downarrow 2)(\downarrow 2)^T$ contains all these dot products, so it must be the identity matrix. The matrix product $(\downarrow 2)^T(\downarrow 2)$ is different, as we see below.

The operator $(\downarrow 2)^T$ is important. ***The transpose of downsampling is upsampling***:

$$\boxed{(\downarrow 2)^T = (\uparrow 2).} \tag{3.4}$$

Upsampling places zeros into the odd-numbered components:

$$\textbf{Definition of } (\uparrow 2): \qquad (\uparrow 2) \begin{bmatrix} \cdot \\ v(-1) \\ v(0) \\ v(1) \\ \cdot \end{bmatrix} = \begin{bmatrix} \cdot \\ v(-1) \\ 0 \\ v(0) \\ 0 \\ v(1) \\ 0 \\ \cdot \end{bmatrix}.$$

We need a two-part formula to describe those components of $\boldsymbol{u} = (\uparrow 2)\boldsymbol{v}$:

$$u(n) = \begin{cases} v(k) & \text{if } n = 2k \\ 0 & \text{if } n = 2k+1. \end{cases} \tag{3.5}$$

To do this with a matrix, insert zero rows between the rows of $\boldsymbol{I}$. You could say that *the identity matrix has been upsampled.* The matrix form of upsampling is

$$\begin{bmatrix} \cdot & & & \\ 1 & 0 & & \\ & 0 & & \\ & 1 & 0 & \\ & & 0 & \\ & & 1 & 0 \\ & & & \cdot \end{bmatrix} \begin{bmatrix} \cdot \\ v(-1) \\ v(0) \\ v(1) \\ \cdot \end{bmatrix} = \begin{bmatrix} \cdot \\ v(-1) \\ 0 \\ v(0) \\ 0 \\ v(1) \\ \cdot \end{bmatrix} \tag{3.6}$$

The matrix in (3.6) is the transpose of the matrix in (3.3). If we multiply them we do get the identity matrix, verifying $(\downarrow 2)(\uparrow 2) = \boldsymbol{I}$:

$$\begin{bmatrix} \cdot & 1 & & & & & \\ & 0 & 0 & 1 & & & \\ & & & 0 & 0 & 1 & \\ & & & & & 0 & \cdot \end{bmatrix} \begin{bmatrix} \cdot & & & \\ 1 & 0 & & \\ & 0 & & \\ & 1 & 0 & \\ & & 0 & \\ & & 1 & 0 \\ & & & \cdot \end{bmatrix} = \begin{bmatrix} 1 & & & \\ & 1 & & \\ & & 1 & \\ & & & \cdot \end{bmatrix}. \tag{3.7}$$

If we upsample and then downsample, we put in zeros and take them out. The original $\boldsymbol{x}$ is recovered. This is not the usual order of the steps!

Normally we downsample and then upsample. *This product is not the identity matrix.* The output $(\uparrow 2)(\downarrow 2)\boldsymbol{x}$ is different from $\boldsymbol{x}$. Watch the steps:

$$\textit{Down:}\ (\downarrow 2)\boldsymbol{x} = \begin{bmatrix} \boldsymbol{x}(-2) \\ \boldsymbol{x}(0) \\ \boldsymbol{x}(2) \\ \cdot \end{bmatrix}. \quad \textit{Then up:}\ (\uparrow 2)(\downarrow 2)\boldsymbol{x} = \begin{bmatrix} \boldsymbol{x}(-2) \\ 0 \\ \boldsymbol{x}(0) \\ 0 \\ \boldsymbol{x}(2) \\ 0 \\ \cdot \end{bmatrix}. \tag{3.8}$$

In this usual order, $(\downarrow 2)$ removes the odd-numbered components and $(\uparrow 2)$ *puts in zeros.* The product $(\uparrow 2)(\downarrow 2)$ replaces the lost components by zeros:

$$(\uparrow 2)(\downarrow 2) = \begin{bmatrix} 1 & & & & & \\ & 0 & & & & \\ & & 1 & & & \\ & & & 0 & & \\ & & & & 1 & \\ & & & & & \cdot \end{bmatrix} \tag{3.9}$$

Thus $(\downarrow 2)$ and $(\uparrow 2)$ are one-sided inverses but not two-sided inverses. This cannot happen for square matrices of finite size. For n by n matrices, $\boldsymbol{AB} = \boldsymbol{I}$ means that $\boldsymbol{BA} = \boldsymbol{I}$. For rectangular matrices and infinite matrices, a *one-sided inverse* is possible. Downsampling is a very important example.

The matrices for $(\downarrow 2)$ and $(\uparrow 2)$ came from downsampling and upsampling the columns of $\boldsymbol{I}$. Here is another way to describe those same matrices:

- The $(\downarrow 2)$ matrix has the rows of $\boldsymbol{I}$ with a **double shift.**
- The $(\uparrow 2)$ matrix has the columns of $\boldsymbol{I}$ with a **double shift.**

Problem Set 3.1

1. Downsampling the alternating vector $\boldsymbol{x} = (\ldots, 1, -1, 1, -1, 1, \ldots)$ produces the vector $(\downarrow 2)\boldsymbol{x} = (\ldots, 1, 1, 1, \ldots)$. If you delay $\boldsymbol{x}$ to $\boldsymbol{Sx}(n) = \boldsymbol{x}(n-1)$, what is $(\downarrow 2)\boldsymbol{Sx}$?
2. Describe the operator $(\uparrow 2)(\downarrow 2)(\uparrow 2)(\downarrow 2)$.
3. Describe the operators $(\downarrow 3)$ and $(\uparrow 3)$ and $(\uparrow 3)(\downarrow 3)$.
4. What are the components of $(\uparrow 3)(\downarrow 2)\boldsymbol{x}$ and $(\downarrow 2)(\uparrow 3)\boldsymbol{x}$?

3.2 Subsampling in the Frequency Domain

For a pure exponential signal $x(k) = e^{ik\omega}$, the effect of downsampling is particularly clear. The kth component of $v = (\downarrow 2)x$ is $v(k) = e^{i2k\omega}$. This is a pure exponential with frequency 2ω. Frequencies are doubled by downsampling.

It looks as if $V(2\omega) = X(\omega)$, in other words $V(\omega) = X\left(\frac{\omega}{2}\right)$. *That is wrong.*

Suppose x' is another pure exponential, but with frequency $\omega + \pi$. Then $x'(k) = e^{ik(\omega+\pi)}$ is downsampled to $v'(k) = e^{i2k(\omega+\pi)}$. The factor $e^{i2k\pi}$ is always 1. This output is the same $e^{i2k\omega}$ as before! Adding π to the frequency only reversed the signs of the *odd*-numbered components of x'. Those components disappear anyway in downsampling, and $v'(k) = v(k)$. The frequency 2ω appears by doubling ω, and it *also* appears by doubling $\omega + \pi$.

In other words, ω enters the downsampled vector v not only by doubling $\frac{\omega}{2}$ but also by doubling $\frac{\omega}{2} + \pi$. There are two sources, not just one, contributing to $V(\omega)$:

Theorem 3.1 *The transform of* $v = (\downarrow 2)x$ *is* $V(\omega) = \frac{1}{2}\left[X\left(\frac{\omega}{2}\right) + X\left(\frac{\omega}{2} + \pi\right)\right]$.

Proof. Downsampling in the time domain is very direct:

$$v = (\downarrow 2)x \text{ means } v(k) = x(2k). \tag{3.10}$$

The Fourier transform of v is

$$V(\omega) = \sum v(k)e^{-ik\omega} = \sum x(2k)e^{-ik\omega}. \tag{3.11}$$

It is natural to set $n = 2k$, but do it carefully. The reverse direction $k = \frac{n}{2}$ is not safe. You might think that our sum is $\sum x(n)e^{-in\omega/2}$, and the frequency variable ω is just halved. *But the sum is only over even* $n = 2k$. So we start again.

Let u be the vector x with its odd-numbered components set to zero:

$$u(n) = \left\{ \begin{array}{ll} x(n), & n \text{ even} \\ 0, & n \text{ odd} \end{array} \right\} \text{ so that } u = (\ldots, x(0), 0, x(2), 0, \ldots). \tag{3.12}$$

Certainly $(\downarrow 2)u$ equals $(\downarrow 2)x$. The transform of u has two terms. The second term includes $(-1)^n$ or $e^{-in\pi}$ so that addition knocks out odd n:

$$U(\omega) = \sum_{n \text{ even}} x(n)e^{-in\omega} = \frac{1}{2}\sum_{\text{all } n} x(n)e^{-in\omega} + \frac{1}{2}\sum_{\text{all } n} x(n)e^{-in(\omega+\pi)}. \tag{3.13}$$

In frequency, this says that

$$U(\omega) = \tfrac{1}{2}\left[X(\omega) + X(\omega + \pi)\right]. \tag{3.14}$$

Now $V(\omega)$ comes safely from $U(\omega)$ by halving the frequency, because $u(n)$ only involves even n:

$$V(\omega) = U\left(\frac{\omega}{2}\right). \tag{3.15}$$

Equations (3.15)–(3.16) mean that the downsampled signal $v = (\downarrow 2)x$ has transform

$$\boxed{V(\omega) = \tfrac{1}{2}\left[X\left(\tfrac{\omega}{2}\right) + X\left(\tfrac{\omega}{2} + \pi\right)\right].} \tag{3.16}$$

This is downsampling in the frequency domain. It is not time-invariant, because $X(\omega)$ is not just multiplied by $C(\omega)$. The rate changes and a new term appears.

Figures 3.1 and 3.2 and 3.3 show extreme aliasing and no aliasing and typical aliasing. You can see the two terms in Figure 3.1, where $X(\omega)$ is a sawtooth function. *The average in $U(\omega)$ is a constant.* That second term is responsible for ***aliasing***: $X\left(\frac{\omega}{2}\right)$ overlaps $X\left(\frac{\omega}{2}+\pi\right)$ in $V(\omega)$. Their sum happens to be constant, so this special input $\boldsymbol{x}$ gives $\boldsymbol{v} = (\downarrow 2)\boldsymbol{x} = \frac{1}{2}\boldsymbol{\delta}$. All its even-numbered samples are zero, except $\boldsymbol{x}(0) = \frac{1}{2}$.

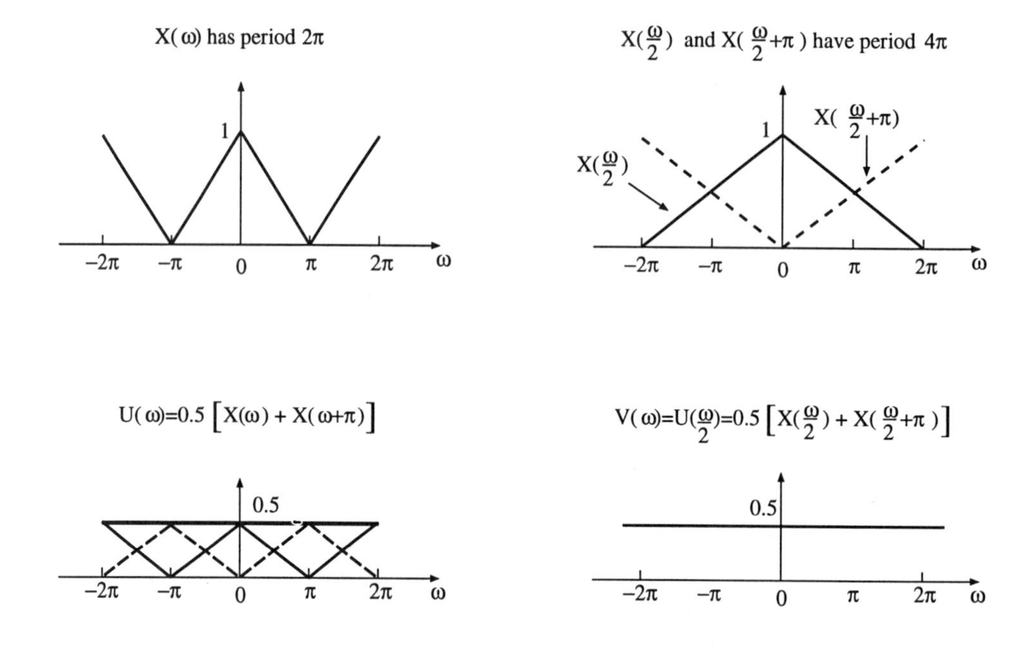

Figure 3.1: Downsampling stretches $X(\omega)$ to $X\left(\frac{\omega}{2}\right)$ and it also creates the alias $X\left(\frac{\omega}{2}+\pi\right)$. Because the alias overlaps, we cannot recover $X(\omega)$ from $U(\omega)$ or from the downsampled $V(\omega)$.

In Figure 3.2, downsampling does not lead to aliasing. The input $X(\omega)$ is properly bandlimited. The stretched transform $X\left(\frac{\omega}{2}\right)$ *does not overlap the aliasing term* $X\left(\frac{\omega}{2}+\pi\right)$. The outputs $V(\omega)$ and $U(\omega)$ from downsampling and upsampling still contain this aliasing part! But a non-overlapping alias can be filtered away.

It is striking that the adjective "halfband" is used in both of these extreme cases, but in *different ways*:

$\boldsymbol{x}$ in Figure 3.1 is a *halfband impulse response* ($\boldsymbol{x}(2n) = 0$ for $n \neq 0$).

$\boldsymbol{x}$ in Figure 3.2 is a *halfband signal* ($X(\omega) = 0$ for $\frac{\pi}{2} \leq |\omega| < \pi$).

Figure 3.3 is somewhere between Figure 3.1 and Figure 3.2. *Aliasing is expected.* The original $\boldsymbol{x}$ cannot be recovered from $(\downarrow 2)\boldsymbol{x}$. But we don't expect the unusual result in Figure 3.1, where downsampling produced a pure impulse.

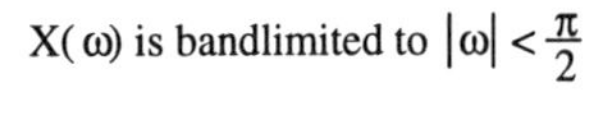

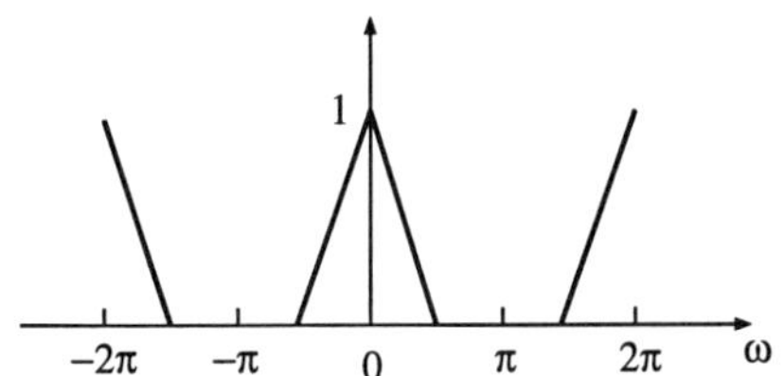

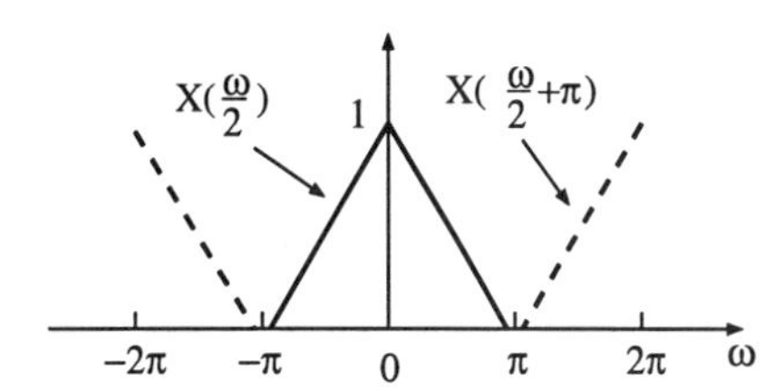

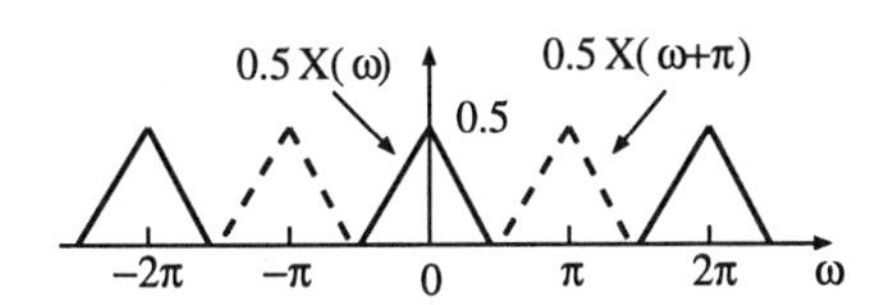

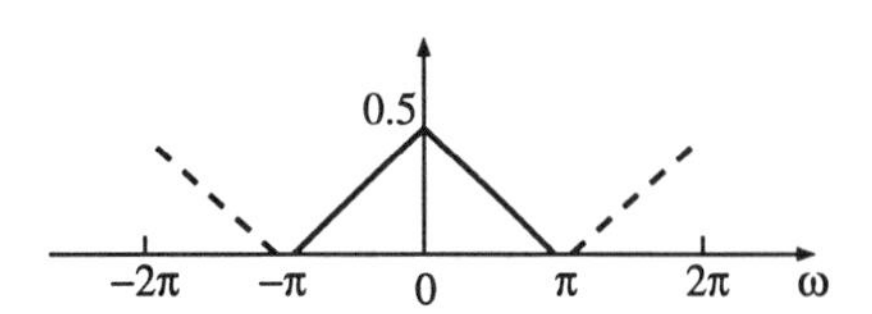

Figure 3.2: Downsampling this bandlimited signal stretches its transform to $[-\pi, \pi]$, no aliasing.

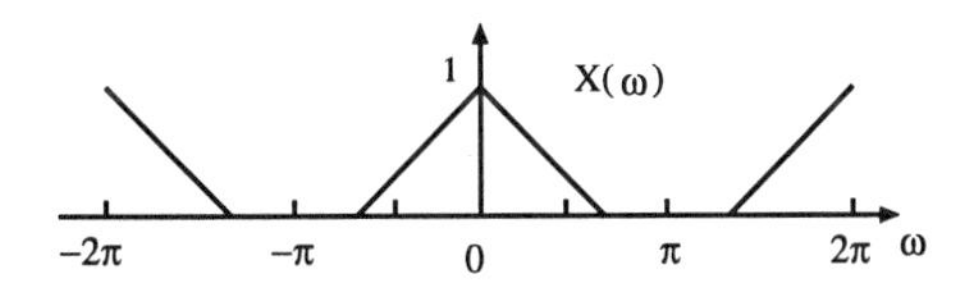

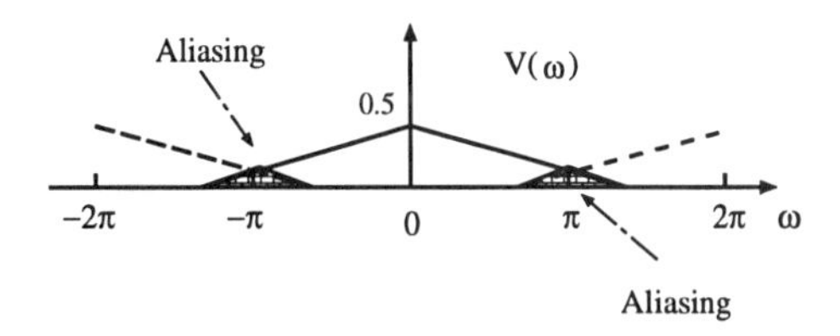

Figure 3.3: Typical aliasing from $(\downarrow 2)$. The stretched $X(\frac{\omega}{2})$ overlaps $X(\frac{\omega}{2} + \pi)$.

Upsampling in the Frequency Domain

In the time domain, upsampling needs separate formulas for even $n = 2k$ and odd $n = 2k + 1$:

$$\boldsymbol{u} = (\uparrow 2)\boldsymbol{v} \text{ means } \begin{cases} \boldsymbol{u}(2k) = \boldsymbol{v}(k) \\ \boldsymbol{u}(2k+1) = 0. \end{cases} \tag{3.17}$$

In the frequency domain this becomes simple. Where downsampling $\boldsymbol{v}(k) = \boldsymbol{x}(2k)$ goes from one term in the time domain to two terms in frequency, upsampling goes from two terms to one. The zeros in $\boldsymbol{u}(2k+1) = 0$ contribute nothing to $U(\omega)$. We now establish the upsampling formula in the ω domain:

Theorem 3.2 *The transform of* $\boldsymbol{u} = (\uparrow 2)\boldsymbol{v}$ *is*

$$U(\omega) = V(2\omega). \tag{3.18}$$

Proof. Only the even indices $n = 2k$ enter because $\boldsymbol{u}(2k+1) = 0$:

$$\begin{aligned} U(\omega) &= \sum \boldsymbol{u}(n)e^{-in\omega} = \sum \boldsymbol{u}(2k)e^{-i2k\omega} \\ &= \sum \boldsymbol{v}(k)e^{-i2k\omega} = V(2\omega). \end{aligned} \tag{3.19}$$

Where $V(\omega)$ has period 2π, this new function $V(2\omega)$ has period π. The original graph is compressed into $|\omega| \leq \frac{\pi}{2}$. *An image of that compressed graph appears next to it.* Upsampling creates an imaging effect! Figure 3.4 shows how two compressed copies of the graph of $V(\omega)$ appear in the spectrum of $U(\omega)$.

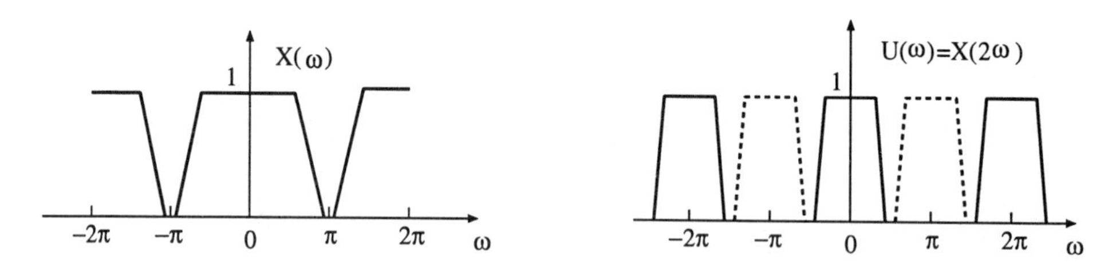

Figure 3.4: The upsampled spectrum $U(\omega) = V(2\omega)$ has compressed images of $V(\omega)$.

This imaging is the *opposite* of aliasing. In aliasing, two input frequencies ω and $\omega + \pi$ give the same output. In imaging, one input frequency ω is responsible for two outputs — at frequency $\frac{\omega}{2}$ and also at $\frac{\omega}{2} + \pi$. Upsampling produces imaging where downsampling produces aliasing.

Upsampling after Downsampling

The analysis bank includes downsampling. Then the synthesis bank includes upsampling. It is very common to see those two operations together:

$$\boldsymbol{v} = (\downarrow 2)\boldsymbol{x} \quad \text{and then} \quad \boldsymbol{u} = (\uparrow 2)\boldsymbol{v} \quad \text{give} \quad \boldsymbol{u} = (\uparrow 2)(\downarrow 2)\boldsymbol{x}.$$

Theorem 3.3 *The transform of* $\boldsymbol{u} = (\uparrow 2)(\downarrow 2)\boldsymbol{x}$ *is*

$$\boxed{U(\omega) = \tfrac{1}{2}\left[X(\omega) + X(\omega + \pi)\right].} \tag{3.20}$$

This was equation (3.14). The aliasing $X(\omega + \pi)$ comes from $(-1)^n \boldsymbol{x}(n)$. When n is odd, this cancels $\boldsymbol{x}(n)$. The average of $\boldsymbol{x}(n)$ and $(-1)^n \boldsymbol{x}(n)$ is

$$\boldsymbol{u}(n) = \begin{cases} \boldsymbol{x}(n), & n \text{ even} \\ 0, & n \text{ odd.} \end{cases} \tag{3.21}$$

In the frequency domain the proof has two steps:

$$\begin{aligned} \text{down:} \quad V(\omega) &= \tfrac{1}{2}\left[X\left(\frac{\omega}{2}\right) + X\left(\frac{\omega}{2} + \pi\right)\right] \\ \text{then up:} \quad U(\omega) &= V(2\omega) = \tfrac{1}{2}\left[X(\omega) + X(\omega + \pi)\right]. \end{aligned}$$

Conclusion: $(\uparrow 2)(\downarrow 2)$ produces *aliasing and also imaging*. Figure 3.5 shows them both; the image of the alias overlaps the original. We can't determine the original $X(\omega)$ from $U(\omega)$.

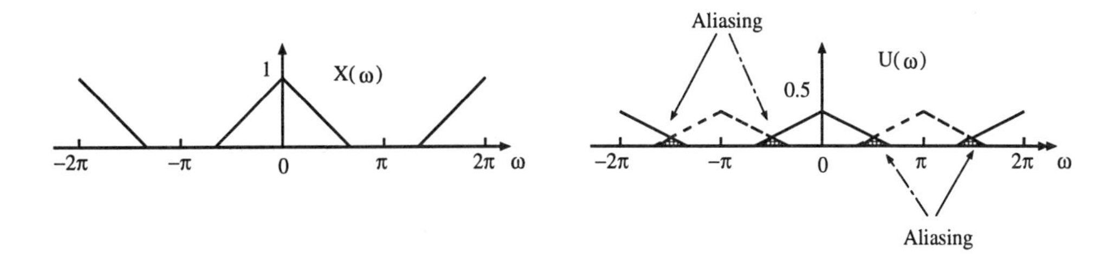

Figure 3.5: The spectrum is stretched by downsampling and compressed by upsampling. The alias from downsampling generally overlaps the image from upsampling.

Filter Bank without Filters

The idea of a filter bank without filters sounds absurd. But this is a very useful example. The filters come from the identity matrix and don't do anything (except possibly a delay). All the activity comes from downsampling and upsampling. It is good to see how the original signal $\boldsymbol{x}$ is reconstructed.

Here is the block form of the simple filter bank, with no filters:

x(n) → $\downarrow 2$ → v → $\uparrow 2$ → u → z^{-1}
z^{-1} → $\downarrow 2$ → $\uparrow 2$ → x(n − 1)

In words, the upper channel downsamples and upsamples. It produces the vector $\boldsymbol{u}$ described above, with all odd components of $\boldsymbol{x}$ replaced by zeros. Then this $\boldsymbol{u}$ is delayed before output.

The lower channel delays $\boldsymbol{x}$ at the start. Then downsampling followed by upsampling removes the *even*-numbered components $\boldsymbol{x}(2k)$. The combined output is just a delay of the input:

$$\begin{matrix}\text{delay of } (\uparrow 2)(\downarrow 2)\boldsymbol{x} \\ + \\ (\uparrow 2)(\downarrow 2)(\text{delay of } \boldsymbol{x})\end{matrix} = \begin{bmatrix} \cdot \\ 0 \\ \boldsymbol{x}(0) \\ 0 \\ \boldsymbol{x}(2) \\ \cdot \end{bmatrix} + \begin{bmatrix} \cdot \\ \boldsymbol{x}(-1) \\ 0 \\ \boldsymbol{x}(1) \\ 0 \\ \cdot \end{bmatrix} = \text{delay of } \boldsymbol{x}.$$

A delay multiplies the transform by $e^{-i\omega}$. *The delay in a filter bank is indicated by* z^{-1}. Instead of the Fourier transform we more often use the z-transform. A delay multiplies by z^{-1}, and a two-step delay multiplies by z^{-2}. An advance (non-causal!) multiplies by z.

We will next express $(\downarrow 2)$ and $(\uparrow 2)$ in the z-domain. And we take this opportunity to do the same for $(\downarrow M)$ and $(\uparrow M)$. Downsampling and upsampling involve every Mth component of the signals in an M-channel filter bank.

Problem Set 3.2

1. Draw $V(\omega)$ and $U(\omega)$ when $X(\omega)$ is a *constant function*. What vector $\boldsymbol{x}$ has this transform? What is the result of downsampling that $\boldsymbol{x}$?

2. What are the usual periods of the functions $X(\omega)$, $X\left(\frac{\omega}{2}\right)$, and $V(\omega)$?

3. For downsampling by 3 instead of 2, write down the equations corresponding to (3.11) through (3.16). In equation (3.10), $v = (\downarrow 3)x$ means $v(k) = x(3k)$.

4. What is the original input x in Figure 3.1, whose transform is the sawtooth function? To invert transforms use $x(n) = \frac{1}{2\pi}\int X(\omega)e^{-in\omega}\,d\omega$.

5. By inverting $X(\omega)$ in Figure 3.2 find the original band-limited signal $x(n)$. What is $v(n) = (\downarrow 2)x(n)$? Compare with the input signal in Figure 3.1.

6. (Review) Verify that the transform of $(\downarrow 2)(\uparrow 2)v$ is still $V(\omega)$. Then $(\downarrow 2)(\uparrow 2) = I$.

7. Draw the block form of an M-channel filter bank without filters. Imitating the $M = 2$ case above, the output equals the input after $M - 1$ delays.

3.3 Sampling Operations in the z-Domain

This section includes a conversion from the frequency variable ω to the complex variable z. The connection between them is always $e^{i\omega} = e^{j\omega} = z$. Real frequencies ω correspond to points on the unit circle $|z| = 1$. The transform is extended beyond this special circle of complex numbers, to include all z in the complex plane.

In the process we overcome one inconvenience. The frequencies ω and $\omega+2\pi$ and $\omega+4\pi$ are impossible to distinguish. In the z-domain we don't attempt to distinguish them. The complex numbers $z = e^{i\omega}$ and $z = e^{i(\omega+2\pi)}$ and $z = e^{i(\omega+4\pi)}$ are absolutely identical. This remains just as true when the "frequency" ω is not real. It depends only on the fact that $e^{2\pi i} = 1$.

We repeat the definition of the z-transform of x:

The signal $x(n)$ is transformed to $X(z) = \sum_n x(n)z^{-n}$.

What we earlier called $X(\omega)$ is now $X\left(e^{j\omega}\right)$. You will know immediately from the letter z that $X(z)$ refers to the z-transform rather than the Fourier transform.

At present, these infinite series are "formal power series". We are not worrying about their convergence. In reality they probably converge for some z and diverge for other z. When positive and negative powers are both included, the region of convergence is essentially an *annulus*. This is the region $r_{\text{inner}} < |z| < r_{\text{outer}}$ between two circles. Convergence may or may not occur on the circles $|z| = r_{\text{inner}}$ and $|z| = r_{\text{outer}}$. There is also the extreme but quite likely possibility that $r_{\text{inner}} = 0$ or $r_{\text{outer}} = \infty$ or $r_{\text{inner}} = r_{\text{outer}}$.

What is the effect on our down and up formulas? The vectors are $v = (\downarrow 2)x$ and $u = (\uparrow 2)v$. In frequency, their transforms are

$$V(\omega) = \tfrac{1}{2}\left[X\left(\frac{\omega}{2}\right) + X\left(\frac{\omega}{2} + \pi\right)\right] \quad \text{and} \quad U(\omega) = V(2\omega).$$

We now derive and explain these rules for converting to the z-transform:

1. Halved frequency $\frac{\omega}{2}$ leads in the z-domain to $z^{1/2}$.

2. Doubled frequency 2ω leads in the z-domain to z^2.

3. Shifted frequency by π leads in the z-domain to $-z$.

These rules follow directly from $z = e^{i\omega}$. Immediately $e^{i\omega/2}$ is $z^{1/2}$ and $e^{i2\omega}$ is z^2 and $e^{i(\omega+\pi)}$ is $-z$. The rules produce the correct down and up formulas in the z-domain:

Theorem 3.4 *The z-transforms of $\boldsymbol{v} = (\downarrow 2)\boldsymbol{x}$ and $\boldsymbol{u} = (\uparrow 2)\boldsymbol{v}$ are*

$$V(z) = \tfrac{1}{2}\left[X(z^{1/2}) + X(-z^{1/2})\right] \quad \textit{and} \quad U(z) = V(z^2). \tag{3.22}$$

The transform of $\boldsymbol{u} = (\uparrow 2)(\downarrow 2)\boldsymbol{x}$ is

$$U(z) = \tfrac{1}{2}\left(X(z) + X(-z)\right). \tag{3.23}$$

The extension from two channels to M channels is straightforward and important. In the time domain, $\boldsymbol{v} = (\downarrow M)\boldsymbol{x}$ and $\boldsymbol{u} = (\uparrow M)\boldsymbol{v}$ have components

$$\boldsymbol{v}(n) = \boldsymbol{x}(Mn) \quad \text{and} \quad \boldsymbol{u}(k) = \begin{cases} \boldsymbol{v}\left(\frac{k}{M}\right), & M \text{ divides } k \\ 0, & \text{otherwise.} \end{cases}$$

Then $(\uparrow M)(\downarrow M)\boldsymbol{x}$ leaves every Mth component unchanged (the components for which M divides k). The other components become zeros.

In the frequency domain there are $M-1$ aliases in downsampling and $M-1$ images from upsampling (it is very instructive to graph $V(\omega)$ and $U(\omega)$):

$$\begin{aligned} \text{down: } V(\omega) &= \frac{1}{M}\left[X\Big(\frac{\omega}{M}\Big) + X\Big(\frac{\omega+2\pi}{M}\Big) + \cdots + X\Big(\frac{\omega+(M-1)2\pi}{M}\Big)\right] \\ \text{up: } U(\omega) &= V(M\omega). \end{aligned} \tag{3.24}$$

Eliminating V leaves $U(\omega) = \frac{1}{M}\,[X(\omega) + X(\omega + \frac{2\pi}{M}) + \cdots]$ for the upsampled $\boldsymbol{u} = (\uparrow M)(\downarrow M)\boldsymbol{x}$. Those aliases overlap the original (not if it is properly bandlimited). To transfer into the z-domain, extend the three rules from $M = 2$ to any M:

1. Dividing ω by M leads to $z^{1/M}$.
2. Multiplying ω by M leads to z^M.
3. Shifting ω by $2\pi/M$ leads to $ze^{2\pi i/M}$.

From this correspondence we can express $(\downarrow M)$ and $(\uparrow M)$ in the z-domain:

Theorem 3.5 *The z-transforms of $\boldsymbol{v} = (\downarrow M)\boldsymbol{x}$ and $\boldsymbol{u} = (\uparrow M)\boldsymbol{v}$ are*

$$V(z) = \frac{1}{M}\sum_{k=0}^{M-1} X\left(z^{1/M}e^{2\pi ik/M}\right) \textit{ and } U(z) = V\left(z^M\right). \tag{3.25}$$

The transform of $\boldsymbol{u} = (\uparrow M)(\downarrow M)\boldsymbol{x}$, down and up, is by eliminating V:

$$U(z) = \frac{1}{M}\left[X(z) + X\left(ze^{2\pi i/M}\right) + \cdots + X\left(ze^{2\pi i(M-1)/M}\right)\right]. \tag{3.26}$$

The *decimator* (downsampling operator) by itself creates aliases. The *expander* (upsampling operator) by itself creates images. To prevent the aliases and to remove the images, we filter the signal. A filter comes before the decimator, to band-limit the input. The aliasing terms become zero. A filter comes *after* the expander, to wipe out the images. This filter removes (or nearly removes) the frequencies beyond $|\omega| = \frac{\pi}{M}$. The block forms are clear:

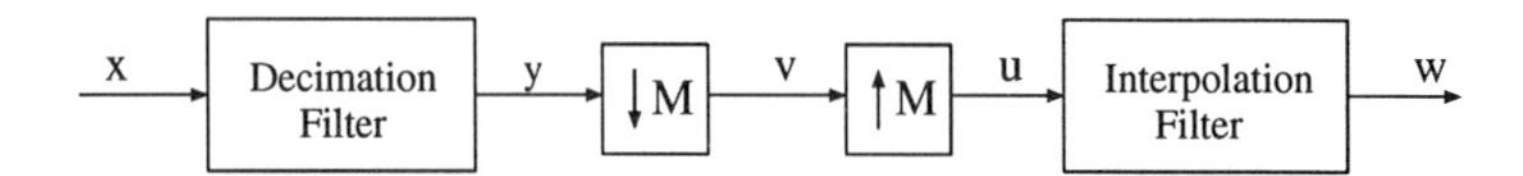

The decimation filter suppresses aliasing. The interpolation filter suppresses imaging.

It is easiest to see the decimation circuit in the frequency domain. The band-limiting filter cuts off $X(\omega)$ outside $\left[-\frac{\pi}{M}, \frac{\pi}{M}\right]$, to give $Y(\omega)$. Downsampling stretches that band to the full interval $[-\pi, \pi]$. But no aliases appear, as shown in Figure 3.6.

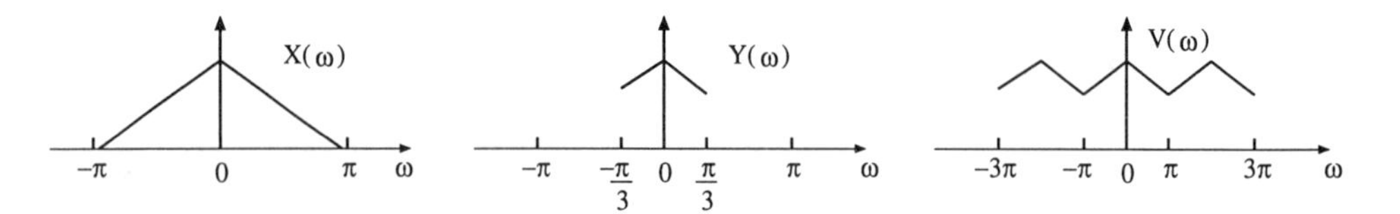

Figure 3.6: Spectra of $X(\omega)$, $Y(\omega)$, and $V(\omega)$ with $M = 3$.

From $V(\omega)$ we can recover $Y(\omega)$. Unless the original signal was band-limited, we cannot recover $X(\omega)$. If it was, we can. The interpolation circuit would do the recovery!

It is important to see the interpolation circuit in the time domain. The upsampling step inserts zeros into the signal. The interpolation filter $\boldsymbol{F}$ changes those zero values to smoother values $\boldsymbol{w}(n)$. Remember that a lowpass filter reduces oscillations to give a smoother output. Here $M = 2$ and the interpolation circuit produces $\boldsymbol{w}$ from $\boldsymbol{v}$:

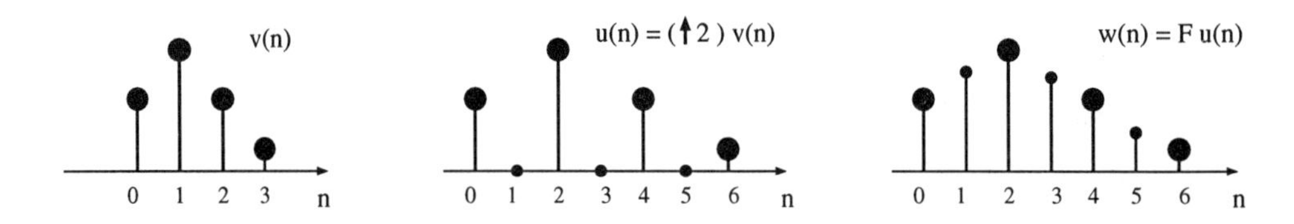

Fractional Sampling Rates

To understand fractional decimation, study the product $(\uparrow 2)(\downarrow 3)$ and also $(\downarrow 3)(\uparrow 2)$. What are the outputs? In some way these should produce an output at a fractional rate, which is $\frac{2}{3}$ of the original rate. The transform $X(\omega)$ should be stretched by a factor $\frac{3}{2}$. Let us see.

The direct approach is in the time domain. When up follows down, we reach $\boldsymbol{x}(0), 0, \boldsymbol{x}(3), 0$:

$$\begin{bmatrix} \boldsymbol{x}(0) \\ \boldsymbol{x}(1) \\ \boldsymbol{x}(2) \\ \boldsymbol{x}(3) \\ \cdot \end{bmatrix} \boxed{\downarrow 3} \begin{bmatrix} \boldsymbol{x}(0) \\ \boldsymbol{x}(3) \\ \cdot \end{bmatrix} \boxed{\uparrow 2} \begin{bmatrix} \boldsymbol{x}(0) \\ 0 \\ \boldsymbol{x}(3) \\ 0 \\ \cdot \end{bmatrix}$$

When down follows up, the remarkable thing is that the result is the same:

$$\begin{bmatrix} x(0) \\ x(1) \\ x(2) \\ x(3) \\ \cdot \end{bmatrix} - \boxed{\uparrow 2} - \begin{bmatrix} x(0) \\ 0 \\ x(1) \\ 0 \\ x(2) \\ 0 \\ x(3) \end{bmatrix} - \boxed{\downarrow 3} - \begin{bmatrix} x(0) \\ 0 \\ x(3) \\ 0 \\ \cdot \end{bmatrix}$$

Thus $(\uparrow 2)$ *commutes with* $(\downarrow 3)$. In either order, every third component appears in every second position. Remember that $(\uparrow 2)$ does *not* commute with $(\downarrow 2)$.

A natural question arises. ***When does*** $(\uparrow L)$ ***commute with*** $(\downarrow M)$? The output will come at $\frac{L}{M}$ times the original rate. It is best to reduce this fraction to lowest terms. The numbers L and M should be *relatively prime*, meaning that they have no common factors. The fraction $\frac{2L}{2M}$ is equal to $\frac{L}{M}$, but it is a mistake to use $(\uparrow 2L)$ and $(\downarrow 2M)$. Those operators do not commute and they are inefficient. The natural choice is the good choice.

Theorem 3.6 *The operators $(\downarrow L)$ and $(\uparrow M)$ commute if and only if L and M are relatively prime. Their greatest common divisor is $(L, M) = 1$. In that case $(\uparrow L)(\downarrow M)\boldsymbol{x} = (\downarrow M)(\uparrow L)\boldsymbol{x}$ has components*

$$\boldsymbol{u}(n) = \begin{cases} \boldsymbol{x}(Mk) & \text{if } \frac{n}{L} = k \text{ is an integer} \\ 0 & \text{if } \frac{n}{L} \text{ is not an integer.} \end{cases} \tag{3.27}$$

This formula is always correct for the order $(\uparrow L)(\downarrow M)\boldsymbol{x}$. With $L = 2$ and $M = 3$ it yields $\boldsymbol{u}(1) = 0$ and $\boldsymbol{u}(2) = \boldsymbol{x}(3)$ as in the example above. With $L = 2$ and $M = 2$ it yields $\boldsymbol{u}(1) = 0$ and $\boldsymbol{u}(2) = \boldsymbol{x}(2)$. This is the familiar output from $(\uparrow 2)(\downarrow 2)\boldsymbol{x}$.

The formula (3.27) can be wrong if down follows up. In that order, the test is whether $\frac{Mn}{L}$ is an integer. For $M = L = 2$ this is true for every n. The output from $(\downarrow 2)(\uparrow 2)$ is the same as the input, and $(\downarrow 2)(\uparrow 2) = \boldsymbol{I}$. But for $M = 3$ and $L = 2$, the fraction $\frac{3n}{2}$ is only an integer when n is even. In this $\frac{2}{3}$ case, up-down agrees with down-up.

When M and L are relatively prime, the output in either order has $\boldsymbol{u}(n) = 0$ unless L divides n. If $\frac{n}{L}$ is not an integer, then $\frac{Mn}{L}$ is not an integer.

Why do we alter the rate by a fraction? The rate may need to change from 60 Hertz to 50 Hertz. The choices $L = 5$ and $M = 6$ will do it. (Not 50 and 60; they are not relatively prime.) We will get five samples instead of six samples, and 50 samples per second instead of 60 per second. There is no aliasing when the input signal is band-limited to $|\omega| < \frac{5\pi}{6}$. Then, stretching by $\frac{6}{5}$ will create no problems:

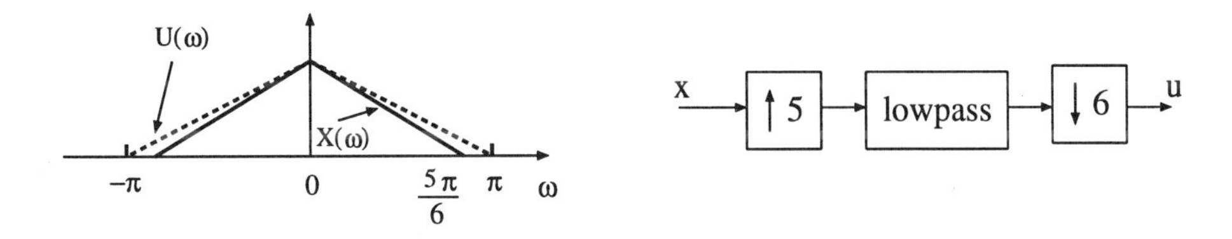

Problem Set 3.3

1. Find $U(\omega)$ from $X(\omega)$ if $\boldsymbol{u} = (\uparrow L)(\downarrow M)\boldsymbol{x}$. Use (3.25) for $(\downarrow M)$. Similarly find $V(\omega)$ if $\boldsymbol{v} = (\downarrow M)(\uparrow L)\boldsymbol{x}$. Show that $U(\omega) \equiv V(\omega)$ if (and only if) L and M are relatively prime.
2. Verify formula (3.27) directly. What happens to the odd-numbered components of $\boldsymbol{x}$ when we compute $(\uparrow 2)(\downarrow 2)\boldsymbol{x}$? What happens to the odd part of $X(z)$ in equation (3.24)?
3. If a band-limited signal $\boldsymbol{v}$ goes through an expander, how is $(\uparrow 2)\boldsymbol{v}$ related to $\boldsymbol{v}$?
4. For a band-limited signal $\boldsymbol{x}$, draw in the frequency domain the steps of decimation and recovery by interpolation. Graph $X(\omega)$, $Y(\omega)$, $V(\omega)$, $U(\omega)$, and $W(\omega)$.
5. Draw $H(\omega)$ for an ideal decimator that prevents aliasing by $(\downarrow 6)$. Draw an ideal $F(\omega)$ that removes images created by $(\uparrow 5)$.
6. What lowpass filter placed between $(\uparrow 5)$ and $(\downarrow 6)$ avoids images and also aliasing?
7. In smoothing $\boldsymbol{u}(n)$ to get the final output $\boldsymbol{w} = \boldsymbol{F}\,\boldsymbol{u}$, which filters $\boldsymbol{F}$ will *interpolate* and not change the even samples: $\boldsymbol{w}(2k) = \boldsymbol{u}(2k)$?

3.4 Filters Interchanged with Samplers

Our notation for the filter coefficients is $\boldsymbol{h}(n)$. When the input $\boldsymbol{x}$ is the impulse $\boldsymbol{\delta}$, the responses are $\boldsymbol{h}(0), \ldots, \boldsymbol{h}(N)$. The vector $\boldsymbol{h}$ is the impulse response (in the time domain). The response in the z-domain (*we are now using this form systematically and often*) is $H(z) = \sum \boldsymbol{h}(n)z^{-n}$.

As an operator, the filter is a combination of delays (and possibly advances). *Note especially that a delay corresponds to* z^{-1}. The filter is linear and time-invariant. Expressed as an infinite matrix, $\boldsymbol{H}$ has the number $\boldsymbol{h}(n)$ along its nth diagonal (counting down from the main diagonal). The i, j entry is $\boldsymbol{h}(i-j)$.

First Noble Identity

In the standard form, downsampling follows filtering. We apply $\boldsymbol{H}$ and then $(\downarrow 2)$ or $(\downarrow M)$. Taken literally, this would be highly inefficient. We are apparently computing all components of $\boldsymbol{Hx}$, and then throwing away many of them. No sensible engineer would do that. The polyphase form of $H(z)$ will tell us how to reorganize the calculation and do the sampling first.

We cannot just reverse the order of $(\downarrow M)$ and $\boldsymbol{H}$ (of course!). Here we note a special and important case, when the sampling can be done first. *The filter has* $M-1$ *zero coefficients between each pair of nonzeros.* The nonzero coefficients $\boldsymbol{h}(0), \boldsymbol{h}(M), \boldsymbol{h}(2M), \ldots$ will be renamed $\boldsymbol{g}(0), \boldsymbol{g}(1), \boldsymbol{g}(2), \ldots$. Then $H(z)$ is the same as $G(z^M)$. For filters of this special type we can move the decimator outside the filter:

$$x \longrightarrow \boxed{\downarrow M} \longrightarrow \boxed{G(z)} \longrightarrow v \quad \equiv \quad x \longrightarrow \boxed{G(z^M)} \longrightarrow \boxed{\downarrow M} \longrightarrow v$$

The new order starting with $(\downarrow M)$ cuts the sampling rate immediately. The old order applies $\boldsymbol{H}$ to the full vector $\boldsymbol{x}$. Here is an example with $M = 2$ and $\boldsymbol{h} = (1, 0, 2)$:

$$\underset{(\downarrow 2)\boldsymbol{x}}{\begin{bmatrix} \boldsymbol{x}(0) \\ \boldsymbol{x}(2) \\ \cdot \end{bmatrix}} \quad \underset{\boldsymbol{G}(\downarrow 2)\boldsymbol{x}}{\begin{bmatrix} \boldsymbol{x}(0) + 2\boldsymbol{x}(-2) \\ \boldsymbol{x}(2) + 2\boldsymbol{x}(0) \\ \cdot \end{bmatrix}} \quad \text{or} \quad \underset{\boldsymbol{Hx}}{\begin{bmatrix} \boldsymbol{x}(0) + 2\boldsymbol{x}(-2) \\ \boldsymbol{x}(1) + 2\boldsymbol{x}(-1) \\ \boldsymbol{x}(2) + 2\boldsymbol{x}(0) \\ \boldsymbol{x}(3) + 2\boldsymbol{x}(1) \\ \cdot \end{bmatrix}} \quad \underset{(\downarrow 2)\boldsymbol{Hx} = \boldsymbol{G}(\downarrow 2)\boldsymbol{x}}{\begin{bmatrix} \boldsymbol{x}(0) + 2\boldsymbol{x}(-2) \\ \boldsymbol{x}(2) + 2\boldsymbol{x}(0) \\ \cdot \end{bmatrix}}$$

In the z-domain, this example yields the *even powers* in $(1+2z^{-2})X(z)$. The odd powers in $X(z)$ don't matter! So the quick way is to pick out the even powers by downsampling first. This "Noble Identity" represents a simple but very useful fact of algebra:

$$\textit{First Noble Identity:}\ G(z)(\downarrow M) = (\downarrow M)G(z^M). \tag{3.28}$$

Proof Downsampling $x(n)$ leaves $x(Mn)$. Then the filter G produces

$$v(n) = \sum g(k)x\big(M(n-k)\big). \tag{3.29}$$

The other way (the slow way), $H(z) = G(z^M)$ uses every Mth sample of the input. The filter gives $\sum g(k)x(n-Mk)$. Then the sampling operator replaces n by Mn. We reach the same $v(n)$.

You may find it useful to follow an impulse $x = \delta$ through both systems, to see that the impulse response is the same. On the left we get the vector $v = g$ of filter coefficients. On the right we keep all the zeros. Then downsampling ends with the same g:

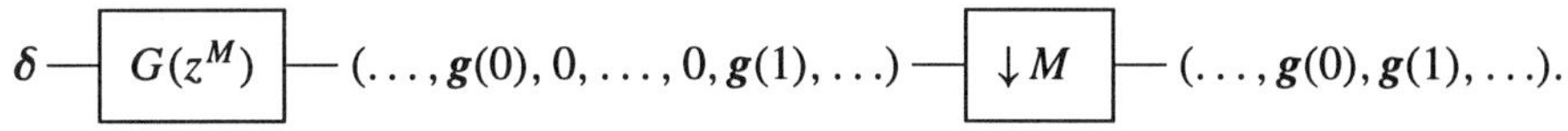

$$\delta \rightarrow \boxed{G(z^M)} \rightarrow (\ldots, g(0), 0, \ldots, 0, g(1), \ldots) \rightarrow \boxed{\downarrow M} \rightarrow (\ldots, g(0), g(1), \ldots).$$

We used G instead of the basic symbol H to emphasize that *the Noble Identities cannot be directly applied to a typical filter*. The identity applies only when $H(z)$ has the form $G(z^M)$. Only filters that have special impulse responses, with $M-1$ zeros between the nonzeros, fit this form. The idea of the polyphase matrix is to decompose a typical H into a set of M filters with these special responses. Then the Noble Identities apply to each polyphase component of H, even if they don't apply to the whole filter.

Second Noble Identity

The second identity concerns *upsampling*. It is a transpose of the first identity. The standard form of a filter bank has $(\uparrow M)$ before a synthesis filter $F(z)$. For filters of the special form $F(z) = G(z^M)$, that order can be reversed. But again we must change $G(z^M)$ to $G(z)$:

$$x \rightarrow \boxed{G(z)} \rightarrow \boxed{\uparrow M} \rightarrow u \quad \equiv \quad x \rightarrow \boxed{\uparrow M} \rightarrow \boxed{G(z^M)} \rightarrow u$$

This time we verify the equivalence in the z-domain. The output $(\uparrow M)Gx$ comes from upsampling $G(z)X(z)$:

$$U(z) = G(z^M)X(z^M).$$

In the other order, the upsampled signal $X(z^M)$ is filtered by the stretched-out $G(z^M)$. Again $U(z) = G(z^M)X(z^M)$, which proves the identity:

$$\textit{Second Noble Identity:}\ (\uparrow M)G(z) = G(z^M)(\uparrow M). \tag{3.30}$$

As before, this identity does not apply to most filters. We still need $M-1$ zeros between the nonzeros. For a simple delay, $G(z^M) = z^{-1}$ is not possible.

The Noble Identities are derived for systems with one input and one output. They also hold for systems with multiple inputs and outputs, as long as the decimation (or interpolation) factors are the same for all channels.

We soon move to the polyphase decomposition of a filter (and a filter bank). The phases have $M-1$ zeros and the special form involving z^M. In the polyphase form, *the downsampling can come first.* And upsampling comes last.

Problem Set 3.4

1. Verify the first Noble Identity $G(z)(\downarrow M) = (\downarrow M)G(z^M)$ in the z-domain.

2. Verify the second Noble Identity $(\uparrow M)G(z) = G(z^M)(\uparrow M)$ in the time domain.

3. Simplify the following system:

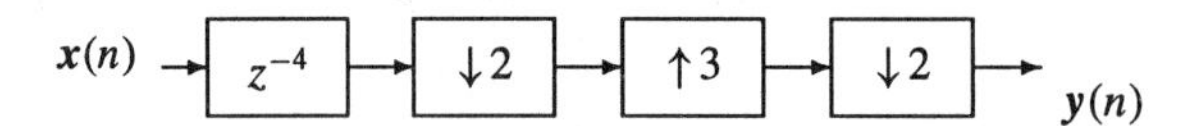

What is $Y(z)$ in terms of $X(z)$? Find $\boldsymbol{y}(n)$ for the following inputs:

$$\boldsymbol{x}(n) = \delta(n), \quad \boldsymbol{x}(n) = (\ldots, 1, 1, 1, 1, \ldots), \quad \boldsymbol{x}(n) = (\ldots, 1, -1, 1, -1, \ldots).$$

4. What is $Y(z)$ in terms of $X(z)$? What is $\boldsymbol{y}(n)$ in terms of $\boldsymbol{x}(n)$?

$\boldsymbol{x}(n) \rightarrow$ [$\uparrow 3$] $\rightarrow$ [z^{-k}] $\rightarrow$ [$\downarrow 3$] $\rightarrow y(n)$

5. If $H(z) = G(z^M)$ has $M-1$ zeros between $\boldsymbol{h}(0), \boldsymbol{h}(M), \ldots,$ find $\boldsymbol{y}(n)$ in terms of $\boldsymbol{x}(n)$ and $\boldsymbol{h}(n)$ for this system:

$\boldsymbol{x}(n) \rightarrow$ [$\uparrow M$] $\rightarrow$ [H] $\rightarrow$ [$\downarrow M$] $\rightarrow y(n)$

6. What identity would you use to compute $\boldsymbol{v} = (\downarrow 2)\boldsymbol{H}\boldsymbol{x}$ efficiently, if $\boldsymbol{H}$ has only *odd-numbered* coefficients $\boldsymbol{h} = (0, \boldsymbol{h}(1), 0, \boldsymbol{h}(3))$?

Chapter 4

Filter Banks

4.1 Perfect Reconstruction

A filter bank is a set of filters, linked by sampling operators and sometimes by delays. The downsampling operators are decimators, the upsampling operators are expanders. In a two-channel filter bank, the analysis filters are normally lowpass and highpass. Those are the filters $\boldsymbol{H}_0$ and $\boldsymbol{H}_1$ at the start of the following filter bank:

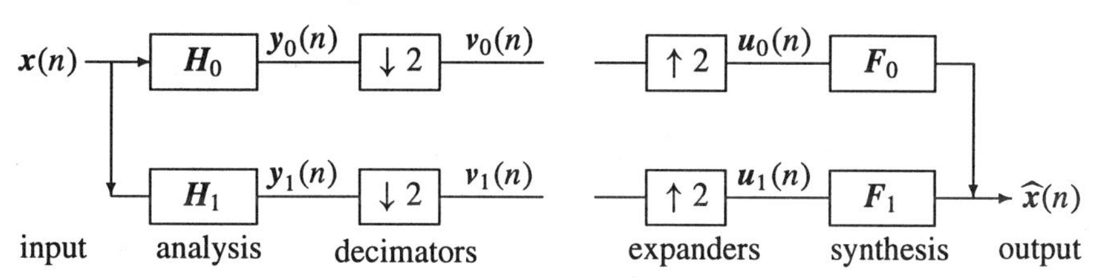

This structure was introduced in the 1980's. It gradually become clear how to choose $\boldsymbol{H}_0$, $\boldsymbol{H}_1$, $\boldsymbol{F}_0$, $\boldsymbol{F}_1$ to get perfect reconstruction: $\widehat{x}(n) = x(n-l)$. The gap in the figure indicates where the downsampled signals might be coded for storage or transmission. At that point we may compress the signal and destroy information. Perfect reconstruction assumes no compression, so the gap is closed.

To indicate that $\boldsymbol{H}_0$ is lowpass and $\boldsymbol{H}_1$ is highpass, we often sketch the frequency responses. Figure 4.1 shows that they are not ideal brick wall filters. The responses overlap. *There is aliasing in each channel.* There is also amplitude distortion and phase distortion (our drawing does not show the phase). The synthesis filters $\boldsymbol{F}_0$ and $\boldsymbol{F}_1$ must be specially adapted to the analysis filters $\boldsymbol{H}_0$ and $\boldsymbol{H}_1$, in order to cancel the errors in this analysis bank.

The goal of this section is to discover the conditions for perfect reconstruction. This means that the filter bank is ***biorthogonal***. The synthesis bank, from $\boldsymbol{F}_0$ and $\boldsymbol{F}_1$ and $\uparrow 2$, is the inverse of the analysis bank. Inverse matrices automatically involve biorthogonality. (The rows of $\boldsymbol{T}$ and the columns of $\boldsymbol{T}^{-1}$ are by definition biorthogonal.) This will extend in Chapter 6 to biorthogonal scaling functions and wavelets.

Perfect reconstruction is a crucial property. If the sampling operators $(\downarrow 2)$ and $(\uparrow 2)$ were not present, a reconstruction without delay would mean that $\boldsymbol{F}_0\boldsymbol{H}_0 + \boldsymbol{F}_1\boldsymbol{H}_1 = \boldsymbol{I}$. A perfect reconstruction with an l-step delay would mean (in the z-domain) that

$$\textit{without } (\downarrow 2) \textit{ and } (\uparrow 2): \quad F_0(z)H_0(z) + F_1(z)H_1(z) = z^{-l}. \tag{4.1}$$

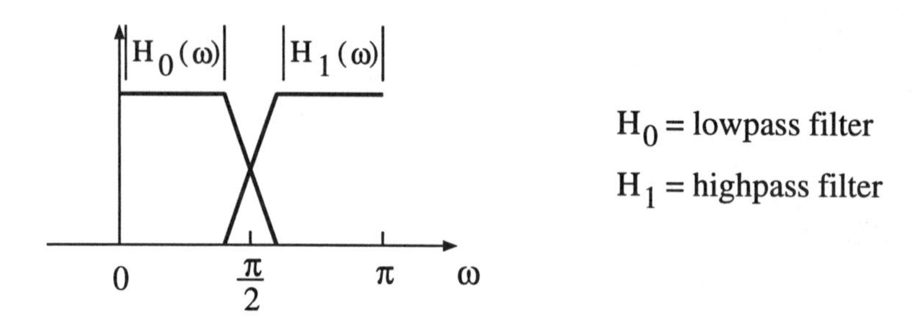

Figure 4.1: Rough sketch of frequency responses. Do not expect $|H_0(\omega)| + |H_1(\omega)| = 1$.

We do expect an overall delay z^{-l}, because each individual filter is causal.

Now take account of the sampling operators, which introduce *aliasing*. We recognize aliasing by the appearance of $-z$ as well as z (and $\omega + \pi$ as well as ω). The combination of $(\downarrow 2)$ followed by $(\uparrow 2)$ zeros out the odd-numbered components. In the z-domain it keeps only the *even powers* of $H_0(z)X(z)$:

$$\text{The transform of } \ (\uparrow 2)(\downarrow 2)\boldsymbol{H}_0\boldsymbol{x} \ \text{ is } \ \tfrac{1}{2}(H_0(z)X(z) + H_0(-z)X(-z)).$$

This is an even function, because the odd components are gone. The aliasing term $H_0(-z)X(-z)$ is multiplied by $F_0(z)$ at the synthesis step. This alias has to cancel the alias $F_1(z)H_1(-z)X(-z)$ from the other channel. So there is an alias cancellation condition in addition to a reconstruction condition:

$$\textit{Alias cancellation } F_0(z)H_0(-z) + F_1(z)H_1(-z) = 0. \tag{4.2}$$

Correction The sampling operators also produce a change in equation (4.1). *The right side has an extra factor* 2. You can see this by considering a simple set of filters: $H_0(z) = 1$ and $H_1(z) = z^{-1}$, $F_0(z) = z^{-1}$ and $F_1(z) = 1$. This satisfies (4.2) and cancels aliasing. The left side of equation (4.1) equals $2z^{-1}$ rather than z^{-1}. The overall delay is $l = 1$ for this filter bank, and its perfect reconstruction comes from

$$\textit{No distortion} \quad F_0(z)H_0(z) + F_1(z)H_1(z) = 2z^{-l}. \tag{4.3}$$

The next page establishes these two conditions (4.2–4.3) for perfect reconstruction.

One further point. In a genuine filter bank, the highpass filter has $H_1 = 0$ at $z = 1$ (or $\omega = 0$). Equation (4.3) becomes $F_0(1)H_0(1) = 2$. That equation is more natural if we include an extra factor $\sqrt{2}$ in the filter coefficients. For a similar reason the highpass filters can be normalized by an extra $\sqrt{2}$. We take this opportunity to assign single letters $\boldsymbol{C} = \sqrt{2}\boldsymbol{H}_0$ and $\boldsymbol{D} = \sqrt{2}\boldsymbol{H}_1$ to the analysis filters, with no need for subscripts. The sum of lowpass coefficients $\boldsymbol{c}(n) = \sqrt{2}\boldsymbol{h}(n)$ is $\sqrt{2}$.

No Aliasing and No Distortion Conditions (4.2) and (4.3) for perfect reconstruction come directly from following a signal through the filter bank. The original signal is $\boldsymbol{x}(n)$. The lowpass analysis filter is $\boldsymbol{H}_0$. In the z-domain this produces $H_0(z)X(z)$. Now downsample and upsample:

$$\text{First } (\downarrow 2) \text{ produces} \quad \tfrac{1}{2}[H_0(z^{\frac{1}{2}})X(z^{\frac{1}{2}}) + H_0(-z^{\frac{1}{2}})X(-z^{\frac{1}{2}})]$$

$$\text{Then } (\uparrow 2) \text{ produces} \quad \tfrac{1}{2}[H_0(z)X(z) + H_0(-z)X(-z)].$$

$(\uparrow 2)\ (\downarrow 2)\ \boldsymbol{H}_0\boldsymbol{x}$ *has zeros in its odd-numbered components.* Those zeros are produced by averaging $\boldsymbol{H}_0\boldsymbol{x}(n)$ with its alternating alias $(-1)^n\boldsymbol{H}_0\boldsymbol{x}(n)$. In the z-domain this is the average of $H_0(z)X(z)$ with $H_0(-z)X(-z)$. The aliasing term has entered the filter bank.

The final filter multiplies by $F_0(z)$. This yields the output from the lowpass channel. Below it we write the corresponding output from the highpass channel (same formula with subscripts changed to 1):

$$\text{lowpass output} = \tfrac{1}{2}F_0(z)\,[H_0(z)X(z) + H_0(-z)\,X(-z)]$$

$$\text{highpass output} = \tfrac{1}{2}F_1(z)\,[H_1(z)X(z) + H_1(-z)\,X(-z)].$$

Now add. The filter bank combines the channels to get $\widehat{\boldsymbol{x}}(n)$. In the z-domain this is $\widehat{X}(z)$. Half the terms involve $X(z)$ and half involve $X(-z)$:

$$\frac{1}{2}[F_0(z)H_0(z) + F_1(z)H_1(z)]X(z) + \frac{1}{2}[F_0(z)H_0(-z) + F_1(z)H_1(-z)]X(-z).$$

For perfect reconstruction with l time delays, this $\widehat{X}(z)$ must be $z^{-l}X(z)$. So the "distortion term" must be z^{-l} and the "alias term" must be zero:

Theorem 4.1 *A 2-channel filter bank gives perfect reconstruction when*

$$F_0(z)H_0(z) + F_1(z)H_1(z) = 2z^{-l} \tag{4.4}$$

$$F_0(z)H_0(-z) + F_1(z)H_1(-z) = 0. \tag{4.5}$$

In vector-matrix form these two conditions involve the *modulation matrix* $\boldsymbol{H}_m(z)$:

$$[F_0(z)\quad F_1(z)]\begin{bmatrix} H_0(z) & H_0(-z) \\ H_1(z) & H_1(-z) \end{bmatrix} = [2z^{-l}\quad 0]. \tag{4.6}$$

This matrix $\boldsymbol{H}_m(z)$ will play a very important role. It involves the responses $H_k(z)$ and their alias terms $H_k(-z)$. For an M-channel bank the matrix will be $M \times M$. But the real problem is clearly identified by the separate conditions (4.4) and (4.5)—*how to design filters that meet those conditions*?

Alias Cancellation and the Product Filter $P_0 = F_0H_0$

At this point we have four filters $\boldsymbol{H}_0, \boldsymbol{H}_1, \boldsymbol{F}_0, \boldsymbol{F}_1$ to design. They must satisfy (4.4) and (4.5). It is almost irresistible to determine some of the filters from the others:

For alias cancellation choose $F_0(z) = H_1(-z)$ *and* $F_1(z) = -H_0(-z)$ (4.7)

Important: This choice automatically satisfies $F_0(z)H_0(-z) + F_1(z)H_1(-z) = 0$. Aliasing is removed; it cancels itself! This relation of $\boldsymbol{F}_0$ to $\boldsymbol{H}_1$ and of $\boldsymbol{F}_1$ to $\boldsymbol{H}_0$ gives the *alternating signs* pattern of a 2-channel filter bank:

$\boldsymbol{H}_0$	$a,\ b,\ c$	$p, -q, r, -s, t$	$F_0(z) = H_1(-z)$
$\boldsymbol{H}_1$	$p,\ q,\ r,\ s,\ t$	$-a,\ b,\ -c$	$F_1(z) = -H_0(-z)$

Now comes a definition that allows us to rewrite equation (4.4) for no distortion:

Define the ***product filter*** *by* $P_0(z) = F_0(z)H_0(z)$.

This is a lowpass filter. The highpass product filter is $P_1(z) = F_1(z)H_1(z)$. These products P_0 and P_1 are exactly the terms in (4.4). The crucial point is the relation between $P_0(z)$ and $P_1(z)$, when the synthesis filters are determined by $F_0(z) = H_1(-z)$ and $F_1(z) = -H_0(-z)$. We substitute directly to find that $P_1(z) = -P_0(-z)$:

$$P_1(z) = -H_0(-z)H_1(z) = -H_0(-z)F_0(-z) = -P_0(-z). \tag{4.8}$$

The reconstruction equation $F_0(z)H_0(z) + F_1(z)H_1(z) = 2z^{-l}$ simplifies to

$$\boxed{F_0(z)H_0(z) - F_0(-z)H_0(-z) = P_0(z) - P_0(-z) = 2z^{-l}.} \tag{4.9}$$

The design of a 2-channel PR filter bank is reduced to two steps:

Step 1. Design a lowpass filter P_0 satisfying (4.9).

Step 2. Factor P_0 into F_0H_0. Then use (4.7) to find F_1 and H_1.

The length of P_0 determines the sum of the lengths of F_0 and H_0. There are many ways to design P_0 in Step 1. And there are many ways to factor it in Step 2. Experiments are going on as this book is written, and undoubtedly they are going on as the book is read, to find the best factors F_0 and H_0 of the best product filter P_0.

Note that (4.9) is a condition on the *odd powers* in $P_0(z) = F_0(z)H_0(z)$. Those odd powers must have coefficient zero, except z^{-l} has coefficient one. The shift to $P(z)$ in equation (4.11) below will remove the even powers except z^0.

A look forward To help the reader find the specific filters that are coming, we point to an outstanding choice for the product filter:

$$P_0(z) = (1 + z^{-1})^{2p} Q(z). \tag{4.10}$$

The polynomial $Q(z)$ of degree $2p-2$ is chosen so that (4.9) is satisfied. There are $2p-1$ odd powers in $P_0(z)$, and $2p-1$ coefficients to choose in $Q(z)$. *Then $Q(z)$ is unique.* This is the Daubechies construction, with a history that we will outline in Section 5.5. Since the construction starts with the special factor $(1+z^{-1})^{2p}$, these filters are called *binomial* or *maxflat*. The binomial factor gives a maximum number of zeros at $z = -1$, which means that the frequency response is maximally flat at $\omega = \pi$. The binomial by itself, without $Q(z)$, represents a "spline filter". $Q(z)$ is needed to give perfect reconstruction.

Splitting P_0 into F_0H_0 can give linear phase filters (symmetry in F_0 and H_0 separately). It can give orthogonal filters (symmetry between F_0 and H_0). *It cannot give both*, except in the Haar case $p = 1$. Section 5.4 discusses the factorization, and Chapter 11 reports some comparisons for image processing.

Simplification The equation $P_0(z) - P_0(-z) = 2z^{-l}$ can be made a little more convenient. The left side is an odd function, so l is odd. Normalize $P_0(z)$ by z^l to center it:

The normalized product filter is $P(z) = z^l P_0(z)$.

Then $P(-z) = (-z)^l P_0(-z)$. Since l is odd, this is $-z^l P_0(-z)$. The reconstruction equation $P_0(z) - P_0(-z) = 2z^{-l}$ takes an extremely simple form when we multiply by z^l. The factor z^{-l} disappears and the minus sign becomes plus:

Perfect Reconstruction Condition. $P(z)$ must be a "halfband filter"

$$P(z) + P(-z) = 2. \tag{4.11}$$

This means that *all even powers in $P(z)$ are zero*, except the constant term (which is 1). The odd powers cancel when $P(z)$ combines with $P(-z)$—so the coefficients of odd powers in $P(z)$ are design variables in 2-channel PR filter banks.

Example 4.1. The maxflat product filter $P_0(z) = (1 + z^{-1})^4 Q(z)$ happens to be

$$P_0(z) = \tfrac{1}{16}(-1 + 9z^{-2} + 16z^{-3} + 9z^{-4} - z^{-6}).$$

The center term is $z^{-3} = z^{-l}$. The requirement $P_0(z) - P_0(-z) = 2z^{-l}$ is quickly verified, because the even powers in P_0 cancel in the difference. The function $Q(z) = -1 + 4z^{-1} - z^{-2}$ was chosen so that the odd powers z^{-1} and z^{-5} are absent from $P_0(z)$.

A centering operation gives the normalized product filter $P(z)$. Multiply by $z^l = z^3$ to symmetrize the polynomial around the constant term z^0:

$$P(z) = \tfrac{1}{16}(-z^3 + 9z + 16 + 9z^{-1} - z^{-3}).$$

This is halfband, because the only even power is z^0 and its coefficient is 1. The perfect reconstruction requirement $P(z) + P(-z) = 2$ is verified. The odd powers in P cancel in the sum.

Notice the variety of factorizations into $P_0(z) = F_0(z)H_0(z)$. The polynomial $P_0(z)$ has *six roots*. The two roots from $Q(z)$ are at $c = 2 - \sqrt{3}$ and $\frac{1}{c} = 2 + \sqrt{3}$. The other four roots from $(1 + z^{-1})^4$ are at $z = -1$. Each factor normally has at least *one* root at $z = -1$. (But the factorization into $H_0 = 1$ and $F_0 = P_0$ is quite interesting.) Thus F_0 or H_0 (*either order is possible*!) could be

(a)	1	degree	$N = 0$
(b)	$(1 + z^{-1})$	degree	$N = 1$
(c)	$(1 + z^{-1})^2$ or $(1 + z^{-1})(c - z^{-1})$	degree	$N = 2$
(d)	$(1 + z^{-1})^3$ or $(1 + z^{-1})^2(c - z^{-1})$	degree	$N = 3$

The number $N + 1$ of filter coefficients is one greater than the degree. Thus (c) can produce a 5/3 filter, with $H_0(z) = \frac{1}{8}(-1 + 2z^{-1} + 6z^{-2} + 2z^{-3} - z^{-4})$ and $F_0(z) = \frac{1}{2}(1 + 2z^{-1} + z^{-2})$. The analysis length is given first. This is a possible choice for compression, with symmetric filters of very low complexity. The binomial $(1 + z^{-1})^2$ in the synthesis filter means that its continuous-time scaling function will be a hat function. Reversing F_0 and H_0 gives a 3/5 pair, not as successful in practice.

The choice (b) is also of interest. One factor is $1 + z^{-1}$ so one scaling function is the box function. Experiments indicate better performance when this short lowpass filter is in the analysis bank. Where $5/3$ was preferred to $3/5$ for odd-length filters, it seems that $2/6$ is preferred to $6/2$. *Five roots go into $F_0(z)$.*

The other outstanding choices are length 4/4 from the factorizations in (d). We get linear phase from $(1+z^{-1})^3$ and $(-1+3z^{-1}+3z^{-2}-z^{-3})$. The orthonormal Daubechies filter comes

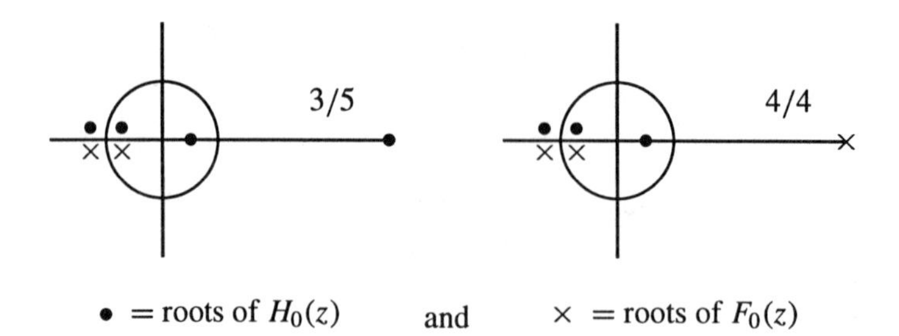

• = roots of $H_0(z)$ and × = roots of $F_0(z)$

from $(1+z^{-1})^2(c-z^{-1})$ and $(1+z^{-1})^2(\frac{1}{c}-z^{-1})$. *These are not linear phase.* (Problem 7 shows that linear phase requires $\frac{1}{c}$ to be a root when c is a root.) For the Daubechies orthogonal choice, the roots $-1, -1, c$ of one factor are the reciprocals of the roots $-1, -1, \frac{1}{c}$ *of the other factor.* Section 5.4 demonstrates that this balanced splitting (*spectral factorization*) of a halfband filter produces an orthogonal filter bank and orthogonal wavelets.

Example 4.2. (Haar filter bank) The average-difference analysis filter has

$$\boldsymbol{H}_m(z) = \begin{bmatrix} H_0(z) & H_0(-z) \\ H_1(z) & H_1(-z) \end{bmatrix} = \frac{1}{\sqrt{2}} \begin{bmatrix} 1+z^{-1} & 1-z^{-1} \\ 1-z^{-1} & 1+z^{-1} \end{bmatrix}.$$

The synthesis filters are

$$\begin{array}{ccccc} F_0(z) & = & H_1(-z) & = & (1+z^{-1})/\sqrt{2} \\ F_1(z) & = & -H_0(-z) & = & -(1-z^{-1})/\sqrt{2} \end{array}.$$

The product filters are

$$\begin{array}{l} P_0(z) = F_0(z)H_0(z) = \frac{1}{2}(1+z^{-1})^2 \\ P_1(z) = F_1(z)H_1(z) = -\frac{1}{2}(1-z^{-1})^2 = -P_0(-z). \end{array}$$

Both P_0 and P_1 contain $+z^{-1}$. Perfect reconstruction $P_0(z) - P_0(-z) = 2z^{-1}$ means that $l = 1$. The normalized product filter (symmetric halfband) is

$$P(z) = z^l P_0(z) = \tfrac{1}{2}z(1+z^{-1})^2 = \tfrac{1}{2}z + 1 + \tfrac{1}{2}z^{-1}.$$

Modulation Matrices

The conditions for perfect reconstruction are expressed in (4.6) by

$$[F_0(z) \quad F_1(z)] \begin{bmatrix} H_0(z) & H_0(-z) \\ H_1(z) & H_1(-z) \end{bmatrix} = [2z^{-l} \quad 0]. \tag{4.12}$$

This displays the *analysis modulation matrix* $\boldsymbol{H}_m(z)$—which is central to filter bank theory. With no extra effort we can also produce the synthesis modulation matrix $\boldsymbol{F}_m(z)$. The two matrices should play matching (and even reversible) roles. This balance between $\boldsymbol{F}_m$ and $\boldsymbol{H}_m$ is achieved by expanding (4.12) into a matrix equation:

$$\begin{bmatrix} F_0(z) & F_1(z) \\ F_0(-z) & F_1(-z) \end{bmatrix} \begin{bmatrix} H_0(z) & H_0(-z) \\ H_1(z) & H_1(-z) \end{bmatrix} = \begin{bmatrix} 2z^{-l} & 0 \\ 0 & 2(-z)^{-l} \end{bmatrix}. \tag{4.13}$$

The second row of equations follows from the first, when $-z$ replaces z. Note that $F_1(z)$ is in the $(1, 2)$ position, but $H_1(z)$ is in the $(2, 1)$ position. This *transpose convention* between analysis and synthesis will appear again for polyphase matrices. The reader sees why it is necessary.

The reconstruction condition (4.13) applies to $\boldsymbol{F}_m(z)\boldsymbol{H}_m(z)$. If we "center" the filter coefficients around the zero position, the right side becomes the identity matrix! This is so desirable and memorable that we do it. It is the same normalization that centered $P_0(z)$ into $P(z)$, and it is especially clear when the filters are linear phase (and l is odd):

Theorem 4.2 *If all filters are symmetric (or antisymmetric) around zero, as in* $H(z) = H(z^{-1})$ *and* $\boldsymbol{h}(k) = \boldsymbol{h}(-k)$*, then the condition for perfect reconstruction becomes a statement about inverse matrices:*

$$\boldsymbol{F}_m(z)\boldsymbol{H}_m(z) = 2\boldsymbol{I}. \tag{4.14}$$

The H's determine the F's. The analysis bank is inverted by the synthesis bank. When we express it that way, equation (4.14) becomes almost obvious.

A Brief History of H_1

The reader understands that the filters $\boldsymbol{H}_0$ and $\boldsymbol{H}_1$ are still to be chosen. These choices are connected. Historically, designers chose the lowpass filter coefficients $\boldsymbol{h}(0), \ldots \boldsymbol{h}(N)$ and then constructed $\boldsymbol{H}_1$ from $\boldsymbol{H}_0$. Here are two possibilities that produce *equal length filters.* $\boldsymbol{H}_1$ will be highpass whenever $\boldsymbol{H}_0$ is lowpass:

Alternating signs : $\quad H_1(z) = H_0(-z)$ comes from $(\boldsymbol{h}(0), -\boldsymbol{h}(1), \boldsymbol{h}(2), -\boldsymbol{h}(3), \ldots)$

Alternating flip : $\quad H_1(z) = -z^{-N}H_0(-z^{-1})$ comes from $(\boldsymbol{h}(N), -\boldsymbol{h}(N-1), \ldots)$.

For convenience we are assuming real coefficients. The number N is odd in the alternating flip. The perfect reconstruction condition is *still to be imposed.* When that is satisfied, the overall system delay is $l = N$.

Early choice. Croisier–Estaban–Galand (1976) chose alternating signs $H_1(z) = H_0(-z)$. The resulting filter bank was called QMF (Quadrature Mirror Filter). The highpass response $|H_1(e^{j\omega})|$ is a mirror image of the lowpass magnitude $|H_0(e^{j\omega})|$ with respect to the middle frequency $\frac{\pi}{2}$ — the quadrature frequency. Note that IIR filters $\boldsymbol{H}_0$ and $\boldsymbol{H}_1$ are allowed (and needed for PR, except for Haar!). This name QMF has since been extended to a larger class of filter banks, allowing M channels.

Better choice. Smith and Barnwell (1984–6) and Mintzer (1985) chose the alternating flip $H_1(z) = -z^{-N}H_0(-z^{-1})$. This leads to orthogonal filter banks, when $\boldsymbol{H}_0$ is correctly chosen. The Daubechies filters will fit this pattern.

General choice. The product $F_0(z)H_0(z)$ is a halfband filter. This gives biorthogonality, when aliasing is cancelled by the relation of $\boldsymbol{F}_0$ to $\boldsymbol{H}_1$ and $\boldsymbol{F}_1$ to $\boldsymbol{H}_0$.

Actually the synthesis bank has little freedom. Alias cancellation requires $F_0(z)H_0(-z) + F_1(z)H_1(-z) = 0$. Croisier-Estaban-Galand wrote each F_k directly in terms of H_k by

$$F_0(z) = H_0(z) \text{ and } F_1(z) = -H_1(z).$$

With alternating signs $H_1(z) = H_0(-z)$ inside the analysis bank, their synthesis construction agrees (as it must) with the anti-aliasing equations

$$F_0(z) = H_1(-z) \text{ and } F_1(z) = -H_0(-z). \tag{4.15}$$

Smith-Barnwell also made the anti-aliasing choice (4.15). With the alternating flip in $H_1(z)$, their synthesis filters (remembering that N is odd) are

$$F_0(z) = H_1(-z) = z^{-N}H_0(z^{-1}) \text{ comes from } (\boldsymbol{h}(N), \boldsymbol{h}(N-1), \ldots, \boldsymbol{h}(0)).$$

$$F_1(z) = -H_0(-z) = z^{-N}H_1(z^{-1}) \text{ comes from } (-\boldsymbol{h}(0), \boldsymbol{h}(1), -\boldsymbol{h}(2), \ldots, \boldsymbol{h}(N)).$$

Notice! Each $\boldsymbol{F}_k$ in Figure 4.1 has become the ordinary flip of the corresponding $\boldsymbol{H}_k$. In matrix language the synthesis matrices are the *transposes* of the analysis matrices. A shift by N delays makes them causal. When we flip to get $\boldsymbol{F}_0$ and then alternate signs to get $\boldsymbol{H}_1$, we have the alternating flip from $\boldsymbol{H}_0$ to $\boldsymbol{H}_1$.

The alternating flip automatically gives double-shift orthogonality between highpass and lowpass (to be explained). *Conclusion*: When the design of $\boldsymbol{H}_0$ leads to perfect reconstruction in the alternating flip filter bank, it also leads to orthogonality.

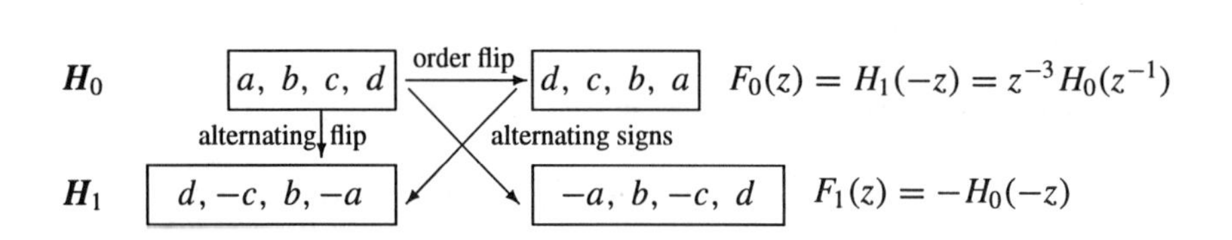

Figure 4.2: Relations between the filters allowing orthogonality when $N = 3$.

With aliasing cancelled, we now look at the PR condition

$$F_0(z)H_0(z) + F_1(z)H_1(z) = 2z^{-l}.$$

The early choice was alternating signs $H_1(z) = H_0(-z)$. With $F_0(z) = H_1(-z)$ and $F_1(z) = -H_0(-z)$, PR requires

$$H_0^2(z) - H_1^2(z) = H_0^2(z) - H_0^2(-z) = 2z^{-l}. \tag{4.16}$$

Therefore $H_0^2(z)$ has exactly one odd power z^{-l}. This is not easy for the square of a polynomial. An FIR filter is restricted to *two coefficients* (not good). Problem 4 asks for the reasoning, which forces the filters to be IIR.

The better choice is alternating flip. Perfect reconstruction is definitely possible. Product filters $\boldsymbol{F}_0\boldsymbol{H}_0$ and $\boldsymbol{F}_1\boldsymbol{H}_1$ become $P_0(z) = z^{-N}H_0(z^{-1})H_0(z)$ and $P_1(z) = -z^{-N}H_0(-z^{-1})H_0(-z)$. Multiply by $z^l = z^N$ to center these filters. The normalized product filter is $P(z)$ and the reconstruction condition is (4.11):

$$P(z) + P(-z) = 2 \text{ with } P(z) = H_0(z^{-1})H_0(z). \tag{4.17}$$

This is spectral factorization of a halfband filter! On the unit circle $z = e^{j\omega}$, the product $H_0(e^{-j\omega})H_0(e^{j\omega})$ is a magnitude squared:

$$P(e^{j\omega}) = \sum_{-N}^{N} p(n)e^{-jn\omega} = \left|\sum_{0}^{N} h(n)e^{-jn\omega}\right|^2. \qquad (4.18)$$

The halfband coefficients are $p(n) = p(-n)$ for odd n and $p(n) = 0$ for even n (except $p(0) = 1$). *We design $P(z)$ and factor to find $H_0(z)$.* This symmetric factorization coincides with the Smith-Barnwell alternating flip. *It yields orthogonal banks with perfect reconstruction.* The flattest $P(z)$ will lead us to the Daubechies wavelets.

A note on biorthogonality (PR) with linear phase

Theorem 4.3 *In a biorthogonal linear-phase filter bank with two channels, the filter lengths are all odd or all even. The analysis filters can be*

(*a*) *both symmetric, of odd length*

(*b*) *one symmetric and the other antisymmetric, of even length.*

Proof: Odd and even lengths behave differently when we alternate signs:

odd length:	$a \quad b \quad c \quad b \quad a$	$\rightarrow \quad a \quad -b \quad c \quad -b \quad a$	(remains symmetric)
even length:	$a \quad b \quad b \quad a$	$\rightarrow \quad a \quad -b \quad b \quad -a$	(becomes antisymmetric).

To cancel aliasing, there is sign alternation in $F_0(z) = H_1(-z)$. There is also alternation in $F_1(z) = -H_0(-z)$. The extra minus sign does not change the symmetry type. The two successful combinations are

$$\begin{array}{ccc} H_0 = \text{symm} & & F_0 = \text{symm} \\ & \times & \\ H_1 = \text{symm} & & F_1 = \text{symm} \end{array} \qquad\qquad \begin{array}{ccc} H_0 = \text{symm} & & F_0 = \text{symm} \\ & \times & \\ H_1 = \text{anti} & & F_1 = \text{anti} \end{array}$$

odd lengths   *even lengths*

The other possibilities are excluded by the PR condition: $F_0(z)\,H_0(z)$ has to be a ***halfband filter***. It must have an odd number of coefficients, and the center coefficient must be 1. For $F_0(z)\,H_0(z)$ to have odd length (which means even degree), the factors $F_0(z)$ and $H_0(z)$ must be both odd length or both even length. If one is symmetric and the other antisymmetric, the product $F_0(z)\,H_0(z)$ will be antisymmetric with zero at the center—not allowed. We conclude that $F_0(z)$ and $H_0(z)$ must match: both odd length or both even length, both symmetric or both antisymmetric.

This leaves the two successful possibilities shown above, and two more: H_0 and F_0 both antisymmetric. But the sum of lowpass coefficients cannot be zero. So antisymmetry of H_0 is ruled out.

Perfect Reconstruction with M Channels In reality a filter bank can have M channels. Although $M = 2$ is standard in many applications, we often see $M > 2$. There are M analysis filters $H_0, H_1, \ldots, H_{M-1}$. The sampling is done at the critical rate by $(\downarrow M)$ and $(\uparrow M)$. There

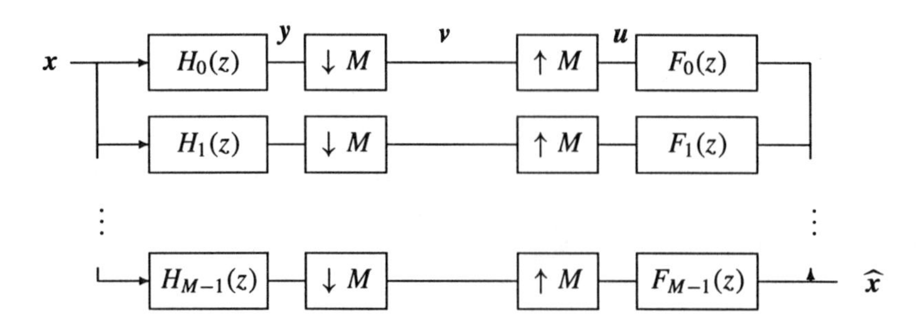

Figure 4.3: Standard form of an M-channel filter bank (maximally decimated).

are M filters $F_0, F_1, \ldots, F_{M-1}$ in the synthesis bank. The outputs from all channels are combined into a single output $\widehat{x}$. Our standard picture of this implementation is Figure 4.3.

For this M-band case, the theory of perfect reconstruction was developed over several years. When we follow each channel from H_k through $(\downarrow M)$ and $(\uparrow M)$ and F_k, and add, we find M conditions for perfect reconstruction. These equations involve the $M \times M$ *modulation matrix* $\boldsymbol{H}_m(z)$ — corresponding exactly to the 2-channel case. The modulations are by multiples of $2\pi/M$ in frequency, and by powers of $W = e^{-2\pi j/M}$ in the z-domain. We put that matrix on record here.

Modulation matrix

$$\boldsymbol{H}_m(z) = \begin{bmatrix} H_0(z) & H_0(zW) & \cdots & H_0(zW^{M-1}) \\ H_1(z) & H_1(zW) & \cdots & H_1(zW^{M-1}) \\ \cdots & \cdots & \cdots & \cdots \\ H_{M-1}(z) & H_{M-1}(zW) & \cdots & H_{M-1}(zW^{M-1}) \end{bmatrix}. \tag{4.19}$$

The last $M-1$ columns represent the $M-1$ aliases created by $(\downarrow M)$, just as $H(-z)$ represented the one alias $(W = -1)$ created when $M = 2$. *The transpose of* $\boldsymbol{H}_m(z)$ *is the* ***alias component matrix***.

Another matrix will play an equally central role — in fact an interchangeable role, because it is very closely linked to $\boldsymbol{H}_m(z)$. This new matrix is the *polyphase matrix* $\boldsymbol{H}_p(z)$. It is developed and explained in the following section, as the natural way to follow the "phases" when a signal is subsampled. We put on record the 2-phase matrix — *this is the polyphase matrix when* $M = 2$:

$$\boldsymbol{H}_p(z) = \begin{bmatrix} H_{0,\text{even}}(z) & H_{0,\text{odd}}(z) \\ H_{1,\text{even}}(z) & H_{1,\text{odd}}(z) \end{bmatrix}.$$

Still looking ahead, we mention especially the orthogonal case. *The polyphase matrix is then unitary for* $|z| = 1$ (this makes it *paraunitary*). $\boldsymbol{H}_m(z)$ is also paraunitary, after dividing by $\sqrt{2}$ (Section 5.1). The analysis of paraunitary matrices was led by Vaidyanathan. He and others built onto the early theory (and nearly indecipherable exposition) of Belevitch. For $M = 2$, the paraunitary matrix leads back to the Smith-Barnwell construction of an orthogonal PR filter bank.

Several perfect reconstruction filter banks deserve special mention. Simplest is the average-difference pair from Chapter 1. This is a useful example but a poor filter. That is the first in a family of "maxflat filters", corresponding to the Daubechies wavelets. The others in the family are orthogonal but not linear phase — since those two properties conflict.

Different factorizations of the product $P_0(z)$ lead to linear phase (not orthogonality). Those filters have become favorites for compression.

For $M > 2$, the design of separate filters $\boldsymbol{H}_0, \boldsymbol{H}_1, \ldots, \boldsymbol{H}_{M-1}$ can become unwieldy. We look for constructions in which these all come from one prototype filter. A particular class is the *cosine-modulated filter banks* in Chapter 9. A phase change (= modulation) is the key to their construction. Those are efficient in every way.

Problem Set 4.1

1. If $(\boldsymbol{H}_m(z))^{-1}$ is also a polynomial, the synthesis bank as well as the analysis bank is FIR. Why must the determinant $H_0(z)H_1(-z) - H_1(z)H_0(-z)$ in the denominator of $\boldsymbol{H}_m^{-1}$ be a monomial cz^{-l}? This determinant is an odd function, $\det \boldsymbol{H}_m(z) = -\det \boldsymbol{H}_m(-z)$. Then the exponent l must be odd.

2. The solution of *both* equations (4.4–5) for F_0 and F_1 involves $(\boldsymbol{H}_m(z))^{-1}$:

$$[F_0(z) \quad F_1(z)] = [2z^{-l} \quad 0](\boldsymbol{H}_m(z))^{-1} = \frac{2z^{-l}}{\det \boldsymbol{H}_m(z)}[H_1(-z) \quad H_0(-z)].$$

If $\det \boldsymbol{H}_m(z) = 2z^{-l}$, as normal, this yields $F_0(z) = H_1(-z)$ and $F_1(z) = -H_0(-z)$.

Extra credit: Verify that the key equation $P(z) + P(-z) = 2$ still holds in the IIR case with the extended definition of the product filter

$$P(z) = \frac{z^l F_0(z) H_0(z)}{\det \boldsymbol{H}_m(z)}.$$

3. Find all filters if $H_0(z) = (\frac{1+z^{-1}}{2})^3$ and $P_0(z) = \frac{1}{16}(-1 + 9z^{-2} + 16z^{-3} + 9z^{-4} - z^{-6})$.

4. If an FIR filter $H_0(z)$ has three or more coefficients, explain why $H_0^2(z)$ has at least two odd powers. Then $H_0^2(z) - H_0^2(-z) = 2z^{-l}$ is impossible. The "alternating signs" construction is not PR. (This is extended in Theorem 5.3: Symmetry prevents orthogonality except with two coefficients.)

5. If H_0 and P_0 are symmetric, why is F_0 symmetric? Why is H_1 linear phase, and when can it be antisymmetric?

6. Prove Theorem 4.3 by observing that an order flip in linear-phase filters $H_0(e^{j\omega})$ and $F_0(e^{j\omega})$ gives $H_{0,R}$ and $F_{0,R}$:

$$\begin{cases} H_0(e^{j\omega}) = e^{-j(N_0/2)\omega} H_{0,R}(e^{-j\omega}), & F_0(e^{j\omega}) = e^{-j(N_1/2)\omega} F_{0,R}(e^{-j\omega}) \\ H_0(e^{j\omega})F_0(e^{j\omega}) + H_0(e^{j(\omega+\pi)})F_0(e^{j(\omega+\pi)}) = 2e^{-j\ell\omega}. \end{cases}$$

N_0 and N_1 are the degrees of $H_0(z)$ and $F_0(z)$, and ℓ is odd.

7. A symmetric filter has $H(z^{-1}) = z^N$ times (___). If $H(c) = 0$ show that also $H(\frac{1}{c}) = 0$.

8. In the example with six roots, show a 4/4 linear phase pair—the roots c and $\frac{1}{c}$ go together. Find the polynomials $H(z)$ and $F(z)$.

9. The 10th degree halfband polynomial $P_0(z) = \left(1 + z^{-1}\right)^6 Q(z)$ has four complex roots r, $\overline{r}$, r^{-1}, $\overline{r}^{-1}$ in the right halfplane (roots of Q). Draw a figure to show the ten roots and how Daubechies 6/6 filters will divide them: r and $\overline{r}$ are separated from r^{-1} and $\overline{r}^{-1}$.

10. For the same 10th degree $P_0(z)$, show how the ten roots can give 6/6 filters with linear phase. One filter has 5 zeros at $z = -1$.

11. (Good problem) Find the actual 4th degree $Q(z)$ that makes $P_0(z)$ halfband. If possible compute its roots.

12. Given a PR filter bank H_0, H_1, F_0, F_1, interchange $H_k(z)$ and $F_k(z)$ (so that the synthesis bank has $H_k(z)$ and analysis bank has $F_k(z)$). Verify that the new system is PR. Define another system $\widehat{H}_k(z) = H_k(-z)$ and $\widehat{F}_k(z) = F_k(-z)$. Is this new system PR?

13. Let $H_0(z)$ be a symmetric lowpass filter with even length and $H_1(z) = H_0(-z)$. Verify that $H_1(z)$ is an antisymmetric highpass filter. Find the synthesis filters $F_k(z)$ that cancel aliasing. Can this system be PR? (Is $P(z)$ a halfband filter?)

4.2 The Polyphase Matrix

This section establishes a key idea and a valuable notation. The word "*polyphase*" has gained a certain mystique in the theory of multirate filters. Perhaps we can begin by explaining the meaning of the word, and also the purpose of the idea. Then the notation and applications will come naturally.

Meaning of polyphase: When a vector is downsampled by 2, its even-numbered components are kept. Its odd-numbered components are lost. Those are the two phases, *even* and *odd.* It is natural to follow the two phases of the input vector, $\boldsymbol{x}_{\text{even}}$ and $\boldsymbol{x}_{\text{odd}}$, as they go through the filter bank. They are acted on by the two phases $\boldsymbol{H}_{\text{even}}$ and $\boldsymbol{H}_{\text{odd}}$ of the filter.

For downsampling by M there are M phases. The ideas still apply to this "several-phase" or "polyphase" decomposition. Instead of even and odd inputs we will have M vectors (phases of $\boldsymbol{x}$). Instead of even and odd filters we will have M filters (phases of $\boldsymbol{H}$). The vector of filter coefficients $\boldsymbol{h}(n)$ is separated into phases, exactly as $\boldsymbol{x}(n)$ is separated. Then we watch those phases during downsampling.

The word "phase" is applied because the even filter with coefficients $\boldsymbol{h}(0), \boldsymbol{h}(2)$ has a different delay (phase shift) from the odd phase with coefficients $\boldsymbol{h}(1), \boldsymbol{h}(3)$.

Purpose of polyphase: The operation $(\downarrow 2)\boldsymbol{Hx}$, taken literally, is not efficient. We are computing all components of $\boldsymbol{Hx}$ and then destroying half of them. If we don't compute them, the system is still working at a fast rate (high bandwidth). The output is at half rate, because of downsampling. Each output component needs N additions and $N+1$ multiplications, to apply all the coefficients $\boldsymbol{h}(0), \ldots, \boldsymbol{h}(N)$.

The polyphase implementation works on the different phases separately. The input vector is separated into $\boldsymbol{x}_{\text{even}}$ and $\boldsymbol{x}_{\text{odd}}$. The operator $(\downarrow 2)$ comes *before the filter*! It changes one input at a high rate to two (or M) inputs at a lower rate. Then the separate phases of the filters act simultaneously (*in parallel*) on separate phases of the input.

The notation has to keep track of each phase. Often we find that "even multiplies even" and "odd multiplies odd". The *Noble Identities* justify an interchange of filtering and sampling. For the whole filter this interchange is forbidden, but it is allowed for each phase.

Polyphase in the time domain (block Toeplitz matrix): We can display the infinite matrix for a 2-channel analysis bank. Recall that $\boldsymbol{c}(n) = \sqrt{2}\,\boldsymbol{h}_0(n)$ and $\boldsymbol{d}(n) = \sqrt{2}\,\boldsymbol{h}_1(n)$. The two filters $\sqrt{2}\boldsymbol{H}_0 = \boldsymbol{C}$ and $\sqrt{2}\boldsymbol{H}_1 = \boldsymbol{D}$ are downsampled by $(\downarrow 2)$. This removes the odd-numbered rows. Then we *interleave the rows* of $\boldsymbol{L} = (\downarrow 2)\boldsymbol{C}$ and $\boldsymbol{B} = (\downarrow 2)\boldsymbol{D}$ to see the analysis bank as a *block Toeplitz matrix*:

$$\textbf{Block Toeplitz } \boldsymbol{H}_b = \begin{bmatrix} \cdot & \cdot & & & & & & \\ \cdot & \cdot & & & & & & \\ c(3) & c(2) & c(1) & c(0) & & & & \\ d(3) & d(2) & d(1) & d(0) & & & & \\ & & c(3) & c(2) & c(1) & c(0) & & \\ & & d(3) & d(2) & d(1) & d(0) & & \\ & & & & \cdot & \cdot & \cdot & \cdot \\ & & & & \cdot & \cdot & \cdot & \cdot \end{bmatrix}.$$

This takes the input in blocks (two samples at a time). It gives the output in blocks. It is time-invariant in blocks! By block z-transform, multiplication by the infinite matrix $\boldsymbol{H}_b$ (which is block convolution) becomes multiplication by the *polyphase matrix*:

$$\textbf{Polyphase matrix} \quad \boldsymbol{H}_p(z) = \begin{bmatrix} c(0) & c(1) \\ d(0) & d(1) \end{bmatrix} + z^{-1} \begin{bmatrix} c(2) & c(3) \\ d(2) & d(3) \end{bmatrix} \tag{4.20}$$

The polyphase matrix is nothing but *the z-transform of a block of filters*. There are 2^2 or M^2 filters, from M phases of M original filters. Here those filters have four coefficients and their phases have two coefficients.

Notice especially how the block matrix $\boldsymbol{H}_b$ relates to the two separate downsampled filters $(\downarrow 2)\boldsymbol{C}$ and $(\downarrow 2)\boldsymbol{D}$:

> The efficient form downsamples the input *first* (to make blocks for $\boldsymbol{H}_b$)
>
> The inefficient form downsamples *last* (after the filters $\boldsymbol{C}$ and $\boldsymbol{D}$)

The Noble Identities prove the equivalence. It is just a removal of useless odd-numbered rows and an interleaving of the remaining rows. Next we discuss the algebra and the implementation.

Key identity in the z-domain: The even part of $X(z)$ is $\frac{1}{2}(X(z) + X(-z))$. The odd part is $\frac{1}{2}(X(z) - X(-z))$. The first has even powers $1, z^2, z^4$; the second has z, z^3, z^5. The original X is the sum of even plus odd (obviously). The same splitting holds for $C(z)$, and furthermore for $C(z)X(z)$. *The key is to find the even part of* $C(z)X(z)$. It is the even coefficients of $\boldsymbol{Cx}$ that survive downsampling and appear in $(\downarrow 2)\boldsymbol{Cx}$. In most of this section the lowpass filter is denoted by $\boldsymbol{C}$, to avoid the subscripts on $\boldsymbol{H}$.

A simple and important identity shows how the even part of $C(z)X(z)$ comes from even times even plus odd times odd:

$$\begin{aligned} \tfrac{1}{2}[C(z)X(z) + C(-z)X(-z)] &= \tfrac{1}{4}[C(z) + C(-z)][X(z) + X(-z)] \\ &\quad + \tfrac{1}{4}[C(z) - C(-z)][X(z) - X(-z)]. \end{aligned} \tag{4.21}$$

In multiplying numbers, odd times odd is odd. But we are *adding exponents*, as in $(z^3)(z^5) = z^8$. So it is really odd *plus* odd, and even *plus* even, that yield the even part of the z-transform. This is the part that downsampling picks out, when $\boldsymbol{Cx}$ is decimated.

The importance of the key identity is this. The left side involves *all* coefficients of $C(z)$ and $X(z)$. Each product on the right involves only *half* the coefficients. The multiplication in the z-domain, which is $(\downarrow 2)\boldsymbol{Cx}$ in the time domain, becomes computationally efficient. We don't want all of $C(z)X(z)$, only the even half. The right side shows how to do half the work. Better still, it shows how even-even and odd-odd can be executed in parallel at half the rate.

Downsampling an even function effectively replaces z by $z^{1/2}$. It "closes the gaps" in $1, z^2, z^4$ by changing to $1, z, z^2$. For an odd function we will need a delay or an advance. We cannot change z and z^3 to $z^{1/2}$ and $z^{3/2}$. You will see how the coefficients of z^{-1}, z^{-3}, z^{-5} in $C(z)$ become coefficients of $1, z^{-1}, z^{-2}$ in the odd phase $C_{\text{odd}}(z)$. There is a delay for the odd phase and a "delay chain" when there are multiple phases.

This chapter works out the polyphase notation. We concentrate most on $M = 2$; the phases are even and odd. Then the polyphase forms of the analysis and synthesis banks lead quickly to a main goal of the theory. We find the perfect reconstruction condition on the polyphase matrices, when the filters are centered:

$$\boxed{\boldsymbol{F}_p(z)\boldsymbol{H}_p(z) = \boldsymbol{I}.}$$

This tells us, clearly and directly, what is required:

1. At a minimum, $\boldsymbol{H}_p(z)$ must be ***invertible***. (biorthogonality)
2. Better than that, its inverse $\boldsymbol{F}_p(z)$ should be a ***polynomial***. (FIR)
3. Better still, $\boldsymbol{F}_p(z)$ might be the ***transpose*** of $\boldsymbol{H}_p(z)$. (orthogonality)

In case **3**, the polyphase matrices are "paraunitary". The analysis and synthesis banks are orthogonal. In the more general case **1**, the banks are "biorthogonal". In case **2**, the synthesis bank is biorthogonal and also FIR.

The rows of a matrix are always biorthogonal to the columns of its inverse. When the rows of one are identical to the columns of the other, the matrix is self-orthogonal. Then it is an *orthogonal* matrix if real, a *unitary* matrix if complex, and a *paraunitary* matrix if it is a function of a complex parameter z.

Polyphase for Vectors

Any input vector $\boldsymbol{x}$ and any filter vector $\boldsymbol{c}$ or $\boldsymbol{h}$ can be separated into even and odd:

$$\boldsymbol{x} = (\ldots, \boldsymbol{x}(0), 0, \boldsymbol{x}(2), 0, \ldots) + (\ldots, 0, \boldsymbol{x}(1), 0, \boldsymbol{x}(3), 0, \ldots).$$

The z-transform is separated into even powers and odd powers, as in

$$X(z) = [\boldsymbol{x}(0) + \boldsymbol{x}(2)z^{-2} + \cdots] + z^{-1}[\boldsymbol{x}(1) + \boldsymbol{x}(3)z^{-2} + \cdots]. \tag{4.22}$$

The even part has powers of z^2. So has the odd part, when we factor out z^{-1}. This is the polyphase decomposition of $\boldsymbol{x}$ in the z-domain:

$$\boxed{X(z) = X_{\text{even}}(z^2) + z^{-1}X_{\text{odd}}(z^2)} \tag{4.23}$$

Each phase has its own z-transform:

$$\boldsymbol{x}_{\text{even}} = \begin{bmatrix} \boldsymbol{x}(0) \\ \boldsymbol{x}(2) \\ \cdot \end{bmatrix} \leftrightarrow X_0(z) = \sum \boldsymbol{x}(2k)z^{-k}$$

$$\boldsymbol{x}_{\text{odd}} = \begin{bmatrix} \boldsymbol{x}(1) \\ \boldsymbol{x}(3) \\ \cdot \end{bmatrix} \leftrightarrow X_1(z) = \sum \boldsymbol{x}(2k+1)z^{-k}.$$

Because of the z^2 in the definition, the in-between zeros are gone from $X_0(z)$ and $X_1(z)$. Please verify that the phases of $X(z) = z^{-1} + z^{-2} + z^{-3}$ are $X_{\text{even}} = z^{-1}$ and $X_{\text{odd}} = 1 + z^{-1}$.

Now reverse the process, to recover $\boldsymbol{x}$. Upsampling puts zeros back into $\boldsymbol{x}_{\text{even}}$ and $\boldsymbol{x}_{\text{odd}}$. *Those zeros change z to z^2.* The odd phase is delayed by z^{-1}, to move $\boldsymbol{x}(1)$ from position 0 to position 1. Then addition reconstructs equation (4.25).

Here is the splitting and the reconstruction in block form. Notice that so far the filters are not included, and $(\downarrow 2)\boldsymbol{x}$ is exactly $\boldsymbol{x}_{\text{even}}$:

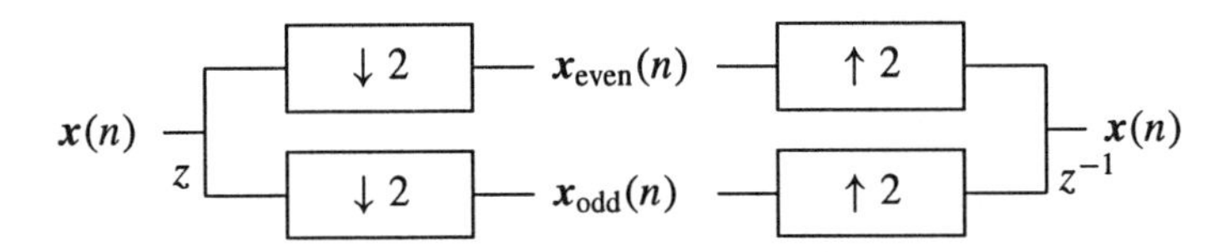

Important! The z at the start of the odd channel is because the odd phase has $\boldsymbol{x}(1)$ in its zeroth position. We have to advance the signal to achieve that. (See Problem 1.) But advances look bad in our flow diagram. So the advance can be replaced by a delay, if we make up for it at the end by delaying the even part too.

Here is the "delay form" that we use in later sections. Please go through that form:

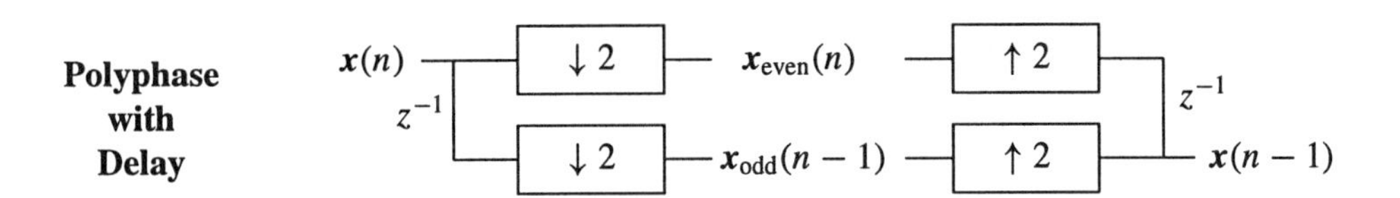

Only delays are involved! This is its advantage. Its disadvantage is that the output $\widehat{\boldsymbol{x}}(n)$ is $\boldsymbol{x}(n-1)$. The whole system equals a delay, where previously the system reproduced $\boldsymbol{x}$. The delay form in the z-domain produces $z^{-1}X(z)$ by delaying the even term:

$$z^{-1}X(z) = z^{-1}X_0(z^2) + z^{-2}X_1(z^2). \tag{4.24}$$

This *polyphase with delay* is just the original definition multiplied by z^{-1}.

Your eye will pick out this delay form. The output $\widehat{\boldsymbol{x}}$ comes later than the input. We still call this perfect reconstruction. Here is the same delay form in the time domain:

$$\begin{bmatrix} x(0) \\ x(1) \\ x(2) \\ x(3) \end{bmatrix} \begin{matrix} \nearrow \\ \searrow \end{matrix} \begin{matrix} \begin{bmatrix} x(0) \\ x(2) \end{bmatrix} \\ \begin{bmatrix} x(-1) \\ x(1) \end{bmatrix} \end{matrix} \begin{matrix} (\uparrow 2) \\ \nearrow \\ \searrow \\ (\uparrow 2) \end{matrix} \begin{matrix} \begin{bmatrix} x(0) \\ 0 \\ x(2) \\ 0 \end{bmatrix} \\ \begin{bmatrix} x(-1) \\ 0 \\ x(1) \\ 0 \end{bmatrix} \end{matrix} \begin{matrix} (z^{-1}) \\ \searrow \\ \\ \nearrow \end{matrix} \begin{bmatrix} x(-1) \\ x(0) \\ x(1) \\ x(2) \end{bmatrix} = \text{delayed } \boldsymbol{x}.$$

Polyphase Matrices for Filters

The polyphase form of a filter $\boldsymbol{C}$ comes directly from the polyphase form of $\boldsymbol{c}$ (the vector of filter coefficients). That vector separates into $\boldsymbol{c}_{\text{even}}$ and $\boldsymbol{c}_{\text{odd}}$. Its z-transform $C(z)$ separates into phases exactly as $X(z)$ did:

$$C(z) = C_0(z^2) + z^{-1}C_1(z^2). \tag{4.25}$$

The filtering step is $C(z)X(z)$. This is ordinary filtering $\boldsymbol{Cx}$, where even mixes with odd. But when downsampling picks out the even part of the product $C(z)X(z)$, it comes from even times even plus odd times odd. The transform of $(\downarrow 2)\boldsymbol{Cx}$ is

$$(C(z)X(z))_{\text{even}} = C_0(z)X_0(z) + z^{-1}C_1(z)X_1(z). \tag{4.26}$$

The direct multiplication of $C(z)$ times $X(z)$ will have even parts from $C_0(z^2)X_0(z^2)$ and from $z^{-2}C_1(z^2)X_1(z^2)$. Those give the even part of $C(z)X(z)$. Downsampling picks out those terms. It changes z^2 to z in their transform. The result is the z-transform of $(\downarrow 2)\boldsymbol{Cx}$.

Example 4.3. The moving average filter, downsampled.
Chapter 1 introduced the lowpass filter $\frac{1}{2}\boldsymbol{x}(n) + \frac{1}{2}\boldsymbol{x}(n-1)$. The coefficients are $\boldsymbol{h}(0) = \frac{1}{2}$ and $\boldsymbol{h}(1) = \frac{1}{2}$. Thus $\frac{1}{2}$ is the leading and only coefficient in H_{even} and H_{odd}. Both matrices have $\frac{1}{2}$ on *one diagonal*—the main diagonal. Here is the two-phase form of $(\downarrow 2)\boldsymbol{Hx}$:

$$\begin{bmatrix} \frac{1}{2} & \frac{1}{2} & & & \\ & & \frac{1}{2} & \frac{1}{2} & \\ & & & & \cdot\ \cdot \end{bmatrix} \begin{bmatrix} \\ \boldsymbol{x} \\ \\ \end{bmatrix} = \begin{bmatrix} \frac{1}{2} & & \\ & \frac{1}{2} & \\ & & \cdot \end{bmatrix} \begin{bmatrix} \boldsymbol{x}(0) \\ \boldsymbol{x}(2) \\ \cdot \end{bmatrix} + \begin{bmatrix} \frac{1}{2} & & \\ & \frac{1}{2} & \\ & & \cdot \end{bmatrix} \begin{bmatrix} \boldsymbol{x}(-1) \\ \boldsymbol{x}(1) \\ \cdot \end{bmatrix}$$

The polyphase components of $\frac{1}{2} + \frac{1}{2}z^{-1}$ are constants: $H_{\text{even}}(z) = H_{\text{odd}}(z) = \frac{1}{2}$.

Polyphase for one filter. The downsampling operator $(\downarrow 2)$ follows the filter $\boldsymbol{C}$. The outputs are $\boldsymbol{Cx}$ and then $(\downarrow 2)\boldsymbol{Cx}$. This is the normal order for filter banks, but it can be made more efficient. A close look at the product $(\downarrow 2)\boldsymbol{C}$ shows that *the filter coefficients* $\boldsymbol{c}(n)$ *are completely separated into even n and odd n.*

Thus $\boldsymbol{C}$ has two parts (or phases), which have their own z-transforms $C_0(z) = C_{\text{even}}(z)$ and $C_1(z) = C_{\text{odd}}(z)$. The 1 by 2 matrix $\boldsymbol{C}_p(z)$ is the two-phase or polyphase form of $C(z)$:

$$\boldsymbol{C}_p(z) = [C_0(z) \quad C_1(z)].$$

Warning. $C_0(z)$ is not $\frac{1}{2}(C(z)+C(-z))$. That involves $1, z^2, z^4, \ldots$ with zeros between. This is $C_0(z^2)$. By definition, $C_0(z)$ closes the zero gaps and has coefficients $\boldsymbol{c}(0), \boldsymbol{c}(2), \boldsymbol{c}(4)$. Similarly $C_1(z)$ has coefficients $\boldsymbol{c}(1), \boldsymbol{c}(3), \boldsymbol{c}(5)$. An advance or delay is involved. We don't keep the zeros from the even powers in $\frac{1}{2}(C(z) - C(-z))$.

The same splitting occurs for the highpass filter $\boldsymbol{D}$. Its even and odd phases are represented by $D_0(z)$ and $D_1(z)$. Those go into the 1 by 2 polyphase matrix $\boldsymbol{D}_p(z)$. Then the whole analysis bank comes together when we combine the polyphase matrices for $\boldsymbol{C}$ and $\boldsymbol{D}$ into a single polyphase matrix $\boldsymbol{H}_p(z)$:

$$\textbf{Polyphase Matrix} \quad \boldsymbol{H}_p(z) = \begin{bmatrix} \boldsymbol{C}_p(z) \\ \boldsymbol{D}_p(z) \end{bmatrix} = \begin{bmatrix} C_0(z) & C_1(z) \\ D_0(z) & D_1(z) \end{bmatrix} \tag{4.27}$$

This shows the matrix that we are aiming for. Now we go back for the close look at $(\downarrow 2)\boldsymbol{C}$. This operator is fundamental in the theory of multirate filters and wavelets.

Polyphase in the time domain. When downsampling follows the filter C, we get the crucial matrix $L = (\downarrow 2)C$. This has to display the separation of even and odd, and I would like to show how this happens. Most of polyphase theory is developed in the z-domain, and we will do that too. But first, look at the filter matrix as it produces $y = Cx$:

$$\begin{bmatrix} \cdot \\ y(0) \\ y(1) \\ y(2) \\ y(3) \\ \cdot \end{bmatrix} = \begin{bmatrix} \cdot & & & & & \\ \cdot & c(0) & & & & \\ \cdot & c(1) & c(0) & & & \\ \cdot & c(2) & c(1) & c(0) & & \\ \cdot & c(3) & c(2) & c(1) & c(0) & \\ & \cdot & \cdot & \cdot & \cdot & \cdot \end{bmatrix} \begin{bmatrix} \cdot \\ x(0) \\ x(1) \\ x(2) \\ x(3) \\ \cdot \end{bmatrix}. \tag{4.28}$$

Downsampling leaves the even-numbered components $y(2n)$. To reach $v = (\downarrow 2)y$, we throw away the odd-numbered rows. This leaves the matrix $L = (\downarrow 2)C$:

$$\begin{bmatrix} \cdot \\ y(0) \\ y(2) \\ y(4) \\ \cdot \end{bmatrix} = \begin{bmatrix} \cdot & & & & & & \\ \cdot & c(1) & c(0) & & & & \\ \cdot & c(3) & c(2) & c(1) & c(0) & & \\ \cdot & c(5) & c(4) & c(3) & c(2) & c(1) & c(0) \\ \cdot & \cdot & \cdot & \cdot & \cdot & \cdot & \cdot \end{bmatrix} \begin{bmatrix} \cdot \\ x(-1) \\ x(0) \\ x(1) \\ x(2) \\ \cdot \end{bmatrix}. \tag{4.29}$$

For polyphase here is the important point. *Only the even-numbered coefficients* $c(2n)$ *are multiplying the even-numbered coefficients* $x(2n)$. The even and odd c's are in separate columns. The even-numbered $x(0)$ is multiplying the column that starts with $c(0)$. The odd-numbered component $x(1)$ is multiplying the column containing $c(1), c(3), \ldots$. We can separate the matrix multiplication $(\downarrow 2)Cx$ into *even times even and odd times odd*:

$$\begin{bmatrix} \cdot \\ y(0) \\ y(2) \\ y(4) \\ \cdot \end{bmatrix} = \begin{bmatrix} \cdot & & & \\ \cdot & c(0) & & \\ \cdot & c(2) & c(0) & \\ \cdot & c(4) & c(2) & c(0) \\ \cdot & \cdot & \cdot & \cdot \end{bmatrix} \begin{bmatrix} \cdot \\ x(0) \\ x(2) \\ \cdot \end{bmatrix} + \begin{bmatrix} \cdot & & & \\ \cdot & c(1) & & \\ \cdot & c(3) & c(1) & \\ \cdot & c(5) & c(3) & c(1) \\ \cdot & \cdot & \cdot & \cdot \end{bmatrix} \begin{bmatrix} \cdot \\ x(-1) \\ x(1) \\ \cdot \end{bmatrix} \tag{4.30}$$

This is a matrix display of equation (4.21):

$$(\downarrow 2)Cx = C_{\text{even}}\, x_{\text{even}} + (\text{delay})\, C_{\text{odd}}\, x_{\text{odd}}. \tag{4.31}$$

The two phases x_{even} and x_{odd} are filtered by the two polyphase components C_{even} and C_{odd}. We need a delay in the odd phase, because $c(1)x(1)$ contributes to $y(2)$ and not to $y(0)$. Then $(\downarrow 2)Cx$ is the sum from the two phases:

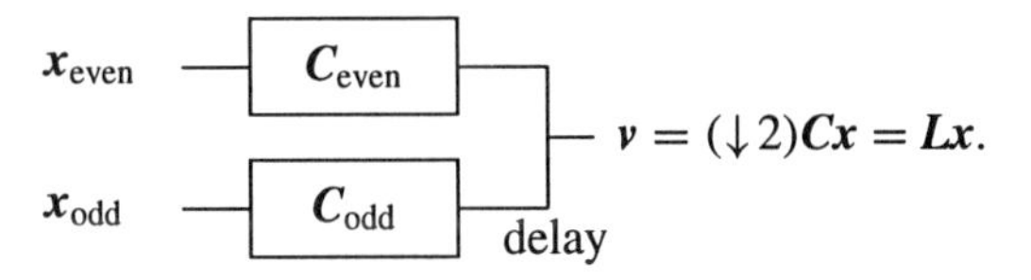

Notice something nice. The two matrices in equation (4.30) have constant diagonals. *The two operators* C_{even} *and* C_{odd} *are time-invariant filters.* They have frequency responses $C_0(z) =$

$C_{\text{even}}(z)$ and $C_1(z) = C_{\text{odd}}(z)$. The delay in the odd channel can go before or after C_1, because it commutes with C_1. (C_1 is time-invariant!) The two filters involve even coefficients $\boldsymbol{c}(2n)$ and odd coefficients $\boldsymbol{c}(2n+1)$, without zeros in between. They can operate in parallel, more efficiently.

Summary: Polyphase form of $L = (\downarrow 2)C$

The matrix C_{even} multiplies $\boldsymbol{x}_{\text{even}}$ in equation (4.31). The matrix C_{odd} multiplies $\boldsymbol{x}_{\text{odd}}$ (with a delay). We repeat this time-domain multiplication so you can compare it with the z-domain:

$$(\downarrow 2)\boldsymbol{C}\boldsymbol{x} = [\boldsymbol{C}_{\text{even}} \quad (delay)\boldsymbol{C}_{\text{odd}}]\begin{bmatrix} \boldsymbol{x}_{\text{even}} \\ \boldsymbol{x}_{\text{odd}} \end{bmatrix}. \tag{4.32}$$

This polyphase form has two ordinary filter matrices side by side. The z-domain polyphase form has two ordinary transfer functions side by side:

$$(C(z)X(z))_{\text{even}} = [C_{\text{even}}(z) \quad z^{-1}C_{\text{odd}}(z)]\begin{bmatrix} X_0(z) \\ X_1(z) \end{bmatrix}. \tag{4.33}$$

If $\boldsymbol{C}$ = identity then $C_{\text{even}}(z) = 1$ and the output is $X_0(z)$. If $\boldsymbol{C}$ = delay then $C_{\text{odd}}(z) = 1$ and the output is $X_1(z)$. All straightforward, but the block form shows something remarkable: *Downsampling comes before filtering* !

The block form with an advance to compute $\boldsymbol{x}_{\text{odd}}$ is ***Polyphase with Advance***:

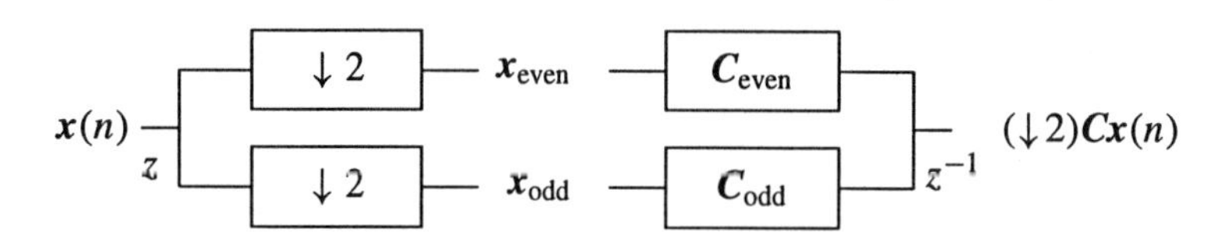

An advance has slipped into this form to get $\boldsymbol{x}(1)$ as the zeroth component of $\boldsymbol{x}_{\text{odd}}$. If we prefer to have only delays (and we do), that is possible. Delay $\boldsymbol{x}$ in the odd channel so that $(\downarrow 2)$ produces $\boldsymbol{x}_{\text{odd}}(n-1)$. Then after filtering, add a delay in the even channel. The result is to delay the whole signal by one time step in ***Polyphase with Delay***:

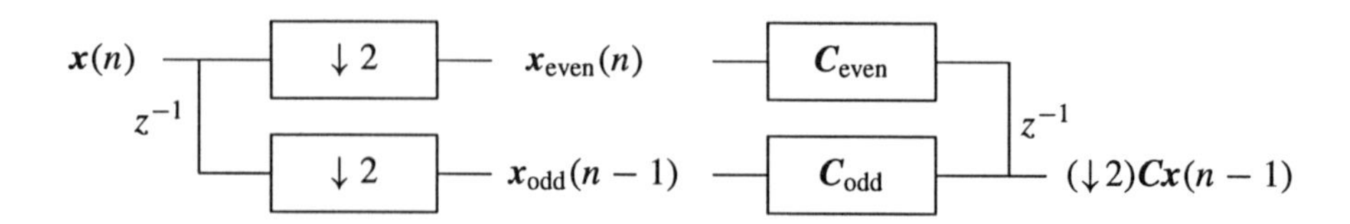

Key point: The polyphase form puts $(\downarrow 2)$ *before* the filters. This order is more efficient. It was possible to use the Noble Identity on each phase separately, because $C_{\text{even}}(z^2)$ and also $C_{\text{odd}}(z^2)$ appeared in the right place:

$$\text{(direct)} \quad (\downarrow 2)C_{\text{even}}(z) = C_{\text{even}}(z^2)(\downarrow 2) \quad \text{(polyphase)}.$$

Note on our convention. It would be very satisfying to avoid the delays completely. We would prefer to have

$$V(z) = [C_0(z) \quad C_1(z)] \begin{bmatrix} X_0(z) \\ X_1(z) \end{bmatrix}. \tag{4.34}$$

To achieve this we can alter the definition of $C_1(z)$. The polyphase decomposition of X would stay the same, but the decomposition of C would change to

$$C(z) = C_0(z^2) + zC_1(z^2). \tag{4.35}$$

Note z in the odd term where we had z^{-1}. This is the convention chosen by [VK]. It is just as good as ours. The zeroth component of C_{odd} becomes $\boldsymbol{c}(-1)$. It multiplies $\boldsymbol{x}(1)$. But I am afraid that in later chapters you would forget (I would too) this convention for $\boldsymbol{c}$, different from $\boldsymbol{x}$. So we keep the same even-odd decomposition of $\boldsymbol{c}$ and $\boldsymbol{x}$, yielding parallel decompositions of $C(z)$ and $X(z)$—and requiring the delay.

Problem Set 4.2

1. Find $X_{\text{even}}(z)$ and $X_{\text{odd}}(z)$ when $X(z) = 1 + 2z^{-5} + z^{-10}$. Verify that $X_{\text{even}}(z^2) = \frac{1}{2}(X(z) + X(-z))$ and $X_{\text{odd}}(z^2) = \frac{z}{2}(X(z) - X(-z))$. The odd definition involves an advance!

2. Express the z-transform of $\uparrow 2(\downarrow 2)\boldsymbol{x}$ in terms of X_{odd} and/or X_{even}.
What operations on $\boldsymbol{x}$ would produce the vector whose transform is $X_1(z^2)$?

3. The phases $\boldsymbol{C}_0, \boldsymbol{C}_1, \boldsymbol{D}_0, \boldsymbol{D}_1$ are all time-invariant, so they commute with delays. Does it follow that
$$\begin{bmatrix} 1 & \\ & \text{delay} \end{bmatrix} \begin{bmatrix} \boldsymbol{C}_0 & \boldsymbol{C}_1 \\ \boldsymbol{D}_0 & \boldsymbol{D}_1 \end{bmatrix} = \begin{bmatrix} \boldsymbol{C}_0 & \boldsymbol{C}_1 \\ \boldsymbol{D}_0 & \boldsymbol{D}_1 \end{bmatrix} \begin{bmatrix} 1 & \\ & \text{delay} \end{bmatrix} ?$$

4. *Polyphase Representation of an IIR Transfer Function*
Let $H(z) = \frac{1}{1-az^{-1}}$ where $0 < a < 1$. Its impulse response is $\boldsymbol{h}(n) = a^n$ for $n \geq 0$ (and zero for negative n). The phases are $\boldsymbol{h}_{\text{even}}(n) = (1, a^2, a^4, \ldots)$ and $\boldsymbol{h}_{\text{odd}}(n) = (a, a^3, a^5, \ldots)$. The z-transforms are $H_{\text{even}}(z) = 1/(1-a^2z^{-1})$ and $H_{\text{odd}}(z) = a/(1-a^2z^{-1})$. This method is very cumbersome. One has to find the impulse response $\boldsymbol{h}(n)$, then its even and odd parts $\boldsymbol{h}_{\text{even}}(n)$ and $\boldsymbol{h}_{\text{odd}}(n)$, then the z-transforms.
An alternate method is to write $H(z) = \frac{1}{1-az^{-1}}$ directly as $H(z) = H_{\text{even}}(z^2) + z^{-1}H_{\text{odd}}(z^2)$. The denominator must be a function of z^2. So multiply above and below by $1 + az^{-1}$:
$$H(z) = \frac{1}{1-az^{-1}} \frac{1+az^{-1}}{1+az^{-1}} = \frac{1+az^{-1}}{1-a^2z^{-2}} = \frac{1}{1-a^2z^{-2}} + z^{-1}\frac{a}{1-a^2z^{-2}}.$$
This displays $H_{\text{even}}(z)$ and $H_{\text{odd}}(z)$. An Nth order filter can be factored as a cascade of first-order sections, and this method applies to each section.
(a) Let $H(z) = \frac{1}{1-\frac{5}{6}z^{-1}+\frac{1}{6}z^{-2}}$. Factor $H(z)$ into two first-order poles. Find the polyphase components of $H(z)$.
(b) Let $H(z) = \frac{1+2z^{-1}+5z^{-2}}{1-\frac{5}{6}z^{-1}+\frac{1}{6}z^{-2}}$. What are its polyphase components?

5. For $M = 4$ channels, we want the four polyphase components of $H(z)$:
$$H(z) = H_0(z^4) + z^{-1}H_1(z^4) + z^{-2}H_2(z^4) + z^{-3}H_3(z^4).$$
(a) What polynomial multiplies $1 - az^{-1}$ to produce $1 - a^4z^{-4}$?
(b) Find the four components for $\frac{1}{1-az^{-1}}$ and $\frac{1+2z^{-1}+5z^{-2}}{1-\frac{5}{6}z^{-1}+\frac{1}{6}z^{-2}}$.

6. If $\boldsymbol{H}$ is a symmetric filter, how many of its phases are symmetric filters?

7. Let $H(z) = 1+2z^{-1}+3z^{-2}+4z^{-3}+4z^{-4}+3z^{-5}+2z^{-6}+z^{-7}$. Find the polyphase components $H_{even}(z)$ and $H_{odd}(z)$. What is the relation between $H_{even}(z)$ and $H_{odd}(z)$ for antisymmetric filters of even length and symmetric filters of odd length?

8. What are the two polyphase components of a symmetric halfband filter? Generalize to an M-th band filter.

4.3 Efficient Filter Banks

Let us repeat the key idea of the polyphase decomposition. The input is $\boldsymbol{x}$, the filter is $\boldsymbol{C}$, and we intend to downsample $\boldsymbol{Cx}$. We want the *even powers of* z in the product $C(z)X(z)$:

$$\text{even powers come from} \begin{cases} \text{even powers in } C(z) \text{ times even powers in } X(z) \\ \text{odd powers in } C(z) \text{ times odd powers in } X(z). \end{cases}$$

The polyphase decomposition is exactly this separation. Even times even is $C_0(z^2)X_0(z^2)$. Odd times odd starts with $\boldsymbol{c}(1)z^{-1}$ times $\boldsymbol{x}(1)z^{-1}$. The product $\boldsymbol{c}(1)\boldsymbol{x}(1)$ enters $\boldsymbol{Cx}(2)$ because indices add. Then $\boldsymbol{Cx}(2)$ after downsampling is $\boldsymbol{v}(1)$. Since $\boldsymbol{c}(1)$ and $\boldsymbol{x}(1)$ are the *zeroth components* of the odd phases, we need a delay to put their product into $\boldsymbol{v}(1)$. The algebra with $X(z) = X_0(z^2) + z^{-1}X_1(z^2)$ and $C(z) = C_0(z^2) + z^{-1}C_1(z^2)$ is

$$\begin{aligned} (CX)_0(z^2) &= C_0(z^2)X_0(z^2) + z^{-2}C_1(z^2)X_1(z^2) \\ (CX)_0(z) &= C_0(z)X_0(z) + z^{-1}C_1(z)X_1(z). \end{aligned} \tag{4.36}$$

We reached this answer by filtering and then subsampling (the step when z^2 becomes z). But now we see a better way. *Sample first* to get $\boldsymbol{x}_0 = \boldsymbol{x}_{\text{even}}$ and $\boldsymbol{x}_1 = \boldsymbol{x}_{\text{odd}}$. Filter those separately (and in parallel) by $\boldsymbol{C}_0$ and $\boldsymbol{C}_1$. Then combine the outputs with a suitable delay — the step down on the right:

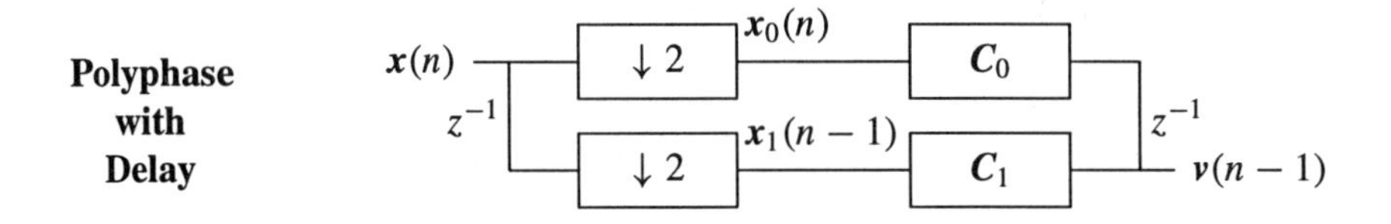

In this polyphase form, *the filters* $\boldsymbol{C}_0$ and $\boldsymbol{C}_1$ *are half as long as* $\boldsymbol{C}$. The output can be computed twice as fast, if all operations are done at the same speed. We could also use two cheaper low bandwidth processors, both working full time. They do $\frac{N}{2}$ multiplications and additions per unit time instead of N. By downsampling first, we reduce the input rate to the filters. The bandwidth is halved.

Example 4.4. The averaging filter in its direct form computes $\frac{1}{2}\boldsymbol{x}(0)+\frac{1}{2}\boldsymbol{x}(-1)$. Then it sits idle for one clock step, before computing $\frac{1}{2}\boldsymbol{x}(2) + \frac{1}{2}\boldsymbol{x}(1)$. The polyphase form has H_{even} and H_{odd} multiplying separately by $\frac{1}{2}$. They do one multiplication each, in parallel. An easy-to-write code will execute the polyphase form.

Polyphase for Filter Banks

The polyphase idea extends from one filter to a bank of filters. The direct form of the analysis bank does the downsampling last. *The polyphase form does the downsampling first.* In the block diagram of the filter bank, the decimators move *outside* the filters. We can write $\boldsymbol{C}$ and $\boldsymbol{D}$ or $\boldsymbol{H}_0$ and $\boldsymbol{H}_1$ for the lowpass and highpass filters (0 and 1 do not mean even and odd!):

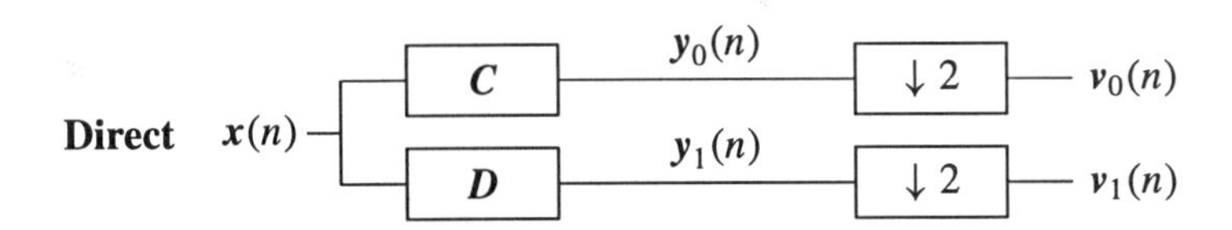

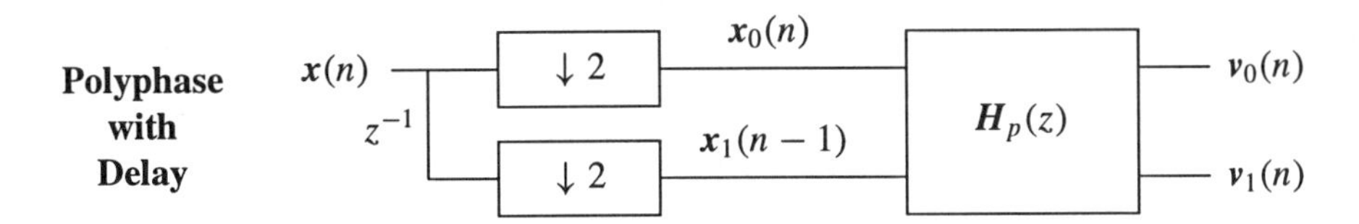

The *polyphase matrix* multiplies $X_0(z)$ and $z^{-1}X_1(z)$ to produce $V_0(z)$ and $V_1(z)$:

$$\begin{bmatrix} V_0(z) \\ V_1(z) \end{bmatrix} = \begin{bmatrix} C_0(z) & C_1(z) \\ D_0(z) & D_1(z) \end{bmatrix} \begin{bmatrix} X_0(z) \\ z^{-1}X_1(z) \end{bmatrix} = \boldsymbol{H}_p(z) \begin{bmatrix} X_0(z) \\ z^{-1}X_1(z) \end{bmatrix} \tag{4.37}$$

This defines and displays $\boldsymbol{H}_p(z)$. For FIR causal filters, the kind we expect to use, the polyphase components are polynomials in z^{-1}. When the input $\boldsymbol{x}$ is also causal, the outputs are causal.

Another point about notation. The indices in X_0 and X_1 refer to even and odd. The indices in V_0 and V_1 refer to the two channels. This is normal for matrix multiplication, when H_{ij} multiplies X_j and contributes to V_i. *Rows of $\boldsymbol{H}_p(z)$ go with channels, and columns of $\boldsymbol{H}_p(z)$ go with phases.*

In an M-channel bank, i is the channel index and j is the phase index in $H_{ij}(z)$. Then V_i is the output from channel i, and X_j is the jth phase of the input. We often reorganize a filter bank into its polyphase form.

Example 4.5. Average-difference filter bank in polyphase form.

The averaging filter $\boldsymbol{H}_0 \boldsymbol{x}(n) = \frac{1}{2}\boldsymbol{x}(n) + \frac{1}{2}\boldsymbol{x}(n-1)$ has polyphase components $\boldsymbol{H}_{0,\text{even}} = \boldsymbol{H}_{0,\text{odd}} = \frac{1}{2}(\text{identity})$. The differencing filter $\boldsymbol{H}_1\boldsymbol{x}(n) = \frac{1}{2}\boldsymbol{x}(n) - \frac{1}{2}\boldsymbol{x}(n-1)$ has components $\boldsymbol{H}_{1,\text{even}} = \frac{1}{2}\boldsymbol{I}$ and $\boldsymbol{H}_{1,\text{odd}} = -\frac{1}{2}\boldsymbol{I}$. Note that $H_0(z)$ and $H_1(z)$ are linear but the polyphase matrix is constant—typical of a *block transform*:

$$\boldsymbol{H}_p(z) = \begin{bmatrix} H_{00}(z) & H_{01}(z) \\ H_{10}(z) & H_{11}(z) \end{bmatrix} = \begin{bmatrix} \frac{1}{2} & \frac{1}{2} \\ \frac{1}{2} & -\frac{1}{2} \end{bmatrix}.$$

Example 4.6. Four-tap filters yield two taps for each phase. The even phase $\boldsymbol{C}_{\text{even}}$ has two coefficients $\boldsymbol{c}(0)$ and $\boldsymbol{c}(2)$. The odd phase has $\boldsymbol{C}_{\text{odd}} = \boldsymbol{c}(1) + \boldsymbol{c}(3)z^{-1}$. The same pattern holds for $\boldsymbol{D}$. The polyphase matrix for the filter bank is

$$\boldsymbol{H}_p(z) = \begin{bmatrix} \boldsymbol{c}(0) + \boldsymbol{c}(2)z^{-1} & \boldsymbol{c}(1) + \boldsymbol{c}(3)z^{-1} \\ \boldsymbol{d}(0) + \boldsymbol{d}(2)z^{-1} & \boldsymbol{d}(1) + \boldsymbol{d}(3)z^{-1} \end{bmatrix}. \tag{4.38}$$

Even is separated from odd. This reflects what happens in the filter bank, when $\boldsymbol{Cx}$ and $\boldsymbol{Dx}$ are downsampled:

$$\begin{bmatrix} v_0 \\ v_1 \end{bmatrix} = \begin{bmatrix} (\downarrow 2)\boldsymbol{Cx} \\ (\downarrow 2)\boldsymbol{Dx} \end{bmatrix} = \begin{bmatrix} \boldsymbol{C}_{\text{even}} & \boldsymbol{C}_{\text{odd}} \\ \boldsymbol{D}_{\text{even}} & \boldsymbol{D}_{\text{odd}} \end{bmatrix} \begin{bmatrix} 1 & \\ & \text{delay} \end{bmatrix} \begin{bmatrix} \boldsymbol{x}_{\text{even}} \\ \boldsymbol{x}_{\text{odd}} \end{bmatrix}$$

In case you like matrices, we are going to write the time-domain filter bank matrix in three ways. Downsampling is included in all three! First comes the matrix $\boldsymbol{H}_d = \boldsymbol{H}_{\text{direct}}$ that multiplies the input vector $\boldsymbol{x}$ in the direct form:

$$\textbf{Direct} \quad \boldsymbol{H}_d = \begin{bmatrix} \cdot & \cdot & & & & & & \\ c(3) & c(2) & c(1) & c(0) & & & & \\ & & c(3) & c(2) & c(1) & c(0) & & \\ & & & & \cdot & \cdot & \cdot & \cdot \\ \cdot & \cdot & & & & & & \\ d(3) & d(2) & d(1) & d(0) & & & & \\ & & d(3) & d(2) & d(1) & d(0) & & \\ & & & & \cdot & \cdot & \cdot & \cdot \end{bmatrix}. \tag{4.39}$$

Downsampling has removed every other row. That leaves this "square" infinite matrix. Each column is completely odd or completely even.

For the second form we rearrange the *rows* of $\boldsymbol{H}_d$. The highpass outputs are interleaved with the lowpass outputs, both downsampled by 2. This produces the block-diagonal form (or *block Toeplitz form*) $\boldsymbol{H}_b = \boldsymbol{H}_{\text{block}}$:

$$\textbf{Block} \quad \boldsymbol{H}_b = \begin{bmatrix} \cdot & \cdot & & & & & & \\ \cdot & \cdot & & & & & & \\ c(3) & c(2) & c(1) & c(0) & & & & \\ d(3) & d(2) & d(1) & d(0) & & & & \\ & & c(3) & c(2) & c(1) & c(0) & & \\ & & d(3) & d(2) & d(1) & d(0) & & \\ & & & & \cdot & \cdot & \cdot & \cdot \\ & & & & \cdot & \cdot & \cdot & \cdot \end{bmatrix}. \tag{4.40}$$

Your eye will divide that matrix into 2 by 2 blocks. It is like an ordinary time-invariant constant-diagonal matrix, but the entries are blocks instead of scalars. The main diagonal block corresponds to the constants in the polyphase matrix. The subdiagonal block produces the z^{-1} terms. There are only two diagonals because the phases of C and D have *two* coefficients. The original C and D had four coefficients.

The third form is the polyphase form $\boldsymbol{H}_p$. We are still in the time domain. For this third form we rearrange the *columns* of the direct form:

$$\textbf{Polyphase} \quad \boldsymbol{H}_p = \begin{bmatrix} \cdot & & & \cdot & & \\ c(2) & c(0) & & c(3) & c(1) & \\ & c(2) & c(0) & & c(3) & c(1) \\ & & \cdot \;\; \cdot & & & \cdot \;\; \cdot \\ \cdot & & & \cdot & & \\ d(2) & d(0) & & d(3) & d(1) & \\ & d(2) & d(0) & & d(3) & d(1) \\ & & \cdot \;\; \cdot & & & \cdot \;\; \cdot \end{bmatrix} = \begin{bmatrix} C_0 & C_1 \\ D_0 & D_1 \end{bmatrix}. \tag{4.41}$$

When the columns are rearranged, the vector $\boldsymbol{x}$ must be rearranged. Here $\boldsymbol{x}_{\text{even}}$ comes above $\boldsymbol{x}_{\text{odd}}$(delayed). The transform of the time-domain matrix $\boldsymbol{H}_p$ is the z-domain polyphase matrix $\boldsymbol{H}_p(z)$. *This 2 by 2 matrix of filters becomes a 2 by 2 matrix of functions.*

The block form $\boldsymbol{H}_b$ is an infinite matrix of 2 by 2 blocks. The polyphase form $\boldsymbol{H}_p$ is a 2 by 2 matrix of infinite blocks. Each block is a time-invariant filter. Either form leads by z-transform to the 2×2 polyphase matrix $\boldsymbol{h}_p(0) + z^{-1}\boldsymbol{h}_p(1)$:

Polyphase matrix $$\boldsymbol{H}_p(z) = \begin{bmatrix} c(0) & c(1) \\ d(0) & d(1) \end{bmatrix} + z^{-1} \begin{bmatrix} c(2) & c(3) \\ d(2) & d(3) \end{bmatrix} \qquad (4.42)$$

Relation Between Modulation and Polyphase

To produce two vectors from $\boldsymbol{x}$, one way is by polyphase. The parts $\boldsymbol{x}_{\text{even}}$ and $\boldsymbol{x}_{\text{odd}}$ are half-length. The other way is by modulation. Both $\boldsymbol{x}$ and $\boldsymbol{x}_{\text{mod}}$ are full length (therefore redundant). *The modulated vector $\boldsymbol{x}_{\text{mod}}$ reverses the sign of odd-numbered components*:

$$\boldsymbol{x}_{\text{mod}}(n) = (-1)^n \boldsymbol{x}(n) \quad \text{and} \quad X_{\text{mod}}(z) = X(-z).$$

In the frequency domain, $-1 = e^{i\pi}$ and ω is modulated by π —which explains the name:

$$X_{\text{mod}}(\omega) = X(\omega + \pi). \qquad (4.43)$$

The vector $\boldsymbol{x}_{\text{mod}}$ appears naturally when upsampling follows downsampling. Remember that $\boldsymbol{u} = (\uparrow 2)(\downarrow 2)\boldsymbol{x}$ is the average of $\boldsymbol{x}$ and $\boldsymbol{x}_{\text{mod}}$:

$$\boldsymbol{u} = \tfrac{1}{2}(\boldsymbol{x} + \boldsymbol{x}_{\text{mod}}) = (\ldots, \boldsymbol{x}(0), 0, \boldsymbol{x}(2), 0, \ldots). \qquad (4.44)$$

In the z-domain, the same fact gives the formula we use constantly:

$$U(z) = \tfrac{1}{2}[X(z) + X(-z)].$$

For downsampling and upsampling by M, the frequency modulation is by multiples of $2\pi/M$. The z-domain equivalent is multiplication by $W = e^{-2\pi i/M}$. There are M terms $X(W^k z)$. We continue with $M = 2$ terms, $X(z)$ and $X(-z)$.

The extra term represents aliasing. This modulated signal can overlap the original signal (in the frequency domain). If it does, we cannot recover $\boldsymbol{x}$ from $\boldsymbol{u}$. Now place this step $(\uparrow 2)(\downarrow 2)$ into a complete filter bank:

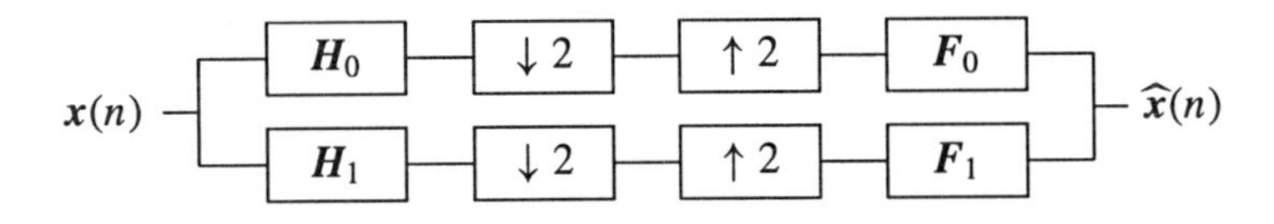

The transform of $\boldsymbol{H}_0\boldsymbol{x}$ is $H_0(z)X(z)$. When this is downsampled and upsampled, its alias appears in $\frac{1}{2}[H_0(z)X(z) + H_0(-z)X(-z)]$. Multiplying by $F_0(z)$ gives the output from the top channel. The lower channel is the same with 0 replaced by 1. The final output is the sum of channels:

$$\widehat{X}(z) = \tfrac{1}{2}[F_0(z) \quad F_1(z)] \begin{bmatrix} H_0(z) & H_0(-z) \\ H_1(z) & H_1(-z) \end{bmatrix} \begin{bmatrix} X(z) \\ X(-z) \end{bmatrix} \qquad (4.45)$$

This is the 2×2 modulation matrix $\boldsymbol{H}_m(z)$. It contains $H_0(z)$ and $H_1(z)$ along with their aliases. *The transpose of* $\boldsymbol{H}_m(z)$ *is also called the alias component matrix.*

A perfect reconstruction filter bank has to avoid aliasing (and other distortions). There is no aliasing when the combination $F_0(z)H_0(-z) + F_1(z)H_1(-z)$ is zero. Section 4.1 presented the synthesis filters $F_0(z) = H_1(-z)$ and $F_1(z) = -H_0(-z)$ that cancel aliasing.

Now turn to polyphase, the separation into even and odd phases:

$$H(z) = \tfrac{1}{2}[H(z) + H(-z)] + \tfrac{1}{2}[H(z) - H(-z)]. \tag{4.46}$$

The first bracket is even. Replacing z by $-z$ leaves it unchanged. The second bracket is odd. Replacing z by $-z$ reverses its sign. Therefore (4.46) must be the polyphase decomposition $H_{\text{even}}(z^2) + z^{-1}H_{\text{odd}}(z^2)$:

$$\begin{aligned}[H_{\text{even}}(z^2) \;\; z^{-1}H_{\text{odd}}(z^2)] &= \tfrac{1}{2}[H(z) + H(-z) \;\; H(z) - H(-z)] \\ &= \tfrac{1}{2}[H(z) \;\; H(-z)]\begin{bmatrix} 1 & 1 \\ 1 & -1 \end{bmatrix}.\end{aligned} \tag{4.47}$$

This connects polyphase to modulation. For one filter $\boldsymbol{H}_p$ and $\boldsymbol{H}_m$ have one row. For a bank of filters $\boldsymbol{H}_p$ and $\boldsymbol{H}_m$ are square matrices. A 2-channel analysis bank has equation (4.47) for H_0 in the top row, and the same equation for H_1 in the lower row:

$$\begin{bmatrix} H_{0,\text{even}}(z^2) & H_{0,\text{odd}}(z^2) \\ H_{1,\text{even}}(z^2) & H_{1,\text{odd}}(z^2) \end{bmatrix}\begin{bmatrix} 1 & \\ & z^{-1} \end{bmatrix} = \frac{1}{2}\begin{bmatrix} H_0(z) & H_0(-z) \\ H_1(z) & H_1(-z) \end{bmatrix}\begin{bmatrix} 1 & 1 \\ 1 & -1 \end{bmatrix} \tag{4.48}$$

Theorem 4.4 *The polyphase matrix* $\boldsymbol{H}_p$ *is connected to the modulation matrix* $\boldsymbol{H}_m$ *by a 2-point DFT and a diagonal delay matrix* $D(z) = diag(1, z^{-1})$.

This pattern extends in Section 9.2 to a bank of M filters. Each filter has M phases, and it has M modulations (by multiples of $2\pi/M$). The phases are associated with time shifts, and the modulations are frequency shifts. So the DFT connects them.

Problem Set 4.3

1. From $H(z) = H_{\text{even}}(z^2) + z^{-1}H_{\text{odd}}(z^2)$, find the corresponding formula for $H(-z)$. Write the two equations as

$$[H(z) \;\; H(-z)] = [H_{\text{even}}(z^2) \;\; H_{\text{odd}}(z^2)]\begin{bmatrix} 1 & \\ & z^{-1} \end{bmatrix}\begin{bmatrix} ? & ? \\ ? & ? \end{bmatrix}.$$

Identify that last matrix and verify that you have inverted equation (4.48). What is $\boldsymbol{H}_m(z)$ in terms of $\boldsymbol{H}_p(z)$?

2. If $\boldsymbol{H}_m^T(z^{-1})\boldsymbol{H}_m(z) = 2\boldsymbol{I}$ show from (4.48) that $\boldsymbol{H}_p^T(z^{-1})\boldsymbol{H}_p(z) = \boldsymbol{I}$.

3. Find a new example of matrices $\boldsymbol{H}_m(z)$ and $\boldsymbol{H}_p(z)$ for Problem 2.

4. Write down the equations for $M = 4$ that are analogous to (4.46)–(4.48).

Problems 5–11 develop an important example of biorthogonal filters.

5. The upper channel has responses $H_0(z) = 1$ and $F_0(z) = \frac{1}{2} + z^{-1} + \frac{1}{2}z^{-2} = \frac{1}{2}(1 + z^{-1})^2$. Follow the signals $\boldsymbol{x}(n) = (-1)^n$ and $\boldsymbol{\delta}(n)$ through this channel by plotting $(\uparrow 2)(\downarrow 2)\boldsymbol{H}_0\boldsymbol{x}(n)$.

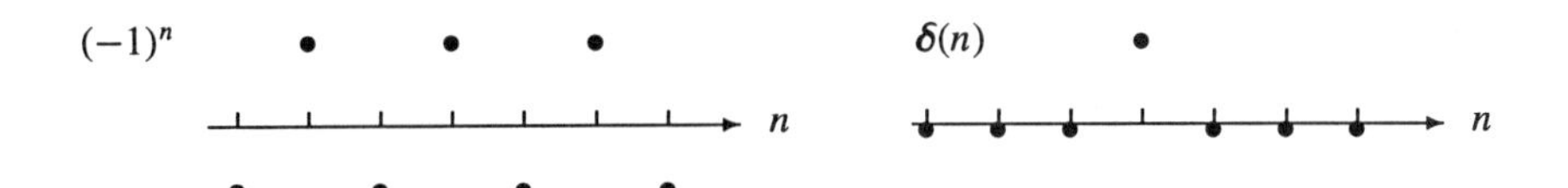

Plot the output $w_0(n) = F_0(\uparrow 2)(\downarrow 2)H_0 x(n)$ on top of these inputs. Describe the output from this channel with any input $x(n)$: The odd $w_0(2k+1)$ are _____ and the even $w_0(2k)$ are _____.

6. The lower channel has responses $H_1(z) = F_0(-z) = \frac{1}{2} - z^{-1} + \frac{1}{2}z^{-2} = \frac{1}{2}(1 - z^{-1})^2$ and $F_1(z) = -H_0(-z) = -1$. Follow the same two inputs $(-1)^n$ and $\delta(n)$ through this channel by plotting $H_1 x(n)$ and $w_1(n) = F_1(\uparrow 2)(\downarrow 2)H_1 x(n)$. Verify that the sum of outputs $w_0(n)+w_1(n)$ is $\widehat{x}(n) = x(n-1)$. In words, the output $w_1(n)$ from the lower channel is _____ for $n = 2k+1$ and _____ for $n = 2k$ for any input $x(n)$.

7. Verify the perfect reconstruction condition on $F_0(z)H_0(z) + F_1(z)H_1(z)$ for this filter bank. What is the delay ℓ? What is the product filter $P_0(z)$? Find the centered product filter $P(z)$, which is halfband.

8. Find the modulation matrix $H_m(z)$ for the analysis bank. Find the synthesis matrix $F_m(z)$, remembering the transpose. Compute the product $F_m(z)\,H_m(z)$.

9. Reverse the filters in the upper channel by $H_0(z) = \frac{1}{2} + z^{-1} + \frac{1}{2}z^{-2}$ and $F_0(z) = 1$. Follow the same signals $x(n) = (-1)^n$ and $x(n) = \delta(n)$ through this new channel.

10. Construct the corresponding $H_1(z) = F_0(-z)$ and $F_1(z) = -H_0(-z)$ and follow the two inputs through the lower channel. Verify that the outputs $w_0(n)+w_1(n)$ reconstruct the impulse $x(n)$. Certainly $F_0(z)H_0(z) + F_1(z)H_1(z)$ is the same as before (still PR). The halfband filter $P(z) = \frac{1}{2}z + 1 + \frac{1}{2}z^{-1}$ was factored into $P(z) \bullet 1$ and then $1 \bullet P(z)$. Which filter bank factors $P(z)$ into $(1+z^{-1})/\sqrt{2}$ times $(1+z)/\sqrt{2}$?

    Big question: Is the more regular filter $\frac{1}{2} + z^{-1} + \frac{1}{2}z^{-2}$ better in analysis (H_0) or in synthesis (F_0)? The output w_0 should be a "good but compressed" copy of x.

11. Show that the linear signal $x(n) = n$ goes entirely through the lowpass channel. The highpass channel has $H_1 x(n) =$ _____. The constant signal $x(n) = 1$ also goes through, so this filter bank (still from Problem 5) has accuracy $p = 2$.

12. Explain the equivalence of these representations before downsampling:

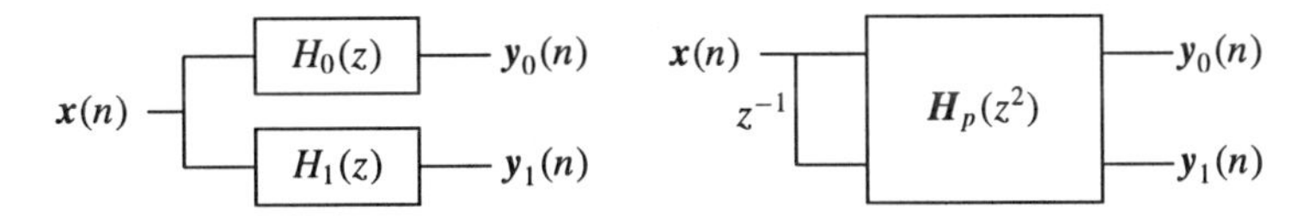

In matrix notation this is $\begin{bmatrix} H_0(z) \\ H_1(z) \end{bmatrix} = \begin{bmatrix} H_{00}(z^2) & H_{01}(z^2) \\ H_{10}(z^2) & H_{11}(z^2) \end{bmatrix} \begin{bmatrix} 1 \\ z^{-1} \end{bmatrix}.$

13. Explain the equivalence of these representations including downsampling:

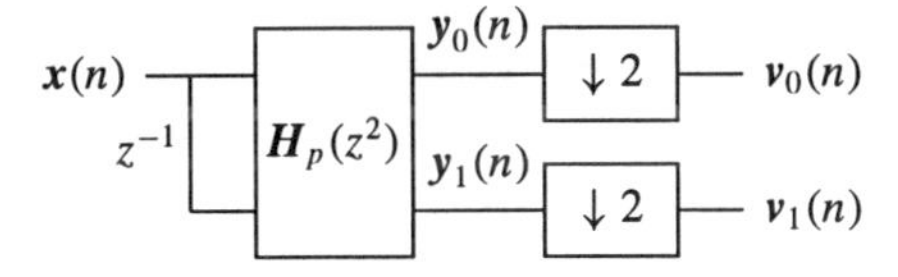

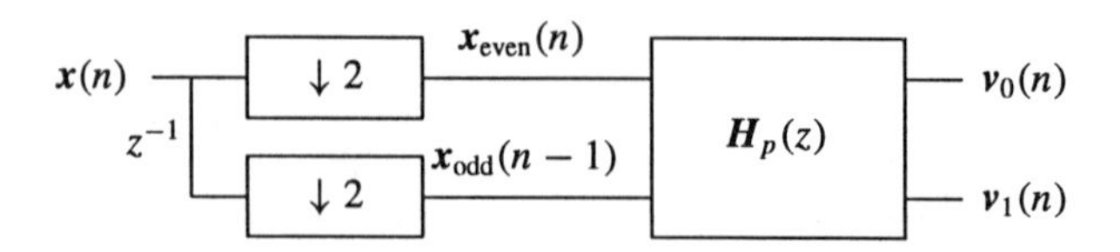

Warning for downsampling by 3. The polyphase components when $M = 3$ are C_0, C_1, C_2 and X_0, X_1, X_2. We want the component Y_0 of the product, because Y_1 and Y_2 are lost in downsampling. *That component Y_0 does not involve C_1 times X_1*. Multiplying z^{-1} by z^{-1} does not give z^{-3}. The exponents $0, 3, 6, \ldots$ come from C_2 times X_1 and C_1 times X_2 (and also C_0 times X_0). To get z^{-3} in the product, we multiply z^{-1} by z^{-2} or z^0 by z^{-3}:

$$Y_0(z^3) = C_0(z^3)X_0(z^3) + z^{-3}C_1(z^3)X_2(z^3) + z^{-3}C_2(z^3)X_1(z^3).$$

Then replace z^3 by z to find the transform of $(\downarrow 3)\boldsymbol{y} = (\downarrow 3)\boldsymbol{Cx}$.

We save this pattern for a later discussion of $(\downarrow M)$. It uses "Type 1 polyphase" components C_0, C_1, C_2. By defining Type 2 components R_0, R_1, R_2 we could achieve that R_i multiplies X_i. (The R's are a permutation of the C's.) Our immediate interest is restricted to $(\downarrow 2)$.

14. Write down the formula for $Y_0(z^3)$ that corresponds to equation (4.46). The polyphase components of C are C_0, C_1, C_2.

15. Show that $X_{\text{even}}(z)$ is the z-transform of the downsampled vector $(\downarrow 2)\boldsymbol{x}$. What vector is transformed to $X_{\text{odd}}(z)$?

16. Write $\boldsymbol{H}_p(z)$ in terms of $H_0(z)$ for these filters of length $N+1$.

$$\text{(a) } H_1(z) = H_0(-z) \quad \text{(b) } H_1(z) = z^{-N}H_0(z^{-1}) \quad \text{(c) } H_1(z) = z^{-N}H_0(-z^{-1}).$$

17. Find the analysis filters $H_0(z)$ and $H_1(z)$ for the following polyphase matrices:

a. $\boldsymbol{H}_p(z) = \begin{bmatrix} 1+2z^{-1}-z^{-2} & 2-z^{-1} \\ z^{-3} & 1+2z^{-1}+z^{-2} \end{bmatrix}$

b. $\boldsymbol{H}_p(z) = \begin{bmatrix} 1+2z^{-2} & 1+z^{-1} \\ 1-z^{-1} & 2+z^{-1} \end{bmatrix} \begin{bmatrix} 2 & 1+z^{-2} \\ 1-z^{-1} & -z^{-3} \end{bmatrix}$

c. $\boldsymbol{H}_p(z) = \begin{bmatrix} c_1 & s_1 \\ -s_1 & c_1 \end{bmatrix} \begin{bmatrix} 1 & 0 \\ 0 & z^{-1} \end{bmatrix} \begin{bmatrix} c_2 & s_2 \\ -s_2 & c_2 \end{bmatrix}$

d. $\boldsymbol{H}_p(z) = \begin{bmatrix} \frac{1}{2+z^{-1}} & 1 \\ z^{-3} & \frac{1+z^{-1}}{3-z^{-1}} \end{bmatrix}$

e. $\boldsymbol{H}_p(z) = \begin{bmatrix} 1 & 1 \\ 1 & -1 \end{bmatrix} \begin{bmatrix} \frac{1+2z^{-1}}{2+z^{-1}} & 0 \\ 0 & \frac{1-2z^{-1}}{2-z^{-1}} \end{bmatrix}$

18. For each system (a-e) in Problem 17, find the modulation matrix $\boldsymbol{H}_m(z)$.

4.4 Polyphase for Upsampling and Reconstruction

The reader can anticipate what is coming. The polyphase form above was for the analysis bank, with downsampling. The same ideas apply to the synthesis bank, with upsampling. The expanders $(\uparrow 2)$ will again move outside the filters in the polyphase implementation. This time "outside" means *after* the synthesis filters.

We start with one filter called $\boldsymbol{G}$, *and move the upsampler*:

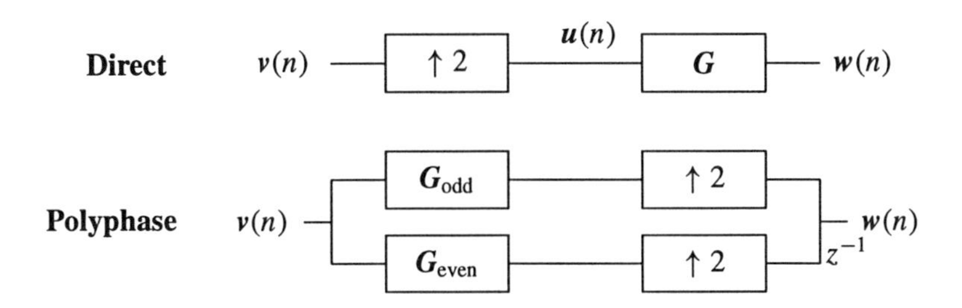

The polyphase form is more efficient, because the half-length filters $\boldsymbol{G}_0 = \boldsymbol{G}_{\text{even}}$ and $\boldsymbol{G}_1 = \boldsymbol{G}_{\text{odd}}$ receive input at the *slow rate* (low bandwidth). *The direct form unnecessarily doubles the rate by* $(\uparrow 2)$ before the filter.

It is always important to verify formulas in the z-domain. In the direct form, upsampling produces $U(z) = V(z^2)$. The only powers to enter $U(z)$ are *even* powers, because upsampling puts zeros in the odd components. Then the filter produces $W(z) = G(z)V(z^2)$. In this multiplication, even powers in G yield even powers in W. The odd phase of G multiplies $V(z^2)$ to give the odd phase of W. That is why polyphase works in the synthesis filters. Even times even is separated from odd times even:

$$W(z) = [1 \quad z^{-1}]\begin{bmatrix} W_{\text{even}}(z) \\ W_{\text{odd}}(z) \end{bmatrix} = [1 \quad z^{-1}]\begin{bmatrix} G_{\text{even}}(z^2) \\ G_{\text{odd}}(z^2) \end{bmatrix} V(z^2).$$

Suppose the filter $\boldsymbol{G}$ is a simple delay. Then its Type 1 polyphase components are $\boldsymbol{G}_{\text{even}} = 0$ and $\boldsymbol{G}_{\text{odd}} =$ identity. The output from this polyphase equation is correct—a delay of $\uparrow 2\boldsymbol{v}$.

Example 4.7. Suppose $\boldsymbol{G}$ has four coefficients $g(0), g(1), g(2), g(3)$. In the time domain, $\boldsymbol{G}$ is a constant-diagonal matrix. When upsampling acts before the filter to produce $\boldsymbol{G}(\uparrow 2)$, it *removes columns* of $\boldsymbol{G}$. (Just as downsampling removed rows of $\boldsymbol{C}$.) The time-domain matrix in direct and inefficient form is $\boldsymbol{G}(\uparrow 2)$:

$$\boldsymbol{G}\,(\uparrow 2) = \begin{bmatrix} g(0) & & & & \\ g(1) & g(0) & & & \\ g(2) & g(1) & g(0) & & \\ g(3) & g(2) & g(1) & g(0) & \\ & \cdot & \cdot & \cdot & \cdot \end{bmatrix}\begin{bmatrix} 1 & \\ 0 & \\ & 1 \\ & 0 \\ & & \cdot \end{bmatrix} = \begin{bmatrix} g(0) & & \\ g(1) & & \\ g(2) & g(0) & \\ g(3) & g(1) & \\ & \cdot & \cdot \end{bmatrix} \tag{4.49}$$

Now comes the polyphase idea. Each row is either all even or all odd. The even rows (*not columns as for* $\boldsymbol{C}$) combine into $\boldsymbol{G}_{\text{even}}$. The odd rows combine into $\boldsymbol{G}_{\text{odd}}$. Each of those is a constant-diagonal matrix (a time-invariant filter!). When they act separately and simultaneously, the process is more efficient:

$$(\uparrow 2)\begin{bmatrix} \boldsymbol{G}_{\text{even}} \\ \boldsymbol{G}_{\text{odd}} \end{bmatrix} = \begin{bmatrix} g(0) & & \\ g(2) & g(0) & \\ & \cdot & \cdot \\ g(1) & & \\ g(3) & g(1) & \\ & \cdot & \cdot \end{bmatrix}$$

In this polyphase form, upsampling comes *after* the two filters. The odd phase is delayed before the even and odd outputs are combined:

$$G(\uparrow 2) \text{ becomes } [1 \quad \text{delay}](\uparrow 2)\begin{bmatrix} G_{\text{even}} \\ G_{\text{odd}} \end{bmatrix}. \tag{4.50}$$

This is the polyphase form in the time domain. Its transform is the polyphase form in the z-domain. We are using the second Noble Identity *on each phase separately*:

$$\textbf{direct} \quad (\uparrow 2)G_{\text{even}}(z^2) \quad = \quad G_{\text{even}}(z)(\uparrow 2) \quad \textbf{polyphase}$$

Synthesis Bank: Direct and Polyphase

Now put *two filters* into a synthesis filter bank. The filters will be G_0 and G_1. Remember that synthesis has two input vectors (at half rate) and one output vector (at full rate). The direct form and the polyphase form are (with our present conventions) as follows:

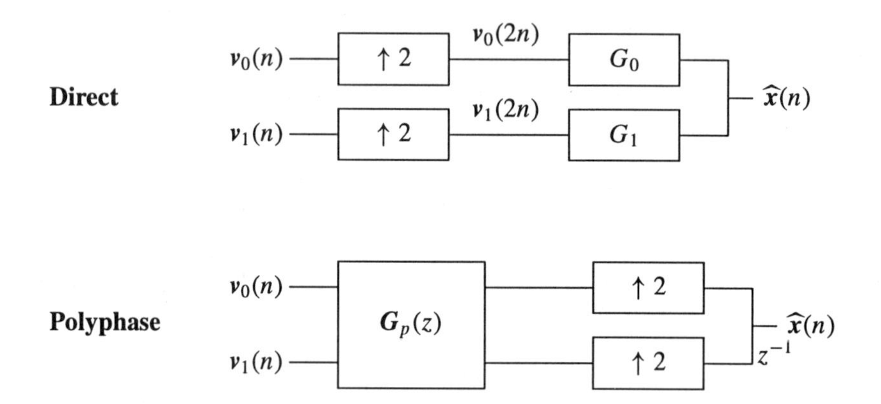

This is a 2-channel synthesis bank. In the z-domain, the *polyphase matrix* $\boldsymbol{G}_p(z)$ multiplies the inputs $V_0(z)$ and $V_1(z)$. The indices 0 and 1 are now channel numbers. The extension of equation (4.41) to two signals coming through two filters G_0 and G_1 will produce $\boldsymbol{G}_p(z)$:

$$\widehat{X}(z) = [1 \quad z^{-1}]\begin{bmatrix} G_{0,\,\text{even}}(z^2) & G_{1,\,\text{even}}(z^2) \\ G_{0,\,\text{odd}}(z^2) & G_{1,\,\text{odd}}(z^2) \end{bmatrix}\begin{bmatrix} V_0(z^2) \\ V_1(z^2) \end{bmatrix} \tag{4.51}$$

Notice! In the analysis polyphase matrix $\boldsymbol{H}_p(z)$, the lower left entry was H_{10}. Now G_{01} is at the lower left. Where we had the even part of $\boldsymbol{H}_1$, we have the odd part of $\boldsymbol{G}_0$. The channel index always comes before the phase index, and this forces a transpose in the polyphase synthesis matrix. Other authors agree with this convention—a necessary evil.

In the M by M case, H_{ij} is the jth polyphase component of the filter in the ith channel. So is G_{ij}. But the entry in row i, column j of the synthesis matrix is G_{ji}.

Useful convention for synthesis: Type 2 polyphase. Vaidyanathan delays channel 0 instead of channel 1, at the end of the synthesis bank. Therefore he reverses the two polyphase components. This produces the Type 2 polyphase decomposition, where 1 *means even and* 0 *means odd*. We write F rather than G to keep this type separate:

$$\textbf{Type 2 is} \quad F(z) = F_1(z^2) + z^{-1}F_0(z^2). \tag{4.52}$$

In the Type 2 polyphase matrix for two filters, the delay z^{-1} goes with 0:

$$\widehat{X}(z) = [z^{-1} \quad 1] \begin{bmatrix} F_{0,0}(z^2) & F_{1,0}(z^2) \\ F_{0,1}(z^2) & F_{1,1}(z^2) \end{bmatrix} \begin{bmatrix} V_0(z^2) \\ V_1(z^2) \end{bmatrix}. \tag{4.53}$$

We just reversed $[1 \quad z^{-1}]$ into $[z^{-1} \quad 1]$, and *reversed the rows of the matrix*. The product is exactly the same. The only difference in the block form is the new position of the delay, now in the upper channel:

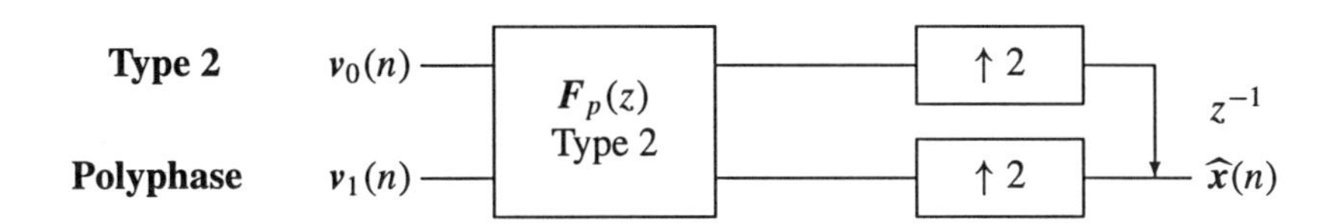

We will follow this arrangement: Type 2 for synthesis and Type 1 for analysis. The main point is that upsampling follows filtering. The phases of the filters are shorter than the filters themselves. Those subfilters operate simultaneously to produce separate phases of the outputs. Then upsampling with delay assembles $\widehat{x}$. We draw an M-channel synthesis bank in both Type 1 and Type 2 polyphase form:

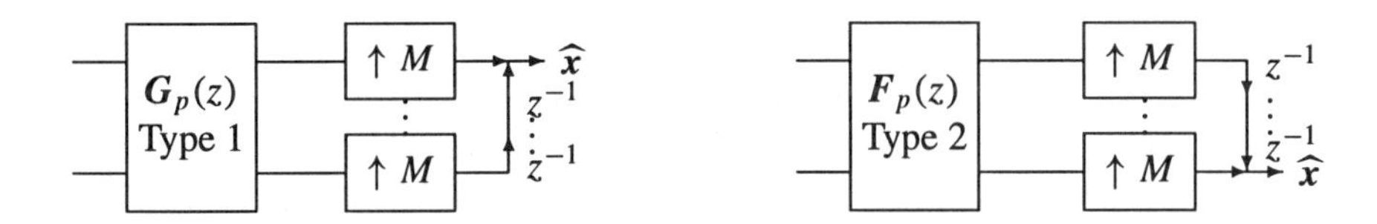

The M polyphase components of a single filter $\boldsymbol{H}$ are

$$\textbf{Type 1:} \quad H(z) = G_0(z^M) + z^{-1}G_1(z^M) + \cdots + z^{-(M-1)}G_{M-1}(z^M)$$

$$\textbf{Type 2:} \quad H(z) = F_{M-1}(z^M) + z^{-1}F_{M-2}(z^M) + \cdots + z^{-(M-1)}F_0(z^M)$$

When the bank has M filters, each with M phases, we have an M by M matrix. This is the polyphase matrix containing M^2 time-invariant filters.

Perfect Reconstruction

One reason for introducing the polyphase form is efficiency. The analysis bank and synthesis bank are faster, when $(\downarrow 2)$ and $(\uparrow 2)$ are moved outside. The even and odd subfilters that appear inside are about 50% shorter. So they can be executed more quickly.

The other reason for polyphase is to simplify the theory. Underlying the whole book is the goal of perfect reconstruction. We have to *connect* the synthesis bank to the analysis bank. One must be the inverse of the other, if the signal $\widehat{\boldsymbol{x}}$ is to agree with the input $\boldsymbol{x}$. When pure filters connect to pure filters, the products are pure filters — and the z-transforms tell all.

Since the analysis bank is represented by a matrix $\boldsymbol{H}_p(z)$, and the synthesis bank is represented by $\boldsymbol{F}_p(z)$, we hope very much that *inverse filter banks* are associated with *inverse matrices*:

Perfect reconstruction should mean that $\boldsymbol{F}_p(z) = \boldsymbol{H}_p^{-1}(z)$.

We note that delays are allowed and expected. The output signal may be $\widehat{x}(n) = x(n-l)$. Then the system delay is l. In the z-domain this is $\boldsymbol{F}_p\boldsymbol{H}_p = z^{-(\ell-1)/2}\boldsymbol{I}$.

The direct connection of analysis bank to synthesis bank has the decimators and expanders inside. This produces the standard QMF bank:

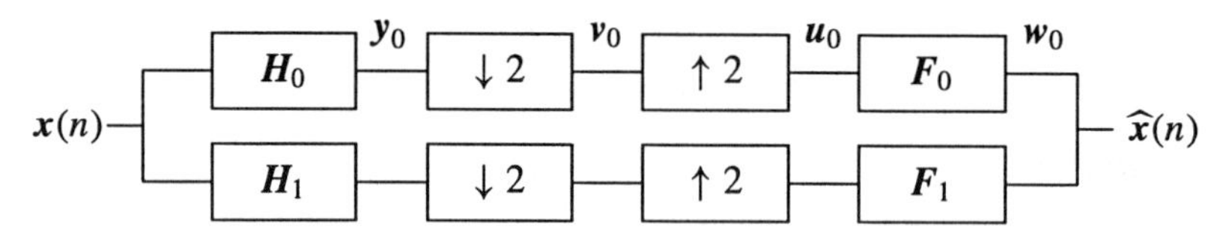

But the standard order is inefficient. The polyphase order is much better, with ($\downarrow$ 2) and ($\uparrow$ 2) moved to the outside. We draw the polyphase form with Type 2 leading to $\boldsymbol{F}_p(z)$. The final delay is moved from channel 2 in the Type 1 form (row 2 of $\boldsymbol{G}_p$) to channel 1 in the Type 2 form (row 1 of $\boldsymbol{F}_p$):

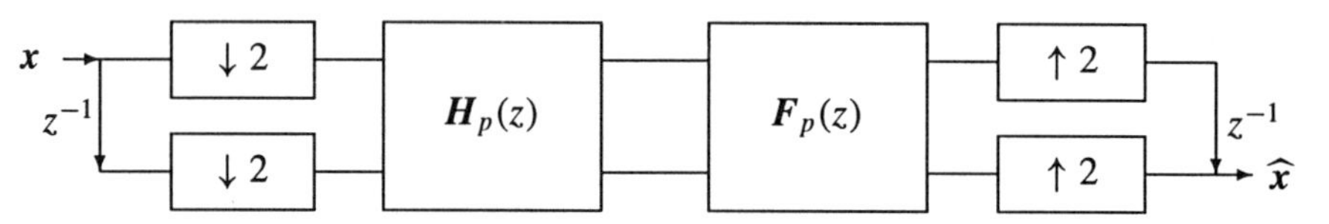

Polyphase is simpler and better because $\boldsymbol{H}_p$ and $\boldsymbol{F}_p$ are now side by side. Those are matrices coming from pure time-invariant filters — the even and odd phases of two analysis filters and two synthesis filters. In between would come compression or transmission. With nothing happening between, $\boldsymbol{F}_p\boldsymbol{H}_p$ disappears into $\boldsymbol{I}$.

Example 4.8. The simplest analysis bank has $H_0(z) = 1$ and $H_1(z) = z^{-1}$. Their polyphase components are $H_{00} = 1$ and $H_{01} = 0$, then $H_{10} = 0$ and $H_{11} = 1$. The polyphase matrix is $\boldsymbol{H}_p = \boldsymbol{I}$.

The corresponding synthesis bank has $G_0(z) = z^{-1}$ and $G_1(z) = 1$ in the two channels. The Type 1 polyphase components of z^{-1} are $G_{00} = 0$ and $G_{01} = 1$, in the first column (not row!) of $\boldsymbol{G}_p$. Then $G_{10} = 1$ and $G_{11} = 0$ go into the second column. These enter the synthesis matrix $\boldsymbol{G}_p$, and in this example the transposing has no effect:

$$\boldsymbol{G}_p = \begin{bmatrix} G_{00} & G_{10} \\ G_{01} & G_{11} \end{bmatrix} = \begin{bmatrix} 0 & 1 \\ 1 & 0 \end{bmatrix}. \tag{4.54}$$

When we account for the delays in the lower channel, the final output is $\widehat{x}(n) = x(n-1)$. The QMF bank is a simple delay chain, with perfect reconstruction.

Now use Type 2 for synthesis. This reverses the rows of $\boldsymbol{G}_p$ to give $\boldsymbol{F}_p$. Thus $\boldsymbol{F}_p = \boldsymbol{I}$. Starting from the two synthesis filters, z^{-1} has Type 2 components $F_{00} = 1$ and $F_{01} = 0$. The second channel yields $F_{10} = 0$ and $F_{11} = 1$ in the second column (not row!) of $\boldsymbol{F}_p$:

$$\boldsymbol{F}_p = \begin{bmatrix} F_{00} & F_{10} \\ F_{01} & F_{11} \end{bmatrix} = \begin{bmatrix} 1 & 0 \\ 0 & 1 \end{bmatrix}. \tag{4.55}$$

You see the advantage of the second form, which is the Type 2 *form* $\boldsymbol{F}_p(z)$. The perfect reconstruction from this delay chain is reflected in

$$\boxed{\boldsymbol{F}_p(z)\,\boldsymbol{H}_p(z) = \boldsymbol{I}.} \tag{4.56}$$

That is the important equation in this section. The example itself is a terrible QMF bank. It has delays but no genuine filters. However the conclusion remains correct when $\boldsymbol{H}_p(z)$ and $\boldsymbol{F}_p(z)$ are polynomials in z^{-1}, from FIR filters. That is the whole point—that useful matrices can be inverses of each other and *both can be polynomials.*

The direct and polyphase forms of a QMF bank are externally equivalent. The observer of $\boldsymbol{x}$ and $\widehat{\boldsymbol{x}}$ does not notice a difference. But the efficiency is improved and the theory is simplified. The theory of perfect reconstruction is a perfect matrix equation:

Theorem 4.5 *QMF banks give perfect reconstruction when $\boldsymbol{F}_p$ and $\boldsymbol{H}_p$ are inverses:*

$$\boldsymbol{F}_p(z)\,\boldsymbol{H}_p(z) = \boldsymbol{I} \quad or \quad z^{-L}\boldsymbol{I}.$$

$\boldsymbol{H}_p(z)$ is Type 1, for analysis, and $\boldsymbol{F}_p(z)$ is Type 2 transposed, for synthesis.

Example 4.9. In the average-difference filter bank all filters have *two taps*. Their even and odd phases have only one tap. Therefore the polyphase matrices are constant. But they are not diagonal, as they were in the simple delay chain. The analysis polyphase matrix is

$$\boldsymbol{H}_p(z) = \tfrac{1}{2}\begin{bmatrix} 1 & 1 \\ 1 & -1 \end{bmatrix}.$$

The synthesis matrices are transposed as always. $\boldsymbol{F}_p$ has "even" in the lower row:

$$\boldsymbol{G}_p(z) = \begin{bmatrix} 1 & -1 \\ 1 & 1 \end{bmatrix} \quad \text{and} \quad \boldsymbol{F}_p(z) = \begin{bmatrix} 1 & 1 \\ 1 & -1 \end{bmatrix} \quad \text{and} \quad \boldsymbol{F}_p\boldsymbol{H}_p = \boldsymbol{I}.$$

Problem Set 4.4

1. Express $X(z) = X_0(z^2) + z^{-1}X_1(z^2)$ as a sum $\boldsymbol{x} = (\uparrow 2)\boldsymbol{x}_0 + (\underline{\quad})$ in the time domain. What are the coefficients $\boldsymbol{x}_0(n)$ and $\boldsymbol{x}_1(n)$ in the polyphase decomposition (Type 1) in terms of the original $\boldsymbol{x}(n)$?

The next three exercises are the synthesis bank equivalents of the time-domain analysis matrices H_d, H_p, H_b in Section 4.3.

2. With filter coefficients $\boldsymbol{c}(0), \ldots, \boldsymbol{c}(3)$ in $\boldsymbol{G}_0$ and $\boldsymbol{d}(0), \ldots, \boldsymbol{d}(3)$ in $\boldsymbol{G}_1$, write down the infinite time-domain matrix $\boldsymbol{G}_d$ for the synthesis bank. The analysis bank had the lowpass $(\downarrow 2)\boldsymbol{C}$ above the highpass $(\downarrow 2)\boldsymbol{D}$. The synthesis bank will have $\boldsymbol{G}_0(\uparrow 2)$ and $\boldsymbol{G}_1(\uparrow 2)$ side by side. Each of these parts looks like equation (4.49).

3. Rearrange the rows of the direct matrix $\boldsymbol{G}_d$ in the previous exercise to give the polyphase matrix $\boldsymbol{G}_p$ in the time domain. This should be a 2 by 2 matrix with time-invariant filters $\boldsymbol{C}_{\text{even}}$, $\boldsymbol{C}_{\text{odd}}$, $\boldsymbol{D}_{\text{even}}$, $\boldsymbol{D}_{\text{odd}}$ as the blocks ($\boldsymbol{C} = \boldsymbol{G}_0$ and $\boldsymbol{D} = \boldsymbol{G}_1$).

4. Rearrange the columns of the direct matrix $\boldsymbol{G}_d$ to produce an infinite matrix $\boldsymbol{G}_b$ of 2 by 2 blocks.

5. Draw a delay chain with M channels. Then $H_0 = 1, H_1 = z^{-1}, \ldots, H_{M-1} = z^{-(M-1)}$ leads to what matrix $\boldsymbol{H}_p(z)$? Choose the synthesis delays to reconstruct $\widehat{\boldsymbol{x}}(n) = \boldsymbol{x}(n-(M-1))$. Create the Type 1 matrix $\boldsymbol{G}_p(z)$ and the Type 2 matrix $\boldsymbol{F}_p(z)$. Check $\boldsymbol{F}_p\boldsymbol{H}_p = \boldsymbol{I}$.

6. Establish the relation between $\boldsymbol{G}_m(z)$ (modulation) and $\boldsymbol{G}_p(z)$ (polyphase Type 1).

7. If $\boldsymbol{G}_m(z)\boldsymbol{H}_m(z) = \begin{bmatrix} 2z^{-l} & 0 \\ 0 & -2z^{-l} \end{bmatrix}$ show that $\boldsymbol{G}_p(z)\boldsymbol{H}_p(z) = \begin{bmatrix} 0 & z^{-L} \\ z^{-L} & 0 \end{bmatrix}$, with $l = 2L+1$. The matrices $\boldsymbol{G}_m$ and $\boldsymbol{G}_p$ include a transpose. Then $\boldsymbol{F}_p$ includes a row exchange for Type 2, and $\boldsymbol{F}_p(z)\boldsymbol{H}_p(z) = z^{-L}\boldsymbol{I}$.

8. Find $F_0(z)$ and $F_1(z)$ for synthesis from the analysis polyphase matrix:

a. $\boldsymbol{H}_p(z) = \begin{bmatrix} 2-4z^{-1} & 1-z^{-1} \\ 3+z^{-1}+2z^{-2} & 1 \end{bmatrix}$

b. $\boldsymbol{H}_p(z) = \begin{bmatrix} c_2 & s_2 \\ -s_2 & c_2 \end{bmatrix} \begin{bmatrix} z^{-1} & 0 \\ 0 & 1 \end{bmatrix} \begin{bmatrix} c_1 & s_1 \\ -s_1 & c_1 \end{bmatrix}$

c. $\boldsymbol{H}_p(z) = \begin{bmatrix} \frac{1+z^{-1}}{5-z^{-1}} & 1 \\ \frac{1}{2-z^{-1}} & z^{-3} \end{bmatrix}$

d. $\boldsymbol{H}_p(z) = \begin{bmatrix} \frac{1+2z^{-1}}{2+z^{-1}} & 0 \\ 0 & \frac{1-3z^{-1}}{3-z^{-1}} \end{bmatrix} \begin{bmatrix} 1 & 1 \\ 1 & -1 \end{bmatrix}$

9. Find the polyphase matrices $\boldsymbol{H}_p(z)$ and $\boldsymbol{F}_p(z)$. Is this system PR?

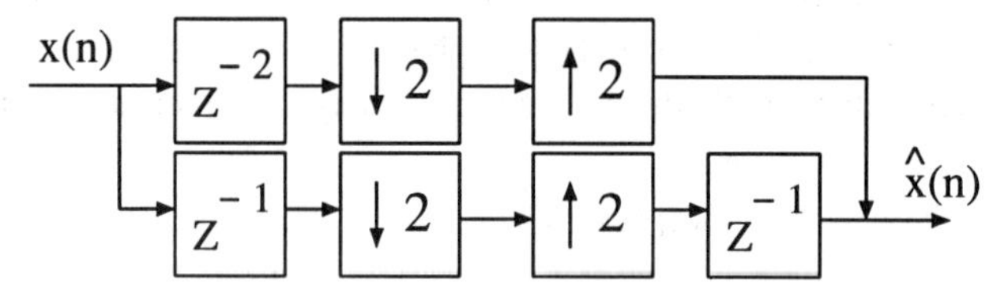

4.5 Lattice Structure

A filter bank is represented by its polyphase matrix. When we know the filters, we know the matrix $\boldsymbol{H}_p(z)$. In the opposite direction, we can choose a suitable $\boldsymbol{H}_p(z)$—often as a product of simple matrices. Then the corresponding filter bank is a "lattice" of simple filters, easy to implement.

Here is a class of polyphase matrices to use as examples. They are linear in z^{-1}. One factor has the coefficients $1, 1$ and $1, -1$ from the Haar filter bank—averages and differences. Another factor has coefficients a, b, c, d that we are free to choose—as long as the matrix is invertible. Between those factors is a diagonal matrix with a delay in the second channel:

$$\boldsymbol{H}_p(z) = \begin{bmatrix} 1 & 1 \\ 1 & -1 \end{bmatrix} \begin{bmatrix} 1 & \\ & z^{-1} \end{bmatrix} \begin{bmatrix} a & b \\ c & d \end{bmatrix} \tag{4.57}$$

$$\boldsymbol{H}_p^{-1}(z) = \frac{1}{2} \begin{bmatrix} a & b \\ c & d \end{bmatrix}^{-1} \begin{bmatrix} 1 & \\ & z \end{bmatrix} \begin{bmatrix} 1 & 1 \\ 1 & -1 \end{bmatrix}. \tag{4.58}$$

First point: $\boldsymbol{H}_p^{-1}$ contains an advance (denoted by z). No problem. Good causal polyphase matrices have anticausal inverses, as this one has. The main point is that $\boldsymbol{H}_p^{-1}$ is a polynomial (FIR not IIR). Add one delay to the whole synthesis bank (multiply by z^{-1}) and it becomes causal:

$$z^{-1}\boldsymbol{H}_p^{-1}(z) = \frac{1}{2}\begin{bmatrix} a & b \\ c & d \end{bmatrix}^{-1}\begin{bmatrix} z^{-1} & \\ & 1 \end{bmatrix}\begin{bmatrix} 1 & 1 \\ 1 & -1 \end{bmatrix}. \tag{4.59}$$

When you multiply out the factors in $\boldsymbol{H}_p(z)$, you see the possibilities:

$$\boldsymbol{H}_p(z) = \begin{bmatrix} a + cz^{-1} & b + dz^{-1} \\ a - cz^{-1} & b - dz^{-1} \end{bmatrix}. \tag{4.60}$$

The top row holds the even and odd phases of the lowpass filter $\boldsymbol{C}$:

$$C(z) = a + bz^{-1} + cz^{-2} + dz^{-3} \quad \text{has phases} \quad a + cz^{-1} \quad \text{and} \quad b + dz^{-1}.$$

The second row is the polyphase form of the highpass filter $\boldsymbol{D}$:

$$D(z) = a + bz^{-1} - cz^{-2} - dz^{-3} \quad \text{has phases} \quad a - cz^{-1} \quad \text{and} \quad b - dz^{-1}.$$

We design these 4-tap filters by choosing a, b, c, d. They can have *linear phase or orthogonality*. Our 2 by 2 matrix can be symmetric or it can be orthogonal:

Linear phase filters (symmetric-antisymmetric). Choose $a = d$ and $b = c$.

The lowpass filter $\boldsymbol{C}$ has coefficients a, b, b, a. The highpass filter $\boldsymbol{D}$ is antisymmetric, with coefficients $a, b, -b, -a$. For the polyphase matrix, just substitute $d = a$ and $c = b$:

$$\textit{Linear phase} \qquad \boldsymbol{H}_p(z) = \begin{bmatrix} a + bz^{-1} & b + az^{-1} \\ a - bz^{-1} & b - az^{-1} \end{bmatrix}. \tag{4.61}$$

How do you recognize the symmetry of $\boldsymbol{C}$ (or $\boldsymbol{H}_0$) from its phases in the top row? One phase is the *flip* of the other phase. Then the whole filter a, b, b, a is the flip of itself, which makes it symmetric.

The synthesis filters are also linear phase. The underlying reason is that the inverse of a symmetric matrix is symmetric. Now change to orthogonal.

Orthogonal filter bank Choose $d = a$ and $c = -b$ and normalize.

The second row $[c\ d] = [-b\ a]$ becomes orthogonal to the first row $[a\ \ b]$. Both rows and both columns have length $\sqrt{a^2 + b^2}$. Each of the three factors in $\boldsymbol{H}_p(z)$ is a unitary matrix, after dividing by that length. The inverses come directly from the conjugate transposes:

$$\begin{bmatrix} a & b \\ -b & a \end{bmatrix}^{-1} = \frac{1}{a^2 + b^2}\begin{bmatrix} a & -b \\ b & a \end{bmatrix}$$

$$\begin{bmatrix} 1 & \\ & z^{-1} \end{bmatrix}^{-1} = \begin{bmatrix} 1 & \\ & \overline{z^{-1}} \end{bmatrix}^T \qquad \text{if } |z| = 1.$$

We use the word *unitary* rather than *orthogonal* because z is complex. The inverse of a unitary matrix $\boldsymbol{U}$ is $\overline{\boldsymbol{U}}^T$. The first matrix is real, so conjugating had no effect. The second is diagonal, so transposing had no effect. We always do both.

Notice that $|z| = 1$. The inverse of $z = e^{-i\omega}$ equals the conjugate: $z^{-1} = e^{i\omega} = \overline{z}$. Section 5.1 will introduce the word *paraunitary* for $\boldsymbol{H}_p(z)$, when it is a unitary matrix on the circle $|z| = 1$. Our present example is paraunitary.

Multiplying the three matrix factors gives $\boldsymbol{H}_p(z)$, with $d = a$ and $c = -b$:

$$\textit{Paraunitary polyphase matrix} \quad \boldsymbol{H}_p(z) = \frac{1}{2}\begin{bmatrix} a - bz^{-1} & b + az^{-1} \\ a + bz^{-1} & b - az^{-1} \end{bmatrix}. \tag{4.62}$$

The impulse response shows the difference between linear phase (earlier) and orthogonal (now). The lowpass $C(z)$ comes from the first row, where a and $-b$ give the even coefficients. The coefficients in $b + az^{-1}$ go with the odd powers z^{-1} and z^{-3}:

$$C(z) = a + bz^{-1} - bz^{-2} + az^{-3}.$$

The highpass response from the second row is

$$D(z) = a + bz^{-1} + bz^{-2} - az^{-3}.$$

We lost symmetry and gained orthogonality. One part of orthogonality is that $(a, b, -b, a)$ is perpendicular to $(a, b, b, -a)$. But this is not all. Another part is that $(a, b, -b, a, 0, 0)$ is perpendicular to $(0, 0, a, b, b, -a)$. We have to consider those *double shifts*, because they enter the time-domain matrix—two filters $\boldsymbol{C}$ and $\boldsymbol{D}$ with downsampling:

$$\boldsymbol{H} = \begin{bmatrix} (\downarrow 2)\boldsymbol{C} \\ (\downarrow 2)\boldsymbol{D} \end{bmatrix} = \frac{1}{2}\begin{bmatrix} a & -b & b & a & & \\ & & a & -b & b & a \\ & & & \cdot & \cdot & \cdot \\ -a & b & b & a & & \\ & & -a & b & b & a \\ & & & \cdot & \cdot & \cdot \end{bmatrix}. \tag{4.63}$$

Linear phase and also orthogonal: ***Not possible for length greater than*** 2.

Later we prove this fact in general. Linear phase FIR pairs that are also orthogonal can have at most two nonzero coefficients in each filter. They are just variations on the average-difference pair. This is a one-step improvement over single filters, which can have only *one* nonzero coefficient. An allpass FIR filter is a delay, with one term $C(z) = cz^{-k}$.

Lattice Structures

The orthogonal bank will now be more general than this example. It will have more coefficients. But $\boldsymbol{H}_p(z)$ can still be produced from simple factors. They will be *cascades* of constant matrices and diagonal matrices. The filter bank has a highly efficient *lattice structure*. We can see already the form of our a, b, c, d factor in the two important cases of linear phase and orthogonality:

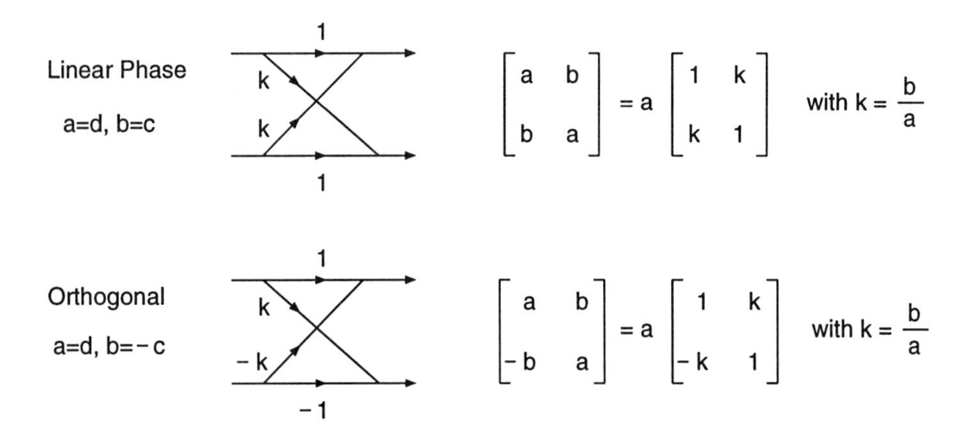

Each lattice involves only *one multiplication* when implemented properly (Problem 7). There is also a single overall multiplying factor, collected from all factors in the cascade. The numbers k can be design parameters. For any k's, a cascade of linear phase filters is linear phase. A cascade of orthogonal filters (including $-k$) is orthogonal. We now study a general lattice built from simple orthogonal factors.

Note. The lattice structure gives long orthogonal filters very easily. They can be good filters. But the k's enter $\boldsymbol{H}_p(z)$ in a complicated nonlinear way. For the ***design*** step, where filter characteristics are optimized, most engineers choose simpler parameters like the coefficients $\boldsymbol{h}(\mathrm{n})$.

For the ***implementation*** step, the lattice has an important advantage. Often the coefficients $\boldsymbol{h}(n)$ will be "quantized"—real numbers are replaced by binary numbers (finitely many digits). In general, this destroys orthogonality. But orthogonality is not lost when implemented as a lattice of simple filters. Each orthogonal factor is determined by a real number k (or by an angle θ). When k or θ is quantized, the factor is not exactly correct—*but it is still orthogonal.* Therefore the whole filter bank remains strictly orthogonal.

We now introduce a product of ***rotation matrices*** and ***delay matrices***. The rotation matrices are constant, exactly as in our examples:

$$\boldsymbol{R} = \begin{bmatrix} \cos\theta & \sin\theta \\ -\sin\theta & \cos\theta \end{bmatrix} = \cos\theta \begin{bmatrix} 1 & k \\ -k & 1 \end{bmatrix}. \tag{4.64}$$

This matrix gives clockwise rotation through the angle θ. When $\cos\theta$ is factored out, it leaves the number $k = \tan\theta$. We will think of the rotation matrix (orthogonal matrix) in terms of θ, but we implement it with k to save multiplications. The delay matrices are used to delay the second channel:

$$\Lambda(z) = \begin{bmatrix} 1 & 0 \\ 0 & z^{-1} \end{bmatrix} \quad \text{has determinant } z^{-1}.$$

It is convenient to have an extra factor of -1 in the lower channel. The matrix $\operatorname{diag}(1, -1)$, which is also $\Lambda(-1)$, accomplishes this. Then the product of $\ell+1$ rotations separated by ℓ delays is the polyphase matrix for the whole lattice:

$$\boldsymbol{H}_p(z) = \Lambda(-1)\,\boldsymbol{R}_\ell\,\Lambda(z)\,\boldsymbol{R}_{\ell-1}\,\Lambda(z)\cdots\boldsymbol{R}_1\,\Lambda(z)\,\boldsymbol{R}_0. \tag{4.65}$$

The rotation R_ℓ is through an angle θ_ℓ. The rotation R_0 is through an angle θ_0. The determinant is $-z^{-\ell}$, because of the delays and the sign change from diag$(1,-1)$. Without the delays in between, the product would be a single rotation through the total angle $\sum \theta_i$. With the ℓ delays, we have something much more important. It is essentially the most general two-channel orthonormal filter bank with l delays.

$H_p(z)$ is the polyphase matrix of an orthonormal filter bank, because each factor is unitary. We show this analysis bank in its efficient form, with the downsampling operators before the filters. R_0 comes first in the structure because it is the right-hand factor in $H_p(z)$.

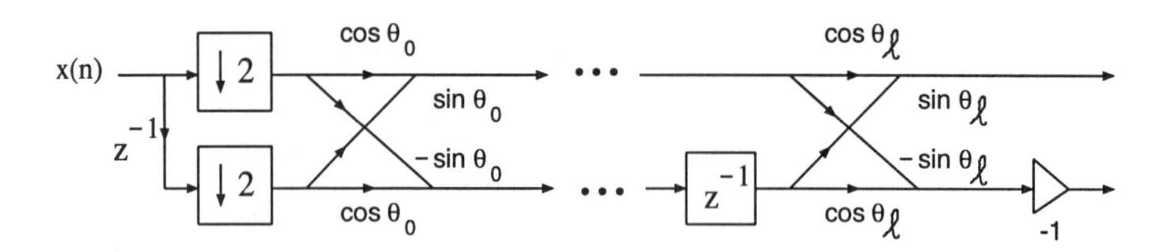

Figure 4.4: The lattice structure for $H_p(z) = \Lambda(-1)\, R_\ell\, \Lambda(z) \cdots R_1\, \Lambda(z)\, R_0$.

The factors $\cos\theta$ that are removed give a multiplication by $\cos\theta_\ell \cdots \cos\theta_0$ at the end. Note that Haar's filter bank comes from ***one*** rotation with angle $\frac{\pi}{4}$:

$$\frac{1}{\sqrt{2}}\begin{bmatrix} 1 & 1 \\ 1 & -1 \end{bmatrix} = \begin{bmatrix} \cos\theta & \sin\theta \\ \sin\theta & -\cos\theta \end{bmatrix} \quad \text{with} \quad \theta = \frac{\pi}{4}.$$

Actually this is a picture of any polyphase matrix $H_p(z)$ at the particular value $z = 1$. The matrix $H_p(1)$ corresponds to zero frequency, because $e^0 = 1$. *The response to direct current is always the Haar matrix, when the lowpass filter has a zero at* $\omega = \pi$.

Theorem 4.6 *If* $H_0 = 0$ *at* $\omega = \pi$ *(which is* $z = -1$*), then the polyphase matrix has*

$$H_p(1) = \frac{1}{\sqrt{2}}\begin{bmatrix} 1 & 1 \\ 1 & -1 \end{bmatrix}.$$

For an orthonormal filter bank the angles in the lattice structure add to $\frac{\pi}{4}$.

Proof: H_0 has a zero at $\omega = \pi$ when the sum of odd-numbered coefficients equals the sum of even-numbered coefficients: $h(0) - h(1) + h(2) - \ldots = 0$. This *first sum rule* means that at $z = 1$, the even phase equals the odd phase:

$$H_{00}(1) = h(0) + h(2) + \cdots = h(1) + h(3) + \cdots = H_{01}(1). \tag{4.66}$$

Row 1 of the polyphase matrix has equal entries at $z = 1$. Row 2 comes from the highpass filter H_1. This has the opposite property $H_1 = 0$ at $\omega = 0$. The odd sum is *minus* the even sum. With the right signs, $H_p(1)$ is the Haar matrix in the Theorem.

Since this is exactly what we get in the product of rotations, when $z = 1$ and the delay matrices become I, the total rotation angle must be $\sum \theta_i = \frac{\pi}{4}$.

The 4-tap Daubechies filter in Section 5.2 has angles $\frac{\pi}{3}$ and $-\frac{\pi}{12}$. Those angles add to $\frac{\pi}{4}$. Looking back at the start of this section, we realize that Haar followed by one rotation is actually useless. That second rotation angle must be zero! The sum rule in the top row of the lowpass filter becomes $a + b = -b + a$, which forces $b = 0$. The good angles are $\frac{\pi}{3}$ and $-\frac{\pi}{12}$, not $\frac{\pi}{4}$ and 0.

The Synthesis Lattice

The synthesis bank inverts the analysis bank. To invert the lattice, reverse the order of rotations. The inverse $\Lambda(-1)\,\boldsymbol{R}_\ell\,\Lambda(z)\,\ldots\,\boldsymbol{R}_1\,\Lambda(z)\,\boldsymbol{R}_0$ is the product of inverses in reverse order:

$$\left(\boldsymbol{H}_p(z)\right)^{-1} = \boldsymbol{R}_0^T\Lambda^{-1}(z)\boldsymbol{R}_1^T \cdots \Lambda^{-1}(z)\boldsymbol{R}_\ell^T\Lambda(-1). \tag{4.67}$$

The inverse of each rotation is the transpose: $\boldsymbol{R}^{-1} = \boldsymbol{R}^T$ = rotation through $-\theta$. These constant matrices are orthogonal. The inverse of a delay matrix is an *advance*:

$$\Lambda^{-1}(z) = \begin{bmatrix} 1 & 0 \\ 0 & z \end{bmatrix} \quad \text{and} \quad z^{-1}\Lambda^{-1}(z) = \begin{bmatrix} z^{-1} & 0 \\ 0 & 1 \end{bmatrix}$$

Since $\boldsymbol{H}_p(z)$ has ℓ delays, its inverse has ℓ advances. Multiplying each advance by z^{-1} puts l delays into the upper channel. The synthesis half has *upsamplers last* (this is the polyphase form in Figure 4.5). An extra advantage is that we can add or remove rotations without losing

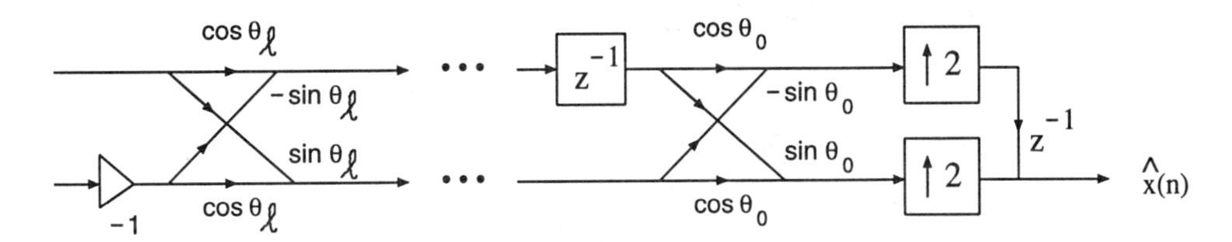

Figure 4.5: Orthogonal synthesis bank in lattice form.

orthogonality. In contrast, the alteration of one filter coefficient $\boldsymbol{h}(n)$ is almost certain to destroy orthogonality and also perfect reconstruction.

What is the synthesis polyphase matrix?

Lattice Coefficients from Filter Coefficients

We now prove that any 2-channel orthonormal filter bank can be expressed in lattice form, with rotations and delays. Thus the lattice structure is "complete". Starting with $\boldsymbol{H}_p(z)$ of degree ℓ, we find a rotation-delay that reduces the degree to $\ell - 1$:

$$\boldsymbol{H}_p(z) = \begin{bmatrix} \cos\theta_\ell & -\sin\theta_\ell \\ \sin\theta_\ell & \cos\theta_\ell \end{bmatrix}\begin{bmatrix} 1 & 0 \\ 0 & z^{-1} \end{bmatrix}\boldsymbol{H}_p^{(\ell-1)}(z). \tag{4.68}$$

The new matrix $\boldsymbol{H}_p^{(\ell-1)}(z)$ is still unitary for $|z| = 1$ and its determinant is $\pm z^{-(\ell-1)}$. So we reduce the degree again. After ℓ steps we reach $\boldsymbol{H}_p(0)$ whose determinant is ± 1. It is unitary for $|z| = 1$ and the only matrices with this property are constants. After properly accounting for the matrix $\Lambda(-1) = \text{diag}(1, -1)$, we have the rotation angle θ_0 that completes the lattice.

Theorem 4.7 *Every lowpass-highpass orthonormal filter bank has a polyphase matrix* $\boldsymbol{H}_p(z)$ *of the lattice form* (4.65).

Proof.. The step (4.68) requires an angle θ_ℓ such that

$$\begin{bmatrix} \cos\theta_\ell & \sin\theta_\ell \\ -\sin\theta_\ell & \cos\theta_\ell \end{bmatrix} \boldsymbol{H}_p(z) = \begin{bmatrix} 1 & 0 \\ 0 & z^{-1} \end{bmatrix} \boldsymbol{H}_p^{(\ell-1)}(z). \tag{4.69}$$

On both sides, the second row must have ***no constant terms***. Then we can factor z^{-1} from that row and the reduction succeeds—we get the next matrix $\boldsymbol{H}_p^{(\ell-1)}(z)$ and continue. If $\boldsymbol{H}_p(z) = \boldsymbol{h}_p(0) + \cdots + \boldsymbol{h}_p(d)z^{-d}$, the constant term on the left side of the equation comes from $\boldsymbol{h}_p(0)$. The second row on that side must give

$$\begin{bmatrix} -\sin\theta_\ell & \cos\theta_\ell \end{bmatrix} \boldsymbol{h}_p(0) = \begin{bmatrix} 0 & 0 \end{bmatrix}. \tag{4.70}$$

This row vector that produces zero must exist if $\boldsymbol{h}_p(0)$ is singular. The whole argument rests on the fact that $\boldsymbol{h}_p^T(d)\boldsymbol{h}_p(0)$ is the ***zero matrix***. The key equation is $\overline{\boldsymbol{H}}_p^T \boldsymbol{H}_p = \boldsymbol{I}$:

$$\left[\boldsymbol{h}_p^T(0) + \cdots + \boldsymbol{h}_p^T(d)z^d\right]\left[\boldsymbol{h}_p(0) + \cdots + \boldsymbol{h}_p(d)z^{-d}\right] = \boldsymbol{I}. \tag{4.71}$$

The coefficient $\boldsymbol{h}_p^T(d)\,\boldsymbol{h}_p(0)$ of the highest order term must be zero. When two nonzero matrices multiply to give zero, $\boldsymbol{h}_p(0)$ and $\boldsymbol{h}_p(d)$ are both singular. There exists a vector $[-\sin\theta_\ell \ \ \cos\theta_\ell]$ that knocks out the constant term in the second row of (4.69). After ℓ steps we reach a final rotation $\boldsymbol{R}_0$ and the lattice is complete.

Lattices for Linear Phase Filters

For orthonormal filters, $\boldsymbol{H}_0$ and $\boldsymbol{H}_1$ have the same (even) length. One is symmetric and the other is antisymmetric. For linear phase filters, those statements are not necessarily true. The great variety of linear phase PR filter banks means a great variety of lattice structures. The equal-length case copies the orthonormal case, but with factors $\boldsymbol{S} = \begin{bmatrix} a & b \\ b & a \end{bmatrix}$ instead of $\boldsymbol{R} = \begin{bmatrix} a & b \\ -b & a \end{bmatrix}$.

Theorem 4.8 *Every linear phase perfect reconstruction filter bank with equal (even) length filters has a lattice factorization*

$$\boldsymbol{H}_p(z) = \begin{bmatrix} 1 & 1 \\ 1 & -1 \end{bmatrix} \boldsymbol{S}_L \Lambda(z) \boldsymbol{S}_{L-1} \Lambda(z) \cdots \boldsymbol{S}_1 \Lambda(z) \boldsymbol{S}_0. \tag{4.72}$$

The filter bank is a cascade of simple two-tap filters:

$$\Lambda(z) = \begin{bmatrix} 1 & 0 \\ 0 & z^{-1} \end{bmatrix} \quad \textit{and} \quad \boldsymbol{S}_i = \begin{bmatrix} a_i & b_i \\ b_i & a_i \end{bmatrix} = a_i \begin{bmatrix} 1 & k_i \\ k_i & 1 \end{bmatrix}.$$

The proof is entirely parallel to the orthogonal case (Theorem 4.7). The implementation just has a sign change. The number of lattice sections is half the length of $\boldsymbol{H}_0$ (and $\boldsymbol{H}_1$). All factors a_i are collected into a single factor $a = \prod a_i$, for efficiency. And the two multiplications by k_i are further reduced to one multiplication (and three additions) in Problem 7.

What is not parallel is that linear phase allows further possibilities: unequal length and symmetric-symmetric pairs. Here are the rules for PR with linear phase:

Even length must differ by a multiple of 4: $\boldsymbol{H}_0 = \text{symm and } \boldsymbol{H}_1 = \text{anti.}$

Odd length must *not* differ by a multiple of 4: $\boldsymbol{H}_0 = \text{symm and } \boldsymbol{H}_1 = \text{symm.}$

An example of the second type is in the "Guide to the Book", with lengths 5 and 3: $\frac{1}{8}(-1, 2, 6, 2, -1)$ in $\boldsymbol{H}_0$ and $\frac{1}{4}(1, -2, 1)$. For symmetric filters whose lengths differ by 2, the (complete) lattice factorization is

$$\boldsymbol{H}_p(z) = \boldsymbol{D} \prod_{k=2}^{L} \mathbf{A}_k(z) \text{ with } \mathbf{A}_k(z) = \begin{bmatrix} 1+z^{-1} & 1 \\ 1+\beta_k z^{-1} + z^{-2} & 1+z^{-1} \end{bmatrix} \begin{bmatrix} \alpha_k & 0 \\ 0 & 1 \end{bmatrix} \tag{4.73}$$

$\boldsymbol{D}$ is diagonal, and a practical system can have only $\beta_2 \neq 0$. This speeds up the implementation. The FBI 9/7 system is an example that requires only two lattice sections.

The general case has a similar (but longer) factorization. The proof is also longer. Our purpose has been to establish the main point: the efficiency and the availability of the lattice structure.

Perfect Reconstruction with FIR

We are interested above all in FIR systems. Then both $\boldsymbol{H}_p$ and $\boldsymbol{F}_p$ have a finite number of terms, from the finite number of filter coefficients. The crucial observation is that with well-chosen matrices, *the product of two polynomials can be a constant or a monomial*:

$$\begin{aligned} \boldsymbol{H}_p^{-1}(z)\boldsymbol{H}_p(z) &= \boldsymbol{I} \\ \boldsymbol{F}_p(z)\,\boldsymbol{H}_p(z) &= z^{-L}\boldsymbol{I} \end{aligned} \qquad \text{are possible for matrix polynomials.}$$

In the scalar case, $1 + \beta z^{-1}$ is a polynomial but $1/(1 + \beta z^{-1})$ is not. The reciprocal of a polynomial scalar is not a polynomial. If an ordinary time-invariant filter is FIR, its inverse is IIR. The only scalar exceptions are trivial cases (delays). *The real exceptions are matrix polynomials — which are polyphase matrices of filter banks.* That is what this book is about.

The inverse of a matrix polynomial can be a matrix polynomial. We ask when.

Theorem 4.9 *An FIR analysis bank has an FIR synthesis bank that gives perfect reconstruction if and only if the determinant of $\boldsymbol{H}_p(z)$ is a monomial cz^{-l} with $c \neq 0$.*

Proof. The entries of $(\boldsymbol{H}_p(z))^{-1}$ are always cofactors divided by the determinant:

$$(\boldsymbol{H}_p^{-1})_{ij} = \frac{(j, i) \text{ cofactor of } \boldsymbol{H}_p(z)}{\text{determinant of } \boldsymbol{H}_p(z)}.$$

The cofactor (the determinant without row j and column i) is certainly a polynomial. If the determinant is cz^{-l} (one term only!) then the division leaves a polynomial: the inverse is FIR. When this inverse $\boldsymbol{H}_p^{-1}(z)$ is delayed by z^{-l}, it becomes causal. This is $\boldsymbol{F}_p(z)$. It is the Type 2 polyphase matrix of the FIR synthesis bank.

The design problem is to maintain a monomial determinant cz^{-l} while building an efficient filter bank. For an M by M matrix the cofactors of $\boldsymbol{H}_p(z)$ may have high degrees. The analysis and synthesis filters generally have different lengths. The design problem is harder, but more is possible. Cosine-modulated filter banks are winners.

Efficiency of Lattice Structure

Consider a symmetric filter bank with length $N = 2J$. The number of multiplications per unit time is $4J$ using the direct form and $2J$ using polyphase. Additions are the same, with two extra at the 2-pt DFT matrix.

On the other hand, there are J lattice sections. Each section requires 2 multiplications and 2 additions. Counting β_1 and β_2 in the figure brings the total complexity to $2J$ multiplications and $2J+2$ additions. The effective complexity (at the input rate) is J multiplications and $J+1$ additions.

Problem 7 discusses an alternative that uses 1 multiplier and 3 adders per lattice section. In summary, the lattice complexity is approximately half of the polyphase complexity. The same is true for orthogonal filters.

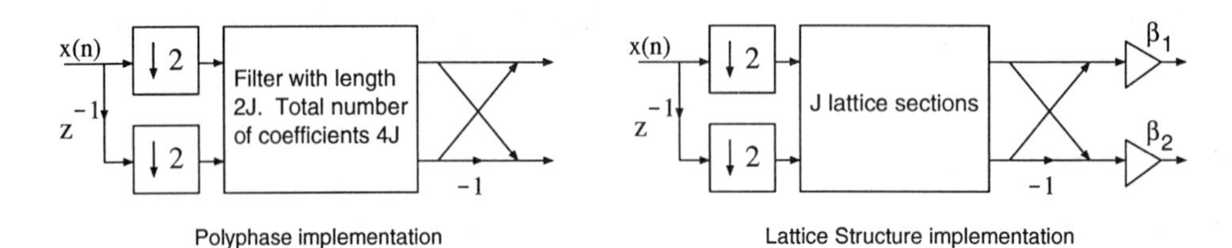

Polyphase implementation Lattice Structure implementation

Problem Set 4.5

1. Show that this matrix gives an orthogonal filter bank. Find a, b, c, d and $\boldsymbol{H}_p^{-1}(z)$:

$$\boldsymbol{H}_p(z) = \frac{1}{2}\begin{bmatrix} \cos\Theta + z^{-1}\sin\Theta & -\sin\Theta + z^{-1}\cos\Theta \\ \cos\Theta - z^{-1}\sin\Theta & -\sin\Theta - z^{-1}\cos\Theta \end{bmatrix}.$$

2. Suppose U and V are unitary matrices (constant but possibly complex). This means that $U^{-1} = \overline{U}^T$ and $V^{-1} = \overline{V}^T$. Show why their product UV is also unitary. *Notice the key point*: Inverses come in reverse order $V^{-1}U^{-1}$ and so do transposes.

3. For a 6-tap antisymmetric filter $a + bz^{-1} + cz^{-2} - cz^{-3} - bz^{-4} - az^{-5}$, show that the flip of one phase is minus the other phase. What if the number of taps is odd?

4. Redraw the lattice cascade with the downsampling operators moved to the right, after the butterfly filters. Be sure to change z^{-1} to z^{-2} (why?).

5. Find the product $\boldsymbol{H}_p(z)$ when the rotation angles are $\theta_0 = 0$, $\theta_1 = \frac{\pi}{2}$, and $\theta_2 = 0$. Check the determinant of $\boldsymbol{H}_p(z)$, remembering the $\ell = 2$ delays. This is an example in which $\boldsymbol{H}_p(z)$ contains no terms in $z^{-\ell}$, although the determinant has degree ℓ. In the notation of Theorem 4.7, $\ell = 2$ but $d = 1$.

   The correct definition of degree of $\boldsymbol{H}_p(z)$ is the *Smith-McMillan degree = minimum number of delays* to realize the system. For orthonormal filter banks, this equals the degree l of the determinant.

6. Find the polyphase matrix $\boldsymbol{H}_p(z)$ for the pair $H_0(z) = 1$ and $H_1(z) = \frac{1}{2} - z^{-1} + \frac{1}{2}z^{-2}$ in Problem Set 4.3. Find the synthesis polyphase matrix $\boldsymbol{F}_p(z)$ for $F_0(z) = \frac{1}{2} + z^{-1} + \frac{1}{2}z^{-2}$ and $F_1(z) = -1$. Remember to transpose and reverse rows for $\boldsymbol{F}_p(z)$, and compute $\boldsymbol{F}_p(z)\boldsymbol{H}_p(z)$.

7. We can multiply by $\boldsymbol{S} = \begin{bmatrix} a & b \\ b & a \end{bmatrix}$ with one multiplication by $c = \frac{a+b}{a-b}$ (and one by $\frac{a-b}{2}$ that can be collected with others at the end of the cascade):

$$\boldsymbol{S} = \frac{a-b}{2}\begin{bmatrix} 1 & 1 \\ 1 & -1 \end{bmatrix}\begin{bmatrix} c & \\ & 1 \end{bmatrix}\begin{bmatrix} 1 & 1 \\ 1 & -1 \end{bmatrix}.$$

Write $\boldsymbol{S}$ with 3 adds using the figure. Find a "one multiplication per rotation $\boldsymbol{R}$".

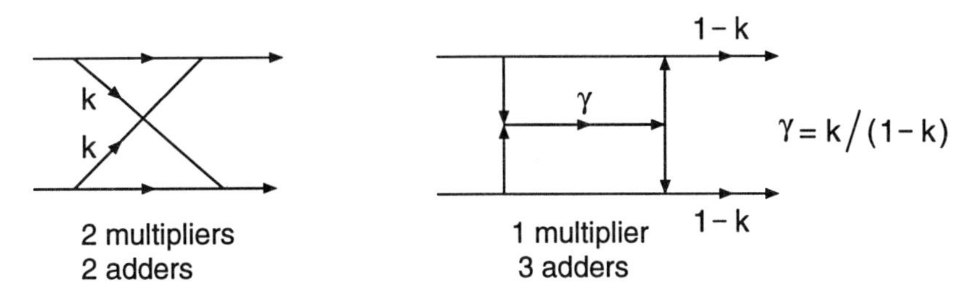

8. Find the polyphase matrix $\boldsymbol{H}_p(z)$ for the analysis pair $H_0(z) = 1$ and $H_1(z) = \frac{1}{16}(-1 + 9z^{-2} - 16z^{-3} + 9z^{-4} - z^{-6})$. Find $\boldsymbol{F}_p(z)$, transposed and row reversed, for the synthesis pair $F_0(z) = \frac{1}{16}(-1+9z^{-2}+16z^{-3}+9z^{-4}-z^{-6})$ and $F_1(z) = -1$. Verify that $\boldsymbol{F}_p(z)\boldsymbol{H}_p(z) = z^{-1}\boldsymbol{I}$. Note $L = 1$ and $\ell = 3$.

9. What is the synthesis polyphase matrix $\boldsymbol{F}_p$ for the lattice structure in the last figure?

10. The rules for linear phase with PR were stated separately for even and odd lengths. Combine them into one rule for the degrees of $H_0(z)$ and $H_1(z)$: $N_0 + N_1 + 2$ *must be a multiple of* 4.

11. Change a given PR filter bank by choosing $\widehat{H}_0(z) = H_0(z)$ and $\widehat{H}_1(z) = z^{-2}H_1(z)$. What is the relation between $\boldsymbol{H}_p(z)$ and $\widehat{\boldsymbol{H}}_p(z)$? Is this new filter bank PR?

12. Find the analysis filters and the relation between $H_0(z)$ and $H_1(z)$:

$$\boldsymbol{H}_p(z) = \begin{bmatrix} 1 & 2 \\ 2 & 1 \end{bmatrix} \begin{bmatrix} 1 & 0 \\ 0 & z^{-1} \end{bmatrix} \begin{bmatrix} 1 & -3 \\ -3 & 1 \end{bmatrix} \begin{bmatrix} 1 & 0 \\ 0 & z^{-2} \end{bmatrix} \begin{bmatrix} 1 & -2 \\ -2 & 1 \end{bmatrix}$$

13. Let $\boldsymbol{H}_p(z) = \begin{bmatrix} z^{-N} & 0 \\ \beta(z) & 1 \end{bmatrix}$, where $\beta(z)$ is a polynomial. Find $H_0(z)$ and $H_1(z)$. Find $\boldsymbol{F}_p(z)$ for a PR system. What are the synthesis filters?

14. Let $\boldsymbol{H}_p(z) = \prod_{k=1}^{L} \begin{bmatrix} z^{-1} & 0 \\ \beta_k(z) & 1 \end{bmatrix}$. Find $\boldsymbol{F}_p(z)$ and all the filters for a PR system.

15. What is the lowpass orthogonal filter with $\theta_1 = \frac{\pi}{3}$, $\theta_2 = -\frac{\pi}{2}$, and $\theta_3 = \frac{\pi}{4}$?

16. What are the symmetric even-length analysis filters with $k_1 = 2$, $k_2 = 0$, $k_3 = -7$, and $k_4 = 5$? What are the PR synthesis filters?

17. If we impose orthogonality on the symmetric factors $\boldsymbol{S}_i$ in (4.72), show that the filters only have two nonzero coefficients. This yields another proof of Theorem 5.3 that symmetry prevents orthogonality.

18. Show that $\begin{bmatrix} 1 & 1 \\ 1 & -1 \end{bmatrix} \begin{bmatrix} 1 & k \\ k & 1 \end{bmatrix}$ is equivalent to $\begin{bmatrix} 1+k & 0 \\ 0 & 1-k \end{bmatrix} \begin{bmatrix} 1 & 1 \\ 1 & -1 \end{bmatrix}$.
This form of the last two blocks in (4.72) reduces the lattice computation time.

Chapter 5

Orthogonal Filter Banks

5.1 Paraunitary Matrices

For an orthogonal matrix, the inverse is the transpose. When the matrix is 2 by 2, this imposes a tremendous constraint. Suppose we choose just one entry of the matrix, in the upper left corner. Our choice should not exceed 1 in absolute value, and we call it $\cos\theta$:

$$H = \begin{bmatrix} \cos\theta & - \\ - & - \end{bmatrix}.$$

The other entries are almost completely determined by this choice. Below $\cos\theta$ we need $\sin\theta$ or $-\sin\theta$, to make the first column a unit vector. The second column must be a unit vector orthogonal to the first one, and we select

$$H = \begin{bmatrix} \cos\theta & -\sin\theta \\ \sin\theta & \cos\theta \end{bmatrix}. \tag{5.1}$$

$\boldsymbol{H}$ rotates every plane vector $\boldsymbol{x}$ by θ. *The length is preserved*: $||\boldsymbol{Hx}|| = ||\boldsymbol{x}||$.

This is not the only orthogonal matrix that starts with $\cos\theta$. We could multiply the second column or the second row by -1. Then the rotation matrix in (5.1) becomes a *reflection* matrix; its determinant changes to -1. If these are *complex* matrices, we could multiply any column and any row by numbers on the unit circle $|z| = 1$. This yields every unitary 2 by 2 matrix. Essentially, one entry determines the whole matrix. The inverse is the conjugate transpose: $\boldsymbol{H}^{-1} = \overline{\boldsymbol{H}}^T$.

An analysis bank is represented by an infinite matrix (in the time domain). But in the frequency domain or z-domain, the matrix is 2 by 2 (from $M = 2$ channels and $M = 2$ phases). This matrix depends on a parameter ω or z. Therefore we stretch the definition from unitary to ***paraunitary***:

Definition 5.1 *The matrix $\boldsymbol{H}(z)$ is paraunitary if it is unitary for all $|z| = 1$:*

$$\boldsymbol{H}^T(e^{-j\omega})\boldsymbol{H}(e^{j\omega}) = \boldsymbol{I} \quad \textit{for all } \omega. \tag{5.2}$$

This extends to all $z \neq 0$ by $\widetilde{\boldsymbol{H}}(z) = \boldsymbol{H}^T(z^{-1})$. Then a paraunitary matrix has

$$\boldsymbol{H}^T(z^{-1})\boldsymbol{H}(z) = \widetilde{\boldsymbol{H}}(z)\,\boldsymbol{H}(z) = \boldsymbol{I} \quad \textit{for all } z. \tag{5.3}$$

When the coefficients $\boldsymbol{h}(k)$ are complex, they are conjugated in $\widetilde{\boldsymbol{H}}(z)$.

The matrix $\boldsymbol{H}$ need not be 2 by 2. If it is 1 by 1, then $|\boldsymbol{H}(e^{j\omega})| = 1$. The corresponding filter is *allpass*. The best allpass examples are ratios of polynomials coming from IIR filters — since only trivial polynomials z^{-l} can have $|\boldsymbol{H}(e^{j\omega})| = 1$.

If $\boldsymbol{H}(z)$ is $M \times M$, it could come from an M-channel filter bank. It might be the polyphase matrix $\boldsymbol{H}_p(z)$ or the modulation matrix $\boldsymbol{H}_m(z)$ (divided by $\sqrt{2}$). We will show that the filter bank is orthogonal if these matrices are paraunitary. That is the important connection for this book.

Equation (5.2) gives the inverse matrix by transposing and conjugating the original. The synthesis bank comes by "reversing" the analysis bank. Note that for a square matrix, $\boldsymbol{H}(z)$ is paraunitary when $\boldsymbol{H}^{-1}(z)$ and $\boldsymbol{H}^T(z)$ and $\widetilde{\boldsymbol{H}}(z)$ are paraunitary. And notice especially what equation (5.3) says about the *determinants* of these matrices:

$$\left(\det \widetilde{\boldsymbol{H}}(z)\right)\left(\det \boldsymbol{H}(z)\right) = 1. \tag{5.4}$$

The determinants are 1 by 1 allpass!

Theorem 5.1 *If a square paraunitary matrix $\boldsymbol{H}(z)$ is FIR (= polynomial), then its determinant must be a delay:*

$$\det \boldsymbol{H}(z) = \pm z^{-l}. \tag{5.5}$$

The determinant of $\boldsymbol{H}_p(z)$ is also a delay for any *bi*-orthogonal filter bank. Orthogonality requires more; the polyphase matrix $\boldsymbol{H}_p(z)$ must be paraunitary.

If $\boldsymbol{H}(z)$ is rectangular, say M by r, then $\widetilde{\boldsymbol{H}}(z)\boldsymbol{H}(z) = \boldsymbol{I}$ is still possible. The identity matrix is now r by r (and necessarily $r \leq M$, since the rank cannot exceed M). The matrix is still called paraunitary. There is no inverse matrix $\boldsymbol{H}^{-1}(z)$ in the rectangular case, but $\widetilde{\boldsymbol{H}}(z)$ is a left-inverse. We hope and expect that $\boldsymbol{H}(z)$ can be completed to a square paraunitary matrix, by adding $M-r$ more columns. In the applications, we are starting with r filters and creating $M - r$ additional filters — while preserving orthogonality.

When $\widetilde{\boldsymbol{H}}(z)\boldsymbol{H}(z) = d\boldsymbol{I}$ with $d > 0$, we could still use the word paraunitary. The chief example is the modulation matrix, which has $d = 2$. Some authors keep this flexibility.

Our chief interest is in the case $M = 2$. A 2 by 2 paraunitary matrix is essentially determined by one entry. For a paraunitary polyphase matrix, the even phase $H_{00}(z)$ of the lowpass filter essentially determines the whole orthogonal filter bank. For filter banks with four taps, there are two free parameters — which can be $\boldsymbol{c}(0)$ and $\boldsymbol{c}(2)$. The design problem is greatly reduced when the filter bank is orthogonal and its polyphase matrix is paraunitary.

2 by 2 Paraunitary Examples

The first example is $\boldsymbol{H}(z) = \boldsymbol{I}$. It corresponds to a "lazy filter bank" without filters. The polyphase matrix is $\boldsymbol{H}_p = \boldsymbol{I}$ when the filters are $H_0(z) = 1$ and $H_1(z) = z^{-1}$. The bank just splits the input vector $\boldsymbol{x}$ into odd and even phases, without filtering it.

The next paraunitary matrix is $\boldsymbol{H}(z) = \frac{1}{2}\begin{bmatrix} 1 & 1 \\ 1 & -1 \end{bmatrix}$. This is still constant. It is the polyphase matrix of the Haar average-difference filter bank. We move on to more interesting examples.

Suppose $\boldsymbol{R}_0, \ldots, \boldsymbol{R}_l$ are constant rotation matrices as in equation (5.1). The rotation angles are $\theta_0, \ldots, \theta_l$. If we multiply those matrices, we get a single rotation through the total angle $\sum \theta_i$. But if we introduce a diagonal matrix $\Lambda(z) = \operatorname{diag}\left(1, z^{-1}\right)$ *between* those rotations, we get something much more general and important:

$$\boldsymbol{H}(z) = \boldsymbol{R}_l\, \Lambda(z)\, \boldsymbol{R}_{l-1}\, \Lambda(z) \cdots \boldsymbol{R}_1\, \Lambda(z)\, \boldsymbol{R}_0. \tag{5.6}$$

This matrix is paraunitary, because it is the product of paraunitary factors. Its determinant is z^{-l} (the product of determinants). The filter bank can be realized as a lattice structure involving l delays. Section 4.5 proved that *this is the most general 2 by 2 paraunitary matrix of degree l.* The only subtle point is the meaning of the word "degree." Here are two small examples to make that point, both with $\boldsymbol{R}_0 = \boldsymbol{R}_2 = \boldsymbol{I}$. Then $H(z)$ is $\Lambda(z)\boldsymbol{R}_1\Lambda(z)$. The only difference is in $\boldsymbol{R}_1$:

$$\boldsymbol{H}(z) = \begin{bmatrix} 1 & \\ & z^{-1} \end{bmatrix} \frac{1}{\sqrt{2}} \begin{bmatrix} 1 & 1 \\ -1 & 1 \end{bmatrix} \begin{bmatrix} 1 & \\ & z^{-1} \end{bmatrix} = \frac{1}{\sqrt{2}} \begin{bmatrix} 1 & z^{-1} \\ -z^{-1} & z^{-2} \end{bmatrix}$$

$$\boldsymbol{H}(z) = \begin{bmatrix} 1 & \\ & z^{-1} \end{bmatrix} \begin{bmatrix} 0 & 1 \\ -1 & 0 \end{bmatrix} \begin{bmatrix} 1 & \\ & z^{-1} \end{bmatrix} = \begin{bmatrix} 0 & z^{-1} \\ -z^{-1} & 0 \end{bmatrix}$$

Both have determinant z^{-2}. Both have degree $l = 2$! But the power z^{-2}, which appears in the first, does not appear in the second. The correct definition of degree is the *Smith-McMillan degree — the minimum number of delays* required to realize the system. For paraunitary systems, but not for all systems, the degree is revealed by the determinant.

For lowpass-highpass filters with four taps, the even and odd phases have two terms each. If $\boldsymbol{H}(z)$ is paraunitary, it fits the general form (5.6) with $l = 1$. There are two parameters θ_0 and θ_1 to be chosen. Again we have two design parameters, for 4-tap orthogonal filters, but this time they are angles.

This chapter constructs a family of *maxflat filters*. Then Chapter 6 shows that these filters give the *Daubechies wavelets*. We mention the connection now, so you will attach importance to maxflat filters. They don't have a sharp transition band — they don't have the quickest transition from passband to stopband — but they produce the most vanishing moments.

Problem Set 5.1

1. Suppose $\boldsymbol{H}(z) = \boldsymbol{h}(0) + \boldsymbol{h}(1)z^{-1} + \cdots + \boldsymbol{h}(N)z^{-N}$ is paraunitary with $N > 0$. Show that $\boldsymbol{h}(0)^T\boldsymbol{h}(N) =$ zero matrix. Deduce that $\boldsymbol{h}(0)$ and $\boldsymbol{h}(N)$ are both singular matrices.

2. Multiply out $\boldsymbol{H}(z) = \boldsymbol{R}_1\Lambda(z)\boldsymbol{R}_0$ for angles θ_1 and θ_0. If this is a polyphase matrix, what filter coefficients $\boldsymbol{d}(k)$ come from the even and odd phases in the lower row of $\boldsymbol{H}(z)$? If that is a highpass filter, with $D = 0$ at $\omega = 0$ (which is $z = 1$), show that $\theta_1 + \theta_0 = -\frac{\pi}{4} + n\pi$. (Haar has $\theta_0 = -\frac{\pi}{4}$ and $\theta_1 = 0$.)

3. Complete this polynomial matrix to have $\det \boldsymbol{H}(z) \equiv 1$. Is it paraunitary?

$$\boldsymbol{H}(z) = \begin{bmatrix} 1+z^{-1} & z^{-1} \\ - & - \end{bmatrix}.$$

4. When can a row be the first row of a paraunitary matrix?

5. Show that $\boldsymbol{H} = \boldsymbol{I} - 2\boldsymbol{v}\boldsymbol{v}^T$ is a unitary matrix, where $\boldsymbol{v}^T\boldsymbol{v} = 1$. Compute $\boldsymbol{H}^{-1}$ and the determinant of $\boldsymbol{H}$. *Construct an example.*

6. Show that $\boldsymbol{H} = \boldsymbol{I} - \boldsymbol{v}\boldsymbol{v}^T + z^{-1}\boldsymbol{v}\boldsymbol{v}^T$ is a paraunitary matrix for a unit-norm vector $\boldsymbol{v}$. Compute its inverse and determinant. The factorization of $\boldsymbol{H}_p(z)$ using these Householder matrices is in [V].

5.2 Orthonormal Filter Banks

This section brings together the requirements for an orthonormal filter bank. We will see those requirements in the *time domain* and the *polyphase domain* and the *modulation domain.* These requirements are conditions on the filter coefficients $\boldsymbol{c}(k)$ and $\boldsymbol{d}(k)$. Then equation (5.19) indicates a simple choice of the $\boldsymbol{d}$'s coming from the $\boldsymbol{c}$'s. If the lowpass filter meets the orthogonality requirements, it is easy to construct a highpass filter to go with it.

The discussion is in terms of a 2-channel FIR filter bank, $M = 2$. But the conditions extend immediately to any M. *The polyphase matrix and the modulation matrix must be paraunitary.* In the M-channel case, the lowpass filter does not immediately determine the $M - 1$ remaining filters (which are bandpass). There is some freedom in their construction, and we come back in Chapter 9 to $M > 2$.

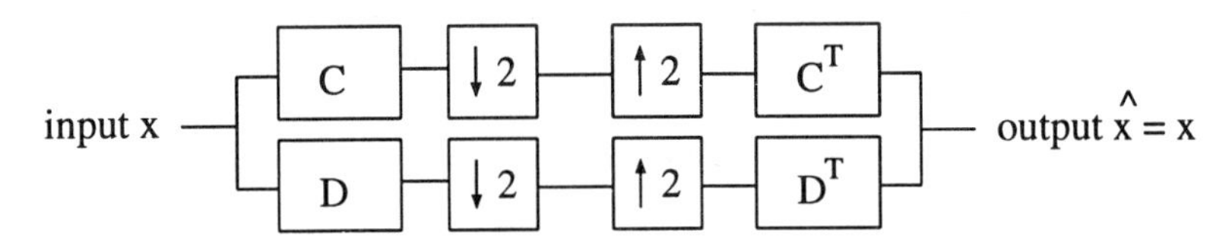

Figure 5.1: An orthogonal filter bank has synthesis bank = transpose of analysis bank.

Figure 5.1 shows the structure of an orthogonal filter bank. We intend to achieve $\widehat{\boldsymbol{x}} = \boldsymbol{x}$, with synthesis filters $\boldsymbol{C}^T$ and $\boldsymbol{D}^T$ that are time-reversals of the analysis filters:

$$\widetilde{\boldsymbol{C}} = \boldsymbol{C}^T \quad \text{and} \quad \widetilde{\boldsymbol{c}}(n) = \boldsymbol{c}(-n) \tag{5.7}$$

$$\widetilde{\boldsymbol{D}} = \boldsymbol{D}^T \quad \text{and} \quad \widetilde{\boldsymbol{d}}(n) = \boldsymbol{d}(-n) \tag{5.8}$$

As it stands, $\widetilde{\boldsymbol{C}}$ and $\widetilde{\boldsymbol{D}}$ are anticausal. At the end we make them causal by N delays. The output $\widehat{\boldsymbol{x}}(n)$ is equally delayed; it is $\boldsymbol{x}(n - N)$. But the algebra is easiest for $\boldsymbol{C}^T$ and $\boldsymbol{D}^T$ with no delays.

This special structure imposes special conditions on the $\boldsymbol{c}$'s and $\boldsymbol{d}$'s for perfect reconstruction. We will call the requirements ***Condition O*** (for orthogonality). This section finds four equivalent forms: Condition O on the infinite matrix, on the lowpass coefficients, and on the polyphase and modulation matrices $\boldsymbol{H}_p$ and $\boldsymbol{H}_m$. We look first at the infinite matrices in the time domain.

Time Domain: Condition O and the Alternating Flip

The key matrix in the time domain is $\boldsymbol{H}_t$. It represents the direct form of the analysis bank, with downsampling. The lowpass part $\boldsymbol{L} = (\downarrow 2)\boldsymbol{C}$ comes above the highpass part $\boldsymbol{B} = (\downarrow 2)\boldsymbol{D}$. We display this infinite matrix for filters of length four:

$$\boldsymbol{H}_t = \begin{bmatrix} \boldsymbol{L} \\ \boldsymbol{B} \end{bmatrix} = \begin{bmatrix} c(3) & c(2) & c(1) & c(0) & & \\ & & c(3) & c(2) & c(1) & c(0) \\ & & & & \cdot & \cdot \\ d(3) & d(2) & d(1) & d(0) & & \\ & & d(3) & d(2) & d(1) & d(0) \\ & & & & \cdot & \cdot \end{bmatrix}. \tag{5.9}$$

The shifts by 2 were created by downsampling, which removed the odd-numbered rows.

With orthogonality, the synthesis filters are to be time-reversals of the analysis filters. The infinite synthesis matrix contains the transposes of $(\downarrow 2)\boldsymbol{C}$ and $(\downarrow 2)\boldsymbol{D}$. Those transposes are $\boldsymbol{C}^T(\uparrow 2)$ and $\boldsymbol{D}^T(\uparrow 2)$, since upsampling is the transpose of downsampling:

$$\boldsymbol{H}_t^T = \begin{bmatrix} \boldsymbol{L}^T & \boldsymbol{B}^T \end{bmatrix} = \begin{bmatrix} c(3) & & & d(3) & & \\ c(2) & & & d(2) & & \\ c(1) & c(3) & & d(1) & d(3) & \\ c(0) & c(2) & & d(0) & d(2) & \\ & c(1) & \cdot & & d(1) & \cdot \\ & c(0) & \cdot & & d(0) & \cdot \end{bmatrix}. \tag{5.10}$$

The shifts by 2 in the columns are created by upsampling, which removes every other column. We require $\widehat{\boldsymbol{x}} = \boldsymbol{x}$. This means that $\boldsymbol{H}_t^T\boldsymbol{H}_t = \boldsymbol{I}$. The matrix $\boldsymbol{H}_t$ is required to be an *orthogonal matrix.* Its columns are orthonormal and so are its rows: $\boldsymbol{H}_t\boldsymbol{H}_t^T = \boldsymbol{I}$. We can express this Condition O in matrix form, and in block form, and in coefficient form.

Condition O An orthogonal filter bank comes from an orthogonal matrix:

$$\boldsymbol{H}_t^T\boldsymbol{H}_t = \boldsymbol{I} \text{ and } \boldsymbol{H}_t\boldsymbol{H}_t^T = \boldsymbol{I}. \tag{5.11}$$

In block form this means that

$$\begin{bmatrix} \boldsymbol{L}^T & \boldsymbol{B}^T \end{bmatrix} \begin{bmatrix} \boldsymbol{L} \\ \boldsymbol{B} \end{bmatrix} = \boldsymbol{L}^T\boldsymbol{L} + \boldsymbol{B}^T\boldsymbol{B} = \boldsymbol{I} \tag{5.12}$$

and

$$\begin{bmatrix} \boldsymbol{L} \\ \boldsymbol{B} \end{bmatrix} \begin{bmatrix} \boldsymbol{L}^T & \boldsymbol{B}^T \end{bmatrix} = \begin{bmatrix} \boldsymbol{L}\boldsymbol{L}^T & \boldsymbol{L}\boldsymbol{B}^T \\ \boldsymbol{B}\boldsymbol{L}^T & \boldsymbol{B}\boldsymbol{B}^T \end{bmatrix} = \begin{bmatrix} \boldsymbol{I} & \boldsymbol{0} \\ \boldsymbol{0} & \boldsymbol{I} \end{bmatrix}. \tag{5.13}$$

For the coefficients $c(k)$ and $d(k)$, equation (5.13) becomes *orthogonality to double shifts*:

$$\boldsymbol{L}\boldsymbol{L}^T = \boldsymbol{I}: \quad \sum c(n)c(n-2k) = \delta(k) \tag{5.14}$$

$$\boldsymbol{L}\boldsymbol{B}^T = \boldsymbol{0}: \quad \sum c(n)d(n-2k) = 0 \tag{5.15}$$

$$\boldsymbol{B}\boldsymbol{B}^T = \boldsymbol{I}: \quad \sum d(n)d(n-2k) = \delta(k). \tag{5.16}$$

Because of (5.14)–(5.16), we refer to Condition O as ***double-shift orthogonality***. Its equivalent in the frequency domain is presented below. This double-shift orthogonality immediately rules out odd length filters! If the length is $N+1=5$, a shift by 4 gives an inner product that cannot be zero:

$$(c(0), c(1), c(2), c(3), c(4)) \cdot (0,0,0,0,c(0)) = c(0)c(4) \neq 0.$$

N cannot be even (the filter length cannot be odd) because $\frac{N}{2}$ double shifts would give a shift by N — and the inner product $c(0)c(N)$ is not zero. So N is odd. The degree $2N$ of the halfband product filter $\boldsymbol{P}$ is 2, 6, 10, ... The clearest examples have $N=3$ and $2N=6$.

Example. Condition O in (5.14) imposes two constraints on four coefficients:

$$c(0)^2 + c(1)^2 + c(2)^2 + c(3)^2 = 1 \quad \text{and} \quad c(0)c(2) + c(1)c(3) = 0. \tag{5.17}$$

Equation (5.16) is an identical condition on the d's, from $\boldsymbol{B}\boldsymbol{B}^T = \boldsymbol{I}$. Equation (5.15) is the orthogonality of the rows of $\boldsymbol{L}$ to the rows of $\boldsymbol{B}$. Those are the fundamental design constraints on an orthogonal filter bank.

The conditions on the c's and d's are independent, but something good happens. *If the c's satisfy equation* (5.14), *it is easy to choose the d's.* We display a choice of d's that automatically gives orthogonality:

$$\begin{bmatrix} L \\ B \end{bmatrix} = \begin{bmatrix} c(3) & c(2) & c(1) & c(0) & & \\ & & c(3) & c(2) & c(1) & c(0) \\ & & & & \cdot & \cdot \\ -c(0) & c(1) & -c(2) & c(3) & & \\ & & -c(0) & c(1) & -c(2) & c(3) \\ & & & & \cdot & \cdot \end{bmatrix}. \tag{5.18}$$

This is the ***alternating flip***. The c's are reversed in order and alternated in sign, to produce the d's. We start with lowpass coefficients $c(0), \ldots, c(N)$, where N is odd. The highpass coefficients are

$$d(k) = (-1)^k c(N-k). \tag{5.19}$$

The essential point can be checked by eye, in the infinite matrix (5.18). *If the **top** rows are orthogonal to each other, then **all** rows are orthogonal.* The zero dot products in LB^T are:

$$\begin{aligned} -c(3)c(0) + c(2)c(1) - c(1)c(2) + c(0)c(3) &= 0 \\ -c(1)c(0) + c(0)c(1) &= 0. \end{aligned} \tag{5.20}$$

Furthermore, the d's are orthonormal within themselves ($BB^T = I$) because the c's are. Equations (5.17) hold for the d's, when they are constructed by flipping the c's. Minus signs cancel in $d(3)d(1)$ which is $(-c(0))(-c(2))$.

Our four-tap example has $N = 3$. The alternating flip gives $LB^T = 0$ for every odd N. The top rows of H_t in (5.18) are *always* orthogonal to the bottom rows. Also (5.16) for the d's follows from (5.14) for the c's. Thus the alternating flip reduces orthogonality to (5.14):

Condition O on the coefficients: $\sum c(n)c(n-2k) = \delta(k)$.

Polyphase Domain: Condition O and the Alternating Flip

The polyphase form separates $(\downarrow 2)C$ and $(\downarrow 2)D$ into even phase and odd phase. In the time domain, we are rearranging the columns of the matrix H_t. All even columns come before the odd columns. The matrix goes into the 2 by 2 block form of Section 4.4, with time-invariant filters as the blocks:

$$H_{\text{block}} = \begin{bmatrix} C_{\text{even}} & C_{\text{odd}} \\ D_{\text{even}} & D_{\text{odd}} \end{bmatrix}.$$

In the z-domain, we are rearranging the response functions $C(z)$ and $D(z)$:

$$\sum c(k) z^{-k} = C_{\text{even}}(z^{-2}) + z^{-1} C_{\text{odd}}(z^{-2}). \tag{5.21}$$

Those phase responses are written $H_{00}(z)$ and $H_{01}(z)$ when $C(z)$ is $H_0(z)$. The highpass response $D(z) = H_1(z)$ decomposes in the same way. *The polyphase matrix is*

$$H_p(z) = \begin{bmatrix} C_{\text{even}}(z) & C_{\text{odd}}(z) \\ D_{\text{even}}(z) & D_{\text{odd}}(z) \end{bmatrix} = \begin{bmatrix} H_{00}(z) & H_{01}(z) \\ H_{10}(z) & H_{11}(z) \end{bmatrix}. \tag{5.22}$$

Now Condition O translates directly into a requirement on $\boldsymbol{H}_p(z)$. A filter bank is **orthogonal** when its polyphase matrix is **paraunitary**:

$$\boldsymbol{H}_p^T(e^{-j\omega})\boldsymbol{H}_p(e^{j\omega}) = \boldsymbol{I} \text{ for all } \omega \text{ and } \widetilde{\boldsymbol{H}}_p(z)\boldsymbol{H}_p(z) = \boldsymbol{I} \text{ for all } z. \tag{5.23}$$

The inverse of $\boldsymbol{H}_p(z)$ is the synthesis polyphase matrix. The matrix and its inverse can be multiplied in either order — here is analysis times synthesis:

$$\begin{bmatrix} C_{\text{even}}(z) & C_{\text{odd}}(z) \\ D_{\text{even}}(z) & D_{\text{odd}}(z) \end{bmatrix} \begin{bmatrix} C_{\text{even}}(z^{-1}) & D_{\text{even}}(z^{-1}) \\ C_{\text{odd}}(z^{-1}) & D_{\text{odd}}(z^{-1}) \end{bmatrix} = \begin{bmatrix} 1 & 0 \\ 0 & 1 \end{bmatrix}. \tag{5.24}$$

On the unit circle, where z^{-1} is $\overline{z}$, row 1 times column 1 becomes

$$|C_{\text{even}}(z)|^2 + |C_{\text{odd}}(z)|^2 = 1 \text{ when } |z| = 1. \tag{5.25}$$

This is the essence of Condition O in the polyphase domain.

For the example with four coefficients, it is helpful to multiply out equation (5.25):

$$\left(c(0) + c(2)z^{-1}\right)\left(c(0) + c(2)z\right) + \left(c(1) + c(3)z^{-1}\right)\left(c(1) + c(3)z\right) =$$
$$c(0)^2 + c(2)^2 + c(1)^2 + c(3)^2 + [c(0)c(2) + c(1)c(3)]\left(z^{-1} + z\right) = 1.$$

Thus (5.25) is equivalent to the explicit statement (5.17). The sum of squares is 1 and the dot product $c(0)c(2) + c(1)c(3)$ is zero.

The other multiplications in (5.24) give answers 1 or 0 in the same way. All these requirements on the d's are automatically satisfied by the flip construction! We write that choice $d(k) = (-1)^k c(N-k)$ in the z-domain:

$$\sum d(k)z^{-k} = \sum c(N-k)(-z)^{-k} = \sum c(n)(-z)^{n-N}.$$

This relation between highpass and lowpass is an *alternating flip*:

$$D(z) = (-z)^{-N} C(-z^{-1}). \tag{5.26}$$

The number N is odd. Because of $(-z)^{-N}$, the even and odd phases in $\boldsymbol{C}$ are reversed to odd and even phases in $\boldsymbol{D}$. We take $N = 3$ as typical. An alternating flip of $c(0), c(1), c(2), c(3)$ yields

$$\begin{aligned} D(z) &= c(3) - c(2)z^{-1} + c(1)z^{-2} - c(0)z^{-3} \\ &= \left(c(3) + c(1)z^{-2}\right) - z^{-1}\left(c(2) + c(0)z^{-2}\right). \end{aligned} \tag{5.27}$$

With this flip in row 2, the multiplication $\boldsymbol{H}_p(z)\boldsymbol{H}_p^T(z^{-1})$ becomes

$$\begin{bmatrix} c(0) + c(2)z^{-1} & c(1) + c(3)z^{-1} \\ c(3) + c(1)z^{-1} & -c(2) - c(0)z^{-1} \end{bmatrix} \begin{bmatrix} c(0) + c(2)z & c(3) + c(1)z \\ c(1) + c(3)z & -c(2) - c(0)z \end{bmatrix} = \boldsymbol{I}.$$

The off-diagonal entries of the product are *automatically* zero. *The alternating flip achieves* $\boldsymbol{L}\boldsymbol{B}^T = \boldsymbol{0}$ *with or without orthogonality.* The 2, 2 entry of the product is the same as the 1, 1 entry, when $|z| = 1$. The orthogonality requirement (5.25) makes the 1, 1 entry equal to 1.

Modulation Domain: Condition O and the Alternating Flip

The function that arises from modulating $C(z)$ is $C(-z)$. The frequency in $z = e^{j\omega}$ changes by π to produce $-z = e^{j(\omega+\pi)}$. This modulation takes $C(\omega)$ to $C(\omega+\pi)$. The frequency response graph is shifted by π.

Our goal is to relate $C(z)$ to $C(-z)$ and $D(z)$ to $D(-z)$ for an orthogonal filter bank. We already know Condition O on the coefficients:

$$c(0)^2 + c(1)^2 + c(2)^2 + c(3)^2 = 1 \quad \text{and} \quad c(0)c(2) + c(1)c(3) = 0.$$

Watch how 1 and 0 appear in $|C(z)|^2$. Stay on the unit circle where $\overline{z}^{-1} = z$:

$$\left(c(0) + c(1)z^{-1} + c(2)z^{-2} + c(3)z^{-3}\right)\left(c(0) + c(1)z + c(2)z^2 + c(3)z^3\right) =$$
$$1 + [c(0)c(1) + c(1)c(2) + c(2)c(3)]\left(z^{-1} + z\right) + c(0)c(3)\left(z^{-3} + z^3\right).$$

Now change z to $-z$. The odd powers z and z^3 change sign. When we add, those odd powers cancel:

$$|C(z)|^2 + |C(-z)|^2 = |C(\omega)|^2 + |C(\omega+\pi)|^2 = 2 \quad \text{for all } z = e^{j\omega}. \tag{5.28}$$

This is the *halfband condition*, also called the *Nyquist condition*: $|C(z)|^2$ is a halfband filter. Those filters we can design! In other words, the lowpass analysis filter $C(z)$ is a spectral factor of a halfband filter.

Condition O on $H_m(z)$. *The modulation matrix of an orthogonal filter bank is a paraunitary matrix times $\sqrt{2}$:*

$$H_m(z)\widetilde{H}_m(z) = 2I \quad \text{for all } z. \tag{5.29}$$

On the circle $z = e^{j\omega}$, the modulation matrix is a unitary matrix times $\sqrt{2}$:

$$\begin{bmatrix} C(\omega) & C(\omega+\pi) \\ D(\omega) & D(\omega+\pi) \end{bmatrix}\begin{bmatrix} \overline{C(\omega)} & \overline{D(\omega)} \\ \overline{C(\omega+\pi)} & \overline{D(\omega+\pi)} \end{bmatrix} = \begin{bmatrix} 2 & 0 \\ 0 & 2 \end{bmatrix}. \tag{5.30}$$

The 1, 1 entry of this matrix product is $|C(\omega)|^2 + |C(\omega+\pi)|^2 = 2$ by (5.28). The other entries, when we multiply them out, follow immediately from (5.15) and (5.16). Thus Condition O on the coefficients is equivalent to Condition O on the modulation matrix H_m. It is also equivalent to Condition O on H_p. It is the statement that the analysis bank followed by its transpose gives perfect reconstruction. We summarize:

Theorem 5.2 *For an orthogonal filter bank the lowpass coefficients must satisfy* Condition O *(we give four equivalent forms).*

Matrix form	$LL^T = (\downarrow 2)CC^T(\uparrow 2) = I$
Coefficient form	$\sum c(n)c(n-2k) = \delta(k)$
Polyphase form	$\lvert C_{\text{even}}(e^{j\omega})\rvert^2 + \lvert C_{\text{odd}}(e^{j\omega})\rvert^2 = 1$
Modulation form	$\lvert C(\omega)\rvert^2 + \lvert C(\omega+\pi)\rvert^2 = 2$

By an alternating flip, $LB^T = 0$ and $BB^T = I$ follow immediately from the lowpass part $LL^T = I$. The real problem is the design of the lowpass filter.

Symmetry Prevents Orthogonality

It is natural to want two good properties at once. Symmetry is good for the eye, and orthogonality is good for the algorithm. But the only filters with both properties are averaging filters (Haar filters) with two coefficients. We are forced to use IIR filters, or M channels, or multifilters with matrix coefficients (Section 7.5). Extra computation is unavoidable, because the next theorem rules out a perfect filter.

Theorem 5.3 *A symmetric orthogonal FIR filter can only have two nonzero coefficients.*

Proof. N is odd for orthogonality. The filter length must be even. With $N = 5$ a symmetric filter of length 6 has the form $(\boldsymbol{c}(0), \boldsymbol{c}(1), \boldsymbol{c}(2), \boldsymbol{c}(2), \boldsymbol{c}(1), \boldsymbol{c}(0))$. This vector must be orthogonal to all its double shifts. The inner product with its shift by *four* must be $2\boldsymbol{c}(0)\boldsymbol{c}(1) = 0$. Therefore $\boldsymbol{c}(1) = 0$. Then the inner product with its shift by *two* gives $2\,\boldsymbol{c}(0)\boldsymbol{c}(2) = 0$. The only nonzero coefficient is $\boldsymbol{c}(0)$ at both ends of the filter. This completes the proof.

By convention $\boldsymbol{c}(0)$ is the first nonzero coefficient. Shift the filter if necessary to achieve this. The only symmetric orthogonal possibilities are $\boldsymbol{c} = (1, 1)/\sqrt{2}$ and $(1, 0, 0, 1)/\sqrt{2}$ and $(1, 0, \ldots, 0, 1)/\sqrt{2}$. Only the Haar coefficients $(1, 1)/\sqrt{2}$ will lead to orthogonal wavelets. Symmetry really conflicts with orthogonality.

A second proof observes that the odd phase is the flip of the even phase:

$$(\boldsymbol{c}(0), \boldsymbol{c}(4), \boldsymbol{c}(2), \boldsymbol{c}(2), \boldsymbol{c}(4), \boldsymbol{c}(0)) \text{ has } |C_{\text{even}}(z)|^2 = |C_{\text{odd}}(z)|^2.$$

Condition O is $|C_{\text{even}}(z)|^2 + |C_{\text{odd}}(z)|^2 = 2$. With symmetry this separates into $|C_{\text{even}}(z)|^2 = 1$ and $|C_{\text{odd}}(z)|^2 = 1$. *The even phase is an allpass filter* ! So is the odd phase. But FIR allpass filters can only have one nonzero coefficient, which completes the second proof.

A third proof is based on the zeros of $C(z)$. This is in Section 5.4 below.

Problem Set 5.2

1. (a) Show that the alternating flip with odd N gives $\overline{D(\omega)} = -e^{iN\omega}C(\omega + \pi)$.
   (b) Then $\overline{D(\omega + \pi)} = e^{iN\omega}C(\omega)$. Verify that $C(\omega)\overline{D(\omega)} + C(\omega + \pi)\overline{D(\omega + \pi)} = 0$.
   (c) Also $|C(\omega)|^2 + |C(\omega + \pi)|^2 = 2$ implies that $|D(\omega)|^2 + |D(\omega + \pi)|^2 = 2$
2. For any four coefficients $\boldsymbol{h}(0), \ldots, \boldsymbol{h}(3)$, verify that
   $$|H_{\text{even}}(z)|^2 + |H_{\text{odd}}(z)|^2 = \tfrac{1}{2}\left(|H(z)|^2 + |H(-z)|^2\right).$$
   Then Condition O for polyphase equals Condition O for modulation.
3. Find the flaw in this construction of the modulation matrix $\boldsymbol{H}_m(z)$. Start with an arbitrary upper left entry $C(z)$. Complete the 2 by 2 matrix to be paraunitary (times $\sqrt{2}$). Then the filter bank is orthogonal.
4. Find $\boldsymbol{d}$ by an alternating flip of $\boldsymbol{c} = (\boldsymbol{c}(0), \ldots, \boldsymbol{c}(5))$. Verify equation (5.15) directly to show that $\boldsymbol{c}$ is double-shift orthogonal to $\boldsymbol{d}$.
5. Verify that $\boldsymbol{c} = \frac{1}{4\sqrt{2}}(1 + \sqrt{3}, 3 + \sqrt{3}, 3 - \sqrt{3}, 1 - \sqrt{3})$ satisfies Condition O (Daubechies).
6. If two lowpass filters $\boldsymbol{C}$ and $\boldsymbol{H}$ satisfy Condition O, does their product satisfy Condition O?
7. If two polyphase matrices $\boldsymbol{H}_p(z)$ and $\boldsymbol{K}_p(z)$ satisfy Condition O (they are paraunitary), does their product satisfy Condition O?

8. If two modulation matrices $\boldsymbol{H}_m(z)$ and $\boldsymbol{K}_m(z)$ satisfy Condition O, show that $\boldsymbol{H}_m(z)\boldsymbol{K}_m(z)/\sqrt{2}$ is also paraunitary. What is the lowpass filter in the product?

9. Why does orthogonality *require* an alternating flip between the lowpass filter $\boldsymbol{C}$ and the highpass filter $\boldsymbol{D}$? Explain why the paraunitary matrix

$$\boldsymbol{H}_p(z) = \begin{bmatrix} C_{\text{even}}(z) & C_{\text{odd}}(z) \\ D_{\text{even}}(z) & D_{\text{odd}}(z) \end{bmatrix} \quad \text{must have} \quad \begin{aligned} |D_{\text{even}}(z)| &= |C_{\text{odd}}(z)| \\ |D_{\text{odd}}(z)| &= |C_{\text{even}}(z)|. \end{aligned}$$

Go further to show that $D_{\text{even}} = \pm z^{-2L}$ (flip of C_{odd}) and $D_{\text{odd}} = \pm z^{-2L}$ (flip of $-C_{\text{even}}$). This gives the alternating flip with any even delay z^{-2L}.

10. Find $H_0(z)$ and $H_1(z)$ and $\boldsymbol{H}_p(z)$. Is this paraunitary? Find the PR synthesis filters. What is $\boldsymbol{F}_p(z)$?

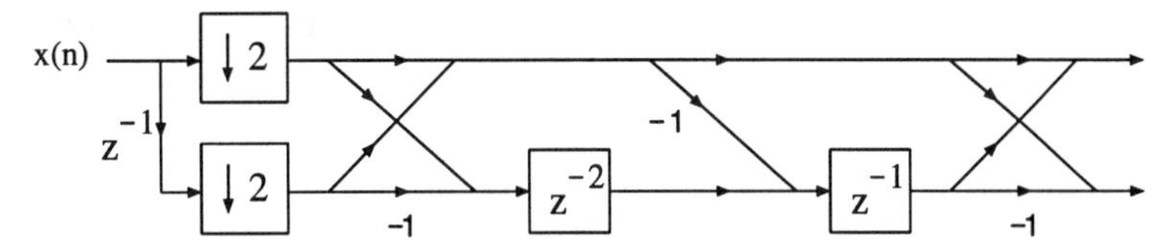

11. For $\boldsymbol{H}_p(z) = \boldsymbol{I} - \boldsymbol{v}\boldsymbol{v}^T + z^{-1}\boldsymbol{v}\boldsymbol{v}^T$ with $\boldsymbol{v} = \frac{1}{3}\begin{bmatrix} 2-1 & 1 & \sqrt{3} \end{bmatrix}^T$, find the four PR filters.

12. Let $\boldsymbol{H}_p(z) = \boldsymbol{R}_1\Lambda(z)\boldsymbol{R}_0$ where $\theta_1 = \frac{\pi}{4}$ and $\theta_0 = -\frac{\pi}{2}$. What are the analysis and synthesis filters? Plot the frequency responses of $H_k(z)$.

5.3 Halfband Filters

Out of all the equations in the previous section, we would like to emphasize one. It came at the end. It applied first of all to the lowpass filter $C(z)$, and then by the alternating flip also to $D(z)$. It was equation (5.28), that the frequency response $C(\omega) = \sum c(k)e^{-jk\omega}$ satisfies

$$|C(\omega)|^2 + |C(\omega+\pi)|^2 = 2. \tag{5.31}$$

The key question is, *what does equation* (5.31) *say about* $|C(\omega)|^2$ *itself?*

We assign the symbol $P(\omega)$ to this important quantity $|C(\omega)|^2$. It is the ***power spectral response***. Because $C(\omega)$ multiplies $\overline{C(\omega)}$, the filter with this response $P(\omega)$ is *symmetric*. It is the "autocorrelation filter":

$$P(\omega) = \sum_{-N}^{N} p(n)e^{-jn\omega} = \left(\sum_{0}^{N} c(k)e^{-jk\omega}\right)\left(\sum_{0}^{N} c(k)e^{jk\omega}\right). \tag{5.32}$$

The function $P(\omega) = |C(\omega)|^2$ is real and nonnegative. It equals its complex conjugate. This verifies the symmetry $\boldsymbol{p}(n) = \boldsymbol{p}(-n)$ that was expected (with real coefficients).

To repeat: When $\sum c(k)e^{-jk\omega}$ multiplies its conjugate $\sum c(l)e^{jl\omega}$, we watch for $n = k - l$ which is $l = k - n$. The coefficient $\boldsymbol{p}(n)$ is the sum of $\boldsymbol{c}(k)$ times $\boldsymbol{c}(k-n)$:

$$p(n) = \sum c(k)c(k-n) = \text{autocorrelation of the sequence } c(k). \tag{5.33}$$

Autocorrelation is $\boldsymbol{p} = \boldsymbol{c} * \boldsymbol{c}^T$. This is the convolution of $\boldsymbol{c} = (\boldsymbol{c}(0), \boldsymbol{c}(1), \boldsymbol{c}(2), \ldots)$ with its time reversal $\boldsymbol{c}^T = (\ldots, \boldsymbol{c}(2), \boldsymbol{c}(1), \boldsymbol{c}(0))$. Replacing $-n$ by n in (5.33) brings no change in $\boldsymbol{p}$.

The reason for the surprising notation $\boldsymbol{c}^T$ is that multiplying $C(\omega)$ by $\overline{C(\omega)}$ corresponds exactly to multiplying the infinite filter matrix $\boldsymbol{C}$ by $\boldsymbol{C}^T$:

$$\boldsymbol{P} = \boldsymbol{C}\boldsymbol{C}^T = \begin{bmatrix} \cdot & & & \\ & \cdot & & \\ \cdot & & \cdot & \\ & \cdot & & c(0) \\ \cdot & & c(1) & & \cdot \\ & c(2) & & \cdot & & \cdot \\ \cdot & & \cdot & & \cdot & & \cdot \end{bmatrix} \begin{bmatrix} \cdot & \cdot & & \cdot & & \cdot \\ & \cdot & \cdot & & c(2) & \\ & & \cdot & c(1) & & \cdot \\ & & c(0) & & \cdot & \\ & & & \cdot & & \cdot \\ & & & & \cdot & \\ & & & & & \cdot \end{bmatrix}. \tag{5.34}$$

$\boldsymbol{P}$ is a symmetric positive definite (or semidefinite) Toeplitz matrix. The transpose of $\boldsymbol{C}\boldsymbol{C}^T$ is $\boldsymbol{C}\boldsymbol{C}^T$. The diagonal $\boldsymbol{p}(0) = \boldsymbol{c}(0)^2 + \boldsymbol{c}(1)^2 + \boldsymbol{c}(2)^2 + \cdots$ is certainly positive. Equation (5.31) says that $\boldsymbol{p}(0) = 1$, when the matrix $\boldsymbol{C}$ comes from an orthonormal filter bank.

What does equation (5.31) say about the other coefficients $\boldsymbol{p}(n)$? In a word, it says *nothing* about the odd coefficients and it assigns *zero* to the even coefficients. $\boldsymbol{C}$ is the start of an orthogonal filter bank if and only if the autocorrelation filter $\boldsymbol{P}$ is a ***halfband filter***:

$$P(z) + P(-z) = 2. \tag{5.35}$$

The *even coefficients* with $n = 2m$ must be $\delta(m)$:

$$\textbf{Halfband filter} \quad p(2m) = \sum c(k)c(k-2m) = \begin{cases} 1 & \text{if } m = 0 \\ 0 & \text{if } m \neq 0. \end{cases} \tag{5.36}$$

The odd coefficients are not necessarily zero! They cancel automatically in equation (5.35). To require $\boldsymbol{C}\boldsymbol{C}^T = \boldsymbol{I}$, with odd coefficients zero, would make $\boldsymbol{C}$ an allpass filter. It could only be a delay. The requirement $P(z) + P(-z) \equiv 2$ is much weaker than $P(z) \equiv 1$. Perfect reconstruction is possible for an FIR filter bank, but essentially impossible for one FIR filter.

The highpass response $D(\omega)$ leads similarly to the autocorrelation $P_1(\omega) = |D(\omega)|^2$. Then $\boldsymbol{D}\boldsymbol{D}^T$ must also be a normalized halfband filter. That result is automatic with the alternating flip, which gives $P_1(\omega) = P(\omega + \pi)$. Then $\boldsymbol{p}_1(2m) = \boldsymbol{p}(2m) = \delta(m)$ for the even coefficients. The odd coefficients change sign, $P_1(z) = P(-z)$, but the halfband condition $P_1(z) + P_1(-z) = 2$ remains true.

The sum $|C(\omega)|^2 + |C(\omega + \pi)|^2$ is $P(\omega) + P(\omega + \pi)$. For a halfband filter, this sum is a constant. The graph of $P(\omega)$ shows a special symmetry with respect to the halfband frequency $\omega = \frac{\pi}{2}$ — hence the name. Notice what happens in downsampling — the even coefficients yield the identity filter:

$$(\downarrow 2)\boldsymbol{P} = \boldsymbol{I} \text{ when } \boldsymbol{P} \text{ is normalized halfband.} \tag{5.37}$$

Example 5.1. The symmetric filter with $\boldsymbol{p}(0) = 1$ and $\boldsymbol{p}(1) = \boldsymbol{p}(-1) = \frac{1}{2}$ is a halfband filter. Its response is

$$P(z) = 1 + \frac{z^{-1} + z}{2} \text{ and equivalently } P(\omega) = 1 + \cos\omega.$$

In the z-domain, $P(z) + P(-z) = 2$. The odd powers cancel and there is no z^2 term. In the ω-domain $1 + \cos\omega + 1 - \cos\omega = 2$. The odd frequency cancels. *Notice that* $P(\omega) = 1 + \cos\omega$

is never negative! It does reach zero at the highest frequency $\omega = \pi$, corresponding to $z = -1$. When $P(\omega)$ touches zero, its spectral factor $C(\omega)$ must also touch zero—since $P = |C|^2$. This leads us to $C(\omega) = (1 + e^{-i\omega})/\sqrt{2}$:

$$P(\omega) = |C(\omega)|^2 = \left(1 + e^{-i\omega}\right)\left(1 + e^{i\omega}\right)/2 = 1 + \cos\omega. \tag{5.38}$$

These coefficients $\boldsymbol{c}(0) = \boldsymbol{c}(1) = 1/\sqrt{2}$ come from the familiar averaging filter. The division by $\sqrt{2}$ gives an ortho*normal* filter. Figure 5.2 shows the lowpass halfband filter $\boldsymbol{P}$ with response $1 + \cos\omega$. Added to $P(\omega + \pi)$ it gives the constant 2.

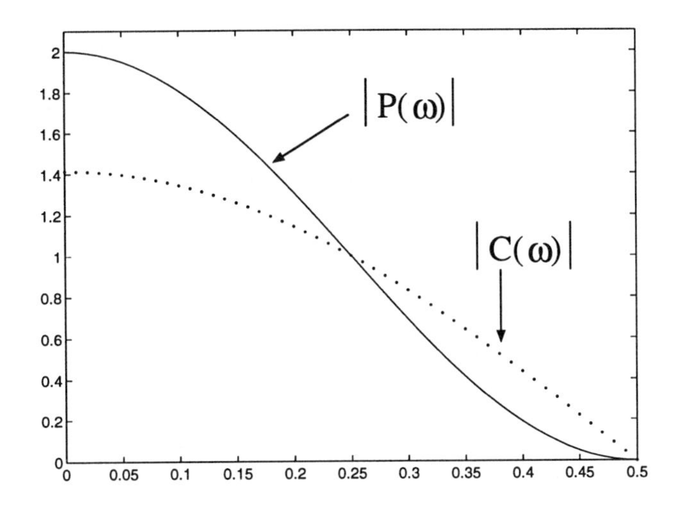

Figure 5.2: The orthonormal filter $C(\omega)$ has $P(\omega) = |C(\omega)|^2$. The normalized frequency 0.5 is $\omega = \pi$.

The requirement $P(\omega) \geq 0$ is crucial. Otherwise we could not factor $P(\omega)$ into $|C(\omega)|^2$. The halfband filter with coefficients $\boldsymbol{p}(-1) = \boldsymbol{p}(0) = \boldsymbol{p}(1) = 1$ could never be $|C(\omega)|^2$. Its response $P(\omega) = e^{i\omega} + 1 + e^{-i\omega}$ is *negative* at $\omega = \pi$.

Example 5.2. The Daubechies 4-tap filter picks out $C(\omega)$ from $P = |C|^2$ when

$$P(\omega) = (1 + \cos\omega)^2 \left(1 - \tfrac{1}{2}\cos\omega\right). \tag{5.39}$$

Note the double zero at $\omega = \pi$, coming from $(1 + \cos\omega)^2$. If we keep only that factor, this $P(\omega)$ would be the square of the previous example. Its factor would be the square of the previous $C(\omega)$, namely $\frac{1}{2} + e^{-i\omega} + \frac{1}{2}e^{-2i\omega}$. Those coefficients $\frac{1}{2}, 1, \frac{1}{2}$ are important—they will lead to the *hat function*, when we study wavelets. But $P(\omega) = (1 + \cos\omega)^2$ does not by itself yield a halfband filter, so the lowpass C with coefficients $\frac{1}{2}, 1, \frac{1}{2}$ cannot go into an orthogonal filter bank. The hat function is not orthogonal to its translates.

To repeat: $(1 + \cos\omega)^2$ includes the term $\cos^2\omega$. This produces an even frequency $\cos 2\omega$. In the z-domain we are squaring $1 + \frac{1}{2}\left(z^{-1} + z\right)$, which produces the even power z^{-2}. This is not halfband! Daubechies' extra factor $1 - \frac{1}{2}\cos\omega$ must be included, to cancel the 2ω term in $P(\omega)$ and the z^2 term in $P(z)$. We will have orthogonality, thanks to that factor.

To see that $P(\omega)$ is halfband, multiply it out. The $\cos 2\omega$ term is missing:

$$P(\omega) = \left(1 + 2\cos\omega + \cos^2\omega\right)\left(1 - \tfrac{1}{2}\cos\omega\right) = 1 + \tfrac{3}{2}\cos\omega - \tfrac{1}{2}\cos^3\omega.$$

In the z-domain, the z^2 term is missing. Its coefficient $\boldsymbol{p}(2)$ is zero:

$$P(z) = -\tfrac{1}{16}z^3 + \tfrac{9}{16}z + 1 + \tfrac{9}{16}z^{-1} - \tfrac{1}{16}z^{-3}. \tag{5.40}$$

We know that $P(\omega) \geq 0$ because $1 - \frac{1}{2}\cos\omega > 0$. Therefore it can be factored into $|C(\omega)|^2$ (*spectral factorization*). This is not a trivial calculation. It is made easier by the fact that we already know the factor for $1 + \cos\omega$, from the first example. This leaves only the linear piece $Q(\omega) = 1 - \frac{1}{2}\cos\omega$ or $Q(z) = 1 - \frac{1}{4}\left(z^{-1} + z\right)$. We factor Q in three steps:

$$1 - \tfrac{1}{4}\left(z^{-1} + z\right) \quad = \quad (\boldsymbol{b}(0) + \boldsymbol{b}(1)z^{-1})\,(\boldsymbol{b}(0) + \boldsymbol{b}(1)z) \tag{5.41}$$

$$1 = \boldsymbol{b}(0)^2 + \boldsymbol{b}(1)^2 \quad \text{and} \quad -\tfrac{1}{4} = \boldsymbol{b}(0)\boldsymbol{b}(1) \tag{5.42}$$

Solving the quadratic equations gives $\boldsymbol{b}(0) = (1 + \sqrt{3})/\sqrt{8}$ and $\boldsymbol{b}(1) = (1 - \sqrt{3})/\sqrt{8}$. The solutions are real because $1 - \frac{1}{2}\cos\omega$ is safely positive.

Another approach, basically the same, is to multiply (5.41) by z to get an ordinary quadratic. The quadratic formula gives its roots as $2 \pm \sqrt{3}$. Since $Q = 0$ at these roots, we have

$$\boldsymbol{b}(0) + \boldsymbol{b}(1)\,(2 \pm \sqrt{3})^{-1} = 0. \tag{5.43}$$

The previous solution is correct because $1+\sqrt{3}+(1-\sqrt{3})(2-\sqrt{3})^{-1} = 0$. This is the *minimum phase solution*, from the root $2 - \sqrt{3}$ inside the unit circle. There is another solution from $2 + \sqrt{3}$, in which $\boldsymbol{b}(0)$ and $\boldsymbol{b}(1)$ are exchanged. This is *maximum phase*. Two more solutions come from reversing signs to $-\boldsymbol{b}(0)$ and $-\boldsymbol{b}(1)$. We are seeing the limited number of possible spectral factors in $|C(\omega)|^2$. The general rule for higher degree polynomials and longer filters is the same:

Minimum phase: Choose roots of $z^N P(z)$ that are on or *inside* $|z| = 1$.

Maximum phase: Choose roots of $z^N P(z)$ that are on or *outside* $|z| = 1$.

Mixed phase is also possible, choosing some roots inside and some outside. That can bring us to linear phase. We will show how linear phase factors of the Daubechies polynomials lead to *biorthogonal filter banks* which are among the current favorites.

Completion of Example. We factored $1 - \frac{1}{2}\cos\omega$ and $1 + \cos\omega$. Multiply to obtain

$$\begin{aligned} C(z) &= \tfrac{1}{4\sqrt{2}}\left(1 + z^{-1}\right)^2\left(\left(1 + \sqrt{3}\right) + \left(1 - \sqrt{3}\right)z^{-1}\right) \\ &= \tfrac{1}{4\sqrt{2}}\left[\left(1 + \sqrt{3}\right) + \left(3 + \sqrt{3}\right)z^{-1} + \left(3 - \sqrt{3}\right)z^{-2} + \left(1 - \sqrt{3}\right)z^{-3}\right]. \end{aligned} \tag{5.44}$$

Those are the four coefficients $\boldsymbol{c}(0), \ldots, \boldsymbol{c}(3)$ ***of the famous Daubechies filter*** $\boldsymbol{D_4}$. In our present normalization, they are divided by $\sqrt{32} = 4\sqrt{2}$. In other normalizations they are divided by 4. Remember their two key properties:

1. The halfband filter has $P(z) + P(-z) = 2$. The factor $C(z)$ goes into an orthonormal filter bank. $D(z)$ comes from $C(z)$ by an alternating flip.

2. The response $C(z)$ has a double zero at $z = -1$. In frequency, $C(\omega)$ has a double zero at $\omega = \pi$. The response is *flat* at π because of $(1 + \cos\omega)^2$.

The double zero at $\omega = \pi$ will produce *two vanishing moments* for the Daubechies wavelets in Chapter 6.

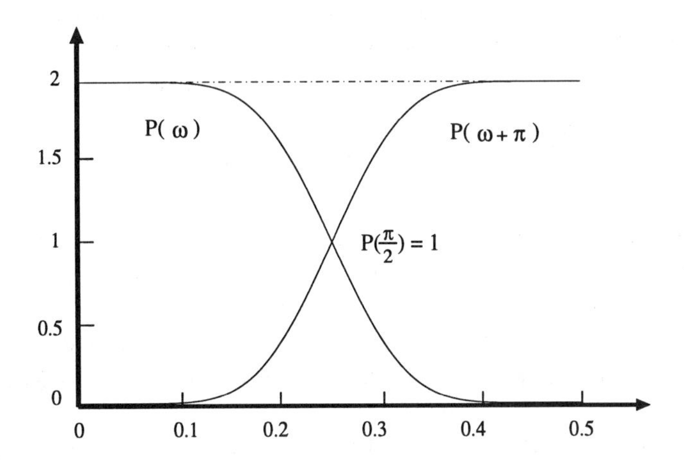

Figure 5.3: A halfband filter $P(\omega)$ and its mirror image $P(\omega+\pi)$. Their sum is constant.

Figure 5.3 shows a graph of $|C(\omega)|^2$, so you can see the halfband property that produces orthogonality. The highpass response in that figure is $|C(\omega+\pi)|^2$, which equals $|D(\omega)|^2$. The sum of the two is constant, so there is no amplitude distortion. The flatness gives great accuracy near $\omega = 0$ and $\omega = \pi$ (not so great in the middle). The filter bank gives perfect reconstruction.

Problem Set 5.3

1. Solve equations (5.42) for $x = b(0)^2$. Confirm the factorization.
2. What is the 2 by 2 polyphase matrix $\boldsymbol{H}_p(z)$ from the Daubechies $C(z)$ and the alternating flip?
3. What is the 2 by 2 modulation matrix $\boldsymbol{H}_m(z)$ for that four-tap Daubechies example? Verify that $\widetilde{\boldsymbol{H}}_m \boldsymbol{H}_m = 2\boldsymbol{I}$.
4. If a linear phase halfband filter satisfies $G(z) + G(-z) = z^{-l}$, what is the relation between l and N? Can $G(z)$ be an antisymmetric?

5.4 Spectral Factorization

In an orthonormal filter bank, $C(z) = \sqrt{2}H(z)$ is a *spectral factor* of a symmetric halfband filter $P(z)$. The factorization is $P(z) = C(z^{-1})C(z)$ and the halfband property is $P(z) + P(-z) = 2$. In frequency, $P(\omega) = |C(\omega)|^2$ achieves the orthogonality condition $|C(\omega)|^2 + |C(\omega+\pi)|^2 = 2$. In the reverse direction, $P(z)$ is the *autocorrelation* of $C(z)$. This intimate relation of spectral factor $C(z)$ and its autocorrelation $P(z)$ is fundamental throughout signal processing.

Two questions arise immediately:

1. (Theory) Can every polynomial with $P(\omega) \geq 0$ be factored into $|C(\omega)|^2$?

2. (Practice) How is this spectral factorization actually done?

The answer to Question 1 is *yes*. This is the Féjer-Riesz Theorem. The answer to Question 2 is not so quick. There are many competing algorithms for spectral factorization. Short filters offer no serious difficulty, but with 100 or even 50 coefficients the weaker algorithms become slow and/or unreliable. When $C(\omega)$ is only approximate, the reconstruction is not perfect.

The trigonometric polynomials $P(\omega)$ and $C(\omega)$ are both of degree N:

$$\sum_{-N}^{N} p(n)e^{-in\omega} = |C(e^{i\omega})|^2 = \left|\sum_{0}^{N} c(n)e^{-in\omega}\right|^2.$$

$P(z)$ has symmetric coefficients $p(n) = p(-n)$. There are $N+1$ independent coefficients in P and the same number in C. They are linked by quadratic equations, when we solve $P(\omega) = |C(\omega)|^2$. Those equations are solvable if and only if $P(\omega) \geq 0$ for all ω.

As an aside, note that *matrix* spectral factorization is also possible where $P(\omega)$ is symmetric positive definite. Both 1 and 2, theory and practice, are nontrivial. The Riccati equation is involved.

We indicate four factorization methods. Three are actually used; Method C is for conversation only. The first method begins by finding the zeros of a polynomial (by a good algorithm!). This proves that spectral factorization is possible, by doing it.

Method A (zeros of a polynomial). With real symmetric coefficients $p(n)$, we have $P(z) = P(1/z)$. ***If*** z_i ***is a root, so is*** $1/z_i$. When z_i is inside the unit circle, $1/z_i$ is outside. The roots z_j *on* the unit circle must have even multiplicity, by the crucial assumption that $P(\omega) \geq 0$. Therefore the polynomial $z^N P(z)$ of degree $2N$, with leading coefficient $p(N) \neq 0$, must have these $2N$ factors:

$$z^N P(z) = p(N) \prod_{i=1}^{M} (z - z_i)\left(z - \frac{1}{z_i}\right) \prod_{j=1}^{N-M} (z - z_j)^2. \tag{5.45}$$

This contains the key point, but we know more. Real coefficients ensure that the complex conjugate $\overline{z}$ is a root when z is a root. The complex roots off the unit circle actually come *four at a time*: z_i and $\overline{z}_i$ inside, $1/z_i$ and $1/\overline{z}_i$ outside. The complex roots on the circle also come four at a time: z_j twice and $\overline{z}_j$ twice. Real roots on the circle come two at a time (even multiplicity).

Now construct $C(z)$ by taking *all* the roots z_i (including $\overline{z}_i$) inside the circle, and also take one out of every double root z_j on the circle:

$$z^N C(z) = |p(N)|^{1/2} \prod_{i=1}^{M} (z - z_i) \prod_{j=1}^{N-M} (z - z_j). \tag{5.46}$$

This is the "minimum phase spectral factor." It has no roots outside the circle. The coefficients of $C(z)$ are still real, because the complex roots are automatically in conjugate pairs: $\overline{z}_i$ and $\overline{z}_j$ came with z_i and z_j.

Example 5.3. The 4-tap Daubechies filter in the previous section led to zeros at $z_i = 2 - \sqrt{3}$ and $z_i^{-1} = 2 + \sqrt{3}$. The other four roots of $z^3 P(z)$ are at $z_j = -1$ (on the unit circle and again real). Two of those roots go into the spectral factor (5.46). Thus $z^3 C(z)$ is a cubic polynomial with roots $2 - \sqrt{3}$, -1, and -1. It is minimum phase.

Every factorization of $P(z)$ into $F(z)H(z)$ must put some roots into $H(z)$ and the remaining roots into $F(z)$. The rules for this separation of roots of $P(z)$ are:

- For F and H to be ***real*** filters, z and $\overline{z}$ must stay together.
- For F and H to be ***symmetric*** filters, z and z^{-1} must stay together.

- For F to be the transpose of H, z and z^{-1} must go separately. This is the spectral factorization $C(z)\,C(z^{-1})$ that gives an orthogonal filter bank when P is halfband.

Figure 5.4a shows a partition of the zeros into circles and squares that makes both factors symmetric. The splitting in Figure 5.4b makes one factor the transpose (coefficients reversed) of the other factor. *To achieve both properties at the same time, all zeros of $P(z)$ – not just the zeros on the unit circle – must be of even multiplicity.* We now show that this is impossible for a halfband filter. This gives another proof of Theorem 5.3, that orthogonality conflicts with symmetry.

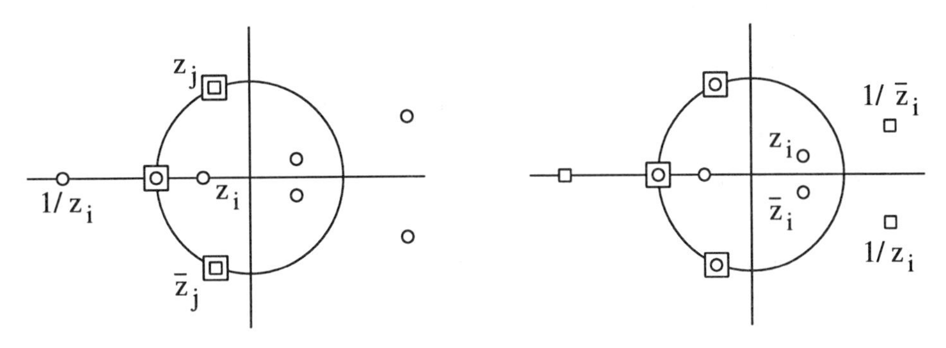

Figure 5.4: Twelve zeros of a halfband filter separate into analysis (○) and synthesis (□).

Theorem 5.4 *A symmetric orthogonal FIR $C(z)$ can only have two nonzero coefficients.*

Proof. An FIR filter is symmetric when $z^N C(z) = C(z^{-1})$. If z_i is a zero of $C(z)$, so is z_i^{-1}. Then $P(z) = C(z^{-1})C(z)$ has a *double root* at z_i. More precisely, all roots of the polynomial $z^N P(z)$ have even multiplicity. This polynomial is a perfect square $[R(z)]^2$.

If the filter is also orthogonal, $P(z)$ must be halfband:

$$z^N P(z) = [\boldsymbol{r}(0) + \cdots + \boldsymbol{r}(N)z^N]^2 \quad \text{has only one odd power } z^N.$$

The first odd power in $R(z)$ produces (when it multiplies $\boldsymbol{r}(0)$) an odd power in $[R(z)]^2$. The last even power in $R(z)$ also produces (when it multiplies $\boldsymbol{r}(N)\,z^N$) an odd power in $[R(z)]^2$. But our halfband filter has only one odd power. *We cannot allow any even powers or any odd powers in the terms $\cdots$ indicated by the three dots.* The polynomial $R(z)$ has only two terms and $P(z)$ has three terms. Then $C(z)$ only has two nonzero coefficients.

All symmetric orthogonal FIR filters have $C(z) = (1+z^{-N})/\sqrt{2}$ with odd N. The halfband product filter is $P(z) = \frac{1}{2}z^N + 1 + \frac{1}{2}z^{-N}$.

Example 5.4. For symmetric filters, the roots $z_i = 2-\sqrt{3}$ and $z_i^{-1} = 2+\sqrt{3}$ must stay together when we factor $P(z)$. The four roots at $z = -1$ can be split between $H(z)$ and $F(z)$. One symmetric splitting is $H(z) = (1 + z^{-1})(-1 + 4z^{-1} - z^{-2})/2\sqrt{2}$ and $F(z) = (1+z^{-1})^3/4\sqrt{2}$. There are several symmetric factorizations, but none of them can be orthogonal.

Now we return to **computation of zeros of polynomials**. For long filters, a good algorithm is needed to find the zeros z_i and z_j of $P(z)$. We quote from the 1994 abstract by Lang and Frenzel [La,Fr]:

> Finding polynomial roots rapidly and accurately is an important problem in many areas of signal processing. We use Müller's method for computing a root of the deflated polynomial. This estimate is improved by applying Newton's method to the original polynomial. Furthermore we give a simple approach to improve the accuracy for spectral factorization when there are double roots on the unit circle.

Müller's method uses three previous estimates of the root of $z^N P(z)$ to find the next estimate. The parabola that interpolates at the three old points has a root at the new point. Since parabolas can have complex roots, Müller's algorithm can find complex roots from a real start — while Newton can only move chaotically on the real line.

Newton's method uses the most recent estimate z_k. For real roots, the tangent line at z_k to the graph of $z^N P(z)$ crosses zero at the new point z_{k+1}. This is the outstanding method for solving nonlinear equations, provided z_0 is close enough — which is the task of Müller's method. The polynomial $z^{10000} - 1$ was one of the tests (not the only one!). The code is on ftp from *cml.rice.edu* under directory *pub/software*.

MATLAB ***uses an eigenvalue method.*** Its subroutine ***roots*** is effective up to quite large degree. The roots of a polynomial $z^N + \cdots$ are the eigenvalues of its $N \times N$ *companion matrix*, which has 1's down a diagonal and minus the polynomial coefficients along a row. For example, $z^3 - 2z^2 - 5z - 9$ is specified by the vector $\boldsymbol{v} = [1 \ \ -2 \ \ -5 \ \ -9]$. The command $\boldsymbol{M} = \text{compan}(\boldsymbol{v})$ produces the matrix $\boldsymbol{M} = \begin{bmatrix} 2 & 5 & 9 \\ 1 & 0 & 0 \\ 0 & 1 & 0 \end{bmatrix}$ with $\det(z\boldsymbol{I} - \boldsymbol{M}) = z^3 - 2z^2 - 5z - 9$. MATLAB finds the eigenvalues (by the QR method) which are the roots of the polynomial.

The next section mentions how rescaling y to $4y$ allowed us to compute Daubechies filters of twice the length achievable without this scaling. The coefficients in $P(z)$ were better controlled. Linear phase filters with extremely good stopband attenuation have many zeros on or near the unit circle. These are the hardest zeros to compute.

Method B (solve quadratic equations). We are looking for $N+1$ numbers $\boldsymbol{c}(0), \ldots, \boldsymbol{c}(N)$. The $N+1$ equations are of second degree, involving $\boldsymbol{c}$'s times $\boldsymbol{c}$'s. The equations come from matching powers of $e^{i\omega}$ in $\overline{C(\omega)}C(\omega) = P(\omega)$:

$$\left(\sum_0^N c(k)e^{ik\omega}\right)\left(\sum_0^N c(k)e^{-ik\omega}\right) = \sum_{-N}^N p(n)e^{-in\omega}. \tag{5.47}$$

One way to make those equations explicit is in matrix form:

$$\begin{bmatrix} c(0) & c(1) & c(2) & \cdot & c(N) \\ & c(0) & c(1) & c(2) & \cdot \\ & & c(0) & c(1) & c(2) \\ & & & c(0) & c(1) \\ & & & & c(0) \end{bmatrix} \begin{bmatrix} c(0) \\ c(1) \\ c(2) \\ \cdot \\ c(N) \end{bmatrix} = \begin{bmatrix} p(0) \\ p(1) \\ p(2) \\ \cdot \\ p(N) \end{bmatrix}. \tag{5.48}$$

The first equation is $\boldsymbol{c}(0)^2 + \cdots + \boldsymbol{c}(N)^2 = \boldsymbol{p}(0)$. This gives the constant term in (5.47). The second equation gives the $e^{-i\omega}$ term. The last equation is $\boldsymbol{c}(0)\boldsymbol{c}(N)e^{-iN\omega} = \boldsymbol{p}(N)\, e^{-iN\omega}$.

Equation (5.48) is not a linear system! It is quadratic, in fact homogeneous of degree 2. It has a real solution if and only if $P(\omega) \geq 0$ for all ω. This is not an easy condition to verify on the coefficients $\boldsymbol{p}(n)$. If $P(\omega) \geq 0$ is not true—in which case our solution methods must fail—we can add enough to the DC term $\boldsymbol{p}(0)$ to make it true.

For orthogonality, the $\boldsymbol{p}(n)$ come from a halfband filter. The even coefficients $\boldsymbol{p}(2), \boldsymbol{p}(4), \ldots$ are all zero. But our discussion is not in any way limited to this halfband case. Spectral factorization applies to all filters with $P(\omega) \geq 0$. It even applies to IIR filters, but those lead to infinitely many equations.

Example 5.5. The previous section factored $P(\omega) = 1 - \frac{1}{4}\left(e^{-i\omega} + e^{i\omega}\right)$. Comparing coefficients of 1 and $e^{-i\omega}$ led us to

$$\begin{aligned} \boldsymbol{c}(0)^2 + \boldsymbol{c}(1)^2 &= 1 \\ \boldsymbol{c}(0)\boldsymbol{c}(1) &= -\tfrac{1}{4} \end{aligned} \quad \text{or} \quad \begin{bmatrix} \boldsymbol{c}(0) & \boldsymbol{c}(1) \\ 0 & \boldsymbol{c}(0) \end{bmatrix} \begin{bmatrix} \boldsymbol{c}(0) \\ \boldsymbol{c}(1) \end{bmatrix} = \begin{bmatrix} 1 \\ -\frac{1}{4} \end{bmatrix}. \tag{5.49}$$

This is our system (5.48) for that particular example with $N = 1$. Eliminating $\boldsymbol{c}(1)$ gave a single quadratic equation. The unknown was $x = \boldsymbol{c}(0)^2$ and the equation was $x + \frac{1}{16x} = 1$ or $16x^2 - 16x + 1 = 0$. For $N > 1$ we *cannot* reduce the $N + 1$ quadratic equations to a single equation for $\boldsymbol{c}(0)$. An approximate solution by method A, B, C, or D (or another method E) is the best we can expect.

To use a nonlinear equation solver, write the kth quadratic equation as $\frac{1}{2}\boldsymbol{c}^T\boldsymbol{Q}(k)\boldsymbol{c} = \boldsymbol{p}(k)$. The symmetric matrix $\boldsymbol{Q}(k)$ has 1's along its kth subdiagonal and superdiagonal (and $\boldsymbol{Q}(0) = 2\boldsymbol{I}$). Here is $\boldsymbol{Q}(2)$ with $N = 3$:

$$\tfrac{1}{2}\begin{bmatrix} \boldsymbol{c}(0) & \boldsymbol{c}(1) & \boldsymbol{c}(2) & \boldsymbol{c}(3) \end{bmatrix} \begin{bmatrix} 0 & 0 & 1 & 0 \\ 0 & 0 & 0 & 1 \\ 1 & 0 & 0 & 0 \\ 0 & 1 & 0 & 0 \end{bmatrix} \begin{bmatrix} \boldsymbol{c}(0) \\ \boldsymbol{c}(1) \\ \boldsymbol{c}(2) \\ \boldsymbol{c}(3) \end{bmatrix} = \boldsymbol{p}(2). \tag{5.50}$$

This is $\boldsymbol{c}(0)\boldsymbol{c}(2) + \boldsymbol{c}(1)\boldsymbol{c}(3) = \boldsymbol{p}(2)$ from matching the $e^{-2i\omega}$ terms in $|C(\omega)|^2 = P(\omega)$. The partial derivatives of that left side $\boldsymbol{L}(2)$ are in the gradient vector $\boldsymbol{Q}(2)\boldsymbol{c}$:

$$\begin{aligned} \partial \boldsymbol{L}(2)/\partial \boldsymbol{c}(0) &= \boldsymbol{c}(2) \\ \partial \boldsymbol{L}(2)/\partial \boldsymbol{c}(1) &= \boldsymbol{c}(3) \\ \partial \boldsymbol{L}(2)/\partial \boldsymbol{c}(2) &= \boldsymbol{c}(0) \\ \partial \boldsymbol{L}(2)/\partial \boldsymbol{c}(3) &= \boldsymbol{c}(1) \end{aligned} \quad \text{agrees with} \quad \begin{bmatrix} 0 & 0 & 1 & 0 \\ 0 & 0 & 0 & 1 \\ 1 & 0 & 0 & 0 \\ 0 & 1 & 0 & 0 \end{bmatrix} \begin{bmatrix} \boldsymbol{c}(0) \\ \boldsymbol{c}(1) \\ \boldsymbol{c}(2) \\ \boldsymbol{c}(3) \end{bmatrix}.$$

The kth quadratic function is $\boldsymbol{L}(k) = \frac{1}{2}\boldsymbol{c}^T\boldsymbol{Q}(k)\boldsymbol{c}$, and its gradient is $\boldsymbol{Q}(k)\boldsymbol{c}$.

The second derivatives are also desired by nonlinear subroutines, and also readily available. They are in the constant matrix $\boldsymbol{Q}(k)$. One successful program using gradients and second derivatives has been the Quadratic Constrained Least Squares (QCLS) optimization code by anonymous ftp from *eceserv0.ece.wisc.edu* under the directory *pub/nguyen/software/QCLS*. We use it in Chapter 9 to design M-band filter banks.

Method C (matrix factorization). Equation (5.48) is an attractive form for the quadratic equations. But there is a more symmetric form. If you like infinite matrices, you will enjoy this. Instead of a finite matrix times a vector, it has an infinite constant-diagonal matrix $\boldsymbol{C}$ times its transpose:

$$C^T C = \begin{bmatrix} c(0) & c(1) & \cdot & c(N) & & & & \\ & c(0) & c(1) & \cdot & c(N) & & & \\ & & c(0) & c(1) & \cdot & c(N) & & \\ & & & \cdot & \cdot & \cdot & \cdot & \end{bmatrix} \begin{bmatrix} c(0) & & & \\ c(1) & c(0) & & \\ \cdot & c(1) & c(0) & \\ c(N) & \cdot & c(1) & \cdot \\ & c(N) & \cdot & \cdot \\ & & c(N) & \cdot \\ & & & \cdot \end{bmatrix} = P.$$

The columns of P have entries $p(-N), \dots, p(0), \dots, p(N)$. In this matrix form of $|C(\omega)|^2 = P(\omega)$, the symmetric matrix P factors into *upper triangular C^T times lower triangular C*. This is only possible if P is *positive semi-definite*—which is exactly our condition $P(\omega) \geq 0$.

Matrices are decomposed into triangular factors every day. This is the matrix statement of ordinary Gaussian elimination. The factorization is usually written $A = LU$. It gives *lower triangular times upper triangular*. Fortunately, infinite constant-diagonal matrices commute; our equation is also $CC^T = P$. The harder problem is to factor infinite matrices in finite time.

Approximate method: take a finite section of P. Keep R rows and columns, where R is larger (perhaps much larger) than N. This finite piece P_R is still symmetric and positive definite. Therefore P_R can be factored into $C_R C_R^T$, where C_R is lower triangular. That is the *Cholesky factorization* of P_R. It is a symmetrized form of $A = LU$, available because P_R is symmetric positive definite.

The finite matrix C_R does not contain the exact $c(k)$. It is not even true that C_R has constant diagonals (although P_R has). The reduction to a finite matrix has chopped the tail ends of the row-column multiplications, either in $C^T C$ or in CC^T. But the rows of the computed factor C_R *do approach the rows of C*. The correct (minimum-phase) coefficients $c(k)$ appear in the limit as $R \to \infty$.

We demonstrate with $P(\omega) = 1 - \frac{1}{2}\cos\omega$. This Daubechies example was solved exactly for $c(0) = (1+\sqrt{3})/\sqrt{8} = 0.9659$ and $c(1) = (1-\sqrt{3})/\sqrt{8} = -0.2588$. Take $R = 4$ and use ***chol*** in MATLAB to factor P_4 into CC^T:

$$P_4 = \begin{bmatrix} 1 & -0.25 & & \\ -0.25 & 1 & -0.25 & \\ & -0.25 & 1 & -0.25 \\ & & -0.25 & 1 \end{bmatrix} \quad \text{and}$$

$$C = \begin{bmatrix} 1 & & & \\ -0.25 & 0.9682 & & \\ & -0.2582 & 0.9661 & \\ & & -0.2588 & 0.9659 \end{bmatrix}.$$

This matrix has $c(0)$ and $c(1)$ correct to four places in the last row. But with long filters this method is very slow.

Method D (Cepstral method: Take logarithms). The idea is to convert the multiplication $P(z) = C(z^{-1})C(z)$ into addition. Formally, $\log\left(\sum p(n)z^{-n}\right)$ is easily separated into positive and negative powers of z. The symmetry $p(n) = p(-n)$ and the positivity $P(\omega) \geq 0$ yield a logarithm $L(z)$ with coefficients $l(n) = l(-n)$:

$$\log P(z) = \sum_{-\infty}^{\infty} l(n)z^{-n} = \left(\frac{l(0)}{2} + \sum_{1}^{\infty} l(n)z^{n}\right) + \left(\frac{l(0)}{2} + \sum_{1}^{\infty} l(n)z^{-n}\right).$$

These are infinite series. The logarithm of a polynomial is not a polynomial. This means that our finite computations can only be approximate. The easy separation into $\log C(z^{-1}) + \log C(z)$ is the key advantage of the method.

The sequence $\boldsymbol{l}(n)$ is the *complex cepstrum* of $\boldsymbol{p}(n)$. The series for $L(z)$ converges in an annulus $|z_i| < |z| < 1/|z_i|$ of the complex plane. Here z_i is the largest root of $P(z)$ inside the unit circle, and $1/z_i$ is the smallest root outside. (The method is in trouble with roots z_j *on* the circle. Best to remove those first. Otherwise $L(z) = \log P(z)$ will be infinite at z_j and the series cannot converge.) The computation of the $\boldsymbol{c}$'s requires an inverse Fourier transform of $\log P(z)$. Then the $\boldsymbol{c}$'s are computed from the $\boldsymbol{l}$'s.

A detailed treatment of the cepstrum $\boldsymbol{l}(n)$ is given by Oppenheim and Schafer [OS]. The constant terms give $\boldsymbol{c}(0) = \exp \frac{1}{2}\boldsymbol{l}(0)$. The next term $\boldsymbol{c}(1)$ is interesting because the z^{-1} terms only involve $\boldsymbol{l}(0)$ and $\boldsymbol{l}(1)$. The recursion for $n = 1, 2, \ldots$ turns out to be

$$\boldsymbol{c}(n) = \boldsymbol{l}(n)\boldsymbol{c}(0) + \tfrac{n-1}{n}\boldsymbol{l}(n-1)\boldsymbol{c}(1) + \cdots + \tfrac{1}{n}\boldsymbol{l}(1)\boldsymbol{c}(n-1).$$

We need only $N+1$ coefficients $\boldsymbol{l}(0), \ldots, \boldsymbol{l}(N)$ in the logarithm to find all $N+1$ coefficients $\boldsymbol{c}(0), \ldots, \boldsymbol{c}(N)$ in the spectral factor. To find those $\boldsymbol{l}(n)$ from the given $\boldsymbol{p}(n)$, we use a large-size FFT in the z-domain. A typical size is $8N$, for acceptable accuracy.

This cepstral method does *not* compute zeros of polynomials. So it doesn't find symmetric filters. It is a good way to find orthogonal filters. A code is available by anonymous ftp at *eceserv0.ece.wisc.edu* under directory *pub/nguyen/software/CEPSTRAL.*

Very optional comment. Spectral factorization also solves *singly infinite* constant-diagonal systems. (This is the genuine Toeplitz problem. Doubly infinite matrices could be named after Laurent — but mostly we still say Toeplitz.) The coefficient matrix $\boldsymbol{P}_+$ has entries $\boldsymbol{p}(i-j)$ only for $i \geq 0$ and $j \geq 0$:

$$\boldsymbol{P}_+\boldsymbol{x}_+ = \boldsymbol{b}_+ \text{ is } \begin{bmatrix} \boldsymbol{p}(0) & \boldsymbol{p}(1) & \boldsymbol{p}(2) & \cdot \\ \boldsymbol{p}(1) & \boldsymbol{p}(0) & \boldsymbol{p}(1) & \cdot \\ \boldsymbol{p}(2) & \boldsymbol{p}(1) & \boldsymbol{p}(0) & \cdot \\ \cdot & \cdot & \cdot & \cdot \end{bmatrix} \begin{bmatrix} \boldsymbol{x}(0) \\ \boldsymbol{x}(1) \\ \boldsymbol{x}(2) \\ \cdot \end{bmatrix} = \begin{bmatrix} \boldsymbol{b}(0) \\ \boldsymbol{b}(1) \\ \boldsymbol{b}(2) \\ \cdot \end{bmatrix}. \tag{5.51}$$

This corresponds in continuous time $t \geq 0$ to a *Wiener-Hopf integral equation*:

$$\int_0^\infty p(s-t)x(t)\,dt = b(s) \text{ for } s \geq 0. \tag{5.52}$$

$\boldsymbol{P}_+$ does not have constant-diagonal factors in $\boldsymbol{P}_+ = \boldsymbol{LU}$. Lower triangular times upper destroys the time-invariant pattern. (Starting at time zero is responsible.) The beautiful Wiener-Hopf idea is that *upper times lower succeeds perfectly*:

$$\boldsymbol{P}_+ = \boldsymbol{C}_+^T\boldsymbol{C}_+ = \begin{bmatrix} \boldsymbol{c}(0) & \boldsymbol{c}(1) & \boldsymbol{c}(2) & \cdot \\ & \boldsymbol{c}(0) & \boldsymbol{c}(1) & \cdot \\ & & \boldsymbol{c}(0) & \cdot \\ & & & \cdot \end{bmatrix} \begin{bmatrix} \boldsymbol{c}(0) & & & \\ \boldsymbol{c}(1) & \boldsymbol{c}(0) & & \\ \boldsymbol{c}(2) & \boldsymbol{c}(1) & \boldsymbol{c}(0) & \\ \cdot & \cdot & \cdot & \cdot \end{bmatrix}. \tag{5.53}$$

Wiener and Hopf computed this spectral factorization of $\boldsymbol{P}_+$ by Method D. One of Norbert Wiener's great theorems is that $\sum |\boldsymbol{l}(n)| < \infty$ when $\sum |\boldsymbol{p}(n)| < \infty$. With no zeros of $P(z)$ on the unit circle, he could take the logarithm even for IIR filters. The solution is

$$\boldsymbol{x}_+ = (\boldsymbol{P}_+)^{-1}\boldsymbol{b}_+ = (\boldsymbol{C}_+)^{-1}(\boldsymbol{C}_+^T)^{-1}\boldsymbol{b}_+. \tag{5.54}$$

The inverses of C_+ and C_+^T are constant-diagonal. So Wiener-Hopf can compute the spectral factorization $P(\omega) = |C(\omega)|^2$ and transform back to the time domain.

Finite constant-diagonal matrices don't have constant-diagonal factors and spectral factorization no longer succeeds. Nevertheless $\boldsymbol{P}_N\boldsymbol{x}_N = \boldsymbol{b}_N$ can be solved quickly by the Levinson algorithm or a "superfast" algorithm or by preconditioned conjugate gradients [ChSt].

We added these comments because transform methods are so central to signal processing. This is the whole underpinning of filter theory.

Problem Set 5.4

1. Suppose $P(z)$ has six zeros at $z = -1$ and four other real zeros at $z = a,\ a^{-1},\ b,\ b^{-1}$. Draw the complex plane and indicate which zeros go into the minimum phase spectral factor $C(z)$.
2. For the same ten zeros, indicate a set of zeros that produces a symmetric (linear phase) $C(z)$. Also indicate a second possibility.
3. Why must all roots of $P(z)$ on the unit circle have even multiplicity, to allow $P(z) = C(z) \times C(z^{-1})$ and $P(\omega) = |C(\omega)|^2$?
4. The coefficients of the Daubechies polynomials $C(z)$ up to order 12 are tabulated at the end of this chapter. Find the zeros using ***roots*** in MATLAB or another algorithm. What were the roots of the halfband polynomial $P(z)$?
5. Show that $P(z) = z^{-N}C(z)C(z^{-1})$ must be symmetric. If C is lowpass with passband and stopband cutoff frequencies ω_p and ω_s and errors δ_p and δ_s, what are the cutoff frequencies and the errors of P?

5.5 Maxflat (Daubechies) Filters

This section is about an important family of filters, which will lead to an outstanding family of wavelets. The same construction yields both. Wavelets come from filters with special properties. Historically, their close relation was not immediately seen — now it is the subject of Chapter 6. The importance of this special construction is in its combination of two key properties:

1. These particular filters (and wavelets) are ***orthogonal.***
2. The frequency responses have ***maximum flatness*** at $\omega = 0$ and $\omega = \pi$.

The lowpass filters will have $p = 1, 2, 3, 4, \ldots$ zeros at π. They have $2p = 2, 4, 6, 8, \ldots$ coefficients, so that $N = 2p - 1$. We use ***boldface p*** for the coefficients of $P(\omega) = |C(\omega)|^2$ and *lightface* p to count the zeros of $C(\omega)$ at $\omega = \pi$. The highpass coefficients $\boldsymbol{d}(k)$ come from an alternating flip. The first member of this family was the subject of Chapter 1: $\boldsymbol{c}(0) = \boldsymbol{c}(1) = 1/\sqrt{2}$. Note the normalization $\boldsymbol{c}(0)^2 + \boldsymbol{c}(1)^2 = 1$. These numbers go into a *unitary matrix*. For each $p = 1, 2, 3, 4, \ldots$ the filter bank is orthonormal. The product filters have degree $2N = 4p - 2$:

$$P_0(z) = \left(\frac{1+z^{-1}}{2}\right)^{2p} Q_{2p-2}(z) \quad \textit{will be halfband by special choice of } Q.$$

In the literature on filters, this family is described as *maxflat*. The coefficients were given by [Herrmann]. They were already in formulas for interpolation, described below. In the history of wavelets, we are reproducing the great 1988 discovery by Ingrid Daubechies. The filters are FIR with $2p$ coefficients. The wavelets are supported on the interval $[0, N] = [0, 2p-1]$. As p increases, the filters are increasingly "regular" and the wavelets are increasingly "smooth."

This section concentrates on filter properties and coefficients. We give a simple derivation of $P(z)$, and new facts about its zeros. The next chapters will concentrate on wavelets and the step into continuous time.

Condition O and Condition A_p

Before starting, it is helpful to count the requirements we must impose. There are $2p$ numbers to be chosen. These can be the coefficients $\boldsymbol{c}(0), \ldots, \boldsymbol{c}(2p-1)$ in the lowpass filter, with frequency response $C(\omega)$. They could equally well be the coefficients $\boldsymbol{p}(0), \ldots, \boldsymbol{p}(2p-1)$ of the centered (even) polynomial $P(\omega) = |C(\omega)|^2$. The $\boldsymbol{c}$'s come from the $\boldsymbol{p}$'s by spectral factorization. The nonnegative polynomial $P(\omega)$ is factored by the methods of the previous section:

$$P(\omega) = \sum_{1-2p}^{2p-1} \boldsymbol{p}(n)e^{-in\omega} \text{ equals } |C(\omega)|^2 = \left|\sum_{0}^{2p-1} \boldsymbol{c}(n)e^{-in\omega}\right|^2. \tag{5.55}$$

Our formulas yield the numbers $\boldsymbol{p}(n) = \boldsymbol{p}(-n)$. Except for the first few filters in the family, there are no simple formulas for $\boldsymbol{c}(n)$.

These $2p$ numbers are determined by p conditions for orthogonality from Condition O, and p conditions for a flat response from Condition A. More precisely, the requirement is "Condition A_p"—the subscript indicates the order of flatness at $\omega = \pi$ (and $\omega = 0$). Here are the $p + p$ conditions:

Condition O $\quad P = |C|^2$ is a normalized halfband filter:

$$\boldsymbol{p}(0) = 1 \text{ and } \boldsymbol{p}(2) = \boldsymbol{p}(4) = \cdots = \boldsymbol{p}(2p-2) = 0. \tag{5.56}$$

Condition A_p $\quad C(\omega)$ has a zero of order p at $\omega = \pi$:

$$C(\pi) = C'(\pi) = \cdots = C^{(p-1)}(\pi) = 0. \tag{5.57}$$

The equation $C(\pi) = 0$ says that $\sum \boldsymbol{c}(n)(-1)^n = 0$. The odd-numbered coefficients have the same sum as the even-numbered coefficients:

$$\text{Condition } A_1 \text{ on } \boldsymbol{c}(n){:} \quad \sum_{\text{odd } n} \boldsymbol{c}(n) = \sum_{\text{even } n} \boldsymbol{c}(n). \tag{5.58}$$

This is the first of the "sum rules." Altogether we can impose the pth order zero in (5.57) as p sum rules on the coefficients:

$$\text{Condition } A_p \text{ on } \boldsymbol{c}(n){:} \quad \sum_{n=0}^{2p-1} (-1)^n n^k \boldsymbol{c}(n) = 0 \quad \text{for } k = 0, 1, \ldots, p-1. \tag{5.59}$$

The factor n^k comes from the kth derivative of $\sum \boldsymbol{c}(n)e^{-in\omega}$. Then $(-1)^n$ comes from substituting $\omega = \pi$. The convention for n^0 is 1.

Note on $C(0) = \sqrt{2}$: The sum rule (5.58) also applies to the coefficients $p(n)$, because $P(\omega) = |C(\omega)|^2$ also vanishes at $\omega = \pi$. The odd sum must be 1, since the only nonzero even-numbered coefficient is $p(0) = 1$:

$$\sum_{\text{odd } n} p(n) = \sum_{\text{even } n} p(n) = p(0) = 1. \tag{5.60}$$

The sum over all n is $P(0) = 2$. Then $P(\omega) = |C(\omega)|^2$ yields $C(0) = \pm\sqrt{2}$. We always choose the plus sign for a lowpass filter, so the DC term at $\omega = 0$ is not reversed in sign:

$$\sum_{\text{all } n} c(n) = C(0) = \sqrt{P(0)} = \sqrt{2}. \tag{5.61}$$

The p zeros at π mean that $C(\omega)$ *has a factor* $\left(1 + e^{-i\omega}\right)^p$:

$$\text{Condition } A_p \text{ on } C(\omega)\text{:} \quad C(\omega) = \left(\frac{1 + e^{-i\omega}}{2}\right)^p R(\omega). \tag{5.62}$$

$R(\omega)$ has degree $p - 1$, to bring the total degree of $C(\omega)$ to $2p - 1$. You could say that the pth order flatness is accounted for by $\left(1 + e^{-i\omega}\right)^p$. Then the p coefficients in $R(\omega)$ are chosen to satisfy the p equations of Condition O.

To repeat: p equations for orthogonality and p equations for flatness. Condition O is applied to $P(\omega)$; it must be halfband. Condition A_p is applied to $C(\omega)$; it must have the factor $(1 + e^{-i\omega})^p$. This is easily converted to a condition on $P(\omega) = |C(\omega)|^2$, when we use $\left|1 + e^{-i\omega}\right|^2 /2 = (1 + \cos\omega)$:

$$\text{Condition } A_p \text{ on } P(\omega)\text{:} \quad P(\omega) \text{ has a factor } \left(\frac{1 + \cos\omega}{2}\right)^p. \tag{5.63}$$

Formulas for $P(\omega)$

We intend to give two formulas for $P(\omega) = |C(\omega)|^2$. The one associated with Ingrid Daubechies has $(1 + \cos\omega)^p$ times a sum of p terms. The formula associated with Yves Meyer gives the derivative of $P(\omega)$ as $-c(\sin\omega)^{2p-1}$. Then integration determines c and $P(\omega)$.

The best starting point is the ordinary polynomial $B_p(y)$. This has degree $p - 1$, with p coefficients. It is the binomial series for $(1 - y)^{-p}$, truncated after p terms:

$$B_p(y) = 1 + py + \frac{p(p+1)}{2}y^2 + \cdots + \binom{2p-2}{p-1} y^{p-1} = (1 - y)^{-p} + O(y^p). \tag{5.64}$$

The coefficient of y^k is $\binom{p+k-1}{k}$. The remainder has order y^p because this is the first term to be dropped. The complex zeros of this polynomial $B_p(y)$ will be all-important for the Daubechies filters.

We combine $B_p(y)$ with the factor $(1 - y)^p$ that has p zeros at $y = 1$. The variable y on $[0, 1]$ will correspond to the frequency ω on $[0, \pi]$. The product $\widetilde{P}(y) = 2(1 - y)^p B_p(y)$ has exactly the flatness we want at $y = 0$:

$$2(1 - y)^p B_p(y) = 2(1 - y)^p[(1 - y)^{-p} + O(y^p)] = 2 + O(y^p). \tag{5.65}$$

This is a polynomial of degree $2p-1$. It is the unique polynomial with $2p$ coefficients that satisfies p conditions at each endpoint:

$\widetilde{P}(y)$ ***and its first*** $p-1$ ***derivatives are zero at*** $y=0$ ***and*** $y=1$, ***except*** $\widetilde{P}(0)=2$.

Two more properties follow quickly. First, the derivative has $p-1$ zeros at both end points. It is a polynomial of degree $2p-2$ and with those zeros it must be

$$\widetilde{P}'(y) = -Cy^{p-1}(1-y)^{p-1} \quad \text{for some } C. \tag{5.66}$$

The second property comes when we add $\widetilde{P}(y)$ to $\widetilde{P}(1-y)$. The sum equals 2 at both ends and is still flat. Its $2p$ coefficients are uniquely determined — it must be the constant polynomial 2:

$$\widetilde{P}(y) + \widetilde{P}(1-y) \equiv 2. \tag{5.67}$$

At $y=\frac{1}{2}$ this gives $\widetilde{P}(\frac{1}{2})=1$. Figure 5.5 shows how $\widetilde{P}(y)$ is odd around its middle value. This "Hermite interpolating polynomial" drops from 2 to 0 with flatness at the ends. Here are the polynomials for $p=2$ and $p=3$:

$$B_2(y) = 1+2y \quad \text{and} \quad \widetilde{P}(y) = 2(1-y)^2(1+2y) = 2-6y^2+4y^3$$
$$B_3(y) = 1+3y+6y^2 \quad \text{and} \quad \widetilde{P}(y) = 2(1-y)^3B_3(y) = 2-20y^3+30y^4-12y^5.$$

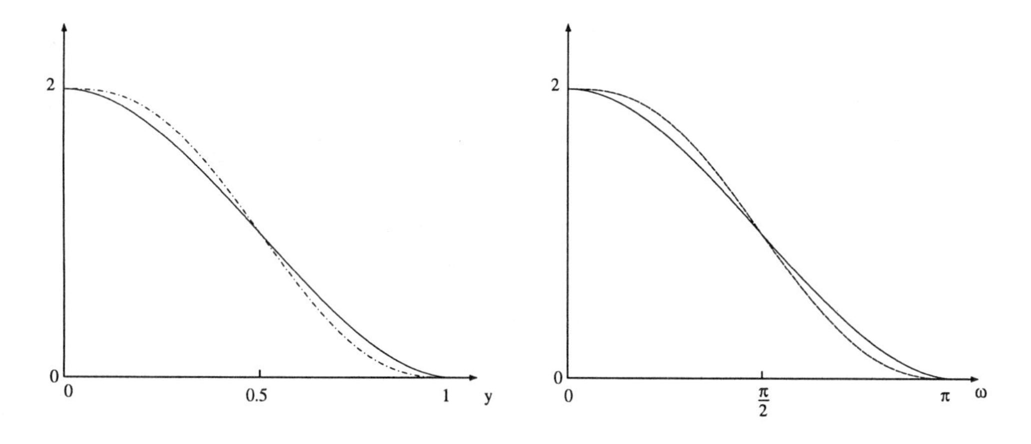

Figure 5.5: $\widetilde{P}(y)$ on the left and $P(\omega)$ on the right, for $p=2$ and $p=3$.

Now we go from ordinary polynomials in y to trigonometric polynomials in ω. The degree stays at $2p-1$. The change that takes $0 \le y \le 1$ into $0 \le \omega \le \pi$ is

$$y = \frac{1-\cos\omega}{2} \quad \text{and} \quad 1-y = \frac{1+\cos\omega}{2}. \tag{5.68}$$

The polynomial $\widetilde{P}(y)$ *becomes our desired* $P(\omega)$. We summarize its properties.

Theorem 5.5 *The polynomial* $2(1-y)^pB_p(y)$ *becomes the halfband response*

$$P(\omega) = 2\left(\frac{1+\cos\omega}{2}\right)^p \sum_{k=0}^{p-1} \binom{p+k-1}{k} \left(\frac{1-\cos\omega}{2}\right)^k. \tag{5.69}$$

This satisfies Conditions O and A_p. Its Meyer form, by integrating $P'(\omega)$ and choosing c to give $P(\pi) = 0$, is

$$P(\omega) = 2 - c\int_0^{\omega} (\sin\omega)^{2p-1} d\omega. \tag{5.70}$$

For $p = 1, 2, 3$ the Daubechies and Meyer forms are

$$\begin{aligned} P(\omega) &= 1 + \cos\omega = 2 - \textstyle\int_0^{\omega} \sin\omega\, d\omega \\ P(\omega) &= (1+\cos\omega)^2(1 - \tfrac{1}{2}\cos\omega) = 2 - \tfrac{3}{2}\textstyle\int_0^{\omega} \sin^3\omega\, d\omega \\ P(\omega) &= (1+\cos\omega)^3(1 - \tfrac{9}{8}\cos\omega + \tfrac{3}{8}\cos^2\omega) = 2 - \tfrac{15}{4}\textstyle\int_0^{\omega} \sin^5\omega\, d\omega \end{aligned}$$

Most authors emphasize the Daubechies form, with its highly visible factor $(1 + \cos\omega)^p$. That immediately ensures a pth order zero for the factors at $\omega = \pi$. Spectral factorization is speeded up, because only a lower-degree polynomial remains. It may not be so clear that (5.69) is a half-band filter. The even powers like $\cos^2\omega$ and $\cos^4\omega$ must disappear and they do. In the explicit formula for $p = 2$, multiplication produces $P(\omega) = 1 + \frac{3}{2}\cos\omega - \frac{1}{2}\cos^3\omega$.

The halfband property is $P(\omega) + P(\omega + \pi) \equiv 2$. This addition cancels the odd powers of $\cos\omega$, and the even powers are not present (except the constant term 1). This identity follows immediately from (5.67) because $1 - y = \frac{1+\cos\omega}{2} = \frac{1-\cos(\omega+\pi)}{2}$:

$$\widetilde{P}(y) + \widetilde{P}(1-y) \equiv 2 \text{ becomes } P(\omega) + P(\omega+\pi) \equiv 2. \tag{5.71}$$

The reader recognizes this "Condition O" as $|C(\omega)|^2 + |C(\omega+\pi)|^2 = 2$.

The halfband property is immediate in the Meyer form, with absolutely no calculations. Replace y by $(1 - \cos\omega)/2$ in (5.66) to find $P'(\omega)\, d\omega$:

$$-Cy^{p-1}(1-y)^{p-1}dy = -C\left(\frac{1-\cos\omega}{2}\right)^{p-1}\left(\frac{1+\cos\omega}{2}\right)^{p-1}\frac{\sin\omega}{2}\, d\omega. \tag{5.72}$$

This is $-c(1 - \cos^2\omega)^{p-1}\sin\omega\, d\omega$, which is also $-c(\sin\omega)^{2p-1}\, d\omega$. Its integral is

$$-c\int \left(1 - \cos^2\omega\right)^{p-1} \sin\omega\, d\omega = \text{odd powers of } \cos\omega.$$

The only even frequency is a constant of integration. The filter is halfband.

The flatness condition requires first of all that $P(\pi) = 0$. The constant c makes this true. The derivative $P'(\omega) = -c(\sin\omega)^{2p-1}$ has a zero of order $2p - 1$ at $\omega = \pi$. Then P itself has a zero of order $2p$. Its factor C has a zero of order p. Condition A_p is satisfied and Meyer's formula is confirmed.

Note that $P(\omega)$ *decreases monotonically* from $P(0) = 2$ to $P(\pi) = 0$. Its derivative $-c(\sin\omega)^{2p-1}$ is everywhere negative between 0 and π. There are no ripples in Figure 5.5. Therefore $P(\omega) \geq 0$ for all ω, and a factorization into $|C(\omega)|^2$ is assured.

The transition from passband (low frequencies) to stopband (high frequencies) becomes steeper and sharper as p increases. The slope at the midpoint $\omega = \frac{\pi}{2}$ is $-c\left(\sin\frac{\pi}{2}\right)^{2p-1}$, which is $-c$. We will show that c increases asymptotically like $\sqrt{p}$ as $p \to \infty$. Thus the transition band has width of order $1/\sqrt{p}$.

The Halfband Filter $P(z)$

Now we change from y and ω to the complex variable z. This will produce the filter coefficients in $P(z)$. That polynomial will be halfband and centered. The shifted polynomial $P_0(z) = z^{-N}P(z) = z^{1-2p}P(z)$ will be halfband and causal. The change of variables comes from $z = e^{i\omega}$:

$$\frac{z+z^{-1}}{2} = \cos\omega = 1-2y. \tag{5.73}$$

Thus $y = 0$ and $\omega = 0$ give $z = 1$. Similarly $y = 1$ and $\omega = \pi$ give $z = -1$.

Notice that the midpoints $y = \frac{1}{2}$ and $\omega = \frac{\pi}{2}$ give $z = \pm i$. *There are two z's for each y*, from $z+z^{-1} = 2-4y$. (This is a quadratic equation for z.) One z is inside the unit circle, the other is $1/z$ outside. This "Joukowski transformation" is also central in fluid flow. The endpoints $z = 1$ and $z = -1$ are really double roots of $z+z^{-1} = 2$ and $z+z^{-1} = -2$.

The change of variable gives $1-y$ and y in factored form:

$$1-y = \frac{1+\cos\omega}{2} = \left(\frac{1+z}{2}\right)\left(\frac{1+z^{-1}}{2}\right) \text{ and } y = \frac{1-\cos\omega}{2} = \left(\frac{1-z}{2}\right)\left(\frac{1-z^{-1}}{2}\right). \tag{5.74}$$

Substituting in $\widetilde{P}(y)$, the maxflat filter in the z-domain becomes $P(z)$:

$$P(z) = 2\left(\frac{1+z}{2}\right)^p\left(\frac{1+z^{-1}}{2}\right)^p\sum_{k=0}^{p-1}\binom{p+k-1}{k}\left(\frac{1-z}{2}\right)^k\left(\frac{1-z^{-1}}{2}\right)^k. \tag{5.75}$$

This factors into $P(z) = C(z)C(z^{-1})$ when $P(\omega)$ factors into $|C(\omega)|^2$. The p zeros at $y = 1$ and $\omega = \pi$ are now $2p$ zeros at $z = -1$. Half of them go into $C(z)$. The $p-1$ complex zeros of the other factor $B_p(y)$ become $2p-2$ zeros of $P(z)$. Half of those (the $p-1$ zeros inside the circle $|z| = 1$, if we want minimum phase) also go into $C(z)$. So the spectral factor $C(z)$ can be computed in two steps:

1. Find the $p-1$ zeros of $B_p(y)$ and the $p-1$ corresponding z's with $|z| < 1$.

2. Include p zeros at $z = -1$. Then $C(z)$ has these $2p-1$ zeros.

Example. $p = 2$ leading to Daubechies D_4 from $B_2(y) = 1+2y$, which is $\frac{1}{2}(-z+4-z^{-1})$.

The zero is at $y = -\frac{1}{2}$. Therefore $z+z^{-1} = 4$. This quadratic equation has roots $z = 2\pm\sqrt{3}$. Then the $2p-1$ roots of $C(z)$ are $-1, -1, 2-\sqrt{3}$. The coefficients of D_4 are approximately 0.4830, 0.8365, 0.2241, and −0.1294:

$$\begin{aligned} C(z) &= \alpha(1+z^{-1})^2(1-(2-\sqrt{3})z^{-1}) \\ &= \left[(1+\sqrt{3}) + (3+\sqrt{3})z^{-1} + (3-\sqrt{3})z^{-2} + (1-\sqrt{3})z^{-3}\right] / 4\sqrt{2}. \end{aligned}$$

Example. $p = 70$ leading to Daubechies D_{140} from $B_{70}(y)$.

From $p = 2$ to $p = 70$ is quite a jump! Figure 5.6 displays the 69 zeros of $B_{70}(y)$. They are close to a limiting curve in the complex y-plane. The equation $|4y(1-y)| = 1$ of that curve is discussed below. In the z-plane the limiting curve is moon-shaped, with the beautiful formula $|z-z^{-1}| = 2$. ***It consists of two circles!*** The roots of the minimum phase factor $C(z)$ are close

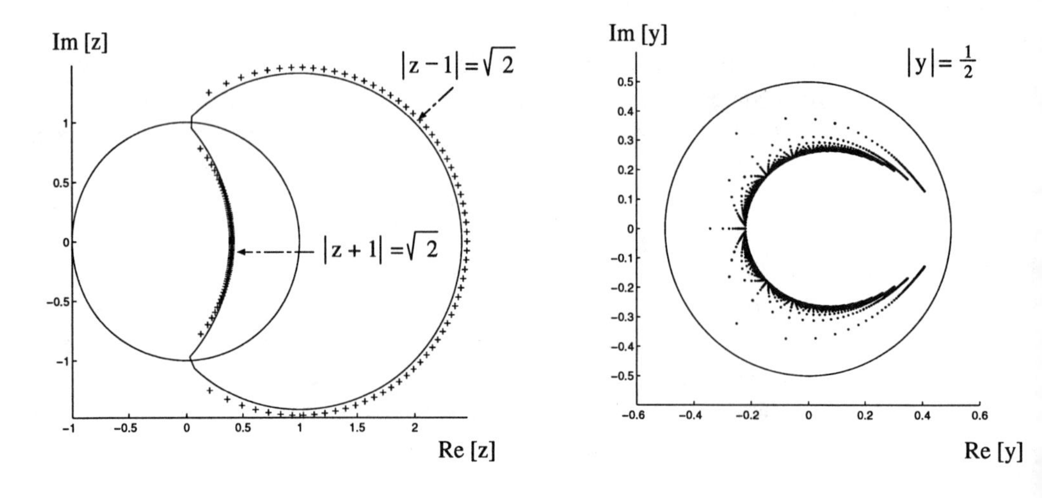

Figure 5.6: The 138 zeros of $P(z)$ and the zeros of $B_p(y)$ up to $p = 60$.

to the inner circle. To those $p - 1 = 69$ inner zeros we add $p = 70$ zeros at $z = -1$. Then the spectral factor of $P(z)$ has 139 roots, and the constant β makes $C(1) = \sqrt{2}$:

$$C(z) = \beta \left(\frac{1+z^{-1}}{2}\right)^{70} \prod_{1}^{69} \left(1 - z^{-1} Z_j\right). \tag{5.76}$$

Same Coefficients in Interpolation

The coefficients in $-\frac{1}{16} + \frac{9}{16}z^{-2} + z^{-3} + \frac{9}{16}z^{-4} - \frac{1}{16}z^{-6}$ appear in many places. This is a typical example, with $p = 2$, of a maxflat halfband filter. Its factor $(1 + z^{-1})^4$ yields $2p = 4$ zeros at $z = -1$. The halfband property means that $\boldsymbol{P}$ is an ***interpolating filter***: $\boldsymbol{Px}$ keeps the even-numbered coefficients of $\boldsymbol{x} = (\ldots, \boldsymbol{x}(0), 0, \boldsymbol{x}(2), 0, \ldots)$. Four zeros at π mean that the four polynomials $1, t, t^2, t^3$ are correctly interpolated in the odd components:

$$\begin{array}{lll} \boldsymbol{x} = (\ldots, 1, 0, 1, 0, 1, \ldots) & \text{gives} & \boldsymbol{Px} = (\ldots, 1, 1, 1, 1, 1, \ldots) \\ \boldsymbol{x} = (\ldots, 0, 0, 2, 0, 4, \ldots) & \text{gives} & \boldsymbol{Px} = (\ldots, 0, 1, 2, 3, 4, \ldots) \\ \boldsymbol{x} = (\ldots, 0, 0, 4, 0, 16, \ldots) & \text{gives} & \boldsymbol{Px} = (\ldots, 0, 1, 4, 9, 16, \ldots) \\ \boldsymbol{x} = (\ldots, 0, 0, 8, 0, 64, \ldots) & \text{gives} & \boldsymbol{Px} = (\ldots, 0, 1, 8, 27, 64, \ldots) \end{array}$$

Expressed differently, $\frac{9}{16}(\boldsymbol{x}(1) + \boldsymbol{x}(-1)) - \frac{1}{16}(\boldsymbol{x}(3) + \boldsymbol{x}(-3))$ is fourth-order accurate at the midpoint $t = 0$. This links wavelet theory to *recursive subdivision* (interpolation to create smooth curves). Starting with equally-spaced values $\boldsymbol{x}(2n)$, $\boldsymbol{P}$ puts new values at the halfway points $t = n$. Then $\boldsymbol{P}$ produces new values at $t = \frac{n}{2}$. The established values do not change. The limit is a smooth curve through the original values. The monograph [CDM] develops this important application, with references. The stability of recursive interpolation is controlled by Condition **E** applied to $\boldsymbol{P}$.

Asymptotics of the Daubechies Filters

Figure 5.6 practically requires us to study the zeros Y and Z as $p \to \infty$. The first steps were taken by [LeKa] and [ShSt]. The truncated binomial series $B_p(y)$ has degree $p-1$. Its $p-1$ zeros yield $2p-2$ zeros in the z-plane, from $Z + Z^{-1} = 2 - 4Y$. The main facts proved so far are:

1. All the zeros have $|Y| \leq \frac{1}{2}$ and $\operatorname{Re} Z > 0$.
2. In the y-plane, the zeros are all outside the limiting curve $|w| = |4y(1-y)| = 1$.
3. In the z-plane, the zeros are all outside the limiting curve $|z - z^{-1}| = 2$.
4. The zeros are near a uniform distribution along the circle of radius $1+ \log(4\pi p)/2p$ in the w-plane.
5. The far left zero is $Z = i - W/\sqrt{p} - iW^2/2p + O(p^{-3/2})$ where $\operatorname{erf}(W) = 1$.

Note that if z lies on that moon-shaped limiting curve $|z - z^{-1}| = 2$, so do $\bar{z}$ and z^{-1} and $\bar{z}^{-1}$. The complex roots in the z-plane come *four at a time*, for finite p and in the limit $p = \infty$. The moon consists of two circles of radius $\sqrt{2}$ (Problem). The outer circle is $|z-1| = \sqrt{2}$ with center at 1. The inner circle is $|z+1| = \sqrt{2}$ with center at -1. The circles meet at $z = \pm i$, and the far left zeros go slowly toward these two points.

Figure 5.6 shows the zeros in the y-plane up to $p = 60$. Their approach to the limiting curve is fascinating. The application of long filters, with large p, is still to be developed. It can use the lattice structure of Section 4.5

Computing the Spectral Factor $C(z)$

For large p, computing the zeros of $B_p(y)$ and the spectral factorization $P(z) = C(z)C(z^{-1})$ are challenging tasks. They are related but not identical. After we find the zeros, the filter coefficients in $C(z)$ are fully determined — but not necessarily in a well-conditioned way. The direct multiplication of 69 or 139 linear factors is not safe. The cepstral method (Section 5.4) is competitive because it goes directly to $C(z)$, without the zeros. It is based on splitting $\log P(z)$ into $\log C(z) + \log C(z^{-1})$. Our experience with the zeros, using MATLAB and also Lang's code from Rice, showed the importance of a simple weighting:

The zeros of $B_p(y)$ were correct up to $p = 34$. The codes fail for $p = 35$.

With weighted variable $4y$, the zeros became correct up to $p = 80$.

The reason for the breakup at $p = 35$ is the wide dynamic range of coefficients in $B_p(y)$. The constant term is 1. The highest term y^{p-1} has coefficient

$$\frac{(2p-2)!}{(p-1)!(p-1)!} \simeq \frac{\sqrt{2\pi(2p-2)}}{2\pi(p-1)} \frac{(2p-2)^{2p-2}}{(p-1)^{2p-2}} = \frac{4^{p-1}}{\sqrt{\pi(p-1)}} \tag{5.77}$$

by Stirling's formula. The leading term 4^{p-1} multiplying y^{p-1} suggests that $4y$ is a better variable than y. This was strongly confirmed by experiment. The recursion

$$b(p) = 1; \quad \text{for } p-1:1:1 \quad b(i) = b(i+1) * (2p-i-1)/4 * (p-i)$$

produces the coefficients in MATLAB order for the command $Y = \textbf{roots}\ (b)/4$.

The zeros are needed in a linear phase factorization $P_0(z) = H_0(z)F_0(z)$, which does not come from the cepstral method. In this case all four zeros $z, \bar{z}, z^{-1}, \bar{z}^{-1}$ go into the same factor (two from inside the unit circle, two from outside). One possibility is to put those quartets alternately in analysis and synthesis, H_0 and F_0, to give filters of nearly equal length. Both get p zeros at $\omega = \pi$. Another possibility is for $H_0(z)$ to be very short, like Haar. Then $F_0(z)$ comes from an exact division $P_0(z)/H_0(z)$. No zeros are needed! Experiment will show which linear-phase factors give the best compression of signals and images.

Transition Band for Maxflat Filters

The equiripple filters from the Remez-Parks-McClellan algorithm have a sharp transition from lowpass to highpass. Their transition band has width of order $\frac{1}{N}$. Their slope at the midpoint $\omega = \frac{\pi}{2}$ is of order N. They minimize the maximum error, giving the best pointwise approximation to 1 in the passband and 0 in the stopband. But they only have one zero (at most) at $\omega = \pi$.

The Daubechies filters, with many zeros at π, have no ripples. $P(\omega)$ and $|C(\omega)|$ decrease monotonically between 0 and π. We now show that the slope at $\omega = \frac{\pi}{2}$ is much smaller, of order $\sqrt{N}$ instead of $N = 2p - 1$.

Theorem 5.6 *The maxflat filter has center slope proportional to $\sqrt{N}$. The transition from $P(\omega) = 0.98$ to $P(\omega) = 0.02$ is over an interval of length $4/\sqrt{N}$.*

Proof. The constant c in Meyer's form is fixed by $c\int_0^\pi (\sin\omega)^N d\omega = 2$. This definite integral is known to be a ratio of Gamma functions (which are factorials $\Gamma(n+1) = n!$). We use Stirling's formula to estimate the integral:

$$\sqrt{\pi}\frac{\Gamma(\frac{N+1}{2})}{\Gamma(\frac{N+2}{2})} \simeq \sqrt{\pi}\left(\frac{N-1}{2e}\right)^{\frac{N-1}{2}}\left(\frac{2e}{N}\right)^{\frac{N}{2}} = \sqrt{\frac{2\pi e}{N-1}}\left(1-\frac{1}{N}\right)^{\frac{N}{2}} \simeq \sqrt{\frac{2\pi}{N}}.$$

The slope of $P(\omega)$ at $\omega = \frac{\pi}{2}$ is $-c \simeq -\sqrt{2N/\pi}$ in Meyer's form. The transition bandwidth is therefore $O(1/\sqrt{N})$ and we make this more precise. Between $\frac{\pi}{2} - \frac{\sigma}{\sqrt{N}}$ and $\frac{\pi}{2} + \frac{\sigma}{\sqrt{N}}$ the drop in $P(\omega)$ is the integral of $c(\sin\omega)^N$. Shift by $\frac{\pi}{2}$ to center the integral, replacing $\sin(\frac{\pi}{2} - \omega)$ by $\cos\omega$:

$$\text{drop} = c\int_{-\sigma/\sqrt{N}}^{\sigma/\sqrt{N}} (\cos\omega)^N d\omega \simeq \frac{c}{\sqrt{N}}\int_{-\sigma}^{\sigma}\left(1-\frac{\theta^2}{2N}\right)^N d\theta \simeq \sqrt{\frac{2}{\pi}}\int_{-\sigma}^{\sigma} e^{-\theta^2/2}d\theta. \qquad (5.78)$$

Here $\theta = \omega\sqrt{N}$. Thus 95% of the drop in $P(\omega)$ comes with $\sigma = 2$ (within two standard deviations of the mean, for the normal distribution in statistics). This transition interval has width $\Delta\omega = 4/\sqrt{N}$, as the theorem predicts. That rule was found experimentally by Kaiser and Reed in 1977, at the beginning of the triumph of digital filters.

Problem Set 5.5

1. The halfband filter $P(\omega) = 1 + \sum p(n)e^{-in\omega}$ (odd n only) satisfies $P(\pi) = 0$. Deduce directly that $P(0) = 2$.

2. Find the zeros of $B_3(y)$ and the Daubechies 6-tap filter D_6.

3. The definite integral $\int_0^\pi (\sin\omega)^N\,d\omega$ equals $2\frac{2\cdot4\cdots(N-1)}{1\cdot3\cdots(N)}$ for odd N. Express this in factorials and use Stirling's formula to rederive the estimate $\sqrt{2\pi/N}$. Then the slope at the center frequency $\frac{\pi}{2}$ is $O(\sqrt{N})$.

4. The points $z = 1+\sqrt{2}e^{i\theta}$ are on the circle $|z-1| = \sqrt{2}$. Substitute for z to show that this circle is one part of our limiting curve:

$$\left|\frac{z-z^{-1}}{2}\right| = \left|\frac{1+\sqrt{2}e^{-i\theta}}{1+\sqrt{2}e^{i\theta}}\right| = 1.$$

5. Difficult problem, *neat identity*: Show that the first moments of the Daubechies wavelets *squared* are (by [Villemoes]) $\int_0^N t(w(t))^2\,dt = N/2$.

Coefficients of the Daubechies Filters

The coefficients of the orthogonal maxflat ***Daubechies filters*** $D_4, D_6, \ldots, D_{12}$ are given below. The normalization is $H(0) = \Sigma h(n) = \sqrt{2}$.

N	n	$h(n)$
3	0	.4829629131445341
	1	.8365163037378077
	2	.2241438680420134
	3	−.1294095225512603
5	0	.3326705529500825
	1	.8068915093110924
	2	.4598775021184914
	3	−.1350110200102546
	4	−.0854412738820267
	5	.0352262918857095
7	0	.2303778133088964
	1	.7148465705529154
	2	.6308807679398587
	3	−.0279837694168599
	4	−.1870348117190931
	5	.0308413818355607
	6	.0328830116668852
	7	−.0105974017850690

N	n	$h(n)$
9	0	.1601023979741929
	1	.6038292697971895
	2	.7243085284377726
	3	.1384281459013203
	4	−.2422948870663823
	5	−.0322448695846381
	6	.0775714930400459
	7	−.0062414902127983
	8	−.0125807519990820
	9	.0033357252854738
11	0	.1115407433501095
	1	.4946238903984533
	2	.7511339080210959
	3	.3152503517091982
	4	−.2262646939654400
	5	−.1297668675672625
	6	.0975016055873225
	7	.0275228655303053
	8	−.0315820393174862
	9	.0005538422011614
	10	.0047772575109455
	11	−.0010773010853085

Chapter 6

Multiresolution

6.1 The Idea of Multiresolution

Our main approach to wavelets is through 2-channel filter banks. Everything develops from the filter coefficients. All constructions are concrete and highly explicit. Choose good coefficients and you get good wavelets. The heart of the theory is to see how conditions on the numbers $\boldsymbol{h}(k)$ and $\boldsymbol{c}(k)$ and $\boldsymbol{d}(k)$ determine properties of $\phi(t)$ and $w(t)$—the scaling function and the basic wavelet. Then the problem is to design filters that achieve those properties.

By iterating the filter bank, Section 6.2 reaches the *dilation equation* for $\phi(t)$ and the *wavelet equation* for $w(t)$. Sections 6.3 and 6.4 study those equations in the time domain and frequency domain. Conditions O and A lead to orthogonality and approximation accuracy. The Daubechies wavelets are "optimal" with respect to those two properties. But these orthogonal wavelets are not and cannot be symmetric (except for Haar). Also the transition from passband to stopband is not sharp. So the design problem is still open. Better wavelets remain to be constructed.

This opening section aims for an overview that brings out the key ideas. Before the construction using discrete time, we describe what is wanted in continuous time. The goal is a decomposition of the whole function space into subspaces. That implies a decomposition of each function—*there is a piece of $f(t)$ in each subspace.* Those pieces (or projections) give finer and finer details of $f(t)$. The signal is "resolved" at scales $\Delta t = 1, 1/2, \ldots, (1/2)^j$.

For audio signals, these scales are essentially *octaves*. They represent higher and higher frequencies. For images and indeed for all signals, the simultaneous appearance of multiple scales is known as ***multiresolution***.

Multiresolution will be described first for subspaces V_j and W_j. The scaling spaces V_j are *increasing*. The wavelet space W_j is the *difference* between V_j and V_{j+1}. *The sum of V_j and W_j is V_{j+1}.* Then these extra conditions involving dilation to $2t$ and translation to $t-k$ define a genuine multiresolution:

> If $f(t)$ is in V_j then $f(t)$ and $f(2t)$ and all $f(t-k)$ and $f(2t-k)$ are in V_{j+1}.

In the end, one wavelet generates a whole basis. The functions $w(2^j t - k)$ come by dilation and translation (all j and all k). There are six steps toward this goal, and we take them one at a time:

1. An increasing sequence of subspaces V_j (complete in L^2)

2. The wavelet subspace W_j that gives $V_j + W_j = V_{j+1}$
3. The dilation requirement from $f(t)$ in V_j to $f(2t)$ in V_{j+1}
4. The basis $\phi(t-k)$ for V_0 and $w(t-k)$ for W_0
5. The basis $\phi(2^j t-k)$ for V_j and $w(2^j t-k)$ for W_j
6. The basis of all wavelets $w(2^j t-k)$ for the whole space L^2.

A shortcut to multiresolution. Before those six steps, may I mention one shortcut step that starts with the filter coefficients $\boldsymbol{h}(k)$. That step is *to solve the dilation equation for the scaling function* $\phi(t)$:

$$\phi(t) = \sum 2\,\boldsymbol{h}(k)\,\phi(2t-k).$$

The first requirements on the coefficients are $\sum \boldsymbol{h}(k) = 1$ and $\sum (-1)^k\,\boldsymbol{h}(k) = 0$. The full requirement is Condition E in Section 7.2. When this is satisfied, $\phi(t)$ can be computed. Then $\{\phi(2^j t-k)\}$ is a basis for V_j. These spaces are automatically increasing and complete and shift-invariant and connected by dilation. Thus multiresolution is achieved.

A Scale of Subspaces

Each V_j is contained in the next subspace V_{j+1}. A function in one subspace is in all the higher (finer) subspaces:

$$V_0 \subset V_1 \subset \cdots \subset V_j \subset V_{j+1} \subset \cdots$$

A function $f(t)$ in the whole space has a piece in each subspace. Those pieces contain more and more of the full information in $f(t)$. The piece in V_j is $f_j(t)$. One requirement on the sequence of subspaces is *completeness*:

$$f_j(t) \to f(t) \qquad \text{as } j \to \infty.$$

The first example will not have the dilation feature required for multiresolution:

Example 6.1. *V_j contains all trigonometric polynomials of degree $\leq j$.*
Certainly V_j is contained in V_{j+1}. The spaces are growing. (Since Daubechies uses $-j$ where we use j, her subspaces are *decreasing*. Most authors now use an increasing sequence, for simpler numbering.) The piece of $f(t)$ in V_j is the partial sum $f_j(t)$ of its Fourier series:

$$f_j(t) = \sum_{|k|\leq j} c_k e^{ikt} \quad \text{is the piece in } V_j.$$

This is the *projection* of $f(t)$ onto V_j. The exponentials e^{ikt} are orthogonal, so the energy in $f_j(t)$ is the sum of $|c_k|^2$ over low frequencies $|k| \leq j$. The energy in $f(t) - f_j(t)$ is the sum over high frequencies $|k| > j$. This approaches zero as $j \to \infty$. Therefore the sequence V_j is complete in the whole 2π-periodic space L^2.

Now we identify the second family of subspaces. *W_j contains the new information* $\Delta f_j(t) = f_{j+1}(t) - f_j(t)$. This is the "detail" at level j. From the viewpoint of individual functions,

$$f_j(t) + \Delta f_j(t) = f_{j+1}(t). \tag{6.1}$$

From the viewpoint of the subspaces they lie in, this is

$$V_j \bigoplus W_j = V_{j+1}. \tag{6.2}$$

Each function in V_{j+1} is the sum of two orthogonal parts, f_j in V_j and Δf_j in W_j. In our example, the new information $\Delta f_j = f_{j+1} - f_j$ is the new term that enters $f_{j+1}(t)$:

$$\Delta f_j(t) = c_{j+1}e^{i(j+1)t} + c_{-j-1}e^{-i(j+1)t}. \tag{6.3}$$

The space W_j contains terms of exact degree $j+1$. Those are orthogonal to all terms of degree $\leq j$. Together, these orthogonal complements W_j and V_j produce V_{j+1}. This is an important part of multiresolution.

> *The spaces W_j are **differences** between the V_j.*
> *The spaces V_j are **sums** of the W_j.*

We can call V_j a partial sum by recognizing how the W's add to V. Start from

$$V_0 \bigoplus W_0 = V_1 \quad \text{and} \quad V_1 \bigoplus W_1 = V_2.$$

Substituting the first into the second, V_2 is the sum of three mutually orthogonal subspaces:

$$V_0 \bigoplus W_0 \bigoplus W_1 = V_2.$$

When you add details up to and including W_j, you have V_{j+1}:

$$V_0 \bigoplus W_0 \bigoplus W_1 \bigoplus \cdots \bigoplus W_j = V_{j+1}. \tag{6.4}$$

For the functions in those subspaces, this equation is just

$$f_0(t) + \Delta f_0(t) + \Delta f_1(t) + \cdots + \Delta f_j(t) = f_{j+1}(t). \tag{6.5}$$

That sum is $f_0 + (f_1 - f_0) + (f_2 - f_1) + \cdots + (f_{j+1} - f_j)$. It "telescopes" into f_{j+1}.

In practice, we can construct the spaces V_j and take differences. Or we can construct the W_j and take sums. It is like differentiating an integral or integrating a derivative. Starting from a basis is like starting from the W_j; then the V_j are partial sums as above. For the exponential basis e^{ikt}, this works fine. It also works for Haar wavelets, which are so simple. But for later wavelets, that direct search for $w(t)$ was very difficult.

Important. The construction of wavelets has succeeded by *finding the V_j first*. We begin with the scaling function $\phi(t)$, not the wavelet! Its translates $\phi(t-k)$ go into V_0. Rescaling to $2^j t$ gives V_j. Then the wavelet spaces W_j are the differences between V_{j+1} and V_j. The functions in W_j are the details at the jth scale.

Similarly for filter banks, we design the lowpass filter by choosing $\boldsymbol{c}(k)$. Then the highpass coefficients $\boldsymbol{d}(k)$ are easy. To maintain this analogy between continuous and discrete, we draw a "logarithmic tree" of filter banks. At each step, the highpass filter (with downsampling as usual) produces the detail Δf_j in W_j. The space V_{j+1} is the sum of $V_0, W_0, \ldots, W_j$.

It is convenient if W_j is orthogonal to V_j. Each W_j is then automatically orthogonal to all other W_k. *Reason:* If $k < j$, then W_k is contained in V_j—which is perpendicular to W_j. The completeness condition can be restated as

$$V_0 \bigoplus \sum_{j=0}^{\infty} W_j = L^2.$$

With orthogonality of each piece $f_j(t)$ to the next detail $\Delta f_j(t)$, these subspaces are orthogonal. But we emphasize now that *orthogonality is not essential.*

A nonorthogonal example comes directly from any nonorthogonal basis $b_0(t)$, $b_1(t)$, ... The piece $f_j(t)$ includes all the terms through $b_j(t)$:

$$\begin{array}{rll}\text{Sum up to } j: & f_j(t)=\sum_0^j c_k b_k(t) & \text{is in } V_j \\ \text{Next term}: & \Delta f_j(t)=c_{j+1}b_{j+1}(t) & \text{is in } W_j.\end{array}$$

The pattern is not lost, just the orthogonality. The new space V_{j+1} is still the "*direct sum*" of V_j and W_j, which intersect only at the zero vector. The angle between subspaces can be less than 90°, as long as every f_{j+1} in V_{j+1} has exactly one splitting into $f_j + \Delta f_j$:

$$\boxed{V_j \cap W_j = \{0\} \quad \text{and} \quad V_j + W_j = V_{j+1}}. \tag{6.6}$$

This nonorthogonal situation applies to *biorthogonal* filters and wavelets (Section 6.5). There W_j is orthogonal to a different subspace $\widetilde{V}_j$. The extra freedom can be put to good use.

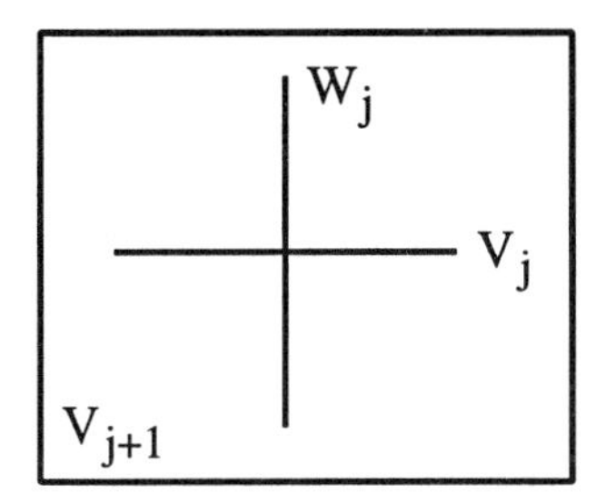

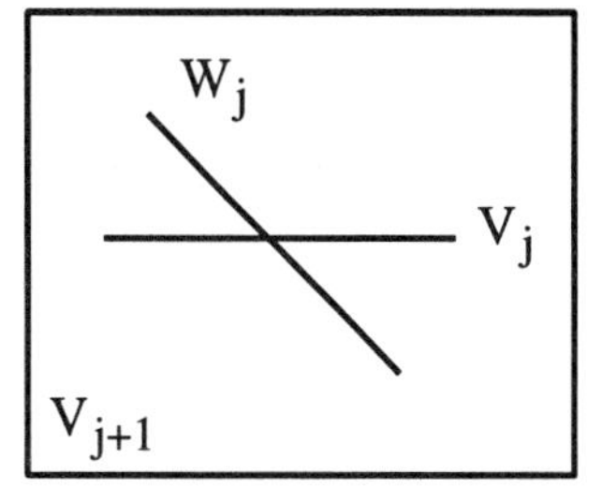

Figure 6.1: An orthogonal sum and a direct sum. Both written $V_j \oplus W_j = V_{j+1}$ and both allowed.

The Dilation Requirement

So far we have an increasing and complete scale of spaces. Each V_j is contained in the next V_{j+1}. For multiresolution, the crucial word *scale* carries an additional meaning. V_{j+1} consists of all rescaled functions in V_j:

$$\textbf{\textit{Dilation:}} \quad f(t) \text{ is in } V_j \quad \Longleftrightarrow \quad f(2t) \text{ is in } V_{j+1}.$$

The graph of $f(2t)$ changes twice as fast as the graph of $f(t)$. On a map, the scale is doubled. At 3,000,000 :1 the state of Utah fills a page. At 6,000,000 :1 its height is a half page. The length that represents a mile is cut in half. This length is Δt or Δx or h.

The example using $f(t) = c_{-j}e^{-ijt} + \cdots + c_j e^{ijt}$ does not meet this rescaling requirement. The highest frequency only increases by one, between V_j and V_{j+1}. But when t is changed to $2t$, the highest frequency becomes $2j$. *The frequencies must double.* The new space V_{j+1} is required to contain all those new frequencies. To satisfy the scaling requirement, the partial sums go an octave at a time. *The sum for f_j should stop at frequency 2^j instead of j.* Then Δf_j contains all frequencies between 2^j and 2^{j+1}:

$$\begin{array}{rlll}\textbf{\textit{Multiresolution example}}: & f_j(t) & = & \sum c_k e^{ikt} \text{ for } |k| \le 2^j \\ \textbf{\textit{Next detail}}: & \Delta f_j(t) & = & \sum c_k e^{ikt} \text{ for } 2^j < |k| \le 2^{j+1}.\end{array}$$

This is a genuine multiresolution, in which V_j and W_j have roughly the same dimension. It is the Littlewood-Paley decomposition of a Fourier series, into octaves instead of single terms.

This is a chief part of the mathematical background. To fit the requirements precisely, when $f(t)$ is defined on the whole line $-\infty < t < \infty$, we should use all frequencies ω and not just integers:

$$f_j(t) = \frac{1}{2\pi} \int_{|\omega| \le 2^j} \widehat{f}(\omega) e^{i\omega t} \, d\omega.$$

Now the spaces V_j go down the scale toward $j = -\infty$, as well as up the scale. The continuous frequency ω can be halved as well as doubled. The basis functions become *sinc functions*, by the sampling theorem. *Continuous frequency but discrete basis*, as is normal for L^2. And the nested spaces include $j < 0$:

$$\cdots \subset V_{-1} \subset V_0 \subset \cdots \subset V_j \subset V_{j+1} \subset \cdots \tag{6.7}$$

In addition to completeness as $j \to \infty$, we require emptiness as $j \to -\infty$:

$$\bigcap V_j = \{0\} \quad \text{and} \quad \overline{\bigcup V_j} = \text{ whole space.} \tag{6.8}$$

Emptiness means that $\|f_j(t)\| \to 0$ as $j \to -\infty$. Completeness still means that $f_j(t) \to f(t)$ as $j \to \infty$. The detail $\Delta f_j = f_{j+1} - f_j$ belongs to W_j and we still have

$$V_j \bigoplus W_j = V_{j+1}. \tag{6.9}$$

This can be an orthogonal sum, with Δf_j orthogonal to f_j. It must be a direct sum, with $V_j \bigcap W_j = \{0\}$. The reconstruction of $f(t)$ from its details Δf_j can start at $j = 0$ as before, or it can start at $j = -\infty$:

$$f(t) = f_0(t) + \sum_0^\infty \Delta f_j(t) \quad \text{or} \quad f(t) = \sum_{-\infty}^\infty \Delta f_j(t).$$

The sum of subspaces can start at $j = 0$ or $j = -\infty$. When the sum stops at $J \ge 0$, we have the subspace V_{J+1}:

$$V_{J+1} = V_0 + \sum_{j=0}^J W_j \quad \text{or} \quad V_{J+1} = \sum_{j=-\infty}^J W_j.$$

The left sum includes the scaling functions in V_0. The sum on the right involves only the wavelets. That form includes all the very large time scales $\Delta t = 2^{-j}$ as $j \to -\infty$.

In practice we use the first sum. Our calculations begin at some unit scale. The scaling functions at $j = 0$ and the wavelets with $j \ge 0$ are the basis. I suppose the scaling functions at level $j = J$ and the wavelets with $j \ge J$ are another basis.

The Translation Requirement and the Basis

Instead of rescaling $f(t)$, we now shift its graph. This is *translation*, and it leads to the fundamental requirement of time-invariance in signal processing. The subspaces are *shift-invariant*:

If $f_j(t)$ is in V_j then so are all its translates $f_j(t-k)$.

Suppose $f(t)$ is in V_0. Then $f(2t)$ is in V_1 and so is $f(2t-k)$. By induction, $f(2^j t)$ is in V_j and so is $f(2^j t - k)$. Dilation and translation are now built in.

With translation we are committed to working on the whole line $-\infty < t < \infty$, or to periodicity. A particular $f(t)$ may have compact support, but the whole space V_0 (all functions together) is shift-invariant. For finite intervals, the requirements have to be (and can be) adjusted. Dilation and translation operate freely on the whole line, and can be studied by Fourier transform.

The final requirement for multiresolution concerns a ***basis*** for each space V_j. If we choose one function $\phi(t)$ in V_0, its translates $\phi(t-k)$ may be independent. These translates may span the whole space V_0. They may even be orthonormal. The starting assumption, to be weakened later, is that V_0 contains such a function:

There exists $\phi(t)$ so that $\{\phi(t-k)\}$ is an orthonormal basis for V_0.

When the functions $\phi(t-k)$ are an orthonormal basis for V_0, the rescaled functions $\sqrt{2}\phi(2t-k)$ will be an orthonormal basis for V_1. At scaling level j, the basis functions $\phi(2^j t-k)$ are normalized by $2^{j/2}$. We collect all the requirements in one place:

Multiresolution Analysis

The subspaces V_j satisfy requirements 1 to 4:

1. $V_j \subset V_{j+1}$ and $\bigcap V_j = \{0\}$ and $\overline{\bigcup V_j} = L^2$ (completeness).
2. *Scale invariance*: $f(t) \in V_j \iff f(2t) \in V_{j+1}$.
3. *Shift invariance*: $f(t) \in V_0 \iff f(t-k) \in V_0$.
4. *Shift-invariant basis*: V_0 has an orthonormal basis $\{\phi(t-k)\}$.

4′. *Shift-invariant basis*: V_0 has a stable basis (Riesz basis) $\{\phi(t-k)\}$.

4 and 4′ are interchangeable. A stable basis can be orthogonalized in a shift-invariant way. This is in Section 6.4, together with the definition: stable = Riesz = uniformly independent. In practice we choose a convenient basis, orthogonal or not. Then V_j has the basis $\phi_{jk}(t) = 2^{j/2}\phi(2^j t-k)$:

$$f_j(t) = \sum_{k=-\infty}^{\infty} a_{jk}\phi_{jk}(t) \quad \text{is the piece in } V_j.$$

In the orthogonal case, the energy in this piece is

$$\|f_j\|^2 = \sum_{k=-\infty}^{\infty} |a_{jk}|^2 . \tag{6.10}$$

Shift-invariance and scale-invariance are built in through the basis $\{2^{j/2}\phi(2^j t-k)\}$. This basis combines requirements 2, 3, and 4!

We have at least three ways to construct or describe a multiresolution:

1. By the spaces V_j

2. By the scaling function $\phi(t)$

3. By the coefficients $2\boldsymbol{h}(k)$ in the dilation equation.

Our next examples use the spaces V_j and their bases. Then we move to description 3 and the dilation equation. Section 6.4 will orthogonalize the basis. The result will be the orthonormal $\phi(t-k)$ that multiresolution originally asks for. What we really need is a good shift-invariant basis.

It is also possible to allow *several* scaling functions $\phi_1, \ldots, \phi_r$, when one function (with its translates) cannot produce the whole space V_0. This occurs in Example 3 below. It corresponds to "*multiwavelets*".

The framework for multiresolution is set by the dilation-translation requirement. Examples come first. Then we study the dilation equation, and construct wavelets.

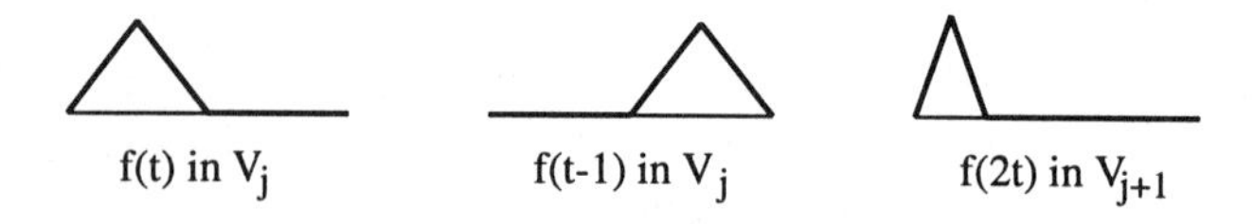

Figure 6.2: Translation stays in V_j. Dilation moves into V_{j+1}. Why is $f(2t)-f(t)$ *not* in W_j?

Examples of Multiresolution

1. Piecewise constant functions. V_0 contains all functions in L^2 that are constant on unit intervals $n \le t < n+1$. These functions are determined by their values $f(n)$ at all integer times $t = n$:

$$f(t) = f(\text{integer part of } t).$$

The function $f(2t)$ in V_1 is then constant on half-intervals. The functions in V_j are constant on intervals of length 2^{-j}. The spaces are increasing, $V_j \subset V_{j+1}$, because any function that is constant on intervals of length 2^{-j} is automatically constant on intervals of half that length. These are *dyadic intervals*, starting at a dyadic number $t = n/2^j$ and ending at $t = (n+1)/2^j$.

These spaces are shift-invariant—the translate of a piecewise constant function is still piecewise constant. The step from j to $j+1$ rescales time by 2 and produces V_{j+1}. What about a basis? The simplest choice is the *box function*:

$$\phi(t) = \begin{cases} 1 & \text{for } 0 \le t < 1 \\ 0 & \text{otherwise} \end{cases} \quad \text{is orthogonal to its translates } \phi(t-k).$$

Every function in V_0 is a combination of boxes $f(t) = \sum f(n)\phi(t-n)$. So requirement 4 is satisfied by the box function $\phi(t)$.

2. Continuous piecewise linear functions. The functions $f(t)$ are now linear between each pair of values $f(n)$ and $f(n+1)$. Notice again the shift-invariance and the scale-invariance:

Shift: If $f(t)$ is piecewise linear, so is $f(t-k)$.
Scale: If $f(t)$ is linear on unit intervals, then $f(2t)$ is linear on half-intervals.

The spaces are right for multiresolution. Is there a shift-invariant basis?

The basis function that comes to mind is the *hat function* $H(t)$, equal to one at $t = 1$, and linear between its values $H(n) = \delta(n-1)$. The translates $H(t-k)$ generate all piecewise linear functions on unit intervals. Any function $f(t)$ in V_0 can be expressed as $\sum f(n-1)H(t-n)$. However $H(t)$ is *not orthogonal* to the neighboring hat $H(t-1)$. The product $H(t)H(t-1)$ is positive on the one interval $1 < t < 2$ where the hats overlap. Its integral (the inner product of the hats) is not zero.

We must work harder to find an orthogonal basis, and the eventual $\phi(t)$ will not have compact support. Or else we keep this non-orthogonal basis.

3. Discontinuous piecewise linear functions. Now $f(t)$ in V_0 may have a jump at each meshpoint $t = n$. There is a value $f(n_-)$ from the left and a value $f(n_+)$ from the right. The hat function is still in the space, but so is the box function! The spaces V_j are clearly shift-invariant and scale-invariant. If $f(t)$ is linear between integers (where it jumps), then $f(2t)$ is linear between half-integers (where it jumps).

There are two degrees of freedom at each meshpoint, the values $f(n_-)$ and $f(n_+)$. Therefore *two scaling functions* $\phi_1(t)$ and $\phi_2(t)$ are required for a shift-invariant basis. They can both be supported on the unit interval, and they can be orthogonal:

$$\phi_1(t) = \textbf{\textit{box function}} \quad \text{and} \quad \phi_2(t) = \textbf{\textit{sloping line}} = 1 - 2t.$$

The union of $\{\phi_1(t-k)\}$ and $\{\phi_2(t-k)\}$ is an orthonormal basis—which illustrates the idea behind "*multiwavelets*". The usual dilation equation for $\phi(t)$ becomes a vector equation for $\phi_1(t)$ and $\phi_2(t)$. The coefficients $\boldsymbol{c}(k)$ in that equation are 2×2 matrices. The associated filter bank in Section 7.5 contains "*multifilters*".

4. Cubic splines. V_0 consists of piecewise cubic polynomials on unit intervals, with $f(t)$ and $f'(t)$ and $f''(t)$ continuous. The third derivative $f'''(t)$ may jump at the integers $t = n$, so the cubics are different in neighboring intervals. We have shift-invariance and scale-invariance, when V_1 contains the cubic splines on half-intervals. *This is the main point*: Approximating subspaces on regular meshes automatically fit the requirements for multiresolution.

The shortest cubic spline is a *B-spline*. It consists of different third-degree polynomials on the four unit intervals within $0 \le t \le 4$. The letter B stands for basis, but not for orthogonal basis. In complete analogy with the hat function, which is a linear spline, the scaling function $\phi(t)$ for the cubic splines cannot have compact support if we insist on orthogonality. An orthogonal basis $\{\phi(t-k)\}$ does exist, but it requires Fourier analysis to find it.

The cubic B-spline satisfies a dilation equation with very simple coefficients, proportional to $1, 4, 6, 4, 1$. But those coefficients do not lead to orthogonal filters. We can stay with these coefficients and go to biorthogonal filters (the best plan). Or we can orthogonalize, losing compact support and reaching a filter with infinitely many coefficients. Section 7.4 develops the theory of splines.

5. Daubechies functions. The search for orthogonal filter banks leads to the four coefficients of a "maxflat" lowpass filter. The response $C(\omega)$ has a double zero at the highest frequency $\omega = \pi$. This is maximal flatness, with four coefficients and orthogonality:

$$\boldsymbol{c}(0),\ \boldsymbol{c}(1),\ \boldsymbol{c}(2),\ \boldsymbol{c}(3) = 1+\sqrt{3},\ 3+\sqrt{3},\ 3-\sqrt{3},\ 1-\sqrt{3} \text{ times } 1/4\sqrt{2}.$$

Their sum is $\sqrt{2}$. Their sum of squares is unity. They are orthogonal to their double shifts, because $c(0)c(2)+c(1)c(3) = 0$. From these coefficients Daubechies constructed $\phi(t)$ by solving the dilation equation

$$\phi(t) = \sqrt{2}\sum_{k=0}^{3} c(k)\,\phi(2t-k).$$

The solution comes in the next section. The zeroth space V_0 contains every $\phi(t-k)$. Those functions are an orthonormal basis. The rescaled functions $\phi(2^j t - k)$ span V_j.

This is our best description of the Daubechies spaces V_j, to give the dilation equation for $\phi(t)$. In Examples 1–4, we started with the spaces. In Example 5, Daubechies started with the coefficients and found $\phi(t)$—which produces the spaces. Either way, we have the scale-invariance and shift-invariance of multiresolution analysis.

The Dilation Equation

The space V_0 is contained in V_1. Therefore $\phi(t)$ is also in V_1. It must be a combination of the basis functions $2^{1/2}\phi(2t-k)$ for that subspace. The coefficients in the combination will be called $c(k)$. Bring the factor $2^{1/2} = \sqrt{2}$ outside:

$$V_0 \subset V_1 \text{ means } \phi(t) = \sqrt{2}\sum_k c(k)\,\phi(2t-k). \tag{6.11}$$

This is the dilation equation. It is a two-scale equation, involving t and $2t$. It is also called a *refinement equation,* because it displays $\phi(t)$ in the refined space V_1. That space has the finer scale $\Delta t = 1/2$, and it contains $\phi(t)$ which has scale $\Delta t = 1$.

To emphasize: The dilation equation is a direct consequence of $V_0 \subset V_1$. It is not an extra requirement! There will be a finite set of coefficients $c(0), \ldots, c(N)$ when $\phi(t)$ is supported on $[0, N]$. In general, $\phi(t)$ has infinite support and we need infinitely many $c(n)$.

To find $c(n)$, multiply the dilation equation (6.11) by $\sqrt{2}\phi(2t-n)$. Integrate and use orthogonality:

$$\sqrt{2}\int_{-\infty}^{\infty} \phi(t)\phi(2t-n)\,dt = c(n). \tag{6.12}$$

If $\phi(t)$ is the unit box and $\phi(2t)$ is the half-box, this gives $c(0) = \sqrt{2}/2$ and $c(1) = \sqrt{2}/2$. The dilation equation for the box function then has coefficients 1 and 1:

$$\phi(t) = \phi(2t) + \phi(2t-1). \tag{6.13}$$

From orthogonality of the basis $\{\phi(t-k)\}$ we have double-shift orthogonality of the dilation coefficients $c(k)$. And unit energy in $\phi(t)$ gives a unit vector of c's:

$$\text{Double-shift: } \sum c(k)c(k-2m) = \delta(m). \quad \text{Unit vector: } \sum |c(k)|^2 = 1 \tag{6.14}$$

For proof, multiply the dilation equations for $\phi(t)$ and $\phi(t-m)$ and integrate. Orthonormality of the ϕ's yields double-shift orthogonality of the c's:

$$\int_{-\infty}^{\infty} \phi(t)\phi(t-m)\,dt = \sum_k c(k)c(k-2m) = \delta(m). \tag{6.15}$$

The coefficients $c(k)$ go into an orthonormal filter bank! Starting with the spaces V_j in a multiresolution, the dilation equation has brought us back to filters—where the key matrix is $L = (\downarrow 2)C$. Double-shift orthogonality becomes $LL^T = I$. The rows of L contain the double shifts $L_{ij} = c(2i - j)$.

Box example:

$$L = \frac{1}{\sqrt{2}} \begin{bmatrix} 1 & 1 & & & & & \\ & & 1 & 1 & & & \\ & & & & 1 & 1 & \\ & & & & & & \cdots \end{bmatrix}$$

Daubechies example:

$$L = \begin{bmatrix} c(3) & c(2) & c(1) & c(0) & & & \\ & & c(3) & c(2) & c(1) & c(0) & \\ & & & & c(3) & c(2) & \cdots \\ & & & & & & \cdots \end{bmatrix}.$$

To end this section, we have to identify the wavelet spaces W_j.

The Wavelet Equation

The scaling functions $\phi(2^j t - k)$ are orthogonal at each scale separately. But $\phi(t)$ is not orthogonal to $\phi(2t)$. They are *not* orthogonal across scales; the level j must be fixed. The function $\phi(t)$ in V_0 is also in V_1 (the dilation equation). Orthogonality *across scales* comes from the wavelet subspaces W_j and their basis functions $w_{jk}(t)$. We study those now, from three starting-points:

1. The spaces W_j. **2.** The wavelets $w(t)$. **3.** The coefficients $d(k)$.

Use Method **1** if you have the V_j. Their differences yield the spaces W_j. Use Method **2** if you can identify the wavelets. Just shift and rescale. Use Method **3** if you have the numbers $c(k)$. The *alternating flip* yields $d(k) = (-1)^k c(N-k)$. Then $w(t)$ comes from the wavelet equation below, and W_j contains the combinations of $w(2^j t - k)$.

The box function gives an example in which all three approaches will work. We construct W_0 and $w(t)$ and the d's:

1. *From the subspaces*: V_0 contains constant functions on unit intervals, and V_1 contains constant functions on *half*-intervals. The space W_0 is in V_1 (therefore constant on half-intervals). It is orthogonal to V_0, so the integral over each full interval is zero. This fact produces the complementary subspace W_0, orthogonal to V_0 inside V_1:

$$W_0 = \{ \text{ constants on half-intervals with } f(n) + f(n + 1/2) = 0 \}.$$

$V_0 \bigoplus W_0$ does give V_1. Combining *equal* values at n and $n + 1/2$ from V_0 with *opposite* values from W_0 gives *any* values $f(n)$ and $f(n + 1/2)$ for V_1.

2. *From the wavelets:* The important function in W_0 is the up-down square wave:

$$\textit{Haar wavelet} \quad w(t) = \begin{cases} 1 & \text{for } 0 \le t < \frac{1}{2} \\ -1 & \text{for } \frac{1}{2} \le t < 1 \\ 0 & \text{otherwise.} \end{cases}$$

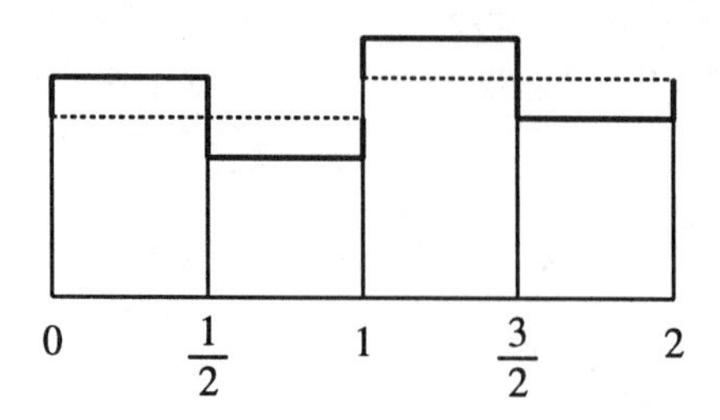

Figure 6.3: Two bases for V_1: Halfsize boxes $\phi(2t-k)$ or full boxes $\phi(t-k)$ plus up-down Haar wavelets $w(t-k)$.

This is orthogonal to the box function $\phi(t)$. It is orthogonal to translates of ϕ and also to its own translates (there is no overlap of $w(t)$ with $w(t-1)$). *More than that, multiresolution says that the wavelet $w(t)$ is orthogonal to rescalings of itself and to translates of rescalings:*

$$\int_{-\infty}^{\infty} w(t)w(2^j t - k)\,dt = 0 \text{ unless } j = k = 0.$$

The translates of $w(t)$ span W_0. The translates of $w(2^j t)$ span W_j. Those wavelet spaces are orthogonal because $W_0 \subset V_j$ and $V_j \perp W_j$. (Exchange j and 0 if j is negative.) From orthogonal spaces we have orthogonal basis functions. Then completeness makes the whole orthonormal system $\{2^{j/2}w(2^j t - k)\}$ a basis for L^2.

3. *From the coefficients* $\boldsymbol{c}(0) = \boldsymbol{c}(1) = 1/\sqrt{2}$: The flip construction gives $\boldsymbol{d}(0) = 1/\sqrt{2}$ and $\boldsymbol{d}(1) = -1/\sqrt{2}$. Those coefficients go into the wavelet equation:

$$\textbf{Wavelet equation} \quad w(t) = \sqrt{2}\sum \boldsymbol{d}(k)\,\phi(2t-k). \qquad (6.16)$$

This equation produces the wavelet directly from the scaling functions—*no equation to solve*! The wavelet is $w(t) = \phi(2t) - \phi(2t-1)$. This is a half-box minus a shifted half-box. It is the up-down square wave, which is Haar's wavelet.

Our final example, from Daubechies, starts with the four $\boldsymbol{c}$'s. Then the flip construction gives the four $\boldsymbol{d}$'s (to normalize, divide again by $4\sqrt{2}$):

$$\boldsymbol{d}(0), \boldsymbol{d}(1), \boldsymbol{d}(2), \boldsymbol{d}(3) = 1-\sqrt{3}, -(3-\sqrt{3}), 3+\sqrt{3}, -(1+\sqrt{3}).$$

Their sum is zero. Their sum of squares (normalized) is 1. They are orthogonal to their double shifts, because the $\boldsymbol{c}$'s are. The wavelet equation gives the Daubechies wavelet $w(t)$, which has no simple formula. *The orthogonality to $w(t-k)$ and $\phi(t-k)$ is only known indirectly*—from the double-shift orthogonality of the $\boldsymbol{d}$'s. The structure of multiresolution gives crucial information that we cannot find in a table of integrals.

The actual construction of $\phi(t)$ and $w(t)$, and the drawing of their graphs, is immediately ahead.

Example 6.2. (Strange but beautiful.) Suppose $\phi(t)$ is the delta function $\delta(t)$. This is not in L^2 but continue anyway. The space V_0 contains combinations $\sum a(n)\delta(t-n)$ of *delta functions at the integers*. What orthogonal wavelet $w(t)$ goes with this scaling function $\delta(t)$?

By scale invariance, V_1 contains $\delta(2t-n)$. The spikes for V_1 are at $t = 0, \pm\frac{1}{2}, \pm 1, \ldots$. Since V_0 holds the delta functions at integers, W_0 contains delta functions *at the midpoints* $t = n + \frac{1}{2}$.

The integers and the midpoints combine to give $V_0 \bigoplus W_0 = V_1$. *The wavelet is the delta function at* $t = \frac{1}{2}$.

Similarly, V_j contains delta functions at $t = n/2^j$. W_j contains delta functions at the midpoints $(n + \frac{1}{2})/2^j$. What is W_{-1}? Its delta functions are at $t = (n + \frac{1}{2})/2^{-1} = 2n + 1$. These are odd integers $\pm 1, \pm 3, \pm 5, \ldots$. The spacing between them is 2 as expected. Then W_{-2} has delta functions at $\pm 2, \pm 6, \pm 10, \ldots$ with spacing 4. The union of all W_j has delta functions at all binary points.

The dilation equation for the delta function is $\delta(t) = 2\delta(2t)$. The only nonzero coefficient is $\boldsymbol{h}(0) = 1$. *The filter is the identity.* The wavelet equation with only one term is $w(t) = 2\delta(2t - 1) = \delta(t - \frac{1}{2})$. This confirms what we found, that W_0 contains delta functions at all midpoints between integers. Notice! An odd number of coefficients (one) means that $N = 0$ (even). *The alternating flip must shift by an odd integer, for double-shift orthogonality.* So the nonzero highpass coefficient was $\boldsymbol{d}(1)$ not $\boldsymbol{d}(0)$.

To linger one last second on this trivial great example, the double-shift matrices from the low and high channels are $\boldsymbol{L} = (\downarrow 2)$ and $\boldsymbol{B} = (\downarrow 2)(\text{ delay })$.

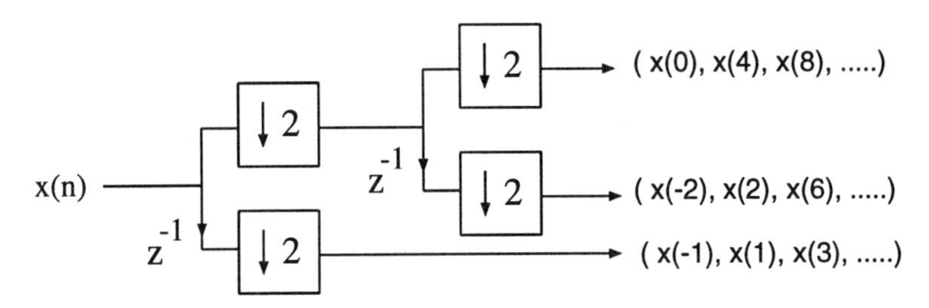

Figure 6.4: The lazy filter $\boldsymbol{H} = \boldsymbol{I}$ leads to delta functions.

The Scaling Function is Supported on $[0, N]$

A remarkable feature of $\phi(t)$ is that it is zero outside the interval $0 \le t \le N$. This could never happen to a ***one-scale*** difference or differential equation (homogeneous). The solutions would be combinations of λ^n and $e^{\lambda t}$, and only occasionally zero. The compact support of $\phi(t)$ comes from the two scales in the dilation equation

$$\phi(t) = \sum_{k=0}^{N} 2\boldsymbol{h}(k)\, \phi(2t - k). \tag{6.17}$$

Theorem 6.1 *The scaling function $\phi(t)$ is supported on the interval* $[0, N]$.

Proof. Suppose we know that the support is a finite interval $[a, b]$. Then $\phi(2t)$ is supported on $[\frac{a}{2}, \frac{b}{2}]$. The shifted function $\phi(2t - k)$ is supported on $[\frac{a+k}{2}, \frac{b+k}{2}]$. The index k goes from zero to N, so the right side of the dilation equation is supported between $\frac{a}{2}$ and $\frac{b+N}{2}$. Comparing with the left side,

$$[a, b] = \left[\frac{a}{2}, \frac{b+N}{2}\right] \quad \text{leads to} \quad a = 0 \text{ and } b = N.$$

How do we know that the support is a finite interval in the first place? From the cascade algorithm. The box function $\phi^{(0)}(t)$ is supported on $[0, 1]$. When this box is substituted into the right side of the dilation equation, the function $\phi^{(1)}(t)$ that comes out has support $[0, \frac{1+N}{2}]$. Then $\phi^{(1)}(t)$ is substituted into the right side and the result $\phi^{(2)}(t)$ is zero outside $[0, \frac{1+3N}{4}]$. The limiting function $\phi(t)$ is certain to be zero outside $[0, N]$. This cascade is studied in the next section.

It will be useful to reach the same conclusion based on the Fourier transform (Section 6.4). That argument can assume less about the filter coefficients. We mention that there are never gaps where $\phi(t)$ is zero on an interval inside $[0, N]$. And if the highpass coefficients $\boldsymbol{h}_1(k)$ run from $k = 0$ to $k = \widetilde{N}$, then the wavelet $w(t) = \sum 2\boldsymbol{h}_1(k)\phi(2t-k)$ has support $[0, \frac{1}{2}(N + \widetilde{N})]$. The last term $\phi(2t - \widetilde{N})$ is zero after $2t - \widetilde{N}$ reaches N.

Problem Set 6.1

1. Explain why the scaling requirement, that $f(t)$ is in V_j if and only if $f(2t)$ is in V_{j+1}, can be restated as $\widehat{f}(\omega)$ is in $\widehat{V}_j$ if and only if $\widehat{f}(2\omega)$ is in $\widehat{V}_{j-1}$. Here $\widehat{V}_j$ is the space of Fourier transforms of functions in V_j.

2. For the space V_0 of piecewise constant functions in Example 1, show that the only shift-invariant basis $\phi(t-k)$ contains box functions. What is the corresponding statement about allpass FIR filters?

3. For piecewise constants, show that $f(t)$ is in L^2 if and only if $f(n)$ is in l^2.

4. Find 2 by 2 matrices $\boldsymbol{c}(0)$ and $\boldsymbol{c}(1)$ so that the box function $\phi_1(t)$ and sloping line $\phi_2(t) = 1-2t$ in Example 3 satisfy

$$\begin{bmatrix} \phi_1(t) \\ \phi_2(t) \end{bmatrix} = \boldsymbol{c}(0)\begin{bmatrix} \phi_1(2t) \\ \phi_2(2t) \end{bmatrix} + \boldsymbol{c}(1)\begin{bmatrix} \phi_1(2t-1) \\ \phi_2(2t-1) \end{bmatrix}.$$

5. If $f(t)$ is in V_0 and $g(t)$ is in V_1, why is it generally false that $g(t) - f(t)$ is in W_1?

6. What multiresolution requirements are violated if W_j consists of all multiples of $\cos(2^j t)$?

6.2 Wavelets from Filters

The previous section reached the dilation equation and the wavelet equation:

$$\phi(t) = \sum \sqrt{2}\,\boldsymbol{c}(n)\,\phi(2t-n) \quad \text{and} \quad w(t) = \sum \sqrt{2}\,\boldsymbol{d}(n)\,\phi(2t-n). \tag{6.18}$$

Those equations are the crucial connections between wavelets and filters. Historically, their development was separate. Now you have to see them together. The lowpass filter $\boldsymbol{c}(0), \ldots, \boldsymbol{c}(N)$ determines the scaling function $\phi(t)$. Then the highpass coefficients produce the wavelets.

Working with $\phi(t)$ and $w(t)$, we really have three basic jobs:

1. Compute the coefficients in $f_j(t) = \sum_k a_{jk}\phi_{jk}(t)$ and $f(t) = \sum_j \sum_k b_{jk} w_{jk}(t)$.

2. Construct $\phi(t)$ by actually solving the dilation equation.

3. Connect the properties of $\phi(t)$ and $w(t)$ to properties of the $\boldsymbol{c}$'s and $\boldsymbol{d}$'s.

This section will do part of each job, the ***recursive part***. This shows how multiresolution (for functions) connects to subband filtering (for vectors). The three parts that we can do immediately are:

1. Compute a_{jk} and b_{jk} recursively from $a_{j+1,k}$ (and vice versa).
2. Set up a recursion (the cascade algorithm) to construct $\phi(t)$.
3. Prove orthogonality for $\phi_{jk}(t)$ and $w_{jk}(t)$ from orthogonality of $\boldsymbol{c}$'s and $\boldsymbol{d}$'s.

Those are the three subsections. Later we have to initialize the recursion in 1, execute and study the cascade algorithm in 2, and derive other properties in 3.

Wavelet Coefficients by Recursion

Suppose $f_1(t)$ is in V_1. It is a combination of the basis functions $\sqrt{2}\,\phi(2t-k)$. These functions $\phi_{1k}(t)$ are at level 1. Multiresolution splits this level into $V_1 = V_0 \oplus W_0$, so $f_1(t)$ is also a combination of the basis functions for V_0 and W_0. Those basis functions are $\phi_{0k}(t) = \phi(t-k)$ and $w_{0k}(t) = w(t-k)$:

$$\begin{aligned} \sum a_{1k}\,\phi_{1k}(t) &= \sum a_{0k}\,\phi_{0k}(t) + \sum b_{0k}\,w_{0k}(t) \\ &= \sum a_{0k}\,\phi(t-k) + \sum b_{0k}\,w(t-k). \end{aligned} \tag{6.19}$$

We are computing a change of basis. Given the coefficients $a_{1k}(t)$ in the V_1 basis, we want the coefficients a_{0k} and b_{0k} in the $V_0 \oplus W_0$ basis. The same step will apply at every level. It takes us from the coefficients $a_{j+1,k}$ in the basis for V_{j+1}, to the coefficients a_{jk} and b_{jk} in the bases for V_j and W_j. This is the recursion that makes the wavelet transform fast.

We will suppose that these bases are ***orthonormal***. Later in this section we prove this property (assuming the cascade algorithm uses orthogonal filters and converges). Orthonormality makes the formulas easy and it makes the inverse easy. Section 6.5 will derive the biorthogonal recursion, when orthogonality is not assumed.

To find the recursion, shift equation (6.18) by k and set $n = \ell - 2k$:

$$\begin{aligned} &\textbf{\textit{Dilation equation:}}\ \phi(t-k) = \sum \sqrt{2}\,\boldsymbol{c}(n)\,\phi(2t-2k-n) = \sum \boldsymbol{c}(\ell-2k)\,\phi_{1\ell}(t) \\ &\textbf{\textit{Wavelet equation:}}\ w(t-k) = \sum \sqrt{2}\,\boldsymbol{d}(n)\,\phi(2t-2k-n) = \sum \boldsymbol{d}(\ell-2k)\,\phi_{1\ell}(t) \end{aligned} \tag{6.20}$$

Multiply by $f_1(t)$ and integrate with respect to t. Since the basis functions are orthonormal, the integral gives the coefficients of $f_1(t)$ in each basis:

$$a_{0k} = \sum \boldsymbol{c}(\ell-2k)\,a_{1\ell} \quad \text{and} \quad b_{0k} = \sum \boldsymbol{d}(\ell-2k)\,a_{1\ell}. \tag{6.21}$$

This is the key recursion. It is the action of a filter bank, which inputs $a_{1\ell}$ and outputs a_{0k} and b_{0k}. But we have to watch indices, because an ordinary convolution would be $\sum \boldsymbol{c}(k-\ell)\,a_{1\ell}$ and downsampling would give $\sum \boldsymbol{c}(2k-\ell)\,a_{1\ell}$. There is a time reversal between this filter $\boldsymbol{C}$ and the *transpose* filter $\boldsymbol{C}^T$ that appears in the recursion (6.21):

$$\boldsymbol{c}^T(n) = \boldsymbol{c}(-n) \quad \text{and} \quad \boldsymbol{d}^T(n) = \boldsymbol{d}(-n). \tag{6.22}$$

Going between levels of a multiresolution is subband filtering with C^T and D^T:

Theorem 6.2 *A function $\sum a_{j+1,\ell}\,\phi_{j+1,\ell}(t)$ in the space $V_{j+1} = V_j \oplus W_j$ has coefficients a_{jk} and b_{jk} in the new orthonormal basis $\{\phi_{jk}(t), w_{jk}(t)\}$:*

$$a_{jk} = \sum_{\ell} \boldsymbol{c}(\ell - 2k)\, a_{j+1,\ell} \quad \text{and} \quad b_{jk} = \sum_{\ell} \boldsymbol{d}(\ell - 2k)\, a_{j+1,\ell}. \tag{6.23}$$

In vector notation this is $\boldsymbol{a}_j = (\downarrow 2)\, \boldsymbol{C}^T\, \boldsymbol{a}_{j+1}$ and $\boldsymbol{b}_j = (\downarrow 2)\, \boldsymbol{D}^T\, \boldsymbol{a}_{j+1}$. The pyramid is

$$\begin{array}{ccccccccc} a_{j+1} & \xrightarrow{C^T} & a_j & \xrightarrow{C^T} & a_{j-1} & \longrightarrow \cdots & a_1 & \xrightarrow{C^T} & a_0 \\ & \searrow D^T & b_j & \searrow D^T & b_{j-1} & \searrow & & \searrow D^T & b_0 \end{array}$$

Proof. For $j = 0$, formula (6.23) is (6.21). The extension to every j comes from the dilation equation. Again $n = \ell - 2k$:

$$2^{j/2}\,\phi(2^j t - k) = 2^{j/2} \sum \sqrt{2}\,\boldsymbol{c}(n)\,\phi(2^{j+1}t - 2k - n) = \sum \boldsymbol{c}(\ell - 2k)\,\phi_{j+1,\ell}(t). \tag{6.24}$$

The wavelet equation has $\boldsymbol{d}$ in place of $\boldsymbol{c}$. The inner products of these equations with $f(t)$ give the recursions (6.23) for the coefficients a_{jk} and b_{jk}.

Now go in the opposite direction. Change from the basis $\{\phi_{jk}(t), w_{jk}(t)\}$ back to the basis $\{\phi_{j+1,\ell}(t)\}$. Since the bases are orthonormal, the inverse operation is given by the transpose.

Theorem 6.3 *$a_{j+1,\ell}$ comes from a_{jk} and b_{jk} by a **synthesis** filter bank:*

$$a_{j+1,\ell} = \sum \boldsymbol{c}(2k - \ell)\, a_{jk} + \boldsymbol{d}(2k - \ell)\, b_{jk}. \tag{6.25}$$

The inverse pyramid is the fast inverse wavelet transform:

$$\begin{array}{cccccccc} a_0 & \xrightarrow{C} & a_1 & \xrightarrow{C} & a_2 & \cdots & a_j & \xrightarrow{C} & a_{j+1} \\ b_0 & \nearrow D & b_1 & \nearrow D & & & b_j & \nearrow D & \end{array}$$

Lowpass Iteration and the Cascade Algorithm

We begin the solution of the dilation equation. Our goal is to construct the scaling function $\phi(t)$. The only inputs are the filter coefficients $\boldsymbol{c}(0), \ldots, \boldsymbol{c}(N)$. The first solution method we propose is the ***cascade algorithm***.

Start the cascade with $\phi^{(0)}(t) =$ box function on $[0, 1]$. *Iterate the lowpass filter:*

$$\phi^{(i+1)}(t) = \sum_n \sqrt{2}\,\boldsymbol{c}(n)\,\phi^{(i)}(2t - n) = \sum_n 2\,\boldsymbol{h}(n)\,\phi^{(i)}(2t - n). \tag{6.26}$$

The algorithm works with functions in continuous time. Those functions are piecewise constant and the pieces become shorter (their length is 2^{-i}). If $\phi^{(i)}(t)$ converges suitably to a limit $\phi(t)$, then this limit function solves the dilation equation.

Notice the two time scales, t and $2t$, which come from the continuous form of downsampling. In place of $(\downarrow 2)\,\phi(n) = \phi(2n)$, we have $(\downarrow 2)\,\phi(t) = \phi(2t)$. The cascade algorithm is really iteration with the filter matrix $\boldsymbol{M} = (\downarrow 2)\,2\boldsymbol{H}$ — as we will see in detail. It is an infinite iteration, and our final formula for $\phi(t)$ will involve an infinite product.

It is easy to associate a continuous-time function $x(t)$ with a discrete-time vector $\boldsymbol{x}(n)$. The function takes the value $\boldsymbol{x}(n)$ over the nth time interval. That is the interval $n \le t < n+1$. Thus the constant vector $\boldsymbol{x} = (\ldots, 1, 1, 1, \ldots)$ produces the constant function $x(t) \equiv 1$. The impulse $\boldsymbol{x} = (\ldots, 0, 1, 0, \ldots)$ produces the standard *box function.* In general $x(t)$ is *piecewise constant:* $x(t) = \boldsymbol{x}(n)$ on the interval $n \le t < n+1$.

The iterations start from the box function $\phi^{(0)}(t)$. There are two steps in each iteration — *filtering* and *rescaling*. Suppose the filter coefficients are $\boldsymbol{h}(0) = 2/3$ and $\boldsymbol{h}(1) = 1/3$. Filtering the input gives $\frac{2}{3}\phi^{(0)}(t) + \frac{1}{3}\phi^{(0)}(t-1)$. Then rescaling t to $2t$ compresses the graph. *To maintain a constant area we multiply the height by* 2:

$$\phi^{(1)}(t) = \tfrac{4}{3}\phi^{(0)}(2t) + \tfrac{2}{3}\phi^{(0)}(2t-1).$$

Filtering and rescaling one box produces two half-width boxes of height $\frac{4}{3}$ and $\frac{2}{3}$. That iteration step preserves the area (=1). Now filter and rescale $\phi^{(1)}(t)$. The two half-boxes become four quarter-boxes, from $\phi^{(2)}(t) = \frac{4}{3}\phi^{(1)}(2t) + \frac{2}{3}\phi^{(1)}(2t-1)$. The first quarter-box has height $\frac{16}{9}$. That height is multiplied by $\frac{4}{3}$ at every iteration!

We wish we could say that the iterations $\phi^{(i)}(t)$ are converging. Their limit $\phi(t)$ would satisfy the dilation equation $\phi(t) = \frac{4}{3}\phi(2t) + \frac{2}{3}\phi(2t-1)$. In some weak sense, this may be true. In a pointwise sense at $t = 0$, the functions $\phi^{(i)}(0)$ diverge because of $(4/3)^i$. The coefficients $2/3$ and $1/3$ illustrate the iteration process, but not its convergence.

We want to see that process also by algebra. It is clearest if we ignore the rescaling and just execute the filtering with coefficients $\boldsymbol{h}(k)$. The heights of the boxes would be $\frac{2}{3}, \frac{1}{3}$, and then $\frac{4}{9}, \frac{2}{9}, \frac{2}{9}, \frac{1}{9}$. In the z-domain, this corresponds to

$$H(z) = \tfrac{2}{3} + \tfrac{1}{3}z^{-1} \quad \text{and} \quad H(z^2)H(z) = \tfrac{4}{9} + \tfrac{2}{9}z^{-1} + \tfrac{2}{9}z^{-2} + \tfrac{1}{9}z^{-3}. \tag{6.27}$$

The actual time intervals go from length 1 to $\frac{1}{2}$ to $\frac{1}{4}$. The actual graph heights are doubled at each step, to preserve area. But the essential point is the product $H(z^2)H(z)$. After three steps, the iteration will produce $H^{(3)}(z) = H(z^4)H(z^2)H(z)$. After i steps we have

$$H^{(i)}(z) = \prod_{k=0}^{i-1} H(z^{2^k}). \tag{6.28}$$

This product is the z-domain equivalent of iterating the lowpass filter $H(z)$. The values of $\phi^{(i)}(t)$ — the heights of the graph after i iterations — are the coefficients of $2^i H^{(i)}(z)$. That factor 2^i accounts for the height-doublings that preserve area, when the time intervals for $\phi^{(i)}(t)$ become 2^{-i}.

You may ask, why not choose the usual averaging filter as a first example? Let me show you why. The averaging coefficients are $\boldsymbol{h}(0) = \boldsymbol{h}(1) = \frac{1}{2}$. The first step of the iteration, with coefficients $2\boldsymbol{h}(0) = 2\boldsymbol{h}(1) = 1$, is

$$\phi^{(1)}(t) = \phi^{(0)}(2t) + \phi^{(0)}(2t-1).$$

From the box function $\phi^{(0)}(t)$ this produces the same box: $\phi^{(1)}(t) = \phi^{(0)}(t)$.

The output equals the input. The iteration process converges immediately. We have found the scaling function! In general $\phi(t)$ is the limit of the sequence $\phi^{(i)}(t)$, when that limit exists as $i \to \infty$. Here $\phi^{(0)} = \phi^{(1)}$ and the box function is a "fixed point" of the iteration. When we filter and rescale $\phi(t)$ we get back $\phi(t)$, because the sum of two half-length boxes is the original box:

$$\textit{Box:} \quad \phi(t) = \phi(2t) + \phi(2t-1). \tag{6.29}$$

The z-domain equivalent is a product built from

$$H(z) = \tfrac{1}{2} + \tfrac{1}{2}z^{-1}.$$

Please notice that we *do not square* this function. $H^{(2)}(z)$ is $H(z^2)H(z)$:

$$H^{(2)}(z) = (\tfrac{1}{2} + \tfrac{1}{2}z^{-2})(\tfrac{1}{2} + \tfrac{1}{2}z^{-1}) = \tfrac{1}{4} + \tfrac{1}{4}z^{-1} + \tfrac{1}{4}z^{-2} + \tfrac{1}{4}z^{-3}. \tag{6.30}$$

After i iterations, $H^{(i)}(z)$ will have 2^i coefficients all equal to 2^{-i}. After rescaling, this still corresponds to the box function.

Now use three filter coefficients $\boldsymbol{h} = \left(\frac{1}{4}, \frac{1}{2}, \frac{1}{4}\right)$. The box $\phi^{(0)}(t)$ produces *three* half-boxes in

$$\phi^{(1)}(t) = \tfrac{1}{2}\phi^{(0)}(2t) + \phi^{(0)}(2t-1) + \tfrac{1}{2}\phi^{(0)}(2t-2).$$

Then there are *seven* quarter-boxes in $\phi^{(2)}(t)$. Rescaling prevents the support interval from becoming long. The limiting interval is $0 \le t < 2$.

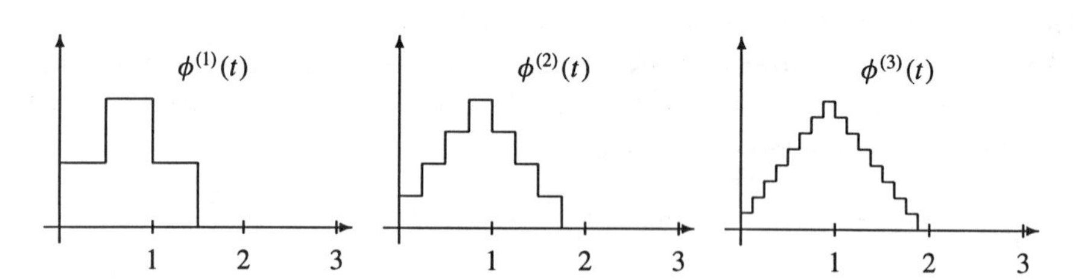

Figure 6.5: The cascade algorithm for $\frac{1}{4}, \frac{1}{2}, \frac{1}{4}$ converges to the hat function.

A reasonable guess for the limiting function $\phi(t)$ is the *hat function*. This is piecewise linear, going up to $\phi(1) = 1$ and down to $\phi(2) = 0$. We verify that the hat function is a fixed point of the iteration. Filtering and rescaling leaves this scaling function $\phi(t)$ unchanged:

$$\phi(t) = \tfrac{1}{2}\phi(2t) + \phi(2t-1) + \tfrac{1}{2}\phi(2t-2). \tag{6.31}$$

Notice how the coefficients $\frac{1}{4}, \frac{1}{2}, \frac{1}{4}$ are doubled. The hat function is a combination of three narrower hats. For future reference, we note the different properties of these examples:

1. $H(z) = \frac{2}{3} + \frac{1}{3}z^{-1}$ is not zero at $z = -1$, corresponding to $\omega = \pi$. The iterations fail to converge.

2. $H(z) = \frac{1}{2} + \frac{1}{2}z^{-1}$ is zero at $z = -1$. The iterations converge. The filter $H(z)H(z^{-1})$ is halfband: no even powers except the constant term. The box function is orthogonal to its translates.

3. $H(z) = \frac{1}{4} + \frac{1}{2}z^{-1} + \frac{1}{4}z^{-2}$ is zero (twice) at $z = -1$. The iterations converge. The filter $H(z)H(z^{-1})$ is *not* halfband. It contains the even powers z^2 and z^{-2}. The hat function $\phi(t)$ is *not* orthogonal to $\phi(t-1)$.

We must quickly emphasize that a zero at $z = -1$ (which is $\omega = \pi$ in the frequency domain) does not guarantee the convergence of $\phi^{(i)}(t)$. But without that zero in the filter response, strong convergence has no chance.

Similarly, a halfband filter does not guarantee that $\phi(t)$ is orthogonal to its translates. But without that halfband property of $H(z)H(z^{-1})$, orthogonality has no chance. Section 7.2 will further indicate those connections; they are not quite two-way implications:

Convergence of $\phi^{(i)}(t)$ to $\phi(t)$ needs $H = 0$ at $z = -1$
Orthogonality of $\phi(t-k)$ needs $H(z)H(z^{-1})$ to be halfband

Orthogonal Functions from Orthogonal Filters

When the filter bank is orthonormal in discrete time, we hope for orthogonal basis functions in continuous time. All wavelets $w(2^j t)$ should be orthogonal to the scaling functions $\phi(t-k)$. Furthermore, the wavelets $w(2^j t - k)$ should be mutually orthogonal and the scaling functions $\phi(t-k)$ should be mutually orthogonal. Note that $\phi(t)$ is *not* orthogonal to $\phi(2t)$.

Theorem 6.4 *Assume that the cascade algorithm converges: $\phi^{(i)}(t) \to \phi(t)$ uniformly in t. If the coefficients $\mathbf{c}(k)$ and $\mathbf{d}(k)$ come from an orthonormal filter bank, so they have double-shift orthogonality, then*

1. *The scaling functions $\phi(t-n)$ are orthonormal to each other:*

$$\int_{-\infty}^{\infty} \phi(t-n)\phi(t-m)\,dt = \delta(m-n).$$

2. *The scaling functions are orthogonal to the wavelets:*

$$\int_{-\infty}^{\infty} \phi(t-m)w(t-n)\,dt = 0.$$

3. *The wavelets $w_{jk}(t) = 2^{j/2}w(2^j t - k)$ at all scales are orthonormal:*

$$\int_{-\infty}^{\infty} w_{jk}(t)w_{JK}(t)\,dt = \delta(j-J)\delta(k-K).$$

Proof of 1: The box functions $\phi^{(0)}(t-k)$ are certainly orthonormal (because nonoverlapping). We will show that when $\phi^{(i)}(t-k)$ are orthogonal, the next iterates $\phi^{(i+1)}(t-k)$ are also orthornormal. Then the limits $\phi(t-k)$ are orthonormal.

The induction step from i to $i+1$ assumes that the $\phi^{(i)}(t-k)$ are orthonormal, and sets $l = m - n$:

$$\begin{aligned}
&\int \phi^{(i+1)}(t-m)\phi^{(i+1)}(t-n)\,dt \\
&= 2\int \left(\sum c(k)\phi^{(i)}(2t-2m-k)\right)\left(\sum c(k)\phi^{(i)}(2t-2n-k)\right)dt \qquad (6.32)\\
&= \int \left(\sum c(k)\phi^{(i)}(2t-2m-k)\right)\left(\sum c(k-2l)\phi^{(i)}(2t-2m-k)\right)2dt \\
&= \sum c(k)c(k-2l) = \delta(l) = \delta(m-n).
\end{aligned}$$

The crucial step came in the last line, when we used the orthogonality of the row $[\boldsymbol{c}(0)\cdots\boldsymbol{c}(N)]$ to its double shifts. These are rows of $\boldsymbol{L} = (\downarrow 2)\boldsymbol{C}$. The orthogonality is in the statement $\boldsymbol{L}\boldsymbol{L}^T = \boldsymbol{I}$. Equivalently, it is in the statement that $\left|\sum c(k)e^{-ik\omega}\right|^2$ is a normalized halfband filter: no even powers except the constant term 1.

Note the important point! Orthogonality of wavelets came from orthogonality of filters. When the infinite iterations converge, the limits retain orthogonality. This holds at each scale level j. In $\int \phi(2^j t-m)\phi(2^j t-n)\,dt$, we replace $2^j t$ by T. Orthogonality *does not hold* between scaling functions at different levels. Certainly, $\phi(t)$ is not orthogonal to all $\phi(2t-n)$, or the dilation equation would require $\phi \equiv 0$.

Proof of 2: Repeat the integration steps above for ϕ times w:

$$\begin{aligned}
&\int \phi(t-m)w(t-n)\,dt \\
&= \int \left(\sum c(k)\sqrt{2}\,\phi(2t-2m-k)\right)\left(\sum d(k)\sqrt{2}\,\phi(2t-2n-k)\right)dt \\
&= \cdots = \sum c(k)d(k-2l) = 0.
\end{aligned}$$

Always, $l = m-n$. The last step uses the orthogonality of the rows of $\boldsymbol{L} = (\downarrow 2)\boldsymbol{C}$ to the rows of $\boldsymbol{B} = (\downarrow 2)\boldsymbol{D}$. Again the double shift is essential. It is false that *all* rows of $\boldsymbol{C}$ and $\boldsymbol{D}$ are orthogonal.

The matrix form of this double-shift orthogonality is $\boldsymbol{L}\boldsymbol{B}^T = \boldsymbol{0}$. It comes from the alternating flip. That choice always produces double-shift orthogonality of $\boldsymbol{d}$'s to $\boldsymbol{c}$'s, but it does not by itself make $w(t)$ orthogonal to $\phi(t)$. To reach the end of part **2**, *we needed part* **1**—orthogonality between the ϕ's.

Proof of 3: The orthogonality of wavelets $w_{jk}(t)$ at the same scale level (the same j) is proved as in parts **1** and **2**:

$$\int_{-\infty}^{\infty} w(t-m)w(t-n)\,dt = \cdots = \sum d(k)d(k-2l) = \delta(l) = \delta(m-n).$$

Again continuous time orthogonality follows from discrete time orthogonality. This is not $\boldsymbol{D}\boldsymbol{D}^T = \boldsymbol{I}$. It is $\boldsymbol{B}\boldsymbol{B}^T = \boldsymbol{I}$, with double shifts in the rows of $\boldsymbol{B} = (\downarrow 2)\boldsymbol{D}$.

The orthogonality of wavelets at different scale levels (different j) is immediate from the rules of multiresolution. Suppose $j < J$. Then W_J is orthogonal to V_J by part **2**. But W_j is contained in V_{j+1} and therefore in V_J. So W_j is orthogonal to W_J. This proves the orthogonality theorem.

Final note: It was convenient to start from the box function $\phi^{(0)}(t)$, which is orthogonal to its translates. Then an orthogonal filter bank maintains this orthogonality to translates. Other starting functions will lead to the same fixed point $\phi(t)$, or at least to a multiple $c\phi(t)$—*if strong convergence holds.*

In general, convergence can be "*weak*" or "*strong*". For weak convergence, the functions $\phi^{(i)}(t)$ can oscillate faster and faster. You would not call this convergence. But the *integral* of $\phi^{(i)}(t)$ converges to the integral of $\phi(t)$, on every fixed interval $[0, T]$. (Integration controls the oscillations.) In the convergence that we assumed, $\phi^{(i)}(t)$ approaches $\phi(t)$ at every point.

There is a better starting function $\phi^{(0)}(t)$ than the box. The constant value $\phi^{(0)}(n)$ on each interval $n \le t < n+1$ can be the *correct* $\phi(n)$. The values of ϕ are filled in at half-integers and quarter-integers by the iterations $\phi^{(1)}(t)$ and $\phi^{(2)}(t)$. The graph of $\phi(t)$ appears, 2^i points at a time. We stop when we have enough points for the printer to connect into a continuous graph.

The next section explains how to start with the correct values of $\phi(n)$ at the integers.

Problem Set 6.2

1. For the filter with $\boldsymbol{h}(0) = \boldsymbol{h}(1) = \frac{1}{2}$ and any $\phi^{(0)}(t)$, describe and draw $\phi^{(i)}(t)$.
2. If $H(z)$ is a polynomial of degree N, what is the degree of $H(z^2)H(z)$? What is the degree of $H^{(i)}(z) = \prod_{k=0}^{i-1} H(z^{2^k})$?
   Rescaling will replace z by $z^{1/2}$. After i steps, the degree is divided by 2^i. Show that the degree of $H^{(i)}(z^{1/2^i})$ approaches N as $i \to \infty$.
3. With coefficients $\boldsymbol{h}(0), \ldots, \boldsymbol{h}(N)$, the support interval of $\phi^{(i)}(t)$ grows to $[0, N]$. What happens if $\phi^{(0)}(t)$ is a box on $[0, 2N]$?
4. The unit area of the box is preserved if and only if $\boldsymbol{h}(0) + \cdots + \boldsymbol{h}(N) = 1$. Are negative coefficients allowed?
5. Suppose the filter coefficients $\boldsymbol{h}(k)$ are $\frac{1}{2}, 0, 0, \frac{1}{2}$. Starting from the box function, take one step of the cascade algorithm and draw $\phi^{(1)}(t)$. Then take a second step and draw $\phi^{(2)}(t)$. Describe $\phi^{(i)}(t)$—on what fraction of the interval $[0, 3]$ does $\phi^{(i)}(t) = 1$?
6. Suppose the only filter coefficient is $\boldsymbol{h}(0) = 1$. Starting from the box function $\phi^{(0)}(t)$, draw the graphs of $\phi^{(1)}(t)$ and $\phi^{(2)}(t)$. In what sense does $\phi^{(i)}(t)$ converge to the delta function $\delta(t)$? To verify the dilation equation $\delta(t) = 2\delta(2t)$, multiply by any smooth $f(t)$ and compare the integrals of both sides.
7. Suppose $\phi^{(0)}(t)$ is a stretched box of unit area: $\phi^{(0)}(t) = 1/2$ for $0 \le t < 2$. Draw the graphs of $\phi^{(1)}(t)$ and $\phi^{(2)}(t)$ when $\boldsymbol{h}(0) = \boldsymbol{h}(1) = 1/2$. On what interval is $\phi^{(i)}(t)$ nonzero? What is the limit $\phi(t)$?
8. Suppose $\phi^{(0)}(t)$ is the Haar wavelet with *zero area*:
   $$\phi^{(0)}(t) = 1 \text{ for } 0 \le t < 1/2 \text{ and } \phi^{(0)}(t) = -1 \text{ for } 1/2 \le t < 1.$$
   With $\boldsymbol{h}(0) = \boldsymbol{h}(1) = 1/2$, draw the graphs of $\phi^{(1)}(t)$ and $\phi^{(2)}(t)$. The sequence $\phi^{(i)}(t)$ converges "weakly" to what multiple $c\phi(t)$?

6.3 Computing the Scaling Function by Recursion

The main point of this section can be stated in three sentences. Then you can follow through on the details, or look ahead for the matrices $\boldsymbol{m}(0)$ and $\boldsymbol{m}(1)$:

> The dilation equation gives $\phi(0), \phi(1), \ldots$ as the eigenvector of a matrix $\boldsymbol{m}(0)$.
> Then $\phi(t)$ at $t =$ half-integers comes from multiplying by a matrix $\boldsymbol{m}(1)$.
> Then $\phi(t)$ at every dyadic t comes by recursion. Each step uses $\boldsymbol{m}(0)$ or $\boldsymbol{m}(1)$.

The scaling function is created recursively. This section gives the rule.

The dilation equation is easiest with only *two coefficients* ($N = 1$). Then $\boldsymbol{m}(0) = 2\boldsymbol{h}(0)$ and $\boldsymbol{m}(1) = 2\boldsymbol{h}(1)$ are scalars not matrices. The two-coefficient dilation equation is

$$\phi(t) = \boldsymbol{m}(0)\,\phi(2t) + \boldsymbol{m}(1)\,\phi(2t-1). \tag{6.33}$$

The solution will be zero outside the interval $0 \le t < 1$. Inside that interval, set $t = 0$ to find $\phi(0)$ and $t = \frac{1}{2}$ to find $\phi(\frac{1}{2})$:

$$\phi(0) = \boldsymbol{m}(0)\,\phi(0) \quad \text{and} \quad \phi\left(\tfrac{1}{2}\right) = \boldsymbol{m}(1)\,\phi(0).$$

Now set $t = \frac{1}{4}$ and $\frac{3}{4}$, then $\frac{1}{8}$ and $\frac{5}{8}$, and onward through all the dyadic points $t = n/2^i$. Directly from the equation you find

$$\begin{aligned} \phi(\tfrac{1}{4}) &= \boldsymbol{m}(0)\phi(\tfrac{1}{2}) \quad \text{and} \quad \phi(\tfrac{3}{4}) = \boldsymbol{m}(1)\phi(\tfrac{1}{2}) \\ \phi(\tfrac{1}{8}) &= \boldsymbol{m}(0)\phi(\tfrac{1}{4}) \quad \text{and} \quad \phi(\tfrac{5}{8}) = \boldsymbol{m}(1)\phi(\tfrac{1}{4}) \\ \phi(\tfrac{3}{8}) &= \boldsymbol{m}(0)\phi(\tfrac{3}{4}) \quad \text{and} \quad \phi(\tfrac{7}{8}) = \boldsymbol{m}(1)\phi(\tfrac{3}{4}). \end{aligned}$$

Each new value comes from multiplying a previous value by $\boldsymbol{m}(0)$ or $\boldsymbol{m}(1)$. At each time t, the right side of equation (6.33) has only *one* nonzero term. Thus $\phi(\frac{3}{4})$ equals $\boldsymbol{m}(1)\phi(\frac{1}{2})$ which is $\boldsymbol{m}(1)\boldsymbol{m}(1)\phi(0)$.

At the next step $\phi(\frac{3}{8})$ equals $\boldsymbol{m}(0)\boldsymbol{m}(1)\boldsymbol{m}(1)\phi(0)$. The key is in the order of $\boldsymbol{m}(0)$ and $\boldsymbol{m}(1)$. *It is the same order as in the binary expansions* $\frac{3}{4} = 0.11$ *and* $\frac{3}{8} = 0.011$. At any point $t = n/2^i$, the solution $\phi(t)$ has i factors:

$$\text{If} \quad t = 0.01101 \quad \text{in base 2, then} \quad \phi(t) = \boldsymbol{m}(0)\boldsymbol{m}(1)\boldsymbol{m}(1)\boldsymbol{m}(0)\boldsymbol{m}(1)\phi(0).$$

We have now solved the two-coefficient dilation equation at all dyadic points.

Admittedly, the restriction to two coefficients looks severe. The pattern is correct and important, but two numbers $\boldsymbol{m}(0)$ and $\boldsymbol{m}(1)$ are not enough. The only normal case is $\boldsymbol{m}(0) = \boldsymbol{m}(1) = 1$, when we get the box function. For $\boldsymbol{m}(0) = \frac{4}{3}$ and $\boldsymbol{m}(1) = \frac{2}{3}$, the first equation becomes $\phi(0) = \frac{4}{3}\phi(0)$. This produces a singularity of $\phi(t)$ at all dyadic points.

We will not pursue that example here, because there is a more valuable application—which reduces $N+1$ coefficients to two. This is the familiar step of reducing a high-order equation to a low-order system. For differential equations that produces a matrix, as in the system $u' = Au$. For dilation equations the reduction will produce *two matrices* $\boldsymbol{m}(0)$ and $\boldsymbol{m}(1)$. The dilation equation will become a ***two-coefficient matrix equation***. *The recursion will not change*, except it has vectors and matrices.

Vector Form of the Dilation Equation

The $N+1$ coefficients in the dilation equation are $\sqrt{2}\boldsymbol{c}(k) = 2\boldsymbol{h}(k)$:

$$\phi(t) \quad = \quad 2\sum_{k=0}^{N} \boldsymbol{h}(k)\,\phi(2t-k). \tag{6.34}$$

Outside the interval $0 \le t < N$, we want and expect $\phi(t) \equiv 0$. Inside that interval, substitute $t = 0, t = 1, \ldots, t = N-1$ to determine $\phi(t)$ at the integers. You will see again the crucial fact that even k goes with even $2t-k$, and odd k goes with odd $2t-k$. (Reason! *The sum of k and* $2t-k$ *is even.*) The right side of (6.34) leads to an N by N matrix $\boldsymbol{m}(0)$ which is displayed for $N=5$:

$$\begin{bmatrix} \phi(0) \\ \phi(1) \\ \phi(2) \\ \phi(3) \\ \phi(4) \end{bmatrix} = 2 \begin{bmatrix} h(0) & & & & \\ h(2) & h(1) & h(0) & & \\ h(4) & h(3) & h(2) & h(1) & h(0) \\ & h(5) & h(4) & h(3) & h(2) \\ & & & h(5) & h(4) \end{bmatrix} \begin{bmatrix} \phi(0) \\ \phi(1) \\ \phi(2) \\ \phi(3) \\ \phi(4) \end{bmatrix} = \boldsymbol{m}(0)\Phi(0). \tag{6.35}$$

This is the dilation equation restricted to the integers. It is an eigenvalue problem for $\boldsymbol{m}(0)$. That matrix has even and odd in separate columns. For a nontrivial solution, this matrix (including the factor 2) must have $\lambda = 1$ as an eigenvalue. Assume this is true. Then the values $\phi(n)$ are in the eigenvector (*which we call* $\Phi(0)$). That eigenvalue problem for $\boldsymbol{m}(0)$ sets the integer values $\phi(n)$, and the recursion starts.

Now look at the vector of half-integer values. Substitute $t = \frac{1}{2}, \frac{3}{2}, \frac{5}{2}, \frac{7}{2}, \frac{9}{2}$ into the dilation equation. This leads to a closely related matrix $\boldsymbol{m}(1)$. The first row comes from $\phi(\frac{1}{2}) = 2\boldsymbol{h}(1)\phi(0) + 2\boldsymbol{h}(0)\phi(1)$. Notice that $2t$ is an *odd* integer, so *the sum of k and* $2t-k$ *is now odd.* The matrix is $\boldsymbol{m}(1)$:

$$\begin{bmatrix} \phi(1/2) \\ \phi(3/2) \\ \phi(5/2) \\ \phi(7/2) \\ \phi(9/2) \end{bmatrix} = 2 \begin{bmatrix} h(1) & h(0) & & & \\ h(3) & h(2) & h(1) & h(0) & \\ h(5) & h(4) & h(3) & h(2) & h(1) \\ & & h(5) & h(4) & h(3) \\ & & & & h(5) \end{bmatrix} \begin{bmatrix} \phi(0) \\ \phi(1) \\ \phi(2) \\ \phi(3) \\ \phi(4) \end{bmatrix} = \boldsymbol{m}(1)\Phi(0). \tag{6.36}$$

As expected, the values at half-integers come from the values at integers. A vector $\Phi(\frac{1}{2})$ comes from a vector $\Phi(0)$. In matrix notation (6.35) was an eigenproblem for $\Phi(0)$ and (6.36) is the step to $\Phi(\frac{1}{2})$:

$$\Phi(0) = \boldsymbol{m}(0)\Phi(0) \quad \text{and} \quad \Phi(\tfrac{1}{2}) = \boldsymbol{m}(1)\Phi(0).$$

This is exactly like the two-coefficient case! Now $\boldsymbol{m}(0)$ and $\boldsymbol{m}(1)$ are $N \times N$ matrices. The beautiful fact is that the same pattern continues to quarter-integers and beyond.

When t is a quarter-integer, the times $2t-k$ are half-integers. The values $\phi(\frac{1}{4}), \phi(\frac{5}{4}), \ldots$ come from $\phi(\frac{1}{2}), \phi(\frac{3}{2}), \ldots$. The dilation equation connects those vectors by the matrix $\boldsymbol{m}(0)$. Similarly the values $\phi(\frac{3}{4}), \phi(\frac{7}{4}), \ldots$ come from multiplying those half-integer values in the vector $\Phi(\frac{1}{2})$ by the matrix $\boldsymbol{m}(1)$:

$$\Phi\left(\tfrac{1}{4}\right) = \boldsymbol{m}(0)\,\Phi\left(\tfrac{1}{2}\right) \quad \text{and} \quad \Phi\left(\tfrac{3}{4}\right) = \boldsymbol{m}(1)\Phi\left(\tfrac{1}{2}\right).$$

Exactly as before, the binary expansion of $t = n/2^i$ reveals the order of the factors $\boldsymbol{m}(0)$ and $\boldsymbol{m}(1)$—as they multiply the initial eigenvector $\Phi(0)$ of values at the integers. We describe the recursion and then prove it is correct.

Theorem 6.5 *The vector form of the dilation equation is*

$$\Phi(t) = \boldsymbol{m}(0)\,\Phi(2t) + \boldsymbol{m}(1)\,\Phi(2t-1). \tag{6.37}$$

The vector $\Phi(t)$ is zero outside the interval $0 \le t \le 1$. Its components are the N slices $\phi(t)$, $\phi(t+1)$, $\phi(t+2)$, ... of the scaling function. Substituting $t = 0$, the values $\phi(n)$ at the integer times $t = 0, 1, \ldots, N-1$ are in the eigenvector of $\boldsymbol{m}(0)$ with $\lambda = 1$:

$$\text{(Fixed point)} \qquad \Phi(0) \quad = \quad \boldsymbol{m}(0)\Phi(0). \tag{6.38}$$

The vector $\Phi(t)$ of values at the dyadic points $t, t+1, \ldots, t+N-1$ comes from i multiplications by $\boldsymbol{m}(0)$ and $\boldsymbol{m}(1)$. Here $t = n/2^i < 1$ with n odd:

$$\text{If } t = \tfrac{3}{8} = 0.011 \text{ then } \Phi(t) = \begin{bmatrix} \phi(t) \\ \phi(t+1) \\ \vdots \\ \phi(t+N-1) \end{bmatrix} = \boldsymbol{m}(0)\boldsymbol{m}(1)\boldsymbol{m}(1)\Phi(0). \tag{6.39}$$

The scalar equation of high order is reduced to a vector equation of low order. It is just a recursion, in which the 0-1 digit t_1 tells whether to use $\boldsymbol{m}(0)$ or $\boldsymbol{m}(1)$:

$$\textbf{\textit{Vector recursion:}} \quad \Phi(.\, t_1\, t_2\, t_3\, \ldots) = \boldsymbol{m}(t_1)\, \Phi(.\, t_2\, t_3\, t_4\, \ldots). \tag{6.40}$$

The vector $\Phi(2t)$ is nonzero on the half-interval $0 \le t < \frac{1}{2}$. The other vector $\Phi(2t-1)$ is nonzero for $\frac{1}{2} \le t < 1$. These two vectors of compressed slices are multiplied by $\boldsymbol{m}(0)$ and $\boldsymbol{m}(1)$. To identify those particular matrices as correct, one way is to substitute t into the dilation equation and watch the numbers $2t - k$. As t crosses $\frac{1}{2}$, those numbers cross an integer—they go from one slice to the next. At that moment $\boldsymbol{m}(0)$ is exchanged for $\boldsymbol{m}(1)$.

The matrix $\boldsymbol{m}(0)$ has the same "*double-shift*" between rows that we saw earlier in filter banks. The earlier matrix was $\boldsymbol{L} = (\downarrow 2)\boldsymbol{C}$, with entries $\boldsymbol{L}_{ij} = \boldsymbol{c}(2i-j)$. Now this matrix $\boldsymbol{L}$ appears in the dilation equation! It is multiplied by the extra factor $\sqrt{2}$ to become $\boldsymbol{M}$. Its entries are $2\boldsymbol{h}(2i-j)$:

$$\boldsymbol{M} = \sqrt{2}\boldsymbol{L} \quad = \quad 2\begin{bmatrix} \cdots & \boldsymbol{h}(0) & & & & \\ \cdots & \boldsymbol{h}(2) & \boldsymbol{h}(1) & \boldsymbol{h}(0) & & \\ \cdots & \boldsymbol{h}(4) & \boldsymbol{h}(3) & \boldsymbol{h}(2) & \boldsymbol{h}(1) & \boldsymbol{h}(0) \\ \cdots & \cdots & \cdots & \cdots & \cdots & \cdots \end{bmatrix}. \tag{6.41}$$

We now show that the dilation equation has an even more compact form $\Phi_\infty(t) = \boldsymbol{M}\Phi_\infty(2t)$ These are infinite vectors and matrices. The nonzero part will be exactly the two-term form of the dilation equation.

Dilation Equation in Infinite Vector Form $\Phi_\infty(t) = \boldsymbol{M}\Phi_\infty(2t)$

$$\text{This form is} \quad \begin{bmatrix} \vdots \\ \phi(t-1) \\ \phi(t) \\ \phi(t+1) \\ \vdots \end{bmatrix} = \boldsymbol{M} \begin{bmatrix} \vdots \\ \phi(2t-1) \\ \phi(2t) \\ \phi(2t+1) \\ \vdots \end{bmatrix} \quad \text{for } -\infty < t < \infty \tag{6.42}$$

Restricted to the interval $0 \le t < 1$, only rows $0, 1, \ldots, N-1$ of $\Phi_\infty(t)$ are nonzero. Those N slices of $\phi(t)$ form the vector $\Phi(t)$ with N components. The dilation equation (6.42) reduces to the vector form (6.37):

$$\Phi(t) = \left\{ \begin{array}{ll} m(0)\Phi(2t) & \text{for } 0 \le t < \frac{1}{2} \\ m(1)\Phi(2t-1) & \text{for } \frac{1}{2} \le t < 1 \end{array} \right\} = m(0)\Phi(2t) + m(1)\Phi(2t-1).$$

The matrices $m(0)$ and $m(1)$ are N by N sections of the infinite matrix M. For $i, j = 0, 1, \ldots, N-1$ the matrix entries are

$$m(0)_{ij} = M_{ij} = 2h(2i-j) \quad \text{and} \quad m(1)_{ij} = M_{i,j-1} = 2h(2i-j+1).$$

Proof. The verification is in three steps, first for M and then $m(0)$ and then $m(1)$. I hope the intuition is already in place, to see the sum $\sum 2h(k)\phi(2t-k)$ as $M\Phi(2t)$ or $\sqrt{2}L\Phi(2t)$ or $2(\downarrow 2)H\Phi(2t)$. We now follow each step:

(Verify M) Row zero of $\Phi_\infty(t) = M\Phi_\infty(2t)$ is $\phi(t) = 2\sum h(k)\phi(2t-k)$.
Row one is $\phi(t+1) = 2\sum h(k)\phi(2t+2-k)$. The double shift works.

(Verify $m(0)$) Restrict to $0 \le t < \frac{1}{2}$ and keep only rows $0, 1, \ldots, N-1$.
Since $\phi(2t-1) = 0$ and $\phi(2t+N) = 0$ we only need columns 0 to $N-1$ of the infinite matrix M. This N by N section is $m(0)$.

(Verify $m(1)$) Restrict to $\frac{1}{2} \le t < 1$ and keep only rows $0, 1, \ldots, N-1$.
Since $\phi(2t-2) = 0$ and $\phi(2t+N-1) = 0$ we only need columns $-1, 0, \ldots, N-2$ of M. This N by N section is $m(1)$.

The change from $m(0)$ to $m(1)$ comes as t crosses $\frac{1}{2}$, because the nonzero entries in $\Phi_\infty(2t)$ appear one component earlier.

We now go back to the eigenvalue problem $\Phi(0) = m(0)\Phi(0)$. Condition A_1 leads to the eigenvalue $\lambda = 1$, and guarantees a solution.

The Fixed Point Equation $\Phi(0) = m(0)\Phi(0)$

The rows of $m(0)$ have a double shift. The columns have entirely even indices or entirely odd indices. The 5 by 5 matrix ($N = 5$) shows this pattern:

$$m(0) \quad = \quad 2\begin{bmatrix} h(0) & & & & \\ h(2) & h(1) & h(0) & & \\ h(4) & h(3) & h(2) & h(1) & h(0) \\ & h(5) & h(4) & h(3) & h(2) \\ & & & h(5) & h(4) \end{bmatrix}.$$

The key requirement on the coefficients $h(n)$ is Condition A_1: *The frequency response $H(\omega)$ has a zero at $\omega = \pi$:*

$$h(0) - h(1) + h(2) - \cdots = 0.$$

Combined with $h(0) + h(1) + h(2) + \cdots = 1$, this means that every column of $m(0)$ and $m(1)$ and M adds to 1.

Theorem 6.6 *Condition A_1 guarantees that $\lambda = 1$ is an eigenvalue of M and $m(0)$ and $m(1)$:*

$$\begin{array}{lcl} h(0) - h(1) + h(2) - \cdots & = & 0 \\ h(0) + h(1) + h(2) + \cdots & = & 1 \end{array} \quad \textit{yields } 2\sum_{\text{even } n} h(n) = 2\sum_{\text{odd } n} h(n) = 1.$$

Any matrix with unit column sums has $\lambda = 1$ as an eigenvalue. Therefore $\Phi(0) = \boldsymbol{m}(0)\Phi(0)$ can be solved to give the scaling function at integer times $\Phi(0) = (\phi(0), \phi(1), \ldots, \phi(N-1))^T$.

Proof. Adding the two equations gives the even part. Subtracting gives the odd part. These are the column sums of the matrix $\boldsymbol{m}(0)$, all equal to 1. A matrix with unit column sums has a *left* eigenvector of ones, $\boldsymbol{em}(0) = \boldsymbol{e}$, because the multiplication just adds up each column:

$$\begin{bmatrix} 1 & 1 & \cdots & 1 \end{bmatrix} \boldsymbol{m}(0) = \begin{bmatrix} 1 & 1 & \cdots & 1 \end{bmatrix}.$$

This means that $\boldsymbol{m}(0) - \boldsymbol{I}$ is not invertible, and $\lambda = 1$ is an eigenvalue. The left eigenvector is $\boldsymbol{e}$. The right eigenvector is $(\phi(0), \phi(1), \ldots, \phi(N-1))$:

$$(\boldsymbol{m}(0) - \boldsymbol{I})\Phi(0) = 0 \text{ which means } \Phi(0) = \boldsymbol{m}(0)\Phi(0).$$

The fundamental fact is that a square matrix and its transpose have the same determinant and same rank and same eigenvalues.

The columns of $\boldsymbol{m}(1)$ and $\boldsymbol{M}$ also add to 1, producing the eigenvalue $\lambda = 1$. Small point! We can safely normalize the eigenvector $\Phi(0)$ by $\sum \phi(n) = 1$. This is the "unit area" requirement that we impose on the function $\phi^{(0)}(t)$ at the start of the iterations. Then the scaling function has $\int \phi(t)dt = 1$ at the end.

Corollary *The sum* $\sum \phi(t+k)$ *is identically* 1.

Proof. Multiply the vector dilation equation $\Phi(t) = \boldsymbol{m}(0)\,\Phi(2t) + \boldsymbol{m}(1)\,\Phi(2t-1)$ on the left by $\boldsymbol{e}$. Use the fact that $\boldsymbol{e}\,\boldsymbol{m}(0) = \boldsymbol{e}$ and $\boldsymbol{e}\,\boldsymbol{m}(1) = \boldsymbol{e}$:

$$\boldsymbol{e}\,\Phi(t) = \boldsymbol{e}\,\Phi(2t) + \boldsymbol{e}\,\Phi(2t-1).$$

This is a dilation equation for $\boldsymbol{e}\,\Phi(t)$ and its solution is the box function! Thus $\boldsymbol{e}\,\Phi(t) = 1$. The "periodized scaling function" $\sum \phi(t+k)$ is identically one.

Example 6.3. The coefficients $2\boldsymbol{h}(k) = \frac{1}{2}, 1, \frac{1}{2}$ lead to the hat function. The 2 by 2 eigenvalue problem for $\boldsymbol{m}(0)$ gives the correct values of $\phi(0)$ and $\phi(1)$:

$$\begin{bmatrix} \frac{1}{2} & 0 \\ \frac{1}{2} & 1 \end{bmatrix} \begin{bmatrix} \phi(0) \\ \phi(1) \end{bmatrix} = \begin{bmatrix} \phi(0) \\ \phi(1) \end{bmatrix} \quad \text{gives} \quad \begin{matrix} \phi(0) = 0 \\ \phi(1) = 1. \end{matrix}$$

The sum of all hat functions $\phi(t+n)$ is identically one. Notice that the first row of the eigenvalue equation is always $2\boldsymbol{h}(0)\phi(0) = \phi(0)$. Then $\phi(0)$ is zero, apart from the exceptional case $\boldsymbol{h}(0) = \frac{1}{2}$ which occurs for the box function. This means that the scaling function $\phi(t)$ is zero up to and *including* $t = 0$. The box function starts with a jump at $t = 0$, because $\boldsymbol{h}(0) = \frac{1}{2}$.

Example 6.4. The Daubechies coefficients have $8\boldsymbol{h}(k) = 1+\sqrt{3}, 3+\sqrt{3}, 3-\sqrt{3}$, and $1-\sqrt{3}$. Dividing by 4 we have $2\boldsymbol{h}(k)$, the numbers that enter $\boldsymbol{m}(0)$:

$$\frac{1}{4}\begin{bmatrix} 1+\sqrt{3} & 0 & 0 \\ 3-\sqrt{3} & 3+\sqrt{3} & 1+\sqrt{3} \\ 0 & 1-\sqrt{3} & 3-\sqrt{3} \end{bmatrix} \begin{bmatrix} \phi(0) \\ \phi(1) \\ \phi(2) \end{bmatrix} = \begin{bmatrix} \phi(0) \\ \phi(1) \\ \phi(2) \end{bmatrix}.$$

The eigenvector gives $\phi(0)$, $\phi(1)$, and $\phi(2)$:

$$\phi(0) = 0 \qquad \phi(1) = \tfrac{1}{2}\left(1 + \sqrt{3}\right) \qquad \phi(2) = \tfrac{1}{2}\left(1 - \sqrt{3}\right).$$

We now know the Daubechies scaling function $\phi(t)$ at the integers. The only nonzeros in the fixed-point eigenvector Φ are $\phi(1)$ and $\phi(2)$. From these values at $t = 1$ and $t = 2$, the recursion produces $\phi(t)$ at any dyadic point.

Practical conclusion Every $\phi(n/2^i)$ comes via $\boldsymbol{m}(0)$ and $\boldsymbol{m}(1)$.

Theoretical conclusion Those dyadic values have a uniform bound if and only if all products of $\boldsymbol{m}(0)$ and $\boldsymbol{m}(1)$ in all orders have a *uniform upper bound.* When this holds, the dilation equation has a bounded solution $\phi(t)$ for all t.

We can propose sufficient conditions so that all products of $\boldsymbol{m}(0)$ and $\boldsymbol{m}(1)$ have a uniform bound. We can also propose necessary conditions. It is not known how to verify the necessary and sufficient condition (Section 7.3).

Derivatives of the Dilation Equation

While working in the time domain, we might as well take the derivative of $\phi(t)$. The result is highly interesting and not fully understood. Part of the problem is that the derivative $\phi'(t)$ may not exist.

The plan is to differentiate each term in the dilation equation for $\phi(t)$:

$$\phi'(t) \quad = \quad 4\sum \boldsymbol{h}(k)\phi'(2t-k).$$

This is another dilation equation, with every coefficient doubled. The equation $\Phi(t) = \boldsymbol{M}\Phi(2t)$ has led to $\Phi'(t) = 2\boldsymbol{M}\Phi'(2t)$. At $t = 0$ this yields the fixed-point equation $\Phi'(0) = 2\boldsymbol{M}\Phi'(0)$. The eigenvector $\Phi'(0)$ contains the derivatives $\phi'(n)$ at the integers $t = n$.

To solve $\Phi'(0) = \boldsymbol{M}\Phi'(0)$, we have a new requirement. *The number $\lambda = \frac{1}{2}$ must also be an eigenvalue of $\boldsymbol{M}$.* Again this applies to the $N \times N$ matrices $\boldsymbol{m}(0)$ and $\boldsymbol{m}(1)$. This new requirement on the entries is stated as Condition A_2 in the following theorem.

Theorem 6.7 *The matrices $\boldsymbol{M}$ and $\boldsymbol{m}(0)$ and $\boldsymbol{m}(1)$ have eigenvalues 1 and $\frac{1}{2}$ if and only if the filter coefficients satisfy Condition A_2 which includes A_1:*

$$\textit{Condition } A_2: \quad \sum_0^N (-1)^k \boldsymbol{h}(k) = 0 \quad \textit{and} \quad \sum_0^N (-1)^k k\boldsymbol{h}(k) = 0.$$

The eigenvector for $\lambda = 1$ is $\Phi(0)$, containing the values $\phi(0), \ldots, \phi(N-1)$. The eigenvector for $\lambda = \frac{1}{2}$ is $\Phi'(0)$, containing the derivatives $\phi'(0), \ldots, \phi'(N-1)$.

In this case $H(\omega)$ has a *double zero* at $\omega = \pi$. This beautiful pattern extends onward to Condition A_p. The matrices have eigenvalues $\lambda = 1, \frac{1}{2}, \ldots, \left(\frac{1}{2}\right)^{p-1}$ if and only if the filter coefficients satisfy p sum rules:

$$\textit{Condition } A_p: \quad \sum (-1)^k k^m \boldsymbol{h}(k) = 0 \quad \text{for } m = 0, \ldots, p-1.$$

The eigenvector for $\lambda = \left(\frac{1}{2}\right)^m$ contains values of the mth derivative of $\phi(t)$ at the integers. Formally, $\Phi^{(m)}(0) = 2^m \boldsymbol{M}\Phi^{(m)}(0)$ comes from differentiating the dilation equation m times at the integers. We mention the frequency domain equivalent:

Condition A_p: The frequency response $H(\omega)$ has a zero of order p at $\omega = \pi$.

We will see this again! And we also begin to uncover the crucial role of the *left eigenvectors* (which are row vectors). Those tell how to produce polynomials from combinations of the translates $\phi(t-k)$. Under Condition A_p, these low order polynomials $1, t, \ldots, t^{p-1}$ are in the lowpass space V_0. They are the keys to approximation of a function $f(t)$ by functions in V_0.

The letters A_p indicate "approximation of order p." The theorem above, with its extension from 2 to p, is absolutely basic to the algebra of downsampled filters. These eigenvalues and eigenvectors control everything in Chapter 7.

You will see that the derivative $\phi'(t)$ is often *one-sided.* Derivatives of $\phi(t)$ may not exist in the usual sense. This subject still contains some mysteries.

Example 6.5. The hat function coefficients $2\boldsymbol{h}(k) = \frac{1}{2}, 1, \frac{1}{2}$ satisfy Condition A_2:

$$\text{First sum rule:} \qquad \tfrac{1}{2} - 1 + \tfrac{1}{2} = 0$$
$$\text{Second sum rule:} \qquad 0\left(\tfrac{1}{2}\right) - 1(1) + 2\left(\tfrac{1}{2}\right) = 0.$$

Therefore $\boldsymbol{m}(0)$ will have eigenvalues 1 and $\frac{1}{2}$:

$$\boldsymbol{m}(0) = \begin{bmatrix} 2\boldsymbol{h}(0) & 0 \\ 2\boldsymbol{h}(2) & 2\boldsymbol{h}(1) \end{bmatrix} = \begin{bmatrix} \frac{1}{2} & 0 \\ \frac{1}{2} & 1 \end{bmatrix}.$$

The eigenvector for $\lambda = 1$ has components 0 and 1. They agree with $\phi(0)$ and $\phi(1)$, the hat function at the integers. The eigenvector for $\lambda = \frac{1}{2}$ has components 1 and -1. They are $\phi'_+(0)$ and $\phi'_+(1)$, the slopes $\phi'(t)$ of the hat function in the two intervals. These are derivatives *from the right* at the points $t = 0$ and $t = 1$. The slopes on the left side of those points are different because the hat function has corners.

The matrix $\boldsymbol{m}(1)$ must also have eigenvalues 1 and $\frac{1}{2}$:

$$\boldsymbol{m}(1) = \begin{bmatrix} 2\boldsymbol{h}(1) & 2\boldsymbol{h}(0) \\ & 2\boldsymbol{h}(2) \end{bmatrix} = \begin{bmatrix} 1 & \frac{1}{2} \\ 0 & \frac{1}{2} \end{bmatrix}.$$

The eigenvector for $\lambda = 1$ has components 1 and 0. Those agree with the hat function at the shifted points $t = 1$ and $t = 2$. The eigenvector for $\lambda = \frac{1}{2}$ has components 1 and -1. Those agree with the slopes of the hat function *from the left* at $t = 1$ and $t = 2$. Remember that $\boldsymbol{m}(0)$ is involved at the start of an interval and $\boldsymbol{m}(1)$ is involved at the end of an interval.

Condition A_3 is not satisfied for the hat function. There is no eigenvalue $\lambda = \frac{1}{4}$. The hat function has no second derivatives at the integers.

Problem Set 6.3

1. If the filter $H(z)$ is halfband, show that the eigenvector in $\boldsymbol{m}(0)\Phi = \Phi$ is an impulse $\boldsymbol{\delta}(n)$. What are the values of $\phi(t)$ at the integers?

2. If $\phi_1(t)$ and $\phi_2(t)$ satisfy dilation equations, does their product $P(t) = \phi_1(t)\phi_2(t)$ satisfy a dilation equation?
3. Show that the convolution $\phi_1(t) * \phi_2(t)$ *does* satisfy a dilation equation with coefficients from $\boldsymbol{h}_1 * \boldsymbol{h}_2$.
4. Find a specific function $f(t)$ that does not satisfy any dilation equation.

6.4 Infinite Product Formula

The scaling function $\phi(t)$ comes from the dilation equation

$$\phi(t) \quad = \quad 2\sum_{k=0}^{N} \boldsymbol{h}(k)\phi(2t-k).$$

Thus $\phi(t)$ is the fixed point, or fixed function, when we iterate with $\boldsymbol{H}$ and rescale. In the time domain, the matrix that filters and rescales is $\boldsymbol{M} = (\downarrow 2)2\boldsymbol{H}$. Now we intend to find the Fourier transform $\widehat{\phi}(\omega)$ in the frequency domain. Just as the time-domain solution involved products of $\boldsymbol{m}(0)$ and $\boldsymbol{m}(1)$, the frequency-domain solution will involve an infinite product of $H(\omega)$'s.

It is quite remarkable that two-scale equations received so little attention for so long. Historically, t and $2t$ were not often seen in the same equation. They began to appear prominently for fractals, which are self-similar. Now, multiple scales seem to be everywhere. We meet them in this book through multirate filters — with two scales. Then the iteration leads to all scales.

If the 2's were removed, the dilation equation would be an ordinary difference equation. The coefficients are constant, so we look for pure exponential solutions $e^{i\omega t}$. When you make that substitution, you are effectively taking the Fourier transform of the equation. The transform turns difference equations and differential equations (*and dilation equations*) into algebraic equations. We do that now for the two-scale equation, and we watch how $2t$ leads to $\omega/2$.

The dilation equation becomes $\widehat{\phi}(\omega) = H(\frac{\omega}{2})\,\widehat{\phi}(\frac{\omega}{2})$. This leads recursively to an infinite product for $\widehat{\phi}(\omega)$. This transform must be a *sinc function* when $\boldsymbol{h}(0) = \boldsymbol{h}(1) = 1/2$ —because the time-domain solution $\phi(t)$ is a box function. That sinc function must be orthogonal to its modulations by $e^{-i\omega k}$, because the box function is orthogonal to its translates $\phi(t-k)$. We have to study orthogonality and also approximation, which is controlled by "zeros at π." These properties are now studied in the frequency domain.

1. Condition **O** for orthogonality:

$\lvert H(\omega)\rvert^2 + \lvert H(\omega+\pi)\rvert^2$	$\equiv$	1 in the frequency domain
$2\sum \boldsymbol{h}(k)\boldsymbol{h}(k-2\ell)$	$=$	$\delta(\ell)$ in the time domain

This is double-shift orthogonality of the lowpass filter coefficients. It connects to orthogonality of the scaling functions $\phi(t-k)$. We use the word "connects" rather than "implies," because a further condition is eventually needed to insure orthogonality in the limit of the iteration. The step from discrete time to continuous time seldom goes wrong, but it can.

This orthogonality will appear in Theorem 6.10 as a neat statement about the Fourier transforms of ϕ and w:

$$\sum_{-\infty}^{\infty} \left|\widehat{\phi}(\omega+2\pi n)\right|^2 = 1 \quad \text{and} \quad \sum_{-\infty}^{\infty} \widehat{\phi}(\omega+2\pi n)\overline{\widehat{w}(\omega+2\pi n)} = 0.$$

The other side of the theory is about approximation. This imposes a very different condition on the $\boldsymbol{h}(k)$ and the polynomial $H(\omega) = \sum \boldsymbol{h}(k)e^{-ik\omega}$.

2. Condition A_p: $H(\omega) = \left(\frac{1+e^{-i\omega}}{2}\right)^p Q(\omega)$.

This factor $\left(1+e^{-i\omega}\right)^p$ means that $H(\omega)$ has a *zero of order* p at $\omega = \pi$. We will prove that this puts the polynomials $1, t, \ldots, t^{p-1}$ in the scaling subspace V_0. They are combinations of $\phi(t-n)$ and they are orthogonal to $w(t-n)$. In the frequency domain, there is again a neat statement about the Fourier transforms of ϕ and w:

$$\widehat{\phi} \text{ has a zero of order } p \text{ at every } \omega = 2\pi n,\ n \neq 0$$
$$\widehat{w} \text{ has a zero of order } p \text{ at zero frequency.}$$

The wavelet coefficients of a smooth function $f(t) = \sum b_{jk} w_{jk}(t)$ decrease faster when p is larger. The estimate is $|b_{jk}| = O\left(2^{-jp}\right)$. This is valuable for compression. This section does the frequency-domain algebra, to solve the dilation equation and to explain Condition O and Condition A_p.

Transform and Solution of the Dilation Equation

To transform the dilation equation, multiply by $e^{-i\omega t}$. Integrate with respect to t:

$$\int_{-\infty}^{\infty} \phi(t)e^{-i\omega t}\,dt \quad = \quad 2\sum_{k=0}^{N} \boldsymbol{h}(k) \int_{-\infty}^{\infty} \phi(2t-k)e^{-i\omega t}\,dt.$$

The left side is $\widehat{\phi}(\omega)$. In the integral on the right, set $u = 2t-k$ and $t = (u+k)/2$:

$$\int_{-\infty}^{\infty} \phi(2t-k)e^{-i\omega t}2\,dt = \int_{-\infty}^{\infty} \phi(u)e^{-i\omega(u+k)/2}\,du = e^{-i\omega k/2}\,\widehat{\phi}\left(\tfrac{\omega}{2}\right). \tag{6.43}$$

Instead of t and $2t$, *the transform involves* ω *and* $\omega/2$. The dilation equation becomes

$$\widehat{\phi}(\omega) = \left(\sum \boldsymbol{h}(k)e^{-i\omega k/2}\right)\widehat{\phi}\left(\tfrac{\omega}{2}\right) = H\left(\tfrac{\omega}{2}\right)\widehat{\phi}\left(\tfrac{\omega}{2}\right). \tag{6.44}$$

This is the result of filtering and rescaling. Filtering multiplies $\widehat{\phi}(\omega)$ by $H(\omega)$. Rescaling changes ω to $\omega/2$. The scaling function (which is unique up to a constant multiple C—this is still to be proved) comes out unchanged:

$$\boxed{\widehat{\phi}(\omega) = H\left(\tfrac{\omega}{2}\right)\widehat{\phi}\left(\tfrac{\omega}{2}\right).}$$

Now iterate this equation. It connects ω to $\omega/2$ and therefore it connects $\omega/2$ to $\omega/4$:

$$\widehat{\phi}(\omega) \quad = \quad H\left(\tfrac{\omega}{2}\right)\left[H\left(\tfrac{\omega}{4}\right)\widehat{\phi}\left(\tfrac{\omega}{4}\right)\right].$$

After N iterations, this becomes

$$\widehat{\phi}(\omega) = H\left(\tfrac{\omega}{2}\right)H\left(\tfrac{\omega}{4}\right)\cdots H\left(\tfrac{\omega}{2^N}\right)\widehat{\phi}\left(\tfrac{\omega}{2^N}\right).$$

In the limit as $N \to \infty$, we have a formula for the solution $\widehat{\phi}(\omega)$. Note that $\omega/2^N$ is approaching zero, and $\widehat{\phi}(0) = \int \phi(t)\,dt$ is the area under the graph of $\phi(t)$. *This equals one.* We impose the normalization $\widehat{\phi}(0) = 1$ in the frequency domain, just as we required unit area in the time domain. Then the formal limit of the iteration leads to the famous infinite product for $\widehat{\phi}$:

$$\widehat{\phi}(\omega) = \prod_{j=1}^{\infty} H\left(\tfrac{\omega}{2^j}\right). \tag{6.45}$$

We note the minimum requirement for convergence of this or any infinite product: the factors $H(\omega/2^j)$ must approach 1 as $j \to \infty$. Thus we need $H(0) = 1$. By periodicity $H(2\pi) = H(4\pi) = 1$. Then the equation $\widehat{\phi}(\omega) = H\left(\frac{\omega}{2}\right)\widehat{\phi}\left(\frac{\omega}{2}\right)$ has a remarkable consequence. *The values $\widehat{\phi}(2\pi)$, $\widehat{\phi}(4\pi)$, $\widehat{\phi}(8\pi)$, ... are all equal. If $H(\pi) = 0$, those values equal zero because $\widehat{\phi}(2\pi) = H(\pi)\,\widehat{\phi}(\pi)$.*

This "zero at π" is a natural requirement on $H(\omega)$ in order that $\widehat{\phi}(\omega)$ may decay and $\phi(t)$ may be a reasonable function.

The infinite product converges for every ω and every $H(\omega)$. We have an explicit formula for $\widehat{\phi}(\omega)$. Whether any function $\phi(t)$ has this Fourier transform is another matter! Convergence follows from a rough bound on $H(\omega)$ in terms of $C = \max|H'(\omega)|$:

$$|H(\omega)| = |1 + H(\omega) - H(0)| \le 1 + C|\omega| \le e^{C|\omega|}.$$

Then the product $\widehat{\phi}(\omega)$ has the same upper bound:

$$|\widehat{\phi}(\omega)| = \left|H\left(\tfrac{\omega}{2}\right)\right| \left|H\left(\tfrac{\omega}{4}\right)\right| \cdots \le e^{C|\omega|/2} e^{C|\omega|/4} \cdots = e^{C|\omega|}.$$

This is a wild overestimate of $\widehat{\phi}(\omega)$, as almost any example will show.

Box example. The coefficients are $\boldsymbol{h}(0) = \boldsymbol{h}(1) = \frac{1}{2}$. Then $H(\omega) = \frac{1}{2}\left(1 + e^{-i\omega}\right)$. The product of the first N factors contains 2^N terms. Looked at correctly, those terms are the first 2^N powers of $e^{-i\omega/2^N}$:

$$\begin{aligned} H^{(N)}(\omega) &= \frac{1}{2^N}\left(1 + e^{-i\omega/2}\right)\left(1 + e^{-i\omega/4}\right)\cdots\left(1 + e^{-i\omega/2^N}\right) \\ &= \frac{1}{2^N}\sum_{k=0}^{2^N-1} e^{-i\omega k/2^N} \quad \text{(geometric series)} \\ &= \frac{1 - e^{-i\omega}}{2^N\left(1 - e^{-i\omega/2^N}\right)} \quad \text{(sum of } 2^N \text{ terms).} \end{aligned} \tag{6.46}$$

Now let $N \to \infty$. The denominator has $1 - e^{-i\theta} = 1 - (1 - i\theta + \cdots) = i\theta + \cdots$ with $\theta = \omega/2^N$. The limit of $2^N i\theta$ is $i\omega$. Therefore the limit of the partial product is the infinite product

$$\widehat{\phi}(\omega) = \prod\left(\tfrac{1}{2} + \tfrac{1}{2}e^{-i\omega/2^j}\right) = (1 - e^{-i\omega})/i\omega. \tag{6.47}$$

This sinc function is the transform of the box function. The integral of $e^{-i\omega t}$ from 0 to 1 agrees with $\widehat{\phi}(\omega)$. Instead of increasing like $e^{C|\omega|}$, as allowed by the general estimate, the transform $\widehat{\phi}(\omega)$ actually decreases to zero as ω becomes large.

Compare the construction of $\phi(t)$ with $\widehat{\phi}(\omega)$. In Section 6.2, we assumed that $\phi^{(i)}(t)$ converged uniformly to $\phi(t)$. Then we studied its properties. In this section, the convergence of the infinite product is cheap (for each separate ω). What we need is sufficient decay of $\widehat{\phi}(\omega)$ as

$|\omega| \to \infty$. Our precise assumption will be continuity of the function $A(\omega)$ in Theorem 6.10 below. Then we can safely study $\widehat{\phi}(\omega)$ in the frequency domain. Note first that for real frequencies, the growth of $\widehat{\phi}(\omega)$ is at most polynomial:

Theorem 6.8 $|\widehat{\phi}(\omega)| \le e^{C|\omega|}$ *for complex* ω *and* $|\widehat{\phi}(\omega)| \le c(1+|\omega|^M)$ *for real* ω.

Brief reason: $H(\omega)$ is periodic. It has a maximum value 2^M. The equation $\widehat{\phi}(2\omega) = H(\omega)\widehat{\phi}(\omega)$ says that $\widehat{\phi}$ grows by at most 2^M when ω is doubled. The bound $|\omega|^M$ has this growth rate. A constant is included to make $c(1+|\omega|^M)$ correct for small $|\omega|$.

The example $H(\omega) = \frac{1}{2} + \frac{1}{2}e^{-i\omega}$ is bounded by 1 for real ω and by $e^{|\omega|}$ for complex ω. Then $M = 0$. The transform $\widehat{\phi}(\omega)$ of the box function has those same bounds:

$$|\widehat{\phi}(\omega)| = \left|H(\frac{\omega}{2})\right| \left|H(\frac{\omega}{4})\right| \cdots \le \begin{cases} 1 & \text{for real } \omega \\ e^{|\omega|/2}\, e^{|\omega|/4} \cdots = e^{|\omega|} & \text{for complex } \omega. \end{cases}$$

Section 6.1 showed that the support interval for $\phi(t)$ is $[0, N]$. This can be proved in the frequency domain too. Our bounds on $\widehat{\phi}(\omega)$ show two fundamental facts about $\phi(t)$:

Theorem 6.9 *Any dilation equation with* $\boldsymbol{h}(0)+\cdots+\boldsymbol{h}(N) = 1$ *has a* unique and compactly supported solution $\phi(t)$. *This solution may be a distribution.*

Compact support comes from $|\widehat{\phi}(\omega)| \le e^{C|\omega|}$. The Paley-Wiener Theorem implies that $\phi(t)$ is supported on the interval $[-C, C]$. With more care [D, p.176] we could find again the exact support interval $[0, N]$.

Uniqueness comes from our formula for the solution! The infinite product converges to $\widehat{\phi}(\omega)$, which is continuous because $\phi(t)$ has compact support:

$$\widehat{\phi}^{(i)}(\omega) = \Big(\prod_{j=1}^{i} H(\omega/2^j)\Big)\, \widehat{\phi}(\omega/2^i) \text{ approaches } \widehat{\phi}(\omega) = \Big(\prod_{1}^{\infty} H(\omega/2^j)\Big)\, \widehat{\phi}(0).$$

In the IIR case, suitable hypotheses will again give uniqueness (of course not compact support). At the other extreme, note how the lazy filter with $\boldsymbol{h}(0) = 1$ leads to $\phi(t) =$ delta function. *The dilation equation* $\phi(t) = 2\phi(2t)$ *is solved by* $\phi(t) = \delta(t)$:

In frequency: $H(\omega) \equiv 1$ so $\widehat{\phi}(\omega) = 1$.

Cascade algorithm: $\phi^{(i)}(t) =$ box function on $[0, 2^{-i}]$ with height 2^i.

Verify directly: $\delta(t) = 2\delta(2t)$ from $\int f(t)\delta(t)dt = f(0) = \int f(t)\delta(2t)2\,dt$.

All these methods show that $\boldsymbol{h}(0) = 1$ produces the best-known distribution $\phi(t) = \delta(t)$.

Orthogonality in the Frequency Domain

The product formula for $\widehat{\phi}(\omega)$ applies with or without Condition O. When that condition holds, we expect orthogonality of the translates $\phi(t-k)$. To establish this orthogonality in the frequency domain, we need to know that the equivalent statement is $A(\omega) \equiv 1$. The function $A(\omega)$ enters naturally into this discussion. It is the transform $\sum \boldsymbol{a}(k)e^{i\omega k}$ of the vector of *inner products* of $\phi(t)$ with $\phi(t-k)$:

Theorem 6.10 *The inner products $\boldsymbol{a}(k)$ are the Fourier coefficients of the 2π-periodic function $A(\omega)$:*

$$\boldsymbol{a}(k) = \int_{-\infty}^{\infty} \phi(t)\overline{\phi(t-k)}\,dt \quad \textit{transforms to} \quad A(\omega) = \sum_{-\infty}^{\infty} \left|\widehat{\phi}(\omega + 2\pi n)\right|^2.$$

The translates $\phi(t-k)$ are orthonormal if and only if $A(\omega) \equiv 1$.

Proof. An inner product in the time domain equals an inner product in the frequency domain, by Parseval's identity. The inner product in the time domain is between $\phi(t)$ and $\phi(t-k)$. The transforms of these functions are $\widehat{\phi}(\omega)$ and $e^{-i\omega k}\widehat{\phi}(\omega)$. Each inner product integrates one function times the complex conjugate of the other:

$$\begin{aligned} \boldsymbol{a}(k) = \int_{-\infty}^{\infty} \phi(t)\overline{\phi(t-k)}\,dt &= \frac{1}{2\pi}\int_{-\infty}^{\infty} \widehat{\phi}(\omega)\overline{\widehat{\phi}(\omega)}e^{i\omega k}\,d\omega \\ &= \frac{1}{2\pi}\int_{0}^{2\pi} \sum_{-\infty}^{\infty} \left|\widehat{\phi}(\omega + 2\pi n)\right|^2 e^{i\omega k}\,d\omega. \end{aligned} \tag{6.48}$$

The last integral split $(-\infty, \infty)$ into an infinite number of 2π-pieces, using the periodicity of $e^{i\omega k}$. This integral defines the kth Fourier coefficient of $A(\omega)$. Thus $A(\omega) = \sum \boldsymbol{a}(k)e^{i\omega k}$.

For an FIR filter, $\phi(t) = 0$ outside the interval $[0, N]$. The inner products are $\boldsymbol{a}(k) = 0$ for $|k| > N$, because $\phi(t)$ and $\phi(t-k)$ have no overlap. *The function $A(\omega) = \sum \boldsymbol{a}(k)e^{i\omega k}$ is a trigonometric polynomial of degree N,* which is not obvious from $\sum |\widehat{\phi}(\omega + 2\pi n)|^2$. In Section 7.3 we will compute $\boldsymbol{a}(k)$ directly from the coefficients $\boldsymbol{h}(n)$. This is always a main point of the theory, to return every calculation to those numbers $\boldsymbol{h}(n)$.

When the translates are orthonormal, all inner products $\boldsymbol{a}(k)$ are zero except for $\boldsymbol{a}(0) = 1$. The function with those coefficients is the constant function $A(\omega) \equiv 1$:

$$\phi(t-k) \quad \text{are orthonormal} \quad \Longleftrightarrow \quad \sum_{-\infty}^{\infty} \left|\widehat{\phi}(\omega + 2\pi n)\right|^2 \equiv 1.$$

We now apply Condition O in the frequency domain to deduce this orthogonality of $\phi(t-k)$. We are repeating in the frequency domain the result of Section 6.2 in the time domain. I believe this is worthwhile! The arguments in the two domains look quite different. Recall the condition on the frequency response $H(\omega)$ to produce an orthonormal filter bank:

$$\text{Condition O:} \quad |H(\omega)|^2 + |H(\omega + \pi)|^2 \equiv 1.$$

This function $H(\omega)$ leads to $\widehat{\phi}(\omega)$ which leads to $A(\omega)$. Somehow, Condition O must imply that $A(\omega) \equiv 1$. The steps are typical of computations in the frequency domain.

Theorem 6.11 *If $A(\omega)$ is continuous, $A(\omega) \equiv 1$ is equivalent to Condition O.*

Proof. We use a very important two-scale identity, proved below:

$$A(2\omega) = |H(\omega)|^2 A(\omega) + |H(\omega + \pi)|^2 A(\omega + \pi). \tag{6.49}$$

If $A(\omega) \equiv 1$, this immediately gives that $|H(\omega)|^2 + |H(\omega + \pi)|^2 \equiv 1$.

For the converse, suppose that Condition O holds. The identity says that $A(2\omega)$ is a weighted average of $A(\omega)$ and $A(\omega+\pi)$. At the point ω_0 where $A(2\omega)$ reaches its maximum, $A(\omega_0)$ and $A(\omega_0+\pi)$ must also reach that maximum. Now repeat the argument at $\omega_0/2$, to show that $A(\omega_0/2)$ shares this same maximum with $A(\omega_0)$. Continuing, the maximum of $A(\omega)$ is achieved at $\omega_0/4$ and $\omega_0/8$ and eventually (by continuity) at $\omega=0$.

By a similar argument, which is due to Tchamitchian, the minimum of $A(\omega)$ is also attained at $\omega=0$. *Therefore $A(\omega)$ is constant.* We verify below that the constant is one: $A(\omega)\equiv 1$. It only remains to prove (6.49).

This valuable identity for $A(2\omega)$ has a nice proof. It uses the dilation equation $\widehat{\phi}(2\omega) = H(\omega)\widehat{\phi}(\omega)$. At the points $2\omega+2\pi n$, this splits into separate cases for even n and odd n:

$$\begin{aligned}\widehat{\phi}(2\omega+2\pi n) &= H(\omega+\pi n)\,\widehat{\phi}(\omega+\pi n)\\ &= \begin{cases} H(\omega)\,\widehat{\phi}(\omega+2k\pi) & \text{if } n=2k\\ H(\omega+\pi)\,\widehat{\phi}(\omega+(2k+1)\pi) & \text{if } n=2k+1.\end{cases}\end{aligned}$$

Now square both sides. Sum from $-\infty$ to ∞ on n and therefore on k. The sum of squares is our function $A(2\omega)$ in the desired identity:

$$\begin{aligned}A(2\omega) &= |H(\omega)|^2\sum_{-\infty}^{\infty}\left|\widehat{\phi}(\omega+2\pi k)\right|^2 + |H(\omega+\pi)|^2\sum_{-\infty}^{\infty}\left|\widehat{\phi}(\omega+\pi+2\pi k)\right|^2\\ &= |H(\omega)|^2A(\omega)+|H(\omega+\pi)|^2A(\omega+\pi).\end{aligned}\tag{6.50}$$

The final step is to confirm that $A(0)=1$. This comes from our other condition on the lowpass filter, not yet used in the frequency domain. Condition A_1 is $H(\pi)=0$. In the time domain, this first sum rule guaranteed an eigenvalue $\lambda=1$ for the matrices $\boldsymbol{M}$ and $\boldsymbol{m}(0)$ and $\boldsymbol{m}(1)$. The fixed-point equation $\phi^{(1)}(n)=\phi^{(0)}(n)$ at the integers could be solved. Condition A_1 is equally essential in the frequency domain. Here we use it to pin down the value $A(0)=1$.

Theorem 6.12 *If $H(\pi)=0$ then $\widehat{\phi}(2\pi n)=0$ for all $n\neq 0$. Therefore*

$$A(0)=\sum_{-\infty}^{\infty}\left|\widehat{\phi}(2\pi n)\right|^2=\left|\widehat{\phi}(0)\right|^2=1.\tag{6.51}$$

Proof. The infinite product for $\widehat{\phi}(2\pi)=H(\pi)H(\pi/2)\cdots$ starts with the factor $H(\pi)$. Immediately this product is zero. For any higher value $n>1$, write $n=2^jm$ with odd m. Then the $(j+1)^{\text{st}}$ factor in the infinite product is zero when $\omega=2\pi n$:

$$\widehat{\phi}(2\pi n)=H(\pi n)H\left(\frac{\pi n}{2}\right)\cdots H\left(\frac{\pi n}{2^j}\right)\cdots=0$$

because $H(\pi n/2^j)=H(\pi m)$. By periodicity this is $H(\pi)=0$. The only nonzero term is $|\widehat{\phi}(0)|^2$. But $\widehat{\phi}(0)=H(0)H(0)H(0)\cdots$ which is 1.

Orthogonalization of the Basis

The condition for an orthonormal basis is $A(\omega)\equiv 1$. When this is not satisfied, there is an easy way to *make* it satisfied. In other words: when the translates $\phi(t-n)$ are not orthonormal, there is an easy way to *make* them orthonormal. Divide $\widehat{\phi}(\omega)$ by the given $A(\omega)$ (or rather, its square root) to get the new orthogonalized function $\widehat{\phi}_{\text{orth}}(\omega)$:

$$\widehat{\phi}_{\text{orth}}(\omega) = \frac{\widehat{\phi}(\omega)}{\sqrt{A(\omega)}}.$$

This immediately gives orthogonality of the new basis $\{\phi_{\text{orth}}(t-n)\}$:

$$A_{\text{orth}}(\omega) = \sum \left| \frac{\widehat{\phi}(\omega + 2\pi n)}{\sqrt{A(\omega)}} \right|^2 = \frac{A(\omega)}{A(\omega)} \equiv 1.$$

That succeeds if $A(\omega)$ is never zero. This is the condition for a ***Riesz basis***.

Theorem 6.13 *The upper and lower bounds on $A(\omega)$ are the Riesz constants B and A for the basis $\{\phi(t-k)\}$ of V_0. Thus $A(\omega) \geq A > 0$ gives a stable basis, and dividing $\widehat{\phi}(\omega)$ by $\sqrt{A(\omega)}$ gives an orthonormal basis.*

When $\phi(t)$ comes from a dilation equation—this is our normal situation—Condition **E** in Chapter 7 gives an equivalent test for a Riesz basis (in terms of eigenvalues). If this test is passed, the wavelets $w_{jk}(t)$ are a Riesz basis for $L^2(\mathbf{R})$.

Proof. To test the linear independence of the functions $\phi(t-k)$, form the matrix $\boldsymbol{A}$ from their inner products. The entries are $\boldsymbol{A}_{ij} = \langle \phi(t-i), \phi(t-j) \rangle$. That number is $\boldsymbol{a}(j-i)$, and $\boldsymbol{A}$ is a Toeplitz matrix! It is the matrix $\boldsymbol{TT}^*$ in Section 2.5. In the frequency domain it becomes multiplication by $A(\omega)$. The upper and lower bounds on $A(\omega)$ determine whether $\{\phi(t-k)\}$ is a Riesz basis.

Orthogonalization is always a basic step in linear algebra. There it is done by the Gram-Schmidt algorithm. We start with independent vectors and produce orthonormal vectors (or functions). This algorithm is not successful here, because it is not time-invariant. The orthogonalized functions will certainly not be translates—when the Gram-Schmidt algorithm works on functions in a definite order like $\phi(t), \phi(t-1), \phi(t+1), \ldots$. To keep a shift-invariant basis, we needed to orthogonalize all these translates at once. The division by $\sqrt{A(\omega)}$ did it.

In matrix language, $\boldsymbol{M}_{\text{orth}} \boldsymbol{M}_{\text{orth}}^T = \boldsymbol{I}$. In the improved factorization by Fourier methods, all rows of $\boldsymbol{M}_{\text{orth}}$ come from the zeroth row by double shifts. In other words, $\boldsymbol{M}_{\text{orth}}$ comes from a filter.

One problem with dividing by $\sqrt{A(\omega)}$. This destroys the finite response of the original filter $\boldsymbol{H}$. The new filter $\boldsymbol{H}_{\text{orth}}$ is IIR, not FIR. The new scaling function $\phi_{\text{orth}}(t)$ that corresponds to $\widehat{\phi}(\omega)/\sqrt{A(\omega)}$ does not have compact support. Vetterli and Herley noticed that this is not as bad as it seems. Since $\phi(t)$ is zero outside the interval $[0, N]$, the inner products $\boldsymbol{a}(k) = \int \phi(t)\overline{\phi(t-k)}\,dt$ are zero for $|k| > N$. The function $A(\omega)$ with these Fourier coefficients is a *real non-negative trigonometric polynomial of degree N*. Its square root $G(\omega) = \sum g(k)e^{-ik\omega}$, by spectral factorization, is also a polynomial of degree N. The frequency response of the new orthogonalized filter is a ratio of polynomials

$$H_{\text{orth}}(\omega) = \frac{H(\omega)}{G(\omega)}.$$

The input-output equation $\boldsymbol{y}(k) = \sum \boldsymbol{h}_{\text{orth}}(k)\boldsymbol{x}(n-k)$ is an implicit difference equation, from an *autoregressive* moving average filter:

$$\sum_0^N \boldsymbol{g}(k)\boldsymbol{y}(n-k) = \sum_0^N \boldsymbol{h}(k)\boldsymbol{x}(n-k).$$

The new filter is IIR but it only involves $2N+2$ parameters $\boldsymbol{g}(k)$ and $\boldsymbol{h}(k)$. Therefore it can be physically realized, and now the basis has been orthogonalized.

Problem Set 6.4

1. Use the identity $\sin 2\theta = 2\sin\theta\cos\theta$ to show that

$$\left(\cos\frac{\omega}{2}\right)\left(\cos\frac{\omega}{4}\right)\cdots\left(\cos\frac{\omega}{2^N}\right)=\frac{1}{2^N}\frac{\sin\omega}{\sin\frac{\omega}{2}}\frac{\sin\frac{\omega}{2}}{\sin\frac{\omega}{4}}\cdots\frac{\sin\frac{\omega}{2^{N-1}}}{\sin\frac{\omega}{2^N}}.$$

Cancel sines and let $N\to\infty$ to find a great infinite product:

$$\prod_1^\infty \cos\left(\frac{\omega}{2^j}\right)=\frac{1}{\omega}\sin\omega.$$

2. The Haar filter has $H(\omega)=\frac{1}{2}(1+e^{-i\omega})=e^{-i\omega/2}\cos\frac{\omega}{2}$. Use $\frac{\omega}{2}$ in Problem 1 to give a new proof for the infinite product (6.47) of $H(\omega/2^j)$:

$$\left(e^{-i\omega/4}\cos\frac{\omega}{4}\right)\left(e^{-i\omega/8}\cos\frac{\omega}{8}\right)\cdots=\left(e^{-i\omega/2}\frac{2}{\omega}\sin\frac{\omega}{2}\right)=\frac{1}{i\omega}(1-e^{-i\omega}).$$

3. Suppose $H(\omega)=\frac{1}{4}(1+e^{-i\omega})^2$. Find $\widehat{\phi}(\omega)$ and $\phi(t)$.

4. If $H(\omega)$ has p zeros at $\omega=\pi$, show that $\widehat{\phi}(\omega)$ has p zeros at $\omega=2\pi n$ for each $n\neq 0$.

6.5 Biorthogonal Wavelets

This chapter has concentrated on orthogonal wavelets, coming from an orthogonal filter bank. The synthesis filters are transposes of the analysis filters. One multiresolution is all we need. The synthesis wavelets are the same, in this self-orthogonal case, as the analysis wavelets. But from ***biorthogonal filters*** we must expect ***biorthogonal wavelets***.

We now meet a new scaling function $\widetilde{\phi}(t)$. Its translates $\widetilde{\phi}(t-k)$ will span a new lowpass space $\widetilde{V}_0$—different from V_0. There is also a wavelet $\widetilde{w}(t)$. The translates $\widetilde{w}(t-k)$ span a complementary highpass space $\widetilde{W}_0$. The sum of those spaces will be $\widetilde{V}_1=\widetilde{V}_0+\widetilde{W}_0$, the next space in the second multiresolution. To a large extent, the theory is achieved by inserting a tilde where appropriate. We want to indicate why this second scale of spaces $\widetilde{V}_j$ is needed.

Biorthogonality comes automatically with inverse matrices. The rows of a 2×2 matrix and the columns of its inverse are biorthogonal:

$$\begin{bmatrix}\text{row} & 1\\ \text{row} & 2\end{bmatrix}\begin{bmatrix}\text{column} & \text{column}\\ 1 & 2\end{bmatrix}=\begin{bmatrix}1 & 0\\ 0 & 1\end{bmatrix}.$$

Notice something pleasant. The product in the other order is still $\boldsymbol{I}$. The right-inverse is also the left-inverse. This order involves *columns times rows*, which are full matrices:

$$\begin{bmatrix}\text{column}\\ 1\end{bmatrix}\begin{bmatrix}\text{row} & 1\end{bmatrix}+\begin{bmatrix}\text{column}\\ 2\end{bmatrix}\begin{bmatrix}\text{row} & 2\end{bmatrix}=\boldsymbol{I}.$$

Those two column-row products are projections, and they add to $\boldsymbol{I}$. These simple facts about 2×2 matrices have important parallels for biorthogonal filters and biorthogonal wavelets.

Filter banks display those parallels immediately. The analysis bank has filters $(\downarrow 2)\boldsymbol{H}_0$ and $(\downarrow 2)\boldsymbol{H}_1$. Those are rows 1 and 2. The synthesis bank has expanders before filters, $\boldsymbol{F}_0\,(\uparrow 2)$ and $\boldsymbol{F}_1\,(\uparrow 2)$. Those are columns 1 and 2. In the orthogonal case $\boldsymbol{F}_0$ is $\boldsymbol{H}_0^T$ and $\boldsymbol{F}_1$ is $\boldsymbol{H}_1^T$. In the biorthogonal case we don't have transposes but we still have inverses. To understand the pattern for wavelets, we absolutely must return to multiresolution. One scale of spaces $V_0 \subset V_1 \subset \cdots \subset V_j$ is too limited. We need two hierarchies of spaces, V_j in synthesis and $\widetilde{V}_j$ in analysis.

Tilde Notation *Does the tilde go on the analysis functions or the synthesis functions?* Both conventions are equally possible. We hope to agree with other authors! More and more, the tilde is going on the ***analysis*** functions. Then $f(t)$ is expanded in synthesis functions, which have no tilde. But the coefficients come from the analysis functions and have tildes:

$$f_0(t) = \sum \tilde{a}_{0k}\,\phi(t-k) \text{ is in } V_0, \text{ with } \tilde{a}_{0k} = \int f(t)\,\tilde{\phi}(t-k)\,dt \tag{6.52}$$

$$f(t) = \sum\sum \tilde{b}_{jk}\,w_{jk} \text{ is in } L^2, \text{ with } \tilde{b}_{jk} = \int f(t)\,\tilde{w}_{jk}(t)\,dt \tag{6.53}$$

What does this mean for the filter banks that process the coefficients? Those filters use the letter $\boldsymbol{H}$ in analysis and $\boldsymbol{F}$ in synthesis. We will stay with $\boldsymbol{H}$ and $\boldsymbol{F}$ (rather than $\boldsymbol{C}$ and $\boldsymbol{D}$) when discussing biorthogonal filters. An important result in this section is the Fast Wavelet Transform in equation (6.70), and its inverse (the biorthogonal IFWT) in Theorem 6.16.

Biorthogonal Multiresolution

This chapter began with orthogonal bases $\{\phi(t-k)\}$ for V_0 and $\{w(t-k)\}$ for W_0. *The equation* $V_0 \oplus W_0 = V_1$ ***started a multiresolution.*** W_j was the orthogonal (!) complement of V_j inside V_{j+1}. All is well if $\phi(t)$ and $w(t)$ come from an orthogonal bank of FIR filters. Their translates are all orthogonal. They span perpendicular spaces and we have an orthogonal multiresolution.

All is *not* so well if $\phi(t)$ and $w(t)$ fail to have compact support. The filters fail to be FIR. Often this means that we have asked for too much! Instead of orthogonal bases, we should be content with stable bases. An outstanding example is the space of piecewise linear functions. The stable basis consists of the hat function $\phi(t)$ and its translates. That basis is *not orthogonal.*

When the basis is not orthogonal, there is no reason to insist that W_0 must be orthogonal to V_0. If we do, the multiresolution is called ***semi-orthogonal*** in Section 7.4, and we have "pre-wavelets". But the important property is a stable basis $\{w(t-k)\}$. The highpass coefficients will construct $w(t)$.

Remember the pattern for perfect reconstruction. Coefficients are chosen so that $F_0(z)H_0(z)$ is halfband. When $\phi(t)$ is the hat function from $F_0(z) = \left(\frac{1+z^{-1}}{2}\right)^2$, the other factor $H_0(z)$ needs *five* coefficients. This means $N = 2$ but $\widetilde{N} = 4$. The wavelet has 3-interval support. Then $\phi(t-k)$ and $w(t-k)$ span V_0 and W_0, without orthogonality.

The new *analysis multiresolution* is the point of this section. The coefficients from $H_0(z)$ go into a different dilation equation, whose solution is the ***analyzing function*** $\tilde{\phi}(t)$:

$$\textbf{Analysis Dilation Equation:} \quad \tilde{\phi}(t) = \sum_{0}^{\widetilde{N}} 2h_0(k)\,\tilde{\phi}(2t-k). \tag{6.54}$$

The coefficients $\boldsymbol{h}_0(k)$ add to 1 as before. The new multiresolution obeys the same conditions as before; just add a tilde. $\widetilde{V}_0$ is spanned by $\{\tilde{\phi}(t-k)\}$. The space $\widetilde{V}_j$ is spanned by $\{\tilde{\phi}(2^j t-k)\}$.

They are clearly shift-invariant. The dilation equation (6.54) says that $\widetilde{V}_0 \subset \widetilde{V}_1$. Then also $\widetilde{V}_j \subset \widetilde{V}_{j+1}$. The highpass coefficients produce the wavelet:

$$\textbf{Analysis Wavelet Equation:} \quad \widetilde{w}(t) = \sum_0^N 2\boldsymbol{h}_1(k)\,\widetilde{\phi}(2t-k). \tag{6.55}$$

This wavelet is supported on $[0, \ell]$, where $2\ell = N + \widetilde{N}$. That sum $N + \widetilde{N}$ is the degree of the product filters $F_0(z)H_0(z)$ and $F_1(z)H_1(z)$. Those are symmetric halfband filters, and in the hat function example the degree is $N + \widetilde{N} = 2+4$. Then $\ell = 3$ is odd. The four functions $\phi(t)$, $w(t)$, $\widetilde{\phi}(t)$, $\widetilde{w}(t)$ are graphed in Figure 6.6.

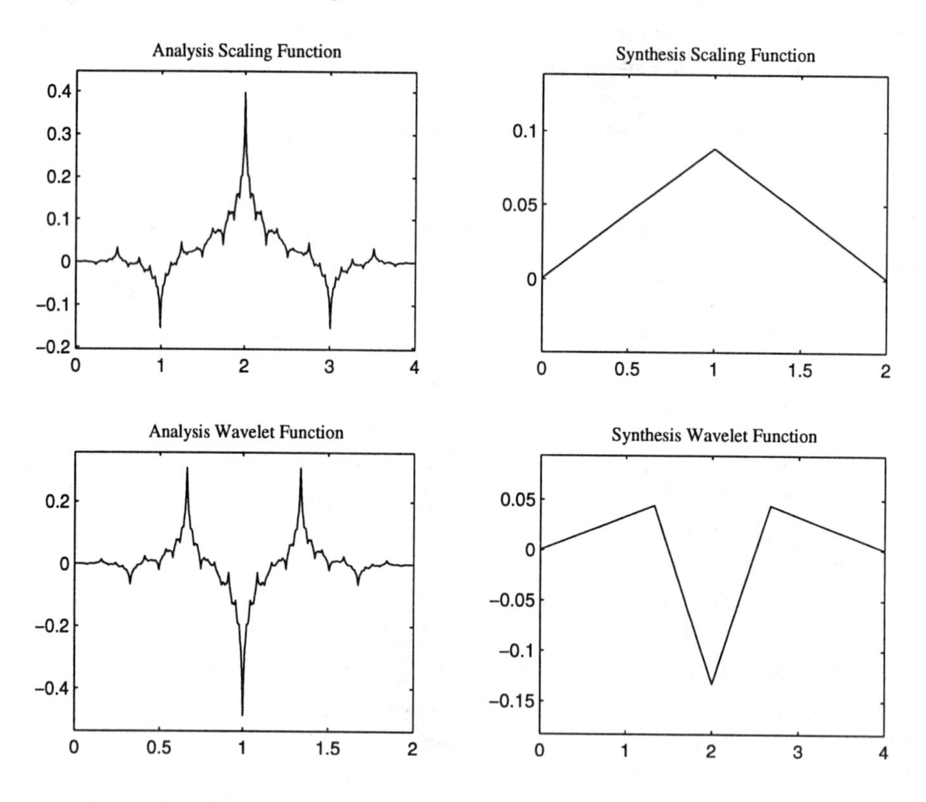

Figure 6.6: Biorthogonal scaling functions and wavelets from a 5/3 PR filter bank.

Biorthogonality in Continuous Time

Our construction of $\phi(t)$, $w(t)$, $\widetilde{\phi}(t)$, $\widetilde{w}(t)$ starts with biorthogonal filters. The lowpass analysis coefficients $\boldsymbol{h}_0(k)$ are *not* double-shift orthogonal to themselves. They are double-shift ***biorthogonal*** to the lowpass synthesis coefficients $\boldsymbol{f}_0(k)$:

$$2\sum \boldsymbol{h}_0(k)\boldsymbol{f}_0(k+2n) = \delta(n). \tag{6.56}$$

This means that $F_0(z)\,H_0(z)$ is halfband. Similarly the highpass filters give $F_1(z)\,H_1(z) =$ halfband:

$$2\sum \boldsymbol{h}_1(k)\boldsymbol{f}_1(k+2n) = \delta(n). \tag{6.57}$$

The other key relation is biorthogonality of highpass to lowpass:

$$\sum \boldsymbol{h}_0(k)\boldsymbol{f}_1(k+2n) = 0 \quad \text{and} \quad \sum \boldsymbol{h}_1(k)\boldsymbol{f}_0(k+2n) = 0. \tag{6.58}$$

The reader knows that all these equations restate perfect reconstruction:

$$F_0(z) = H_1(-z), \quad F_1(z) = -H_0(-z), \quad F_0(z)H_0(z) \textit{ is a halfband filter.}$$

Our question is, *how does biorthogonality appear in continuous time*? The functions $\phi(t)$ and $\widetilde{\phi}(t)$ come from iterating the lowpass filters $\boldsymbol{F}_0$ and $\boldsymbol{H}_0$ (with rescaling!). Start the cascade of iterations from $\phi^{(0)}(t) = \widetilde{\phi}^{(0)}(t) =$ box function on $[0, 1]$. Their translates have biorthogonality. The box $\phi^{(0)}(t-k)$ has no overlap with $\widetilde{\phi}^{(0)}(t-\ell)$ when $k \neq \ell$. This biorthogonality is preserved at every iteration step, when we use equation (6.56). This is exactly parallel to the earlier proof (6.25) that orthogonality is preserved at each iteration. When $\phi^{(i)}(t)$ and $\widetilde{\phi}^{(i)}(t)$ converge in L^2 to the scaling functions $\phi(t)$ and $\widetilde{\phi}(t)$, those limit functions inherit the same biorthogonality:

$$\int_{-\infty}^{\infty} \phi(t-k)\widetilde{\phi}(t-\ell)\,dt = \delta(k-\ell). \tag{6.59}$$

With $i \to \infty$ in the cascade algorithm, these limits $\phi(t)$ and $\widetilde{\phi}(t)$ solve the synthesis and analysis dilation equations. Now bring in the wavelet equations:

$$\int_{-\infty}^{\infty} \phi(t)\widetilde{w}(t)\,dt = \int_{-\infty}^{\infty} \left(\sum 2h_0(k)\phi(2t-k)\right)\left(\sum 2f_1(\ell)\widetilde{\phi}(2t-\ell)\right)dt. \tag{6.60}$$

That right side is zero because of (6.58) and (6.59). And biorthogonality extends to the translates for the same reason:

$$\int_{-\infty}^{\infty} \phi(t-k)\widetilde{w}(t-\ell)\,dt = 0 \quad \text{for all } k \text{ and } \ell. \tag{6.61}$$

Finally the wavelets w and $\widetilde{w}$ are biorthogonal from the wavelet equations and (6.57):

$$\int_{-\infty}^{\infty} w(t-k)\widetilde{w}(t-\ell)\,dt = \delta(k-\ell). \tag{6.62}$$

All this is routine, *provided the cascade algorithms for $\phi(t)$ and $\widetilde{\phi}(t)$ both converge in L^2*. Wavelet theory gives the requirements in Section 7.2, as tests on eigenvalues of two matrices $\boldsymbol{T}$ and $\widetilde{\boldsymbol{T}}$. Suppose those tests are passed (not at all automatic!). The basis functions are biorthogonal. What does this say about the subspaces they span?

Theorem 6.14 *Suppose the filter coefficients satisfy* (6.56)–(6.58) *and also Condition* E *(for L^2 convergence of the cascade algorithm). Then the synthesis functions $\phi(t-k)$, $w(t-k)$ are biorthogonal to the analysis functions $\widetilde{\phi}(t-\ell)$, $\widetilde{w}(t-\ell)$ as in* (6.59)–(6.62). *Each scaling space is orthogonal to the dual wavelet space:*

$$V_j \perp \widetilde{W}_j \quad \textit{and} \quad W_j \perp \widetilde{V}_j. \tag{6.63}$$

V_0 and $\widetilde{W}_0$ are perpendicular because their bases $\phi(t-k)$ and $\widetilde{w}(t-k)$ are perpendicular. When t is replaced by $2^j t$, the zero inner products are still zero. (Change variables back to $T = 2^j t$.) So at each scale j we have perpendicular subspaces.

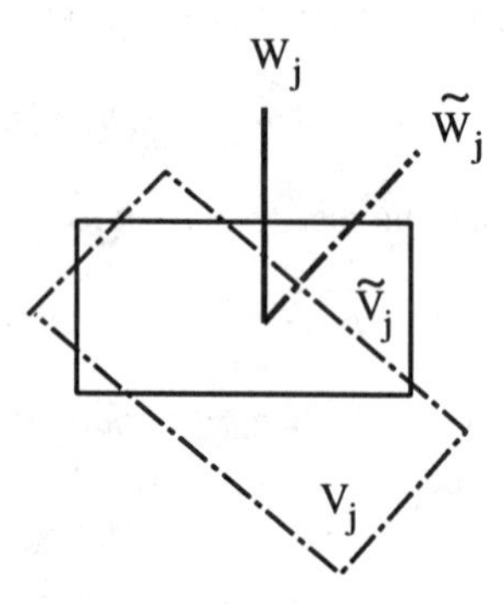

Figure 6.7: V_j is perpendicular to $\widetilde{W}_j$, while $\widetilde{V}_j$ is perpendicular to W_j.

The two multiresolutions are intertwining. As always we have

$$V_j + W_j = V_{j+1} \quad \text{and} \quad \widetilde{V}_j + \widetilde{W}_j = \widetilde{V}_{j+1}. \tag{6.64}$$

These are direct sums but generally not orthogonal sums. The subspaces V_j and W_j have zero intersection, but they are not perpendicular. Instead V_j is perpendicular to $\widetilde{W}_j$. All the subspaces $W_{j-1}, W_{j-2}, \ldots$ are then perpendicular to $\widetilde{W}_j$. Similarly all the subspaces $\widetilde{W}_{j-1}, \widetilde{W}_{j-2}, \ldots$ are perpendicular to W_j. Therefore we have *biorthogonal bases* (*dual bases*):

Corollary *The wavelets $w_{jk}(t) = 2^{j/2}w(2^jt-k)$ and $\widetilde{w}_{jk}(t) = 2^{j/2}\widetilde{w}(2^jt-k)$ are **biorthogonal bases** for L^2:*

$$\int_{-\infty}^{\infty} w_{jk}(t)\widetilde{w}_{JK}(t)\,dt = \delta(j-J)\,\delta(k-K). \tag{6.65}$$

Representing $f(t)$ in a Wavelet Series

If we have a wavelet basis, we have a wavelet series. Any square-integrable (finite energy) function $f(t)$ can be expanded in wavelets:

$$f(t) = \sum_{-\infty}^{\infty}\sum_{-\infty}^{\infty} \widetilde{b}_{jk}\, w_{jk}(t). \tag{6.66}$$

The synthesis wavelets are used to synthesize the function (of course). But the coefficients $\widetilde{b}_{jk}$ come from inner products with the analysis wavelets. *This is why $\widetilde{b}_{jk}$ has a tilde.* Multiply (6.66) by the analysis wavelet $\widetilde{w}_{JK}(t)$ and integrate over t. Biorthogonality yields

$$\widetilde{b}_{JK} = \int_{-\infty}^{\infty} f(t)\,\widetilde{w}_{JK}(t)\,dt. \tag{6.67}$$

Equations (6.66) and (6.67) are the biorthogonal wavelet transform and its inverse. The transform takes function to coefficients, the inverse transform synthesizes the function. We show below how the coefficients can be computed recursively (or pyramidally). This is the *fast* wavelet transform.

Note that Parseval's equality between $\int |f(t)|^2\,dt$ and $\sum\sum |b_{jk}|^2$ is *not true*. That required orthogonality — and not biorthogonality. When we square and integrate the series for $f(t)$, nonzero inner products come from the products $w_{jk}(t)w_{JK}(t)$. We do have an inequality

$$A\int_{-\infty}^{\infty} |f(t)|^2\,dt \le \sum\sum \left|\widetilde{b}_{jk}\right|^2 \le B\int_{-\infty}^{\infty} |f(t)|^2\,dt. \tag{6.68}$$

With $B \ge A > 0$ this says we have a stable basis or ***Riesz basis***. This is true for wavelets, subject to the same Condition **E** [Cohen-Daubechies].

Fast Biorthogonal Wavelet Transform

The reader knows that in practice the wavelet expansion cannot go all the way back to $j = -\infty$. We do not use arbitrarily low frequencies (longer and longer wavelets). A more practical expansion starts with V_0 and $\widetilde{V}_0$, at a scale normalized to $\Delta t = 1$, and goes to V_J and $\widetilde{V}_J$, where the finer scale is 2^{-J}. Enough high-frequency details are included to reproduce an accurate signal.

Thus we work primarily with the subspaces $V_J = V_0 + W_0 + \cdots + W_{J-1}$. There are two important bases for V_J. One is $\phi_{Jk}(t) = 2^{J/2}\phi(2^J t - k)$ for $-\infty < k < \infty$. These scaling functions are *at level* J. The other basis consists of $\phi_{0k}(t)$ from V_0 and $w_{jk}(t)$ for $0 \le j < J$. Since life is recursive, we are interested first of all in $J = 1$. Then the two ways to expand a signal in V_1 are the scaling basis (fine scale) or scaling functions plus wavelets (coarse scale):

$$\sum \widetilde{a}_{1k}\,\sqrt{2}\,\phi(2t-k) = \sum \widetilde{a}_{0k}\,\phi(t-k) + \sum \widetilde{b}_{0k}\,w(t-k). \tag{6.69}$$

The key is the pyramid algorithm. This connects $\widetilde{a}_{1k}$ to $\widetilde{a}_{0k}$ and $\widetilde{b}_{0k}$. We are using ϕ and w because this is synthesis of the signal. But the coefficients come from *analysis* of the signal, which uses $\widetilde{\phi}$ and $\widetilde{w}$:

$$\widetilde{a}_{0k} = \int f(t)\,\widetilde{\phi}(t-k)\,dt = \int f(t)\sum_{\ell} \boldsymbol{h}_0(\ell-2k)\,\widetilde{\phi}_{1\ell}(t)\,dt.$$

For $\widetilde{b}_{0k}$ we use the wavelet equation with $\boldsymbol{h}_1(\ell-2k)$ instead of the dilation equation with $\boldsymbol{h}_0(\ell-2k)$. The pyramid has filtering and downsampling:

$$\textbf{\textit{Fast Wavelet Transform}}\quad \widetilde{a}_{0k} = \sum \boldsymbol{h}_0(\ell-2k)\,\widetilde{a}_{1\ell} \;\text{ and }\; \widetilde{b}_{0k} = \sum \boldsymbol{h}_1(\ell-2k)\,\widetilde{a}_{1\ell}. \tag{6.70}$$

This includes a time-reversal! The same filters $\boldsymbol{H}_0^T$ and $\boldsymbol{H}_1^T$ operate at level j.

That change of basis was not orthogonal. Going backwards, the inverse will not be the transpose. The synthesis filters $\boldsymbol{F}_0$ and $\boldsymbol{F}_1$ must do their part.

Theorem 6.15 **Inverse Fast Wavelet Transform:** *The coefficients* $\widetilde{a}_{1\ell}$ *in the basis* $\phi_{1\ell}(t)$ *can be computed from* $\widetilde{a}_{0k}$ *and* $\widetilde{b}_{0k}$ *by time-reversed synthesis filters:*

$$\widetilde{a}_{1\ell} = \textstyle\sum_k \boldsymbol{f}_0(\ell-2k)\,\widetilde{a}_{0k} + \sum_k \boldsymbol{f}_1(\ell-2k)\,\widetilde{b}_{0k}. \tag{6.71}$$

Proof. Perfect reconstruction operates when (6.70) is substituted into (6.71):

$$\widetilde{a}_{1\ell} = \sum_k \boldsymbol{f}_0(\ell-2k)\sum_m \boldsymbol{h}_0(\ell-2m)\widetilde{a}_{1m} + \sum_k \boldsymbol{f}_1(\ell-2k)\sum_m \boldsymbol{h}_1(\ell-2m)\widetilde{a}_{1m} \tag{6.72}$$

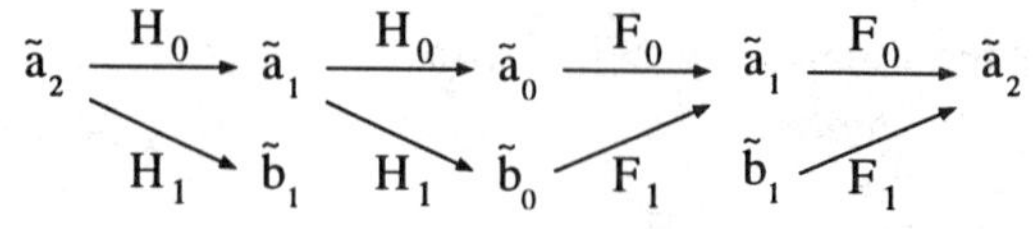

Figure 6.8: The pyramid algorithm from $\tilde{a}_2$ back to $\tilde{a}_2$.

The double-shift biorthogonality in (6.56) and (6.57) makes this correct. The beauty of this Mallat algorithm (Figure 6.8 goes up and back down) is the way it connects continuous-time multiresolution to discrete-time filters.

Notice that the whole pyramid operates equally well if $\boldsymbol{F}_0$ and $\boldsymbol{F}_1$ are exchanged with $\boldsymbol{H}_0$ and $\boldsymbol{H}_1$. The dual expansion of $f(t)$ is:

$$f(t) = \sum \sum b_{jk} \, \widetilde{w}_{jk}(t) \quad \text{and} \quad b_{jk} = \int f(t) w_{jk}(t) \, dt. \tag{6.73}$$

All products have a tilde multiplying a non-tilde! This starts with the inverse relation of synthesis to analysis. Tildes can be exchanged with non-tildes throughout (if we want to do it). We will select ϕ and w to be effective in synthesis.

We emphasize one final point, often ignored. The ***coefficients*** in the expansion of $f(t)$ are really different from ***samples*** of $f(t)$. They are inner products, not point values. This distinction is made in Section 7.1.

Filter Construction by Lifting

Herley and Sweldens have proposed (independently) a systematic way to construct biorthogonal filter banks. Only one lowpass filter changes at each step. Starting from short filters, he quickly builds longer ones. In all cases the highpass filters remove aliasing in the standard way: $H_1(z) = F_0(-z)$ and $F_1(z) = -H_0(-z)$. Then equation (4.9) on H_0 and F_0 is the remaining condition for perfect reconstruction. *We drop the subscript zero on these lowpass filters*, and recall the condition (4.9) that removes distortion:

$$F(z)H(z) - F(-z)H(-z) = 2z^{-l} \quad (\text{odd } l). \tag{6.74}$$

Suppose this is satisfied by F and H; the filter bank is PR. *Keeping F fixed, what are the other possible choices for H?* The answer is simple and important:

Theorem 6.16 **(Lifting)** *For fixed $F(z)$, the solutions $H^{\#}(z)$ to (6.74) are*

$$H^{\#}(z) = H(z) + F(-z)S(z^2) \quad \textit{for any} \quad S(z). \tag{6.75}$$

Proof. Substitute $H^{\#}(z)$ to show that equation (6.74) is still satisfied. The new terms are $F(z)F(-z)S(z^2) - F(-z)F(z)S(z^2) = 0$. This is in [HerVet] and [Swel].

Note that with F fixed, *the equation is linear in H*. We are starting from a particular solution (right side $= 2z^{-l}$). To this particular H we are adding solutions $F(-z)S(z^2)$ to the homogeneous equation (right side $=$ zero). Thus the even $S(z^2)$ displays the degrees of freedom that remain when the PR condition is satisfied (Problem 5). That freedom is used in the Daubechies construction to achieve zeros at $z = -1$.

We still want and need zeros at -1, which is $\omega = \pi$. We also need stable bases. (Section 7.2 will state the stability requirement as Condition E. There is not yet a simple way to decide which $S(z^2)$ are permitted by this condition. In practice, we construct a potentially useful $H^\#(z)$ and test it for Condition E.) This section will build in other important properties — *linear phase, interpolating scaling functions, binary* (dyadic) *filter coefficients*. Those come at the expense of higher-order zeros at π.

Dual lifting is also useful. In this case we fix the analysis filter $H(z)$. The PR condition (6.74) becomes linear in $F(z)$. The lifted solutions are

$$F^\#(z) = F(z) + H(-z)T(z^2) \quad \text{for any} \quad T(z). \tag{6.76}$$

Starting from the Haar filter or even from the "Lazy filter", we alternate lifting and dual lifting to construct high-order biorthogonal filter banks with good properties. All these filters would be attainable directly from the (second-degree) PR equations. Lifting is a way to solve them as a sequence of linear equations, with F or H fixed at each step. Then we control more closely the final result.

Sweldens also emphasizes that lifting yields a *faster implementation* of the wavelet transform and its inverse. The Mallat filter tree, which is subband filtering, is climbed in smaller steps. This is related to a lattice factorization.

Example 6.6. The Lazy filter has $H(z) = 1$ and $F(z) = z^{-1}$. There is no true filtering, only subsampling from $(\downarrow 2)$. Section 4.3 displayed block diagrams of this filter bank. Its polyphase matrix is $\boldsymbol{H}_p = \boldsymbol{I}$.

Suppose we keep $H(z) = 1$ and apply dual lifting to the synthesis filter:

$$F^\#(z) = z^{-1} + T(z^2). \tag{6.77}$$

$F^\#(z)$ can be any halfband filter, with one odd power $z^{-l} = z^{-1}$. We are allowing S and T to contain powers of z as well as z^{-1}. This is needed in order to create symmetric filters.

Earlier we centered the product $H(z)F(z)$, multiplying by z^l. In this case centering gives $zF^\#(z) = 1 + zT(z^2)$. Then D_4 comes from $T(z) = (-z + 9 + 9z^{-1} - z^2)/16$. Every maxflat halfband filter D_{2p} can be lifted and centered from the Lazy filter. This section will create symmetric biorthogonal filters, by lifting $H = 1$ while $F = D_{2p}$.

Note the scaling functions for this important example. In analysis we have the delta function $\widetilde{\phi}(t) = \delta(t)$ from $H(z) = 1$. This solves the dilation equation $\delta(t) = 2\delta(2t)$. In synthesis $F^\#$ yields an ***interpolating scaling function***, with $\phi(n) = \delta(n)$. This is certainly biorthogonal to the analysis functions!

$$\int \phi(t)\widetilde{\phi}(t-n)\,dt = \int \phi(t)\delta(t-n)\,dt = \phi(n) = \delta(n). \tag{6.78}$$

You can see in another way that $\phi(n) = \delta(n)$, because the values of ϕ at the integers come from the $\lambda = 1$ eigenvector of $(\downarrow 2)2\boldsymbol{F}^\#$. When the filter is halfband, the center column of the matrix $2\boldsymbol{F}^\#$ is the vector $\boldsymbol{\delta}$. This is an eigenvector with $\lambda = 1$. Assuming a stable basis, there are no other eigenvectors for $\lambda = 1$ by Condition E. So $\phi(n)$ agrees with $\delta(n)$.

This interpolating property is highly useful in several applications. But the analysis filter with $H(z) = 1$ and $\widetilde{\phi}(t) = \delta(t)$ is generally not acceptable. *Therefore we now lift H. That produces a new pair $(H^\#, F^\#)$ which seems extremely promising.*

Biorthogonal Filters with Binary Coefficients

A *binary coefficient* or *dyadic coefficient* is an integer divided by a power of 2. The maxflat halfband filters $D_{2p}(z)$ all have binary coefficients. This is clear from the Daubechies formula (5.75), where the binomial coefficients are integers. Multiplication by a binary number can be executed entirely by *shifts* and *adds*. Roundoff error is eliminated. And on some architectures, the filter needs less time and less space.

We are therefore highly interested in binary filters. Most factorizations of D_{2p} — this has been our route to orthogonal and biorthogonal filters — are *not* binary. There are zeros at irrational points like $z = 2-\sqrt{3}$. But we can certainly move zeros at $z = -1$ between analysis and synthesis. This operation we call ***balancing***.

Moving $(\frac{1+z^{-1}}{2})$ *from* $F(z)$ *to* $H(z)$ *maintains binary coefficients and symmetry*:

$$\boldsymbol{h}_{new}(n) = \tfrac{1}{2}[\boldsymbol{h}_{old}(n) + \boldsymbol{h}_{old}(n-1)] \quad \text{and} \quad \boldsymbol{f}_{new}(n) = 2\boldsymbol{f}_{old}(n) - \boldsymbol{f}_{new}(n-1). \tag{6.79}$$

Note $\boldsymbol{f}_{new}$ at the end. We are dividing $F(z)$ by $\left(\frac{1+z^{-1}}{2}\right)$. The product $F_{new}H_{new}$ equals $F_{old}H_{old}$, so biorthogonality is preserved. The scaling function $\widetilde{\phi}_{new}(t)$ is $\widetilde{\phi}_{old}(t)$ convolved with the box function. Therefore it has exactly one more derivative than $\widetilde{\phi}_{old}(t)$ (Section 7.3). Similarly $\phi_{new}(t)$ from F_{new} has one less derivative than $\phi_{old}(t)$ from F_{old}. In a filter bank we avoid the destructive factors $\sqrt{2}$, by putting both of them into synthesis. Our convention below is $H(1) = 1$ and $F(1) = 2$.

Our example $\boldsymbol{h1} = [1]$ and $\boldsymbol{f7} = [-1\ \ 0\ \ 9\ \ 16\ \ 9\ \ 0\ \ -1]/16$ is binary. This 1/7 filter bank has denominator 16. *Balancing will produce* 2/6 *and* 3/5, *still binary and symmetric*:

$$\boldsymbol{h2} = [1\ \ 1]/2 \ \text{ and } \ \boldsymbol{f6} = [-1\ \ 1\ \ 8\ \ 8\ \ 1\ \ -1]/8 \ \text{ with 1/3 zeros at } \pi$$

$$\boldsymbol{h3} = [1\ \ 2\ \ 1]/4 \ \text{ and } \ \boldsymbol{f5} = [-1\ \ 2\ \ 6\ \ 2\ \ -1]/4 \ \text{ with 2/2 zeros at } \pi.$$

These are very effective short filters. They are probably the best — after reversing the second pair to 5/3. In the experiments of Chapters 10 and 11 they are comparable and quite effective. As factors of the maxflat D_6 filter, we have seen them before. An interesting feature of $\boldsymbol{f5}$ is that its scaling function $\phi(t)$ is infinite at all binary points! Section 7.3 confirms that this $\phi(t)$ nevertheless has finite energy (and even 0.44 derivatives in the energy sense). By removing two zeros at $z = -1$ from $\boldsymbol{f7}$, we have stolen its smoothness. Enough is left to make it good among short filters (but moved to the analysis side).

For serious compression we need more zeros — which means longer filters. Our previous route was to factor a long maxflat filter. When one factor is $F(z) = \left(\frac{1+z^{-1}}{2}\right)^p$ the pair is still binary and symmetric. The scaling function $\phi(t)$ for this factor is a ***spline***. It has maximum smoothness for its length (p intervals) coming from a maximum number of zeros at π (p zeros). These are outstanding filters, when we keep enough smoothness in the other factor. Taking out three zeros from D_6 would leave a filter $[-1\ 3\ 3\ -1]/2$ which has one zero but negative smoothness — too risky to iterate. Taking three zeros from D_8 is allowed. Now, instead of factoring a long filter, we will lift a short filter.

Lifting will not maximize the number of zeros at π (although we like those zeros). Our first lifting will go from 1/7 to 9/7, and we choose $S(z^2)$ in (6.75) to give two zeros at π. Here is the result of lifting $H = 1$ (all filters are symmetric):

$$\boldsymbol{h9} = [1\ \ 0\ \ -8\ \ 16\ \ 46\ \ldots]/64 \ \text{ and } \ \boldsymbol{f7} = [-1\ \ 0\ \ 9\ \ 16\ \cdots]/16 \ \text{ (2/4 zeros).}$$

These are binary filters. Balancing the zeros by (6.79) yields 10/6 from 9/7:

$$\boldsymbol{h10} = [1 \;\; 1 \;\; -8 \;\; 8 \;\; 62 \;\cdots]/128 \;\text{ and }\; \boldsymbol{f6} = [-1 \;\; 1 \;\; 8 \;\; 8 \;\; 1 \;\; -1]/8 \;\; \text{(3/3 zeros)}.$$

This 10/6 pair gives better compression as 6/10 — reversing analysis and synthesis. So does 5/11 from 11/5, after another balancing (or unbalancing!) step:

$$\boldsymbol{h11} = [1 \;\; 2 \;\; -7 \;\; 0 \;\; 70 \;\; 124 \;\cdots]/256 \;\text{ and }\; \boldsymbol{f5} \text{ above } \;\text{(4/2 zeros)}.$$

Figure 6.9 shows scaling functions $\widetilde{\phi}(t)$ and $\phi(t)$ for analysis and synthesis. You can see how the zeros affect the smoothness. You cannot easily see which pair is best in compression — that depends on the image.

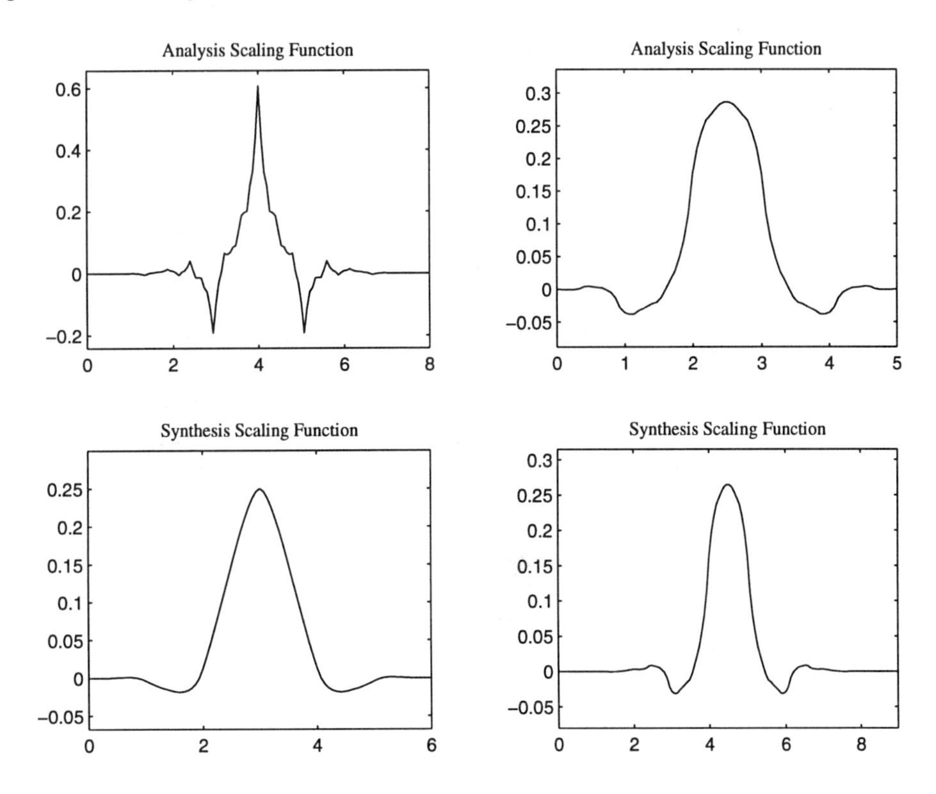

Figure 6.9: Scaling functions for ***h9/f7*** and ***h6/f10***.

Note 1 The reader might be interested in the construction of these new (1995) binary filters. The first author found the 9/7 pair in September, but not by lifting. With $\boldsymbol{f7}$ fixed, he solved the halfband equation $(\downarrow 2)(\boldsymbol{f7} * \boldsymbol{h9}) = \delta$ for the symmetric filter $\boldsymbol{h9}$ with two zeros at π. (The first zero determines the middle coefficient from the others; the second zero is automatic by symmetry.) When reporting this result for the *Wavelet Digest*, he learned that Wim Sweldens had created a whole family of *binary symmetric filters* earlier in 1995 by lifting. We propose to call them "binlets". Here are the next filters $\boldsymbol{h13}/\boldsymbol{f7}$ and $\boldsymbol{13h}/\boldsymbol{f11}$. All signs indicate that $\boldsymbol{h13}/\boldsymbol{f7}$ is the right choice:

$$\boldsymbol{h13} = [-1 \;\; 0 \;\; 18 \;\; -16 \;\; -63 \;\; 144 \;\; 348 \;\cdots]/512 \;\text{ with }\; \boldsymbol{f7} \;\; \text{(4/4 zeros)}$$

$$\mathbf{13h} = [-3\ 0\ 22\ 0\ -125\ 256\ 724\ 256\ -125\ 0\ 22\ 0\ -3]/1024 \quad \text{with}$$

$$\boldsymbol{f11} = D_6 = [3\ 0\ -25\ 0\ 150\ 256\ 150\ 0\ -25\ 0\ 3]/256 \quad \text{(2/6 zeros)}.$$

Extra length gives more zeros and higher compression, up to a point. *Then ringing destroys the image quality.* See Sections 10.1 and 11.2 for the artifacts that plague image compression. The boats in Figure 7.4 offer a visual comparison.

Note 2 [Majani2] emphasizes the importance of a *reversible integer implementation.* Integer inputs are reconstructed exactly. Orthogonal transforms seem not to be reversible (except Haar for $M = 2$ channels and Hadamard for certain $M > 2$). The 2/6 biorthogonal transform with $\boldsymbol{h2} = [1\ 1]$ is reversible and very useful. Lowpass components y_{2i} come first:

$$y_{2i} = x_{2i} + x_{2i+1} \quad \text{and then} \quad y_{2i+1} = x_{2i+1} - \lfloor y_{2i}/2 \rfloor + \lfloor (y_{2i-2} - y_{2i+2})/16 \rfloor.$$

The inverse also has "even = even + f(odd)" and "odd = odd + g(even)":

$$x_{2i} = y_{2i} - x_{2i+1} \quad \text{and then} \quad x_{2i+1} = y_{2i+1} + \lfloor y_{2i}/2 \rfloor - \lfloor (y_{2i-2} - y_{2i+2})/16 \rfloor.$$

Majani has shown that the new binary 9/7 and 13/7 transforms have reversible forms (lossless in integers). The normalized DCT is not reversible for integer data.

Note 3 The maxflat Daubechies filters with $4p-1$ coefficients and $2p$ zeros are also known as Deslauriers-Dubuc filters [DesDub, CDM]. They interpolate because they are halfband. They leave the values $\boldsymbol{x}(n)$ unchanged and produce midpoint values $\boldsymbol{x}(n+\frac{1}{2})$. Section 5.5 confirmed that all polynomials of degree less than $2p$ are interpolated exactly. Recursive subdivision starting from $\boldsymbol{x}(n) = \boldsymbol{\delta}(n)$ converges to the scaling function $\phi(t)$ by the cascade algorithm!

Problem Set 6.5

1. Double-shift orthogonality of lowpass filters is $2\sum \boldsymbol{h}_0(k)\boldsymbol{f}_0(k+2n) = \boldsymbol{\delta}(n)$. Show that in frequency this becomes

$$H_0(\omega)F_0(\omega) + H_0(\omega+\pi)F_0(\omega+\pi) = 1.$$

Write the same equation in the z-domain.

2. Problem **1** involves a row and column of the modulation matrices $\boldsymbol{F}_m$ and $\boldsymbol{H}_m$:

$$\boldsymbol{F}_m(z)\boldsymbol{H}_m(z) = \begin{bmatrix} F_0(z) & F_1(z) \\ F_0(-z) & F_1(-z) \end{bmatrix} \begin{bmatrix} H_0(z) & H_0(-z) \\ H_1(z) & H_1(-z) \end{bmatrix}.$$

Which row-column multiplications correspond to which equations (6.56)–(6.58)?

3. Suppose $H_0(\omega)F_0(\omega) + H_0(\omega+\pi)F_0(\omega+\pi) = 1$ as required. By alternating flip

$$H_1(\omega) = e^{-i\omega}F_0(\omega+\pi) \quad \text{and} \quad F_1(\omega) = -e^{-i\omega}H_0(\omega+\pi).$$

Show that the other entries of $\boldsymbol{F}_m(z)\boldsymbol{H}_m(z) = \boldsymbol{I}$ are then correct.

4. What wavelets come from the biorthogonal filters with $H_0 = 1$, $F_0 = \frac{1}{2}z + 1 + \frac{1}{2}z^{-1}$, $H_1 = \frac{1}{2}z - 1 + \frac{1}{2}z^{-1}$, $F_1 = -1$? Recognize the delta and hat:

$$\widetilde{\phi}(t) = 2\widetilde{\phi}(2t) \quad \text{and} \quad \phi(t) = \tfrac{1}{2}\phi(2t+1) + \phi(2t) + \tfrac{1}{2}\phi(2t-1).$$

Then construct wavelets from $\widetilde{w}(t) = -\frac{1}{2}\widetilde{\phi}(2t+1) + \widetilde{\phi}(2t) - \frac{1}{2}\widetilde{\phi}(2t-1)$ and $w(t) = 2\phi(2t-1)$. Check the biorthogonality conditions

$$\int \phi(t)\widetilde{\phi}(t-k)\,dt = \int w(t)\widetilde{w}(t-k)\,dt = \delta(k) \quad \text{and}$$
$$\int \phi(t)\widetilde{w}(t-k)\,dt = \int \widetilde{\phi}(t)w(t-k)\,dt = 0.$$

5. Lifting from the Haar filters $H(z) = \frac{1}{2}(1+z^{-1})$ and $F(z) = 1+z^{-1}$, show that all PR solutions have the form (6.75). The difference $D = H^{\#} - H$ must satisfy $D(z)F(z) = D(-z)F(-z)$ from (6.74). Substitute $D(z) = a + bz^{-1} + cz^{-2} + dz^{-3} + \cdots$ to show that it has the form $F(-z)S(z^2)$.

6. Lifting (H, F) to $(H^{\#}, F)$ does not change the scaling function $\phi(t)$. Show that the new wavelet is $w^{\#}(t) = w(t) - \sum s(k)\phi(t-k)$.

7. The *fast wavelet transform* is subband filtering of the inner products $a_{jk} = \langle f(t), \widetilde{\phi}_{jk}(t)\rangle$. The highpass channel produces $b_{jk} = \langle f(t), \widetilde{w}_{jk}(t)\rangle$ by (6.70):

$$a_{jk} = \sum \boldsymbol{h}(l-2k)a_{j+1,l} \quad \text{and} \quad b_{jk} = \sum \boldsymbol{h}_1(l-2k)a_{j+1,l}.$$

Suppose H is lifted to $H^{\#}$. *Show that the lifted $a^{\#}_{jk}$ are* $a^{\#}_{jk} = a_{jk} + \sum \boldsymbol{s}(l-k)b_{jl}$. The fast *inverse* transform unlifts by $-\boldsymbol{s}(l-k)$ in the same way. Then it inverts the (H, H_1) transform as usual by $a_{j+1,k} = \sum \boldsymbol{f}(k-2l)a_{jl} + \sum \boldsymbol{f}_1(k-2l)b_{jl}$.

8. From the input $\boldsymbol{x}(n) = a_{1k}$ compute even samples $a_{0k} = a_{1,2k}$ and odd $b_{0k} = a_{1,2k+1} - (a_{0k} + a_{0,k+1})$. Then lift to $a^{\#}_{0k} = a_{0k} + (b_{0,k-1} + b_{0k})/4$. Combine to recognize the 5/3 filter bank, computed more efficiently and in place—no auxiliary memory.

9. Which filter $\boldsymbol{h}$ gives linear interpolation at each step of recursive subdivision? What is $\phi(t)$?

Chapter 7

Wavelet Theory

This chapter follows through on important questions about scaling functions and wavelets. We study their smoothness, their inner products, their accuracy of approximation, and the number of vanishing moments. Even more basic are the existence and the construction of $\phi(t)$. *Does the dilation equation have a solution* $\phi(t)$ *with finite energy*? *Does the cascade algorithm converge to this solution*? We answer those questions in terms of the filter coefficients $\boldsymbol{h}(k)$ (and the answers are not always yes).

Our overall aim is understanding, with a minimum of technical details. We point immediately to two fundamental operators in wavelet theory. They both have a double shift between rows coming from $(\downarrow 2)$:

$$\boldsymbol{M} = (\downarrow 2)\, 2\boldsymbol{H} \quad \text{and} \quad \boldsymbol{T} = (\downarrow 2)\, 2\boldsymbol{H}\boldsymbol{H}^T. \tag{7.1}$$

Apart from its extra factor 2, $\boldsymbol{M}$ is completely familiar: filter by $\boldsymbol{H}$ and then downsample. This is the decimated lowpass analysis filter, with coefficients $\boldsymbol{h}(k)$ adding to 1. In the time domain, those are on the diagonals of $\boldsymbol{H}$.

The symmetric product $\boldsymbol{H}\boldsymbol{H}^T$ is also a Toeplitz matrix. Its entries are the coefficients in $|H(\omega)|^2$. The rows are double-shifted by $(\downarrow 2)$. In frequency, downsampling produces an aliasing term from modulation by π:

$$\begin{aligned} (\boldsymbol{M}f)(\omega) &= H(\tfrac{\omega}{2})\, f(\tfrac{\omega}{2}) + H(\tfrac{\omega}{2}+\pi)\, f(\tfrac{\omega}{2}+\pi) \\ (\boldsymbol{T}f)(\omega) &= |H(\tfrac{\omega}{2})|^2\, f(\tfrac{\omega}{2}) + |H(\tfrac{\omega}{2}+\pi)|^2\, f(\tfrac{\omega}{2}+\pi) \end{aligned} \tag{7.2}$$

The properties of $\boldsymbol{M}$ and $\boldsymbol{T}$ hold the answers to our questions. Iterating the lowpass filter, with subsampling, involves powers of matrices. The convergence depends on the eigenvalues. That will be the message in this chapter: ***Watch the eigenvalues.***

The "transition operator" or "transfer operator" $\boldsymbol{T}$ turns out to be simpler than $\boldsymbol{M}$ — because $|H(\omega)|^2 \geq 0$ and $\boldsymbol{H}\boldsymbol{H}^T$ is symmetric positive definite. $\boldsymbol{T}$ enters when we compute inner products and energies (L^2 norms). After i iterations of $\boldsymbol{M}$ in the cascade algorithm, starting from $\phi^{(0)}(t)$ and reaching $\phi^{(i)}(t)$, the inner products are

$$\boldsymbol{a}^{(i)}(k) = \int_{-\infty}^{\infty} \phi^{(i)}(t)\, \phi^{(i)}(t+k)\, dt. \tag{7.3}$$

The key point will be that $\boldsymbol{T}\boldsymbol{a}^{(i)}(k)$ gives the inner products $\boldsymbol{a}^{(i+1)}(k)$. The powers of $\boldsymbol{T}$ (and therefore the eigenvalues of $\boldsymbol{T}$!) decide whether the cascade algorithm converges, when we use the L^2 energy norm. The properties of $\boldsymbol{M}$ give *pointwise* answers, and the properties of $\boldsymbol{T}$ give *mean square* answers.

We can summarize in a few lines approximately what those answers are:

1. Combinations of $\phi(t-k)$ can exactly produce the polynomials $1, t, \ldots, t^{p-1}$ if $\boldsymbol{M}$ has eigenvalues $1, \frac{1}{2}, \ldots, \left(\frac{1}{2}\right)^{p-1}$; $\boldsymbol{T}$ has eigenvalues $1, \frac{1}{2}, \ldots, \left(\frac{1}{2}\right)^{2p-1}$.

2. The wavelets are orthogonal to $1, \ldots, t^{p-1}$ so they have p vanishing moments. The functions $\phi(t-k)$ give pth order approximation from the space V_0.

3. The cascade algorithm converges to $\phi(t)$ in L^2 if the other eigenvalues of $\boldsymbol{T}$ satisfy $|\lambda| < 1$.

4. $\phi(t)$ and $w(t)$ have s derivatives in L^2 if the other eigenvalues of $\boldsymbol{T}$ satisfy $|\lambda| < 4^{-s}$.

In **1**, the special eigenvalues (powers of $\frac{1}{2}$) give a new form of Condition A_p. It is equivalent to p zeros at π, from a factor $(1+z^{-1})^p$ in $H(z)$.

In **3**, the requirement $|\lambda| < 1$ on the other eigenvalues will be called **Condition E**. This is new to the book. It gives convergence of the cascade algorithm to a Riesz basis $\{\phi(t-k)\}$. Condition E *is the key to iteration of filters, and thus to wavelets.*

In **4**, the smoothness s is very likely not an integer (but we work with integers for simplicity). The number s_{max} of derivatives of $\phi(t)$ is never greater than the number p of vanishing moments. The smoothness of $\phi(t)$ is important in image processing, and p is more important.

This chapter begins with the *accuracy of approximation*. That has the most direct answer. The approximation order is p when the frequency response $H(\omega)$ has a pth order zero at $\omega = \pi$. *Zeros at π are the heart of wavelet theory!* This multiple zero is produced by the factor $(1+e^{-i\omega})^p$ that appears throughout the design of filters. $\boldsymbol{T}$ is associated with $|H(\omega)|^2$ for which the order of the zero becomes $2p$. This is reflected in the $2p$ special eigenvalues of $\boldsymbol{T}$.

Sections 7.2–7.3 analyze the cascade algorithm $\phi^{(i+1)}(t) = \sum 2\boldsymbol{h}(k)\,\phi^{(i)}(2t-k)$. Section 7.4 explains splines and spline wavelets. Section 7.5 introduces multiwavelets.

We note here the MATLAB commands to construct $\boldsymbol{M}$ and $\boldsymbol{T}$ from the vector $\boldsymbol{h}$:

```
function M = down(h)
n = length(h); M = zeros(n,n); for i = 1:n, for j = 1:n,
if (0 < 2 * i - j) & (2 * i - j) <= n) M(i,j) = 2 * h(2 * i - j);
end
end, end
autocor = conv(h, fliplr(h)); T = down(autocor)
```

7.1 Accuracy of Approximation

In applications of wavelets, a function $f(t)$ is projected onto a scaling space V_j. This index j gives the time scale $\Delta t = 2^{-j}$ in the calculation. The scaling functions are $2^{j/2}\phi(2^j t)$ and its translates by $k\Delta t$. Those represent one basis for the space V_j. The projection $f_j(t)$ is the piece of $f(t)$ in that subspace, so it is a combination of those basis functions:

$$\text{For each } j, \quad f_j(t) = \sum_{k=-\infty}^{\infty} a_{jk}\, 2^{j/2}\phi(2^j t - k). \tag{7.4}$$

This is all at level j. In contrast, wavelets come from splitting functions into several scales. ***Multiresolution*** combines the details at levels zero through $j-1$ and the coarse average at level zero. For subspaces this is $V_j = V_0 \bigoplus W_0 \bigoplus \cdots \bigoplus W_{j-1}$. Except for V_0, the basis functions are now wavelets:

$$f_j(t) = \sum_k a_{0k}\,\phi(t-k) + \sum_k b_{0k}\,w(t-k) + \sum_k b_{1k}\,2^{1/2}w(2t-k) + \cdots \tag{7.5}$$

In practice, the level j is determined by balancing *accuracy* with *cost*. In other words, we balance *distortion* with *rate*. The cost and the bit rate are approximately doubled between one level and the next—there are twice as many basis functions and twice as many coefficients. This section will estimate the improvement in accuracy (the drop in distortion).

The accuracy depends partly on the filter bank coefficients and partly on $f(t)$. A smooth function and a smooth signal will be easier to approximate and send. Part of our goal is to separate the two contributions to the error (the distortion), one part from the properties of $\boldsymbol{h}(k)$ and the other from the properties of $f(t)$. We choose $\boldsymbol{h}(k)$. The application presents us with $f(t)$. A typical form of the error estimate will involve the pth derivative of $f(t)$:

$$\|f(t) - f_j(t)\| \approx C(\Delta t)^p\,\|f^{(p)}(t)\|. \tag{7.6}$$

The constant C and the exponent p depend on our choice of $\boldsymbol{h}(k)$. That determines $\phi(t)$ and the subspaces. The step from $\Delta t = 2^{-j}$ to $\Delta t = 2^{-(j+1)}$ divides the error by about 2^p. Thus the number p is critical, when we reach the level at which this "asymptotic" error estimate is accurate. Usually the constant C is less critical, but for wavelets it is an order of magnitude larger than for splines.

The smoothness or roughness of $f(t)$ may be outside our control. This contributes to the error through the norm $\|f^{(p)}(t)\|$ of the pth derivative. The global error estimate (7.6) can be made *local*, if there are regions where $f^{(p)}(t)$ is small and also regions of sudden change. The error will be small in one region and large in the other—unless we increase j in the region of sudden change. Then we have an *adaptive mesh*.

A major advantage of wavelets over Fourier methods is this possibility of local refinement. This is the multigrid idea for finite differences, and it is a key virtue of finite elements. Adaptivity holds also for wavelets—mesh refinement is relatively convenient. (The refinement is usually by factors of 2. Irregular meshes are anathema to Fourier.) We add the word "relatively" because adaptivity has an overhead that practitioners have come to respect.

The error estimate (7.6) is completely typical of numerical analysis. It appears in finite elements, where Δt becomes the element size [SF]. It appears in difference methods for initial-value problems ($p = 1$ for Euler's method, $p = 2$ for centered leapfrog, $p = 4$ for Runge-Kutta, etc.). The form of (7.6) is already set by the basic problem of numerical integration:

$$\int_0^1 f(t)dt - \sum c_k f(t_k) \approx \begin{cases} (\Delta t)^1: & \text{rectangle rule} \\ (\Delta t)^2: & \text{midpoint or trapezoid rule} \\ (\Delta t)^4: & \text{Simpson's rule}\ldots \end{cases} \tag{7.7}$$

The exponent is p, the scale length is Δt, and the symbol $\approx$ hides a constant C and a factor $\|f^{(p)}(t)\|$. All these examples, which extend to other numerical approximations too, have the same form because they are based on the same idea.

That key idea is: ***Watch the polynomials***. The rectangular rule is exact for $f(t) = $ constant. The midpoint rule is exact for $f(t) = $ linear. Simpson's rule is exact for $f(t) = $ cubic polynomial.

The exponent p is *the degree of the first polynomial that gives an error*. Locally, every smooth function looks like a polynomial. This is the essential idea of the Taylor series (and of calculus!). The lowest degree term $C(\Delta t)^p f^{(p)}(t)$ in the error will dominate. We determine the order of accuracy p by computing with polynomials.

An important point is developed at the end of this section. In digital signal processing, the input is not a function $f(t)$. It is a discrete vector $\boldsymbol{x}(n)$. This vector can come from sampling $f(t)$. But if you use the sampled values as the coefficients a_{jk} that enter the filter bank, you are doing violence to the projection. The theory requires a preprocessing step, to transform sampled values to wavelet coefficients. Then postprocessing converts coefficients back to function values. These two steps can be approximated; they should not be ignored.

Note. The constant C is much smaller for splines than for other well-known scaling functions. Compare these asymptotic constants for $p = 2, 3, \ldots, 9$:

Splines	**0.07**	**0.03**	**0.02**	**0.02**	**0.02**	**0.02**	**0.02**	**0.03**
Daubechies	**0.22**	**0.30**	**0.56**	**1.3**	**3.8**	**13.0**	**49.0**	**216.0**

The leading error term is $C(\Delta t)^p f^{(p)}(t)/p!$ with these constants [SwPi,Unser]. This error is $f(t)$ minus its projection onto V_j as $\Delta t = 2^{-j} \to 0$. The growth in the Daubechies constants means roughly that approximation at scale j is only as close as splines at the coarser scale $j-1$ (*half the resolution and a fraction of the work*). There is a price after all for the irregularity of $\phi(t) = D_{2p}(t)$, even though it can reproduce polynomials and achieve the exponent in $(\Delta t)^p$.

Determination of the Accuracy p

Before developing the theory, we compute the number p. The approximation is by translates of $\phi(t)$ and $w(t)$, which can be difficult and intricate functions. *The computation of p always goes back to the lowpass coefficients* $\boldsymbol{h}(k)$. We can determine p from the $\boldsymbol{h}(k)$, or from $H(\omega)$. What we cannot do, and wish to avoid, is the exact projection in continuous time based on $\phi(t)$.

The test for accuracy p is Condition A_p. We recognize it in the time domain, the frequency domain, and now also in the eigenvalue domain.

Theorem 7.1 *The accuracy is p if the lowpass filter coefficients* $\boldsymbol{h}(k)$ *satisfy these three equivalent forms of Condition* A_p*:*

1. *p sum rules on the coefficients:* $\displaystyle\sum_{n=0}^{N}(-1)^n n^j \boldsymbol{h}(n) = 0$ *for* $j = 0, 1, \ldots, p-1$.

2. *p zeros at* π*:* $H(\omega) = \left(\frac{1+e^{-i\omega}}{2}\right)^p Q(\omega)$ *and* $H(z) = (\frac{1+z^{-1}}{2})^p Q(z)$.

3. *p eigenvalues* $1, \frac{1}{2}, \ldots, \left(\frac{1}{2}\right)^{p-1}$ *of the matrix* $\boldsymbol{M} = (\downarrow 2)2\boldsymbol{H} = \{2\boldsymbol{h}(2i-j)\}$:

$$\boldsymbol{M}\Phi^{(j)} = \left(\tfrac{1}{2}\right)^j \Phi^{(j)} \quad \textit{for} \quad j = 0, 1, \ldots, p-1. \tag{7.8}$$

Proof. The equivalence of **1** and **2** is straightforward. Substituting $\omega = \pi$ in the frequency response yields the alternating sum of coefficients:

$$\sum \boldsymbol{h}(n)e^{-in\pi} = \boldsymbol{h}(0) - \boldsymbol{h}(1) + \boldsymbol{h}(2) - \cdots \tag{7.9}$$

Therefore $H = 0$ at $\omega = \pi$ when the first sum rule holds. For the next rule, when $j = 1$, the derivative of $\boldsymbol{h}(n)e^{-in\omega}$ brings down a factor $-in$:

$$\sum \boldsymbol{h}(n)(-in)e^{-in\pi} = -i(0\boldsymbol{h}(0) - 1\boldsymbol{h}(1) + 2\boldsymbol{h}(2) - \cdots). \tag{7.10}$$

Then $H' = 0$ at $\omega = \pi$ is equivalent to the second sum rule. A similar reasoning applies for higher order zeros and higher sum rules. A p-fold zero at π is equivalent to p sum rules. The notation has assumed an FIR filter but it could be IIR. We turn now to statement **3** about eigenvalues, and we start with examples.

Suppose $H(\omega)$ is the pth power of $\frac{1}{2}(1 + e^{-i\omega})$. It has p zeros at π. The other factor in this $H(\omega)$ is simply $Q = 1$. In that particular case ***all*** the eigenvalues are powers of $\frac{1}{2}$. These examples for $p = 2, 3, 4$ come from double shifts of 1, 2, 1 and 1, 3, 3, 1 and 1, 4, 6, 4, 1:

$$\boldsymbol{m}_2 = \frac{1}{2}\begin{bmatrix} 1 & 0 \\ 1 & 2 \end{bmatrix} \qquad \boldsymbol{m}_3 = \frac{1}{4}\begin{bmatrix} 1 & 0 & 0 \\ 3 & 3 & 1 \\ 0 & 1 & 3 \end{bmatrix} \qquad \boldsymbol{m}_4 = \frac{1}{8}\begin{bmatrix} 1 & 0 & 0 & 0 \\ 6 & 4 & 1 & 0 \\ 1 & 4 & 6 & 4 \\ 0 & 0 & 1 & 4 \end{bmatrix}$$

$$\lambda = 1, \tfrac{1}{2} \qquad\qquad \lambda = 1, \tfrac{1}{2}, \tfrac{1}{4} \qquad\qquad \lambda = 1, \tfrac{1}{2}, \tfrac{1}{4}, \tfrac{1}{8}$$

The N by N matrix $\boldsymbol{m}$ has entries $2\boldsymbol{h}(2i-j)$ for $i, j = 0, \ldots, N-1$. We use the letter $\boldsymbol{m}$, because this is a submatrix of $\boldsymbol{M}$. The eigenvalues of $\boldsymbol{m}$ are also eigenvalues of the infinite matrix $\boldsymbol{M}$, with the eigenvectors extended in both directions by zeros. The submatrix $\boldsymbol{m}$ is called $\boldsymbol{m}(0)$ in Section 6.3. The other N by N submatrix is $\boldsymbol{m}(1) = 2\boldsymbol{h}(2i - j)$ for $i, j = 1, \ldots, N$. This matrix $\boldsymbol{m}(1)$ also has the special eigenvalues $(\frac{1}{2})^k$.

To establish that $\boldsymbol{m}$ has these eigenvalues, we increase the number of "zeros at π" one step at a time. At each step, we watch the eigenvectors and eigenvalues. It will be very convenient to work in the z-domain, where each additional zero at $z = -1$ (which is $\omega = \pi$) comes from another factor $(\frac{1+z^{-1}}{2})$. The new eigenvalues of $\boldsymbol{m}$, when $H(z)$ has that extra factor, are ***half*** the old eigenvalues. There is also one new eigenvalue at $\lambda = 1$. Theorem 7.2 describes the new eigenvalues, and its proof completes Theorem 7.1.

Theorem 7.2 *When $H(z)$ is multiplied by $(\frac{1+z^{-1}}{2})$, $\boldsymbol{m}_{\text{new}}$ is one size larger:*

(a) *The eigenvalues λ_{new} are $\frac{1}{2}\lambda_{\text{old}}$. There is an extra eigenvalue $\lambda_{\text{new}} = 1$.*

(b) *The eigenvectors $\boldsymbol{x}_{\text{new}}$ are the differences of the eigenvectors $\boldsymbol{x}_{\text{old}}$:*

$$\boldsymbol{x}_{\text{new}}(k) = \boldsymbol{x}_{\text{old}}(k) - \boldsymbol{x}_{\text{old}}(k-1) \;\; \textit{and} \;\; X_{\text{new}}(z) = (1 - z^{-1})X_{\text{old}}(z). \tag{7.11}$$

The new $\lambda = 1$ has left eigenvector $\boldsymbol{e}_0 = \begin{bmatrix} 1 & 1 & \cdots & 1 \end{bmatrix}$. The right eigenvector for $\lambda = 1$ gives the values $\phi_{\text{new}}(n)$ of the scaling function at the integers.

Proof. We will write the eigenvalue equation $\boldsymbol{m}_{\text{old}}\boldsymbol{x}_{\text{old}} = \lambda_{\text{old}}\boldsymbol{x}_{\text{old}}$ in the z-domain. Then multiplication by $(\frac{1+z^{-1}}{2})$ gives the corresponding equation for $\boldsymbol{m}_{\text{new}}$. Notice the aliasing term from $(\downarrow 2)$:

$$\boldsymbol{m}\boldsymbol{x} = \lambda\boldsymbol{x} \;\; \text{is} \;\; H(z)X(z) + H(-z)X(-z) = \lambda X(z^2). \tag{7.12}$$

This is for X_{old}. Now give $H(z)$ the extra $(\frac{1+z^{-1}}{2})$ and $X(z)$ the factor $(1 - z^{-1})$:

$$\left(\frac{1+z^{-1}}{2}\right)H(z)(1-z^{-1})X(z)+\left(\frac{1-z^{-1}}{2}\right)H(-z)(1+z^{-1})X(-z) = \tfrac{1}{2}\lambda(1-z^{-2})X(z^2). \tag{7.13}$$

This is the eigenvalue equation $\boldsymbol{m}_{\text{new}}\boldsymbol{x}_{\text{new}} = \lambda_{\text{new}}\boldsymbol{x}_{\text{new}}$. The eigenvalue is multiplied by $\frac{1}{2}$ and the eigenvectors obey (7.11). The whole proof is in $(\frac{1+z^{-1}}{2})(1-z) = \frac{1}{2}(1-z^{-2})$. In the examples $\boldsymbol{m}_2, \boldsymbol{m}_3, \boldsymbol{m}_4$ above, the eigenvectors for $\lambda = 1, \frac{1}{2}, \frac{1}{4}, \frac{1}{8}$ are in the columns of these matrices:

$$\begin{bmatrix} 0 & 1 \\ 1 & -1 \end{bmatrix} \quad \begin{bmatrix} 0 & 0 & 1 \\ \frac{1}{2} & 1 & -2 \\ \frac{1}{2} & -1 & 1 \end{bmatrix} \quad \begin{bmatrix} 0 & 0 & 0 & 1 \\ \frac{1}{6} & \frac{1}{2} & 1 & -3 \\ \frac{4}{6} & 0 & -2 & 3 \\ \frac{1}{6} & -\frac{1}{2} & 1 & -1 \end{bmatrix}$$

Take differences of the eigenvectors $0, 1, 0, \ldots$ and $1, -1, 0, \ldots$ in the 2×2 matrix. Those differences $0, 1, -1, \ldots$ and $1, -2, 1, \ldots$ are in the second matrix. They are eigenvectors of $\boldsymbol{m}_3$ for $\lambda = \frac{1}{2}$ and $\frac{1}{4}$. The new eigenvector for $\lambda = 1$ gives the values $\phi(1) = \phi(2) = \frac{1}{2}$ of the scaling function for $H(z) = (\frac{1+z^{-1}}{2})^3$. Section 6.3 explained how the dilation equation at the integers is exactly $\boldsymbol{m}\Phi = \Phi$ (with $\lambda = 1$). In this example $\phi(t)$ is a *spline*—linear, quadratic, and cubic for $p = 2, 3, 4$.

The 4×4 matrix $\boldsymbol{m}_4$ comes from $H(z) = (\frac{1+z^{-1}}{2})^4$. Its eigenvectors in the last three columns are differences of the columns in the 3×3 matrix. The first column, for $\lambda = 1$, holds the values $\frac{1}{6}, \frac{4}{6}, \frac{1}{6}$ of the cubic spline at the integers. Section 7.4 develops the special properties of splines.

Also of importance are the ***left eigenvectors***. For the special eigenvalues $1, \frac{1}{2}, \frac{1}{4}, \cdots$ those eigenvectors are "discrete polynomials". This means that the left eigenvector for $\lambda = 2^{-k}$ is a combination of the row vectors $\boldsymbol{e}_0, \boldsymbol{e}_1, \ldots, \boldsymbol{e}_k$:

$$\boldsymbol{e}_0 = [1 \;\; 1 \;\; \cdots \;\; 1], \boldsymbol{e}_1 = [0 \;\; 1 \;\; \cdots \;\; N-1], \boldsymbol{e}_k = \left[0^k \;\; 1^k \;\; \cdots \;\; (N-1)^k\right].$$

The all-ones vector satisfies $\boldsymbol{e}_0 = \boldsymbol{m}\boldsymbol{e}_0$, as we have seen before. It is the left eigenvector $\boldsymbol{y}_0$ of $\boldsymbol{m}$ corresponding to $\lambda = 1$:

$$[1 \;\; 1 \;\; 1 \;\; \cdots \;\; 1] \begin{bmatrix} 2\boldsymbol{h}(0) & & & \\ 2\boldsymbol{h}(2) & 2\boldsymbol{h}(1) & 2\boldsymbol{h}(0) & \\ & 2\boldsymbol{h}(3) & 2\boldsymbol{h}(2) & \cdots \\ \vdots & \vdots & \vdots & \cdots \end{bmatrix} = [1 \;\; 1 \;\; 1 \;\; \cdots \;\; 1]. \tag{7.14}$$

This just says that the even sum $\sum 2\boldsymbol{h}(2k)$ and the odd sum $\sum 2\boldsymbol{h}(2k+1)$ are equal to 1. That comes from adding and subtracting the first sum rule $\boldsymbol{h}(0) - \boldsymbol{h}(1) + \cdots = 0$ and the lowpass rule $\boldsymbol{h}(0) + \boldsymbol{h}(1) + \cdots = 1$.

When we multiply $H(z)$ by $(\frac{1+z^{-1}}{2})$, the other left eigenvectors of $\boldsymbol{m}_{\text{new}}$ come from the left eigenvectors of $\boldsymbol{m}_{\text{old}}$. Where the right eigenvectors take differences of $\boldsymbol{x}_{\text{old}}$, the left eigenvectors take ***sums***. Summing increases the polynomial degree by 1. In the z-domain, this sum corresponds to multiplication by $\frac{1}{1-z} = \frac{z^{-1}}{z^{-1}-1}$:

Theorem 7.3 *The left eigenvector for $\lambda = 1$ is always $\boldsymbol{e}_0$. The other left eigenvectors in $\boldsymbol{y}_{\text{new}}\boldsymbol{m}_{\text{new}} = \lambda_{\text{new}}\boldsymbol{y}_{\text{new}}$ come from $\boldsymbol{y}_{\text{old}}$ by summing and adding $C\boldsymbol{e}_0$:*

$$Y_{\text{new}}(z) = \frac{1}{1-z} Y_{\text{old}}(z) + C E_0(z). \tag{7.15}$$

Proof. Left eigenvectors of $\boldsymbol{m}$ are right eigenvectors of $\boldsymbol{m}^T$. We transpose the operations of $(\downarrow 2)$ and multiplication by $2H(z)$. The transposes are $(\uparrow 2)$ and multiplication by $2H(z^{-1})$. The eigenvalue equation $\boldsymbol{m}^T\boldsymbol{y}^T = \lambda\boldsymbol{y}^T$ has z^2 because of $(\uparrow 2)$:

$$2H(z^{-1})Y_{\text{old}}(z^2) = \lambda Y_{\text{old}}(z). \tag{7.16}$$

Now give $H(z^{-1})$ the extra factor $(\frac{1+z}{2})$, and divide $Y(z)$ by $1-z$:

$$\left(\frac{1+z}{2}\right)H(z^{-1})\frac{1}{1-z^{-2}}Y(z^2) = \tfrac{1}{2}\lambda\frac{1}{1-z}Y(z). \tag{7.17}$$

This is $\boldsymbol{m}^T_{\text{new}}\boldsymbol{y}^T_{\text{new}} = \lambda_{\text{new}}\boldsymbol{y}^T_{\text{new}}$, which completes the proof. The eigenvalue is multiplied by $\frac{1}{2}$ (of course, since $\boldsymbol{m}^T$ has the same eigenvalues as $\boldsymbol{m}$). The summation in the left eigenvectors (*row vectors*) can be seen in the same three examples $\boldsymbol{m}_2, \boldsymbol{m}_3, \boldsymbol{m}_4$:

$$\begin{matrix}\lambda = 1\\ \lambda = \frac{1}{2}\end{matrix}\begin{bmatrix}1 & 1\\ 1 & 0\end{bmatrix} \qquad \begin{matrix}\lambda = 1\\ \frac{1}{2}\\ \frac{1}{4}\end{matrix}\begin{bmatrix}1 & 1 & 1\\ \frac{3}{2} & \frac{1}{2} & -\frac{1}{2}\\ 1 & 0 & 0\end{bmatrix} \qquad \begin{matrix}\lambda = 1\\ \frac{1}{2}\\ \frac{1}{4}\\ \frac{1}{8}\end{matrix}\begin{bmatrix}1 & 1 & 1 & 1\\ 2 & 1 & 0 & -1\\ \frac{11}{6} & \frac{2}{6} & -\frac{1}{6} & \frac{2}{6}\\ 1 & 0 & 0 & 0\end{bmatrix}$$

These left eigenvector matrices are the inverses of the previous right eigenvector matrices! The left eigenvectors are always biorthogonal to the right eigenvectors. The diagonalization by $\boldsymbol{S}^{-1}\boldsymbol{M}\boldsymbol{S}$ has the right eigenvectors in the columns of $\boldsymbol{S}$ and the left eigenvectors in the rows of $\boldsymbol{S}^{-1}$.

The left eigenvector for $\lambda = 1$ is always $\boldsymbol{e}_0$, the row of ones. The other left eigenvectors of $\boldsymbol{m}_3$ come from *minus* the sum of the $\boldsymbol{m}_2$ eigenvectors, *plus constant*:

$$\begin{matrix}[\ 1 & 1\ ] & \to & [\ 0 & -1 & -2\ ] + [\ \frac{3}{2} & \frac{3}{2} & \frac{3}{2}\ ] & = & [\ \frac{3}{2} & \frac{1}{2} & -\frac{1}{2}\ ]\\ [\ 1 & 0\ ] & \to & [\ 0 & -1 & -1\ ] + [\ 1 & 1 & 1\ ] & = & [\ 1 & 0 & 0\ ].\end{matrix}$$

The vector after the arrow has a minus sign and delay. This is because $\frac{1}{1-z}$ equals $-z^{-1}$ times the summing operator $1 + z^{-1} + z^{-2} + \cdots$.

Extension to Infinite Matrices

For the infinite matrix $\boldsymbol{M}$, these left eigenvectors are ***not*** extended by zeros. The finite eigenvector $\boldsymbol{e}_0$ becomes an infinite all-ones eigenvector. The linear vector $\boldsymbol{e}_1 = [0\ 1\ \cdots\ N-1]$ becomes infinite too: $\boldsymbol{e}_1(n) = n$ for all n. All "polynomial vectors" $\boldsymbol{e}_k$ extend as polynomials. Then the combination of the $\boldsymbol{e}$'s that gives the left eigenvector of $\boldsymbol{m}$ also gives the extended left eigenvector of the infinite $\boldsymbol{M}$.

The eigenvector $[\ \frac{3}{2}\ \ \frac{1}{2}\ \ -\frac{1}{2}\ ]$ of $\boldsymbol{m}_3$ extends to a linear left eigenvector of $\boldsymbol{M}_3$:

$$[\cdots\ \tfrac{5}{2}\ \tfrac{3}{2}\ \tfrac{1}{2}\ -\tfrac{1}{2}\ \cdots]\,\tfrac{1}{4}\begin{bmatrix}\cdots & 0 & 0 & 0 & 0 & \\ & 3 & 1 & 0 & 0 & \\ & 1 & 3 & 3 & 1 & \\ & 0 & 0 & 1 & 3 & \cdots\end{bmatrix} = \tfrac{1}{2}[\cdots\ \tfrac{5}{2}\ \tfrac{3}{2}\ \tfrac{1}{2}\ -\tfrac{1}{2}\ \cdots].$$

Being linear, it has $\lambda = \frac{1}{2}$. The other eigenvector $\begin{bmatrix} 1 & 0 & 0 \end{bmatrix}$ of $\boldsymbol{m}_3$ extends to a "*quadratic*" left eigenvector (for $\lambda = \frac{1}{4}$) of $\boldsymbol{M}_3$:

$$\begin{bmatrix} 1 & 0 & 0 \end{bmatrix} = \begin{bmatrix} 1 & 1 & 1 \end{bmatrix} - \tfrac{3}{2}\begin{bmatrix} 0 & 1 & 2 \end{bmatrix} + \tfrac{1}{2}\begin{bmatrix} 0 & 1 & 4 \end{bmatrix} = \boldsymbol{e}_0 - \tfrac{3}{2}\boldsymbol{e}_1 + \tfrac{1}{2}\boldsymbol{e}_2.$$

The nth component of the eigenvector will be $1 - \frac{3}{2}n + \frac{1}{2}n^2$. Now we explain why these left eigenvectors are important in wavelet theory.

Theorem 7.4 *The left eigenvector in $\boldsymbol{y}_k\boldsymbol{M} = (\frac{1}{2})^k \boldsymbol{y}_k$ gives the combination of scaling functions $\phi(t+n)$ that equals t^k:*

$$\sum y_k(n)\phi(t+n) = t^k \quad \textit{for} \quad k = 0, 1, \ldots, p-1. \tag{7.18}$$

Thus the space V_0 spanned by $\{\phi(t+n)\}$ contains all polynomials of degree less than p.

Proof. We are to show that the inner product $G(t) = \boldsymbol{y}_k \Phi_\infty(t) = \sum y_k(n)\phi(t+n)$ equals a multiple of t^k. Here $\boldsymbol{y}_k$ is a left eigenvector of $\boldsymbol{M}$ and $\Phi_\infty(t) = \boldsymbol{M}\Phi_\infty(2t)$ solves the dilation equation. Put those two facts together:

$$\boldsymbol{y}_k\, \Phi_\infty(t) = \boldsymbol{y}_k \boldsymbol{M}\, \Phi_\infty(2t) = \left(\tfrac{1}{2}\right)^k \boldsymbol{y}_k\, \Phi_\infty(2t). \tag{7.19}$$

The left side is $G(t)$ and the right side is $(\frac{1}{2})^k G(2t)$. Since those are equal, $G(t)$ is a multiple of t^k. That is what we really wanted to prove. More details are in [HeStSt].

The all-ones eigenvector $\boldsymbol{e}_0$ says that $\sum \Phi(t+n) = 1$. This constant function assures at least $p = 1$. Since the wavelet is orthogonal to 1, we have $\int w(t)\,dt = 0$—the first vanishing moment. Hopefully there are more, and $p > 1$.

For $\boldsymbol{M}_2$, it is no surprise that 1 and t can be produced from translates of the hat function. More important is that 1 and t can be produced from the Daubechies scaling function $\phi(t) = D_4(t)$. This is a typical case in which $H(z)$ has an extra factor $\frac{1}{2}\left[1+\sqrt{3}+(1-\sqrt{3})z^{-1}\right]$ for double-shift orthogonality. The matrix $\boldsymbol{m}$ is 3 by 3 but only two eigenvalues 1 and $\frac{1}{2}$ are special (with their constant and linear left eigenvectors $\boldsymbol{y}_0$ and $\boldsymbol{y}_1$):

$$\frac{1}{4}\begin{bmatrix} 1+\sqrt{3} & & \\ 3-\sqrt{3} & 3+\sqrt{3} & 1+\sqrt{3} \\ & 1-\sqrt{3} & 3-\sqrt{3} \end{bmatrix} \quad \text{has} \quad \begin{matrix} \lambda = 1 & \boldsymbol{y}_0 = [1 \;\; 1 \;\; 1] \\ \lambda = \frac{1}{2} & \boldsymbol{y}_1 = [3-\sqrt{3} \;\; 1-\sqrt{3} \;\; -1-\sqrt{3}]/2 \\ \lambda = \frac{1+\sqrt{3}}{4} & \boldsymbol{y}_2 = [1 \;\; 0 \;\; 0] \end{matrix}$$

The sum $\sum \phi(t-n)$ is identically 1. The combination $\sum \boldsymbol{y}_1(n)\phi(t+n)$ equals a multiple of t. Thus the Daubechies space V_0 contains 1 and t. Those are orthogonal to the wavelets in W_0. This orthogonality $\int w(t)\,dt = 0$ and $\int t\,w(t)\,dt = 0$ says that the Daubechies wavelet has two vanishing moments.

Corollary (7.4) *When $H(\omega)$ has p zeros at π, the wavelets orthogonal to $\phi(t-n)$ have p vanishing moments. Those are the synthesis wavelets $\widetilde{w}(t)$:*

$$\int_{-\infty}^{\infty} \widetilde{w}(t)\,dt = 0, \quad \int_{-\infty}^{\infty} t\widetilde{w}(t)\,dt = 0, \quad \ldots, \quad \int_{-\infty}^{\infty} t^{p-1}\widetilde{w}(t)\,dt = 0. \tag{7.20}$$

Reason: $1, \ldots, t^{p-1}$ are combinations of $\phi(t-n)$. Orthogonality to these polynomials means p vanishing moments for the wavelets.

Remember that V_0 is orthogonal to $\widetilde{W}_0$ rather than W_0. Thus it is $\widetilde{w}(t)$, not $w(t)$, that has p vanishing moments! For an orthogonal example like a Daubechies filter, $\widetilde{W}_0 = W_0$ and $\widetilde{w}(t) = w(t)$. In the biorthogonal case, the *analysis* wavelet $w(t)$ has $\widetilde{p}$ vanishing moments when the lowpass *synthesis* filter has $\widetilde{p}$ zeros at π.

Note. The polynomials $1, \ldots, t^{p-1}$ are not actually in the subspace V_0. They are indeed combinations of the translates of $\phi(t)$. But polynomials have infinite energy, $\int_{-\infty}^{\infty} (t^j)^2 dt = \infty$. If V_0 is defined as a subspace of L^2, it cannot contain polynomials.

This point is only formal, not essential. The eigenvectors $\boldsymbol{y}$ have infinitely many nonzero components, which multiply all translates of $\phi(t)$. This maintains the polynomial for all time. Figure 7.1 shows how a finite combination maintains the polynomials 1 and t for finite time. This figure uses the Daubechies scaling function $\phi = D_4$ which has $p = 2$.

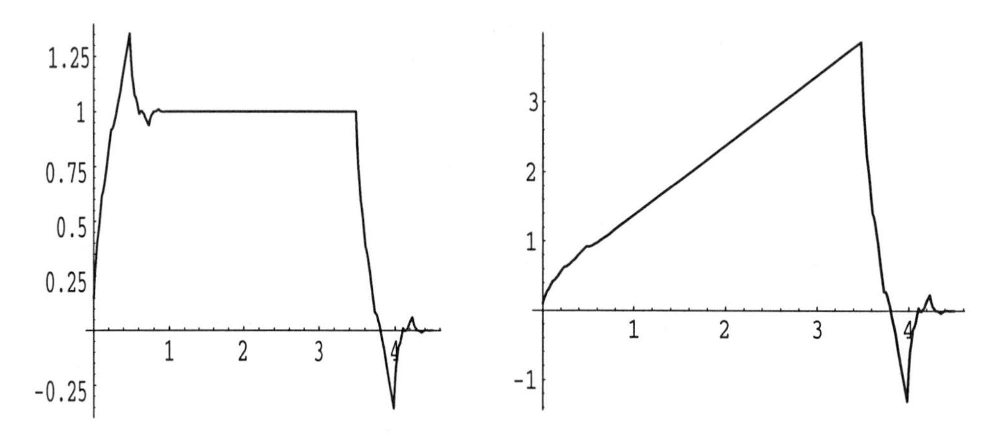

Figure 7.1: A combination of $D_4(t+n)$ can exactly reproduce 1 and t on any interval.

Approximation by Functions in V_j

In continuous time, vectors and matrices are replaced by functions of t. The lowpass filter coefficients $\boldsymbol{h}(k)$ have done their part. By iteration they led to $\phi(t)$. Its translates can reproduce $1, \ldots, t^{p-1}$ — and thus all polynomials of degree less than p. Its support interval is $[0, N]$ and we can estimate its smoothness. The theory in continuous time is now a job for "*harmonic analysis*".

Harmonic analysis is the study of function spaces and transforms. It takes its name from the all-important Fourier example, which analyzes $f(t)$ into a sum of harmonics $e^{i\omega t}$. The key problem is *to connect the properties of $f(t)$ with the properties of the transform* (especially the size of the Fourier coefficients). Because $\sin \omega t$ and $\cos \omega t$ and $e^{i\omega t}$ have nonlocal support, cancellation is crucial. The size of the Fourier coefficients tells a lot, but not everything. Only in the energy norm do we have a perfect match:

$$\text{The energy } \int |f(t)|^2 \, dt \text{ equals the energy } \frac{1}{2\pi} \int |\widehat{f}(\omega)|^2 \, d\omega.$$

In other L^p norms, and in other function spaces, the magnitude $|\widehat{f}(\omega)|$ does not completely decide whether $f(t)$ belongs to the space. We need the phase, which is more difficult. The theory using magnitudes is important. It can never be complete.

For a local wavelet basis, this situation is radically different. The magnitudes are enough! We can match the space of functions $f(t)$ to a space of wavelet coefficients b_{jk}. For $f(t)$ in L^p the coefficients are in the discrete space ℓ^p:

$$A \int |f(t)|^p \, dt \leq \sum_{j,k} |b_{jk}|^p \leq B \int |f(t)|^p \, dt. \tag{7.21}$$

In the language of harmonic analysis, wavelets are an ***unconditional basis*** when $p > 1$. The magnitudes $|b_{jk}|$ give sufficient information without phases. For L^2, which is always the simplest and clearest, an unconditional basis is a ***Riesz basis***:

$$A \int \left|\sum a_n \, \phi(t-n)\right|^2 dt \leq \sum |a_n|^2 \leq B \int \left|\sum a_n \, \phi(t-n)\right|^2 dt.$$

Then the translation invariance of the basis $\phi(t-n)$ yields the requirement (Section 6.5) on $A(\omega) = \sum \boldsymbol{a}(k) \, e^{ik\omega}$ in the frequency domain:

$$0 < A \leq A(\omega) = \sum_{-\infty}^{\infty} \left|\widehat{\phi}(\omega + 2\pi k)\right|^2 \leq B \quad \text{for all } \omega. \tag{7.22}$$

Exact numbers A and B will come from the components $\boldsymbol{a}(k)$ of the eigenvector $\boldsymbol{a} = \boldsymbol{Ta}$. So will a similar inequality for the wavelet basis $w_{jk}(t)$ and the coefficients b_{jk}.

Approximation by Wavelets: Errors $f(t) - f_j(t)$

The number p of zeros at π tells how many basis functions are needed to approximate $f(t)$. The smoother the function, and the higher the order p, the faster the expansion coefficients go to zero and the fewer we need to keep.

We are touching here on the central problem of transform analysis—*to find a convenient basis that yields accurate approximation of the signal with few basis functions.* The best basis depends on the signal (of course). We have to choose a basis for a class of signals. For smooth signals, the Fourier basis is usually satisfactory. Perhaps the essential message of wavelet theory can be captured in a sentence:

For piecewise smooth functions, a wavelet basis is better.

These functions may have jumps. They may be smooth and suddenly rough. A wavelet basis, which is local, can separate those pieces. We keep more coefficients in the rough neighborhoods, by going to a smaller scale 2^{-j}. The mesh adapts to $f(t)$ in a way that Fourier finds difficult.

Here is the fundamental theorem on approximation by scaling functions and/or wavelets. The space of approximating functions is V_j, so the scale is $\Delta t = 2^{-j}$. This space is spanned by the scaling functions $\phi(2^j t - k)$, and it is also spanned by the wavelets $w_{jk}(t)$ at all scales below j. We may choose either basis, since multiresolution says that $V_j = V_0 \oplus W_0 \oplus \cdots \oplus W_{j-1}$. The basis is not important at this point, because we are looking for the best function in the space.

Theorem 7.5 *When $H(\omega)$ has p zeros at π, any $f(t)$ with p derivatives is approximated to order $(\Delta t)^p = 2^{-jp}$ by its projection $f_j(t)$ in V_j:*

$$\|f(t) - f_j(t)\| \le C(\Delta t)^p \|f^{(p)}(t)\|$$

$$\|f(t) - \sum_k a_k 2^{j/2}\phi(2^j t - k)\| \le C 2^{-jp} \|f^{(p)}(t)\|. \tag{7.23}$$

This follows the expected pattern. For approximation by box functions or Haar wavelets, the error is of order Δt because $p = 1$. *Try this piecewise constant approximation on the linear function $f(t) = t$.* The closest constant on the interval $[0, \Delta t]$ is the halfway choice $a_1 = \frac{\Delta t}{2}$. The error $f(t) - f_0(t)$ on this interval is $t - \frac{\Delta t}{2}$. The maximum error is $\frac{\Delta t}{2}$ at $t = 0$. The L^2 error is

$$\left\|t - \frac{\Delta t}{2}\right\| = \left[\frac{1}{\Delta t}\int_0^{\Delta t}\left(t - \frac{\Delta t}{2}\right)^2 dt\right]^{1/2} = \frac{\Delta t}{2\sqrt{3}}. \tag{7.24}$$

The first derivative $f^{(1)}(t)$ on the right side of (7.23) is just 1. The constant is $C = \frac{1}{2}$ in the maximum norm. It is $C = 1/2\sqrt{3}$ if we use the L^2 norm. The important part is the power of Δt. But if a different $\phi(t)$ gives a much larger constant C, that is important too.

The main point is that the basis $\{\phi(t-k)\}$ can locally produce $1, \ldots, t^{p-1}$. In each interval we can "essentially" match the start of the Taylor series. The error is the first Taylor series term we ***cannot*** match. This produces $(\Delta t)^p f^{(p)}(t)$ in the error bound. A detailed proof would lead far into approximation theory. The theorem was known earlier in the particular case of splines. Then it was extended to include ***finite elements*** [SF]. Those are local basis functions — normally piecewise polynomials. When the approximation theorem was first proved, nobody was thinking of wavelets! Who would use irregular functions to approximate smooth functions? It defies common sense, but wavelets come from iterations of simple filters. The computations are quick (if done recursively). The theorem applies because those irregular functions $\phi(t-k)$ can exactly reproduce polynomials.

The requirement on $H(\omega)$ is p zeros at π. It is interesting to note the corresponding requirements on $\widehat{\phi}(\omega)$. These are the so-called Strang-Fix conditions:

$$\widehat{\phi}(\omega) \text{ must have zeros of order } p \text{ at all frequencies } \omega = 2\pi n, n \neq 0. \tag{7.25}$$

The connection to zeros of $H(\omega)$ is through the infinite product $\widehat{\phi}(\omega) = \prod_1^\infty H(\omega/2^j)$. At frequency $\omega = 2\pi n$, we write $n = 2^j q$ with q odd. Then the $(j+1)$st factor in the infinite product is $H(2\pi n/2^{j+1}) = H(q\pi)$. By periodicity this is $H(\pi)$. A pth order zero of H at $\omega = \pi$ yields a pth order zero of $\widehat{\phi}$ at $\omega = 2\pi n$. Thus the Strang-Fix condition on $\phi(t)$ becomes Condition A_p on $H(\omega)$.

The goal is always to find conditions on $H(\omega)$ that make $\widehat{\phi}(\omega)$ do what we want. For good approximation, we require zeros at π. This same condition improves the smoothness of $\phi(t)$ and stabilizes the lowpass filter under iteration. It is an open question ***how many*** zeros to ask for. Enough to stabilize the iterations, but not so many that we overconstrain the lowpass filter. Designers are often satisfied with two derivatives for $\phi(t)$, which occurs for $p \approx 4$. Other designers accept a smaller p.

Decay of the Wavelet Coefficients

The order p enters in another way, to improve compression. It allows the wavelet coefficients to do what Fourier coefficients do automatically: *they decrease rapidly for a smooth function.* For Fourier, $f(t)$ has to be smooth everywhere. One small jump, and the coefficients decrease no faster than $\frac{1}{k}$. For wavelets, the slow decrease will only apply around the jump. In smooth regions the coefficients drop off quickly. Multiresolution offers a wavelet basis in which the coefficients are directly linked to local properties of $f(t)$.

Theorem 7.6 *If $f(t)$ has p derivatives, its wavelet coefficients decay like 2^{-jp}:*

$$|b_{jk}| = \left|\int f(t)\, w_{jk}(t)\, dt\right| \le C\, 2^{-jp} \left\| f^{(p)}(t) \right\|. \tag{7.26}$$

Proof. We plan to integrate by parts in $\int f(t) w_{jk}(t)\, dt$. That gives the derivative of f and the integral of w. (For biorthogonal wavelets replace w by $\widetilde{w}$.) The first vanishing moment means that the integral of $w(t)$ from $-\infty$ to ∞ is zero. *The indefinite integral has compact support*:

$$I_1(t) = \int_{-\infty}^{t} w(u)\, du \quad \text{is nonzero only on } [0, N].$$

$I_1(t)$ is bounded, with finite energy. It will be a hat function, when $w(t)$ is Haar's up-down square wave. Integrate by parts in (7.26) to produce f' times a rescaled I_1:

$$b_{jk} = 2^{-j} \int_{-\infty}^{\infty} f'(t) 2^{j/2} I_1(2^j t - k) dt = O(2^{-j}). \tag{7.27}$$

The factor 2^{-j} comes because we are integrating $w_{jk}(t)$ instead of $w(t)$:

$$\int_{-\infty}^{t} w_{jk}(t) dt = \int_{-\infty}^{t} 2^{j/2} w(2^j t - k) dt = 2^{-j} \int_{-\infty}^{2^j t - k} w(u)\, du.$$

Now repeat this step p times. Each integration by parts brings an extra derivative of $f(t)$ and an extra integral of $w_{jk}(t)$ — with its factor 2^{-j}. If successive integrals of $w(t)$ are $I_1(t)$ and $I_2(t)$ and finally $I_p(t)$, we end with

$$|b_{jk}| = \left| 2^{-jp} \int_{-\infty}^{\infty} f^{(p)}(t) 2^{j/2} I_p(2^j t - k) dt \right| \le C 2^{-jp} \left\| f^{(p)}(t) \right\|.$$

Even when $f(t)$ has more derivatives, we cannot continue beyond p. The integral of $I_p(t)$ from $-\infty$ to ∞ is not zero! If it were, p integrations by parts would bring us back to $\int t^p w(t) dt = 0$ — but this pth moment of $w(t)$ does ***not*** vanish. $I_{p+1}(t)$ is a nonzero constant when t is large. Its energy is infinite and (7.26) breaks down for $p+1$. Look again at Haar wavelets with $p = 1$. They have a square wave up and a square wave down, at scale 2^{-j}. The differences $f(t) - f(t - 2^{-j})$ are of order 2^{-j}. Integration gives a direct estimate $|b_{jk}| = O(2^{-j})$ in this particular case. The general method integrates by parts to get the derivative $f'(t)$ times the hat function I_1 times the key factor 2^{-j}.

Sample Values vs. Expansion Coefficients

Start with a function $x(t)$. Its samples $\boldsymbol{x}(n)$ are often the inputs to the filter bank. Is this legal? ***No. It is a wavelet crime.*** Some can't imagine doing it, others can't imagine not doing it. Is this crime convenient? ***Yes.*** We may not know the whole function $x(t)$, it may not be a combination of $\phi(t-k)$, and computing the true coefficients in $\sum a(k)\phi(t-k)$ may take too long. But the crime cannot go unnoticed—we have to discuss it.

When the samples $\boldsymbol{x}(n)$ are direct inputs to the filter bank (at unit scale $j=0$), you effectively assume a particular continuous-time function. The pyramid algorithm (filter and downsample) acts on the numbers $\boldsymbol{x}(n)$ as if they were expansion coefficients of its underlying function $x_s(t)$:

$$\textbf{\textit{Samples as coefficients:}} \quad x_s(t) = \sum \boldsymbol{x}(n)\phi(t-n). \tag{7.28}$$

Does $x_s(t)$ have the correct sample values $\boldsymbol{x}(n)$? This seems a minimum requirement. It holds when $\phi(k) = \delta(k)$. Then the only term in (7.28) at $t=n$ is the correct $\boldsymbol{x}(n)$. A centered hat function has this property but most $\phi(t)$ do not. One possible solution is to adjust the coefficients in $\sum a(k)\phi(t-k)$ to produce the known samples $\boldsymbol{x}(n)$:

$$\textit{Determine } a_{int}(k) \textit{ from } \quad \boldsymbol{x}(n) = \sum a_{int}(k)\phi(k-n). \tag{7.29}$$

This linear system has a constant-diagonal Toeplitz matrix. The n,k entry is $\phi(k-n)$. We are inverting an FIR filter that has the response $\sum \phi(k)e^{-ik\omega}$. Then the underlying function that interpolates the samples $\boldsymbol{x}(n)$ is $x_{int}(t) = \sum a_{int}(k)\phi(t-k)$.

We believe that generally the samples $\boldsymbol{x}(n)$ should be pre-filtered, before they enter the filter bank. Solving (7.29) puts the samples through an IIR filter. A different approach yields an FIR filter, by approximating the "correct" coefficients $a(k)$. Those are inner products of $x(t)$ with the analyzing function $\widetilde{\phi}(t-k)$. The pre-filter replaces this integral by a sum:

$$\text{Replace} \quad a(k) = \int x(t)\widetilde{\phi}(t-k)\,dt \quad \text{by} \quad a_q(k) = \sum \boldsymbol{x}(n)\widetilde{\phi}(n-k). \tag{7.30}$$

In the Daubechies examples, $\widetilde{\phi} = \phi$ from orthogonality and the tilde disappears.

The pre-filter that gives $a_q(k)$ is normally FIR. Except for sinc wavelets and the duals to splines, our basis functions have compact support. The ideal function underlying this pre-filter is $x_q(t) = \sum a_q(k)\phi(t-k)$, a sensible choice.

Important: In continuous time, the synthesis $x(t) = \sum a(k)\phi(t-k)$ is consistent with the analysis $a(k) = \int x(t)\widetilde{\phi}(t-k)\,dt$. In discrete time those are *not consistent.* When t changes to n and the integral changes to a sum, inverse operators do not become inverse matrices (unfortunately). But the discrete approximations are exactly correct for polynomials up to order p or $\widetilde{p}$:

$$\sum_{-\infty}^{\infty} n^r\phi(n) = \int_{-\infty}^{\infty} t^r\phi(t)\,dt \quad \text{for} \quad r < p. \tag{7.31}$$

The right side is the rth derivative of $\widehat{\phi}(\omega) = \int \phi(t)e^{-i\omega t}dt$ at $\omega = 0$, times i^r. For the left side we use Poisson's summation formula (Chapter 2). This gives the same rth derivative at $\omega = 0$ and at points $\omega = 2\pi k$:

$$\sum_{-\infty}^{\infty} n^r\phi(n) = \sum_{-\infty}^{\infty} i^r\widehat{\phi}^{(r)}(2\pi k).$$

But all terms on the right are zero except for $k = 0$, by the Strang-Fix condition (7.25) on the function $\phi(t)$. Thus the equality (7.31) holds for any $\phi(t)$ whose translates can reproduce polynomials to degree p. Similarly $a_q(k) = a(k)$, sum equals integral, when $x(t)$ is a low-degree polynomial.

Summary. We recommend that the samples $\boldsymbol{x}(n)$ be converted to coefficients $a_q(k)$ by (7.30). Those enter the filter bank, not $\boldsymbol{x}(n)$. The output $\widehat{a}(k)$ can be post-filtered to recover sample values.

Other pre-filters are also reasonable. [Flandrin] proposed that the underlying $x(t)$ should be band-limited. The sampling theorem gives $x(t)$ as a sum of $\text{sinc}(t-n)\boldsymbol{x}(n)$. Projecting this band-limited $x(t)$ onto V_0 gives $\sum a_{bl}(k)\phi(t-k)$, and those coefficients a_{bl} can enter the filter bank.

The samples $\boldsymbol{x}(n)$ could be regarded as *averages* instead of point values. Then $x(t)$ is assumed piecewise constant, a combination of box functions $B(t-n)$. Projecting $x(t)$ onto V_0 gives $\sum a_{ave}(k)\phi(t-k)$. The filters can operate on $a_{ave}(k)$.

There is no unique answer to our question of how to process the sample values. We may choose $a_{int}(k)$ or $a_q(k)$ or $a_{bl}(k)$ or $a_{ave}(k)$. We should not send samples automatically through the filter bank.

Problem Set 7.1

1. Find the accuracy p from the sum rules for these filter coefficients:

   (a) $\boldsymbol{h} = (\ \frac{1}{4},\quad \frac{1}{2},\quad \frac{1}{4}\ )$

   (b) $\boldsymbol{h} = \frac{1}{16}(\ 1,\quad 4,\quad 6,\quad 4,\quad 1\ )$

   (c) $\boldsymbol{h} = \frac{1}{8}(\ 1-\sqrt{3},\quad 3-\sqrt{3},\quad 3+\sqrt{3},\quad 1+\sqrt{3}\ )$ = Daubechies in reverse.

   (d) $\boldsymbol{h} = D_6$

   (e) $\boldsymbol{h} = (\ \frac{1}{3},\quad \frac{1}{3},\quad \frac{1}{3}\ )$

2. Factor the frequency response for filters (a) to (e) into $H(\omega) = (\frac{1+e^{-i\omega}}{2})^p R(\omega)$.

3. Find the eigenvalues of the N by N matrix $\boldsymbol{m}$ for each of the filters (a) to (e). The number of taps is $N+1$; cases (b) and (d) are less convenient by hand.

4. Which 5 by 5 matrix $\boldsymbol{m}_5$ comes from $H(z) = (\frac{1+z^{-1}}{2})^5$? Find the right eigenvectors for $\lambda = \frac{1}{2}, \frac{1}{4}, \frac{1}{8}, \frac{1}{16}$ from differences of the eigenvectors of $\boldsymbol{m}_4$ in the text. Find directly the eigenvector for $\lambda = 1$.

5. Verify that the left eigenvectors given in the text for $\boldsymbol{m}_4$ are sums of the left eigenvectors for $\boldsymbol{m}_3$, plus a constant $[c\ \ c\ \ c\ \ c]$. Find the five left eigenvectors of $\boldsymbol{m}_5$.

6. $\boldsymbol{M} = (\downarrow 2)2\boldsymbol{H}$ transforms $\widehat{f}(\omega)$ into $\widehat{\boldsymbol{M}f}(\omega) = H\left(\frac{\omega}{2}\right)\widehat{f}\left(\frac{\omega}{2}\right) + H\left(\frac{\omega}{2}+\pi\right)\widehat{f}\left(\frac{\omega}{2}+\pi\right)$. The first sum rule is $H(\pi) = 0$. If $\widehat{f}(0) = 0$, show that $\widehat{\boldsymbol{M}f}(0) = 0$, and explain what this means in the time domain.

7. The first two sum rules are $H(\pi) = 0$ and $H'(\pi) = 0$. Suppose that $\widehat{f}(0) = \widehat{f'}(0) = 0$. Show that $\widehat{\boldsymbol{M}f}(0) = (\widehat{\boldsymbol{M}f})'(0) = 0$.

8. If $w(t)$ has p vanishing moments, show that its Fourier transform has a pth order zero at $\omega = 0$. Then factoring $(i\omega)^p$ from $\widehat{w}$ gives the transform $\widehat{I_p}$ of the p-fold integral of $w(t)$.

9. Find the Fourier transform of the Haar wavelet and factor out $i\omega$ to obtain the transform of $I_1(t)$. Show that $I_1(t)$ is a hat function.

10. The left eigenvector $\begin{bmatrix} 1 & 0 \end{bmatrix}$ for $\boldsymbol{m}_2$ extends to $\begin{bmatrix} \cdots & 2 & 1 & 0 & -1 & \cdots \end{bmatrix}$ for the infinite matrix $\boldsymbol{M}_2$. Write out $\boldsymbol{y}\boldsymbol{M}_2 = \frac{1}{2}\boldsymbol{y}$ and circle the 2 by 2 subvector and submatrix in the middle. Verify that $\cdots + 2\phi(t-1) + \phi(t) + 0 - \phi(t-1) - \cdots$ equals t, when $\phi(t)$ is the hat function from this filter.

11. How would you compute $a_{ave}(k)$ so that $\sum a_{ave}(k)\phi(t-k)$ is the projection onto V_0 of the piecewise constant $x(t)$ with *average* $\boldsymbol{x}(k)$ over the kth subinterval?

7.2 The Cascade Algorithm for the Dilation Equation

In the theory of wavelets, the two-scale dilation equation $\phi(t) = \sum 2\boldsymbol{h}(k)\phi(2t-k)$ is central. Its solution is the scaling function, which leads to wavelets. The equation arises in the limit of the *cascade algorithm*

$$\phi^{(i+1)}(t) = \sum 2\boldsymbol{h}(k)\phi^{(i)}(2t-k). \tag{7.32}$$

This is an iteration (with rescaling) of the lowpass filter. We find a simple proof of the necessary and sufficient condition on the $\boldsymbol{h}(k)$ for $\phi^{(i)}(t)$ to converge in L^2. The cascade normally begins from $\phi^{(0)}(t)$ = box function. Problem 4 finds all other $\phi^{(0)}(t)$ that yield convergence to $\phi(t)$. Thus there are two conditions for convergence, one on the filter (this is the important one) and a condition on $\phi^{(0)}(t)$.

Our method is to compute the inner products $\boldsymbol{a}^{(i)}(k) = \int \phi^{(i)}(t)\phi^{(i)}(t+k)dt$ at each step of the algorithm. The vectors $\boldsymbol{a}^{(i+1)}$ and $\boldsymbol{a}^{(i)}$ are connected by a *transition matrix* $\boldsymbol{T}$ formed from the $\boldsymbol{h}(k)$. The cascade algorithm for $\phi(t)$ becomes the power method $\boldsymbol{a}^{(i+1)} = \boldsymbol{T}\boldsymbol{a}^{(i)}$ for the equation $\boldsymbol{a} = \boldsymbol{T}\boldsymbol{a}$. "**Condition E**" for convergence is that all eigenvalues of $\boldsymbol{T}$ satisfy $|\lambda| < 1$ except for a simple eigenvalue at $\lambda = 1$.

Recall that $\lambda = 1$ is an eigenvalue of $\boldsymbol{M} = (\downarrow 2)2\boldsymbol{H}$ by the *first sum rule*:

$$\sum_{\textit{even } k} 2\boldsymbol{h}(k) = \sum_{\textit{odd } k} 2\boldsymbol{h}(k) = 1.$$

These even k and odd k appear in separate columns of $\boldsymbol{M}$. Each column adds to 1. Therefore $\lambda = 1$ is an eigenvalue, and the left eigenvector is $\boldsymbol{e} = [1 \ 1 \ \cdots \ 1]$. This is necessary for convergence, pointwise or L^2, but far from sufficient. It means that $H(\omega) = \sum \boldsymbol{h}(k)e^{-ik\omega}$ has a zero at $\omega = \pi$. Other things being equal, every zero at π gives a boost to convergence. This is a double zero in the nonnegative function $P(\omega) = |H(\omega)|^2$. The key to L^2 convergence is the matrix $\boldsymbol{T}$ associated with $|H(\omega)|^2$ in the same way that $\boldsymbol{M}$ is associated with $H(\omega)$:

$$\boldsymbol{T} = (\downarrow 2)2\boldsymbol{H}\boldsymbol{H}^T.$$

When $\boldsymbol{H}$ and $\boldsymbol{H}^T$ are infinite Toeplitz matrices, $\boldsymbol{M}$ and $\boldsymbol{T}$ are also infinite. They are block Toeplitz matrices, with 1 by 2 blocks because of the operator $(\downarrow 2)$. $\boldsymbol{T}$ is illustrated in equation (7.38) below. The important action is in the finite matrix at the center. The central submatrix of $\boldsymbol{T}$ has order $2N-1$:

$$\boldsymbol{T}_{jk} = 2\boldsymbol{p}(2j-k) \quad \text{if} \quad |H(\omega)|^2 = \sum \boldsymbol{p}(k)e^{-ik\omega}, \quad -N < j, k < N.$$

The columns of $\boldsymbol{T}$ are still even or odd, containing coefficients $\boldsymbol{p}(2n)$ or else $\boldsymbol{p}(2n+1)$. Those columns add to 1, since $|H(\omega)|^2$ has a zero at π. The all-ones vector $\boldsymbol{e}$ is still a left eigenvector for $\lambda = 1$:

$$\boldsymbol{e}\boldsymbol{T} = \boldsymbol{e}\boldsymbol{M}\boldsymbol{H}^T = \boldsymbol{e}\boldsymbol{H}^T = \boldsymbol{e}. \tag{7.33}$$

The crucial question is the significance of the *right eigenvector* in $\boldsymbol{Ta} = \boldsymbol{a}$. It gives the inner products $\langle\phi(t), \phi(t+k)\rangle$. *Convergence of the cascade algorithm in L^2 becomes convergence of the power method* $\boldsymbol{a}^{(i+1)} = \boldsymbol{Ta}^{(i)}$.

We include this convergence proof in the text because the argument is straightforward. Condition E on $\boldsymbol{T}$ also determines [Cohen-Daubechies] whether the translates $\phi(t+k)$ form a Riesz basis. [Eirola] and [Villemoes] use the same matrix $\boldsymbol{T}$ in a different way, to find the smoothness of $\phi(t)$ (next section).

We work with one-dimensional filters, but the analysis is the same in higher dimensions [Lawton-Lee-Shen]. In a sense our approach completes the convergence analysis of [CDM], by working with $|H(\omega)|^2$—the autocorrelation of the filter (or mask). The key is to identify $\boldsymbol{T}$ and watch its eigenvalues.

Examples of Divergence and Weak Convergence

Example 7.1. The coefficients $\boldsymbol{h}(k) = 1, \frac{1}{2}, -\frac{1}{2}$ have even sum = odd sum = $\frac{1}{2}$. The dilation equation is

$$\phi(t) = 2\phi(2t) + \phi(2t-1) - \phi(2t-2). \tag{7.34}$$

This looks innocent, but in a few steps the cascade algorithm is a disaster. The value at $t = 0$ is doubled at every iteration. Each step gives $\phi^{(i+1)}(0) = 2\phi^{(i)}(0)$.

This blowup at a point does not by itself rule out finite energy. The function $f(t) = t^{-1/3}$ also blows up at $t = 0$, but on the interval $[0, 1]$ its energy is $\int t^{-2/3}dt = 3$. Therefore this example is continued below, to prove that the energy in $\phi^{(i)}(t)$ does become infinite as $i \to \infty$. When the first coefficient in (7.34) is between $\frac{1}{2}(1-\sqrt{3})$ and $\frac{1}{2}(1+\sqrt{3})$, which allows the possibility of blowup at $t = 0$, the cascade algorithm converges in L^2.

Example 7.2. The coefficients $\boldsymbol{h}(k) = \frac{1}{2}, 0, 0, \frac{1}{2}$ have even sum = odd sum = $\frac{1}{2}$ and also double-shift orthogonality. All the shifted vectors $0, 0, \frac{1}{2}, 0, 0, \frac{1}{2}, 0, \ldots$ and $0, 0, 0, 0, \frac{1}{2}, 0, 0, \frac{1}{2}, \ldots$ are orthogonal. The expected solution of the dilation equation is a *stretched box*:

$$\phi(t) = \phi(2t) + \phi(2t-3) \text{ leads to } \phi(t) = \begin{cases} \frac{1}{3} & 0 \le t < 3 \\ 0 & \text{else.} \end{cases}$$

The box area is still one. Two half-boxes still fit into a whole box. But when the cascade algorithm starts with the unit box, it converges only *weakly* to $\phi(t)$.

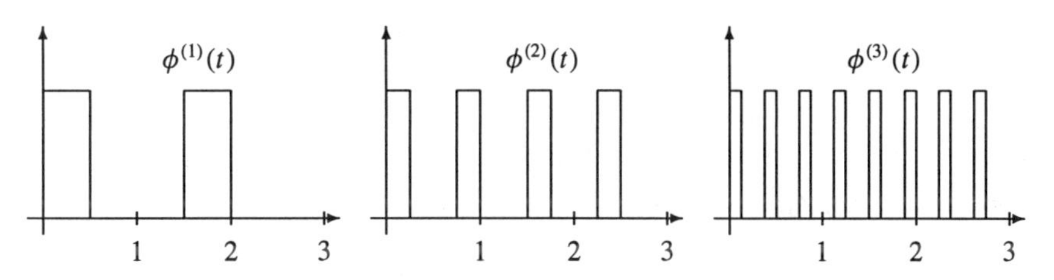

Figure 7.2: Three cascade steps $\phi^{(i+1)}(t) = \phi^{(i)}(2t) + \phi^{(i)}(2t-3)$.

The area between 0 and 3 is one-third filled, more and more densely. At no point in Figure 7.2 do the functions $\phi^{(i)}(t)$ actually equal $\frac{1}{3}$; they always equal 0 or 1. But the area above every interval Δt approaches $\frac{1}{3}\Delta t$. This is weak convergence to $\phi(t) = \frac{1}{3}$, stated for any smooth function $f(t)$:

$$\lim_{i\to\infty}\int_0^3 \phi^{(i)}(t)f(t)\,dt = \int_0^3 \tfrac{1}{3}f(t)\,dt. \tag{7.35}$$

The adjective "weak" means that integrals converge, even if functions themselves do not. Fast oscillations in $\phi^{(i)}(t)$ are averaged out by the integration.

Note that this shifted box $\phi(t)$ is *not orthogonal* to its translates. Furthermore, its energy $\int(\phi(t))^2\,dt$ is $\frac{1}{3}$ instead of 1. Those facts seem surprising, because at every step $\phi^{(i)}(t)$ has unit energy and is orthogonal to all other $\phi^{(i)}(t+k)$. This illustrates the weakness of convergence—inner products of $\phi^{(i)}(t)$ with fixed $f(t)$ converge but inner products with $\phi^{(i)}(t+k)$ do not converge to $\int\phi(t)\phi(t+k)\,dt$.

Our inner product and energy formulas will involve $\boldsymbol{T}$. This matrix always has $\lambda = 1$ as an eigenvalue, because of the zero at π. Explosive failure as in Example 1 is associated with an eigenvalue that has $|\lambda| > 1$. Weak convergence as in Example 2 is associated with a repeated $\lambda = 1$ and/or other eigenvalues with $|\lambda| = 1$ (with a full set of eigenvectors). Total success occurs when all other eigenvalues of $\boldsymbol{T}$ have $|\lambda| < 1$; this will be Condition E. The smaller those other eigenvalues, the smoother the function $\phi(t)$.

The Inner Product Formula

The energy in a real function $\phi(t)$ (its L^2 norm) is its inner product with itself. We also want its inner product with its translates $\phi(t+k)$. This inner product is $\boldsymbol{a}(k)$:

$$\boldsymbol{a}(k) = \int_{-\infty}^{\infty}\phi(t)\phi(t+k)\,dt, \quad -\infty < k < \infty. \tag{7.36}$$

Note that $\boldsymbol{a}(k) = \boldsymbol{a}(-k)$. The number $\boldsymbol{a}(0)$ is the energy $||\phi||^2$. When $\phi(t)$ is zero outside the interval $0 \le t < N$, the inner products $\boldsymbol{a}(k)$ are all zero for $|k| \ge N$. The translated $\phi(t+N)$ does not overlap $\phi(t)$. Only the $2N-1$ central components of $\boldsymbol{a}$ can be nonzero.

The cascade algorithm starts each step with $\phi^{(i)}(t)$. Suppose we know the inner products $\boldsymbol{a}^{(i)}(k)$ between that function and its translates. Iteration produces a new function $\phi^{(i+1)}(t)$, and we want the new inner products $\boldsymbol{a}^{(i+1)}(k)$. The new function is a combination $\sum 2\boldsymbol{h}(k)\phi^{(i)}(2t-k)$, *with t rescaled to* $2t$. The new inner products depend on the old ones, and on the numbers $\boldsymbol{h}(k)$. We now find the formula that the inner products obey.

Lemma. To find the vector $\boldsymbol{a}^{(i+1)}$ of inner products, multiply $\boldsymbol{a}^{(i)}$ by the Toeplitz matrix $2\boldsymbol{H}\boldsymbol{H}^T$ (which gives $4N-1$ components) and downsample:

$$\boldsymbol{T}\boldsymbol{a}^{(i)} = \boldsymbol{a}^{(i+1)} = (\downarrow 2)2\boldsymbol{H}\boldsymbol{H}^T\boldsymbol{a}^{(i)}. \tag{7.37}$$

Example 7.1 (continued) With $2\boldsymbol{h}(k) = 2, 1, -1$, the numbers along the diagonals of the symmetric matrix $2\boldsymbol{H}\boldsymbol{H}^T$ are $\frac{1}{2}(2,1,-1)*(-1,1,2) = \frac{1}{2}(-2,1,6,1,-2)$. This comes from convolution and also from matrix multiplication: $2\boldsymbol{H}\boldsymbol{H}^T$ is

$$\frac{1}{2}\begin{bmatrix} \cdot & & & \\ 1 & 2 & & \\ -1 & 1 & 2 & \\ & -1 & 1 & 2 \end{bmatrix}\begin{bmatrix} \cdot & 1 & -1 & \\ & 2 & 1 & -1 \\ & & 2 & 1 \\ & & & 2 \end{bmatrix} = \frac{1}{2}\begin{bmatrix} \ddots & \ddots & \ddots & \ddots & \\ -2 & 1 & 6 & 1 & -2 \\ & \ddots & \ddots & \ddots & \ddots \end{bmatrix}.$$

$(\downarrow 2)$ removes the odd-numbered rows to leave double shifts in $\boldsymbol{T}$:

$$\boldsymbol{T} = \frac{1}{2}\begin{bmatrix} \cdot & -2 & & & & & \\ \cdot & 6 & 1 & -2 & & & \\ & -2 & 1 & 6 & 1 & -2 & \\ & & & -2 & 1 & 6 & \cdot \\ & & & & & -2 & \cdot \end{bmatrix}. \tag{7.38}$$

All columns of $\boldsymbol{T}$ add to 1! At the first cascade step, the box function $\phi^{(0)}(t)$ has inner products $\boldsymbol{a}^{(0)} = (\ldots, 0, 1, 0, \ldots)$. The Lemma says that the inner products of $\phi^{(1)}(t)$ with its translates are

$$\boldsymbol{a}^{(1)} = \boldsymbol{T}\boldsymbol{a}^{(0)} = \frac{1}{2}\begin{bmatrix} \cdot \\ 0 \\ -2 \\ 6 \\ -2 \\ 0 \\ \cdot \end{bmatrix} = \begin{bmatrix} \cdot \\ 0 \\ -1 \\ 3 \\ -1 \\ 0 \\ \cdot \end{bmatrix}.$$

Since $N = 2$ for this filter $\boldsymbol{H}$, all inner products are zero for $|k| \geq 2$. We only need the center submatrix of order $2N - 1 = 3$, and we iterate again:

$$\boldsymbol{a}^{(2)} = \frac{1}{2}\begin{bmatrix} 1 & -2 & 0 \\ 1 & 6 & 1 \\ 0 & -2 & 1 \end{bmatrix}\begin{bmatrix} -1 \\ 3 \\ -1 \end{bmatrix} = \frac{1}{2}\begin{bmatrix} -7 \\ 16 \\ -7 \end{bmatrix}. \tag{7.39}$$

Clearly the numbers are growing. The cascade algorithm is diverging. $\boldsymbol{T}$ has an eigenvalue larger than 1. It is $\frac{5}{2}$, and this filter is a disaster in iteration.

Example 7.2 (continued) With coefficients $2\boldsymbol{h}(k) = 1, 0, 0, 1$, the matrix $2\boldsymbol{H}\boldsymbol{H}^T$ has rows containing $\frac{1}{2}(1, 0, 0, 2, 0, 0, 1)$. Downsampling removes every other row to leave double shifts in $\boldsymbol{T}$. Since $N = 3$ and $2N - 1 = 5$, we look only at $\boldsymbol{T}_5$:

$$T_5 = \frac{1}{2}\begin{bmatrix} 0 & 1 & & & \\ 2 & 0 & 0 & 1 & \\ 0 & 0 & 2 & 0 & 0 \\ & 1 & 0 & 0 & 2 \\ & & & 1 & 0 \end{bmatrix}.$$

Again all columns add to 1. Starting again with the box function $\phi^{(0)}(t)$, its inner products are $\boldsymbol{a}^{(0)} = (0, 0, 1, 0, 0)$. *Multiplying by T_5 produces this same vector $\boldsymbol{a}^{(1)} = \boldsymbol{a}^{(0)}$*. Therefore $\phi^{(1)}(t)$ is also orthogonal to its translates.

That conclusion is no surprise. The $\boldsymbol{h}(k)$ have double-shift orthogonality. The center column of $\boldsymbol{T}$ agrees with the identity matrix. At every step $\boldsymbol{a}^{(i)} = \boldsymbol{\delta}$.

Still there is weak trouble. The reason is that $\boldsymbol{T}_5$ has a repeated eigenvalue at $\lambda = 1$. Its other eigenvalues are $-1, -\frac{1}{2}, \frac{1}{2}$. A second eigenvector for $\lambda = 1$ is $\frac{1}{9}(1, 2, 3, 2, 1)$. Those are the actual inner products of the stretched ϕ_{box} on the interval $0 \leq t < 3$. The inner products $\boldsymbol{a}^{(i)} = \boldsymbol{\delta}$ do not approach the inner products of ϕ_{box}. The cascade algorithm does not converge in L^2 to this stretched box.

Example 7.3. (Convergence to the hat function) With coefficients $2\boldsymbol{h}(k) = \frac{1}{2}, 1, \frac{1}{2}$, the matrix $2\boldsymbol{H}\boldsymbol{H}^T$ has entries $\frac{1}{8}(1, 4, 6, 4, 1)$. Downsampling leaves $\boldsymbol{T}_{2N-1} = \boldsymbol{T}_3$:

$$\boldsymbol{T}_3 = \frac{1}{8}\begin{bmatrix} 4 & 1 & 0 \\ 4 & 6 & 4 \\ 0 & 1 & 4 \end{bmatrix} \quad \text{has } \lambda = 1, \tfrac{1}{2}, \tfrac{1}{4}. \tag{7.40}$$

This example is successful but not orthogonal. Each multiplication of $\boldsymbol{a}^{(0)} = (0, 1, 0)$ by $\boldsymbol{T}_3$ gives the inner products at the next cascade step:

$$\boldsymbol{a}^{(1)} = \frac{1}{8}\begin{bmatrix} 1 \\ 6 \\ 1 \end{bmatrix} \quad \boldsymbol{a}^{(2)} = \frac{1}{64}\begin{bmatrix} 10 \\ 44 \\ 10 \end{bmatrix} \quad \cdots \quad \boldsymbol{a}^{(\infty)} = \frac{1}{6}\begin{bmatrix} 1 \\ 4 \\ 1 \end{bmatrix}. \tag{7.41}$$

We jumped to the limit $\boldsymbol{a}^{(\infty)}$ because it is the eigenvector of $\boldsymbol{T}_3$ for $\lambda = 1$.

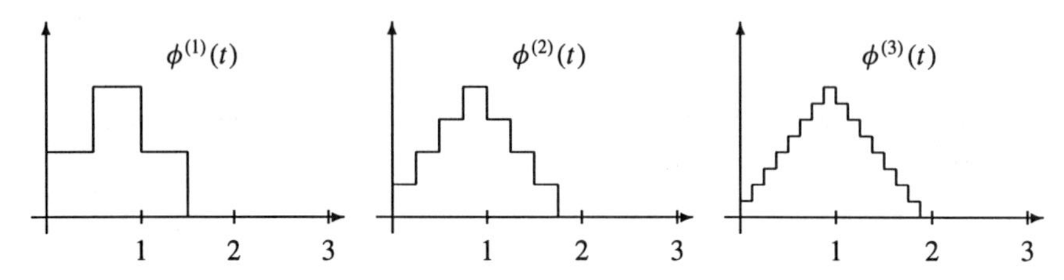

Figure 7.3: Three cascade steps $\phi^{(i+1)}(t) = \frac{1}{2}\phi^{(i)}(2t) + \phi^{(i)}(2t-1) + \frac{1}{2}\phi^{(i)}(2t-2)$.

The limit vector $\boldsymbol{a}^{(\infty)} = \boldsymbol{a}$ contains the inner products of the hat function with its translates. The $\phi^{(i)}(t)$ in Figure 7.3 are converging to $\phi^{(\infty)}(t) =$ hat function, and the inner products $\boldsymbol{a}^{(i)}(k)$ are converging to $\boldsymbol{a}^{(\infty)}(k) = \frac{1}{6}, \frac{4}{6}, \frac{1}{6}$. We are seeing the **power method** in operation, for the functions and also for the vectors. The hat function is the steady-state fixed point of the operator in the cascade. The vector $\boldsymbol{a}$ is the steady-state eigenvector of $\boldsymbol{T}$ with $\lambda = 1$.

Proof of the Lemma To show that the inner products are $\boldsymbol{a}^{(i+1)} = \boldsymbol{T}\boldsymbol{a}^{(i)}$, it is very convenient to compute all $\boldsymbol{a}^{(i+1)}(k)$ at once. *Thus we work with vectors*:

$$\Phi^{(i)}(t) = \begin{bmatrix} \cdot \\ \phi^{(i)}(t-1) \\ \phi^{(i)}(t) \\ \phi^{(i)}(t+1) \\ \cdot \end{bmatrix} \quad \text{and} \quad \Phi(t) = \begin{bmatrix} \cdot \\ \phi(t-1) \\ \phi(t) \\ \phi(t+1) \\ \cdot \end{bmatrix}. \tag{7.42}$$

The next vector in the cascade is $\Phi^{(i+1)}(t) = (\downarrow 2)2\boldsymbol{H}\Phi^{(i)}(2t)$. The rescaling to $2t$ is accounted for by $(\downarrow 2)$. This is the way to a vector-based calculation:

$$\begin{aligned} \boldsymbol{a}^{(i+1)} &= \int_{-\infty}^{\infty} \phi^{(i+1)}(t)\, \Phi^{(i+1)}(t)\, dt \\ &= \int_{-\infty}^{\infty} \left[2\sum \boldsymbol{h}(k)\phi^{(i)}(2t-k)\right]\left[(\downarrow 2)2\boldsymbol{H}\Phi^{(i)}(2t)\right] dt. \end{aligned} \tag{7.43}$$

Bring the operator $(\downarrow 2)2\boldsymbol{H}$ outside the integral. Change variables in the kth term to $u = 2t - k$. That kth term of the integral becomes (with $du = 2\,dt$)

$$\int_{-\infty}^{\infty} \boldsymbol{h}(k)\phi^{(i)}(u)\Phi^{(i)}(u+k)du = \boldsymbol{h}(k)\boldsymbol{S}^{-k}\boldsymbol{a}^{(i)}. \tag{7.44}$$

The k-step shift $\boldsymbol{S}^{-k}$ allowed us to write $\Phi^{(i)}(u+k)$ as $\boldsymbol{S}^{-k}\Phi^{(i)}(u)$. Then the integration with respect to u produced $\boldsymbol{a}^{(i)}$. Now sum equation (7.44) on k to reach the matrix $\sum \boldsymbol{h}(k)\boldsymbol{S}^{-k}$, which is $\boldsymbol{H}^T$ as the Lemma requires:

$$\boldsymbol{a}^{(i+1)} = (\downarrow 2)2\boldsymbol{H}\boldsymbol{H}^T\boldsymbol{a}^{(i)} = \boldsymbol{T}\boldsymbol{a}^{(i)}. \tag{7.45}$$

Corollary *The inner products $\boldsymbol{a}(k)$ of the scaling function $\phi(t)$ with $\phi(t+k)$ are in the eigenvector of $\boldsymbol{T}$ corresponding to $\lambda = 1$:*

$$\boldsymbol{a} = \boldsymbol{T}\boldsymbol{a}. \tag{7.46}$$

This assumes that the scaling function exists in L^2, which we will prove. To reach $\boldsymbol{a} = \boldsymbol{T}\boldsymbol{a}$, repeat the calculation above without the superscripts.

The power method $\boldsymbol{a}^{(i)} = \boldsymbol{T}^i\boldsymbol{a}^{(0)}$ converges when $\boldsymbol{T}$ has a non-repeated eigenvalue $\lambda = 1$ and all other eigenvalues have $|\lambda| < 1$. This "Condition E" gives $\boldsymbol{a}^{(i)} \to \boldsymbol{a}$. (Note! All vectors are normalized by $\sum \boldsymbol{a}(k) = 1$.) Convergence for the functions $\phi^{(i)}(t)$ is still to prove, but convergence for their inner products $\boldsymbol{a}^{(i)}$ is easier — just linear algebra.

Theorem 7.7 *The infinite matrix $\boldsymbol{T} = (\downarrow 2)2\boldsymbol{H}\boldsymbol{H}^T$ and its submatrix $\boldsymbol{T}_{2N-1}$ always have $\lambda = 1$ as an eigenvalue. The power iteration $\boldsymbol{a}^{(i+1)} = \boldsymbol{T}\boldsymbol{a}^{(i)}$ converges to the eigenvector $\boldsymbol{a} = \boldsymbol{T}\boldsymbol{a}$ if and only if $\boldsymbol{T}_{2N-1}$ satisfies*

Condition E: *$\boldsymbol{T}_{2N-1}$ has all $|\lambda| < 1$ except for a simple eigenvalue $\lambda = 1$.*

Proof. Suppose the starting $\boldsymbol{a}^{(0)}$ is expanded as a combination $\boldsymbol{a} + c_2\boldsymbol{v}_2 + c_3\boldsymbol{v}_3 + \cdots$ of eigenvectors of $\boldsymbol{T}_{2N-1}$. Every time we multiply by that matrix, each $\boldsymbol{v}_j$ is multiplied by the corresponding λ_j. Since $|\lambda_j| < 1$ by Condition E, those components get smaller. In the limit as $i \to \infty$, the powers $\boldsymbol{T}^i\boldsymbol{a}^{(0)}$ converge to the eigenvector $\boldsymbol{a}$ — whose coefficient stays at 1 (because it has $\lambda = 1$).

This proof only works if $\boldsymbol{T}_{2N-1}$ has a full set of eigenvectors, to expand $\boldsymbol{a}^{(0)}$. To cover all cases we use the Jordan form of $\boldsymbol{T}_{2N-1}$. It has $\lambda = 1$ alone in a 1×1 block. All other blocks have $|\lambda| < 1$ and their powers approach zero.

Convergence of the Cascade Algorithm in L^2

The convergence proof will be easy if we know that the dilation equation $\phi(t) = \sum 2\boldsymbol{h}(k)\phi(2t-k)$ has a finite energy solution. We now prove that $\phi^{(i)}(t)$ converges to this scaling function $\phi(t)$. Properly speaking, we must also show the existence of $\phi(t)$ itself. This existence step is logically first but it will come later for simplicity.

Theorem 7.8 *Assume that $\phi(t)$ is in L^2. The cascade sequence $\phi^{(i)}(t)$ converges to $\phi(t)$ if and only if $\boldsymbol{T}$ satisfies Condition* E. *Then*

$$||\phi^{(i)} - \phi||^2 = ||\phi^{(i)}||^2 - 2 < \phi^{(i)}, \phi > + ||\phi||^2 = \boldsymbol{a}^{(i)}(0) - 2\boldsymbol{b}^{(i)}(0) + \boldsymbol{a}(0) \tag{7.47}$$

$$\textbf{\textit{converges to}} \quad \boldsymbol{a}(0) - 2\boldsymbol{a}(0) + \boldsymbol{a}(0) = 0.$$

Proof. The numbers $\boldsymbol{a}^{(i)}(0)$ and $\boldsymbol{a}(0)$ are the energies $||\phi^{(i)}(t)||^2$ and $||\phi(t)||^2$:

$$\int \phi^{(i)}(t)\phi^{(i)}(t+0)\,dt = \boldsymbol{a}^{(i)}(0) \quad \text{and} \quad \int \phi(t)\phi(t+0)\,dt = \boldsymbol{a}(0).$$

We know already that $\boldsymbol{a}^{(i)}$ converges to $\boldsymbol{a}$. This was the preceding theorem. Equation (7.47) also contains the inner product of $\phi^{(i)}(t)$ with $\phi(t)$. This is the zeroth component $\boldsymbol{b}^{(i)}(0)$ of a new vector of inner products $\boldsymbol{b}^{(i)}(k) = \int \phi^{(i)}(t)\phi(t+k)\,dt$.

Our main calculation found each vector $\boldsymbol{a}^{(i+1)}$ from the previous $\boldsymbol{a}^{(i)}$. The rule was to multiply by $\boldsymbol{T}$. Condition E gave convergence to $\boldsymbol{a}$. In the same way, we now show that $\boldsymbol{b}^{(i+1)} = \boldsymbol{T}\boldsymbol{b}^{(i)}$. Then the vectors $\boldsymbol{b}^{(i)}$ converge (this is the power method again) to the same eigenvector $\boldsymbol{a}$. Therefore $-2\boldsymbol{b}^{(i)}(0)$ in (7.47) converges to $-2\boldsymbol{a}(0)$, which completes the proof that $||\phi^{(i)} - \phi||^2$ converges to zero.

For the new calculation $\boldsymbol{b}^{(i+1)} = \boldsymbol{T}\boldsymbol{b}^{(i)}$, it is again convenient to work with vectors:

$$\boldsymbol{b}^{(i+1)} = \int_{-\infty}^{\infty} \phi^{(i+1)}(t) \begin{bmatrix} \cdot \\ \phi(t-1) \\ \phi(t) \\ \phi(t+1) \\ \cdot \end{bmatrix} dt = \int_{-\infty}^{\infty} \phi^{(i+1)}(t)\Phi(t)\,dt.$$

Substitute the cascade formula for $\phi^{(i+1)}(t)$ and the dilation equation for $\Phi(t)$:

$$\boldsymbol{b}^{(i+1)} = \int_{-\infty}^{\infty} \Big[2\sum_k \boldsymbol{h}(k)\phi^{(i)}(2t-k)\Big]\,[(\downarrow 2)2\boldsymbol{H}\Phi(2t)]\,dt. \tag{7.48}$$

This matches equation (7.43) when $\Phi^{(i)}$ is replaced by Φ. Change variables in the kth term to $u = 2t - k$, and that term matches (7.44)—with $\boldsymbol{a}$ replaced by $\boldsymbol{b}$. Then sum on k to match equation (7.45). These same steps give the new answer, with $\boldsymbol{a}$ changed to $\boldsymbol{b}$:

$$\boldsymbol{b}^{(i+1)} = (\downarrow 2)2\boldsymbol{H}\boldsymbol{H}^T\boldsymbol{b}^{(i)} = \boldsymbol{T}\boldsymbol{b}^{(i)}. \tag{7.49}$$

The normalization $\sum \boldsymbol{b}^{(0)}(k) = 1$ is still true, because $\int \phi^{(i)}(t)\sum \phi(t+k)\,dt = \int \phi^{(i)}\,dt = 1$. So the power method starting from $\boldsymbol{b}^{(0)}$ converges to the same vector $\boldsymbol{a}$ (not $\boldsymbol{b}$!). By equation (7.47) the sequence $\phi^{(i)}(t)$ converges to $\phi(t)$.

The converse is also straightforward [Strang3]. Convergence of the functions $\phi^{(i)}$ to ϕ implies convergence of their inner products $\boldsymbol{a}^{(i)}$ to $\boldsymbol{a}$. Thus the power method always goes to the same limiting vector $\boldsymbol{a}$. Condition E must hold.

Existence of the Scaling Function

This is not a book about proofs. But the dilation equation is so fundamental that we must be sure it has a solution. Formally the infinite product $\prod H(\omega/2^j)$ yields the Fourier transform $\widehat{\phi}(\omega)$. The real question is when this product converges strongly to a finite energy solution, and the answer to that question is Condition E.

We base existence of $\phi(t)$ on our calculation of inner products, which is the key to this section. The space L^2 is *complete*, so when we prove that the energies $\left\|\phi^{(m)} - \phi^{(n)}\right\|^2$ approach zero, there is guaranteed to exist a limit function $\phi(t)$ in L^2.

Theorem 7.9 *If Condition E holds then* $\left\|\phi^{(m)} - \phi^{(n)}\right\|^2$ *approaches zero as* $m, n \to \infty$. *Therefore the sequence* $\phi^{(0)}(t), \phi^{(1)}(t), \ldots$ *converges to a limit* $\phi(t)$ *in* L^2.

Proof. The energy $\left\|\phi^{(m)} - \phi^{(n)}\right\|^2$ is $\left\|\phi^{(m)}\right\|^2 - 2\langle\phi^{(m)}, \phi^{(n)}\rangle + \left\|\phi^{(n)}\right\|^2$. The first and third terms are $\boldsymbol{a}^{(m)}(0)$ and $\boldsymbol{a}^{(n)}(0)$, both approaching the limit $\boldsymbol{a}(0)$. We must show that the inner product $\langle\phi^{(m)}, \phi^{(n)}\rangle$ also approaches $\boldsymbol{a}(0)$.

Suppose $m = i + n$ with fixed $i > 0$. The same inner product calculation, following the pattern (7.43)–(7.44)–(7.45) and multiplying by $\boldsymbol{T}$ at each step, yields

$$\langle\phi^{(m)}, \phi^{(n)}\rangle = \boldsymbol{T}^n \langle\phi^{(i)}, \phi^{(0)}\rangle = \boldsymbol{T}^n \boldsymbol{c}^{(i)}(0). \tag{7.50}$$

The vector $\boldsymbol{c}^{(i)}$ contains the inner products $\boldsymbol{c}^{(i)}(k) = \int \phi^{(i)}(t)\phi^{(0)}(t+k)\,dt$. For each fixed i, the power method $\boldsymbol{T}^n\boldsymbol{c}^{(i)}$ approaches $\boldsymbol{a}$. The limit of (7.50) is $\boldsymbol{a}(0)$ as desired. But the difficulty (since $i = m - n$ is arbitrary) is that this convergence must hold *uniformly for all starting vectors* $\boldsymbol{c}^{(i)}$. We know two facts about the components $\boldsymbol{c}^{(i)}(k)$. They are uniformly bounded and they add to 1:

$$|\boldsymbol{c}^{(i)}(k)| \le \left\|\phi^{(i)}\right\| \left\|\phi^{(0)}\right\| \le \left\|\boldsymbol{a}^{(i)}\right\| \le C \tag{7.51}$$

$$\sum_k \boldsymbol{c}^{(i)}(k) = \int_{-\infty}^{\infty} \phi^{(i)}(t) \sum \phi^{(0)}(t+k)\,dt = \int_{-\infty}^{\infty} \phi^{(i)}(t)\,dt = 1. \tag{7.52}$$

Condition E and the Jordan form $\boldsymbol{J}$ of $\boldsymbol{T}$ give uniform convergence $\boldsymbol{T}^n\boldsymbol{c} \to \boldsymbol{a}$:

$$\boldsymbol{T}^n\boldsymbol{c} = \boldsymbol{S}\boldsymbol{J}^n\boldsymbol{S}^{-1}\boldsymbol{c} = \begin{bmatrix} \boldsymbol{a} & -- \end{bmatrix} \begin{bmatrix} 1 & \\ & \boldsymbol{B}^n \end{bmatrix} \begin{bmatrix} \boldsymbol{e} \\ --- \end{bmatrix} \begin{bmatrix} \boldsymbol{c} \end{bmatrix}. \tag{7.53}$$

The left eigenvector $\boldsymbol{e} = [1 \;\; 1 \;\; \ldots \;\; 1]$ is in row 1, and $\boldsymbol{a}$ in column 1 is the right eigenvector. These eigenvector matrices $\boldsymbol{S}$ and $\boldsymbol{S}^{-1}$ are fixed, and the block $\boldsymbol{B}$ has eigenvalues $|\lambda| < 1$. Therefore $\boldsymbol{B}^n \to \boldsymbol{0}$ and we have uniform convergence to $\boldsymbol{aec} = \boldsymbol{a}\sum c(k) = \boldsymbol{a}$. Equation (7.50) approaches $\boldsymbol{a}(0)$ as m and n get large, completing the proof that $||\phi^{(m)} - \phi^{(n)}||^2 \to 0$.

Remark 1 The convergence of the cascade algorithm can be interpreted in the frequency domain, which is illuminating. The convergence of $\phi^{(i)}(t)$ to $\phi(t)$ becomes convergence of the infinite product to $\widehat{\phi}(\omega)$:

$$\widehat{\phi}^{(i)}(\omega) = \Big[\prod_{j=1}^{i} H(\omega/2^j)\Big]\widehat{\phi}^{(0)}(\omega/2^i) \;\; \text{converges in } L^2 \text{ to } \;\; \widehat{\phi}(\omega) = \prod_{j=1}^{\infty} H(\omega/2^j).$$

Remark 2 For filters with double-shift orthogonality, there is no danger that $\boldsymbol{T}$ has an eigenvalue with $|\lambda| > 1$. The norm of $\boldsymbol{T}$ is $sup(|H(\omega)|^2 + |H(\omega+\pi)|^2)$ and this is 1. Condition E reduces to the Cohen-Lawton condition that $\lambda = 1$ is a simple eigenvalue of $\boldsymbol{T}$. Then $\{\phi(t+k)\}$ is an *orthonormal* basis and $\boldsymbol{a} = \boldsymbol{\delta}$.

Condition E holds in this orthogonal case if $H(\omega) \neq 0$ for $|\omega| \le \frac{\pi}{3}$ [Mallat1, JiaWang].

Remark 3 The same method (7.43)-(7.44)-(7.45) that gives inner products of scaling functions also gives inner products with wavelets. The highpass operator $\boldsymbol{H}_1$ replaces the lowpass $\boldsymbol{H}$ in the appropriate places:

$$\langle\phi(t), w(t+k)\rangle = (\downarrow 2)2\boldsymbol{H}\boldsymbol{H}_1^T\boldsymbol{a} \tag{7.54}$$

$$\langle w(t), w(t+k)\rangle = (\downarrow 2)2\boldsymbol{H}_1\boldsymbol{H}_1^T\boldsymbol{a} \tag{7.55}$$

These are derived in Section 11.6. *We find inner product formulas for any functions that satisfy two-scale equations.*

Remark 4 For "multifilters" the coefficients $\boldsymbol{h}(k)$ are $r \times r$ matrices. The dilation equation determines a *vector* of r scaling functions. The inner product $\int \boldsymbol{\phi}(t)\boldsymbol{\phi}^T(t+k)\,dt$ is an $r \times r$ matrix $\boldsymbol{a}(k)$. The equation $\boldsymbol{Ta} = \boldsymbol{a}$ for a vector of matrices now involves a matrix convolution:

$$\boldsymbol{Ta} = (\downarrow 2)2\boldsymbol{h} * \boldsymbol{a} * \boldsymbol{h}^T. \tag{7.56}$$

In the frequency domain this *matrix transition operator* $\boldsymbol{T}$ becomes

$$\boldsymbol{TA}(\omega) = \boldsymbol{H}\left(\frac{\omega}{2}\right)\boldsymbol{A}\left(\frac{\omega}{2}\right)\boldsymbol{H}^*\left(\frac{\omega}{2}\right) + \boldsymbol{H}\left(\frac{\omega}{2}+\pi\right)\boldsymbol{A}\left(\frac{\omega}{2}+\pi\right)\boldsymbol{H}^*\left(\frac{\omega}{2}+\pi\right). \tag{7.57}$$

The theory develops on the same lines [Cohen-Daubechies-Plonka] to give the existence in L^2, the smoothness, the approximation properties, and the stability of the basis $\{\phi_i(t+k)\}$. The eigenvalues of $\boldsymbol{T}$ are still in control.

Problem Set 7.2

1. If $\boldsymbol{h}(k)$ has double-shift orthogonality (Condition O), show that the central column of $\boldsymbol{T}$ is $\boldsymbol{\delta}$. This is an eigenvector of $\boldsymbol{T}$ for $\lambda = 1$.
2. Construct the finite matrices $\boldsymbol{T}$ for $\boldsymbol{h} = \frac{1}{3}(1, 1, 1)$ and $\boldsymbol{h} = (c, \frac{1}{2}, \frac{1}{2}, -c)$.
3. Draw the output from one Haar cascade step $\phi^{(1)}(t) = \phi^{(0)}(2t) + \phi^{(0)}(2t-1)$ with
   (a) $\phi^{(0)}(t) =$ unit box on the interval $[1, 2]$
   (b) $\phi^{(0)}(t) =$ hat function on the interval $[0, 2]$
   (c) $\phi^{(0)}(t) =$ hat function on the interval $[0, 1]$.

   Two of those $\phi^{(0)}(t)$ lead to convergence. Which one doesn't?
4. Prove that $P^{(1)}(t) = \sum \phi^{(1)}(t-n)$ equals $P^{(0)}(2t) = \sum \phi^{(0)}(2t-n)$ for any starting function $\phi^{(0)}(t)$ and any coefficients $\boldsymbol{h}(k)$ with a zero at π: even sum = odd sum = $\frac{1}{2}$.
   It follows that $P^{(i)}(t) = P^{(0)}(2^i t)$ will oscillate faster and faster. *There is no convergence unless $P^{(0)}(t) = \sum \phi^{(0)}(t-n)$ is identically one.* This is the condition on $\phi^{(0)}(t)$ in the cascade algorithm.
5. Find the eigenvalues of $\boldsymbol{T}$ in Problem 2 (depending on c).
6. Show that the filter $\boldsymbol{h} = [-1\ \ 3\ \ 3\ \ -1]/4$ does *not* satisfy Condition E, and $\boldsymbol{T}$ has $\lambda = 2.1712$. This bad $\boldsymbol{h}$ is dual to the good spline filter $\boldsymbol{f} = [1\ \ 3\ \ 3\ \ 1]/8$; their product $\boldsymbol{h} * \boldsymbol{f}$ is Daubechies maxflat halfband.

7.3 Smoothness of Scaling Functions and Wavelets

The previous section established Condition **E** for the convergence of the cascade algorithm. The eigenvalues of $\boldsymbol{T}$ are required to be less than 1 (except the simple eigenvalue $\lambda = 1$). The limit function $\phi(t)$ is then in L^2—which assures some minimal smoothness, but not much. If the eigenvalues of $\boldsymbol{T}$ are less than 4^{-s}, apart from the special eigenvalues that are powers of $\frac{1}{2}$, *we now show that $\phi(t)$ and $w(t)$ have s derivatives.*

This conclusion is true whether s is an integer or not. The proof for integers s is very direct, so we include it. The proof for noninteger s needs more space and effort, so we refer for example to [Villemoes]. These are derivatives in the L^2 sense because we work with the matrix $\boldsymbol{T}$. In the frequency domain, each derivative is a multiplication by $i\omega$. Therefore $\phi(t)$ in L^2 has s derivatives in L^2 when

$$||\phi^{(s)}(t)||^2 = \frac{1}{2\pi}\int_{-\infty}^{\infty} |\omega|^{2s}|\widehat{\phi}(\omega)|^2 d\omega \text{ is finite.}$$

This definition allows s to be a fraction (or negative) with no difficulty. Since $w(t)$ is a combination of $\phi(2t-k)$, we only need to study the smoothness of $\phi(t)$.

The basic idea is simple. *Each new factor* $\left(\frac{1+z^{-1}}{2}\right)$ *in $H(z)$ has four effects*:

1. All eigenvalues of $\boldsymbol{T}$ are divided by 4.
2. The old $\phi(t)$ is convolved with the box function.
3. The old $\widehat{\phi}(\omega)$ is multiplied by $\left(1-e^{-i\omega}\right)/i\omega$.
4. The new $\phi(t)$ has *one more derivative* than the old $\phi(t)$.

When we check these facts, our desired result is proved. You might think that the final $\phi(t)$ has the full p derivatives, because $H(z)$ has p factors of $\frac{1+z^{-1}}{2}$. But some of those factors are needed to get the eigenvalues of $\boldsymbol{T}$ below 1 (always excluding the special eigenvalues $1, \frac{1}{2}, \ldots$). If s (integer) is less than the number $s_{\max}$ below, there will be s factors still left after this Condition **E** is met. Then $\phi(t)$ has s derivatives. The non-special eigenvalues are below 4^{-s}.

The word "regularity" has been applied to s and also to p. *Those are different numbers* (we will prove $s < p$). So we avoid that word, and refer to smoothness s and accuracy p. We state the conclusions for any s; our proof was for s = integer.

Theorem 7.10 *Each new factor* $\frac{1+z^{-1}}{2}$ *has the effects* **1, 2, 3, 4**. *Then $\phi(t)$ and $w(t)$ have s derivatives in L^2 when the non-special eigenvalues of $\boldsymbol{T}$ have $|\lambda| < 4^{-s}$. The supremum $s_{\max}$, when $\phi(t)$ comes from $H(z) = \left(\frac{1+z^{-1}}{2}\right)^p Q(z)$, is*

$$s_{\max} = p - \log_4\left|\lambda_{\max}\left(\boldsymbol{T}_Q\right)\right| \quad \textit{with} \quad \boldsymbol{T}_Q = (\downarrow 2)2QQ^T. \tag{7.58}$$

A short MATLAB code will construct $\boldsymbol{T}$ or $\boldsymbol{T}_Q$ from $\boldsymbol{h}$ and find its eigenvalues. In practice, we exclude the special $\lambda = 1, \ldots, (\frac{1}{2})^{2p-1}$ directly from the eigenvalues of $\boldsymbol{T}$, and then $s_{\max} = -\log_4(|\lambda_{\max}(\boldsymbol{T})|)$.

Actually **1, 2, 3** are already proved. The effect on the eigenvalues of $\boldsymbol{T}$ was established in Section 7.2. Each eigenvalue is divided by 2 twice, because $\boldsymbol{T}$ comes from $H(z)H(z^{-1})$. So all eigenvalues of $\boldsymbol{T}_{\text{old}}$ are divided by 4. Then $\boldsymbol{T}_{\text{new}}$ also has the special eigenvalues 1 and $\frac{1}{2}$. The extra $\frac{1+z^{-1}}{2}$ in $H(z)$ changes the infinite product to $\widehat{\phi}_{\text{new}}(\omega)$:

$$\widehat{\phi}_{\text{new}}(\omega) = \prod_{j=1}^{\infty}\left(\tfrac{1}{2}+\tfrac{1}{2}e^{-i\omega/2^j}\right)\widehat{\phi}_{\text{old}}(\omega) = \left(\frac{1-e^{-i\omega}}{i\omega}\right)\widehat{\phi}_{\text{old}}(\omega). \tag{7.59}$$

The extra "Haar factors" multiply to give the "box factor." This is the infinite product in Section 6.4 that yields the box function $B(t)$. In the time domain $\phi_{\text{new}}(t)$ is the box convolved with $\phi_{\text{old}}(t)$. This adds one more derivative for ϕ_{new}.

Lemma 7.1 *The convolution $\phi_{\text{new}}(t) = B(t) * \phi_{\text{old}}(t)$ has $s+1$ derivatives if and only if $\phi_{\text{old}}(t)$ has s derivatives.*

Proof. In frequency we are multiplying $\widehat{\phi}_{\text{old}}(\omega)$ by $\left(1 - e^{-i\omega}\right)/i\omega$. This has magnitude at most $2/|\omega|$. Therefore $\widehat{\phi}_{\text{new}}$ decreases at least one order faster than $\widehat{\phi}_{\text{old}}$:

$$\int_{-\infty}^{\infty} |\omega|^{2(s+1)} \left|\widehat{\phi}_{\text{new}}(\omega)\right|^2 d\omega \le 4 \int_{-\infty}^{\infty} |\omega|^{2s} \left|\widehat{\phi}_{\text{old}}(\omega)\right|^2 d\omega.$$

The factor 4 comes from $(2/|\omega|)^2$ times $|\omega|^2$. The last integral is finite when ϕ_{old} has s derivatives. So the first integral is finite and ϕ_{new} has $s+1$ derivatives. In the time domain, the derivative of the convolution $\phi_{\text{new}}(t) = B(t) * \phi_{\text{old}}(t)$ is a *difference*:

$$\phi'_{\text{new}}(t) = \frac{d}{dt}\int_0^1 \phi_{\text{old}}(t-s)\,ds = \int_0^1 \phi'_{\text{old}}(t-s)\,ds = \phi_{\text{old}}(t) - \phi_{\text{old}}(t-1). \tag{7.60}$$

Again ϕ_{new} has one more derivative than ϕ_{old}. This is true in the pointwise sense as well as the L^2 sense. The smoothness increases by one from the extra zero at π in the filter.

For completeness we prove the converse, following [Villemoes]. If ϕ_{new} has $s+1$ derivatives then certainly ϕ'_{new} has s derivatives. From (7.60) this means that $\phi_{\text{old}}(t) - \phi_{\text{old}}(t-1)$ has s derivatives. Use this fact N times:

$$\phi_{\text{old}}(t) - \phi_{\text{old}}(t-N) = [\phi_{\text{old}}(t) - \phi_{\text{old}}(t-1)] + [\phi_{\text{old}}(t-1) - \phi_{\text{old}}(t-2)] + \cdots$$

Each difference on the right has s derivatives. So does the difference on the left. But $\phi(t)$ does not overlap $\phi(t-N)$, so $\phi_{\text{old}}(t)$ by itself must have s derivatives.

Example 7.4. For any $s < \frac{1}{2}$, the box function has s derivatives in L^2.

Reason: $\left|(1-e^{-i\omega})/i\omega\right|$ is below and often near $\frac{2}{|\omega|}$. The integral $\int |\omega|^{2s} \left|\frac{2}{\omega}\right|^2 d\omega$ is finite for $s < \frac{1}{2}$ but infinite for $s = \frac{1}{2}$. Therefore the value $s_{\max} = \frac{1}{2}$ is not actually achieved, which is typical. Normally $\phi(t)$ has s derivatives for all $s < s_{\max}$.

We can check formula (7.58). The box filter $H(z) = \frac{1+z^{-1}}{2}$ has $Q(z) \equiv 1$. Then $\boldsymbol{Q} = \boldsymbol{I}$ and $\boldsymbol{T}_Q = (\downarrow 2)2\boldsymbol{I}$ has $\lambda_{\max} = 2$. The logarithm of 2 to base 4 is $\frac{1}{2}$. Formula (7.58) correctly gives $s_{\max} = 1 - \frac{1}{2} = \frac{1}{2}$.

The splines of degree $p-1$, which come from $p-1$ additional convolutions of the box, have $s_{\max} = p - \frac{1}{2}$. *This is the largest possible* $s_{\max}$! With p zeros at π, $\phi(t)$ cannot have more than $p - \frac{1}{2}$ derivatives in L^2. We will stay with integers to prove that $s \le p-1$, after noting how the smoothness of splines drops by $\frac{1}{2}$ when we change from L^2 to pointwise.

Pointwise, the box function has zero smoothness. The hat function has one derivative (only one-sided, because the slope jumps). The spline of degree $p-1$ has $p-1$ one-sided derivatives. The general theory says that pointwise smoothness for every function is between $s_{\max} - \frac{1}{2}$ and $s_{\max}$ (this is a *Sobolev inequality*). For splines the pointwise smoothness is at the low end of that range, which is $p-1$.

Theorem 7.11 *If $\phi(t)$ has s derivatives in L^2* (integer s) *then $s < p$. Allowing fractions, $s_{\max}$ cannot exceed $p - \frac{1}{2}$. Pointwise, the smoothness cannot exceed $p-1$.*

Proof. For $\phi(t)$ in L^2 we need at least one zero at π ($H(\pi) = 0$ by Chapter 6). For $\phi'(t)$ in L^2 we need at least two zeros at π (use the Lemma). Continuing, s derivatives in L^2 (*integer s*) require at least $s+1$ zeros at π. Therefore $p \geq s+1$ and $s < p$.

These conclusions are consistent with the sth derivative of the dilation equation:

$$\phi^{(s)}(t) = 2^s \sum 2\boldsymbol{h}(k)\phi^{(s)}(2t-k). \tag{7.61}$$

The values $\phi^{(s)}(n)$ at the integers come from the eigenvalue problem $\Phi^{(s)} = 2^s \boldsymbol{M}\Phi^{(s)}$. We know that $\boldsymbol{M}$ has eigenvalue 2^{-s} provided $s < p$. For higher derivatives we cannot even find values at the integers.

We caution that $s < p$ only means that equation (7.61) *might* produce an sth derivative in L^2. It might not! Splines have the highest smoothness that p allows. Compare splines with other scaling functions when $p = 2$. The hat function has $s_{\max} = \frac{3}{2}$, the Daubechies function only has $s_{\max} = 1$. It has .55 derivatives in the pointwise sense [Daubechies-Lagarias]. Smoother wavelets have been constructed (orthogonal or biorthogonal). The neat fact is that the eigenvalues of $\boldsymbol{T}$ immediately give $s_{\max}$.

The Daubechies coefficients are $1+\sqrt{3}, 3+\sqrt{3}, 3-\sqrt{3}, 1-\sqrt{3}$ (divided by 8). The product $H(z)H(z^{-1})$ has coefficients $-1, 0, 9, 16, 9, 0, -1$ (divided by 16). This halfband property gives $(\ldots, 0, 1, 0, \ldots)$ in the center column of $\boldsymbol{T}$:

$$\boldsymbol{T}_5 = \frac{1}{16}\begin{bmatrix} 0 & -1 & 0 & 0 & 0 \\ 16 & 9 & 0 & -1 & 0 \\ 0 & 9 & 16 & 9 & 0 \\ 0 & -1 & 0 & 9 & 16 \\ 0 & 0 & 0 & -1 & 0 \end{bmatrix} \quad \text{has eigenvalues} \quad 1, \tfrac{1}{2}, \tfrac{1}{4}, \tfrac{1}{4}, \tfrac{1}{8}.$$

The eigenvector $\boldsymbol{a} = (0, 0, 1, 0, 0)$ says that all $\phi(t-n)$ are orthogonal to $\phi(t)$. The approximation order is $p = 2$, since $H(\omega)$ has a double zero at π and $(\frac{1}{2})^4$ is not an eigenvalue of $\boldsymbol{T}$. *The smoothness index is $s_{\max} = 2 - 1 = 1$ because the largest "other eigenvalue" is the repeated $\lambda = 4^{-1}$.* The function $\phi(t)$ has *almost* one derivative with finite energy. The integral of $|\omega|^2|\widehat{\phi}(\omega)|^2$ is not finite, but any smaller power of $|\omega|$ will make it finite.

The smoothness of D_6, with a triple zero at π, turns out to be $s_{\max} = 3 - \frac{\log 3}{\log 2}$. With accuracy p, the number of Daubechies coefficients is $2p$. Eirola computed the smoothness of all Daubechies functions up to D_{40} which has $p = 20$, and also found the asymptotic formula:

p	$s_{\max}$	p	$s_{\max}$	Asymptotically
1	0.5	6	2.388	$s_{\max} \approx 0.2075\, p + \text{constant}$
2	1.0	7	2.658	
3	1.415	8	2.914	
4	1.775	9	3.161	
5	2.096	10	3.402	

Cascade Algorithm in the Frequency Domain

When t is rescaled to $2t$, the Fourier transform rescales to $\frac{\omega}{2}$. The shift by k that usually produces $e^{-ik\omega}$ now produces $e^{-ik\omega/2}$. The cascade equation $\phi^{(i+1)}(t) = \sum 2\boldsymbol{h}(k)\phi^{(i)}(2t-k)$ transforms into

$$\widehat{\phi}^{(i+1)}(\omega) = \left(\sum \boldsymbol{h}(k)e^{-ik\omega/2}\right)\widehat{\phi}^{(i)}\left(\frac{\omega}{2}\right) = H\left(\frac{\omega}{2}\right)\widehat{\phi}^{(i)}\left(\frac{\omega}{2}\right). \tag{7.62}$$

The first cascade step multiplies by $H\left(\frac{\omega}{2}\right)$ and the next step by $H\left(\frac{\omega}{4}\right)$. Thus $\widehat{\phi}^{(2)}(\omega)$ is $H\left(\frac{\omega}{4}\right)H\left(\frac{\omega}{2}\right)\widehat{\phi}^{(0)}\left(\frac{\omega}{4}\right)$. The output at step i involves $H^{(i)}$ with i factors:

$$\widehat{\phi}^{(i)}(\omega) = \left[\prod_{j=1}^{i} H\left(\frac{\omega}{2^j}\right)\right]\widehat{\phi}^{(0)}\left(\frac{\omega}{2^i}\right) = H^{(i)}(\omega)\,\widehat{\phi}^{(0)}\left(\frac{\omega}{2^i}\right). \tag{7.63}$$

We expect that this i-term product approaches the infinite product

$$\widehat{\phi}(\omega) = \text{ limit of } \widehat{\phi}^{(i)}(\omega) = \prod_{j=1}^{\infty} H\left(\frac{\omega}{2^j}\right). \tag{7.64}$$

The question is: *When does this limit exist and how smooth is $\phi(t)$?* At each separate frequency ω, the limit exists. That requires only $H(0) = 1$ and a bound C on the derivative $|H'(\omega)|$. Then

$$\left|H\left(\frac{\omega}{2^j}\right)\right| \le 1 + C\frac{|\omega|}{2^j} \le e^{C|\omega|/2^j}.$$

If we take logarithms, to look at a sum instead of a product, that sum converges like $\sum C|\omega|/2^j$. The sum $\log|\widehat{\phi}(\omega)|$ is less than $C|\omega|$. Therefore the product $|\widehat{\phi}(\omega)|$ is less than $e^{C|\omega|}$.

Such a bound is useless for large ω! For $\phi(t)$ to be a reasonable function, we need $\widehat{\phi}(\omega)$ to decay rather than grow as $|\omega| \to \infty$. Working with the energy $\int |\widehat{\phi}(\omega)|^2 d\omega$, there is no doubt that each iteration $\widehat{\phi}(\omega)$ retains finite energy. From step i to $i+1$ the energy grows by no more than

$$\left\|\widehat{\phi}^{(i+1)}\right\| \le \|T\|\,\left\|\widehat{\phi}^{(i)}\right\| \quad \text{where} \quad \|T\|^2 = \max\left(|H(\omega)|^2 + |H(\omega+\pi)|^2\right). \tag{7.65}$$

The orthonormal case has $\|T\| = 1$, by Condition O. The energy is the same at every iteration of the cascade algorithm. For the biorthogonal case we expect $\|T\| > 1$ and the bound in (7.65) also becomes useless. To find the new energy $\|\widehat{\phi}^{(i+1)}\|^2$ from $\|\widehat{\phi}^{(i)}\|^2$, we look again at the operator $\boldsymbol{T}$—which is absolutely fundamental to the theory of wavelets.

In the time domain, $\boldsymbol{T}$ is a double-shift Toeplitz matrix. That double shift corresponds to picking out *even frequencies* $0, 2\omega, 4\omega$. In the z-domain it corresponds to picking out *even powers* of z:

Definition 7.1 The transition operator $\boldsymbol{T}$ in the three domains is:

$$\begin{aligned}
\boldsymbol{Ta}(k) &= (\downarrow 2)2\boldsymbol{HH}^T\boldsymbol{a}(k) \\
TA(2\omega) &= |H(\omega)|^2A(\omega) + |H(\omega+\pi)|^2A(\omega+\pi) \\
&= \text{even frequencies in } 2|H(\omega)|^2A(\omega) \\
TA(z^2) &= H(z)A(z)H(z^{-1}) + H(-z)A(-z)H(-z^{-1}) \\
&= \text{even powers in } 2H(z)A(z)H(z^{-1}).
\end{aligned}$$

$A(\omega)$ is the function $\sum \boldsymbol{a}(k)e^{-ik\omega}$ and $A(z)$ is the function $\sum \boldsymbol{a}(k)z^{-k}$. In our application, the $\boldsymbol{a}(k)$ are inner products and $A(\omega)$ is taken from Section 6.4:

$$\boldsymbol{a}(k) = \int \phi(t)\phi(t+k)\,dt \quad \text{and} \quad A(\omega) = \sum \left|\widehat{\phi}(\omega + 2\pi\ell)\right|^2.$$

We give an example immediately, and ask for more in the problem set.

Example 7.5. The coefficients $\boldsymbol{h}(k) = \frac{1}{4}, \frac{1}{2}, \frac{1}{4}$ give

$$H(\omega) = \tfrac{1}{4}(1 + 2e^{-i\omega} + e^{-2i\omega}) \quad \text{and} \quad H(z) = (1 + 2z^{-1} + z^{-2}).$$

Suppose $\phi^{(0)}(t)$ is the box function, so its inner products are $\mathbf{a}^{(0)}(k) = (\ldots, 0, 1, 0, \ldots)$. The corresponding $A^{(0)}(\omega)$ is the constant function 1. The first step of the cascade algorithm produces $\phi^{(1)}(t)$ as *three half-boxes* with heights $\frac{1}{2}, 1, \frac{1}{2}$. The new energy $\|\phi^{(1)}\|^2$ is $\frac{1}{2}(\frac{1}{4} + 1 + \frac{1}{4}) = \frac{6}{8}$. This should agree with $\boldsymbol{a}^{(1)}(0)$ after the action of $\boldsymbol{T}$:

$$\text{Time:} \qquad \boldsymbol{T}\boldsymbol{a}^{(0)} = \frac{1}{8}\begin{bmatrix} 4 & 1 & 0 \\ 4 & 6 & 4 \\ 0 & 1 & 4 \end{bmatrix}\begin{bmatrix} 0 \\ 1 \\ 0 \end{bmatrix} = \frac{1}{8}\begin{bmatrix} 1 \\ 6 \\ 1 \end{bmatrix}$$

$$\begin{aligned} \text{Frequency:} \quad A^{(1)}(2\omega) &= \text{even frequencies in } 2|H(\omega)|^2 A^{(0)}(\omega) \\ &= \text{even frequencies in } \tfrac{1}{8}\left(e^{2i\omega} + 4e^{i\omega} + 6 + 4e^{-i\omega} + e^{-2i\omega}\right) \\ A^{(1)}(\omega) &= \tfrac{1}{8}\left(e^{i\omega} + 6 + e^{-i\omega}\right) \end{aligned}$$

$$\begin{aligned} z\text{-domain:} \quad A^{(1)}(z^2) &= \text{even powers in } 2H(z)A^{(0)}(z)H(z^{-1}) \\ &= \text{even powers in } \tfrac{1}{8}(z^2 + 4z + 6 + 4z^{-1} + z^{-2}) \\ A^{(1)}(z) &= \tfrac{1}{8}(z + 6 + z^{-1}). \end{aligned}$$

Summary: The main point is to connect L^2 convergence of the infinite product for $\widehat{\phi}(\omega)$ with the number p of vanishing moments of the wavelets, and the smoothness s of $\phi(t)$. Everything depends on the eigenvalues of the matrix $\boldsymbol{T}_{2N-1}$:

Cascade convergence requires all $|\lambda| < 1$ except for a simple $\lambda = 1$.

Approximation of order p requires eigenvalues $\lambda = 1, \frac{1}{2}, \frac{1}{4}, \ldots, (\frac{1}{2})^{2p-1}$.

Smoothness (s derivatives in L^2) requires all other $|\lambda| < 4^{-s}$.

Continuity of the Scaling Function

Functions in L^2 have finite energy. They may or may not be continuous. Continuity is a "pointwise" property, not revealed by inner products and not automatic under Condition **E**. (Haar satisfies this condition.) We describe now the test for continuity of $\phi(t)$. In borderline cases it is not always easy to apply, because it involves *two matrices*.

Those matrices are $\boldsymbol{m}(0)$ and $\boldsymbol{m}(1)$. Remember from Section 6.3 that these are $N \times N$ submatrices of the double-shift lowpass matrix $\boldsymbol{M} = (\downarrow 2)2\boldsymbol{H}$. The columns of $\boldsymbol{m}(0)$ and $\boldsymbol{m}(1)$ add to 1. If $\boldsymbol{e} = [1 \ \ldots \ 1]$ is the all-ones row vector then $\boldsymbol{em}(0) = \boldsymbol{e}$ and $\boldsymbol{em}(1) = \boldsymbol{e}$. The dilation equation in vector form is on the interval $[0, 1)$:

$$\boldsymbol{\Phi}(t) = \boldsymbol{m}(0)\ \boldsymbol{\Phi}(2t) + \boldsymbol{m}(1)\ \boldsymbol{\Phi}(2t-1). \tag{7.66}$$

The vector $\boldsymbol{\Phi}(t) = [\phi(t)\ \phi(t+1)\ \cdots]^T$ stacks the N slices of $\phi(t)$. Because equation (7.66) has only two coefficients, it gives a simple recursion (Section 6.3). The first digit t_1 in $t = .t_1t_2t_3\cdots$ tells whether we use $\boldsymbol{m}(0)$ or $\boldsymbol{m}(1)$:

$$\boldsymbol{\Phi}(t) = \boldsymbol{m}(t_1)\ \boldsymbol{\Phi}(.t_2t_3\cdots). \tag{7.67}$$

The next 0–1 digit t_2 tells whether the next step uses $\boldsymbol{m}(0)$ or $\boldsymbol{m}(1)$:

$$\boldsymbol{\Phi}(t) = \boldsymbol{m}(t_1)\boldsymbol{m}(t_2)\ \boldsymbol{\Phi}(.t_3t_4\cdots). \tag{7.68}$$

The matrices $\boldsymbol{m}(0)$ and $\boldsymbol{m}(1)$ can come in any order (determined by the digits in t). A nearby point T will *begin* with the same digits. At some later point the digits will differ. If $T = .t_1t_2T_3T_4\cdots$ then

$$\boldsymbol{\Phi}(t) - \boldsymbol{\Phi}(T) = \boldsymbol{m}(t_1)\boldsymbol{m}(t_2)[\ \boldsymbol{\Phi}(.t_3t_4\cdots) - \boldsymbol{\Phi}(.T_3T_4\cdots)]. \tag{7.69}$$

To prove continuity is to show that $\boldsymbol{\Phi}(t)$ is close to $\boldsymbol{\Phi}(T)$ when the neighbors t and T share many digits $t_1t_2\cdots t_K$. This will be true if the product of $\boldsymbol{m}$'s in every order is small. Actually we work with matrices of order $N-1$, after removing $\lambda = 1$.

Theorem 7.12 *The scaling function $\phi(t)$ is continuous if all products of $\boldsymbol{m}_{N-1}(0)$ and $\boldsymbol{m}_{N-1}(1)$ approach zero as the number of factors increases.*

The matrices $\boldsymbol{m}(0)$ and $\boldsymbol{m}(1)$ have the eigenvalue 1, with the all-ones left eigenvector $\boldsymbol{e}$. All products of the $\boldsymbol{m}$'s will have this eigenvalue and eigenvector. They won't go to zero! But in equation (7.69) these products multiply a vector that is *orthogonal* to $\boldsymbol{e}$:

$$\boldsymbol{e}\ \boldsymbol{\Phi}(t) \equiv 1 \text{ implies that } \boldsymbol{e}[\ \boldsymbol{\Phi}(.t_3t_4\cdots) - \boldsymbol{\Phi}(.T_3T_4\cdots)] = 1 - 1 = 0. \tag{7.70}$$

Restricted to vectors orthogonal to $\boldsymbol{e}$, $\boldsymbol{m}(0)$ becomes $\boldsymbol{m}_{N-1}(0)$ and $\boldsymbol{m}(1)$ becomes $\boldsymbol{m}_{N-1}(1)$. If long products of these matrices are small, then (7.69) says that $\Phi(t)$ is close to $\Phi(T)$. This means that $\Phi(t)$ is continuous.

Continuity requires longer and longer products of $A = \boldsymbol{m}_{N-1}(0)$ and $B = \boldsymbol{m}_{N-1}(1)$ to approach zero. This may be easy to test, or hard. Any eigenvalue with $|\lambda| \geq 1$ guarantees failure. To have $\|A\| < 1$ and $\|B\| < 1$ guarantees success. But the right norm can be very difficult to find. The problem is the possibility of small eigenvalues and dangerous products:

$$A = \begin{bmatrix} \epsilon & 1 \\ 0 & 0 \end{bmatrix} \quad \text{and} \quad B = \begin{bmatrix} \epsilon & 0 \\ 1 & 0 \end{bmatrix} \quad \text{give} \quad AB = \begin{bmatrix} 1+\epsilon^2 & 0 \\ 0 & 0 \end{bmatrix}.$$

The powers A^n and B^n go to zero. But $ABABAB\cdots$ blows up because $1+\epsilon^2 > 1$.

Problem 2 shows how to compute $\boldsymbol{m}_{N-1}$ from $\boldsymbol{m}_N$. We cannot show how to test all products in all orders. [Heil-Colella] and many others discuss this problem but it has no complete solution.

Smoothness of Binary Filters

A striking example of the difference between pointwise smoothness and derivatives in L^2 is the filter $\boldsymbol{h} = [-1\ 2\ 6\ 2\ -1]/8$. Its scaling function is infinite at all dyadic points. Pointwise, $\phi(t)$ is a failure. The matrix $\boldsymbol{M} = (\downarrow 2)2\boldsymbol{H}$ has a double eigenvalue at $\lambda = 1$, with only one eigenvector. The powers of $\boldsymbol{M}$ are therefore unbounded. But the eigenvalues of $\boldsymbol{T}$—its MATLAB construction in Section 6.5 is followed by $\boldsymbol{eig(T)}$—are $\lambda = 1, \frac{1}{2}, \frac{1}{4}, \frac{1}{8}, 0.5428, \ldots$. The smoothness of $\phi(t)$ is $s_{\max} = -\log(.5428)/\log 4 = .44$. Thus $\phi(t)$ and $w(t)$ have 0.44 derivatives with finite energy, in spite of the fact that $\phi(t)$ blows up on a dense set!

The maxflat Daubechies filters of length $4p-1$ have $2p$ zeros at π. Their scaling functions interpolate at the integers because D_{2p} is halfband: $\phi(n) = \delta(n)$ is the eigenvector of $\boldsymbol{M}$ for $\lambda = 1$. We expect high smoothness for $\phi(t)$ because of all the zeros:

$$2p = 0, 2, 4, 6 \quad \text{leads to} \quad s_{\max} = -0.5, 1.5, 2.44, 3.17.$$

We were also interested in the smoothness of the new binary filters (the *binlets*, dual to these symmetric Daubechies halfband filters). A new 9/7 pair was constructed by lifting in Section 6.5. There we also balanced its zeros to 10/6 by moving $1+z^{-1}$ between analysis and synthesis. (This just changes $s_{\max}$ by 1.) The new filters are good for text plus image:

$$\boldsymbol{h9} = [1 \;\; 0 \;\; -8 \;\; 16 \;\; 46 \;\; 16 \;\; -8 \;\; 0 \;\; 1]/64 \quad \text{with 2 zeros and} \quad s_{\max} = 0.59$$

$$\boldsymbol{h13} = [-1 \;\; 0 \;\; 18 \;\; -16 \;\; -63 \;\; 144 \;\; 348 \;\; \cdots]/512 \quad \text{with 4 zeros and} \quad s_{\max} = 1.18.$$

Compare with the standard symmetric biorthogonal FBI 9/7 pair, which has 16 nonzero coefficients. Two digits are inadequate but here they are:

$$\boldsymbol{h}\mathbf{FBI} = [0.03 \;\; -0.02 \;\; -0.08 \;\; 0.27 \;\; 0.60 \;\; \cdots] \text{ and } \boldsymbol{f}\mathbf{FBI} = [-0.05 \;\; -0.03 \;\; 0.30 \;\; 0.56 \;\; \cdots].$$

These have $s_{\max} = 1.4$ in analysis and 2.1 in synthesis. The FBI pair has higher coding gain and 4/4 zeros but no interpolating property. It gives higher PSNR and lower error on Lena and Barbara. But the perceptual quality of the new pair seems sharper (to our eyes). The analysis function $\widetilde{\phi}_{\text{new}}(t)$ is more peaked and the synthesis $\phi_{\text{new}}(t)$ is smoother. Our latest test on the "*boats*" image at 0.32 bpp was a tie in objective measures (PSNR, MSE, Max error), but we see more in the new image (Figure 7.4). The cable at the upper right is lost by the standard 9/7 pair, and the ship name PICARDIE becomes unreadable. You see that the reality of filter comparison is not totally precise!

Figure 7.4: Original of *boats* and two competing 9/7 reconstructions: FBI and binlet.

Problem Set 7.3

1. Suppose $eA = e$. If x is a vector perpendicular to e, show that Ax is perpendicular to e. (If $ex = 0$ prove that $e(Ax) = 0$.)

2. Suppose $eA = e$. Show that multiplying $S^{-1}AS$ by blocks gives

$$\begin{bmatrix} I_{N-1} & 0_{N-1} \\ e_{N-1} & 1 \end{bmatrix} \begin{bmatrix} a_{N-1} & b_{N-1} \\ c_{N-1} & d \end{bmatrix} \begin{bmatrix} I_{N-1} & 0_{N-1} \\ -e_{N-1} & 1 \end{bmatrix} = \begin{bmatrix} A_{N-1} & b_{N-1} \\ 0_{N-1} & 1 \end{bmatrix}.$$

The first $N-1$ columns of S are perpendicular to e. The matrix $A_{N-1} = a_{N-1} - b_{N-1}e_{N-1}$ is the restriction of A to those vectors. Compute this restriction $m_{N-1}(0)$ for the matrix $m(0)$ from the hat coefficients $\frac{1}{4}, \frac{1}{2}, \frac{1}{4}$.

3. For the transform $\widehat{\phi}(\omega) = (1-e^{-i\omega})^2/i^2\omega^2$ of the hat function show that $\int |\omega|^{2s}|\widehat{\phi}(\omega)|^2\,d\omega < \infty$ if and only if $s < \frac{3}{2}$.

4. Find by hand the matrix T and its eigenvalues and $s_{\max}$, starting from the filters $h = \left(\frac{1}{4}, \frac{1}{2}, \frac{1}{4}\right)$ and $h = \left(\frac{1}{6}, \frac{1}{2}, \frac{1}{3}\right)$.

5. Find by MATLAB the matrix T and its eigenvalues and $s_{\max}$ for the "D_6 filter" that has $h * h^T = \left(\frac{1+z^{-1}}{2}\right)^6 Q(z) =$ halfband of degree 10.

6. Show from their lengths that h and f and $p = h * f$ cannot all be symmetric halfband.

7. Dual to $f = [-1\ \ 0\ \ 9\ \ 16\ \ 9\ \ 0\ \ -1]/32$ is another *halfband but unsymmetric* filter $h^\# = [1\ \ 0\ \ 23\ \ 16\ \ -9\ \ 0\ \ 1]/32$. Verify that $p^\# = f * h^\#$ is halfband to give PR. This is our first unsymmetric product filter. What is the system delay l? What is $\lambda_{\max}(T^\#)$ coming from $h^\#$? Why is $\widetilde{\phi}^\#(t)$ interpolating (equal to $\boldsymbol{\delta}$ at the integers)? How smooth is it?

8. Explain why the smoothness of the delta function is $s_{\max} = -\frac{1}{2}$.

7.4 Splines and Semiorthogonal Wavelets

Splines are piecewise polynomials, with a smooth fit between the pieces. They are older than wavelets. The "two-scale equation" or dilation equation was at first not particularly noticed. Now we will see that the numbers $h(k)$ are binomial coefficients, directly from Pascal's triangle.

For a cubic spline the coefficients are 1, 4, 6, 4, 1 divided by 16. The transfer function is $H(z) = \left(1+z^{-1}\right)^4/16$. All four zeros are at $z = -1$. The filter is lowpass, the spline is as smooth as possible, and it has the highest accuracy $p = 4$ that is possible with $N = 4$. Almost every formula in this book comes out neatly and explicitly for splines.

One application of splines is to ***interpolation***, when data points need to be connected by a smooth curve. To put one high-degree polynomial through all the points is very unwise. A small movement of a single point produces an extreme change in the polynomial (which oscillates violently between the interpolation points). It is much better to use short pieces of low-degree polynomials—often cubic splines.

A cubic spline has degree 3, for any number of interpolation points. It has ***two continuous derivatives***, at the points where two different cubics meet. Thus $\phi_+(t)$ and $\phi'_+(t)$ and $\phi''_+(t)$ on one side of the meeting point agree with $\phi_-(t)$ and $\phi'_-(t)$ and $\phi''_-(t)$ on the other side. When $t = 0$ is the meeting point, only the coefficient of t^3 can change. The curve looks smooth and its coefficients are easy to find from the data points—but not trivial. There is a system of linear equations to solve, because all data points influence all coefficients. To say this in another way, the spline that matches the data values $0, 0, 1, 0, 0, \ldots$ is not zero in the intervals between those values. It decays exponentially as $|t| \to \infty$ but this "cardinal spline" does not give the best basis for computations.

The good function is the "B-spline" with compact support. It is our scaling function $\phi(t)$. This function matches the data values $0, \frac{1}{6}, \frac{4}{6}, \frac{1}{6}, 0$, at the integers. It has unit area $\int \phi(t)dt = 1$ and a smooth fit (two derivatives). The spline is nonzero on ***four intervals*** (Figure 7.5). Outside this range $\phi(t)$ is identically zero.

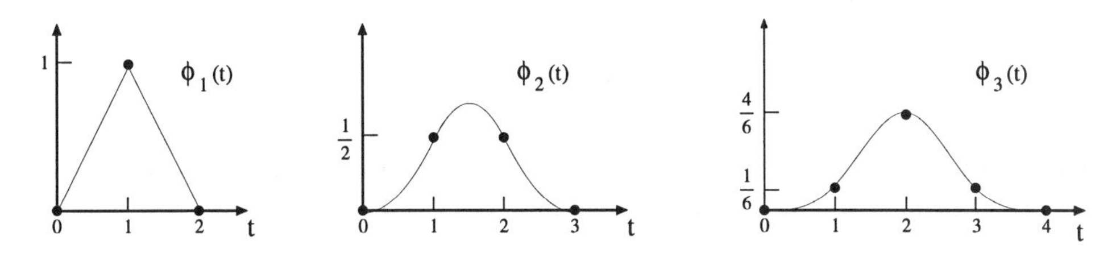

Figure 7.5: The spline $\phi_{N-1}(t)$ of degree $N-1$ is the convolution of N box functions. It is supported on $[0, N]$. Its filter $H(z) = (\frac{1+z^{-1}}{2})^N$ has a zero of order $p = N$ at $z = -1$.

A note on interpolation at the integers $t = n$. To match a set of values $f(n)$, we look for a combination of B-splines. The equations are $\sum F(k)\phi(n-k) = f(n)$. This is a constant-diagonal (Toeplitz) system, with $\frac{1}{6}, \frac{4}{6}, \frac{1}{6}$ on the diagonals coming from $\phi(n)$. To compute $F(k)$ we are inverting an FIR filter, and the inverse is IIR (a recursive filter). This explains why all $F(k)$ are nonzero when we interpolate an impulse $f(n) = \delta(n)$. It is like Shannon's Sampling Theorem, with cardinal splines instead of sinc functions. *In fact the splines converge to* $\frac{\sin \pi t}{\pi t}$ *as* $p = N \to \infty$.

One major advantage: Splines do not require equally spaced data points. At the limit of unequal spacing, pairs of nodes come together. The spline becomes a ***finite element*** with ***one*** continuous derivative. The cubic finite element with values 0, 1, 0, and slopes 0, 0, 0, is nonzero only on ***two intervals*** around the center point. A second cubic element interpolates 0, 0, 0, with slopes 0, 1, 0. This short support (Figure 7.7) makes finite elements very popular in solving differential equations — much more popular than splines.

When there are two data values (function and slope) at the mesh points, each input $x(n)$ to the corresponding filter is a *pair of numbers*. We have a ***multifilter*** and it leads to ***multiwavelets***. Everything is FIR. The values and slopes give the cubic in between. Multiwavelets are developed in the next section — we now return to the cubic spline $\phi(t)$.

Splines from Box Functions

Before two-scale equations, there was a direct approach to splines. This is still the fastest way. ***The cubic B-spline is the convolution of four box functions***:

$$\phi_3(t) = (B * B * B * B)(t). \tag{7.71}$$

The results from each convolution step are in Figure 7.5. These are linear splines $\phi_1(t)$, quadratic splines $\phi_2(t)$, and cubic splines $\phi_3(t)$. They are in the continuity classes C^0, C^1, and C^2, since they have 0, 1, and 2 continuous derivatives. *The box function is* $\phi_0(t)$. The convolution of N box functions has degree $N-1$ in each piece, with $N-2$ continuous derivatives between pieces.

We want to show that the fourth derivative of a cubic spline is *a sequence of delta functions*. The coefficients of these delta functions are $1, -4, 6, -4, 1$. These are the jumps in the third derivative, and this binomial pattern applies for every N. The first derivative of the hat function has jumps $1, -2, 1$.

To take the fourth derivative of a cubic spline, or the Nth derivative of a convolution of N box functions, we can work in time or frequency. We do both. First is the pleasant computation of $\widehat{\phi}_{N-1}(\omega)$ from convolving N box functions:

$$\widehat{\phi}_{N-1}(\omega) = (\widehat{\phi}_0(\omega))^N = \left(\frac{1}{i\omega}\right)^N (1 - e^{-i\omega})^N. \tag{7.72}$$

The box function has $\widehat{\phi}_0(\omega) = \int_0^1 e^{-i\omega t}\,dt = \frac{1}{i\omega}(1 - e^{-i\omega})$. Convolution in t is multiplication in ω, so the convolution of N box functions has transform $[\widehat{\phi}_0(\omega)]^N$.

Theorem 7.13 *The convolution of N boxes is a piecewise polynomial $\phi_{N-1}(t)$ of degree $N-1$. The jumps in the $(N-1)$st derivative at $t = 0, 1, \ldots, N$ are the alternating binomial coefficients $(-1)^t \binom{N}{t}$.*

Proof in the frequency domain: Each derivative multiplies the transform by $i\omega$. The Nth derivative cancels the denominator in $\widehat{\phi}_{N-1}(\omega)$ and has transform $(1 - e^{-i\omega})^N$. This is the transform of a sequence of delta functions at the points $t = 0, 1, \ldots, N$. Since the Nth derivative is zero between those points, the spline $\widehat{\phi}_{N-1}(t)$ must be a piecewise polynomial of degree $N-1$.

The fourth derivative of $\phi_3(t)$ has transform $(1 - e^{-i\omega})^4$. So the third derivative has jumps $1, -4, 6, -4, 1$.

Proof in the time domain: The derivative of $f(t) * g(t)$ is $f'(t) * g(t)$ or equally it is $f(t) * g'(t)$. The fourth derivative of the cubic $\phi = B * B * B * B$ has four factors:

$$\frac{d^4\phi}{dt^4} = (B' * B' * B' * B')(t). \tag{7.73}$$

Each factor $B'(t)$ is $\delta(t) - \delta(t-1)$. This is the derivative of the box function, which jumps up at $t = 0$ and down at $t = 1$. The convolution (7.73) becomes

$$\frac{d^4\phi}{dt^4} = \delta(t) - 4\delta(t-1) + 6\delta(t-2) - 4\delta(t-3) + \delta(t-4). \tag{7.74}$$

The third derivative has jumps $1, -4, 6, -4, 1$ at $t = 0, 1, 2, 3, 4$. Otherwise $d^4\phi/dt^4 = 0$ and $\phi(t)$ is an ordinary cubic polynomial.

The Coefficients $h(n)$ for Splines

We know that the filter coefficients $\frac{1}{2}, \frac{1}{2}$ lead to the box function $\phi_0(t)$. The coefficients $\frac{1}{4}, \frac{1}{2}, \frac{1}{4}$ lead to the hat function $\phi_1(t)$. These numbers appear in the lowpass filter $\boldsymbol{H}$. We suspect that the coefficients $\boldsymbol{h}(n) = \frac{1}{16}(1, 4, 6, 4, 1)$ lead to the cubic spline $\phi_3(t)$. The pattern for the convolution of N box functions could be described in two ways:

1. $\boldsymbol{h}(n)$ are the coefficients in the special polynomial $H(z) = \left(\frac{1+z^{-1}}{2}\right)^N$.

2. $\boldsymbol{h}(n)$ is the convolution $\left(\frac{1}{2}, \frac{1}{2}\right) * \left(\frac{1}{2}, \frac{1}{2}\right) * \cdots * \left(\frac{1}{2}, \frac{1}{2}\right)$.

Those two patterns are equivalent. Each multiplication by $(\frac{1+z^{-1}}{2})$ in the z-domain is convolution by $(\frac{1}{2}, \frac{1}{2})$ in the time domain. The question is, *what is the effect on the scaling function*? Apparently it is convolved with the box function.

This suggests a more general pattern. The transfer function $F(z)G(z)$ and the filter coefficients $\boldsymbol{f}(n) * \boldsymbol{g}(n)$ correspond to the scaling function $\phi_f(t) * \phi_g(t)$. *Multiplication of filters gives convolution of scaling functions!* This is not hard to prove. It applies immediately to the special case $F(z) = \frac{1+z^{-1}}{2}$. Convolution with the box function gives the next spline scaling function, one degree higher.

Lemma 7.2 *The scaling function $\phi_h(t)$ corresponding to $\boldsymbol{H} = \boldsymbol{FG}$ and to $\boldsymbol{h}(n) = \boldsymbol{f}(n) * \boldsymbol{g}(n)$ is the convolution of the scaling functions for $\boldsymbol{F}$ and $\boldsymbol{G}$:*

$$\phi_h(t) = \phi_f(t) * \phi_g(t) \quad \text{and} \quad \widehat{\phi}_h(\omega) = \left(\widehat{\phi}_f(\omega)\right)\left(\widehat{\phi}_g(\omega)\right). \tag{7.75}$$

Proof The dilation equations for ϕ_f and ϕ_g involve ω and $\omega/2$:

$$\widehat{\phi}_f(\omega) = F(\tfrac{\omega}{2})\,\widehat{\phi}_f(\tfrac{\omega}{2}) \quad \text{and} \quad \widehat{\phi}_g(\omega) = G(\tfrac{\omega}{2})\,\widehat{\phi}_g(\tfrac{\omega}{2}).$$

Multiply to get $\widehat{\phi}_h(\omega) = H(\frac{\omega}{2})\widehat{\phi}_h(\frac{\omega}{2})$. This is the dilation equation for $\boldsymbol{H}$. If you like infinite products, multiply $\prod F(\omega/2^i)$ and $\prod G(\omega/2^i)$ to get $\widehat{\phi}_h(\omega) = \prod H(\omega/2^i)$.

Example 7.6. If the coefficients in $\boldsymbol{f}$ and $\boldsymbol{g}$ are both $\frac{1}{2}, \frac{1}{2}$, the scaling functions $\phi_f(t)$ and $\phi_g(t)$ are the box function. Then $\boldsymbol{h} = \frac{1}{4}, \frac{1}{2}, \frac{1}{4}$ produces the hat function:

$$\begin{array}{ccccc} (\frac{1}{2}, \frac{1}{2}) * (\frac{1}{2}, \frac{1}{2}) & = & (\frac{1}{4}, \frac{1}{2}, \frac{1}{4}) \text{ because } \left(\frac{1+z^{-1}}{2}\right)\left(\frac{1+z^{-1}}{2}\right) & = & \frac{1+2z^{-1}+z^{-2}}{4}. \\ B(t) * B(t) & = & \text{Box function} * \text{Box function} & = & \text{Hat function.} \end{array}$$

Every extra factor $(\frac{1+z^{-1}}{2})$ means a convolution with the box function. The new $\phi(t)$ has one more derivative. For splines, each new $\phi_N(t) = \phi_{N-1}(t) * B(t) = \int_0^1 \phi_{N-1}(t-x)\,dx$ is one degree higher. This gives a nice formula for the time derivative:

$$\phi_N'(t) = \int_0^1 \phi_{N-1}'(t-x)\,dx = \phi_{N-1}(t) - \phi_{N-1}(t-1). \tag{7.76}$$

The spline of degree $N-1$ is the convolution of N box functions, and the corresponding filter has $H(z) = (\frac{1+z^{-1}}{2})^N$. We can verify in one step that this spline satisfies the correct dilation equation! Its transform $\widehat{\phi}_{N-1}(\omega)$ agrees with $H\left(\frac{\omega}{2}\right)\widehat{\phi}_{N-1}\left(\frac{\omega}{2}\right)$;

$$\left(\frac{1-e^{-i\omega}}{i\omega}\right)^N = \left(\frac{1+e^{-i\omega/2}}{2}\right)^N \left(\frac{1-e^{-i\omega/2}}{i\omega/2}\right)^N. \tag{7.77}$$

We started with $\phi(t)$ and determined $\boldsymbol{h}(k)$. Normally we do the opposite — solve the dilation equation for $\phi(t)$. Here we have found the dilation equation for a spline:

$$\phi_{N-1}(t) = 2^{1-N} \sum_0^N \binom{N}{k} \phi_{N-1}(2t-k). \tag{7.78}$$

The lowpass space V_0 contains all smooth splines of degree $N-1$ on unit intervals. The B-splines are its basis (not orthogonal!). The space V_1 contains piecewise polynomials on *half*-intervals, and therefore contains V_0. The dilation equation (7.78) expresses $\phi_{N-1}(t)$ in V_0 as a combination of basis functions of V_1. It remains to find the other space W_0 which contains the wavelets.

Summary: $H(z)$ has a zero of order $p = N$ at $z = -1$. *There are N zeros at* $\omega = \pi$. Correspondingly, the polynomials $1, t, t^2, \ldots, t^{N-1}$ are combinations of the splines of degree $N-1$. Those polynomials are all in the lowpass space V_0. The accuracy is $p = N$. The wavelets will have N vanishing moments. ***What are those spline wavelets*?**

Inner Products and Riesz Bounds

The inner products of splines with their translates give the values of higher splines. The inner products for the hat function are $\boldsymbol{a} = \frac{1}{6}, \frac{4}{6}, \frac{1}{6}$:

$$\boldsymbol{a}(0) = 2\int_0^1 t^2\,dt = \tfrac{4}{6} \text{ and } \boldsymbol{a}(1) = \boldsymbol{a}(-1) = 2\int_0^1 t(t-1)\,dt = \tfrac{1}{6}.$$

Those numbers agree with the cubic B-spline at $t = 1, 2, 3$. The formula is:

$$\textit{Spline Inner Products} \quad \boldsymbol{a}(k) = \int_{-\infty}^{\infty} \phi_{N-1}(t)\,\phi_{N-1}(t+k)\,dt = \phi_{2N-1}(N+k). \tag{7.79}$$

The integral is convolving N box functions with N more box functions—and shifting by k. The $2N$ boxes produce the higher spline ϕ_{2N-1}.

The vector $\boldsymbol{a}$ of inner products solves $\boldsymbol{a} = \boldsymbol{T}\boldsymbol{a}$. The operator $\boldsymbol{T}$ comes from the product $H(z)H(z^{-1})$, which for splines is $(\frac{1+z^{-1}}{2})^N(\frac{1+z}{2})^N$. This is z^N times the function $(\frac{1+z^{-1}}{2})^{2N}$ that gives the higher spline $\phi_{2N-1}(t)$. In other words, *the matrix* $\boldsymbol{T}$ *for the lower spline is identical to the matrix* $\boldsymbol{M}$ *for the higher spline.*

The eigenvector of $\boldsymbol{M}$ gives the values of the higher spline at the integers. The same eigenvector (of $\boldsymbol{T}$!) gives the inner products of the lower splines in (7.79). All inner products have $\boldsymbol{a}(k) \geq 0$, because all splines have $\phi(t) \geq 0$. The inner products sum to 1. The maximum value of $A(\omega) = \sum \boldsymbol{a}(k)e^{-ik\omega}$ is $B = 1$, which occurs at $\omega = 0$. It is not hard to show that the minimum value of $A(\omega)$ occurs at $\omega = \pi$:

$$A_{\min} = \sum (-1)^k \phi_{2N-1}(N+k). \tag{7.80}$$

Thus the translates $\phi_{N-1}(t-k)$ form a Riesz basis for V_0 with bounds $A_{\min}$ and 1. For the hat function ($N = 2$) the lower bound is $A_{\min} = -\frac{1}{6} + \frac{4}{6} - \frac{1}{6} = \frac{1}{3}$. In frequency this means that

$$\tfrac{1}{3} \leq A(\omega) = \sum_{-\infty}^{\infty} \left|\widehat{\phi}_1(\omega + 2\pi k)\right|^2 = \tfrac{4}{6} + \tfrac{2}{6}\cos\omega \leq 1. \tag{7.81}$$

The basis of hat functions is well conditioned. Of course it is not orthogonal. If we orthogonalize, following the shift-invariant method of Section 6.5, we get the Battle-Lemarié $\phi(t)$ and $w(t)$—smooth, quickly decaying, infinite support.

Spline Wavelets

There are several important possibilities for the wavelets. One family is FIR and biorthogonal, the other is IIR and "semiorthogonal." Semiorthogonal wavelets are perpendicular to the spline $\phi(t)$, but they are not orthogonal among themselves.

The FIR biorthogonal construction follows the usual rules. We need a halfband product filter $P(z)$. One factor is $(\frac{1+z^{-1}}{2})^N$, to produce the spline. This can be $H(z)$ in the analysis bank, or (better) it can be $F(z)$ in the synthesis bank. The other bank must contain an extra factor to make the product halfband. The natural choices for this second filter are $(\frac{1+z^{-1}}{2})^{2p-N}Q(z)$, which brings the product filter back to the standard Daubechies polynomial (where $Q(z)$ has degree $2p-2$). We need $2p > N$ to have a zero at $z=-1$. For the smallest p, the second filter may not satisfy Condition E for a stable basis in L^2; *more zeros may be needed.* The low degree spline filters and dual filters are the best known:

$$F(z) = (\tfrac{1+z^{-1}}{2})^2 \quad \text{goes with} \quad H(z) = (\tfrac{1+z^{-1}}{2})^2(-1+4z^{-1}-z^{-2}) \ : \ \textbf{\textit{stable}}$$

$$F(z) = (\tfrac{1+z^{-1}}{2})^3 \quad \text{goes with} \quad H(z) = (\tfrac{1+z^{-1}}{2})(-1+4z^{-1}-z^{-2}) \ : \ \textbf{\textit{unstable}}$$

$$F(z) = (\tfrac{1+z^{-1}}{2})^4 \quad \text{goes with} \quad H(z) = (\tfrac{1+z^{-1}}{2})^4 Q_6(z) \ : \ \textbf{\textit{stable.}}$$

The new book [CR] is a very good reference for biorthogonal FIR filters.

Example 7.7. Hat function from $F(z) = (\frac{1+z^{-1}}{2})^2$ gives the 5/3 filter bank.

If the analysis filter has $H(z) \equiv 1$, the product with $F(z)$ is halfband. The analysis scaling function $\widetilde{\phi}(t)$ is the delta function. Its translates $\delta(t-k)$ are biorthogonal to the hat functions $H(t-k)$. But the delta function is not acceptable.

If $H(z)$ is given two zeros at π, it needs the extra $Q(z)$:

$$H(z) = \left(\frac{1+z^{-1}}{2}\right)^2\left(\frac{-1+4z^{-1}-z^{-2}}{2}\right) = \frac{-1+2z^{-1}+6z^{-2}+2z^{-3}-z^{-4}}{8}.$$

Those coefficients $-1, 2, 6, 2, -1$ appeared in the Guide to the Book. This is the analysis part of the biorthogonal $5/3$ pair. The synthesis part is the linear spline (the hat). The product $P(z) = F(z)H(z)$ has four zeros at $z=-1$. Instead of the orthogonal D_4 factors of $P(z)$, the spline $5/3$ factorization gives linear phase.

In these biorthogonal examples, one scaling function is a spline. Best if this is ***synthesis***. The other scaling function is not a spline or a combination of splines or a piecewise polynomial. The space $\widetilde{V}_0$ spanned by $\widetilde{\phi}(t-k)$ is different from V_0. This is normal, but spline people expect to live and work exclusively in spline spaces. They want $V_0 = \widetilde{V}_0 =$ all splines of degree $N-1$. We now achieve this, but an IIR filter appears in the analysis half of the filter bank.

Semiorthogonal Wavelets

Start with a basis $\{\phi(t-k)\}$ for V_0. Suppose this basis is not orthogonal. The hat function and the cubic B-spline are examples of $\phi(t)$. We want to find wavelets $w(t-k)$ that are *orthogonal* to $\phi(t)$. Thus we maintain what was true in the fully orthogonal case:

$$V_0 \perp W_0 \quad \text{and} \quad V_0 \oplus W_0 = V_1. \tag{7.82}$$

The spaces are orthogonal but the bases within those spaces are not orthogonal.

Multiresolution in the biorthogonal case always has

$$V_0 \perp \widetilde{W}_0 \text{ and } W_0 \perp \widetilde{V}_0 \text{ and } V_0 + W_0 = V_1 \text{ and } \widetilde{V}_0 + \widetilde{W}_0 = \widetilde{V}_1. \tag{7.83}$$

Compare with (7.82) to see that $V_0 = \widetilde{V}_0$ and $W_0 = \widetilde{W}_0$. At every scale we will have $V_j \perp W_j$ and $V_j = \widetilde{V}_j$ and $W_j = \widetilde{W}_j$. There is only *one* multiresolution in the semiorthogonal case, one family $V_0 \subset V_1 \subset V_2$, as in the orthogonal case. The difference is that we have two bases for V_0, the given basis $\phi(t-k)$ and the biorthogonal basis $\widetilde{\phi}(t-k)$. This applies at every scale j. The tilde is needed for the dual basis, even if it is not needed for the space.

Semiorthogonality has an important property, directly from (7.83):

Semiorthogonal wavelets $w\left(2^j t-k\right)$ and $w\left(2^J t-l\right)$ are perpendicular if $j \neq J$.

At the same scale $j = J$, semiorthogonal wavelets are not generally perpendicular. But because W_j is orthogonal to V_j which contains all previous $W_{j-1}, W_{j-2}, \ldots$, we are guaranteed that W_j is perpendicular to all wavelets at other scales.

To construct this new wavelet $w(t)$, we need the highpass coefficients $f_1(k)$. In the orthogonal case, they come from $f_0(k)$ by an alternating flip. $F_1(z)$ is $-z^{-N}F_0\left(-z^{-1}\right)$, where z^{-1} gives the flip and $-z^{-1}$ makes it alternating. In that orthogonal case, the inner products $\langle\phi(t), \phi(t+k)\rangle$ are $\boldsymbol{a}(k) = \delta(k)$. The polynomial $A(z) = \sum \boldsymbol{a}(k)z^{-k}$ is identically 1. In the semiorthogonal case, when $\{\phi(t-k)\}$ is not orthonormal, *this inner product polynomial enters the highpass coefficients.*

The highpass function $F_1(z)$ becomes an alternating flip of $F_0(z)A(z)$. The analysis filters become IIR. Here is the general rule for semiorthogonality:

Theorem 7.14 *Suppose the lowpass $F_0(z)$ leads to scaling functions $\phi(t-k)$ whose inner products are the coefficients in $A(z)$. Then the highpass $F_1(z)$ that yields semiorthogonal wavelets $w(t-k)$ is the alternating flip of $F_0(z)A(z)$:*

$$F_1(z) = -z^{1-2N}F_0\left(-z^{-1}\right)A\left(-z^{-1}\right). \tag{7.84}$$

Proof. We want $\int \phi(t)w(t-n)\,dt = 0$ for all n. Use the dilation equation for $\phi(t)$ and the wavelet equation for $w(t)$:

$$\int_{-\infty}^{\infty}\left[\sum 2f_0(k)\phi(2t-k)\right]\left[\sum 2f_1(k)\phi(2t-2n-k)\right]dt = 0. \tag{7.85}$$

Change the second sum to $\sum 2f_1(\ell-2n)\phi(2t-\ell)$. The inner product of $\phi(2t-k)$ with $\phi(2t-\ell)$ is $\boldsymbol{a}(\ell-k)$. The orthogonality requirement (7.89) becomes

$$\sum_{\ell}\sum_{k} f_0(k)\boldsymbol{a}(\ell-k)f_1(\ell-2n) = 0. \tag{7.86}$$

The highpass filter is double-shift orthogonal, but not to the lowpass filter. The double-shift orthogonality is to the sequence $\sum f_0(k)\boldsymbol{a}(\ell-k)$ which corresponds to $F_0(z)A(z)$. Therefore $F_1(z)$ comes from $F_0(z)A(z)$ by an alternating flip. This completes the proof.

The sequence $f_0(0), \ldots, f_0(N)$ gives a scaling function supported on $[0, N]$. The inner product $\boldsymbol{a}(N) = \int \phi(t)\phi(t+N)\,dt = \boldsymbol{a}(-N)$ is automatically zero. At most, the symmetric polynomial $A(z)$ has terms $\boldsymbol{a}(k)z^{-k}$ and $\boldsymbol{a}(k)z^{k}$ for $k < N$. Note that $A(z) = A\left(z^{-1}\right)$. The degree of $F_0(z)A(z)$ is at most $2N-1$.

Example 7.8. The hat function has $\boldsymbol{a}(0) = \frac{4}{6}$ and $\boldsymbol{a}(1) = \boldsymbol{a}(-1) = \frac{1}{6}$. Find $F_1(z)$.

The product $F_0(z)A(z)$ is $\frac{1}{24}\left(1 + 2z^{-1} + z^{-2}\right)\left(z + 4 + z^{-1}\right)$. By alternating flip

$$\begin{aligned} F_1(z) &= \frac{-z^{-3}}{24}\left(1 - 2z + z^2\right)\left(-z + 4 - z^{-1}\right) \\ &= \tfrac{1}{24}\left(1 - 6z^{-1} + 10z^{-2} - 6z^{-3} + z^{-4}\right). \end{aligned}$$

The highpass coefficients $2f_1(k)$ yield $w(t)$ as drawn in Figure 7.6:

$$w(t) = \tfrac{1}{12}\left[\phi(2t) - 6\phi(2t-1) + 10\phi(2t-2) - 6\phi(2t-3) + \phi(2t-4)\right]. \qquad (7.87)$$

This is orthogonal to all hat functions $\phi(t-k)$. It is *not* orthogonal to all $w(t-k)$. But it *is* orthogonal to every $w\left(2^j t - k\right)$ for $j \neq 0$. That is semiorthogonality.

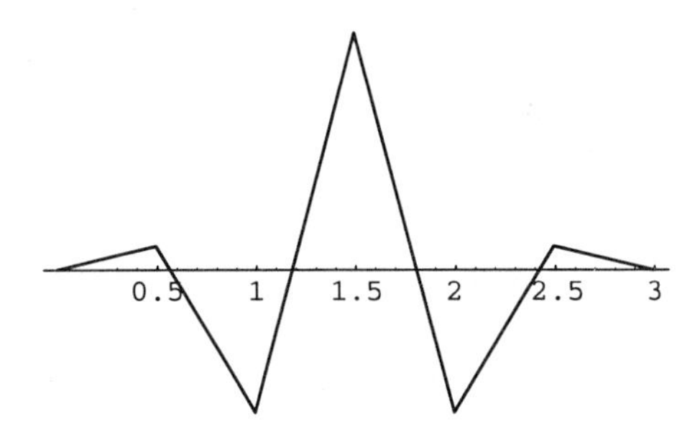

Figure 7.6: The linear wavelet $w(t)$ orthogonal to the hat functions $\phi(t-k)$.

The IIR Half of the Semiorthogonal Filter Bank

The analysis functions $H_0(z)$ and $H_1(z)$ must come from the perfect reconstruction condition $\boldsymbol{H}_m(z)\boldsymbol{F}_m(z) = 2z^{-\ell}\boldsymbol{I}$. We know the modulation matrix $\boldsymbol{F}_m(z)$. Its inverse times $2z^{-\ell}$ gives $\boldsymbol{H}_m(z)$:

$$\begin{bmatrix} F_0(z) & F_1(z) \\ F_0(-z) & F_1(-z) \end{bmatrix} = \begin{bmatrix} F_0(z) & -z^{1-2N}F_0\left(-z^{-1}\right)A\left(-z^{-1}\right) \\ F_0(-z) & z^{1-2N}F_0\left(z^{-1}\right)A\left(z^{-1}\right) \end{bmatrix}$$

$$\begin{bmatrix} H_0(z) & H_0(-z) \\ H_1(z) & H_1(-z) \end{bmatrix} = \frac{2z^{-\ell}}{D(z)}\begin{bmatrix} z^{1-2N}F_0\left(-z^{-1}\right)A\left(z^{-1}\right) & z^{1-2N}F_0\left(-z^{-1}\right)A\left(-z^{-1}\right) \\ -F_0(-z) & F_0(z) \end{bmatrix}$$

The key is always in the determinant $D(z)$. For an FIR filter bank, $D(z)$ is the delay $z^{-\ell}$. For these semiorthogonal filters, we don't get a delay but we do get a remarkable formula:

$$\begin{aligned} D(z) &= z^{1-2N}\left[F_0(z)F_0\left(z^{-1}\right)A\left(z^{-1}\right) + F_0(-z)F_0\left(-z^{-1}\right)A\left(-z^{-1}\right)\right] \\ &= z^{1-2N}A\left(z^{-2}\right). \end{aligned} \qquad (7.88)$$

This is the equation $\boldsymbol{Ta} = \boldsymbol{a}$ for the eigenvector of the transition matrix! The matrix $\boldsymbol{T}$ in the time domain is $(\downarrow 2)2\boldsymbol{F}\boldsymbol{F}^T$. In the z-domain this is exactly what appears in (7.88); the aliasing term with $-z$ comes from $(\downarrow 2)$. The determinant $D(z)$ is identified. It is an odd function, and we take $\ell = 2N-1$. The analysis filters can be read from the matrix $\boldsymbol{H}_m(z)$:

$$H_0(z) = \frac{2z^{-\ell} F_0\left(z^{-1}\right) A\left(z^{-1}\right)}{A\left(z^{-2}\right)} \quad \text{and} \quad H_1(z) = \frac{2F_0(-z)}{A\left(z^{-2}\right)}. \tag{7.89}$$

The division by $A\left(z^{-2}\right)$ yields IIR filters except in the orthogonal case when $A \equiv 1$.

Example 7.9. (Hats continued) The modulation matrix from F_0 and F_1 is

$$\frac{1}{4} \cdot \frac{1}{24} \begin{bmatrix} 1+2z^{-1}+z^{-2} & 1-6z^{-1}+10z^{-2}-6z^{-3}+z^{-4} \\ 1-2z^{-1}+z^{-2} & 1+6z^{-1}+10z^{-2}+6z^{-3}+z^{-4} \end{bmatrix}.$$

The determinant is verified to be $\frac{1}{6}\left(z^{-1}+4z^{-3}+z^{-5}\right) = z^{-3}A\left(z^{-2}\right)$.

The synthesis coefficients require a division by $A\left(z^{-2}\right)$. This has zeros at $z^2 = 2 \pm \sqrt{3}$. We can express $1/A\left(z^{-2}\right)$ by partial fractions (Problem 7). The power series for these fractions will have $\left(2-\sqrt{3}\right)^n$ in the coefficients of z^{2n} and z^{-2n}. That gives the decay rate of the (IIR!) filter coefficients $\boldsymbol{h}(k)$. The scaling functions $\widetilde{\phi}(t-k)$ are biorthogonal to the hats.

Summary: Splines open the possibility of maintaining $V_j \perp W_j$ in the biorthogonal case. Then the wavelets are semiorthogonal. The highpass coefficients involve the inner products $\boldsymbol{a}(k)$ of the scaling functions. The associated polynomial $A(z)$ is called the "*Euler-Frobenius polynomial*" in spline theory, but there is no restriction to splines. For every lowpass filter that yields a scaling function $\phi(t)$, the theory produces semiorthogonal wavelets.

For splines of degree $p-1$, the inner products $\boldsymbol{a}(k)$ are the values at integers of the B-splines of degree $2p-1$. For all cases, the inner products are in the eigenvector $\boldsymbol{Ta} = \boldsymbol{a}$.

We emphasize that the special properties of splines are attracting a lot of attention in signal processing. They have maximum regularity (and symmetry) with minimum support and complexity. ***Splines are outstanding in synthesis***. They give approximation of high order p with low constant C in the error $C(\Delta t)^p$. The dual analysis filter has to be longer, but no construction will ever be perfect.

Problem Set 7.4

1. Prove that the spline $\phi_{N-1}(t)$ is symmetric about the center point $t = N/2$.
2. Explain the formula $\phi_{N-1}(t) = \frac{1}{(N-1)!}\sum(-1)^k \binom{N}{k}(t-k)_+^{N-1}$. Here $(t-k)_+ = \max(t-k, 0)$. One proof uses the jumps in the $(N-1)$st derivative.
3. Take derivatives in Problem 2 to verify $\phi_N'(t) = \phi_{N-1}(t) - \phi_{N-1}(t-1)$.
4. (Challenge) Prove that $\phi_N(t) = \frac{t}{N}\phi_{N-1}(t) + \frac{N+1-t}{N}\phi_{N-1}(t-1)$. This recursion gives a quick stable computation of $\phi_N(t)$. Hint: take out a factor to get $\phi_N(t) = \frac{t}{N}\phi_N'(t) + \frac{N+1}{N}\phi_{N-1}(t-1)$. Use Problems 2–3.
5. Suppose $P(z) = F(z)H(z)$ is centered halfband. Why is $\phi_f(t) * \phi_h(t)$ equal to $\boldsymbol{\delta}(n)$ at $t = n$?
6. Find the polyphase matrix for $\boldsymbol{h}_0(z) = \frac{1}{4}(1, 2, 1)$ and $\boldsymbol{h}_1(z) = \frac{1}{24}(1, -6, 10, -6, 1)$. Connect its determinant to $A(z)$.

7. Find A and B in the partial fraction expansion

$$\frac{6}{x^2+4x+1} = \frac{A}{x+2-\sqrt{3}} + \frac{B}{x+2+\sqrt{3}}.$$

Expand the last two fractions in powers of $\frac{2-\sqrt{3}}{x}$ and $\frac{x}{2+\sqrt{3}}$. Combine into a power series for $6/(x+4+x^{-1})$. The coefficients of x^n and x^{-n} should be equal.

8. Show that the inner products $\boldsymbol{a}(k) = \int \phi(t)\phi(t+k)\,dt$ for quadratic splines are $\boldsymbol{a} = \frac{1}{120}(1, 26, 66, 26, 1)$. You could verify that this is the eigenvector in $\boldsymbol{Ta} = \boldsymbol{a}$, or evaluate the 5th-degree spline at the integers. What is the support interval of the semiorthogonal quadratic $w(t)$?

9. For splines show that $A(\omega) = \sum |\widehat{\phi}_{N-1}(\omega+2\pi k)|^2$ equals $\left(2\sin\frac{\omega}{2}\right)^{2N} \sum(\omega+2\pi k)^{-2N}$. Verify $A'(\pi) = 0$. The minimum is at $\omega = \pi$.

7.5 Multifilters and Multiwavelets

This brief section describes a recent development—to allow the filter coefficients $\boldsymbol{h}(k)$ to be $r \times r$ matrices. Each input sample $\boldsymbol{x}(n)$ is a vector with r components. So is each output $\widehat{\boldsymbol{x}}(n)$. The bank of multifilters has the same structure as an ordinary filter bank, with the extra freedom that comes with matrix coefficients.

In continuous time, the dilation equation will have matrix coefficients $\boldsymbol{h}(k)$. The solution gives r scaling functions $\phi_1(t), \dots, \phi_r(t)$. Then the wavelet equation with highpass matrices $\boldsymbol{h}_1(k)$ yields r corresponding wavelets. *Properly chosen, all these functions can have symmetry as well as orthogonality!* They and all their translates are orthogonal when the polyphase matrix (of order $2r$) is paraunitary. Multiresolution produces an orthonormal basis of wavelets—the translates of r functions at all scales $-\infty < j < \infty$:

$$w_{ijk}(t) = 2^{-j/2} w_i\left(2^j t - k\right), \quad 1 \le i \le r.$$

In some situations $r > 1$ is quite reasonable. When sampling a function $x(t)$, we may also sample its slope $x'(t)$. Velocity may be involved as well as displacement. The pair $\left(x(n), x'(n)\right)$ is a vector with $r = 2$. Those samples could go through two separate scalar filter banks, or through one bank of multifilters. This example already shows, because $x'(t)$ is dimensionally different from $x(t)$, that scaling is important for the r inputs.

The cubic scaling functions in this $\left(x(n), x'(n)\right)$ example are "*finite elements.*" They are drawn in Figure 7.7, with support $[0, 2]$. Like splines, these cubics $\phi_1(t)$ and $\phi_2(t)$ have linear phase—but they are not orthogonal to their translates. They have *one* continuous derivative, not two. The space of C^1 cubics has a more local basis (but with $r = 2$ functions per interval), while the spline space of C^2 cubics has one basis function per interval (the B-spline).

Figure 7.7 also displays the wavelets $w_1(t)$ and $w_2(t)$ in the semiorthogonal case. They are supported on $[0, 3]$. They and their translates span W_0 and are orthogonal to $\phi_1(t)$ and $\phi_2(t)$. They are piecewise cubic on half-intervals, and they come from a wavelet equation $\boldsymbol{w}(t) = \sum 2\boldsymbol{h}_1(k)\boldsymbol{\phi}(2t-k)$. Those coefficients $\boldsymbol{h}_1(k)$ are 2×2 matrices!

The finite element spaces V_0 of degree $1, 3, 5, 7$ are spanned by $r = 1, 2, 3, 4$ scaling functions. Degree 1 has $\phi(t) =$ hat function which has C^0 smoothness (no continuous derivatives). From the polynomials contained in V_0 we know the accuracy $p = 2, 4, 6, 8$. The smoothness is C^0, C^1, C^2, C^3 (and the pattern continues). The r semiorthogonal multiwavelets are always

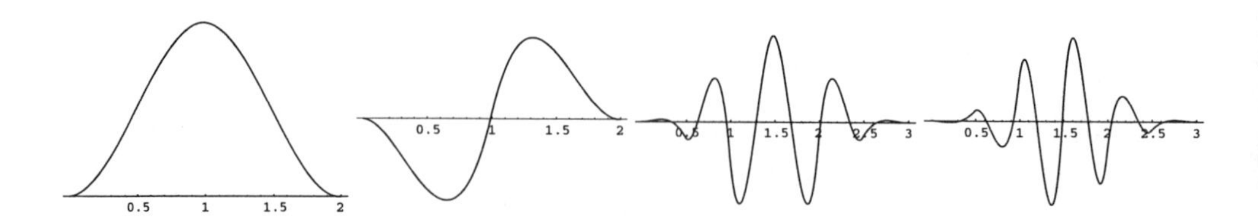

Figure 7.7: Cubic multiwavelets $w_1(t)$ and $w_2(t)$ orthogonal to cubic finite elements $\phi_1(t-k)$ and $\phi_2(t-k)$.

supported on $[0,3]$. These finite element scaling functions are cousins to the splines — which have more smoothness but longer support. Here is a piecewise linear construction with even less smoothness and even shorter support.

Multiwavelet example: "Haar's Hat" These functions are *linear* on each interval but *not continuous* at the ends. Each piece is a straight line between the end values. That piece is determined by $r=2$ numbers, its average and its slope. Its average goes with the box function $\phi_1(t)$, its slope goes with $\phi_2(t) = 2t-1$. Those scaling functions on $[0,1]$ are combinations of the same functions on half-intervals:

Box: $\phi_1(t) = \phi_1(2t) + \phi_1(2t-1)$ (***usual Haar equation***)

Slope: $\phi_2(t) = \frac{1}{2}[\phi_2(2t) + \phi_2(2t-1) - \phi_1(2t) + \phi_1(2t-1)]$ (***zero outside*** $[0,1]$)

This is a ***matrix dilation equation***. The coefficients are 2×2 matrices $\boldsymbol{h}(0)$ and $\boldsymbol{h}(1)$:

$$\begin{bmatrix} \phi_1(t) \\ \phi_2(t) \end{bmatrix} = \begin{bmatrix} 1 & 0 \\ -\frac{1}{2} & \frac{1}{2} \end{bmatrix} \begin{bmatrix} \phi_1(2t) \\ \phi_2(2t) \end{bmatrix} + \begin{bmatrix} 1 & 0 \\ \frac{1}{2} & \frac{1}{2} \end{bmatrix} \begin{bmatrix} \phi_1(2t-1) \\ \phi_2(2t-1) \end{bmatrix} \tag{7.90}$$

You see the eigenvalues 1 and $\frac{1}{2}$ that always accompany linear functions in V_0. The accuracy is $p=2$. The support is short! This is a block transform, not overlapping the next pair of samples — the average and slope on the next interval $[1,2]$. We get good accuracy, but the price is complete lack of smoothness.

It is an exercise to find the wavelets for Haar's Hat. They will be linear on half-intervals (and still discontinuous). Another example of two scaling functions: the real and imaginary parts of a complex scaling function. Symmetry with orthogonality is possible [Lena]. We turn to a more magical construction that combines orthogonality and symmetry and short support and continuity.

This special construction by Geronimo, Hardin, and Massopust gave a strong impetus to multiwavelet theory. The functions $\phi_1(t)$ and $\phi_2(t)$ in Figure 7.8 were found by a recursive interpolation process. The 2×2 coefficients $\boldsymbol{h}_0(k)$ in their dilation equation came later. So did the highpass coefficients and the wavelets [StSt**1**, GHM**2**]. All these functions have linear phase and short support and orthogonality. The only earlier example with these properties was Haar's. Now the accuracy is $p=2$, because the scaling functions can reproduce a hat:

$$\phi_1(t) + \phi_1(t-1) + \phi_2(t) = \text{hat function.}$$

The space V_0 has an FIR orthogonal basis $\{\phi_1(t-k)\}$ joined with $\{\phi_2(t-k)\}$:

$$f_0(t) = \sum a_{10k}\,\phi_1(t-k) + a_{20k}\,\phi_2(t-k) \ \text{ is in } \ V_0. \tag{7.91}$$

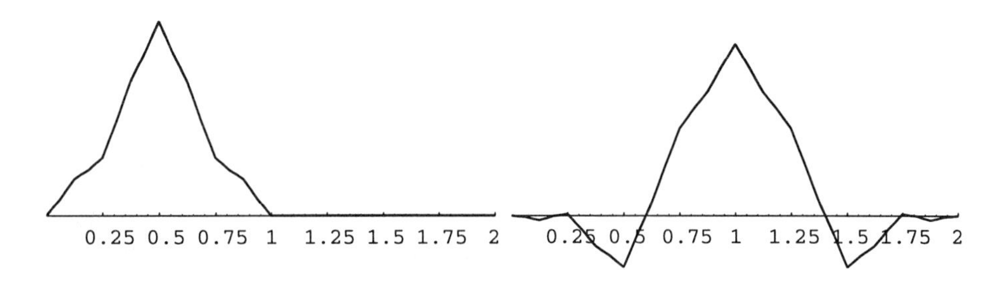

Figure 7.8: GHM scaling functions combine orthogonality of all translates and linear phase.

There is a serious problem for multifilters and multiwavelets. Everyone in signal processing asks about it immediately. We have twice as many filters (or r times as many filters) as usual. One input stream $\boldsymbol{x}(n)$ produces $2r$ half-length outputs from the analysis bank. This means extra computation (but shorter filters). It also means that the stream $\boldsymbol{x}(n)$ has to be *vectorized*—so r inputs go together. It is not at all satisfactory just to take the original stream in blocks of r samples. This does violence to the whole idea. The $\phi_i(t)$ are not time shifts of one function, they are r functions *at each time*.

The $\big(x(n), x'(n)\big)$ example comes naturally in blocks of two at each time. So does Haar's Hat. For other multifilters, one possibility is to repeat $(x(n), cx(n))$ with a suitable scalar c. What we really want is a discrete form of (7.91). The recent papers [Heller**2**, XiaG] show how to convert a single input stream to multi-inputs which give samples of the data at half-intervals (thus $r = 2$ samples per interval). These Geronimo-Hardin-Massopust multiwavelets have given the first experimental results in compression. In competition against D_4 wavelets, with the same accuracy $p = 2$, the multiwavelets gave better compression and required more computations.

Perfect Reconstruction and Orthogonality and Accuracy

The theory of multiwavelets looks completely familiar up to one point, where everything changes. The multifilters still have transfer functions $\boldsymbol{H}(z) = \sum \boldsymbol{h}(k)z^{-k}$. These are now matrices. The perfect reconstruction conditions are the same as before:

$$\boldsymbol{F}_0(z)\boldsymbol{H}_0(z) + \boldsymbol{F}_1(z)\boldsymbol{H}_1(z) = 2z^{-\ell}\boldsymbol{I} \tag{7.92}$$

$$\boldsymbol{F}_0(z)\boldsymbol{H}_0(-z) + \boldsymbol{F}_1(z)\boldsymbol{H}_1(-z) = \boldsymbol{0}. \tag{7.93}$$

The big difference is that the anti-aliasing equation (7.93) is no longer satisfied by $\boldsymbol{F}_0(z) = \boldsymbol{H}_1(-z)$ and $\boldsymbol{F}_1(z) = -\boldsymbol{H}_0(-z)$. *The matrices $\boldsymbol{H}_0$ and $\boldsymbol{H}_1$ need not commute!* Probably they don't. A satisfactory construction method for PR (= biorthogonal) multifilters is not yet available.

For orthogonality the situation is similar. We want the synthesis filters to be the transposes $\boldsymbol{F}_0(z) = \boldsymbol{H}_0^T(z^{-1})$ and $\boldsymbol{F}_1(z) = \boldsymbol{H}_1^T(z^{-1})$ of the analysis filters (times a delay to make them causal). Omitting that delay, the PR conditions (7.92) require the modulation matrix to be paraunitary: $\boldsymbol{H}_m(z)\boldsymbol{H}_m^T(z^{-1}) = 2\boldsymbol{I}$. These matrices have order $2r$. The first block yields the famous Condition **O**:

$$\boldsymbol{H}_0(z)\boldsymbol{H}_0^T(z^{-1}) + \boldsymbol{H}_0(-z)\boldsymbol{H}_0^T(-z^{-1}) = 2\boldsymbol{I}. \tag{7.94}$$

Suppose this is achieved—it is the hard part. Then in the scalar case, $\boldsymbol{H}_1$ is the alternating flip of $\boldsymbol{H}_0$. That no longer works. The lowpass and highpass rows

$$\begin{bmatrix} \boldsymbol{h}(N) & \boldsymbol{h}(N-1) & \cdots & \boldsymbol{h}(1) & \boldsymbol{h}(0) \\ -\boldsymbol{h}(0) & \boldsymbol{h}(1) & \cdots & -\boldsymbol{h}(N-1) & \boldsymbol{h}(N) \end{bmatrix}$$

are not orthogonal, if the $r \times r$ matrices $\boldsymbol{h}(k)$ do not commute. We need to compute, from scratch, highpass coefficients $\boldsymbol{h}_1(k)$ that will complete the paraunitary matrix $\boldsymbol{H}_m/\sqrt{2}$.

This completion is possible [StSt1]. If we have r paraunitary rows of length $2r$, the factorization in equation (9.49) still exists. The constant matrix $\boldsymbol{Q}$ in that equation is $r \times 2r$. Complete it to a square constant orthogonal matrix. Then the factors multiply to give a square paraunitary matrix. Its last r rows contain the highpass coefficients we need.

Finally, we mention the accuracy p. As always, combinations of $\phi_i(t-k)$ must produce the polynomials $1, t, \ldots, t^{p-1}$. In the scalar case, $H(z)$ will have a factor $\left(1+z^{-1}\right)^p$. The matrix factorization is not so simple [CoDaPl]. Also in the scalar case, $\boldsymbol{M} = (\downarrow 2)2\boldsymbol{H}$ has eigenvalues $1, \frac{1}{2}, \ldots, \left(\frac{1}{2}\right)^{p-1}$. This is still the correct Condition A when the entries of $\boldsymbol{M}$ are $r \times r$ matrices:

$$\boldsymbol{M} = 2\begin{bmatrix} \boldsymbol{h}(N) & \boldsymbol{h}(N-1) & \cdots & \cdots & & \\ & & \boldsymbol{h}(N) & \boldsymbol{h}(N-1) & \cdots & \cdots \end{bmatrix}.$$

The column sums $2\sum \boldsymbol{h}(2k)$ and $2\sum \boldsymbol{h}(2k+1)$ were 1 in the scalar case. They need not be $\boldsymbol{I}$ in the matrix case. That would be far too restrictive. To achieve $p = 1$, the sums must have $\lambda = 1$ as an eigenvalue with the same left eigenvector. The condition for higher p is recursive [StSt2,HeStSt]. In some way the scalar case is understood more deeply, when the theory of multiwavelets forces us into the matrix case.

Problem Set 7.5

1. **(Haar's Hat)** Find wavelets $w_1(t)$ and $w_2(t)$ on $[0, 1]$ that are orthogonal to each other and to $\phi_1(t) = $ box and $\phi_2(t) = 2t - 1$. They will be linear on $\left[0, \frac{1}{2}\right]$ and $\left[\frac{1}{2}, 1\right]$. Draw graphs of $w_1(t)$ and $w_2(t)$.

2. Express the wavelets in Problem **1** as combinations of $\phi_1(2t), \phi_1(2t-1), \phi_2(2t), \phi_2(2t-1)$. What are the highpass matrix coefficients $\boldsymbol{h}_1(0)$ and $\boldsymbol{h}_1(1)$?

3. Display a 4×4 block of the analysis bank and verify orthogonality:

$$\text{block} = \begin{bmatrix} \boldsymbol{h}(0) & \boldsymbol{h}(1) \\ \boldsymbol{h}_1(0) & \boldsymbol{h}_1(1) \end{bmatrix} = \begin{bmatrix} \text{lowpass} \\ \text{highpass} \end{bmatrix}.$$

4. Find the matrix dilation equation for the C^1 cubics $\phi_1(t)$ and $\phi_2(t)$ drawn in Figure 7.8. They are combinations of C^1 cubics on half-intervals.

5. The Fourier transform gives what product formula to solve the matrix dilation equation $\phi(t) = \sum 2\boldsymbol{h}(k)\phi(2t-k)$?

Chapter 8

Finite Length Signals

8.1 Circular Convolution and the DFT

Our signals have so far had infinite length. We expected $\boldsymbol{x}(n)$ and $f(t)$ to be defined forever—for all integers n and all real numbers t. The data streams for audio signals are so long that this model is very reasonable. Other applications (like image processing) have *data streams of a definite length* L. Then the finiteness of the signal $\boldsymbol{x}(0), \ldots, \boldsymbol{x}(L-1)$ must be taken into account.

The immediate problem is filtering a finite signal. The computation of $\sum \boldsymbol{h}(k)\boldsymbol{x}(n-k)$ may ask for $\boldsymbol{x}(-1)$ which is not defined. The purpose of extension is to define it. The difficulty is not in the middle of the signal (unless the filter is IIR). The problem is near the ends. One possibility is to *change to a one-sided boundary filter*. The end values of $\boldsymbol{x}$ are processed without crossing the boundary. The other possibility is to *extend the signal beyond the boundary.*

We concentrate first on extending the signal, to define $\boldsymbol{x}(n)$ for every n. Filtering this extension is equivalent to one particular set of boundary filters. Section 8.5 discusses boundary filters in general.

A finite-length filter bank uses L inputs to produce L outputs. An L by L matrix must be present. It is conceptually easiest to extend to an infinite signal, followed by ordinary time-invariant filtering. The question is how to extend, and there are many possibilities:

1. Extend by zeros (*zero-padding*) (*not studied*)

2. Extend by periodicity (*wraparound*) (*this section*)

3. Extend by reflection (*symmetric extension*) (*next section*).

These three methods are applied in Figure 8.1. You notice immediately a very important point:

> Zero-padding and wraparound generally introduce a *jump in the function.*
>
> Reflection generally introduces a *jump in the first derivative.*

Section 10.1 shows the three methods applied to a filtered image. The sky at the top is degraded by zero-padding. Wraparound is better but a thin error layer is visible. Symmetric extension is the best.

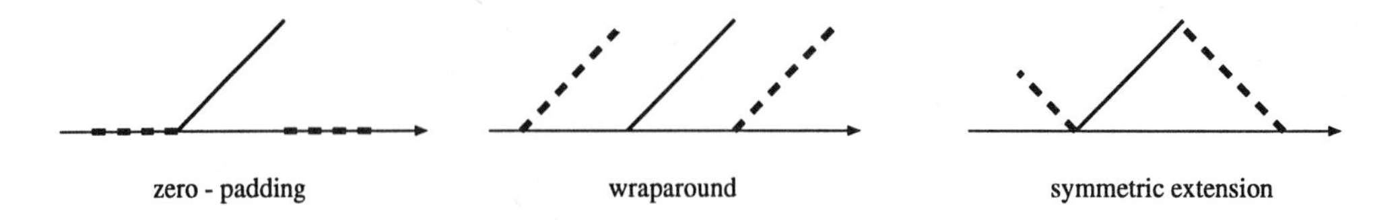

Figure 8.1: Three extension methods: zero-padding, wraparound, reflection.

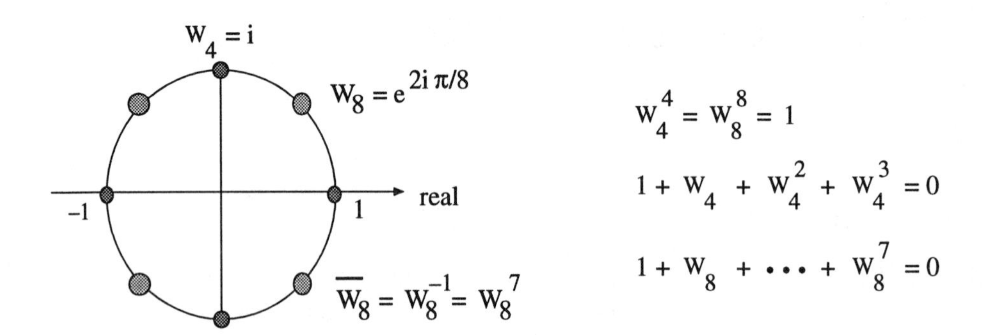

Figure 8.2: The fourth roots of 1 and the eighth roots of 1. They add to zero.

We do not pursue zero-padding in the text. It chops off the infinite Toeplitz matrix $\boldsymbol{H}$, with no adjustment at the boundary. Wraparound is good if the original problem is genuinely periodic or close to it. The filter matrix becomes a *circulant*. All indices are interpreted "*modulo L*". The kth diagonal of the matrix contains $\boldsymbol{h}(k)$, and it wraps around to continue as the $(k - L)$th diagonal.

All circulant matrices are diagonalized by the DFT. Therefore multiplications and inversions are very fast. Convolution $\boldsymbol{h} * \boldsymbol{x}$ is still multiplication of transforms $H(z)X(z)$ (with $z^L = 1$). In frequency this is the component-by-component multiplication $\widehat{\boldsymbol{h}}(k)\widehat{\boldsymbol{x}}(k)$. The inverse matrix exists if all $\widehat{\boldsymbol{h}}(k) \neq 0$, and it is also a circulant.

These circulant matrices are so simple and useful that we take time out to explain them (and also the DFT). Together they are a perfect match. Circulants have constant diagonals in the time domain. They are built out of (circular) shift matrices. The transform to frequency domain is executed at high speed by the Fast Fourier Transform, and we will put the FFT in a matrix form. The Fourier matrix is a product of very sparse matrices.

The next section returns to the critical question of symmetric extension. *Do we want period* $2(L-1)$ *or* $2L$? The answer depends on the length of the signal and the length of the filter! We hope those details will be useful.

Note 1 Other extensions are possible and interesting. Zero-padding could change to constant-padding (almost as simple and more accurate). A more careful extension could have constant slope. Section 8.5 shows how to fit a polynomial and extrapolate. In all cases we only need to provide the filter with a few values (at most N values) beyond the endpoints 0 and $L - 1$. Symmetric extension finds those values entirely by data addressing, without extra computations.

When it helps the understanding, we are free to assume that the periodic signals exist at all times $-\infty < n < \infty$. The essential problem is to implement a filter bank on a finite length signal, *without expanding that length.* Circular extension allows a probable jump at the endpoints.

Properties of the Circular Shift

The building blocks for time-invariant filters are the shift matrices $\boldsymbol{S}$ and $\boldsymbol{S}^{-1}$. All components of a vector are delayed one step (by $\boldsymbol{S}$) and advanced one step (by $\boldsymbol{S}^{-1}$). The building blocks for circulants (*cyclic filters*) are cyclic shifts. Every component is delayed by $\boldsymbol{S}_L$, except the last component which comes around to be the first component.

The 4 by 4 delay takes components $0, 1, 2, 3$ *forward* into positions $1, 2, 3, 4$ – except that 4 becomes 0. Algebraically, we work "modulo 4". Intuitively, our problem has period 4, so that $\boldsymbol{x}(4)$ is the same as $\boldsymbol{x}(0)$. Here is $\boldsymbol{S}_4$ multiplying the vector $\boldsymbol{x}$:

$$\boldsymbol{S}_4\boldsymbol{x} = \begin{bmatrix} 0 & 0 & 0 & 1 \\ 1 & 0 & 0 & 0 \\ 0 & 1 & 0 & 0 \\ 0 & 0 & 1 & 0 \end{bmatrix} \begin{bmatrix} \boldsymbol{x}(0) \\ \boldsymbol{x}(1) \\ \boldsymbol{x}(2) \\ \boldsymbol{x}(3) \end{bmatrix} = \begin{bmatrix} \boldsymbol{x}(3) \\ \boldsymbol{x}(0) \\ \boldsymbol{x}(1) \\ \boldsymbol{x}(2) \end{bmatrix}. \tag{8.1}$$

$\boldsymbol{S}_4$ is a *cyclic permutation matrix.* It has the same rows as $\boldsymbol{I}$, in a different order. Those rows are mutually orthogonal and they are unit vectors – so $\boldsymbol{S}_4$ is an orthogonal matrix. The shift matrices have very special properties:

1. The inverse equals the transpose. The Lth power of $\boldsymbol{S}_L$ is $\boldsymbol{I}$.

2. The eigenvalues of $\boldsymbol{S}_L$ are the L roots of 1 in the complex plane (on the unit circle in Figure 8.2). They are the powers of the number $W = e^{2\pi i/L}$. They are also the powers of $\overline{W} = W^{-1} = e^{-2\pi i/L}$.

   It is convenient to list the eigenvalues in the clockwise order $1, \overline{W}, \overline{W}^2, \ldots,$ because $\overline{W}^k = e^{-2\pi ik/L}$ fits into the Discrete Fourier Transform.

3. Corresponding to $\lambda = \overline{W}^k$ is the eigenvector $\boldsymbol{x} = (1,\ W^k,\ W^{2k},\ \ldots)$. These eigenvectors are the columns of the L by L ***Fourier matrix* = *DFT matrix*** $= \boldsymbol{F}_L$. The entries of $\boldsymbol{F}_L$ are $W^{nk} = e^{2\pi ink/L}$. This matrix diagonalizes the shift (and all circulants):

$$\boldsymbol{F}_L^{-1}\boldsymbol{S}_L\boldsymbol{F}_L = \boldsymbol{D}_L = \text{diag}\,(1,\ \overline{W},\ \overline{W}^2,\ \ldots). \tag{8.2}$$

We can quickly establish Properties **1–3**, using $\boldsymbol{S}_4$ as the example to work with:

$$\boldsymbol{S}_4^T = \begin{bmatrix} 0 & 1 & 0 & 0 \\ 0 & 0 & 1 & 0 \\ 0 & 0 & 0 & 1 \\ 1 & 0 & 0 & 0 \end{bmatrix} \quad \text{times} \quad \boldsymbol{S}_4 = \begin{bmatrix} 0 & 0 & 0 & 1 \\ 1 & 0 & 0 & 0 \\ 0 & 1 & 0 & 0 \\ 0 & 0 & 1 & 0 \end{bmatrix} \quad \text{is} \quad \boldsymbol{S}_4^T\boldsymbol{S}_4 = \boldsymbol{I}.$$

An advance times a delay is the identity. This is true and useful on the whole line, with times $-\infty < n < \infty$. It is true here *on the circle.* The times are $0, \ldots, L-1$, and they repeat. On an infinite half-line, $\boldsymbol{S}_+^T\boldsymbol{S}_+ = \boldsymbol{I}$ is true but $\boldsymbol{S}_+\boldsymbol{S}_+^T = \boldsymbol{I}$ is not true! An advance $\boldsymbol{S}_+^T$ will wipe out the first component $\boldsymbol{x}(0)$, and a delay $\boldsymbol{S}_+$ cannot bring it back. Real boundaries are distinctly harder than periodic boundaries.

The fourth power of S_4 is the identity matrix. When we shift forward four times, the cycle completes – we come back to the start. Always $(S_L)^L = I$. Here are S_4, S_4^2, $S_4^3 = S_4^{-1}$, and $S_4^4 = I$:

$$\begin{bmatrix} 0 & 0 & 0 & 1 \\ 1 & 0 & 0 & 0 \\ 0 & 1 & 0 & 0 \\ 0 & 0 & 1 & 0 \end{bmatrix}, \quad \begin{bmatrix} 0 & 0 & 1 & 0 \\ 0 & 0 & 0 & 1 \\ 1 & 0 & 0 & 0 \\ 0 & 1 & 0 & 0 \end{bmatrix}, \quad \begin{bmatrix} 0 & 1 & 0 & 0 \\ 0 & 0 & 1 & 0 \\ 0 & 0 & 0 & 1 \\ 1 & 0 & 0 & 0 \end{bmatrix}, \quad \begin{bmatrix} 1 & 0 & 0 & 0 \\ 0 & 1 & 0 & 0 \\ 0 & 0 & 1 & 0 \\ 0 & 0 & 0 & 1 \end{bmatrix}.$$

Properties **2** and **3** are about eigenvalues and eigenvectors. To find the eigenvalues, we normally solve $\det(S - \lambda I) = 0$. For the L by L shift matrix, this determinant (see Problem 1) yields the equation $\lambda^L = 1$. The eigenvalues are roots of unity, in agreement with the fact that $S^L = I$.

$$\textit{The determinant of} \quad \begin{bmatrix} -\lambda & 0 & 0 & 1 \\ 1 & -\lambda & 0 & 0 \\ 0 & 1 & -\lambda & 0 \\ 0 & 0 & 1 & -\lambda \end{bmatrix} \quad \textit{is} \quad \lambda^4 - 1.$$

The four eigenvalues are $1, -i, -1, i$. Those fourth roots of 1 are equally spaced around the unit circle, at angles $\frac{8\pi}{4}, \frac{6\pi}{4}, \frac{4\pi}{4}, \frac{2\pi}{4}$. In general the Lth roots are at multiples of the angle $\frac{2\pi}{L}$. Going clockwise around the circle, we have the L powers of $\overline{W} = e^{-2\pi i/L}$ in Figure 8.2.

For the circular shift, we can announce the eigenvectors at the same time. The eigenvector for $\lambda = 1$ is just $\boldsymbol{x} = (1, 1, \ldots, 1)$. The shift leaves it unchanged: $S\boldsymbol{x} = 1\boldsymbol{x}$. The second eigenvalue $\overline{W}$ has eigenvector $\boldsymbol{x} = (1, W, W^2, \ldots, W^{L-1})$. Shifting brings $W^{L-1} = W^{-1}$ to the top:

$$\textit{Circular shift of} \quad \begin{bmatrix} 1 \\ W \\ W^2 \\ \vdots \\ W^{L-1} \end{bmatrix} \quad \textbf{is} \quad \begin{bmatrix} W^{L-1} \\ 1 \\ W \\ \vdots \\ W^{L-2} \end{bmatrix} = W^{-1}\boldsymbol{x} = \overline{W}\boldsymbol{x}.$$

Thus $S\boldsymbol{x} = \overline{W}\boldsymbol{x}$ as required. Similarly the kth eigenvalue $\lambda = W^{-k} = \overline{W}^k$ has the eigenvector $\boldsymbol{x}_k = (1, W^k, W^{2k}, \ldots, W^{(L-1)k})$. Here is $S\boldsymbol{x} = \lambda\boldsymbol{x}$:

$$\textit{Circular shift of} \quad \begin{bmatrix} 1 \\ W^k \\ W^{2k} \\ \vdots \end{bmatrix} \quad \textbf{is} \quad \begin{bmatrix} W^{(L-1)k} \\ 1 \\ W^k \\ \vdots \end{bmatrix} = W^{-k}\boldsymbol{x} = \overline{W}^k\boldsymbol{x}.$$

We are making liberal use of the fact that $W^L = 1$. Now summarize:

Diagonalization of S by F. The eigenvectors of the shift S are the columns of the Fourier matrix F. The matrix $F^{-1}SF$ is diagonal:

$$F^{-1}SF = D \quad \text{and} \quad SF = FD \quad \text{and} \quad S = FDF^{-1}. \tag{8.3}$$

D is the diagonal matrix of eigenvalues $1, \overline{W}, \overline{W}^2, \ldots$. For $L = 4$ these matrices are displayed in the multiplication $S_4F_4 = F_4D_4$, whose columns are $Sx = \lambda x$:

$$\begin{bmatrix} 0 & 0 & 0 & 1 \\ 1 & 0 & 0 & 0 \\ 0 & 1 & 0 & 0 \\ 0 & 0 & 1 & 0 \end{bmatrix} \begin{bmatrix} 1 & 1 & 1 & 1 \\ 1 & i & i^2 & i^3 \\ 1 & i^2 & i^4 & i^6 \\ 1 & i^3 & i^6 & i^9 \end{bmatrix} = \begin{bmatrix} 1 & 1 & 1 & 1 \\ 1 & i & i^2 & i^3 \\ 1 & i^2 & i^4 & i^6 \\ 1 & i^3 & i^6 & i^9 \end{bmatrix} \begin{bmatrix} 1 & & & \\ & \bar{i} & & \\ & & \bar{i}^2 & \\ & & & \bar{i}^3 \end{bmatrix} \tag{8.4}$$
$$\quad S \qquad\qquad F \qquad = \qquad F \qquad\qquad D$$

The columns of F are ***orthogonal eigenvectors*** of S, with length $\sqrt{L}$.

Why are the eigenvectors orthogonal? The answer from linear algebra is: every real matrix with the property $SS^T = S^TS$ has orthogonal eigenvectors. These are the "normal matrices." In our case $SS^T = S^TS = I$.

A second answer is to verify orthogonality directly. The eigenvectors of S (the columns of F) are complex even though S is real. Permutations like S generally have complex eigenvalues and eigenvectors. Take the inner product:

$$\bar{x}_k^T x_\ell = \begin{bmatrix} 1 & \overline{W}^k & \overline{W}^{2k} & \cdots \end{bmatrix} \begin{bmatrix} 1 \\ W^\ell \\ W^{2\ell} \\ \vdots \end{bmatrix} = 1 + z + z^2 + \cdots + z^{L-1} = 0.$$

Here $z = \overline{W}^k W^\ell$. The sum is $(1 - z^L)/(1 - z)$. This is zero because $z^L = 1$ (remember that $W^L = 1$). So the eigenvectors of S are orthogonal to each other: $\overline{F}F = LI$. We are ready to move forward to circulants, via the DFT.

Discrete Fourier Transform and Circulants

An ordinary filter H is a combination of powers of the (infinite) shift matrix S. A circular filter H_L is a combination of powers (with $N < L$) of the finite shift matrix S_L:

Ordinary filter $H = \sum_{-\infty}^{\infty} h(k)S^k$ Circular filter $H_L = \sum_{n=0}^{N} h(n)(S_L)^n$.

H is a Toeplitz matrix and H_L is a ***circulant matrix***. Multiplication by H is convolution. Multiplication by H_L is circular convolution.

For $L = 4$, the four powers of S_4 were displayed earlier. They combine into a typical 4×4 circulant matrix $H_4 = h(0)I + h(1)S_4 + h(2)S_4^2 + h(3)S_4^3$:

$$H_4 = \begin{bmatrix} h(0) & h(3) & h(2) & h(1) \\ h(1) & h(0) & h(3) & h(2) \\ h(2) & h(1) & h(0) & h(3) \\ h(3) & h(2) & h(1) & h(0) \end{bmatrix}. \tag{8.5}$$

Notice how the diagonals wrap around. We still have $h(n)$ along diagonal n. In this circular world 3 is the same as -1, so $h(3)$ is also on diagonal -1. With periodicity we work modulo L. Two numbers are the same ($m \equiv n \bmod L$) when $m - n$ is a multiple of L.

Usually N is much smaller than L. We have a band of nonzeros below the diagonal, wrapping around to the upper right corner (as in S). Otherwise H_L is zero. We give the properties of

H_L and then the applications. The powers of S all have the same eigenvectors as S, so the key to a circulant is its *eigenvectors* and *eigenvalues*.

Diagonalization of H_L by F_L. The eigenvectors of H_L are the columns of the Fourier matrix F_L. The eigenvalues appear in the discrete Fourier transform $(\widehat{h}(0), \widehat{h}(1), \widehat{h}(2), \ldots)$ of the coefficients $(h(0), h(1), h(2), \ldots)$:

$$F_L^{-1} H_L F_L = \widehat{H}_L = \operatorname{diag}\big(\widehat{h}(0), \widehat{h}(1), \ldots, \widehat{h}(L-1)\big). \tag{8.6}$$

Proof. Every time we multiply $x = (1, W^k, W^{2k}, \ldots)$ by the shift matrix S, we get one more power of $\lambda = \overline{W}^k$. A combination of powers S^n produces the same combination of powers $\overline{W}^{kn}$:

$$H_L x = \Big(\sum_0^N h(n) S^n\Big) x = \Big(\sum_0^N h(n) \overline{W}^{kn}\Big) x = \widehat{h}(k) x. \tag{8.7}$$

The eigenvalues of H_L are exactly the components $\widehat{h}(k)$ of the DFT:

The eigenvalue for $x_0 = (1, 1, \ldots)$ is $h(0) + h(1) + \cdots = \widehat{h}(0)$
The eigenvalue for $x_k = (1, W^k, \ldots)$ is $h(0) + h(1)\overline{W}^k + \cdots = \widehat{h}(k)$.

The matrix form $HF = F\widehat{H}$ just gives these L equations all at once.

Example 8.1. Choose $h(0) = 4$, $h(1) = 1$ and $h(2) = 1$. Take $L = 3$:

$$H = \begin{bmatrix} 4 & 1 & 1 \\ 1 & 4 & 1 \\ 1 & 1 & 4 \end{bmatrix} \quad \text{has eigenvalues} \quad \begin{aligned} \widehat{h}(0) &= 4 + 1 + 1 = 6 \\ \widehat{h}(1) &= 4 + \overline{W} + \overline{W}^2 = 3 \\ \widehat{h}(2) &= 4 + \overline{W}^2 + \overline{W}^4 = 3. \end{aligned}$$

The eigenvalues are real because H is symmetric. The eigenvectors are orthogonal (of course). The first eigenvector is $(1, 1, 1)$. The other eigenvectors are $(1, e^{2\pi i/3}, e^{4\pi i/3})$ and its complex conjugate. We could change to real eigenvectors $(1, \cos\frac{2\pi}{3}, \cos\frac{4\pi}{3})$ and $(0, \sin\frac{2\pi}{3}, \sin\frac{4\pi}{3})$ if we wanted to (both with eigenvalue 3). This is the option in the real case of using sines and cosines. Everywhere we use $W^3 = 1$ and $1 + W + W^2 = 0$.

Example 8.2. Choose $h(0) = 4$, $h(1) = 1$, $h(2) = 0$, $h(3) = 1$ and $L = 4$:

$$H = \begin{bmatrix} 4 & 1 & 0 & 1 \\ 1 & 4 & 1 & 0 \\ 0 & 1 & 4 & 1 \\ 1 & 0 & 1 & 4 \end{bmatrix} \quad \text{has eigenvalues} \quad \begin{aligned} \widehat{h}(0) &= 4 + 1 + 0 + 1 = 6 \\ \widehat{h}(1) &= 4 + i^3 + 0 + i = 4 \\ \widehat{h}(2) &= 4 + i^2 + 0 + i^6 = 2 \\ \widehat{h}(3) &= 4 + i + 0 + i^3 = 4. \end{aligned}$$

Again H is symmetric and the eigenvalues are real. Note the eigenvector $(1, -1, 1, -1)$ corresponding to $\widehat{h}(2) = 2$. The DFT matrix diagonalizes H_L:

For infinite signals $y = Hx$ becomes $Y(z) = H(z)X(z)$
For length L signals $y = Hx$ becomes $\widehat{y} = \widehat{H}\widehat{x}$.

Hx is a matrix-vector multiplication. $\widehat{H}\widehat{x}$ is a pointwise multiplication.

We now explain the finite Fourier transform $\widehat{x}$ and circular convolution. The vector x has length L and so does $\widehat{x}$:

Definition 8.1 *The Discrete Fourier Transform of* $\boldsymbol{x}$ *is* $\widehat{\boldsymbol{x}} = \overline{\boldsymbol{F}}_L \boldsymbol{x}$. *Its components are*

$$\widehat{x}(k) = \sum_{n=0}^{L-1} x(n)\overline{W}^{nk} = \sum_{n=0}^{L-1} x(n)e^{-2\pi ink/L}. \tag{8.8}$$

The index n corresponds to time and k corresponds to frequency:

$$\text{DFT}[x(n)] = X(k) = \widehat{x}(k) \qquad \text{IDFT}[\widehat{x}(k)] = \text{IDFT}[X(k)] = x(n).$$

The inverse transform is $\boldsymbol{x} = \overline{\boldsymbol{F}}_L^{-1}\widehat{\boldsymbol{x}} = \frac{1}{L}\boldsymbol{F}_L\boldsymbol{x}$. Its components are

$$x(n) = \frac{1}{L}\sum_{k=0}^{L-1}\widehat{x}(k)e^{2\pi ink/L}. \tag{8.9}$$

The DFT analyzes the signal into pure frequencies. The IDFT synthesizes it again. The exponent is $-2\pi ink/L$ in (8.8) and $2\pi ink/L$ in (8.9), just as the Fourier integral in continuous time has $e^{-i\omega t}$ and $e^{i\omega t}$. Do not forget why the complex conjugate gives the inverse! It is because the columns of $\boldsymbol{F}_L = \boldsymbol{F}_L^T$ are orthogonal. The inner products give a multiple of the identity matrix:

$$\overline{\boldsymbol{F}}_L\boldsymbol{F}_L = L\boldsymbol{I} \quad \text{and} \quad \boldsymbol{F}_L\overline{\boldsymbol{F}}_L = L\boldsymbol{I} \quad \text{and} \quad \overline{\boldsymbol{F}}_L^{-1} = \tfrac{1}{L}\boldsymbol{F}_L. \tag{8.10}$$

Dividing by L gives the inverse, like dividing by 2π in continuous time. The transform preserves energy (apart from this factor L). This is the discrete version of Parseval's theorem $\int |x(t)|^2 dt = \frac{1}{2\pi}\int |X(\omega)|^2 d\omega$:

$$\textit{Discrete Parseval theorem:} \quad \|\boldsymbol{x}\|^2 = \tfrac{1}{L}\,\|\widehat{\boldsymbol{x}}\|^2. \tag{8.11}$$

This is $\sum |x(n)|^2 = \frac{1}{L}\sum |\widehat{x}(k)|^2$. It is based on orthogonality of the columns:

$$\langle \boldsymbol{x}, \boldsymbol{x}\rangle = \langle \tfrac{1}{L}\boldsymbol{F}_L\widehat{\boldsymbol{x}}, \tfrac{1}{L}\boldsymbol{F}_L\widehat{\boldsymbol{x}}\rangle = \tfrac{1}{L}\langle \widehat{\boldsymbol{x}}, \widehat{\boldsymbol{x}}\rangle \quad \text{because} \quad \overline{\boldsymbol{F}}_L^T\boldsymbol{F}_L = L\boldsymbol{I}.$$

The Circular Convolution Rule

Diagonalization by the Fourier matrix is true for all circulants. Therefore any multiplication $\boldsymbol{Hx}$ can be done two ways. A direct matrix multiplication is a "circular convolution". The indirect way is multiplication by $\boldsymbol{F}_L^{-1}$, then the diagonal matrix $\boldsymbol{D}_L$, then $\boldsymbol{F}_L$. This is a transform to the frequency domain, where $\boldsymbol{H}$ is diagonalized. This ***convolution rule*** underlies the whole theory of filters: convolution in time $\Leftrightarrow$ multiplication after DFT.

The circular filter $\boldsymbol{y} = \boldsymbol{Hx}$ becomes $\widehat{\boldsymbol{y}} = \widehat{\boldsymbol{H}\boldsymbol{x}}$ in frequency. This comes from the diagonalization $\boldsymbol{F}^{-1}\boldsymbol{HF} = \widehat{\boldsymbol{H}}$ and the fact that $\boldsymbol{F}\overline{\boldsymbol{F}} = L\boldsymbol{I}$:

$$\widehat{\boldsymbol{y}} = \widehat{\boldsymbol{H}\boldsymbol{x}} \quad \text{is} \quad \overline{\boldsymbol{F}}\boldsymbol{y} = (\boldsymbol{F}^{-1}\boldsymbol{HF})\overline{\boldsymbol{F}}\boldsymbol{x} \quad \text{which is} \quad \boldsymbol{y} = \boldsymbol{Hx}.$$

In words, the diagonalization is a change of basis from time to frequency. In the frequency domain, the filter multiplies each $\widehat{x}(k)$ by $\widehat{h}(k)$. We have agreement between a convolution $\boldsymbol{Hx}$ and a pointwise multiplication $\widehat{\boldsymbol{H}\boldsymbol{x}}$:

$$\begin{bmatrix} 4 & 1 & 1 \\ 1 & 4 & 1 \\ 1 & 1 & 4 \end{bmatrix}\begin{bmatrix} x(0) \\ x(1) \\ x(2) \end{bmatrix} \quad \text{transforms to} \quad \begin{bmatrix} 6 & & \\ & 3 & \\ & & 3 \end{bmatrix}\begin{bmatrix} \widehat{x}(0) \\ \widehat{x}(1) \\ \widehat{x}(2) \end{bmatrix}.$$

For a specific example take

$$\boldsymbol{x} = \begin{bmatrix} 2 \\ 1 \\ 1 \end{bmatrix} \quad \text{and} \quad \widehat{\boldsymbol{x}} = \begin{bmatrix} 1 & 1 & 1 \\ 1 & \overline{W} & \overline{W}^2 \\ 1 & \overline{W}^2 & \overline{W}^4 \end{bmatrix} \begin{bmatrix} 2 \\ 1 \\ 1 \end{bmatrix} = \begin{bmatrix} 4 \\ 1 \\ 1 \end{bmatrix}.$$

Parseval's identity says that $2^2 + 1^2 + 1^2$ should be $\frac{1}{L}$ times $4^2 + 1^2 + 1^2$. (6 is $\frac{1}{3}$ of 18.) The matrix multiplication $\boldsymbol{Hx}$ and the pointwise $\widehat{\boldsymbol{H}\boldsymbol{x}}$ give

$$\boldsymbol{y} = \boldsymbol{Hx} = \begin{bmatrix} 10 \\ 7 \\ 7 \end{bmatrix} \quad \text{and} \quad \widehat{\boldsymbol{y}} = \widehat{\boldsymbol{H}\boldsymbol{x}} = \begin{bmatrix} 6 \cdot 4 \\ 3 \cdot 1 \\ 3 \cdot 1 \end{bmatrix} = \begin{bmatrix} 24 \\ 3 \\ 3 \end{bmatrix}.$$

Again Parseval's identity checks: $10^2 + 7^2 + 7^2 = 198$ is $\frac{1}{3}$ of $24^2 + 3^2 + 3^2$.

Finally we display $\boldsymbol{Hx}$ as a ***circular convolution***. One way is to extend $\boldsymbol{x}$ periodically. Then ordinary convolution has periodic output 10, 7, 7 (repeated):

Periodic signal	2	1	1	2	1	1	2	1	1	2	1	1
Normal filter		1	1	4		1	1	4		1	1	4
			10				7				7	

Another way is to think of convolution as multiplication $H(z)X(z)$. But in this circular case $z^3 = 1$ and $z^4 = z$. Therefore $H(z)$ times $X(z)$ is

$$\begin{aligned} (4 + z + z^2)(2 + z + z^2) &= 8 + 6z + 7z^2 + 2z^3 + z^4 && \text{(ordinary)} \\ &= 10 + 7z + 7z^2 && \text{(wraparound).} \end{aligned}$$

This is $Y(z)$. Therefore $\boldsymbol{y} = (10, 7, 7)$. We summarize the essential points:

> The circular convolution of $\boldsymbol{h}$ with $\boldsymbol{x} = (\boldsymbol{x}(0), \boldsymbol{x}(1), \ldots)$ is $\boldsymbol{h} \circledast \boldsymbol{x} = \boldsymbol{Hx}$.
>
> All vectors have length L. The transform of $\boldsymbol{y} = \boldsymbol{h} \circledast \boldsymbol{x}$ is $\widehat{\boldsymbol{y}}(k) = \widehat{\boldsymbol{h}}(k)\widehat{\boldsymbol{x}}(k)$.

This convolution rule replaces the matrix multiplication $\boldsymbol{Hx}$ by the pointwise $\widehat{\boldsymbol{H}\boldsymbol{x}}$. We have to transform $\boldsymbol{h}$ and $\boldsymbol{x}$ to the Fourier domain! And then transform $\widehat{\boldsymbol{y}}$ back to $\boldsymbol{y}$. So three DFT's and the L multiplications $\widehat{\boldsymbol{h}}(k)\widehat{\boldsymbol{x}}(k)$ replace *one* circular convolution—which is multiplication by $\boldsymbol{H}$:

$$\boldsymbol{h} = \begin{bmatrix} 4 \\ 1 \\ 1 \end{bmatrix} \to \widehat{\boldsymbol{h}} = \begin{bmatrix} 6 \\ 3 \\ 3 \end{bmatrix} \searrow$$

$$\widehat{\boldsymbol{y}} = \widehat{\boldsymbol{h}}\widehat{\boldsymbol{x}} = \begin{bmatrix} 24 \\ 3 \\ 3 \end{bmatrix} \to \boldsymbol{y} = \begin{bmatrix} 10 \\ 7 \\ 7 \end{bmatrix}$$

$$\boldsymbol{x} = \begin{bmatrix} 2 \\ 1 \\ 1 \end{bmatrix} \to \widehat{\boldsymbol{x}} = \begin{bmatrix} 4 \\ 1 \\ 1 \end{bmatrix} \nearrow$$

In the time domain $\boldsymbol{Hx}$ needs about NL multiplications. The N filter coefficients are used L times each. Compare with three DFT steps, which are executed by the ***Fast Fourier Transform***. Each FFT requires only $\frac{1}{2}L \log_2 L$ individual steps. Transforming is worthwhile when $N > \frac{3}{2} \log_2 L$.

In matrix notation, $\boldsymbol{F}$ is the product of $\log_2 L$ very sparse matrices, involving only $L/2$ multiplications each. This FFT recursion connects the L-point transform to two copies of the $\frac{L}{2}$-point transform. It produces a matrix that is *half zero* and two very simple matrices around it. Here is the key to the FFT when $L = 1024$:

$$\textbf{\textit{FFT factors}}: \quad F_{1024} = \begin{bmatrix} I_{512} & D_{512} \\ I_{512} & -D_{512} \end{bmatrix} \begin{bmatrix} F_{512} & \\ & F_{512} \end{bmatrix} \begin{bmatrix} \text{even} \\ \text{odd} \end{bmatrix}. \tag{8.12}$$

I_{512} is the identity matrix. D_{512} is the diagonal matrix with entries $(1, W, \ldots, W^{511})$. The key is the two copies of F_{512} (which use W^2, the 512th root of 1). The permutation at the end separates the incoming vector into its even part $(\downarrow 2)x = (x(0), x(2), \ldots)$ and its odd part $(x(1), x(3), \ldots)$.

This reduction from F_L to two copies of $F_{L/2}$ almost cuts the work in half. We only need an extra 512 multiplications for D_{512}, plus additions and permutations. The full FFT algorithm keeps going, recursively. Each Fourier matrix F_{512} is replaced by two copies of F_{256}, with diagonal matrices D_{256} and a permutation that starts with $(\downarrow 4)$. Since 1024 is 2^{10}, it takes only 10 steps to go from F_{1024} to F_1! Maybe we don't go that far, but we could:

F_{1024} is the product of 10 matrices with I's and D's and a permutation.

The $10 = \log_2 L$ *matrices each require* $512 = \frac{1}{2}L$ *multiplications for the* D*'s. So the total count for the FFT is* $\frac{1}{2}L\log_2 L$.

This simple idea changes $(1024)^2$ multiplications for the full F_{1024} into $5(1024)$ multiplications for its ten sparse factors. That is a savings ratio of 200. Very rarely, perhaps never, has a single algorithm produced such a revolution as the FFT.

Problem Set 8.1

1. Find the determinant of $S_4 - \lambda I$ and the eigenvalues of S_4:

$$\det(S_4 - \lambda I) = \det \begin{bmatrix} -\lambda & 0 & 0 & 1 \\ 1 & -\lambda & 0 & 0 \\ 0 & 1 & -\lambda & 0 \\ 0 & 0 & 1 & -\lambda \end{bmatrix}.$$

2. Each term in an L by L determinant has one factor from each row and column. Why are $(-\lambda)^L$ and 1^L the only possibilities in the determinant of $S_L - \lambda I$? Why not products of $-\lambda$ with 1?

3. If $h(0), \ldots, h(3)$ is the first column of the circulant matrix H and $g(0), \ldots, g(3)$ is the first column of the circulant matrix G, show that the product GH is a circulant matrix with first column $g \circledast h$. One approach is to diagonalize $H = F\widehat{H}F^{-1}$ and $G = F\widehat{G}F^{-1}$.

4. Find the eigenvalues and eigenvectors of

$$H_3 = \begin{bmatrix} 6 & -1 & 1 \\ 1 & 6 & -1 \\ -1 & 1 & 6 \end{bmatrix} \quad \text{and} \quad H_4 = \begin{bmatrix} 0 & 1 & 2 & 1 \\ 1 & 0 & 1 & 2 \\ 2 & 1 & 0 & 1 \\ 1 & 2 & 1 & 0 \end{bmatrix}.$$

5. If $x = (1, 1, 1)$ then $H_3 x = (6, 6, 6)$. Find that answer from the convolution rule using $\widehat{h}(0)\widehat{x}(0)$ and $\widehat{h}(1)\widehat{x}(1)$ and $\widehat{h}(2)\widehat{x}(2)$.

6. Suppose the decimation operator $(\downarrow M)$ is applied to a periodized signal of length L. How many independent samples do we get if M divides L? If this occurs in each of M channels, find the total number of subsamples. Why is M required to divide L?

7. On the interval $0 \le t \le 1$, draw the four Haar wavelets and the four (periodized) Daubechies D_4 wavelets at scale level $\frac{1}{4}$. These are the basis functions for the 4-dimensional space W_2.

8. When is a circulant matrix invertible? The entries in its first column are $\boldsymbol{h}(0), \ldots, \boldsymbol{h}(L-1)$. The transform of that column vector is $(\widehat{\boldsymbol{h}}(0), \ldots, \widehat{\boldsymbol{h}}(L-1))$. The answer is in the transform, since $\boldsymbol{H}$ becomes a multiplication.

9. Write out explicitly the factorization (8.12) for the Fourier matrix $\boldsymbol{F}_4$.

8.2 Symmetric Extension for Symmetric Filters

Now we come to the heart of the finite length problem. A signal of length L will be extended in a symmetric way. The extended signal is $\boldsymbol{Ex}$ and the filter produces $\boldsymbol{HEx}$. If the filter is symmetric (or antisymmetric), then the output $\boldsymbol{HEx}$ will be symmetric (or antisymmetric). Downsampling yields $(\downarrow 2)\boldsymbol{HEx}$. ***The extension must be chosen so that this subsample is symmetric.*** Then we restrict attention to a piece of that subband signal, of length approximately $L/2$. (The exact integer will be crucial.) This restriction $\boldsymbol{R}$ yields the output from that channel:

$$\boldsymbol{R}(\downarrow 2)\boldsymbol{HEx} = (\textbf{\textit{restriction}})(\downarrow 2)(\textbf{\textit{filter}})(\textbf{\textit{extension of }}\boldsymbol{x}).$$

To repeat: A linear phase signal convolved with a linear phase filter has a linear phase output. You can see this in the time domain by simple algebra. In the frequency domain, the linear exponents in the phase add and stay linear. The key question is whether the *subsample* has linear phase, and what phase it has. We need to know the *point of symmetry* at every step of $\boldsymbol{R}(\downarrow 2)\boldsymbol{HEx}$.

The advantage of symmetric extension is this. *We may introduce a corner but we don't introduce a jump.* Wraparound creates a jump because generally $\boldsymbol{x}(L)$ does not equal $\boldsymbol{x}(0)$. With the extension, $\boldsymbol{x}(2L)$ *does* equal $\boldsymbol{x}(0)$. So you should extend before wraparound.

The Discrete Cosine Transform (DCT) is an example of symmetric extension. But there is no filter. It extends a block at a time, period. The DFT of the double block has only cosine terms, by symmetry. The DCT of Type II is the restriction back to the original block.

The DCT of Type IV extends even further, to period $4L$ instead of $2L$. After computing the DFT of a full period, we always restrict back to L coefficients. (More would be redundant, since we started with L samples.) Section 8.3 will describe both types of DCT. The present section is about extension and restriction of ordinary signals, before and after ordinary filtering.

Symmetric Extension and Symmetric Subsamples

One decision becomes crucial for multirate filter banks. Do we repeat the first and last samples, or do we not repeat? The results after the sampling operator $(\downarrow 2)$ are very different. We face a question that does not arise for the DCT. *Is the subsampled signal still symmetric?* That requirement will govern our extension of the input signal. We will take up in order the key decisions and the reasons behind them:

1. When to repeat the first and last samples (depending on N).

2. Where to start the subsampling (depending on L).

3. How many subsamples to keep from each channel.

The input is a set of L numbers $x(0), \ldots, x(L-1)$. The output is to be a total of L samples from the lowpass and highpass channels. We must choose a method that gives exactly L nonredundant outputs. Then we have a nonexpansive transform (a square matrix).

Repeating (H Extension) or Not Repeating (W Extension)

There are two symmetric ways to extend a signal that starts with $x(0)$. Everything depends on the choice of $x(-1)$. We may take $x(-1) = x(1)$, and then continue with $x(-2) = x(2)$. The "point of symmetry" is $t = 0$. When we do this at both ends, $t = 0$ and $t = L - 1$, the period is $2L - 2$. This is called ***whole-point symmetry***, and it is indicated by the letter **W**—not in italic.

This **W** extension corresponds to $x(-t) = x(t)$ in continuous time. In discrete time, it means that the number $x(0)$ is not repeated. At the other end, $x(L-1)$ also appears only once. All other samples $x(n)$ reappear as $x(-n)$.

The other possibility is to choose $x(-1) = x(0)$. The value at $t = 0$ is repeated at $t = -1$. The point of symmetry is halfway between, at $t = -\frac{1}{2}$. The extension continues with $x(-2) = x(1)$. Figure 8.3 shows how this *half-point symmetry* produces a signal that has period $2L$. It is sometimes referred to as a (2, 2) extension, to indicate the repetition at both ends. The extension is symmetric about $t = -\frac{1}{2}$ and $t = L - \frac{1}{2}$ and $t = 2L - \frac{1}{2}$. This half-point symmetry is indicated by the non-italic letter **H**.

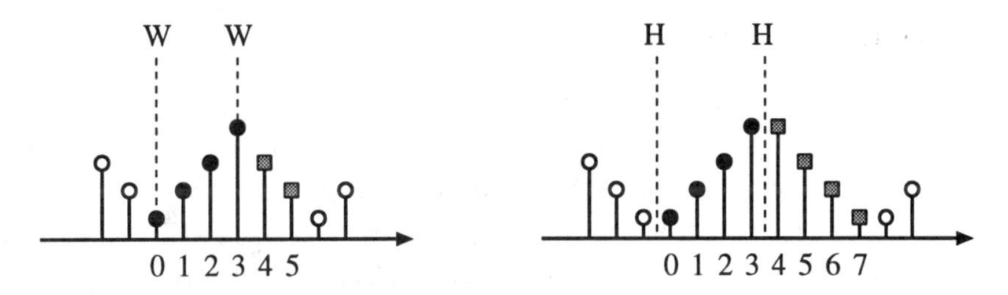

Figure 8.3: Whole-point symmetry (**W**) and half-point symmetry (**H**). The input has $L = 4$. The period is $2(L-1) = 6$ or $2L = 8$.

Which extension to choose, **W** *or* **H**? The answer depends on the filter length. *The right choice depends on whether the filter itself is* **W** *or* **H**. A symmetric filter with an odd number of coefficients, say $h(0)$, $h(1)$, $h(0)$, is a **W** filter. The point of symmetry is the center point 1. A symmetric filter with an even number of coefficients, say $h(0)$, $h(1)$, $h(1)$, $h(0)$, is an **H** filter. It is symmetric about the half-point $\frac{3}{2}$. The **H** filter has repetition, $h(1) = h(2)$, and the **W** filter has whole-point symmetry with no repetition of the center coefficient.

The goal is to have a symmetric extension after downsampling. Then restriction step will succeed. The subsampled signal $(\downarrow 2)y = (\downarrow 2)HEx$ is always periodic, but it is symmetric only if we follow the rule:

Use a **W** (no repeat) *extension for a* **W** (odd length) *filter.*

Use an **H** (repeat) *extension for an* **H** (even length) *filter.*

Those rules both give a **W** *extension for* HEx. Downsampling preserves symmetry.

A mixture of **H** and **W** would give an **H** extension for $y = HEx$. Downsampling an **H** extension goes wrong! Try it for $y(2), y(1), y(0), y(0), y(1), y(2)$.

The clearest way to justify this rule is to try to break it. Figure 8.4 shows the filtered outputs $\boldsymbol{y} = \boldsymbol{HEx}$ when the extension and filter are of opposite types (one is **W**, the other is **H**). Downsampling $\boldsymbol{y}$ will not produce a symmetric signal. Then the restriction of $(\downarrow 2)\boldsymbol{y}$ will fail. We show only failure here because there are many variations of success still to consider. The signal length L is going to enter soon (and $L = 4$ does not show all variations of failure either).

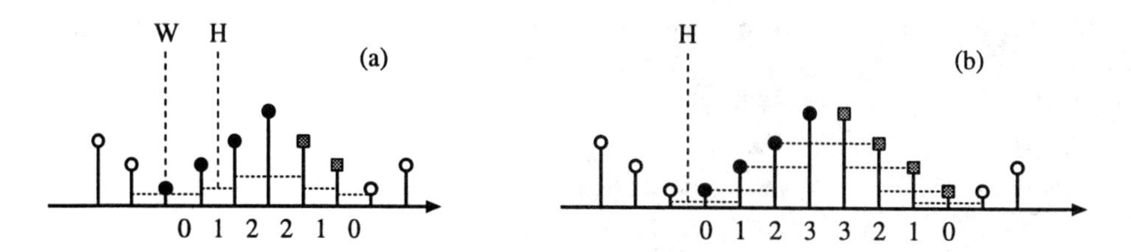

Figure 8.4: **H** and **W** do not mix. (a) **W** extension and **H** filter (two taps) give **H** extension. Downsampling yields 0, 2, 1 repeated. (b) **H** extension and **W** filter (three taps) give **H** extension. Downsampling yields 0, 2, 3, 1 repeated. *These subsamples are not symmetric and restriction fails.*

Length of Signal and Center of Filter

W extension will be chosen for **W** filters (odd length, symmetric or antisymmetric). **H** extension will be used for **H** filters. The next question is where to "center" the filter, and that depends on the signal length L. The goal is a total of L nonredundant samples from the two channels.

We do not study a $(1, 2)$ or $(2, 1)$ extension, with one endpoint repeated, because the period is then odd. This cannot be used with $(\downarrow 2)$. For an M-channel filter bank with odd M, new possibilities appear: **W** can mix with **H**. A complete table of admissible extensions is in the valuable paper by [Brislawn], which has been our guide. Earlier references are [SmEd, KiYaIw].

Start with the **W** extension (no repeat of end values). The filter length is odd, and both analysis filters (lowpass and highpass) are symmetric. The extended signal $\boldsymbol{Ex}$ has period $2(L-1)$. Figure 8.5 shows $L = 4$ on the left with an extension of period 6. The filtered values $\boldsymbol{y} = \boldsymbol{HEx}$ also have period 6. Those $\boldsymbol{y}$'s display a **W** extension (no repeats). Downsampling gives a symmetric extension. (It has to be $(1, 2)$ because the half-period $L - 1$ is odd.) Then restriction produces $L/2 = 2$ independent samples from each channel.

On the right side of Figure 8.5 is an odd-length signal ($L = 5$). The filter has odd length 3. So a **W** extension of $\boldsymbol{x}$ leads to a **W** extension for $\boldsymbol{y}$, after filtering. There are 5 independent $\boldsymbol{y}$'s. By correct centering (in other words, correct phase) we get $\frac{1}{2}(L+1)$ subsamples from the lowpass channel and $\frac{1}{2}(L-1)$ subsamples from the highpass channel. The analysis bank is nonexpansive, producing L outputs.

In the time domain this analysis bank is expressed by an $L \times L$ matrix. Then the synthesis bank (from biorthogonal filters) is expressed by the inverse matrix.

Now we go to an **H** extension (with repeat) for an **H** filter (even length). The extension has period $2L$. The lowpass filter will be symmetric and the highpass filter will be antisymmetric. For simplicity we use the average – difference pair with coefficients $\frac{1}{2}$, $\frac{1}{2}$ and $\frac{1}{2}$, $-\frac{1}{2}$. We still aim for L nonredundant subband samples.

L = 4
x(3)
x(2)
x(1)
x(0)
y = H E x
y(0) y(2) y(2) y(0)
y(1) y(3) y(1)
L = 5
x(4)
x(3)
x(2)
x(1)
x(0)
y = H E x
y(0) y(2) y(4) y(2) y(0)
y(1) y(3) y(3) y(1)
R (↓ 2) H E x = y(0) y(2)
R (↓ 2) H E x = y(1) y(3)
(low channel)
(high channel)
R (↓ 2) H E x = y(0) y(2) y(4)
R (↓ 2) H E x = y(1) y(3)

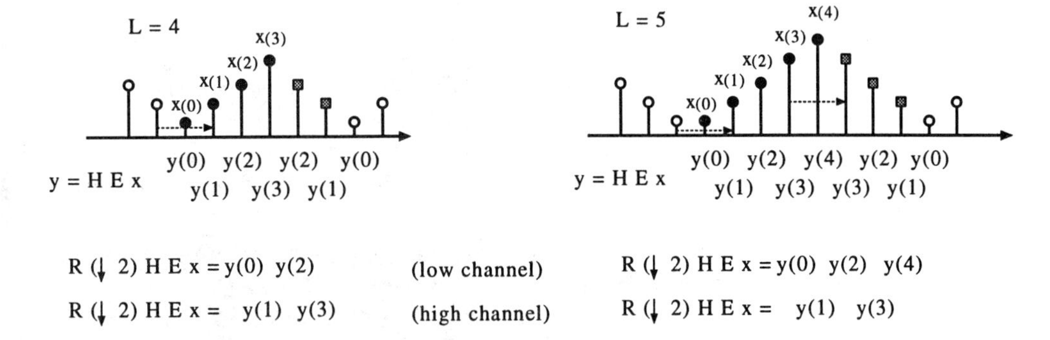

Figure 8.5: **W** extension and **W** filter yield $\boldsymbol{y} = \boldsymbol{HEx}$ with a **W** extension. The samples $(\downarrow 2)\boldsymbol{y}$ are symmetric and the restriction $\boldsymbol{R}$ succeeds.

On the left of Figure 8.6 is the case of even length $L = 4$ (four solid bullets). The extended period is 8. The lowpass filter produces $\boldsymbol{y}$'s and the antisymmetric highpass filter produces $\boldsymbol{z}$'s. (This combination is required for biorthogonality with even length.) Now the centering makes a difference. We can get subsamples of the original size from downsampling:

$$\boldsymbol{R}\,(\downarrow 2)\,\boldsymbol{HEx} = \boldsymbol{y}(0),\ \boldsymbol{y}(2),\ \boldsymbol{y}(4)\ \text{and}\ \boldsymbol{z}(2) \qquad \text{(total } L = 4)$$

Or, as we prefer, a change of center gives equal size samples from the two channels:

$$\boldsymbol{R}\,(\downarrow 2)\,\boldsymbol{HEx} = \boldsymbol{y}(1),\ \boldsymbol{y}(3)\ \text{and}\ \boldsymbol{z}(1),\ \boldsymbol{z}(3) \qquad \text{(total } L = 4)$$

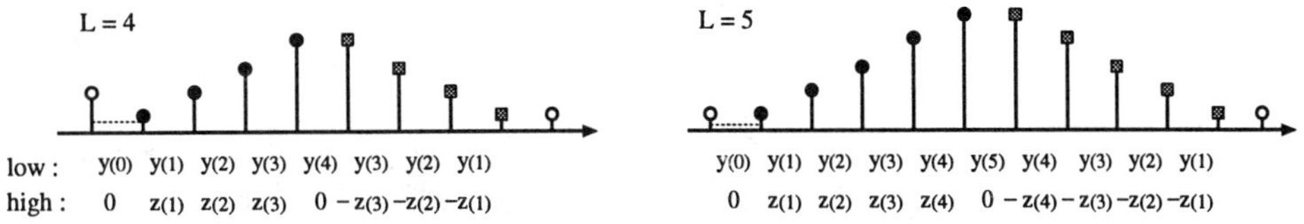

L = 4
low : y(0) y(1) y(2) y(3) y(4) y(3) y(2) y(1)
high : 0 z(1) z(2) z(3) 0 −z(3) −z(2) −z(1)
L = 5
y(0) y(1) y(2) y(3) y(4) y(5) y(4) y(3) y(2) y(1)
0 z(1) z(2) z(3) z(4) 0 −z(4) −z(3) −z(2) −z(1)

Figure 8.6: Even-length **H** filters with **H** extension. The filtered and subsampled signals have L independent samples: $L/2$ from each channel on the left, $(L+1)/2$ and $(L-1)/2$ on the right.

On the right of Figure 8.6 is a signal with odd length $L = 5$. The lowpass filter gives 6 independent $\boldsymbol{y}$'s. The highpass filter gives 4 independent $\boldsymbol{z}$'s. There will be 3 and 2 good subsamples. It is natural to keep the same centering for the two channels. We may select $\boldsymbol{y}(0),\ \boldsymbol{y}(2),\ \boldsymbol{y}(4)$ and $\boldsymbol{z}(2),\ \boldsymbol{z}(4)$. Or select $\boldsymbol{y}(1),\ \boldsymbol{y}(3),\ \boldsymbol{y}(5)$ and $\boldsymbol{z}(1),\ \boldsymbol{z}(3)$. In all cases L independent subsamples are returned by the analysis bank.

Problem Set 8.2

1. Show that the symmetric extension of $f(t)$ from $[0, \pi]$ to $[-\pi, 0]$ gives a function $Ef(t)$ that has no sine terms: $\int_{-\pi}^{\pi} Ef(t)\,\sin kt\,dt = 0$. Give a formula for $Ef(t)$.
2. Extend a function $x(t)$ to have period L (wraparound instead of symmetric extension). Draw the output from the centered filter $\boldsymbol{h}(-1) = \boldsymbol{h}(1) = \frac{1}{2}$. Rescale to $2t$ (period $L/2$) and draw the output $v(t)$.

3. Summarize the reason for choosing **H** extensions for **H** filters and **W** extensions for **W** filters: *We want the filtered values* $\boldsymbol{y} = \boldsymbol{HEx}$ *to be* **W**. If the filtered values are **H**, why does symmetry fail for $(\downarrow 2)\boldsymbol{y}$?
4. What is the difference in the z-domain between a **W**-extended signal and an **H**-extended signal? One of the extensions has $X(z) = X(z^{-1})$ and the other doesn't.
5. Draw the **W** and **H** extensions of a signal $\boldsymbol{x}$ of length $L = 6$. Apply a four-coefficient filter to each, and indicate the full period of the filtered signal $\boldsymbol{y}$. Which yields a satisfactory subsample $(\downarrow 2)\boldsymbol{y}$?
6. Draw the **W** and **H** extensions of a signal $\boldsymbol{x}$ of length $L = 7$. Apply a symmetric 5-coefficient filter and indicate a full period of the filtered signal $\boldsymbol{y}$. Which subsamples would you choose?
7. Suppose the decimation operator is $(\downarrow 3)$ instead of $(\downarrow 2)$. How many independent samples if $\boldsymbol{x}$ has length L, with **W** extension and then with **H** extension? The main requirement is that the period ($2L$ or $2L - 2$ or even $2L - 1$) should be divisible by M: ***Why?***
8. Show that $Ex(t) = \sum_{-\infty}^{\infty} [f(t - 2n) + f(2n - t)]$ gives the symmetric extension with period 2 of a function $x(t)$ defined on $[0, 1]$.

8.3 Cosine Bases and the DCT

The Discrete Cosine Transform (DCT) improves on the DFT for the same reason that symmetric extension improves on periodic extension. *The symmetric extension is continuous.* The periodic extension generally has a jump. This section develops the DCT as an extension followed by the DFT of size $2L$ followed by a restriction back to length L. There are *four types of* DCT coming from four types of extension.

We start in continuous time, by extending $f(t)$ from $[0, \pi]$ to $[-\pi, \pi]$. The symmetric extension is $Ef(t) = f(-t)$ on $[-\pi, 0]$ and then 2π-periodic. This even function $Ef(t)$ is a combination of cosines:

$$Ef(t) = \sum_{0}^{\infty} a_k \cos kt.$$

The discrete analogue is the DCT of Type I or II. There is a choice between **W** extension and **H** extension! We may define $\boldsymbol{x}(-1)$ to equal $\boldsymbol{x}(1)$ or to equal $\boldsymbol{x}(0)$. This choice doesn't arise in continuous time, but the previous section showed its importance in discrete time.

There is another extension in continuous time. It is surprising and very important. We call it E^{IV} because it leads to the DCT of Type IV. Start again with $f(t)$ on $[0, \pi]$. The extension is *even* across the left endpoint $t = 0$, as before, but it is *odd* across the right endpoint $t = \pi$. The extended (unfolded) function $E^{\mathrm{IV}} f(t)$ is drawn in Figure 8.7. *It has period* 4π. There is a jump at $t = \pi$ where the extension is odd. From the left side the graph approaches $f(\pi)$, from the right side the graph approaches $-f(\pi)$. This even-odd extension E^{IV} is remarkably useful.

Notice that $E^{\mathrm{IV}} f(t)$ involves cosines of period 4π (they are $\cos \frac{\ell t}{2}$). But it requires only *half of those cosines*, because the even frequencies $\ell = 2k$ have zero coefficients. These even frequencies yield $\cos kt$, which is an even function around $t = \pi$. The extension is an *odd* function around $t = \pi$, so the integral of $\cos kt$ times $E^{\mathrm{IV}} f(t)$ over a period is zero. We are left with the odd frequencies $\ell = 2k + 1$ and the basis functions $\cos \frac{\ell t}{2} = \cos\left(k + \frac{1}{2}\right) t$:

$$E^{\mathrm{IV}} f(t) = \sum_{0}^{\infty} a_k \cos\left(k + \tfrac{1}{2}\right) t. \tag{8.13}$$

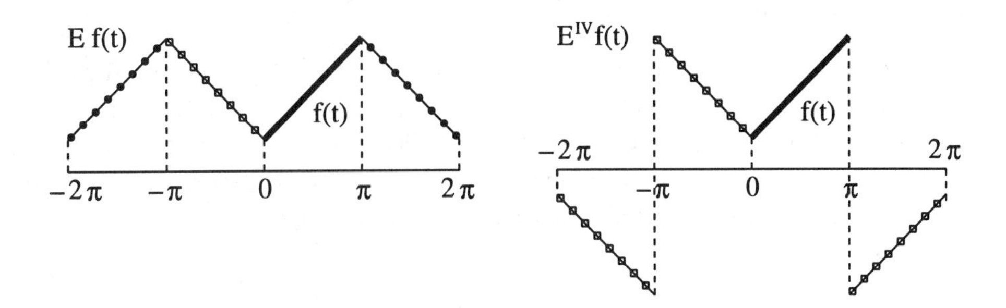

Figure 8.7: The symmetric extension $Ef(t)$ is continuous with period 2π. The even-odd extension $E^{\mathrm{IV}} f(t)$ is discontinuous with period 4π.

Discrete Extension and the DCT-II

The function $f(t)$ on $[0, \pi]$ is now replaced by a vector $(\boldsymbol{x}(0), \ldots, \boldsymbol{x}(L-1))$. We will extend that vector and apply the DFT (the Fourier matrix $\boldsymbol{F}_{2L}$) of *doubled length* $2L$. ***The whole-sample* W *extension is not satisfactory***. By symmetry it creates $\boldsymbol{x}(-n) = \boldsymbol{x}(n)$, extending the definition to $|n| \leq L-1$. This is $2L-1$ samples, and we do not want an odd number. It is better to use the half-sample **H** extension, which repeats $\boldsymbol{x}(0)$ as $\boldsymbol{x}(-1)$. The point of symmetry is $-\frac{1}{2}$, but this inconvenience is small. The symmetric **H** extension has period $2L$, and we call it $\boldsymbol{E}^{\mathrm{II}}$:

$$\boldsymbol{E}^{\mathrm{II}}\boldsymbol{x}(-n) = \boldsymbol{x}(n-1) \quad \text{for} \quad n = 1, \ldots, L. \tag{8.14}$$

The Fourier expansion of $\boldsymbol{E}^{\mathrm{II}}\boldsymbol{x}$ uses an ordinary DFT of length $2L$. That has $2L$ basis functions indexed by frequency k. The nth component of the kth basis vector is normally $e^{2\pi i k n/2L}$. But we shift n to $n+\frac{1}{2}$ in order to move the point of symmetry. The shifted signal is even around zero, and can be expressed by L cosines. The shift brings out a factor $e^{i\pi k/2L}$, and the expansion of the $2L$-periodic signal is

$$\boldsymbol{E}^{\mathrm{II}}\boldsymbol{x}(n) = \sum_{k=0}^{L-1} e^{i\pi k/2L} \widehat{\boldsymbol{x}}^{\mathrm{II}}(k) \cos \frac{\pi k\left(n+\frac{1}{2}\right)}{L}. \tag{8.15}$$

Orthogonality of the exponentials leads to orthogonality of the cosines (not trivial to see). The modulation factors $e^{i\pi k/2L}$ of absolute value 1 just multiply the separate basis functions (indexed by k).

The $L \times L$ DCT-II matrix in this cosine expansion has entries

$$\boldsymbol{C}^{\mathrm{II}}(k, n) = b(k)\sqrt{\frac{2}{L}} \cos \frac{\pi k\left(n+\frac{1}{2}\right)}{L}.$$

The square root of $\frac{2}{L}$ is included to make the rows of $\boldsymbol{C}^{\mathrm{II}}$ into unit vectors. So is the factor

$$b(k) = \begin{cases} 1/\sqrt{2} & \text{if } k = 0 \\ 1 & \text{if } k = 1, \ldots, L-1. \end{cases}$$

This is the familiar factor for the zero frequency $k = 0$ when the cosine stays at 1.

The key points about $\boldsymbol{C}^{\mathrm{II}}$ are its *orthogonality* and *fast execution*. Those are proved by the factorization (8.18) below. They follow from the connection to the DFT (and therefore the FFT).

Even–Odd Extension and the DCT-IV

The even-odd extension E^{IV} was applied to $f(t)$ in continuous time. Its discrete analogue leads to the DCT matrix of Type IV. This is the L by L matrix $\boldsymbol{E}^{\mathrm{IV}}$, with remarkable properties.

Again the original signal has L components $\boldsymbol{x}(n)$. This is extended evenly across $n = -\frac{1}{2}$, with **H** symmetry, to a signal with $2L$ components. *Then comes an antisymmetric* **H** *extension to* $4L$ *components.* Figure 8.8 shows the basic period of $4L$, which repeats.

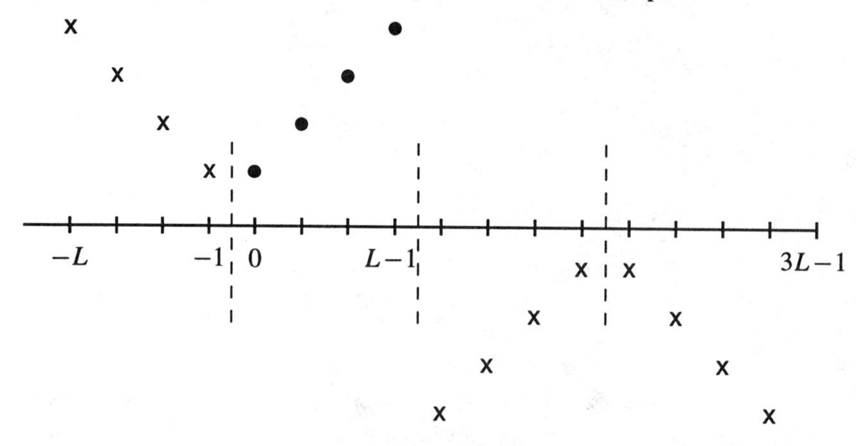

Figure 8.8: $\boldsymbol{x}^{\mathrm{IV}}(n)$ is even across $n = -\frac{1}{2}$ and odd across $n = L - \frac{1}{2}$. Here $L = 4$.

The basis functions for $E^{\mathrm{IV}} f(t)$ in continuous time were $\cos(k+\frac{1}{2})t$. Those came from odd frequencies $\ell = 2k+1$ and from period 4π. The basis functions in discrete time have period $4L$ and the same odd frequencies $\ell = 2k+1$. There is also the shift from n to $n+\frac{1}{2}$, which moves the point of symmetry to zero and the point of antisymmetry to L. Then the basis functions appear in the $L \times L$ DCT-IV matrix

$$C^{\mathrm{IV}}(k,n) = \sqrt{\frac{2}{L}}\,\cos\frac{\pi\left(k+\frac{1}{2}\right)\left(n+\frac{1}{2}\right)}{L}. \tag{8.16}$$

Notice the good properties of this matrix. It is symmetric. It needs no $b(k)$ to insert $\sqrt{2}$ into a zero-frequency term. Its columns are the even-odd basis vectors, and Figure 8.9 shows the first four vectors in the $\boldsymbol{C}^{\mathrm{IV}}$ basis of length $L = 32$. The DCT matrix also has *orthogonality* and *fast execution*, because of its direct link to the DFT matrix of order $2L$—which we now explain.

Matrix Factorizations and Orthogonality: DCT-II and DCT-IV

By slightly adapting Wickerhauser's very neat exposition [W, pp. 90–96], we can achieve three useful purposes at the same time:

1. Include the discrete *sine* transforms: DST-II and DST-IV

2. Prove the *orthogonality* of all four transforms: DCT and DST, II and IV

3. Show explicitly how these four transforms connect by extremely sparse matrices to the DFT matrix $\boldsymbol{F}_{2L}$.

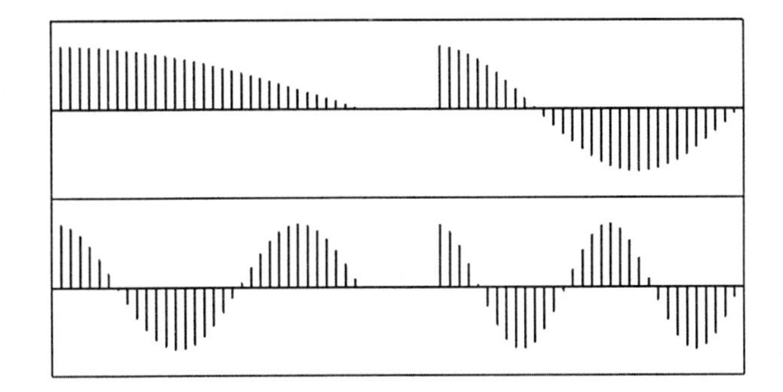

Figure 8.9: The first four basis functions of the DCT-IV, for $M = 32$.

These results come directly from two matrix factorizations, which we will display but not prove. A MATLAB exercise verifies their correctness for each L. You will notice that the DCT-IV and DST-IV formulas are cleaner, except for the multiplier $e^{-\pi i/4L}$ from the shift by $\frac{1}{2}$. Type II must assign $1/\sqrt{2}$ to the zero frequency DC term. Here are the formulas that prove everything.

$$\textbf{\textit{Factorization IV:}} \quad e^{-\pi i/4L}\boldsymbol{R}^T\boldsymbol{F}_{2L}\boldsymbol{R} = \begin{bmatrix} \boldsymbol{C}^{\text{IV}} & \\ & -i\boldsymbol{S}^{\text{IV}} \end{bmatrix} \tag{8.17}$$

$$\textbf{\textit{Factorization II:}} \quad \overline{\boldsymbol{P}}^T\boldsymbol{F}_{2L}\boldsymbol{Q} = \begin{bmatrix} \boldsymbol{C}^{\text{II}} & \\ & \boldsymbol{S}^{\text{II}} \end{bmatrix} \tag{8.18}$$

All matrices on the left are square and orthogonal of size $2L$. Then each block on the right is square and orthogonal of size L. The sine and cosine transforms preserve length (discrete Parseval identity as for the DFT). Note that $\boldsymbol{R}$ is only transposed while $\boldsymbol{P}$ is also conjugated.

It remains to display the sparse matrices, with at most two entries in each column and row. $\boldsymbol{J}$ is the reverse identity, with 1's on the antidiagonal. The combination of $\boldsymbol{I}$ and $\boldsymbol{J}$ gives the Gray ordering in (8.21). The matrices $\boldsymbol{R}$ and $\boldsymbol{P}$ clearly have orthogonal columns, and the division by $\sqrt{2}$ yields unit vectors:

$$\boldsymbol{R} = \frac{1}{\sqrt{2}}\begin{bmatrix} 1 & & & & 1 & & & \\ & \overline{W} & & & & \overline{W} & & \\ & & \ddots & & & & \ddots & \\ & & & \overline{W}^{L-1} & & & & \overline{W}^{L-1} \\ & & & W^L & & & & -W^L \\ & & \cdot^{\cdot^{\cdot}} & & & & \cdot^{\cdot^{\cdot}} & \\ & W^2 & & & & -W^2 & & \\ W & & & & -W & & & \end{bmatrix}, \quad W = e^{\pi i/2L}$$

$$\boldsymbol{P} = \frac{1}{\sqrt{2}}\begin{bmatrix} \sqrt{2} & & & & 0 & & & \\ & W & & & W & & & \\ & & \ddots & & & \ddots & & \\ & & & W^{L-1} & & & W^{L-1} & 0 \\ & & & 0 & & & & i\sqrt{2} \\ & & & \overline{W}^{L-1} & & & -\overline{W}^{L-1} & \\ & & \cdot^{\cdot^{\cdot}} & & & \cdot^{\cdot^{\cdot}} & & \\ 0 & \overline{W} & & & -\overline{W} & & & \end{bmatrix}, \quad \boldsymbol{Q} = \frac{1}{\sqrt{2}}\begin{bmatrix} \boldsymbol{I} & \boldsymbol{I} \\ \boldsymbol{J} & -\boldsymbol{J} \end{bmatrix}.$$

Summary: We now have two discrete cosine transforms, II and IV. There are two further types, I and III, from **W** extensions. Those are definitely less useful. If we omit the $\sqrt{2/L}$ normalization for simplicity, the DCT-II and DCT-IV transform a length L input signal into vectors X^{II} and X^{IV}:

$$X^{\mathrm{II}}(k) = b(k)\sum_{n=0}^{L-1} x(n)\cos\frac{\left(n+\frac{1}{2}\right)k\pi}{L} \tag{8.19}$$

$$X^{\mathrm{IV}}(k) = \sum_{n=0}^{L-1} x(n)\cos\frac{\left(n+\frac{1}{2}\right)\left(k+\frac{1}{2}\right)\pi}{L}. \tag{8.20}$$

Malvar describes the DCT-II as better for transform coding. The DCT-IV is good for spectrum estimation and adaptive filtering. For us the key point is the efficiency of the algorithms.

Several implementations are possible and we describe them briefly, for even L. First we connect each DCT to the DFT. Then we connect DCT-II to DCT-IV.

DCT-II via DFT. The trick is in this reordering of $x(n)$ (the Gray order):

$$y(n) = x(2n) \text{ and } y(L-n-1) = x(2n+1) \text{ for } n = 0, 1, \ldots, \frac{L}{2}-1. \tag{8.21}$$

Now take the DFT of $y(n)$. The DCT-II coefficients are

$$X^{\mathrm{II}}(k) = \cos\frac{\pi k}{2L}\,\mathrm{Re}[\widehat{y}(k)] - \sin\frac{\pi k}{2L}\,\mathrm{Im}[\widehat{y}(k)]. \tag{8.22}$$

The numbers $X^{\mathrm{II}}(k)$ and $X^{\mathrm{II}}(L-k)$ come from a plane rotation by $\frac{\pi k}{2L}$. Each step (permutation, DFT, rotation) preserves vector length. Again, an *orthogonal matrix*.

DCT-IV via DFT. The same reordering and modulation will create $\frac{L}{2}$ complex numbers

$$c(n) = (x(2n) + ix(L-1-2n))\,e^{-i\left(n+\frac{1}{4}\right)\pi/L}. \tag{8.23}$$

Now follows the half-length DFT. The kth output is multiplied by $e^{-ik\pi/L}$. Then the real and imaginary parts of these $\frac{L}{2}$ complex numbers give the DCT-IV coefficients $X^{\mathrm{IV}}(k)$.

Since codes are available, we imitate the simplified block form of Malvar:

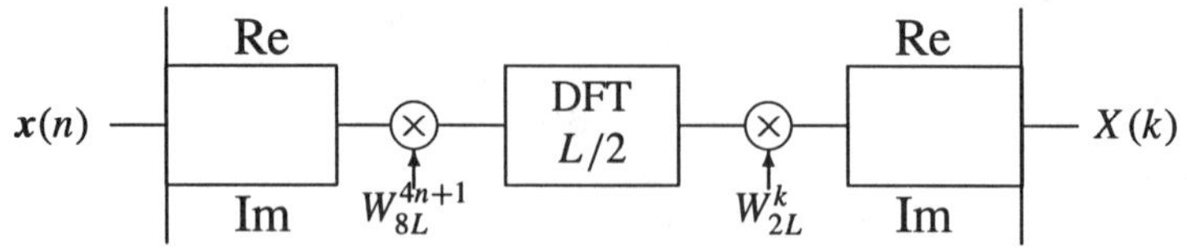

Re Re
$x(n)$ — [ ] — ⊗ — DFT $L/2$ — ⊗ — [ ] — $X(k)$
Im W_{8L}^{4n+1} W_{2L}^{k} Im

The algebra behind this algorithm is expressed in two lines, which require superhuman patience. The first line separates even and odd components $x(n)$:

$$X^{\mathrm{IV}}(k) = \sum_{n=0}^{\frac{L}{2}-1}\left[x(2n)\cos\frac{\left(2n+\frac{1}{2}\right)\left(k+\frac{1}{2}\right)\pi}{L} + x(L-1-2n)\cos\frac{\left(L-1-2n+\frac{1}{2}\right)\left(k+\frac{1}{2}\right)\pi}{L}\right].$$

Replace cosines by exponentials and $\boldsymbol{x}$'s by $\boldsymbol{c}$'s. Then separate into the real and imaginary parts of an $\frac{L}{2}$-point DFT times a modulation, for a fast DCT-IV:

$$\begin{matrix} X^{\mathrm{IV}}(2k) \\ X^{\mathrm{IV}}(L-1-2k) \end{matrix} = \begin{matrix} \mathrm{Re} \\ -\mathrm{Im} \end{matrix} \left[e^{-ik\pi/L} \sum_{n=0}^{\frac{L}{2}-1} c(n) e^{-4i\pi kn/L} \right] \tag{8.24}$$

DCT-II via DCT-IV. Finally we show that X^{II} of length L is immediately available from a DCT-II and a DCT-IV of length $\frac{L}{2}$. Recursively, this produces X^{II} from a sequence of DCT-IV's — whose fast computation was just given. The reduction of X^{II} comes by rewriting (8.19) in the form II plus IV:

$$\begin{aligned} X^{\mathrm{II}}(k) = b(k) \sum_{n=0}^{\frac{L}{2}-1} [\boldsymbol{x}(n) + \boldsymbol{x}(L-1-n)] \cos \frac{\left(n+\frac{1}{2}\right)k\pi}{L/2} \\ + \sum_{n=0}^{\frac{L}{2}-1} [\boldsymbol{x}(n) - \boldsymbol{x}(L-1-n)] \cos \frac{\left(n+\frac{1}{2}\right)\left(k+\frac{1}{2}\right)\pi}{L/2}. \end{aligned} \tag{8.25}$$

This recursion to a shorter DCT-II and DCT-IV was known early. It requires only human patience to verify. When a fast DCT-IV was not available, the connection was useless. Now the DCT-IV algorithm gives a unified approach to the two most important versions of the discrete cosine transform.

In closing we emphasize two points. First, orthogonality of the DCT comes from extension followed by DFT followed by restriction. Second, there are still blocking effects from the limited smoothness. Symmetric extension is a big improvement on periodicity. The next section shows how to smooth the basis functions much more.

Problem Set 8.3

1. Find the DCT-II and DCT-IV transforms of the signals $\boldsymbol{x} = (1, 1, 1, 1)$ and $\boldsymbol{x} = (1, -1, 1, -1)$.
2. Find the DCT-II and DCT-IV transforms of the signals $\boldsymbol{x} = (1, 0, 0, 0)$ and $\boldsymbol{x} = (0, 1, 0, 0)$.
3. Show that the Gray reordering $\boldsymbol{y}(n)$ in equation (8.21) transforms (8.19) into
$$\boldsymbol{x}^{\mathrm{II}}(k) = b(k) \sum_{n=0}^{L-1} \boldsymbol{y}(n) \cos \frac{\pi(4n+1)k}{2L}.$$
4. Write out explicitly the 2×2 and 3×3 DCT-II matrices and show that they are orthogonal.
5. The DCT-IV matrix has entries $\boldsymbol{C}_{kn} = \cos[\frac{\pi}{L}(k+\frac{1}{2})(n+\frac{1}{2})]$. Write down this symmetric $L \times L$ matrix for $L = 2$ and 3. Verify that columns 0 and 1 are orthogonal.
6. Execute this MATLAB verification of (8.17–8.18), connecting the DCT-DST-DFT:

```
L=16; i=eye(L); k=[0:L-1]; n=k';
dctiv=sqrt(2/L)*cos((k+1/2)*(n+1/2)*pi/L);   % orthogonal dctiv matrix
dctii=sqrt(2/L)*cos(k*(n+1/2)*pi/L);
dctii(1,:)=sqrt(1/2)*dctii(1,:);             % orthogonal dctii matrix
dft=sqrt(1/(2*L))*fft(eye(2*L));             % orthogonal dft matrix

Q=sqrt(1/2)*[i;i(L:-1:1,:)];                 % Q matrix
W=exp(sqrt(-1)*pi/(2*L)); Wb=conj(W);        % W=2L-th root of one
```

```
p1=diag([sqrt(2)*W.^ (1:L-1)]);         % P matrix
p2=diag([Wb.^ (1:L-1)]);
P=sqrt(1/2)*[p1;zeros(1,L);[zeros(L-1,1) p2(L-1:-1:1,:)]];
c=P'*dft*Q;                             % dft to dctii mapping

r1=diag(Wb.^ (0:L-1)); r2=diag(W.^ (1:L));
R=sqrt(1/2)*[r1;r2(L:-1:1,:)];          % R matrix
z=exp(-sqrt(-1)*pi/(4*L));              % z=conjugate of 4L-th root of one
d=real(z*conj(R)'*dft*R);               % dft to dctiv
                                       % conj(x') gives x' without conjugate!
max(max(abs(c-dctii)))                  % peak error, should be VERY small
max(max(abs(d-dctiv)))                  % peak error, should be VERY small
```

8.4 Smooth Local Cosine Bases

To "localize" a cosine, just multiply it by a real window function $g(t)$. Windowing is the central idea of the Short Time Fourier Transform (STFT). When we use $e^{-i\omega t}$ rather than cosines, this is the representation pioneered by Gabor:

$$\text{The windowed transform of } f(t) \text{ is } F(\omega, a) = \int_{-\infty}^{\infty} f(t)g(t-a)e^{-i\omega t}\,dt. \tag{8.26}$$

Note the two variables ω = oscillation frequency and a = window position. We have a *time-frequency plane*, in the same way that k and j for the wavelet $w\left(2^j t - k\right)$ place us on a *time-scale plane*. With all frequencies ω and all positions a, we expect redundancy in $F(\omega, a)$ and cannot expect orthogonality. Nevertheless $f(t)$ can be recovered:

$$\text{The inverse windowed transform is } f(t) = C_g \iint F(\omega, a)g(t-a)e^{i\omega t}\,dt\,da.$$

The constant C_g depends on the window. This STFT is competition for the continuous wavelet transform. Both are described in Section 2.6.

With discrete frequencies and positions, we have an oscillation (a cosine) inside a modulating envelope (the window). When the window is a simple box, each block is independent. A compressed image looks as if it is built of tiles. This blocking effect is reduced by filtering and by smooth windows, but ringing often persists and the result is not perfect. The DCT was established as the JPEG standard in 1992, after six years of work by the Joint Photographic Experts Group. But the technology continues to move forward.

This whole construction has recently been made sharper, in order to obtain an ***orthonormal basis***. Specific frequencies and specific windows are chosen, and this is what we want to explain. The frequencies involve the factor $k + \frac{1}{2}$ that is responsible for the even-odd symmetry of the extension E^{IV} and the discrete DCT-IV. Requirements are imposed on the windows to give orthogonality where one window overlaps its translate. This is not orthogonality of the windows alone or the cosines alone, but orthogonality of their products $g_{nk}(t)$ for $k = 0, 1, \ldots$ and $-\infty < n < \infty$:

$$g_{nk}(t) = g(t-n)\cos\left[\left(k+\tfrac{1}{2}\right)\pi(t-n)\right]. \tag{8.27}$$

If $g(t)$ is the box function, then $g(t-n) = 1$ on the time interval $[n, n+1]$. We have the basis functions (8.13) of the even-odd extension. They are the continuous analog of the DCT-IV basis. Moving to smoother windows $g(t-n)$ will reduce the blocking effects at the box edges.

A more general form, very useful in applications, allows windows of *varying length.* The time line is divided into unequal intervals $\left[t_n, t_{n+1}\right]$. There is a window function $g_n(t)$ for each interval. It equals one on an inner interval, and it drops toward zero as it passes t_n and t_{n+1}. The window reaches zero well before t_{n-1} on the left and t_{n+2} on the right. This function $g_n(t)$ overlaps only its two neighbors $g_{n-1}(t)$ and $g_{n+1}(t)$. We mention immediately one key requirement for orthogonality of the basis:

$$\sum (g_n(t))^2 \equiv 1. \tag{8.28}$$

The cosines within the window still have the even-odd symmetry associated with DCT-IV. They are *even* when continued across the left end t_n of the window, and they are *odd* across the right end t_{n+1}. This symmetry leads to orthogonality, when the windows have even symmetry at both ends and satisfy $\sum (g_n(t))^2 = 1$. The more general form of the basis functions is

$$g_{nk}(t) = g_n(t) \cos \frac{\left(k+\frac{1}{2}\right)\pi\,(t-t_n)}{t_{n+1}-t_n}. \tag{8.29}$$

This reduces to the previous form when $t_n = n$ and the window lengths are equal and the windows are translates of one window $g(t)$.

Varying window lengths is an extremely important advantage in speech processing. We attack and hold phonemes for very different lengths of time—and the waveforms in those intervals are very different. We indicate below how the windows lead to an orthogonal basis $g_{nk}(t)$ in this nonuniform case also. For simplicity of exposition, we concentrate first on the equal-interval construction of (8.27), which is important in discrete time too.

This section analyzes cosine windows in continuous time. The next chapter will study ***cosine-modulated filter banks*** in discrete time.

The Window Functions

Start with a single window $g(t)$ and its translates $g(t-n)$. This window function is supported on an interval $[-a, 1+a]$ that stretches beyond $[0, 1]$. If the extension length is $a < \frac{1}{2}$ at both ends, $g(t)$ will overlap only its nearest neighbors $g(t-1)$ and $g(t+1)$.

The window rises from zero at $t=-a$ to one at $t=a$. Here is a continuous rise:

$$g(t) = \sin\left[\frac{\pi}{4}\left(1+\frac{t}{a}\right)\right] \quad \text{for } -a \le t \le a.$$

At $t=-a$ this is $g(-a) = \sin 0 = 0$. At $t=a$ the window reaches $g(a)=1$. The window continues with $g(t) \equiv 1$ on the inner interval $[a, 1-a]$. At the right end $g(t)$ drops to zero in the same way that it rose. Then the window is symmetric around its center point $t=\frac{1}{2}$, and the reflection $t \to 1-t$ leaves it unchanged:

$$g(t) = \sin\left[\frac{\pi}{4}\left(1+\frac{1-t}{a}\right)\right] \quad \text{for } 1-a \le t \le 1+a.$$

We have drawn this $g(t)$ in Figure 8.10. The overlap with $g(t-1)$ is crucial. The key property in that overlap region is

$$(g(t-1))^2 + (g(t))^2 = \sin^2\left[\frac{\pi}{4}\left(1-\frac{t}{a}\right)\right] + \sin^2\left[\frac{\pi}{4}\left(1+\frac{t}{a}\right)\right] = 1.$$

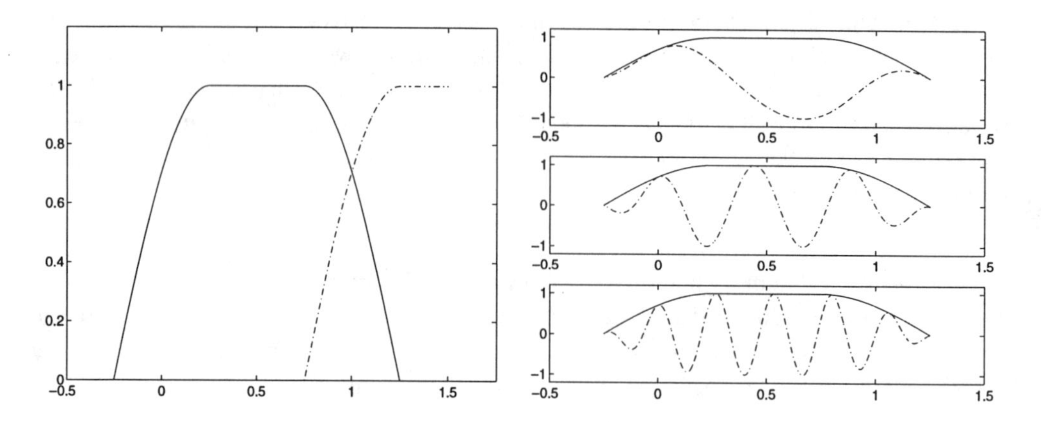

Figure 8.10: The window functions $g(t)$ and $g(t-1)$ for $a = 0.25$ (left). The basis functions of the local cosine transform for $k = 1, 4$ and 7.

This is really $(\text{cosine})^2 + (\text{sine})^2 = 1$. It leads directly to $\sum (g(t-n))^2 \equiv 1$ for all time. Away from the overlap intervals, a single window has unit height and all other windows are at zero.

This particular window can be made smoother. Replace t in its definition by $\sin \frac{\pi t}{2a}$. The derivative $g'(t)$ has a factor $\cos \frac{\pi t}{2a}$ from the chain rule, and that factor is zero at $t = \pm a$. Each repetition of this step introduces one more continuous derivative in $g(t)$. (We do not know of a careful search for an optimal a and an optimal window.) Allowing for unequal spacings and varying windows, the requirements at the overlap around $t = t_n$ are

$$(g_{n-1}(t))^2 + (g_n(t))^2 = 1 \quad \text{and} \quad g_{n-1}(t) = g_n(2t_n - t) \text{ for } |t - t_n| \le a_n. \tag{8.30}$$

In the time-invariant case, with uniform spacing, a fixed window $g(t)$ is shifted to $g_n(t) = g(t - n)$. The conditions (8.30) around $n = 0$ and $t_0 = 0$ apply at every $t_n = n$. These are the conditions that we work with first:

$$(g(t+1))^2 + (g(t))^2 = 1 \quad \text{and} \quad g(t+1) = g(-t) \text{ for } |t| \le a. \tag{8.31}$$

Local Orthogonal Bases

Suppose the window $g(t)$ is a box on [0, 1]. The basis functions $\cos(k + \frac{1}{2})\pi t$ are orthogonal within that box. They are even functions around $t = 0$, where the cosine equals one. They are odd functions around $t = 1$, where the cosine of $(k + \frac{1}{2})\pi$ equals zero. The main point is that *we can smooth those windows and retain orthogonality.* The first author learned the proof from lectures at MIT by Stéphane Mallat, which will lead to a book on time-frequency resolutions including wavelets [Mt].

Theorem 8.1 *The functions $g_{nk}(t) = g(t-n)\cos\left[(k + \frac{1}{2})\pi(t - n)\right]$ are an orthogonal basis for $L^2(\mathbf{R})$.*

Proof. The function $g_{0k}(t)$ is centered on [0, 1] and extends over $[-a, 1 + a]$. On the interval to the left, $g_{-1\ell}(t)$ is centered on $[-1, 0]$ and reaches as far right as $t = a$. The two functions overlap on the interval $[-a, a]$, where

$$g_{0k}(t)\, g_{-1\ell}(t) = g(t)\, g(t+1) \cos\left[(k + \tfrac{1}{2})\pi t\right] \cos\left[(\ell + \tfrac{1}{2})\pi(t+1)\right]. \tag{8.32}$$

We must show that the integral is zero and "the tails are orthogonal".

The product $g(t)g(t+1) = g(t)g(-t)$ is an even function around $t = 0$. So is the cosine of $(k+\frac{1}{2})\pi t$. The second cosine is odd, because the addition formula for cosines gives a sine:

$$\cos\left[(\ell+\tfrac{1}{2})\pi(t+1)\right] = -\sin\left[(\ell+\tfrac{1}{2})\pi t\right]\sin\left[(\ell+\tfrac{1}{2})\pi\right].$$

The reader notices that the key is in $\cos(\ell+\frac{1}{2})\pi = 0$! Then the product (8.32) is odd and its integral over $[-a, a]$ is zero. This proves orthogonality of any two g_{nk}'s with different n's.

Now consider $g_{0k}(t)g_{0\ell}(t)$. Both factors are centered on $[0, 1]$, with different frequencies. Define the product of cosines $C_{k\ell}(t)$ as $\cos\left[\left(k+\frac{1}{2}\right)\pi t\right]\cos\left[\left(\ell+\frac{1}{2}\right)\pi t\right]$. Then $C_{k\ell}(t)$ is even around $t = 0$ and also around $t = 1$. (Odd times odd is even.) The other factor is $(g(t))^2$. At the left end, the key is that $(g(t))^2 + (g(-t))^2 = 1$:

$$\begin{aligned}\int_{-a}^{a}(g(t))^2\,C_{k\ell}(t)\,dt &= \int_0^a (g(-t))^2\,C_{k\ell}(t)\,dt + \int_0^a (g(t))^2\,C_{k\ell}(t)\,dt\\ &= \int_0^a C_{k\ell}(t)\,dt. \end{aligned} \tag{8.33}$$

Around the right endpoint we use $(g(t))^2 + (g(t-1))^2 = 1$:

$$\int_{1-a}^{1+a}(g(t))^2\,C_{k\ell}(t)\,dt = \int_{1-a}^{1} C_{k\ell}(t)\,dt. \tag{8.34}$$

The center interval $[a, 1-a]$ has $g(t) = 1$. Again we are integrating $C_{k\ell}(t)$, the product of cosines. Orthogonality of the cosines on $[0, 1]$ gives orthogonality of $g(t)$ times cosines on the larger interval. Combine (8.33) and (8.34) to see that $g(t)$ disappears:

$$\int_{-a}^{1+a}(g(t))^2\,C_{k\ell}(t)\,dt = \int_0^1 C_{k\ell}(t)\,dt = \tfrac{1}{2}\delta(k-\ell). \tag{8.35}$$

This proves that the local cosines $g_{nk}(t)$ are **orthogonal**. The integral on the left can go from $-\infty$ to ∞, since $g(t)$ is zero on the rest of the line. It remains to prove that the local cosines are a **basis** for $L^2(\mathbf{R})$. How do we express an arbitrary $f(t)$ as a combination of $g_{nk}(t)$?

The ordinary cosines $\cos\left[\left(k+\frac{1}{2}\right)\pi t\right]$ are an orthogonal basis on $[0, 1]$. We "fold" the function $f(t)$ into this interval, creating $h(t)$:

$$\textbf{Folding} = h(t) = \begin{cases} g(t)f(t) + g(-t)f(-t) & \text{on } [0, a]\\ f(t) & \text{on } [a, 1-a]\\ g(t)f(t) - g(2-t)f(2-t) & \text{on } [1-a, 1]\end{cases}$$

Within $[0, 1]$ expand $h(t)$ in the basis $\cos\left(k+\frac{1}{2}\right)\pi t$. Those cosines extend $h(t)$ outside $[0, 1]$, keeping it even around $t = 0$ and odd around $t = 1$. These symmetries are built into the definition of $h(t)$, so there is no change in formula on the larger interval $[-a, 1+a]$. Multiplying by $g(t)$ turns $h(t) =$ *combination of cosines* into $g(t)h(t) =$ *combination of* $g_{0k}(t)$. Here is $g(t)h(t)$:

$$g(t)h(t) = \begin{cases} (g(t))^2\,f(t) + g(t)g(-t)f(-t) & \text{on } [-a, a]\\ f(t) & \text{on } [a, 1-a]\\ (g(t))^2\,f(t) - g(t)g(2-t)f(2-t) & \text{on } [1-a, 1+a]\end{cases}$$

A similar step takes place on the interval $[-1, 0]$. Replace t by $t+1$ in all formulas. Then $h_{-1}(t)$ is defined from $f(t)$ and expressed as a combination of cosines. Multiplication by the shifted window gives $g(t+1)h_{-1}(t)$ as a combination of windowed cosines $g_{-1\ell}$, while $g(t)h(t)$ is produced out of $g_{0k}(t)$. All we have to check is that on the overlap interval $[-a, a]$, the sum $g(t)h(t) + g(t+1)h_{-1}(t)$ adds to $f(t)$. Since $g(t+1) = g(-t)$ in the overlap, we have it:

$$(g(t))^2 f(t) + g(t)g(-t)f(-t) + (g(-t))^2 f(t) - g(-t)g(t)f(-t) = f(t). \tag{8.36}$$

Thus every $f(t)$ is a combination of local cosines. Those are an orthogonal basis. *Theorem proved.*

Note that $g(t)h(t)$ is the even part of $f(t)$ around $t = 0$, and $g(t+1)h_{-1}(t)$ is the odd part. Equation (8.36) is simply (even part) + (odd part) $= f(t)$. An early construction by Wilson alternated $\cos k\pi t$ in one interval with $\sin k\pi t$ in the next interval. The cosines are even at both ends, the sines are odd at both ends, and each overlap interval again has even and odd—which gives orthogonality and reproduces $f(t)$ as above.

The proof also yields a ***fast algorithm***. To expand $f(t)$ in the local cosines, we expanded $h(t)$ and $h_{-1}(t)$ in ordinary cosines on $[0, 1]$ and $[-1, 0]$. So compute those coefficients b_{0k} and $b_{-1\ell}$ by a fast DCT-IV. The same computation on other unit intervals gives all coefficients in $f(t) = \sum b_{nk} g_{nk}(t)$. We record this fact for the basic interval $[0, 1]$ and the coefficients b_{0k}:

$$2b_{0k} = \int_{-\infty}^{\infty} f(t)\, g_{0k}(t)\, dt = \int_0^1 h(t) \cos\left[\left(k+\tfrac{1}{2}\right)\pi t\right] dt. \tag{8.37}$$

The local cosine coefficients of $f(t)$ are the ordinary cosine coefficients of the "folded" function $h(t)$. The normalization by 2 comes from (8.35).

Unequal Spacing and Different Windows

The analysis was simplified above, by assuming all windows to be translates $g(t-n)$ of a basic window. This is not necessary. The construction allows different windows $g_n(t)$ on intervals of different lengths. The time variable $t - n$ in the cosines can become $(t - t_n)\,/\,(t_{n+1} - t_n)$. The functions $g_{nk}(t)$ still provide an orthonormal basis.

The changes are easy to indicate, and their confirmation is a good exercise. The window $g_n(t)$ in Figure 8.11 extends from $t_n - a_n$ to $t_{n+1} + a_{n+1}$. The previous window $g_{n-1}(t)$ ends at $t_n + a_n$, and on the overlap interval the requirements are (8.30). The condition $a_n + a_{n+1} \le t_{n+1} - t_n$ avoids any overlap of $g_{n-1}(t)$ with $g_{n+1}(t)$. This is the non-uniform equivalent of $2a \le 1$. Then we have $\sum (g_n(t))^2 \equiv 1$ as desired.

The local cosine coefficients b_{nk} of an arbitrary $f(t)$ are still the ordinary cosine coefficients of *folded functions* $h_n(t)$:

$$\int_{-\infty}^{\infty} f(t) g_n(t) \cos\left[\left(k+\tfrac{1}{2}\right)\pi \frac{t-t_n}{t_{n+1}-t_n}\right] dt = \int_{t_n}^{t_{n+1}} h_n(t) \cos\left[\left(k+\tfrac{1}{2}\right)\pi \frac{t-t_n}{t_{n+1}-t_n}\right] dt. \tag{8.38}$$

The latter integral moves to the interval $[0, 1]$ and is quickly computed by the DCT-IV. The folded function is

$$h_n(t) = \begin{cases} g_n(t)f(t) + g_{n-1}(t)f(2t_n - t) & \text{on } [t_n, t_n + a_n] \\ f(t) & \text{on } \left[t_n + a_n, t_{n+1} - a_{n+1}\right] \\ g_n(t)f(t) - g_{n+1}(t)f(2t_{n+1} - t) & \text{on } \left[t_{n+1} - a_{n+1}, t_{n+1}\right] \end{cases}$$

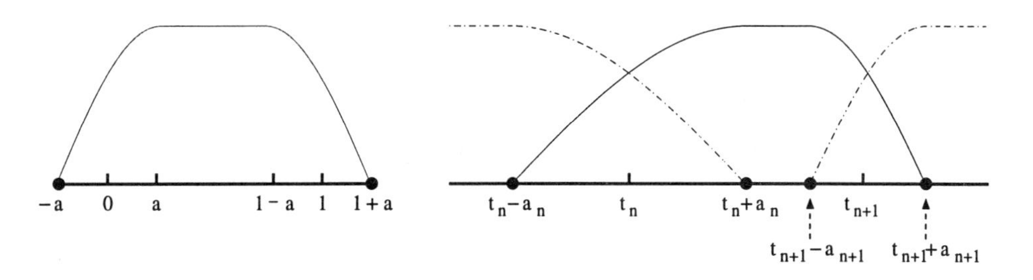

Figure 8.11: Window $g(t)$ and its non-uniform translates at t_n and t_{n+1}.

This even-odd folding matches the even-odd property of $\cos(k+\frac{1}{2})\pi t$. An even-even extension would fail! (Unless it is alternated with odd-odd.) The construction is neat and efficient, and must be regarded as strong competition for wavelets. A fully detailed description is given by Wickerhauser [W]. It has a close relation to wavelets, as we now indicate.

Sinc wavelets and Meyer wavelets The local cosines give a splitting of the time axis. The sinc wavelets are also local cosines, but on the frequency axis. Remember from Section 2.3 that the sinc wavelet and its Fourier transform are

$$\begin{aligned} w(t) &= \frac{\sin 2\pi t - \sin \pi t}{\pi t} = \phi(2t) - \phi(t) \\ \widehat{w}(\omega) &= \begin{cases} 1 & \text{for } \pi \le |\omega| < 2\pi \\ 0 & \text{otherwise.} \end{cases} \end{aligned}$$

When the wavelet is dilated and shifted to $w\left(2^j t - k\right)$, its Fourier transform is moved to the interval $2^j\pi \le |\omega| < 2^{j+1}\pi$. It is modulated by $\exp\left(-i\omega k/2^j\right)$. The transform becomes a local cosine, or rather a local exponential. It is not smooth in ω because the wavelet $w(t)$ does not decay quickly in time.

Note that the order of accuracy is $p = \infty$! The transform is *infinitely flat* at $\omega = 0$ and $\omega = \pi$. (The scaling function is a box function in ω, the ideal lowpass filter.) This corresponds to the ***spectral method*** in the numerical solution of partial differential equations, which also achieves $p = \infty$. Finite differences and finite elements have finite p, like FIR filters. An infinite value of p can only be achieved by functions $\phi(t)$ and $w(t)$ with infinite support. They are nonzero on the whole line, but the simplest sinc construction has slow decay.

Smoothing the local cosines (in ω) will produce faster decay (in t). This leads to the Meyer wavelets, which are *windowed in frequency*. The window can have infinite smoothness in ω, so the time decay can be faster than any power of t. The orthogonality of the sincs is maintained by the conditions on the windows. But windowing in time (to get smooth local cosines) is more practical.

Biorthogonal Local Cosines

Every wavelet construction becomes more general when the orthogonality requirement is lifted. The same is true of local cosines. For *biorthogonal* bases, the sum of squares $S(t) = \sum_{-\infty}^{\infty}(g(t-n))^2$ of window functions is not necessarily one. Then a ***dual window*** is defined by

$$\widetilde{g}(t) = \frac{g(t)}{S(t)}. \tag{8.39}$$

The dual basis functions are the local cosines using the dual window:

$$\widetilde{g}_{nk}(t) = \widetilde{g}(t-n)\cos\left[\left(k+\tfrac{1}{2}\right)\pi\,(t-n)\right]$$

[Matv] gives a nice discussion of biorthogonality of the dual bases. This follows directly from orthogonality with window functions $g_{\text{orth}}(t-n) = g(t-n)/\sqrt{S(t)}$ on the interval $\left[-\frac{1}{2}, \frac{1}{2}\right]$. The reader sees immediately that the sum of $g^2_{\text{orth}}(t-n)$ is $S(t)/S(t) = 1$.

The condition number of the dual bases governs stability. This number is one for an orthonormal basis (using the earlier windows $g(t-n)$) and greater than one for a biorthogonal basis (using $g(t)$ and $\widetilde{g}(t)$). The condition number naturally depends on the distance of $S(t)$ from 1. Matviyenko shows how to construct windows with $S_{\max}(t) \le 2S_{\min}(t)$. They give better compression than the orthogonal local cosines, and stability is well in control. He carries out an approximate optimization of $g(t)$ for the compression of sinusoids $\cos(\omega t + \alpha)$ of arbitrary frequency and phase, and tabulates the resulting windows. The graphs of $\widetilde{g}(t)$ show a double peak which looks more alarming than it is.

The main point is *freedom versus constraints*, in the choice of window as in the choice of wavelets. Freedom gives the possibility of smooth windows. Constraints make those windows successful. A possible constraint is ***accuracy***: to reproduce exactly a constant function (or all polynomials of degree less than p). We can further limit the search to functions $g(t) = \sum c_k \cos\left[\left(k+\frac{1}{2}\right)\pi t\right]$ around $t=0$, and ask for maximum smoothness at $t = \pm\frac{1}{2}$. This leads to "spline windows." The real tests of coding gain and compression are still ahead.

Note. The ***folding operator*** is defined above by $h(t) = \boldsymbol{F}f(t)$. There is also an ***unfolding operator***. When they use the same window, these operators $\boldsymbol{F}$ and $\boldsymbol{U}$ are transposes. The plus sign and minus sign on the first and third lines of $h(t)$ are reversed for the unfolded $u(t) = Uh(t)$. When $\boldsymbol{F}$ and $\boldsymbol{U}$ use dual windows $g(t)$ and $\widetilde{g}(t)$, these operators are inverses. When they use the self-dual windows that have $S(t) = \sum g^2(t-n) \equiv 1$, the operators are both transposes and inverses. Therefore they are unitary operators. The families $\{\boldsymbol{F}_n\}$ and $\{\boldsymbol{U}_n\}$ which fold and unfold around $t=n$ (or more generally $t = t_n$) are the foundation for a full theory of smooth local cosines.

Jawerth, Liu, and Sweldens point out that in image compression, *folding can be seen as a preprocessing step for JPEG*. It is a generalized extension. The extension can be even or odd, symmetric or antisymmetric, and it changes the signal also *inside* the basic interval. The overlapping intervals allow perfect reconstruction—orthogonal or biorthogonal—of the signal.

Problem Set 8.4

1. Why does the linear rise $g(t) = \frac{1}{2}\left(1+\frac{t}{a}\right)$ in the interval $[-a, a]$ not lead to orthogonality? The problem is not lack of smoothness.

2. A rise function can be written as $g(t) = \sin\left(\theta\left(\frac{t}{a}\right)\right)$. Here $\theta(-1) = 0$ and $\theta(1) = \frac{\pi}{2}$. Show that $\theta(t) + \theta(-t) = \frac{\pi}{2}$ assures $(g(t))^2 + (g(-t))^2 = 1$.

3. Compute a cubic spline $\theta(t)$, increasing from $\theta(-1) = 0$ to $\theta(1) = \frac{\pi}{2}$, that meets the requirement in Problem 2.

4. (Thesis project) Experiment with several uniform windows $g(t)$ and several overlap lengths a to optimize compression. Compare with the blocking effects from a window box.

5. Find the parity (even or odd?) of the function $\sin\left(k+\frac{1}{2}\right)\pi t$ around the endpoints of $[0, 1]$. This is the sine-IV basis used in the DST-IV transform.

6. Show that the conditions on the uniform windows $g(t-n)$ can be written as

$$\begin{array}{lccccc} g^2(t)+g^2(-t)=1 & -\frac{1}{2} & \leq & x & \leq & \frac{1}{2} \\ g(t)=g(1-t) & \frac{1}{2} & \leq & x & \leq & \frac{3}{2} \\ g(t)=0 & \text{elsewhere.} & & & & \end{array}$$

This yields $g(t)\equiv 1$ in $[a, 1-a]$ where there is no overlap.

7. (a) Draw a linear function $f(t)$ on $[0, 1]$ and its folding $h(t)=Ff(t)$.

(b) Draw a linear function $h(t)$ on $[0, 1]$ and its unfolding $u(t)=Uh(t)$.

8. Explain (8.34) by following the proof of (8.33).

8.5 Boundary Filters and Wavelets

The striking fact about FIR filter banks is that ***the banded analysis matrix has a banded inverse***. That inverse is the synthesis matrix.

Banded matrices correspond to FIR filters and compactly supported wavelets. The basis functions have finite length. The matrices may even be orthogonal. The question is how to maintain these properties at a boundary, by adding new rows and new functions:

Boundary filters are the "end rows" *of an $L\times L$ filter matrix $\boldsymbol{H}_L$.*

Boundary scaling functions and wavelets are the "end functions" *of a basis.*

If we construct boundary filters, the dilation equation yields boundary functions. The reverse direction also succeeds. The functions have dilation coefficients that go into the filters. All constructions are in the time domain, because transform methods have major difficulty at boundaries. We begin with boundary filters (completing a matrix) and then create boundary functions (completing a basis). Those are really equivalent.

The matrices begin with the lowpass $\boldsymbol{H}_0$ and the highpass $\boldsymbol{H}_1$, subsampled. These are infinite 1×2 block Toeplitz matrices. The rows have a double shift from $(\downarrow 2)$. *We interleave the filters* $(\downarrow 2)\boldsymbol{H}_0$ *and* $(\downarrow 2)\boldsymbol{H}_1$ *to produce* 2×2 *blocks.* Then there is an ordinary Toeplitz shift of *one block* between rows of $\boldsymbol{H}_b$:

$$\boldsymbol{H}_b=\begin{bmatrix} \cdots & \cdots & & & & & & \\ h_0(N) & h_0(N-1) & \cdots & \cdots & h_0(1) & h_0(0) & & \\ h_1(N) & h_1(N-1) & \cdots & \cdots & h_1(1) & h_1(0) & & \\ & & h_0(N) & h_0(N-1) & \cdots & \cdots & h_0(1) & h_0(0) \\ & & h_1(N) & h_1(N-1) & \cdots & \cdots & h_1(1) & h_1(0) \\ & & & & & & \cdots & \cdots \end{bmatrix}$$

Chapter 9 will have M filters and $(\downarrow M)$. Then $\boldsymbol{H}_b$ has $M\times M$ blocks.

The synthesis filters have $F_0(z)=H_1(-z)$ and $F_1(z)=-H_0(-z)$. In the orthogonal case, F_0 is the ordinary flip (the transpose) of H_0. Therefore H_1 is the alternating flip. The key point is that both polyphase matrices, $\boldsymbol{H}_p(z)$ in analysis and $\boldsymbol{H}_p^{-1}(z)$ in synthesis, are *polynomial.* The determinant of $\boldsymbol{H}_p(z)$ is z^{-N}, by the halfband condition on $H_0(z)H_1(-z)$ that makes everything work.

When signals have finite length L, the infinite matrix $\boldsymbol{H}_b$ must change to $L \times L$. We assume that $L > N$ and probably $L \gg N$. Then the "middle" of the matrices is not affected, but the "ends" will be new. We have to choose those end rows — the *boundary filters* in the $L \times L$ matrix $\boldsymbol{H}$. The first question is whether FIR boundary filters can yield $\boldsymbol{H}^{-1}\boldsymbol{H} = \boldsymbol{I}$.

The answer is *yes*. The inverse matrix stays banded. Then the problem is to choose among boundary filters. We will indicate some reasonable choices, but more experiments and experience are needed. This is a research problem of great interest.

In the orthogonal case, the boundary filters are to preserve $\boldsymbol{H}^T\boldsymbol{H} = \boldsymbol{I}$. The middle filters (interior filters) have this orthogonal property — the middle is unchanged from $\boldsymbol{H}_b^T\boldsymbol{H}_b = \boldsymbol{I}$ in the infinite case. We have to be sure that the new end rows of $\boldsymbol{H}$ and the new end columns of $\boldsymbol{H}^{-1}$ have limited length $\leq N$ and not full length L.

Wavelets have similar difficulties. When $\phi(t-n)$ and $w(t-n)$ fall across the boundary, they must change. This happens at each scale, and the multiresolution $V_0 \subset V_1 \subset \cdots$ should be maintained. The boundary functions are to be combinations of internal functions and boundary functions at scale $2t$. The shapes are the same at all scales. The coefficients in that boundary dilation equation and boundary wavelet equation give the boundary filters!

Filter Bank Completions

Our goal is an $L \times L$ analysis matrix $\boldsymbol{H}_L$. The synthesis matrix will be its inverse. In the orthogonal case, which we emphasize most, this inverse is $\boldsymbol{H}_L^T$.

$\boldsymbol{H}_L$ begins with rows from the infinite filter matrix. For the interior part $\boldsymbol{H}_{\text{in}}$, we only save *complete rows*. Our example will be the Daubechies 4-tap filter with coefficients $(a, b, c, d) = \left(1+\sqrt{3}, 3+\sqrt{3}, 3-\sqrt{3}, 1-\sqrt{3}\right)/4\sqrt{2}$. For convenience we save only *one* lowpass and highpass row:

$$\boldsymbol{H}_{\text{in}} = \begin{bmatrix} 0 & d & c & b & a & 0 \\ 0 & -a & b & -c & d & 0 \end{bmatrix}.$$

These rows are orthonormal. Therefore $(\boldsymbol{H}_{\text{in}})(\boldsymbol{H}_{\text{in}}^T) = \boldsymbol{I}$. The task is to add four new orthonormal rows — the boundary filters — to achieve $(\boldsymbol{H}_L)(\boldsymbol{H}_L^T) = \boldsymbol{I}$ with square matrices.

Those four filters will separate into left end and right end, and into lowpass and highpass. The square matrix will have the form

$$\boldsymbol{H}_L = \begin{bmatrix} \boldsymbol{H}_{\text{left}} \\ \boldsymbol{H}_{\text{in}} \\ \boldsymbol{H}_{\text{right}} \end{bmatrix} = \begin{bmatrix} r & s & t & 0 & 0 & 0 \\ u & v & w & 0 & 0 & 0 \\ 0 & d & c & b & a & 0 \\ 0 & -a & b & -c & d & 0 \\ 0 & 0 & 0 & e & f & g \\ 0 & 0 & 0 & x & y & z \end{bmatrix}. \tag{8.40}$$

Notice the freedom that has been removed and the freedom that remains. The boundary filters are allowed only three coefficients. At first this seems difficult, because the short rows (r, s, t) and (u, v, w) must be orthogonal to $(0, d, c)$ and also $(0, -a, b)$. Fortunately those vectors are parallel! The product $bd + ac$ is zero, from the key property of the Daubechies coefficients — that (a, b, c, d) is orthogonal to its double shift. The Gram-Schmidt process can easily produce (r, s, t) and (u, v, w). Similarly we find the two short boundary filters at the right end, which go in the last two rows of $\boldsymbol{H}_L$.

What freedom remains? We can premultiply the first two rows by any 2×2 orthogonal matrix. We use that freedom to make $u + v + w = 0$; that boundary filter becomes highpass. Similarly $x + y + z = 0$ makes the last row orthogonal to DC inputs (constant inputs), which therefore go through the lowpass channel. The actual coefficients are

$$(r, s, t) = (\ 0.93907,\ 0.29767,\ -0.17186) \qquad (e, f, g) = (0.40345,\ 0.69879,\ 0.59069)$$
$$(u, v, w) = (-0.34372,\ 0.81326,\ -0.46954) \qquad (x, y, z) = (0.25535,\ 0.51155,\ -0.80690).$$

In practice the interior matrix $\boldsymbol{H}_{\text{in}}$ has many more rows ($L - 4$ rows). These inner rows are not truncated and remain safely orthogonal. They separate the left end from the right end.

We now show that a similar construction of boundary filters will succeed for other (longer) interior filters. It is almost simple enough to do by hand. Herley gives a MATLAB code. The starting point is $(\boldsymbol{H}_{\text{in}})(\boldsymbol{H}_{\text{in}}^T) = \boldsymbol{I}$. Remember that $\boldsymbol{H}_{\text{in}}$ is rectangular. The product $(\boldsymbol{H}_{\text{in}}^T)(\boldsymbol{H}_{\text{in}})$ in the opposite order cannot equal $\boldsymbol{I}$. But the discrepancy is all near boundaries:

$$\boldsymbol{P} = \boldsymbol{I} - (\boldsymbol{H}_{\text{in}}^T)\,(\boldsymbol{H}_{\text{in}}) = \begin{bmatrix} \boldsymbol{P}_{\text{left}} & 0 & 0 \\ 0 & 0 & 0 \\ 0 & 0 & \boldsymbol{P}_{\text{right}} \end{bmatrix}. \tag{8.41}$$

This matrix satisfies $\boldsymbol{H}_{\text{in}}\boldsymbol{P} = \boldsymbol{H}_{\text{in}} - (\boldsymbol{H}_{\text{in}})(\boldsymbol{H}_{\text{in}}^T)(\boldsymbol{H}_{\text{in}}) = \boldsymbol{0}$. ***The columns of*** $\boldsymbol{P}_{\text{left}}$ ***and*** $\boldsymbol{P}_{\text{right}}$ ***are orthogonal to the rows of*** $\boldsymbol{H}_{\text{in}}$. These columns with short vectors (boundary filters) can complete a full set of orthogonal rows in $\boldsymbol{H}_L$. The rank of $\boldsymbol{P}$ is necessarily the difference between the number of columns and rows in $\boldsymbol{H}_{\text{in}}$. This is the correct number to complete a square matrix.

For the 2×6 example $\boldsymbol{H}_{\text{in}}$ we find a 6×6 matrix $\boldsymbol{P}$:

$$\boldsymbol{P} = \begin{bmatrix} 1 & 0 & 0 & & & \\ 0 & d^2+a^2 & cd-ab & & & \\ 0 & cd-ab & c^2+b^2 & & & \\ & & & b^2+c^2 & ab-cd & 0 \\ & & & ab-cd & a^2+d^2 & 0 \\ & & & 0 & 0 & 1 \end{bmatrix}.$$

The double-shift orthogonality $ac + bd = 0$ gave the zero blocks. Here $\boldsymbol{P}_{\text{left}}$ is 3×3 but its rank must be 2. The first two columns (or rows) of $\boldsymbol{P}$ are orthogonal to the rows of $\boldsymbol{H}_{\text{in}}$. ***They give two left boundary filters***. After normalization to unit length, and rotation to make the second one highpass, they become (r, s, t) and (u, v, w).

Some patience is needed to find the right number of zero columns for $\boldsymbol{H}_{\text{in}}$. For D_4, one zero column at each end gave two boundary filters at each end. [HerVet] propose a way to keep this number even for the Daubechies filters D_{2p}. We illustrate for D_6 a slightly different approach, which keeps four rows of $\boldsymbol{H}_{\text{in}}$ with no zero columns:

$$\boldsymbol{H}_{\text{in}} = \begin{bmatrix} f & e & d & c & b & a & & \\ -a & b & -c & d & -e & f & & \\ & & f & e & d & c & b & a \\ & & -a & b & -c & d & -e & f \end{bmatrix}. \tag{8.42}$$

Thus $\boldsymbol{H}_{\text{in}}$ is 4×8 (and $8 - 4$ is a multiple of 4). Using $(\boldsymbol{H}_{\text{in}})(\boldsymbol{H}_{\text{in}}^T) = \boldsymbol{I}$, we find 4×4 blocks in $\boldsymbol{P}$:

$$\boldsymbol{P} = \boldsymbol{I} - (\boldsymbol{H}_{\text{in}}^T)(\boldsymbol{H}_{\text{in}}) = \begin{bmatrix} \boldsymbol{P}_{\text{left}} & 0 \\ 0 & \boldsymbol{P}_{\text{right}} \end{bmatrix}. \tag{8.43}$$

$\boldsymbol{P}$ has rank 4 and each block has rank 2. We obtain two 4-tap boundary filters at each end. When $\boldsymbol{H}_L$ has more inner filters, for signal lengths $L > 8$, they start and end with at least four zeros. This makes all inner filters orthogonal to our boundary filters. The $L \times L$ orthogonal analysis bank is complete.

General case: Biorthogonal filters have $(\boldsymbol{H}_{\text{in}})(\boldsymbol{F}_{\text{in}}) = \boldsymbol{I}$ but not $(\boldsymbol{F}_{\text{in}})(\boldsymbol{H}_{\text{in}}) = \boldsymbol{I}$. Again these are rectangular matrices with no truncation of the interior filters. We need boundary filters to complete square matrices. Here is the key:

$$\boldsymbol{P} = \boldsymbol{I} - (\boldsymbol{F}_{\text{in}})\,(\boldsymbol{H}_{\text{in}}) \quad \text{has} \quad (\boldsymbol{H}_{\text{in}})\,\boldsymbol{P} = \boldsymbol{H}_{\text{in}} - \boldsymbol{H}_{\text{in}}\boldsymbol{F}_{\text{in}}\boldsymbol{H}_{\text{in}} = \boldsymbol{0}.$$

The columns of $\boldsymbol{P}$ contain boundary filters that can be included with the columns of $\boldsymbol{F}_{\text{in}}$. Similarly $\boldsymbol{P}(\boldsymbol{F}_{\text{in}}) = \boldsymbol{0}$. The rows of $\boldsymbol{P}$ contain boundary filters that are included with the rows of $\boldsymbol{H}_{\text{in}}$. Finally we biorthogonalize these boundary filters. A useful identity (to put it mildly) is $\boldsymbol{P}^2 = \boldsymbol{P}$.

Time-varying Filter Banks: A Remark

Changing from length 4 filters to length 6 will create an internal boundary, at the moment of change. One way to maintain orthogonality is to use boundary filters just before the change and just after. Such a transition has no overlap. A smoother approach is to create *transition filters* that cross the internal boundary. In this case $\boldsymbol{H}_{\text{in}}$ has filters purely to the left and purely to the right.

Again $(\boldsymbol{H}_{\text{in}})(\boldsymbol{H}_{\text{in}}^T) = \boldsymbol{I}$. Now $\boldsymbol{P} = \boldsymbol{I} - \boldsymbol{H}_{\text{in}}^T\boldsymbol{H}_{\text{in}}$ contains the transition filters. See [HeKoRaVe] for details of this important construction.

Direct Extrapolation at the Boundary

We briefly mention another approach to boundary filters. It is closer to the usual treatment of boundary conditions for differential equations. *The governing principle is to fit a polynomial to the data and extend that polynomial.* The job of extrapolation is to define functions and signals beyond the boundary — so the filter can be applied.

Zero-padding and symmetric extension are basic extrapolation methods. Higher degree polynomials give higher degree extrapolation and higher accuracy.

To be consistent with multiresolution, the input $\boldsymbol{x}(n)$ should give two half-length outputs. If the inputs are coefficients of $\phi(2t-n)$ in a function $f(t)$, the output are coefficients of $\phi(t-n)$ and $w(t-n)$ for $f(t)$ at the next scale. In the interior all is normal. *At the boundary we determine a polynomial $F_{p-1}(t)$ of degree $p-1$ that models $f(t)$.* Using only the p numbers $\boldsymbol{x}(0), \ldots, \boldsymbol{x}(p-1)$ to determine the p coefficients in the polynomial gives a square matrix — but this is dangerous. The preference in [WiAm] is to use more inputs $\boldsymbol{x}(0), \ldots, \boldsymbol{x}(N)$ and determine the polynomial coefficients by least squares.

By extending the polynomial we extrapolate the signal. Then filter and downsample as usual. This corresponds to a splitting of V_1 into $V_0 + W_0$. It also corresponds to a completion of the filter matrix near the boundary. For the lowpass Daubechies D_4, new coefficients appear in the first row of $\boldsymbol{H}_L$. The second lowpass row contains the usual d, c, b, a:

$$\begin{bmatrix} 0.8924 & 0.5174 & 0.0129 & -0.0085 & 0 & 0 \\ -0.1294 & 0.2241 & 0.8365 & 0.4830 & 0 & 0 \\ & & \cdots & \cdots & \cdots & \cdots \end{bmatrix}.$$

Orthogonality is lost; accuracy is preserved; highpass coefficients are in [WiAm]. Also important in differential equations is the fact that *boundary conditions can be built into the extrapolation.*

There is a third "quick and dirty" approach that deals entirely with the matrix entries. We follow [KwTa] by considering a biorthogonal example with attractive coefficients:

$$H = \frac{1}{4}\begin{bmatrix} & & & & & \\ -1 & 3 & 3 & -1 & & \\ -1 & 3 & -3 & 1 & & \\ & & -1 & 3 & 3 & -1 \\ & & -1 & 3 & -3 & 1 \\ & & & & & \cdots \end{bmatrix} \quad \text{and} \quad H^{-1} = \frac{1}{4}\begin{bmatrix} 1 & 1 & & & \\ 3 & 3 & & & \\ 3 & -3 & 1 & 1 & \\ 1 & -1 & 3 & 3 & \\ & & 3 & -3 & \cdots \\ & & 1 & -1 & \cdots \end{bmatrix}. \tag{8.44}$$

The column coefficients $1, 3, 3, 1$ in H^{-1} come from $(1+z^{-1})^3$. The synthesis scaling function is a quadratic spline (accuracy $p = 3$). The row coefficients $-1, 3, 3, -1$ come from $(1+z^{-1})(-1+4z^{-1}-z^{-2})$. This single zero gives $\widetilde{p} = 1$. The product filter is the familiar halfband $P(z)$ of degree 6. This is a linear phase choice, not to be iterated too often! We noted in Section 7.2 that Condition **E** is violated. The cascade algorithm will not converge for the filter H.

Nevertheless it is a good example. Suppose the second column shown for H corresponds to the sample $x(0)$. The first column would normally multiply $x(-1)$, *which does not exist.* Extrapolation could create $x(-1)$. Equivalently, this external column $c = -\frac{1}{4}, -\frac{1}{4}$ can be folded into the internal columns by one of these rules:

Zero-padding: Delete the external column c.

Constant extrapolation: Add c to the zeroth column $(3/4, 3/4)$.

Linear extrapolation: Add $2c$ to column 0 and $-c$ to column 1.

Quadratic extrapolation: Add $3c, -3c, c$ to columns 0, 1, 2.

The key point is that H and H^{-1} both maintain their stacked double-shift form (rows of H and columns of H^{-1}). The boundary filters based on constant extrapolation $x(-1) = x(0)$, which adds c to column zero, keep their lowpass and highpass character:

$$H_L = \frac{1}{4}\begin{bmatrix} 2 & -3 & 1 & & \\ 2 & -3 & 1 & & \\ & -1 & 3 & 3 & -1 \\ & -1 & 3 & -3 & 1 \\ & & & & \cdots \end{bmatrix} \quad \text{and} \quad H_L^{-1} = \frac{1}{4}\begin{bmatrix} 4 & 4 & & & \\ 3 & -3 & 1 & 1 & \\ 1 & -1 & 3 & 3 & \\ & & 3 & -3 & \cdots \\ & & 1 & -1 & \cdots \end{bmatrix}. \tag{8.45}$$

For longer filters the extrapolation rules are similar, but care is needed. There is room for experiment. ***Boundary filters are now more easily constructed by** "lifting"* (see the new Appendix 4).

Lowpass Cascade and the Boundary Dilation Equation

Scaling functions come from lowpass filters. The lowpass coefficients go into the dilation equation. This equation is solved for $\phi(t)$ by iteration — which means cascade. A good example is the synthesis filter bank H_L^{-1} given above. Its interior lowpass columns contain $(\frac{1}{4}, \frac{3}{4}, \frac{3}{4}, \frac{1}{4})$ and those coefficients lead to the quadratic spline $\phi(t)$ on $[0, 3]$:

$$\phi(t) = \tfrac{1}{4}\phi(2t) + \tfrac{3}{4}\phi(2t-1) + \tfrac{3}{4}\phi(2t-2) + \tfrac{1}{4}\phi(2t-3). \tag{8.46}$$

This spline has an explicit formula, which we mention but do not need. $2\phi(t)$ equals t^2 and $-2t^2+6t-3$ and $(t-3)^2$ on the intervals $[0,1]$ and $[1,2]$ and $[2,3]$.

The boundary filter $(1, \frac{3}{4}, \frac{1}{4})$ in the first column of $\boldsymbol{H}^{-1}$ leads to a boundary scaling function $\phi_b(t)$. The boundary dilation equation uses those coefficients:

$$\phi_b(t) = \phi_b(2t) + \tfrac{3}{4}\phi(2t) + \tfrac{1}{4}\phi(2t-1) \quad \text{for} \quad t \geq 0. \tag{8.47}$$

Notice how ϕ_b is included with the functions at scale $2t$. Since $\phi(2t)$ and $\phi(2t-1)$ are already known, this is an "*inhomogeneous dilation equation*" for the boundary function $\phi_b(t)$. It can be solved in at least three ways:

1. By creating a general method for inhomogeneous two-scale equations.
2. By an inspired guess.
3. By iteration, cascading the lowpass filter.

We choose Method 2, because the insight is important. The boundary filter comes from constant extrapolation, which preserves minimum accuracy $p=1$. The constant function will belong to V_0. In the interior we do have $\sum \phi(t-n) \equiv 1$. But near $t=0$, the pieces from $\phi(t+1)$ and $\phi(t+2)$ are missing—those functions cross the boundary. It is natural to suspect that the boundary function $\phi_b(t)$ takes their place—but only on the right side $t \geq 0$:

$$\text{Inspired guess:} \quad \phi_b(t) = \phi(t+1) + \phi(t+2) \quad \text{for} \quad t \geq 0. \tag{8.48}$$

The support is $[0,2]$. We try this function in the boundary dilation equation (8.47). On the right side it contributes $\phi(2t+1)+\phi(2t+2)$. On the left side, replace $\phi(t+1)$ and $\phi(t+2)$ using the interior dilation equation (8.46). Compare only for $t \geq 0$, so terms involving $\phi(2t+3)$ and $\phi(2t+4)$ can be and must be ignored. Equation (8.47) is satisfied!

This function $\phi_b(t) = \sum_{n<0} \phi(t-n)$ will appear again, as a direct construction in continuous time. Figure 8.12 shows how it builds $f(t) \equiv 1$ into V_0.

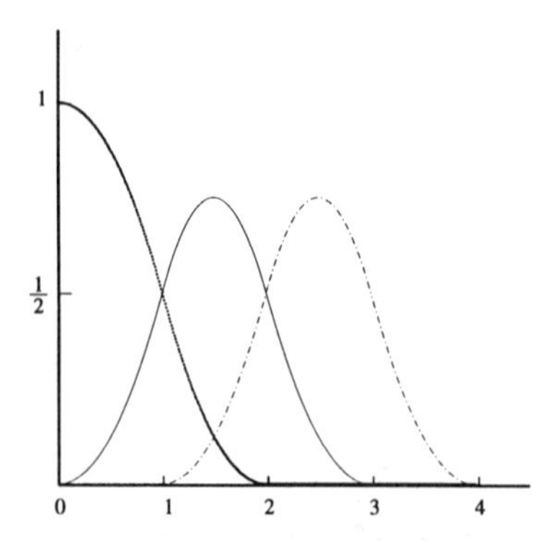

Figure 8.12: Quadratic splines and $\phi_b(t)$ add to 1.

Its dilation equation reveals the boundary filter. In that order, $\phi_b(t)$ produces the filter matrix rather than vice versa.

It is helpful also to see the lowpass cascade (Method **3**). This means powers of the double-shift lowpass synthesis matrix:

$$L = \frac{1}{4}\begin{bmatrix} 4 & 3 & 1 & & & & & \\ & 1 & 3 & 3 & 1 & & & \\ & & & 1 & 3 & 3 & 1 & \cdots \end{bmatrix}.$$

Squaring leaves the unit column sums, and produces a *quadruple shift*:

$$L^2 = \frac{1}{16}\begin{bmatrix} 16 & 15 & 13 & 10 & 6 & 3 & 1 & \\ & 1 & 3 & 6 & 10 & 12 & 12 & \cdots \\ & & & & & 1 & 3 & \cdots \end{bmatrix}.$$

The boundary filter affects only the first row. Elsewhere, convolving 1, 3, 3, 1 with 1, 0, 3, 0, 3, 0, 1 gives the interior sequence 1, 3, 6, 10, 12, 12, 10, 6, 3, 1. This corresponds to $H(z)H(z^2)$ —which is not $[H(z)]^2$. *The cascade includes rescaling.* L relates to integer spacing $\Delta t = 1$, while L^2 relates to $\Delta t = \frac{1}{2}$. The fact that all column sums are 1 means that, at convergence, the scaling functions $\phi_b(t), \phi(t), \phi(t-1), \ldots$ add to 1. This brings back our formula (8.9) for $\phi_b(t)$.

It is unusual to solve a dilation equation so explicitly. But convergence of the cascade is not difficult to establish. Condition **E** holds for the interior filter. The boundary requires Condition $\mathbf{E}_b$ on a small square matrix in the corner of L. That 2×2 matrix has eigenvalues 1 and $\frac{1}{4}$ and the cascade converges.

For this example, the powers L^i are the slow way to solve the dilation equation. *For most examples L^i is the only way.* The cascade algorithm is also called the "graphical algorithm," because a finite power of L yields an approximate graph of the scaling functions. Then one application of the other submatrix B (from the highpass filters) produces the boundary and interior wavelets.

Basis Completion by Boundary Functions

For a direct construction of boundary functions, suppose $\phi(t)$ is supported on $[0, 3]$. Our overall interval is finite, say $[0, 4]$. Then we can only shift once, to $\phi(t-1)$. All other translates of $\phi(t)$ cross the boundary and have to be changed. We are looking for four basis functions in V_0 and $4 \cdot 2^j$ basis functions in V_j. There are three important ways to construct them [CoDaVi]:

1. Periodic Extension: The shifted function $\phi(t-2)$ is supported on $[2, 5]$. The segment on $[4, 5]$ is wrapped back to $[0, 1]$. Similarly, the piece of $\phi(t-3)$ on $[4, 6]$ is wrapped back to $[0, 2]$. We are just periodizing the original functions.

This corresponds in the discrete case to *circulant matrices.* The lowpass analysis filter is a 4×4 circulant containing the four filter coefficients. It stretches to 4×8 with double shift from $(\downarrow 2)$. The highpass filter, similarly stretched, completes the 8×8 block circulant H_L.

2. Symmetric Extension: This would apply to the linear phase example based on the coefficients $(1, 3, 3, 1)$. An even-length symmetric filter (**H** filter) requires an **H** extension, centered at half-samples. The detailed discussion is in Section 8.2. Our point here is that the extension is implicitly creating boundary filters.

Symmetric extension is used for symmetric scaling functions. Those are not self-orthogonal; we are in the biorthogonal case. Our 1, 3, 3, 1 example has an even number of taps, equal in analysis and synthesis. The scaling function $\phi(t+1)$ on $[-1,2]$ extends outside the basic interval $[0,4]$, and is ***folded*** back inside by a symmetric extension to the whole line:

$$f^{\text{fold}}(t) = \sum_{-\infty}^{\infty} [f(t-8n) + f(8n-t)]. \tag{8.49}$$

The terms with $n=0$ are $f(t)+f(-t)$, even across $t=0$. The other terms keep this symmetry and produce the double period 8. Thus any folded function is even with period 8:

$$f^{\text{fold}}(-t) = f^{\text{fold}}(t) \quad \text{and} \quad f^{\text{fold}}(t+8) = f^{\text{fold}}(t). \tag{8.50}$$

Figure 8.13 shows an idealized (!) picture of a symmetric $\phi(t)$ extended to the one-period in-

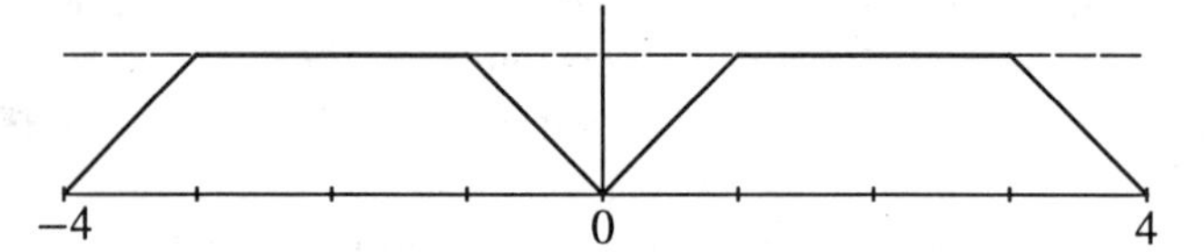

Figure 8.13: Symmetric folding of four symmetric scaling functions.

terval $[-4,4]$. Their restrictions to $[0,4]$ span the space V_0 with the correct dimension $4 \cdot 2^j = 4$. These are the four scaling functions and there are four corresponding wavelets. Similarly there will be four synthesis functions $\widetilde{\phi}(t)$ and wavelets $\widetilde{w}(t)$. *Furthermore biorthogonality is preserved* (Problems 1-2).

The original scaling functions $\phi(t+n)$ on the whole line add to the constant function 1, because the lowpass filter has the required zero at π. This good property is preserved by symmetric extension. *The folded functions still add to* 1. In our idealized figure the flat middle segments have height $\frac{1}{2}$. This piecewise linear $\phi(t)$ comes from a simple filter that cannot be part of an FIR filter bank (Problems 3–4). In general all the pieces of scaling functions that add to 1 are folded back into the interval so constants are in V_0.

Even if $H(z)$ has extra zeros at π, we do not expect V_0 to give accuracy beyond $p=1$. Symmetric extension (folding) generally produces jumps in the slope at the boundary. We turn to general methods that can achieve $p>1$.

3. Boundary Functions: We need a completion method for the orthogonal case, when symmetric extension is unnatural. We also want to maintain the accuracy p of the internal functions. The following method will apply to all cases, orthogonal or not and linear phase or not. *We focus on completing the lowpass synthesis part* (the space V_0 and the filter $\boldsymbol{F}_0$). A parallel construction applies to the analysis space $\widetilde{V}_0$ and the filter $\boldsymbol{H}_0$. Then multiresolution makes W_0 and $\widetilde{W}_0$ orthogonal to $\widetilde{V}_0$ and V_0.

The plan is to start with internal scaling functions supported inside the given interval, and add new functions at each end. To start, the new ones are linearly independent "*boundary pieces.*" We insist on short support and a dilation equation, so that $V_0 \subset V_1$. Additional properties lead to full accuracy. ***Those short pieces are orthogonalized against the existing scaling function and wavelets***. In the orthogonal case those are the internal $\phi(t-n)$ and $w(t-n)$; in general they are the synthesis functions. The algorithm is just Gram-Schmidt or its biorthogonal extension to "bi-Gram-Schmidt."

The key point is that this orthogonalization produces functions supported *near the boundary.* Each boundary piece, however chosen, is already orthogonal to all scaling functions that are far inside the interval. There is no overlap. One source of boundary pieces is the tails of existing $\phi(t+k)$ and $w(t+k)$. Numerically this is a poor choice. The tails are very small for longer wavelets, which leads to severe ill-conditioning.

A better construction [CoDaVi] is to build polynomials up to degree $p-1$ into V_0. On the whole line we have $\sum \phi(t-n) \equiv 1$. Then the first boundary piece ($t \geq 0$ only) is

$$\phi_b(t) = 1 - \textstyle\sum_{n\geq 0} \phi(t-n) = \sum_{n<0} \phi(t-n). \tag{8.51}$$

This is just a *sum of tails.* It is orthogonal to all internal $\phi(t-n)$, $n \geq 0$. Its support is $[0, N-1]$ because $\phi(t), \phi(t-1), \ldots$ already add to 1 beyond $t = N-1$.

The boundary function $\phi_b(t)$ satisfies a dilation equation. Replacing t by $2t$ gives $\phi_b(2t) = 1 - \sum_{n\geq 0} \phi(2t-n)$. Use this to substitute for 1 in (8.51):

$$\phi_b(t) = \phi_b(2t) + \sum_{n\geq 0} [\phi(2t-n) - \phi(t-n)]. \tag{8.52}$$

Replace each $\phi(t-n)$ using its ordinary dilation equation. Thus $\phi_b(t)$ is a combination of $\phi_b(2t)$ and internal $\phi(2t-n)$. It belongs to V_1 and we are on our way.

Higher Accuracy. Suppose the original scaling function $\phi(t-n)$ can reproduce linear functions. The filter has at least two zeros at π. The wavelets have two vanishing moments. The combination $\sum n\phi(t-n)$ equals $\alpha t + \beta$; the constants $\alpha \neq 0$ and β are determined by the filter coefficients. Our task is to ensure that $\alpha t + \beta$ is in V_0 when we remove all $\phi(t-n)$ for $n < 0$. In their place goes $\phi_b(t)$ and a second boundary piece $\phi_{bb}(t)$:

$$\phi_{bb}(t) = \alpha t + \beta - \sum_{n\geq 0} n\phi(t-n) = \sum_{n<0} n\phi(t-n). \tag{8.53}$$

The sum over $n < 0$ shows that the support is $[0, N-1]$. This piece need not be orthogonal to $\phi_b(t)$ or to the internal $\phi(t-n)$. We must orthogonalize it (or biorthogonalize it). Also we check that $\phi_{bb}(t)$ satisfies a dilation equation, so the multiresolution hierarchy $V_0 \subset V_1$ is still secure. Replacing t by $2t$ in (8.53) gives

$$\phi_{bb}(2t) = 2\alpha t + \beta - \sum_{n\geq 0} n\phi(2t-n). \tag{8.54}$$

Divide by 2 and substitute for αt in (8.54):

$$\phi_{bb}(t) = \tfrac{1}{2}\phi_{bb}(2t) + \tfrac{1}{2}\beta + \sum_{n\geq 0} n\left[\tfrac{1}{2}\phi(2t-n) - \phi(t-n)\right]. \tag{8.55}$$

Replace each $\phi(t-n)$ using its dilation equation, and replace $\frac{1}{2}\beta$ by $\left[\frac{1}{2}\beta\phi_b(2t) + \sum \phi(2t-n)\right]$. Then (8.55) guarantees that $\phi_{bb}(t)$ is in the next space V_1.

It is important to count the functions! With p zeros at π, we are choosing p boundary pieces at each end to build the polynomials $1, \ldots, t^{p-1}$ into V_0. An interval of length L will have $L - 2p + 2$ internal Daubechies functions. For $L = 4$ and $p = 2$ we had $\phi(t)$ on $[0, 3]$ and $\phi(t-1)$ on $[1, 4]$. *We only need $2p-2$ additional functions, but we have $2p$.*

To make space for $2p$ boundary pieces, we throw out the perfectly good $\phi(t)$ at the left end and the outermost scaling function at the right end. The term with $n = 0$ moves to the other side of (8.51). This only means that $\phi_b(t)$ has support $[0, N]$. A similar replacement at the other end assures that constant functions are in V_0.

The simplest highpass construction starts with scaling functions $\phi(2t - n)$ that cross the boundary. Subtract their projections on V_0 to get *boundary wavelets* [CoDaVi, p. 73]. Together with the internal wavelets this produces $W_0 \perp V_0$ and $W_0 \subset V_1$. We indicate the D_4 case with $p = 2$ and normal Daubechies coefficients (a, b, c, d):

$$\boldsymbol{F}_{\text{low}} = \begin{bmatrix} \times & \times & \times & & & & & & & \\ \times & \times & \times & \times & \times & & & & & \\ & & & d & c & b & a & & & \\ & & & & & \times & \times & \times & \times & \times \\ & & & & & & & \times & \times & \times \end{bmatrix}. \tag{8.56}$$

The 5×10 highpass matrix completes the 10×10 synthesis bank.

After orthogonalization, the filter coefficients are computed once and for all by [CoDaVi]. *Correction* to line 6 on page 291: $x = 0.29535$.

Problem Set 8.5

1. Verify from (8.49) that folding plus restriction preserves inner products:

$$\int_0^4 f^{\text{fold}}(t) g^{\text{fold}}(t)\, dt = \int_{-\infty}^{\infty} f(t) g(t)\, dt.$$

2. If the functions $\phi(t-n)$ and $\widetilde{\phi}(t-n)$ are biorthogonal on the whole line, verify that their folded versions are still biorthogonal on the finite interval. This interval is $[0, 4]$ when the folding in (8.49) has period 8.

3. What symmetric 4-tap filter produces $\phi(t)$ that is linear on $[0, 1]$ and $[2, 3]$ and constant on the middle interval $[1, 2]$? How many zeros at π in $H(\omega)$?

4. Show that the filter with $\boldsymbol{h} = (\frac{1}{4}, \frac{1}{4}, \frac{1}{4}, \frac{1}{4})$ has $H(i) = H(-i) = 0$. It cannot be part of a PR filter bank! Why can no synthesis filter make $P(z) = F(z)H(z)$ into a halfband filter with $P(z) + P(-z) = z^l$?

5. Take two rows of (8.44) in $\boldsymbol{H}_{\text{in}}$ and two columns of the inverse matrix in $\boldsymbol{F}_{\text{in}}$. Verify $\boldsymbol{H}_{\text{in}}\boldsymbol{F}_{\text{in}} = \boldsymbol{I}_{2\times 2}$ and compute $\boldsymbol{P} = \boldsymbol{I} - \boldsymbol{F}_{\text{in}}\boldsymbol{H}_{\text{in}}$ (which is 6×6). Determine a set of boundary filters.

6. From the double-shift orthogonality of Daubechies D_6, verify that the zero blocks of $\boldsymbol{P}$ in equation (8.43) really are zero.

7. Suppose $\boldsymbol{H}_{\text{in}}$ is $K \times L$ and $\boldsymbol{H}_{\text{in}}\boldsymbol{H}_{\text{in}}^T = \boldsymbol{I}_{K\times K}$. Verify that $\boldsymbol{Q} = \boldsymbol{H}_{\text{in}}^T\boldsymbol{H}_{\text{in}}$ and $\boldsymbol{P} = \boldsymbol{I} - \boldsymbol{Q}$ satisfy $\boldsymbol{Q}^2 = \boldsymbol{Q}$ and $\boldsymbol{P}^2 = \boldsymbol{P}$ and $\boldsymbol{Q}\boldsymbol{H}_{\text{in}}^T = \boldsymbol{H}_{\text{in}}^T$. $\boldsymbol{Q}$ is the projection onto the column space of $\boldsymbol{H}_{\text{in}}^T$ and $\boldsymbol{P}$ is the projection onto the nullspace of $\boldsymbol{H}_{\text{in}}$.

8. (MATLAB) Zero-padding removes the first column $\boldsymbol{c}$ of $\boldsymbol{H}$ in the $-1, 3, 3, 1$ matrix in (8.44). Find the inverse of the resulting matrix.

9. (MATLAB) Linear extrapolation adds $2\boldsymbol{c}$ and $-\boldsymbol{c}$ to the next columns of $\boldsymbol{H}_L$, when the first column $\boldsymbol{c}$ is removed. Find the upper left corner of the resulting $\boldsymbol{H}_L^{-1}$. What are the boundary dilation equations in analysis and synthesis?

Chapter 9

M-Channel Filter Banks

9.1 Freedom versus Structure

An M-channel filter bank has very considerable freedom for $M > 2$. That is true when we require perfect reconstruction. It is still true when we also require orthogonality. Those properties can be expressed directly as conditions on the $M \times M$ analysis polyphase matrix $\boldsymbol{H}_p(z)$:

1. Perfect reconstruction requires $\boldsymbol{H}_p(z)$ to be invertible for all z.

2. The synthesis filters are FIR when the determinant of $\boldsymbol{H}_p(z)$ is a delay z^{-l}.

3. The filter bank is orthogonal when $\boldsymbol{H}_p^T(z^{-1})\boldsymbol{H}_p(z) = \boldsymbol{I}$.

In the orthogonal case, each synthesis filter F_k is the transpose (or flip, or time-reversal) of the analysis filter H_k. Then delays make F_k causal.

How to design a good M-band filter bank? For $M = 2$, orthogonality is already restrictive (and effectively prevents linear phase). The PR condition is less strict; linear phase can be achieved. In both cases the product filters $F_0(z)H_0(z)$ and $F_1(z)H_1(z)$ are halfband. Furthermore, one comes from the other by alternating signs. All four filters are created by factoring *one halfband filter*.

For $M > 2$, the number of freedoms grows faster than the number of restrictions. The choice of one M-th band filter (or one lowpass filter $H_0(z)$) does not determine the other choices. This is attractive at first — we can have linear phase with orthogonality. But decisions are still needed. We can make those decisions late or early! The range of designs can be left very wide (late decision). Or we can restrict the filter bank to have a structure that we know is desirable (early decision, simplifying the design). This chapter studies both possibilities. In all cases we absolutely want *fast implementation.*

In practice, fast algorithms come by cascading simple functions. At the lowest level they are based on 2×2 butterflies. At a higher level three structures are quick:

Rotations and delays or ***DFT banks*** or ***DCT banks***.

All constant orthogonal matrices are products of $M(M-1)/2$ plane rotations. All paraunitary matrices are products of rotations and delays. (Householder would use reflections and delays — this factorization is an important theorem.) Under quantization and roundoff, an orthogonal butterfly remains orthogonal. The difficulty will be the large number of rotation angles, approximately $NM(M-1)/2$ for filters of lengths N. An optimal design may have excellent properties but it will not be easy to find.

This chapter will pursue all three structures. We develop the polyphase approach in Section 9.2 and the time domain approach (to LOT and GenLOT) in Section 9.3. Those sections extend to M channels the earlier ideas from two channels. Then Section 9.4 focuses on cosine modulation — not seen for two channels but now important.

It may assist the reader if we now briefly highlight DFT and DCT filter banks. These are *modulated* filter banks because all M filters come from frequency shifts of one prototype filter. It is usual to refer to DCT filter banks as *cosine-modulated filter banks*. In many applications, the DFT loses and the DCT wins.

Block Transforms

The simplest filter banks use only the M-point DFT or the M-point DCT. The signal is split into blocks of length M. These blocks are *separately* transformed. There is no overlapping or interaction between blocks, and no filter design is involved. The polyphase matrix can be the Fourier DFT matrix or the DCT matrix (this is the JPEG standard). In a block transform, $\boldsymbol{H}_p$ is just a constant matrix — and there is no smoothing between blocks.

The analysis half can be drawn with the modulators first or the decimators first. Figure 9.1 shows the direct form of the block DFT and the more efficient polyphase form ($W_M = e^{-j2\pi/M}$). Section 9.2 discusses the block generation step in detail, because a serial to parallel S/P converter is extremely important — as a way to start the filter bank.

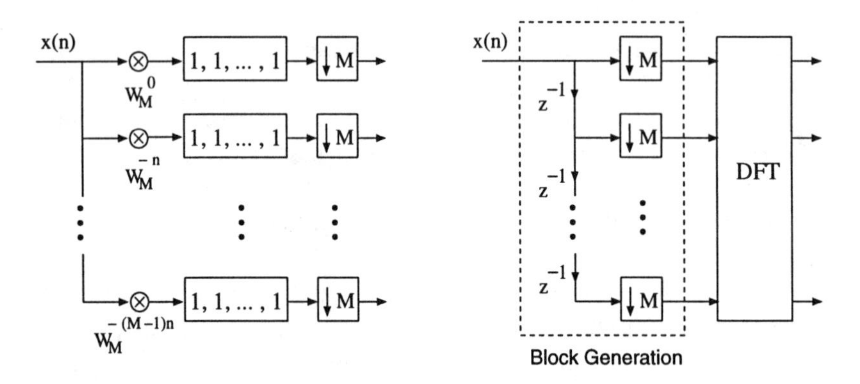

Figure 9.1: The DFT block transform: M samples at a time. Using DCT gives JPEG.

DFT Filter Banks

Now we include filters. The two-channel DFT filter bank appeared early in Chapter 4 (not with that name). The highpass frequency response shifted $H_0(e^{j\omega})$ by π. Consequently, $\boldsymbol{h}_1(n)$ was the ***alternating sign*** version of $\boldsymbol{h}_0(n)$. The reader will remember that $H_1(z) = H_0(-z)$ did not go well with perfect reconstruction. For two channels, the best arrangement is to alternate signs between H_0 and F_1 and between H_1 and F_0. This is not DFT.

For an M-channel DFT bank, the frequency response from H_k is $H_0(e^{j\omega})$ shifted by $\frac{2\pi k}{M}$. The corresponding z-transform and time-domain relations are

$$H_k(z) = H_0(ze^{-j2\pi k/M}) \quad \text{and} \quad \boldsymbol{h}_k(n) = \boldsymbol{h}_0(n)e^{j2\pi k/M}.$$

The synthesis filters are modulated in the same way. Note that the coefficients become *complex* for $M > 2$. This is a disadvantage. The great advantage is the simplicity of design and the speed of implementation, when the whole analysis bank is based on one filter H_0 (and the DFT).

Figure 9.2 shows the direct form and polyphase form of the DFT analysis bank. It differs from Figure 9.1 only by including filters — which come before $(\downarrow M)$ in the direct form and after $(\downarrow M)$ in the polyphase form. We are free to use the DFT or the IDFT in analysis, and reverse this choice in synthesis. The transfer functions $E_k(z)$ are the polyphase components of $H_0(z)$.

Figure 9.2 reduces to Figure 9.1 when $\boldsymbol{h}_0 = (1, 1, \ldots, 1)$. Then the product of DFT and IDFT gives perfect reconstruction trivially, a block at a time. Section 9.2 derives the PR condition for a DFT bank that involves filters. The requirements on those filters are quite restrictive. DFT banks are generally superseded by filter banks based on the DCT, if reconstruction is desired.

We saw the same for continuous time in Chapter 8. *Cosine modulation is the best.*

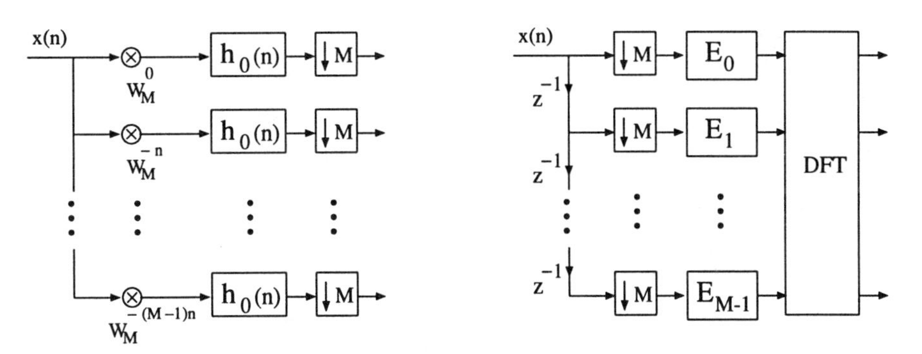

Figure 9.2: Direct form and polyphase form: DFT bank built from one filter.

Cosine-modulated Filter Banks

Cosine modulation replaces the complex DFT by the real DCT. Remember that there are four different cosine transforms. They use different extensions; Types II and IV are appropriate for filter banks. (Types I and III give undesirable bandwidths for the first and last filters. For the same reason, the filter length is constrained to be $2M$ or more generally $2KM$.) We use Type IV with the factors $k + \frac{1}{2}$ and $n + \frac{1}{2}$ in the frequencies.

The DCT matrix $\boldsymbol{C}^{\text{IV}}$ is symmetric and orthogonal. We shift the frequencies to achieve perfect reconstruction, starting from the lowpass prototype $\boldsymbol{p}(n)$:

$$\boldsymbol{h}_k(n) = \boldsymbol{f}_k(n) = \boldsymbol{p}(n)\sqrt{\frac{2}{M}}\cos\left[\left(k + \tfrac{1}{2}\right)\left(n + \frac{M+1}{2}\right)\frac{\pi}{M}\right]. \tag{9.1}$$

The shift of n by $\frac{M}{2}$ has the effect of centering $\boldsymbol{p}$. That simplifies the conditions on this prototype filter — "*the window*" — to achieve perfect reconstruction. The PR conditions are beautiful for filter length $2M$. We write them now without proof:

$$\textbf{Even symmetry} \quad p(n) = p(N-n) \tag{9.2}$$

$$\textbf{Orthogonality} \quad p^2(n) + p^2(n+M) = 1. \tag{9.3}$$

These are *precisely analogous* to the conditions on the continuous-time cosines in Chapter 8. There the window was subject to $g^2(t) + g^2(t+1) = 1$. Section 9.4 will extend the discrete-time requirements to filter lengths $2KM$, and establish orthogonality and perfect reconstruction. There are analogous conditions in continuous time — when each window overlaps several other windows.

We can already see the major advantages of cosine modulation:

- Simplicity of design: one filter $p(n)$ only
- Symmetry and orthogonality
- Very fast implementation.

The simplicity is crucial when M is large. That is the central point — to accept and indeed to welcome restrictions that simplify the structure. The implementation is fast because the DCT is fast. Ultimately this is because the FFT is fast.

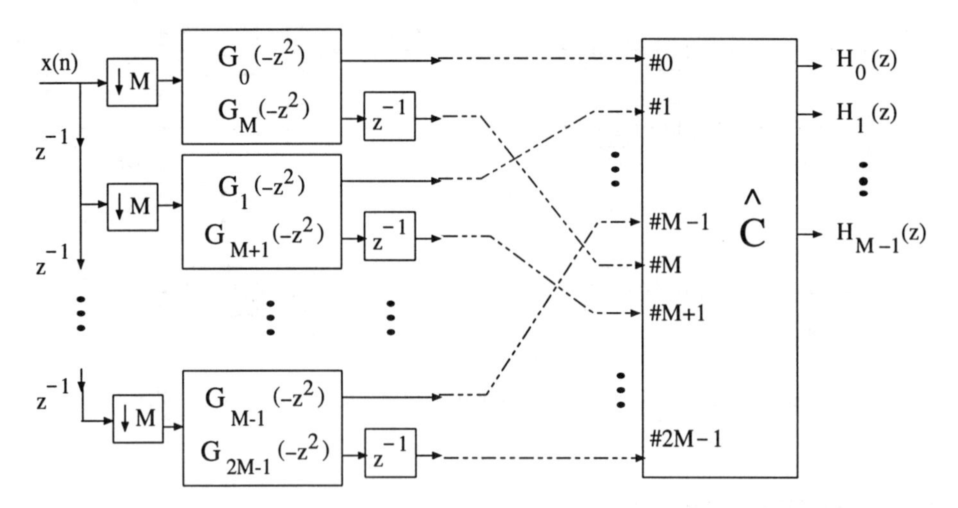

Figure 9.3: Cosine modulation by the DCT: k and $M+k$ are twins.

Remember that the M-point DCT is created from the $2M$-point DFT. In our applications the DCT has real outputs, reducing the full (complex) computation by half. The analysis bank will involve

- $2M$ polyphase filters
- $2M$ complex modulators with real inputs
- one $2M$-point DFT with real outputs.

Figure 9.3 is directly based on the DCT, and emphasizes how phases k and $M+k$ are twinned. Each pair of phases is like a 2-channel orthogonal filter bank! (On which the reader is by now an expert.) The good implementations have a delay chain of length $2M$ and decimators $(\downarrow M)$

coming first. They generate *two* blocks of length M in Figure 9.3, as the double-length DFT requires.

The cosine-modulated basis functions are *not* linear phase. They are even around $\omega = 0$ and odd around $\omega = \pi$, which produces Type-IV orthogonality. The first four basis functions when $M = 32$ are drawn in Section 8.3.

Figure 9.4a shows the idealized frequency response $P(e^{j\omega})$ of the lowpass prototype filter—the window $\boldsymbol{p}$. Then Figure 9.4b shows the phase shifts from cosine modulation. Notice especially how each cosine (the sum of two exponentials) produces two copies of $P(e^{j\omega})$. One copy is shifted left and one copy is shifted right. A good prototype will have passband approximately $|\omega| \le \frac{\pi}{2M}$. The design of that symmetric window subject to (9.2) gives a good M-channel filter bank—structured by cosine modulation.

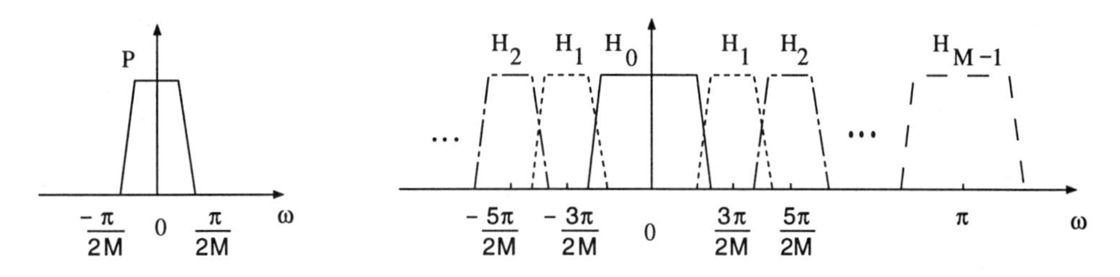

Figure 9.4: Idealized frequency response of the prototype and the cosine-modulated filters.

The original references for the discrete-time filter bank are in [Mr]. The key to the continuous-time construction was found by [Coifman-Meyer]. More recent references are in Section 9.4, where we also pursue the possibility of *biorthogonality*. In that case the analysis window $\widetilde{p}(n)$ is dual to the synthesis window $p(n)$:

$$\widetilde{p}(n) = \frac{p(n)}{p^2(n) + p^2(n+M)}.$$

This corresponds to $\widetilde{g}(t) = g(t)/(g^2(t) + g^2(t+1))$ in continuous time.

It seems fair to say that the discrete-time construction is always a little more delicate. For orthogonality, we check sums instead of integrals. The sums often depend on equal spacing in time or frequency. Where time-varying windows and window lengths were no problem in Section 8.4, they are more difficult for discrete filters. But their potential value is so great that the effort is worth making.

Problem Set 9.1

1. Draw the DCT basis functions $\boldsymbol{h}_0, \boldsymbol{h}_1, \boldsymbol{h}_2, \boldsymbol{h}_3$ when $M = 8$. They are drawn in Section 8.3 for $M = 32$.

2. Suppose the prototype $\boldsymbol{p}$ in Figure 9.4 is an ideal brick wall. What are the coefficients $p(n)$ and what is $P(\omega)$?

3. Invent a prototype $p(n)$ that satisfies the PR conditions (9.2) and (9.3), with $M = 2$ and then general M. The filter has $N + 1$ coefficients.

4. Verify that the two systems below are equivalent where $H_1(z) = H(ze^{-j2\pi/M})$.

→ ⊗ → H(z) → ↓M → (multiplier input W_M^{-1}) → H₁(z) → ↓M →

9.2 Polyphase Form: *M* Channels

Digital filter banks divide the signal into M subbands, then process the subbands and reconstruct. The analysis bank splits the input signal and the synthesis bank recombines it. The essential information is extracted from the subband signals in the *processing block*. Its form varies and depends on the applications. In an audio/video compression system, the spectral contents of the subband signals are coded depending on their energies. In a radar system, the subband signals might be used to null out a narrow-band interference adaptively. Other applications are image analysis and enhancement, robotics, computer vision, echo-cancellation, voice privacy and communications.

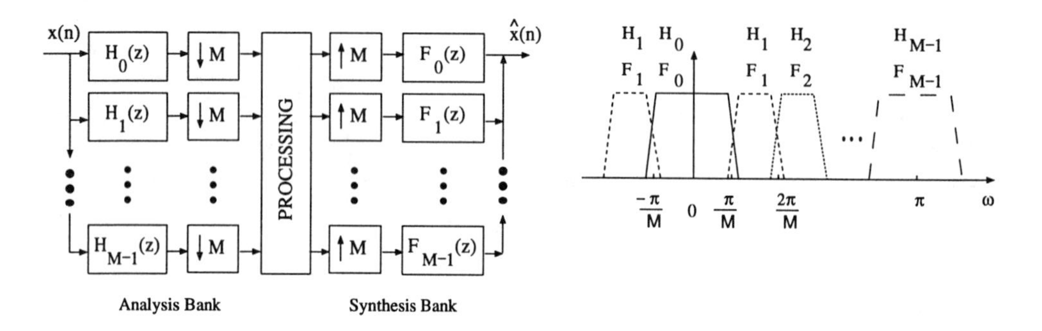

Figure 9.5: A maximally-decimated uniform filter bank with M channels.

Figure 9.5 illustrates a maximally decimated M-channel filter bank. The frequency responses $H_k(e^{j\omega})$ and $F_k(e^{j\omega})$ have passbands as shown. The analysis filters $H_k(z)$ channelize the input signal $x(n)$ into M subband signals, which are downsampled (decimated) by a factor M. At the receiving end, the M subband signals are decoded, interpolated and recombined by the synthesis filters $F_k(z)$. The decimator, which decreases the sampling rate, and the expander, which increases the rate, are denoted by $(\downarrow M)$ and $(\uparrow M)$.

Reconstruction Error

Since the filters $H_k(z)$ are not ideal, the filtered signals are not bandlimited to π/M. Therefore the aliases from downsampling will overlap. They depend on the stopband attenuation of $H_k(e^{j\omega})$ and their transition bands. The interpolated (upsampled) signals have M images of the compressed spectrum (assuming that no processing has been done after the analysis bank). These images are filtered by the synthesis filters $F_k(z)$.

There are two types of errors in the reconstructed output signal $\widehat{x}(n)$: *distortions* (magnitude and phase) and *aliasing*. The nonideal filtering characteristics of $H_k(z)$ and $F_k(z)$ contribute to both distortion and aliasing. The changes in sampling rates (downsampling and upsampling) contribute to the aliasing error. A system with no aliasing error is "alias-free". We compute the z-transform relation between input and output for $M = 3$:

- The analysis filters produce $H_0(z)X(z)$ and $H_1(z)X(z)$ and $H_2(z)X(z)$.
- $(\downarrow 3)$ and $(\uparrow 3)$ then produce two aliases in each channel $k = 0, 1, 2$:

$$\tfrac{1}{3}\left(H_k(z)X(z) + H_k(zW)X(zW) + H_k(zW^2)X(zW^2)\right) \quad \text{with } W = e^{-j2\pi/3}.$$

- The synthesis filters multiply those z-transforms by $F_k(z)$. Adding the nine terms, and collecting the terms for $X(z)$ and its aliases $X(zW)$ and $X(zW^2)$, the total output from the filter bank is

$$\widehat{X}(z) = T_0(z)X(z) + T_1(z)X(zW) + T_2(z)X(zW^2). \tag{9.4}$$

The perfect reconstruction conditions are $T_0(z) = z^{-\ell}$ (no amplitude distortion) and $T_1(z) = T_2(z) = 0$ (no aliasing). These three equations are written out explicitly in (9.10) below. For general M, the input-output relation is

$$\widehat{X}(z) = \sum_{k=0}^{M-1} T_k(z)X(zW^k) \qquad \text{where} \qquad T_k(z) = \frac{1}{M}\sum_{\ell=0}^{M-1} F_\ell(z)H_\ell(zW^k). \tag{9.5}$$

Since the transfer functions $T_1(z), \quad T_2(z), \quad \ldots, \quad T_{M-1}(z)$ multiply the shifted versions of the input spectrum, they are the *aliasing transfer functions. The distortion function* $T_0(z)$ multiplies the original spectrum. Then $T_0(z)X(z)$ is the output when all aliasing is cancelled. The objective is to find a set of PR filters $H_k(z)$ and $F_k(z)$:

$$\textbf{\textit{Perfect Reconstruction:}} \quad \begin{cases} T_0(z) = z^{-\ell} & \text{(no distortion for } k = 0) \\ T_k(z) = 0 & \text{(alias-free for } 1 \le k < M). \end{cases} \tag{9.6}$$

Paraunitary and biorthogonal filter banks (with additional properties such as linear phase and cosine modulation) satisfy all conditions in (9.6). The conventional Pseudo-QMF bank cancels aliasing at adjacent bands ($T_1(z) = 0$). The "Near PR" banks have $T_0(z) = z^{-\ell}$ and $T_k(z)$ near zero. The DFT filter bank cancels all aliasing components, but suffers from distortions! It satisfies the last $M - 1$ conditions in (9.6) but not the first. In discussing a specific M-channel filter bank, one has to keep in mind its reconstruction properties.

Note: For a two-channel filter bank, perfect reconstruction requires

$$\begin{cases} T_0(z) = \frac{1}{2}[F_0(z)H_0(z) + F_1(z)H_1(z)] = z^{-\ell} \\ T_1(z) = \frac{1}{2}[F_0(z)H_0(-z) + F_1(z)H_1(-z)] = 0. \end{cases} \tag{9.7}$$

Solving for $F_k(z)$ yields the synthesis filters from the analysis filters:

$$F_0(z) = \frac{2z^{-\ell}}{\Delta(z)}H_1(-z) \quad \text{and} \quad F_1(z) = -\frac{2z^{-\ell}}{\Delta(z)}H_0(-z) \tag{9.8}$$

where $\Delta(z) = H_0(z)H_1(-z) - H_0(-z)H_1(z)$. An FIR system has ℓ delays:

$$\text{Determinant} \quad \Delta(z) = 2z^{-\ell}, \quad F_0(z) = H_1(-z), \quad F_1(z) = -H_0(-z). \tag{9.9}$$

This simple relationship ***does not extend*** to filter banks with more channels. The perfect reconstruction equations for a three-channel bank are

$$\begin{cases} T_0(z) &= F_0(z)H_0(z) &+ F_1(z)H_1(z) &+ F_2(z)H_2(z) &= 3z^{-\ell} \\ T_1(z) &= F_0(z)H_0(zW) &+ F_1(z)H_1(zW) &+ F_2(z)H_2(zW) &= 0 \\ T_2(z) &= F_0(z)H_0(zW^2) &+ F_1(z)H_1(zW^2) &+ F_2(z)H_2(zW^2) &= 0 \end{cases} \tag{9.10}$$

where $W = e^{-j2\pi/3}$. Writing this system in matrix form, the solution for $F_k(z)$ is

$$\begin{bmatrix} F_0(z) \\ F_1(z) \\ F_2(z) \end{bmatrix} = \begin{bmatrix} H_0(z) & H_1(z) & H_2(z) \\ H_0(zW) & H_1(zW) & H_2(zW) \\ H_0(zW^2) & H_1(zW^2) & H_2(zW^2) \end{bmatrix}^{-1} \begin{bmatrix} 3z^{-\ell} \\ 0 \\ 0 \end{bmatrix}. \tag{9.11}$$

This inverse transpose of the modulation matrix $\boldsymbol{H}_m(z)$ gives

$$\begin{bmatrix} F_0(z) \\ F_1(z) \\ F_2(z) \end{bmatrix} = \frac{3z^{-\ell}}{\Delta(z)} \begin{bmatrix} H_1(zW)H_2(zW^2) - H_1(zW^2)H_2(zW) \\ H_2(zW)H_0(zW^2) - H_2(zW^2)H_0(zW) \\ H_0(zW)H_1(zW^2) - H_0(zW^2)H_1(zW) \end{bmatrix} \tag{9.12}$$

where $\Delta(z)$ is the determinant of $\boldsymbol{H}_m(z)$. The synthesis filter $F_k(z)$ depends on *two* filters $H_j(z)$ for $j \neq k$. This relationship complicates the design for M channels.

The number of parameters grows linearly with M. To simplify the design and implementation, explicit relations between the filters are often imposed:

Paraunitary Synthesis is time-reversed from analysis: $F_k(z) = z^{-N}H_k(z^{-1})$.

Linear Phase $H_k(z) = \pm z^{-N}H_k(z^{-1})$ (symmetric or antisymmetric).

DFT Filter Bank $H_k(z)$ comes from $H_0(zW^k)$.

Cosine Modulation $H_k(z)$ and $F_k(z)$ are DCT modulations of ***one*** prototype.

Pairwise Mirror Image $H_{M-1-k}(z) = H_k(-z)$ for odd M and $z^{-N}H_k(-z^{-1})$ for even M. The frequency responses are symmetric about $\frac{\pi}{2}$.

Each relation reduces the number of parameters by 2 (or by M, for DFT and cosine modulation). But the requirements might prevent our design methods from finding a good solution. The only two-channel linear-phase paraunitary filter bank is Haar's $H_0(z) = 1+z^{-1}$ and $H_1(z) = 1-z^{-1}$. By imposing both properties, we have limited our solution.

The choices depend on the applications. Cosine modulation is used in audio compression and telecommunications. It is efficient and has high stopband attenuation. Linear phase is important in image compression and signal detection.

Before developing filter banks for $M > 2$, we generalize the delay chain, the serial-parallel and parallel-serial converter, and the polyphase form.

Serial-Parallel and Parallel-Serial Converters

The figure below shows the block diagram of a delay chain. The transfer function from the input to the kth output is z^{-k}. By itself, the delay chain is not very interesting. However, cascading the delay chain with downsamplers or upsamplers will yield serial-parallel or parallel-serial converters.

Delay chain cascaded with downsamplers	=	Serial to Parallel (S/P) converter
Upsamplers cascaded with delay chain	=	Parallel to Serial (P/S) converter

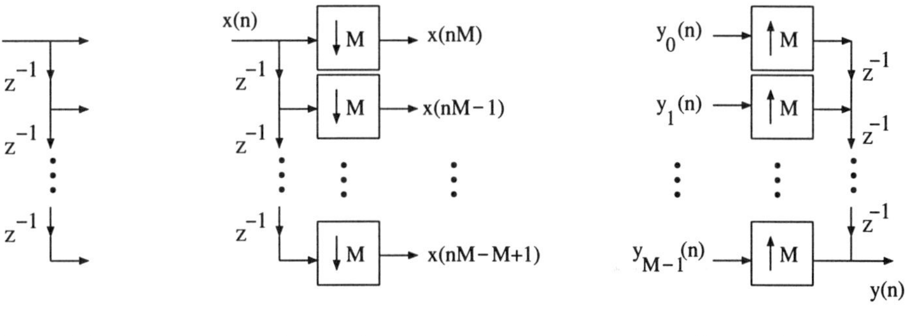

The output at the kth branch of the S/P converter is $x(nM-k)$, which implies that the input sequence is selected in a *counter-clockwise* fashion. The branches select signals in the order $0,\ M-1,\ M-2,\ \ldots,\ 2,\ 1,\ 0,\ M-1,\ldots$. When $x(n)$ is a causal signal, the output of the S/P converter ($M=4$) is:

branch						
0	$x(0)$	$x(4)$	$x(8)$	$x(12)$	$\cdots$	
1	0	$x(3)$	$x(7)$	$x(11)$	$\cdots$	output rate = (input rate)$/M$
2	0	$x(2)$	$x(6)$	$x(10)$	$\cdots$	
3	0	$x(1)$	$x(5)$	$x(9)$	$\cdots$	

A P/S converter is a cascade of expanders and a *reverse-ordered* delay chain. The output $y(n)$ is an interleaved combination of the signals $y_k(n)$. Thus its rate is M times the rate of $y_k(n)$. The signals $y_k(n)$ are selected in a *clockwise* fashion. Assuming that $y_k(n)$ are causal, and $M=3$, the serial output is

$$y_2(0)\ y_1(0)\ y_0(0)\ y_2(1)\ y_1(1)\ y_0(1)\ y_2(2)\ y_1(2)\ y_0(2)\ \cdots$$

Polyphase Representation of a Filter Bank

An analysis filter $H_k(z)$ is the sum of M phases $H_{kl}(z)$:

$$H_k(z)=\sum_{\ell=0}^{M-1} z^{-\ell}\, H_{k,\ell}(z^M), \qquad h_{k,\ell}(n)=h_k(Mn+\ell). \tag{9.13}$$

The four phases ($M=4$) of $1+3z^{-1}-4z^{-2}+7z^{-3}+6z^{-4}-3z^{-5}+z^{-6}$ are

$$\left\{\begin{array}{l} H_{k,0}(z)=1+6z^{-1} \\ H_{k,1}(z)=3-3z^{-1}\end{array}\right. \qquad \left\{\begin{array}{l} H_{k,2}(z)=-4+z^{-1} \\ H_{k,3}(z)=7.\end{array}\right. \tag{9.14}$$

Consider a lowpass $H(z)$ of length 81. Figure 9.6(a) shows $|H(e^{j\omega})|$ with center frequency at $\frac{\pi}{4}$. The magnitudes (nearly constant) and phases (nearly linear) of the four components $H(z)$ are plotted in 9.6(b) and (c). The phase responses are such that $|H|\approx 1$ in the passband and $H\approx 0$ in the stopband. Since magnitudes are approximately equal, the phase angles accomplish this task. Figure 9.6(d) plots the offsets $\Delta_\ell(\omega)=-10\omega-\phi_\ell(\omega)$. We observe that $\Delta_\ell(\omega)$ is nearly $-\ell\omega/4$ (in general $-\ell\omega/M$). The polyphase components provide *fractional delays* so that $H(z)$ is a good lowpass filter.

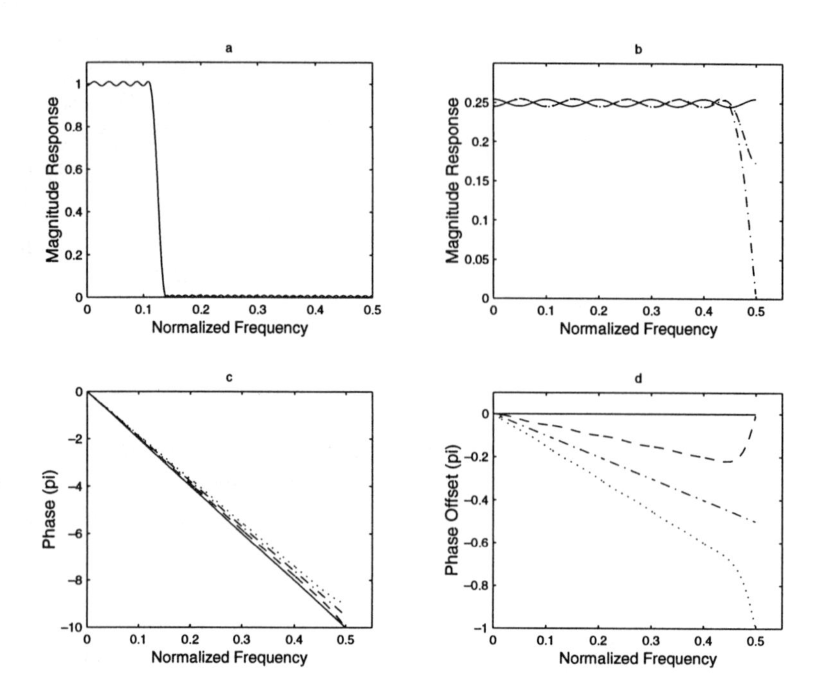

Figure 9.6: (a) Magnitude response of a lowpass filter, (b) Magnitude response of the polyphase filters, (c) Phase responses of the polyphase filters, (d) Phase offset.

There are M^2 polyphase components since each $H_k(z)$ has M components. Grouping these components in row k, the analysis transfer function is

$$\begin{bmatrix} H_0(z) \\ H_1(z) \\ \vdots \\ H_{M-1}(z) \end{bmatrix} = \underbrace{\begin{bmatrix} H_{0,0}(z^M) & H_{0,1}(z^M) & \cdots & H_{0,M-1}(z^M) \\ H_{1,0}(z^M) & H_{1,1}(z^M) & \cdots & H_{1,M-1}(z^M) \\ \vdots & \vdots & \ddots & \vdots \\ H_{M-1,0}(z^M) & H_{M-1,1}(z^M) & \cdots & H_{M-1,M-1}(z^M) \end{bmatrix}}_{\boldsymbol{H}_p(z^M)} \begin{bmatrix} 1 \\ z^{-1} \\ \vdots \\ z^{-(M-1)} \end{bmatrix}. \tag{9.15}$$

$\boldsymbol{H}_p(z)$ is the **polyphase matrix** of the analysis bank. The mapping between $\boldsymbol{h}_k(n)$ and $\boldsymbol{H}_p(z)$ is one to one. The impulse response of $H_{k,\ell}(z)$ is $\boldsymbol{h}_{k,\ell}(n) = \boldsymbol{h}_k(Mn+\ell)$.

Example 9.1. Suppose that the analysis filters of a 3-channel filter bank are

$$\begin{array}{lcl} H_0(z) & & 1+2z^{-1}+3z^{-2}+4z^{-3}+5z^{-4}+6z^{-5}+7z^{-6} \\ H_1(z) & = & 1-2z^{-1}+3z^{-2}-4z^{-3}+5z^{-4}-6z^{-5}+7z^{-6} \\ H_2(z) & & 1+2z^{-1}-3z^{-2}+4z^{-3}+5z^{-4}-6z^{-5}+7z^{-6} \end{array}$$

The corresponding polyphase transfer matrix is

$$\boldsymbol{H}_p(z) = \begin{bmatrix} 1+4z^{-1}+7z^{-2} & 2+5z^{-1} & 3+6z^{-1} \\ 1-4z^{-1}+7z^{-2} & -2+5z^{-1} & 3-6z^{-1} \\ 1+4z^{-1}+7z^{-2} & 2+5z^{-1} & -3-6z^{-1} \end{bmatrix}.$$

The synthesis filter has a Type-II representation (phase ℓ before channel k):

$$F_k(z) = \sum_{\ell=0}^{M-1} z^{-(M-1-\ell)} F_{\ell,k}(z^M) \qquad f_{\ell,k}(n) = f_k(Mn + M - 1 - \ell). \tag{9.16}$$

For $F_k(z) = 1 + 3z^{-1} - 4z^{-2} + 7z^{-3} + 6z^{-4} - 3z^{-5} + z^{-6}$, the four components that enter column k of the Type-II polyphase matrix $\boldsymbol{F}_p(z)$ are

$$F_{0,k}(z) = 7 \quad F_{1,k}(z) = -4 + z^{-1} \quad F_{2,k}(z) = 3 - 3z^{-1} \quad F_{3,k}(z) = 1 + 6z^{-1}.$$

The corresponding synthesis bank can be rewritten as

$$\begin{bmatrix} F_0(z) \\ F_1(z) \\ \vdots \\ F_{M-1}(z) \end{bmatrix} = \underbrace{\begin{bmatrix} F_{0,0}(z^M) & F_{1,0}(z^M) & \cdots & F_{M-1,0}(z^M) \\ F_{0,1}(z^M) & F_{1,1}(z^M) & \cdots & F_{M-1,1}(z^M) \\ \vdots & \vdots & \ddots & \vdots \\ F_{0,M-1}(z^M) & F_{1,M-1}(z^M) & \cdots & F_{M-1,M-1}(z^M) \end{bmatrix}}_{\boldsymbol{F}_p^T(z^M)} \begin{bmatrix} z^{-(M-1)} \\ z^{-(M-2)} \\ \vdots \\ 1 \end{bmatrix}. \tag{9.17}$$

Transposing, $\boldsymbol{F}_p(z)$ is the polyphase transfer matrix of the synthesis bank. Using $\boldsymbol{H}_p(z)$ (Type-I polyphase) and $\boldsymbol{F}_p(z)$ (Type-II polyphase), one can redraw Figure 9.5(a) as Figure 9.7(a). Then decimators move to the left of $\boldsymbol{H}_p(z^M)$ by the *Noble Identity*. Similarly the expanders move to the right of $\boldsymbol{F}_p(z^M)$.

A few words on the implementation efficiency of Figure 9.7(b). The input is blocked into M vectors by a Serial/Parallel converter (implemented as cascade of delay chain and decimators). The blocks are filtered by $\boldsymbol{F}_p(z)\boldsymbol{H}_p(z)$ and then recombined using a Parallel/Serial converter. The total number of nonzero coefficients in $\boldsymbol{H}_p(z)$ and $\boldsymbol{F}_p(z)$ is the same as that in $H_k(z)$ and $F_k(z)$. The main difference is a more efficient rate of operation. The filtering in the polyphase form is done at the input rate divided by M.

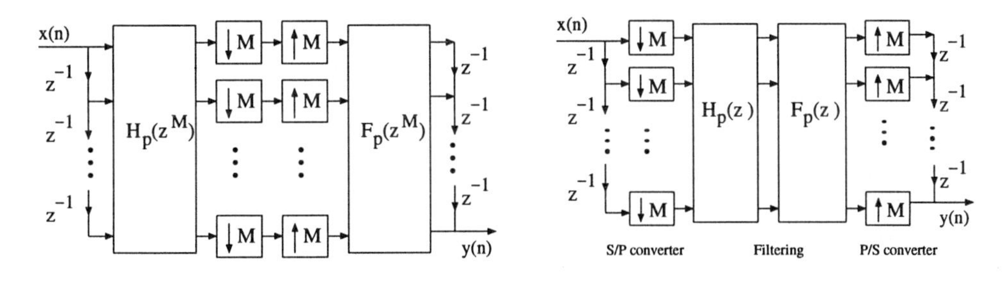

Figure 9.7: Polyphase representations of an M-channel uniform filter bank.

Modulation Matrix

Writing the PR conditions (9.6) in matrix-vector form displays the aliasing component matrix, which is the transposed modulation matrix $\boldsymbol{H}_m^T(z)$:

$$\begin{bmatrix} H_0(z) & H_1(z) & \cdots & H_{M-1}(z) \\ H_0(zW) & H_1(zW) & \cdots & H_{M-1}(zW) \\ \vdots & \vdots & \ddots & \vdots \\ H_0(zW^{M-1}) & H_1(zW^{M-1}) & \cdots & H_{M-1}(zW^{M-1}) \end{bmatrix} \begin{bmatrix} F_0(z) \\ F_1(z) \\ \vdots \\ F_{M-1}(z) \end{bmatrix} = M \begin{bmatrix} T_0(z) \\ T_1(z) \\ \vdots \\ T_{M-1}(z) \end{bmatrix}$$

The term ***modulation*** comes from the fact that row k of $\boldsymbol{H}_m$ is obtained by modulating row zero (shift in frequency by $2\pi k/M$). The term ***aliasing component*** comes from the fact that row k of $\boldsymbol{H}_m^T$ determines the aliasing transfer function $T_k(z)$—which should be zero for $k > 0$. The theory of filter banks can be derived using either $\boldsymbol{H}_m(z)$ or $\boldsymbol{H}_p(z)$. Although they are equivalent, $\boldsymbol{H}_p(z)$ is preferred because it is used in the implementation.

To connect $\boldsymbol{H}_m(z)$ to $\boldsymbol{H}_p(z)$ we need two matrices. One is the diagonal delay matrix $\boldsymbol{D}(z) = diag(1, z^{-1}, \ldots, z^{-(M-1)})$. The other is the M-point DFT matrix $\boldsymbol{F}_M$ which gives the modulations:

$$\text{Modulation and Polyphase} \quad \boldsymbol{H}_m(z) = \boldsymbol{H}_p(z^M)\boldsymbol{D}(z)\boldsymbol{F}_M. \tag{9.18}$$

Proof. The first row of $\boldsymbol{H}_m(z)$ contains the responses $H_0(zW^k)$ of the first filter. The first row of $\boldsymbol{H}_p(z^M)$ contains the phases H_{0j} of that same filter. To assemble phases of any function, we multiply by delays:

$$\begin{bmatrix} H_0(z) & H_0(zW) & \cdots \\ \cdots & \cdots & \cdots \\ \\ \cdots & \cdots & \cdots \end{bmatrix} = \begin{bmatrix} H_{00}(z^M) & H_{01}(z^M) & \cdots \\ \cdots & \cdots & \cdots \\ \\ \cdots & \cdots & \cdots \end{bmatrix} \begin{bmatrix} 1 & 1 & \cdots \\ z^{-1} & (zW)^{-1} & \cdots \\ \vdots & \vdots & \vdots \end{bmatrix}$$

Factoring the delays $1, z^{-1}, \ldots, z^{-(M-1)}$ from the rows of the last matrix leaves $\boldsymbol{F}_M$. The whole matrix multiplication becomes exactly $\boldsymbol{H}_m(z) = \boldsymbol{H}_p(z^M)\boldsymbol{D}(z)\boldsymbol{F}_M$.

The diagonal delay matrix is clearly paraunitary. The Fourier matrix $\boldsymbol{F}_M/\sqrt{M}$ is unitary (thus paraunitary, but complex). Then (9.18) shows that $\boldsymbol{H}_p(z)$ is paraunitary when $\boldsymbol{H}_m(z)/\sqrt{M}$ is paraunitary.

Theorem 9.1 *The equivalent conditions in the z-domain for an orthogonal M-channel filter bank are*

1. *The polyphase matrix is paraunitary:* $\boldsymbol{H}_p^T(z^{-1})\boldsymbol{H}_p(z) = \boldsymbol{I}$.

2. *The modulation matrix divided by* $\sqrt{M}$ *is paraunitary:* $\overline{\boldsymbol{H}}_m^T(z^{-1})\boldsymbol{H}_m(z) = M\boldsymbol{I}$.

It is satisfying that each condition has a direct proof, on its own. The direct proof for polyphase is to see the analysis bank ending with $\boldsymbol{H}_p(z)$, and the synthesis bank starting with $\boldsymbol{H}_p^T(z^{-1})$. That meeting produces $\boldsymbol{I}$ at the center of the filter bank. In polyphase form, the whole bank generates blocks of M samples, filters with $\boldsymbol{I}$, and reconstructs the signal from the blocks. This is perfect reconstruction.

The direct proof of the modulation condition comes from following each channel instead of each phase. When we did this with synthesis filters F_k, the perfect reconstruction condition $T_k(z) = z^{-l}\delta(k)$ was

$$[F_0(z) \;\; F_1(z) \;\; \cdots \;\; F_{M-1}(z)]\,\boldsymbol{H}_m(z) = M[z^{-l} \;\; 0 \;\; \cdots \;\; 0]. \tag{9.19}$$

This is the top row of $\boldsymbol{F}_m(z)\boldsymbol{H}_m(z)$. The other rows come from modulating the F's:

$$\begin{bmatrix} F_0(z) & F_1(z) & \cdots \\ F_0(zW) & F_1(zW) & \cdots \\ \cdots & \cdots & \cdots \end{bmatrix} \begin{bmatrix} H_0(z) & H_0(zW) & \cdots \\ H_1(z) & H_1(zW) & \cdots \\ \cdots & \cdots & \cdots \end{bmatrix} = M \begin{bmatrix} z^{-l} & & \\ & (zW)^{-l} & \\ & & \ddots \end{bmatrix}. \tag{9.20}$$

This is perfect reconstruction with causal filters. It applies to all cases—orthogonal or bi-orthogonal. In the orthogonal case the filters are $F_k(z) = z^{-l}H_k(z^{-1})$. So bring the diagonal matrix in (9.20) to the other side to find $\overline{\boldsymbol{H}}_m^T(z^{-1})\boldsymbol{H}_m(z) = M\boldsymbol{I}$. Note that $\boldsymbol{H}_m$ has complex coefficients because of the powers of W. The polyphase matrix $\boldsymbol{H}_p$ has real coefficients, because all filters are real.

Perfect Reconstruction: Polyphase Form

We have just given the perfect reconstruction condition in modulation form. It was written explicitly for $M = 3$ in equation (9.10), by following the signal through three channels. Equation (9.19) is the same statement for M channels. Equation (9.20) expresses that PR condition in terms of the matrices $\boldsymbol{H}_m(z)$ and $\boldsymbol{F}_m(z)$. Now we express perfect construction in terms of $\boldsymbol{H}_p(z)$ and $\boldsymbol{F}_p(z)$.

The reader will not be surprised by the result. The beauty of the polyphase form is the way it handles the algebra, for any number of filters.

Theorem 9.2 *An M-channel filter bank gives perfect reconstruction if*

$$\boldsymbol{F}_p(z)\boldsymbol{H}_p(z) = z^{-L}\boldsymbol{I}. \tag{9.21}$$

The overall delay of the system is $l = M - 1 + LM$, so that $T_0(z) = cz^{-l}$. The analysis and synthesis filters are ***biorthogonal****.*

A first proof begins with the modulation matrices in (9.20). Then the identity (9.18) converts to polyphase matrices. The result is the PR condition (9.21).

For a second proof, imagine the analysis-synthesis cascade *with the polyphase matrices in the middle*. The bank begins with an S/P converter and ends with a P/S converter. When the product of polyphase matrices is $z^{-L}\boldsymbol{I}$, the whole filter bank is a simple delay. This is perfect reconstruction.

Note: The most general case allows reordering of the channels. The product of the Type-I analysis matrix $\boldsymbol{H}_p(z)$ and the Type-II synthesis matrix $\boldsymbol{F}_p(z)$ is

$$\boldsymbol{F}_p(z)\boldsymbol{H}_p(z) = z^{-L}\begin{bmatrix} \boldsymbol{0} & \boldsymbol{I}_{M-r} \\ z^{-1}\boldsymbol{I}_r & \boldsymbol{0} \end{bmatrix}. \tag{9.22}$$

The overall delay is increased by r. We mention that z^{-1} is present below the diagonal whenever the system is alias-free; this gives the "pseudo-circulant" of Problem 4. For $r = 0$ we return to the fundamental case (9.21), when the product is $z^{-1}\boldsymbol{I}_M$.

Example 9.2. Suppose $\boldsymbol{v}$ is any unit column vector: $\boldsymbol{v}^T\boldsymbol{v} = 1$. Then choose

$$\boldsymbol{H}_p(z) = \boldsymbol{I} - \boldsymbol{v}\boldsymbol{v}^T + z^{-1}\boldsymbol{v}\boldsymbol{v}^T \quad \text{with} \quad \boldsymbol{H}_p^{-1}(z) = z\boldsymbol{v}\boldsymbol{v}^T + \boldsymbol{I} - \boldsymbol{v}\boldsymbol{v}^T. \tag{9.23}$$

You can verify immediately that the product is $\boldsymbol{I}$. The matrix $\boldsymbol{H}_p(z)$ gives an orthogonal analysis bank, and the causal matrix $\boldsymbol{F}_p(z) = z^{-1}\boldsymbol{H}_p^{-1}(z)$ gives the synthesis bank. These degree-one matrices are the building blocks for *all* paraunitary matrices. This is the great factorization theorem began by Belevitch and completed by Vaidyanathan [V, p. 273]:

$$\textit{Every paraunitary } \boldsymbol{H}_p(z) \textit{ factors into } \left(\prod_{j=1}^{L}\left[\boldsymbol{I} - \boldsymbol{v}_j\boldsymbol{v}_j^T + \boldsymbol{v}_j\boldsymbol{v}_j^T z^{-1}\right]\right)\boldsymbol{Q}. \tag{9.24}$$

Example 9.3. Suppose $\boldsymbol{w}^T\boldsymbol{v} = 1$. Choose the analysis polyphase matrix

$$\boldsymbol{H}_p(z) = \boldsymbol{I} - \boldsymbol{v}\boldsymbol{w}^T + z^{-1}\boldsymbol{v}\boldsymbol{w}^T \quad \text{with} \quad \boldsymbol{H}_p^{-1}(z) = z\boldsymbol{v}\boldsymbol{w}^T + \boldsymbol{I} - \boldsymbol{v}\boldsymbol{w}^T. \tag{9.25}$$

Again the product is $\boldsymbol{I}$. But $\boldsymbol{H}_p^{-1}$ is not $\boldsymbol{H}_p^T$. The causal matrix $\boldsymbol{F}_p(z) = z^{-1}\boldsymbol{H}_p^{-1}(z)$ gives the *biorthogonal* synthesis bank (not orthogonal unless $\boldsymbol{v} = \boldsymbol{w}$). These degree-one matrices do *not* give building blocks for the most general biorthogonal filter banks. [PhVaid] have identified the correct subclass.

DFT Filter Banks

The analysis filters are all modulations $H_k(z) = H_0(zW^k)$ of the lowpass filter. The idealized responses are shown in Figure 9.8. Notice that the frequency allocations are very different from Figure 9.5. Similarly the synthesis filters are $F_k(z) = F_0(zW^k)$. All filters have linear phase if $\boldsymbol{H}_0$ and $\boldsymbol{F}_0$ have linear phase. But the filter coefficients are complex for $M > 2$, because $W = e^{-j2\pi/M}$ is complex.

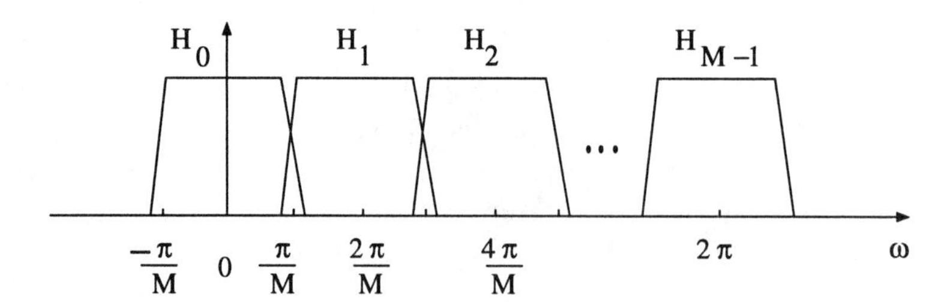

Figure 9.8: Idealized frequency response of the DFT filter bank.

This DFT filter bank is an excellent example for the use of polyphase matrices. Figure 9.2 showed the result of reordering the steps of subsampling and filtering. The lowpass filter coefficients $\boldsymbol{h}_0(n)$ on the left were separated into M phases on the right: $H_0(z) = E_0(z^M) + z^{-1}E_1(z^M) + \cdots$. The polyphase matrix for the whole analysis bank is

$$\boldsymbol{H}_p(z) = [DFT]\,\mathrm{diag}(E_0(z), E_1(z), \ldots, E_{M-1}(z)). \tag{9.26}$$

The implementation of a DFT filter bank is shown in Figure 9.9. The input signal $\boldsymbol{x}(n)$ is blocked into vectors of M components by the delay chain and downsamplers. The subband signals are then filtered by the polyphase components $E_\ell(z)$ of $H_0(z)$ and passed through an M-point DFT. The polyphase filters have real coefficients, so the inputs to the DFT are real. If the

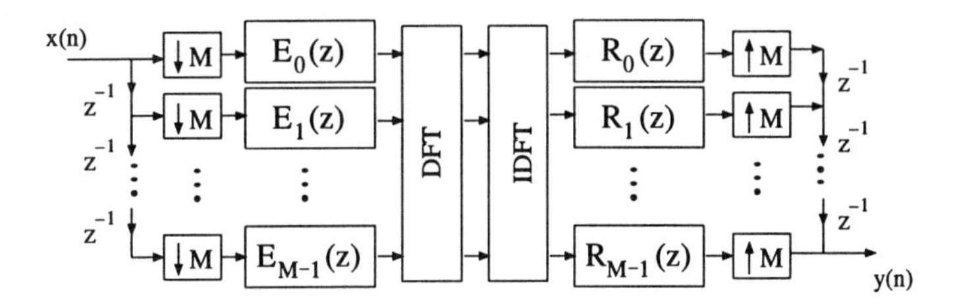

Figure 9.9: Polyphase representation of M-channel uniform DFT filter bank.

length of the lowpass filter is N, the computation load for the analysis bank is N multiplications and an M-point DFT for every M input samples.

The synthesis bank reverses these steps. We use the Type-II polyphase form of the lowpass filter $F_0(z) = R_{M-1}(z^M)+\cdots+z^{-(M-1)}R_0(z^M)$. Then the polyphase matrix $\boldsymbol{F}_p(z)$ has the IDFT multiplied by the diagonal matrix with entries $R_0(z), \ldots, R_{M-1}(z)$. The product of analysis and synthesis is a *diagonal matrix* when the DFT and IDFT cancel:

$$\boldsymbol{F}_p(z)\boldsymbol{H}_p(z) = \operatorname{diag}(R_0(z)E_0(z), R_1(z)E_1(z), \ldots, R_{M-1}(z)E_{M-1}(z)). \tag{9.27}$$

This filter bank is alias-free if all diagonal elements are equal. It gives perfect reconstruction if those equal elements are pure delays:

$$\textbf{\textit{PR condition}} \quad R_k(z)E_k(z) = z^{-L} \quad \text{for all } k. \tag{9.28}$$

But these filters $R_k(z) = z^{-L}E_k^{-1}(z)$ are IIR rather than FIR. For stability we would want each $E_k(z)$ to be minimum phase (zeros inside the unit circle). We also hope for linear phase. It is impossible to have both.

Theorem 9.3 [Nguyen-Vaidyanathan] *The polyphase components of a linear phase filter $H_0(z)$ cannot all have minimum phase.*

The anti-aliasing option is to choose FIR synthesis filters $R_k(z)$:

$$R_k(z) = E_0(z)E_1(z)\cdots E_{M-1}(z) \quad \text{omitting } E_k(z). \tag{9.29}$$

These filters are very long, about $M-1$ times as long as the analysis filters. The products $R_k(z)E_k(z)$ are all equal and aliasing is cancelled. The distortion function $T_0(z)$ is the product with a delay:

$$T_0(z) = z^{-l}E_0(z)E_1(z)\cdots E_{M-1}(z). \tag{9.30}$$

Perfect reconstruction is not possible. The natural question is whether one can design $H_0(z)$ such that either the magnitude or phase distortion is cancelled. Is there any choice of $H_0(z)$ such that T_0 is an allpass function (thus, no magnitude distortion) or a linear-phase function (thus, no phase distortion)? The answers are direct and beautiful:

If $H_0(z)$ has linear phase then $T_0(z)$ has linear phase.

If the components $E_k(z)$ are allpass then $T_0(z)$ is allpass.

In either case the output needs to be equalized. The equalizer for the allpass case should be an allpass filter that equalizes the phase distortion. The equalizer for the linear phase case should be a linear phase filter that equalizes the magnitude distortion.

For these alias-free filter banks, with synthesis filters from (9.29), we note that $H_0(z)$ and $F_0(z)$ cannot both be good lowpass filters. If they were good, there would be almost no overlap between $H_k(z)$ and $H_{k+2}(z)$, and between $F_k(z)$ and $F_{k+2}(z)$. Then the products $F_k(z)H_k(zW)$ do not overlap and their sum $T_1(z)$ cannot be zero! A typical $T_1(z)$ from good filters is drawn below:

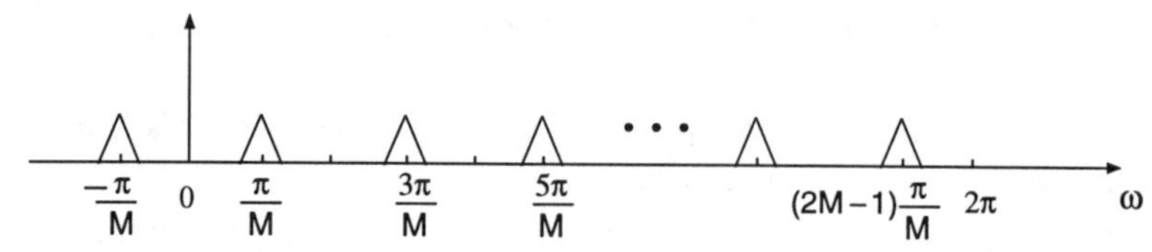

Summary The polyphase method gives a direct approach to the analysis of DFT filter banks. Unfortunately, the results are almost all negative. We cannot even cancel aliasing. The DFT bank is fast, but not good in reconstruction. The DCT bank in Section 9.4 is also fast, and it reconstructs perfectly.

Properties of Paraunitary Filter Banks and Matrices

Power Complementary $\sum_{k=0}^{M-1} \left|H_k(e^{j\omega})\right|^2 = 1$ and $\sum_{k=0}^{M-1} H_k(z^{-1})H_k(z) = 1$.

This assures that there is no magnitude distortion: $T_0(z) = 1$. It does not determine the aliasing transfer functions. Thus paraunitary implies power complementary but power complementary does not imply paraunitary.

Time Reversal $F_k(z) = z^{-N}H_k(z^{-1})$

The magnitude responses satisfy $|F_k(z)| = |H_k(z)|$. $F_k(z)$ has maximum phase if $H_k(z)$ has minimum phase. If $H_k(z)$ is linear phase, then so is $F_k(z)$. The modulation matrix $\boldsymbol{H}_m(z)$ and the AC matrix $\boldsymbol{H}_m^T(z)$ are paraunitary. This follows from $\boldsymbol{H}_m(z) =$ (DFT)$\boldsymbol{D}(z)\boldsymbol{H}_p^T(z^M)$.

Spectral Factor of M-th Band Filter The rows of $\boldsymbol{H}_m(z)$ are paraunitary: $P_k = \widetilde{H}_k H_k$ has

$$\sum_{l=0}^{M-1} \widetilde{H}_k(zW^l)H_k(zW^l) = \sum_{l=0}^{M-1} P_k(zW^l) = 1.$$

Then P_k is an M-th band filter. Notice that $P_k(z)$ is linear phase by definition. Thus, $H_k(z)$ is a spectral factor of an M-th band linear phase filter.

Columnwise orthogonality The kth and ℓth columns are orthogonal. The elements in column k are power complementary: $\sum_m H_{mk}(z^{-1})\, H_{mk}(z) = 1$.

Determinant is an allpass function Let $\boldsymbol{H}(z)$ be a square paraunitary matrix and let $A(z)$ denote its determinant. Then $\widetilde{\boldsymbol{H}}(z)\,\boldsymbol{H}(z) = \boldsymbol{I}$ implies that $\widetilde{A}(z)A(z) = 1$. Consequently, $A(z)$ is an allpass function. For an FIR paraunitary system, $A(z)$ is a delay.

Submatrices and cascades Any columns of a paraunitary matrix $\boldsymbol{H}(z)$ give a paraunitary submatrix. Any cascade $\boldsymbol{H}_1(z)\boldsymbol{H}_2(z)$ is also paraunitary.

Problem Set 9.2

1. Show that $\boldsymbol{H} = \boldsymbol{I} - 2\boldsymbol{v}\boldsymbol{v}^T$ is a symmetric orthogonal matrix if $\boldsymbol{v}^T\boldsymbol{v} = 1$. This is a *Householder reflection* (det $H = -1$) in the plane perpendicular to $\boldsymbol{v}$. Show that $\boldsymbol{H} = \boldsymbol{I} - \boldsymbol{v}\boldsymbol{v}^T + z^{-1}\boldsymbol{v}\boldsymbol{v}^T$ is a paraunitary matrix.

2. What is the entry in row k, column l, of the modulation matrix $\boldsymbol{H}_m(z)$?

3. Verify that $\boldsymbol{F}_p(z)\boldsymbol{H}_p(z) = z^{-1}\boldsymbol{I}$ and find the coefficients in all four filters:

$$\boldsymbol{F}_p(z) = \begin{bmatrix} 3+4z^{-1} & -3-2z^{-1} \\ -2-2z^{-1} & 2+z^{-1} \end{bmatrix} \quad \boldsymbol{H}_p(z) = \begin{bmatrix} 2+z^{-1} & 3+2z^{-1} \\ 2+2z^{-1} & 3+4z^{-1} \end{bmatrix}.$$

Is this an orthogonal bank? Is $\boldsymbol{H}_p(z)$ paraunitary? What are the determinants?

4. A two-channel bank is ***alias-free*** if $T_1(z) = F_0(z)H_0(-z) + F_1(z)H_1(-z) = 0$. Verify that $\boldsymbol{F}_m(z)\boldsymbol{H}_m(z)$ is diagonal. Then substitute (9.18) to prove that $\boldsymbol{F}_p(z)\boldsymbol{H}_p(z)$ is a *pseudo-circulant matrix*:

$$\boldsymbol{F}_m(z)\boldsymbol{H}_m(z) = \begin{bmatrix} T_0(z) & \\ & T_0(-z) \end{bmatrix} \quad \boldsymbol{F}_p(z)\boldsymbol{H}_p(z) = \begin{bmatrix} T_{0,\text{even}} & T_{0,\text{odd}} \\ z^{-1}T_{0,\text{odd}} & T_{0,\text{even}} \end{bmatrix}.$$

This pattern extends to all M-channel alias-free filter banks. When the aliasing functions $T_1, \ldots, T_{M-1}$ are zero, the product $\boldsymbol{F}_m(z)\boldsymbol{H}_m(z)$ is a diagonal matrix with entries $T_0(z)$, $T_0(zW)$,.... The product $\boldsymbol{F}_p(z)\boldsymbol{H}_p(z)$ is again a pseudocirculant. The prefix "pseudo" comes from the extra factor z^{-1} multiplying all entries below the main diagonal.

5. For the following analysis filters, find $\boldsymbol{H}_p(z)$ and its determinant. Find synthesis filters $F_k(z)$ such that the overall system is PR.

   (a) $H_0(z) = 1 + 3z^{-1} - 2z^{-2}$, $H_1(z) = z^{-1} + 2z^{-2}$, $H_2(z) = z^{-2}$
   (b) $H_0(z) = 1 + z^{-1} + z^{-2} + z^{-3} + 2z^{-4} - z^{-5}$, $H_1(z) = 1 + z^{-1} + z^{-2} + z^{-3} + 3z^{-4} + z^{-6} + 2z^{-7} - z^{-8}$, $H_2(z) = z^{-1} + z^{-2} - 2z^{-4} + z^{-5}$

6. Let $H_0(z) = 1 + z^{-1} + z^{-2} - 0.5z^{-3}$, $H_1(z) = z^{-1} + z^{-2} - 0.25z^{-4}$, and $H_2(z) = z^{-2}$. Find the PR synthesis filters in polyphase form $\boldsymbol{F}_p(z)$ and modulation form $\boldsymbol{F}_m(z)$ using (9.12).

7. Find the relation between $Y(z)$ and $X(z)$ below. Is it LTI? What is the system if $H(z)$ is an odd-length linear phase M-th band filter?

x(n) → ↑M → H(z) → ↓M → y(n)

8. What is the polyphase matrix for a Serial/Parallel converter?

9. Let $\boldsymbol{H}_1(z)$ and $\boldsymbol{H}_2(z)$ be paraunitary. Show that the cascade system $\boldsymbol{H}_1(z)\boldsymbol{H}_2(z)$ is paraunitary. How about the system $\frac{1}{2}(\boldsymbol{H}_1(z) + \boldsymbol{H}_2(z))$?

10. Show that the 3×3 polyphase matrix is paraunitary, given that $\boldsymbol{B}_k$ are orthogonal matrices. Suppose the delays z^{-1} are replaced by the first-order allpass section $A(z) = \frac{a^* + z^{-1}}{1 + az^{-1}}$. Find $\boldsymbol{H}_p(z)$ and its determinant. Find the synthesis polyphase matrix $\boldsymbol{F}_p(z)$ that cancels aliasing. Does PR synthesis exist?

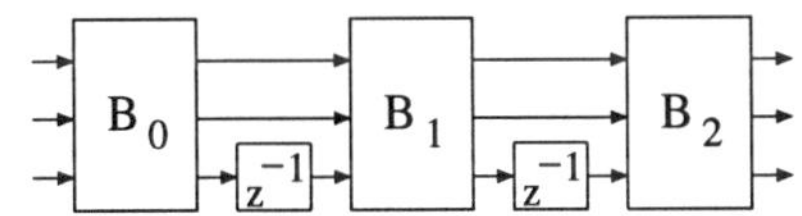

11. A *transmultiplexer* is the reverse of a filter bank (synthesis first). Two or more signals $x_k(n)$ are multiplexed and sent over a high bandwidth channel. At the receiving end, the signal is demultiplexed. The outputs $\widehat{x}_k(n)$ suffer from distortion and crosstalk. A PR transmultiplexer cancels crosstalk and reconstructs the signals $x_k(n)$ exactly.

(a) Express $\widehat{X}_k(z)$ in terms of $X(z)$ and the filters. What are the conditions on $H_k(z)$ and $F_k(z)$ such that the transmultiplexer is PR?

(b) Suppose $(H_k'(z),\ F_k'(z))$ yield a PR filter bank, and a transmultiplexer is constructed by $H_k(z) = H_k'(z),\ F_k(z) = z^{-1}F_k'(z)$. Show that this choice yields a PR transmultiplexer.

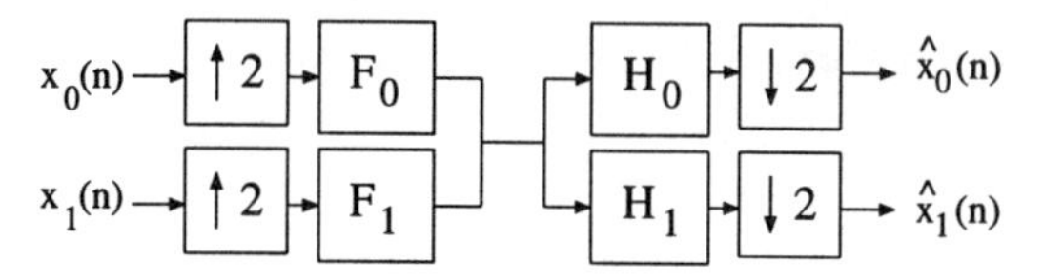

12. Let $(H_k(z),\ F_k(z))$ be an M-channel PR filter bank. What are the exact conditions on L such that $(H_k(z^L),\ F_k(z^L))$ is also PR?

13. Let $(H_k(z),\ F_k(z))$ be the filters in a paraunitary filter bank. Define $H_k'(z) = H_k(ze^{j\theta})$ and show that the filter bank is still paraunitary.

9.3 Perfect Reconstruction, Linear Phase, Orthogonality

There is much to recommend the time domain. The filter matrix $\boldsymbol{H}_b$ in block form — all M analysis channels together — is a *block Toeplitz matrix.* These matrices are easy to understand and analyze, if you remember that each entry is an $M \times M$ block. With filters of length $2M$, we have two blocks on each row as shown:

$$\boldsymbol{H}_b = \begin{bmatrix} \cdot & \cdot & \cdot & & \\ & \underline{h}(1) & \underline{h}(0) & & \\ & & \underline{h}(1) & \underline{h}(0) & \\ & & & \underline{h}(1) & \underline{h}(0) \\ & & & \cdot & \cdot \end{bmatrix} \tag{9.31}$$

This is the Lapped Orthogonal Transform (LOT) of Malvar. It is simply a filter bank with some overlapping but not much. We could have less overlapping or more:

1. No overlapping: $\boldsymbol{H}_b$ is block diagonal = block transform (DFT and DCT).

2. One overlap: $\boldsymbol{H}_b$ is block 2-diagonal = Lapped Orthogonal Transform (LOT).

3. Filter lengths BM: $\boldsymbol{H}$ has B block diagonals = general case.

We are not introducing new filter banks. These are standard M-channel analysis banks, including the downsampling operation $(\downarrow M)$. The only novelty is to interleave the channels, so that we watch all M channels at once. Previously $(\downarrow 2)\boldsymbol{H}_0$ from the lowpass channel was written above $(\downarrow 2)\boldsymbol{H}_1$ from the highpass channel:

$$\begin{bmatrix} (\downarrow 2)\boldsymbol{H}_0 \\ (\downarrow 2)\boldsymbol{H}_1 \end{bmatrix} = \begin{bmatrix} h_0(3) & h_0(2) & h_0(1) & h_0(0) & & \\ & & h_0(3) & h_0(2) & h_0(1) & h_0(0) \\ & & & & \cdots & \cdots \\ h_1(3) & h_1(2) & h_1(1) & h_1(0) & & \\ & & h_1(3) & h_1(2) & h_1(1) & h_1(0) \\ & & & & \cdots & \cdots \end{bmatrix} \tag{9.32}$$

Interleaving rows gives our block Toeplitz matrix, with block size $M = 2$:

$$H_b = \begin{bmatrix} h_0(3) & h_0(2) & h_0(1) & h_0(0) & & \\ h_1(3) & h_1(2) & h_1(1) & h_1(0) & & \\ & & h_0(3) & h_0(2) & h_0(1) & h_0(0) \\ & & h_1(3) & h_1(2) & h_1(1) & h_1(0) \\ & & & & \cdots & \cdots \\ & & & & \cdots & \cdots \end{bmatrix} \tag{9.33}$$

This is the time domain block form when $M = 2$ and $B = 2$. The filter lengths are $BM = 4$. The z-transform is the polyphase matrix

$$H_p(z) = \underline{h}(0) + z^{-1}\underline{h}(1) = \begin{bmatrix} h_0(0) & h_0(1) \\ h_1(0) & h_1(1) \end{bmatrix} + z^{-1} \begin{bmatrix} h_0(2) & h_0(3) \\ h_1(2) & h_1(3) \end{bmatrix}.$$

Note The block $h(k)$ contains the k-th coefficient from each phase of each filter. The underbar in $\underline{h}(k)$ is used to emphasize that this $M \times M$ block is the coefficient of z^{-k} in $H_p(z)$. When we form blocks, *the column order inside each block is to be reversed.* Thus $h_0(0)$ is to the left of $h_0(1)$ in the block, where it was to the right in (9.33). The orthogonality conditions on $\underline{h}(0)$ and $\underline{h}(1)$ are immediate in the time and z-domains:

Theorem 9.4 *The Lapped Orthogonal Transform* (LOT) *requires*

$$\begin{aligned} \underline{h}(0)^T\underline{h}(0) + \underline{h}(1)^T\underline{h}(1) &= I_{M\times M} \\ \textit{(orthogonality of tails)} \quad \underline{h}(1)^T\underline{h}(0) &= 0_{M\times M} \end{aligned} \tag{9.34}$$

Also $H_bH_b^T = I$. *This moves the transposes to the second factors in* (9.34).

The general case with B blocks per row (and filter lengths BM) will produce B block equations from $H_b^TH_b = I$. You could say that these equations are "Condition O" in the time domain. The paraunitary requirements were "Condition O" in the polyphase domain and modulation domain. Notice the difference from the two-channel case. There Condition O was applied only to one filter H_0. The highpass filter H_1 was determined by an alternating flip. Here, unless we impose a special structure on the bandpass filters $H_1, \ldots, H_{M-1}$, we must include them all in Condition O.

The DFT filter bank does impose such a structure—but it makes orthogonality impossible (except in the block transform case $B = 1$ which is pure DFT with no filters). The DCT filter bank also imposes a structure. It produces all M filters from the knowledge of one filter. But this time, for the DCT bank, *orthogonality is possible.* In that case Condition O is no longer double-shift orthogonality, it is M-shift orthogonality.

For the most general LOT, multiply (9.34) by $\underline{h}(0)$ to find $\underline{h}(0)\underline{h}(0)^T\underline{h}(0) = \underline{h}(0)$. We freely use $\underline{h}(0)\underline{h}(1)^T = 0$ and $\underline{h}(0)^T\underline{h}(1) = 0$ to obtain

$$\begin{aligned} \underline{h}(0) &= \underline{h}(0)\underline{h}(0)^T(\underline{h}(0) + \underline{h}(1)) &= PQ \\ \underline{h}(1) &= \underline{h}(1)\underline{h}(1)^T(\underline{h}(0) + \underline{h}(1)) &= (I - P)Q. \end{aligned} \tag{9.35}$$

The matrix $Q = \underline{h}(0) + \underline{h}(1)$ is orthogonal because (9.34) gives $Q^TQ = I$. The matrix $P = \underline{h}(0)\underline{h}(0)^T$ is a projection matrix. It is symmetric and $P^2 = \underline{h}(0)\underline{h}(0)^T\underline{h}(0)\underline{h}(0)^T =$

$\underline{h}(0)\underline{h}(0)^T = P$. Similarly, $\underline{h}(1)\underline{h}(1)^T$ is a projection, and the two projections add to I. The general solution (9.35) for the blocks in H_b was found by Heller and Tolimieri:

$$\underline{h}(0) = PQ \text{ and } \underline{h}(1) = (I - P)Q \quad \text{with } P = \text{projection}, Q = \text{orthogonal}.$$

The special case $P = I$ gives *one* block. Then H_b is block diagonal and it represents a simple orthogonal block transform. The general LOT can choose any projection matrix P and orthogonal matrix Q. Notice that $Q = H_p(1)$.

Note 1. The polyphase matrix has the convenient form $[P + (I - P)z^{-1}]Q$. This is fast if Q and P are fast. Below, we choose Q = DCT matrix:

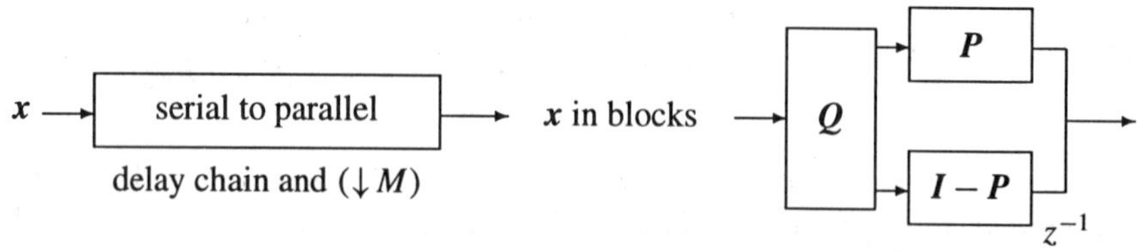

Note 2. The *biorthogonal* lapped transform with $B = 2$ has a very similar pattern (Problem 2). The matrix $P = \underline{h}(0)\underline{f}(0)$ still equals P^2. In this case, $P = P^T$ and $Q^TQ = I$ are not required. This gives the most general PR bank with filter lengths $2M$. The biorthogonal BOLT [PhVaid] extends this design to $B > 2$.

Note 3. For a paraunitary matrix $H_p(z)$, the degree of the determinant is the *Smith-McMillan degree*. When P has rank r, the degree is $L = M - r$. For the very special case of diagonal P, with r ones and $M - r$ zeros, the factors are clear:

$$H_p(z) = \begin{bmatrix} I_r & 0 \\ 0 & I_{M-r}z^{-1} \end{bmatrix} Q \text{ has determinant } z^{-(M-r)}.$$

Note 4. *For fast implementation, Q often starts with the DCT matrix* (the matrix C of cosines). The order M is even. Half the DCT rows and columns are symmetric, half are antisymmetric. It is natural to think of separating those parts and maintaining linear phase. Figure 9.10 does this separation by reordering the DCT rows to put C_{even} above C_{odd}. Then Malvar uses Haar butterflies and delays to separate $\underline{h}(1)$ from $\underline{h}(0)$. At the end we may apply different orthogonal matrices U and V. The blocks $\underline{h}(0)$ and $\underline{h}(1)$ are then

$$\begin{bmatrix} \underline{h}(0) & \underline{h}(1) \end{bmatrix} = \begin{bmatrix} U & V \end{bmatrix} \begin{bmatrix} C_e - C_o & (C_e - C_o)J \\ C_o - C_e & (C_e - C_o)J \end{bmatrix}. \tag{9.36}$$

The free parameters in this LOT are U and V. In practice, those must also allow fast multiplication. Malvar [Mr, p. 167] suggests $U = I$ and V = product of plane rotations or V = product of DST-IV and transposed DCT-II. The LOT is traditionally chosen to be linear phase and to start with the DCT-II.

Figure 9.11 shows the frequency responses of the LOTs. They are obtained by optimizing the coding gain (part a) and the stopband attenuation (part b). We show four basis functions (impulse responses) of the LOT that has high coding gain. Notice that they are symmetric and antisymmetric from the DCT. In summary, LOT is a linear phase paraunitary filter bank where the filter length is $2M$.

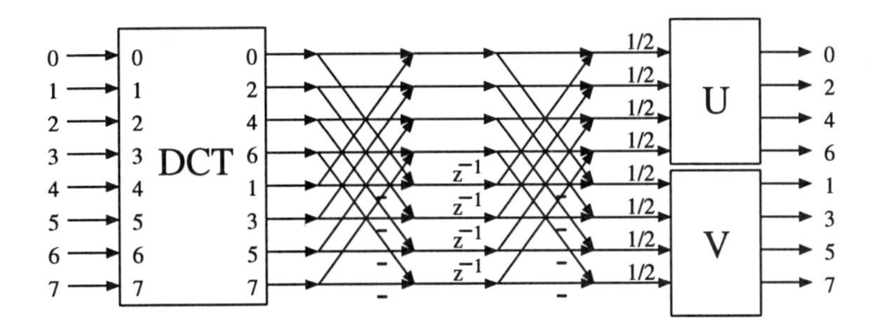

Figure 9.10: Polyphase transfer matrix of the Lapped Orthogonal Transform.

Longer Filters and GenLOT

You recognize that the two blocks ($B = 2$) of the LOT cannot give a sharp cutoff in frequency with great attenuation. The filters need to be longer. The time domain matrix H_b is always block Toeplitz when the input signal is blocked into M samples at a time. If the filter lengths are BM, there are B blocks in each row of H_b (and $B = 2$ for LOT):

$$H_b = \begin{bmatrix} \cdots & \cdots & & \cdots & & & \\ & \underline{h}(B-1) & & \cdots & \underline{h}(0) & & \\ & & \underline{h}(B-1) & & \cdots & \underline{h}(0) & \\ & & & & \cdots & \cdots & \cdots \end{bmatrix}. \tag{9.37}$$

The polyphase matrix for this FIR analysis bank with decimators ($\downarrow M$) is

$$H_p(z) = \underline{h}(0) + \underline{h}(1)z^{-1} + \cdots + \underline{h}(B-1)z^{-(B-1)}. \tag{9.38}$$

Note again that all these blocks are $M \times M$. The conditions for orthogonality come directly from $H_b^T H_b = I$ and from $H_p^T(z^{-1})H_p(z) = I$:

Theorem 9.5 *The analysis bank with filter length BM is orthogonal if*

$$\sum_{k=0}^{B-1-l} \underline{h}(k)^T \underline{h}(k+l) = \delta(l)I. \tag{9.39}$$

A block transform has $B = 1$ and only diagonal blocks in H_b. The constant polyphase matrix is $H_p(z) = \underline{h}(0)$. This gives bad results at block boundaries after compression. Overlapping blocks are much smoother but the filters have BM coefficients, which gives a huge design problem until we narrow it by structural decisions. One good decision is to start the filter bank with the DCT-II.

The simplest filters to follow the DCT are *block Haar and block diagonal and block delay*:

$$W = \frac{1}{\sqrt{2}}\begin{bmatrix} I & I \\ I & -I \end{bmatrix} \text{ and } Q_i = \begin{bmatrix} U_i & 0 \\ 0 & V_i \end{bmatrix} \text{ and } D(z) = \begin{bmatrix} I & 0 \\ 0 & z^{-1}I \end{bmatrix}.$$

The blocks are of order $\frac{M}{2}$ and these matrices are orthogonal. Therefore, all products are paraunitary. The product also has linear phase, if Q at the beginning of the filter bank separates even and odd rows of the cosine matrix C^{II} (symmetric and antisymmetric cosine basis functions). The GenLOT is a cascade of these special filters:

$$\textit{The GenLOT has } H_p(z) = \left(Q_{B-1}WD(z)W\right)\cdots\left(Q_1 WD(z)W\right)Q. \tag{9.40}$$

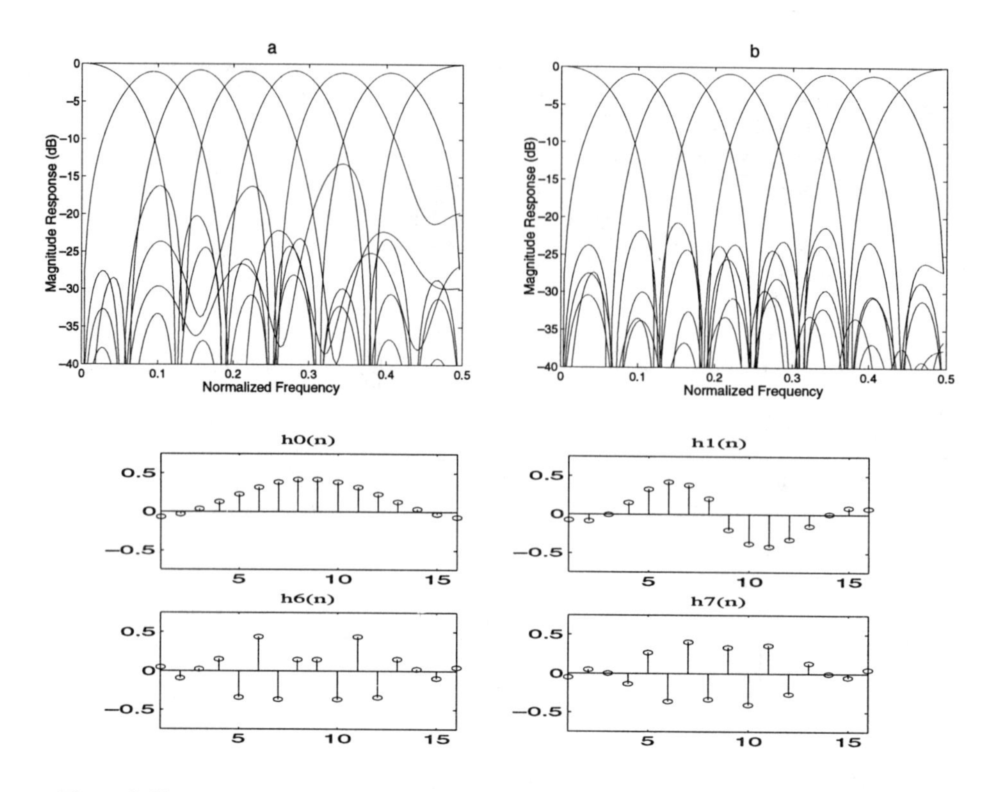

Figure 9.11: Frequency responses of the LOT. (a) High coding gain. (b) High stopband attenuation. Four of the eight symmetric-antisymmetric basis functions in (a).

The implementation flow graph for the analysis bank is in Figure 9.12 (see *GenLOT* in the Glossary for the details of K_1). The DCT-based LOT is the case with $B = 2$. The DCT itself is the case with $B = 1$ and $\boldsymbol{H}_p = \boldsymbol{Q}$. This framework covers all orthogonal filter banks with $\frac{M}{2}$ symmetric and antisymmetric channels — linear phase! The design problem is always to choose fast $\boldsymbol{Q}_i$ that also achieve sharp frequency discrimination.

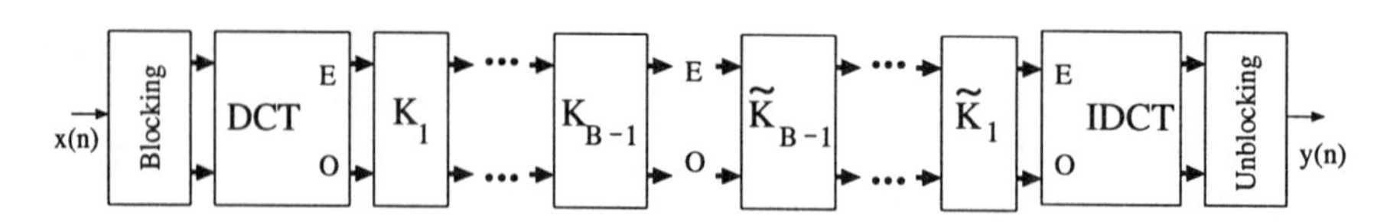

Figure 9.12: Polyphase transfer matrix of GenLOT.

Figure 9.13 shows the frequency responses of $H_k(z)$. The coding gain is 9.36 dB (left) and 23 dB attenuation (right). GenLOT design is included in

http://saigon.ece.wisc.edu/~waveweb/QMF.html

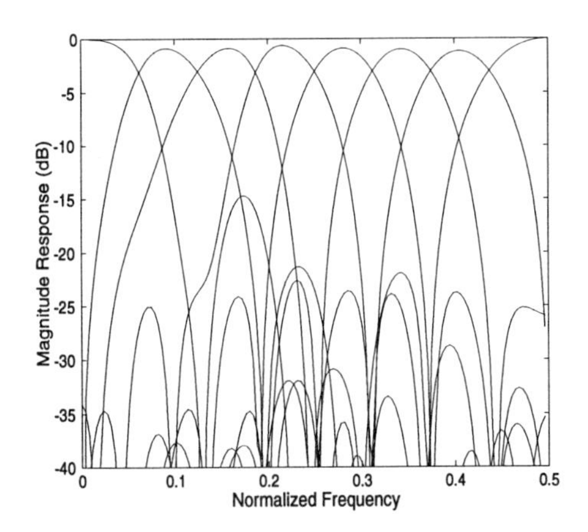

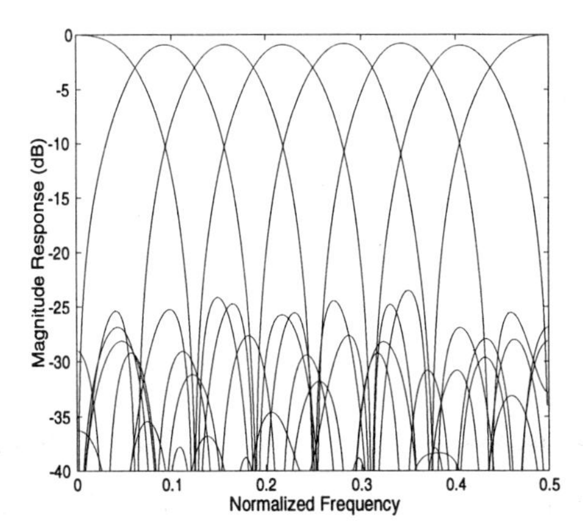

Figure 9.13: Frequency response plots for the 8-channel GenLOT (length 32) with high coding gain and high stopband attenuation.

Products of Rotations and Delays

Every $M \times M$ orthogonal matrix Q is a product of $\frac{1}{2}M(M-1)$ plane rotations (*Givens rotations*). R_{ij} gives rotation by θ_{ij} in the plane of axes i and j:

$$R_{ij} = \begin{bmatrix} 1 & & & & \\ & c & & s & \\ & & 1 & & \\ & -s & & c & \\ & & & & 1 \end{bmatrix} \qquad \begin{array}{l} c = \cos(\theta_{ij}) \text{ and } s = \sin(\theta_{ij}) \\ \text{in rows and columns } i \text{ and } j. \end{array}$$

The angles θ_{ij} can be design variables (but strongly nonlinear). The rotations in the GenLOT stay within even channels (to give U_i) or odd channels (to give V_i). The *Givens factorization* of any causal FIR paraunitary matrix is parallel to the *Householder factorization* in (9.24). Where Householder uses reflections, Givens uses a sequence of rotations:

$$H_p(z) = R_L D(z) \cdots R_2 D(z) R_1 D(z) Q. \tag{9.41}$$

Here $D(z) = \text{diag}(1, 1, \ldots, z^{-1})$ delays only the last channel. Its Smith-McMillan degree is 1, so the degree of $H_p(z)$ is L. Q is any unitary matrix. The matrices R_k can be chosen as *products of $M-1$ rotations only*, in neighboring channels $(1, 2), (2, 3), \ldots, (M-1, M)$. The total number of parameters (plane rotations) is then $L(M-1)$ plus $\frac{1}{2}M(M-1)$ for Q. This is the same total as in the product (9.24) from Householder reflections.

Example 9.4. Consider a three-channel orthogonal filter bank whose polyphase transfer matrix has McMillan degree 2. Figure 9.14 shows the corresponding lattice structure. Each rotation is a simple butterfly connecting channels i and j. The total number of angles is $4+3=7$, and the filters have length 9.

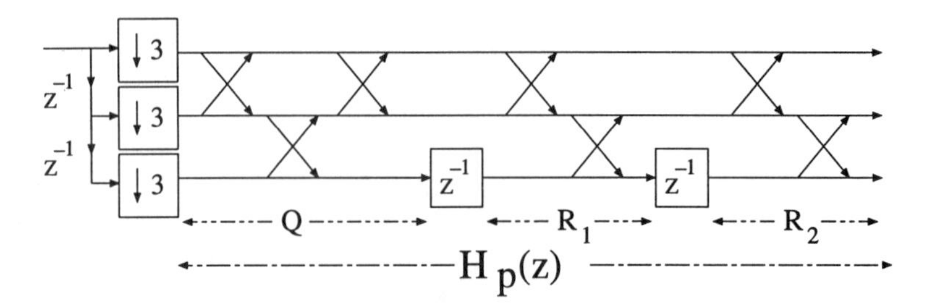

Figure 9.14: A three-channel paraunitary filter bank with McMillan degree 2.

Pairwise Mirror Image (PMI) Filter Banks

A further simplification of the GenLOT and other filter banks is to make each filter H_{M-1-k} a "mirror image" of H_k with respect to $\omega = \frac{\pi}{2}$:

$$|H_{M-1-k}(\omega)| = |H_k(\pi - \omega)| \,. \tag{9.42}$$

This reduces the number of design parameters by a factor of 2. When M is odd, say $M = 3$, the middle filter is its own mirror image and $H_1(z)$ is a function of z^2. The other two channels have $H_2(z) = H_0(-z)$. There is a convenient lattice structure [NgVa**3**], and the PMI property is structurally imposed.

In the GenLOT this pairwise property connects the orthogonal matrices $\boldsymbol{U}_i$ and $\boldsymbol{V}_i$. The polyphase matrix satisfies $\boldsymbol{J}\boldsymbol{H}_p(z) = \boldsymbol{H}_p(z)\,\boldsymbol{\Gamma}$, where $\boldsymbol{\Gamma} = \text{diag}(1, -1, \ldots, 1, -1)$. The matrices $\boldsymbol{V}_i$ are equal to $\boldsymbol{\Gamma}\boldsymbol{U}_i\boldsymbol{\Gamma}$, except the last of the $\boldsymbol{V}$'s is $\boldsymbol{J}\boldsymbol{U}_i\boldsymbol{\Gamma}$.

We compare the frequency responses of two 8-channel GenLOT's with $B = 3$ and filter length 24. The second set of responses has the pairwise mirror image (PMI) property. It displays better stopband attenuation (measured in dB), which was the objective function in the design process.

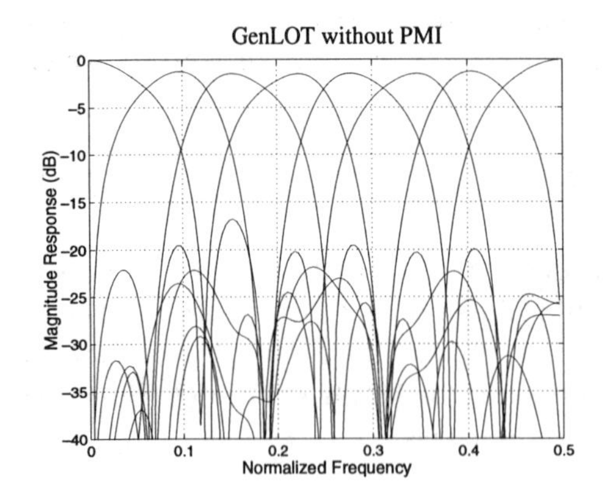

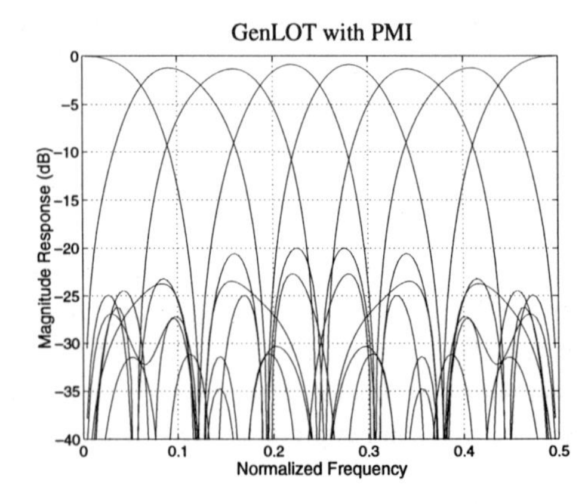

Permissible Lengths and Symmetry for Linear Phase

The filter lengths in GenLOT are the same and the number of symmetric and antisymmetric filters are equal. The conditions on the filter lengths $K_l M + \beta$ and the symmetries for *biorthogonal* filter banks are summarized below [TranNg]:

Case	*Symmetry/Antisymmetry*	*Lengths*	*Sum of Lengths*
M even, β even	$\frac{M}{2}$ S & $\frac{M}{2}$ A	$\sum K_\ell$ is even	$2mM$
M even, β odd	$(\frac{M}{2}+1)$ S & $(\frac{M}{2}-1)$ A	$\sum K_\ell$ is odd	$2mM$
M odd, β even	$(\frac{M+1}{2})$ S & $(\frac{M-1}{2})$ A	$\sum K_\ell$ is odd	$(2m+1)M$
M odd, β odd	$(\frac{M+1}{2})$ S & $(\frac{M-1}{2})$ A	$\sum K_\ell$ is even	$(2m+1)M$

Tree-structured Filter Banks

A popular method to design M-channel filter banks is to cascade smaller systems. Wavelet packets use two-channel systems. The six-channel filter bank below is obtained from cascading a two-channel system with two three-channel systems:

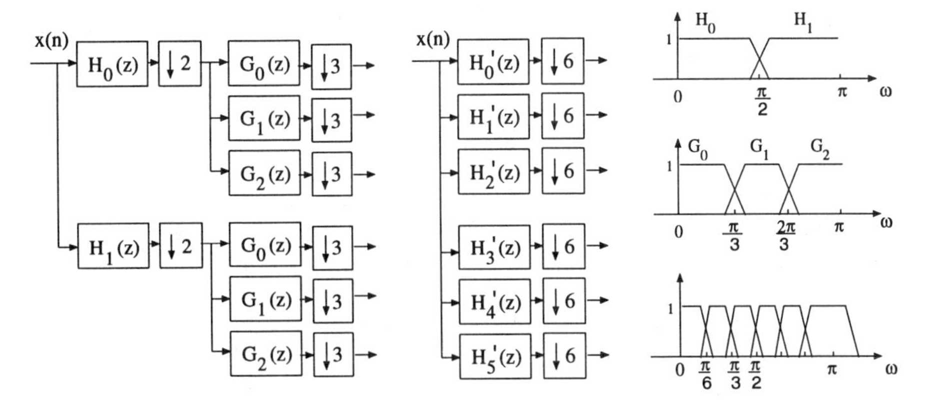

Typical relations between $H'_k(z)$ and $H_k(z)$ and $G_k(z)$ are $H'_0(z) = H_0(z)G_0(z^2)$ and $H'_5(z) = H_1(z)G_2(z^2)$. The 6-channel filter bank is PR if and only if the 2-channel and 3-channel filter banks are PR.

One obtains *nonuniform* filter banks by cascading systems with different decimation factors [HoVaid]. A nonuniform bank with decimation factors (6, 6, 6, 4, 4) uses a 3-channel and a 2-channel system at the second tree level. This cascade is drawn in the Glossary under *Tree-Structured Filter Banks*. We emphasize the simplicity of these designs.

Summary The analysis bank is a block Toeplitz multiplication in the time domain. It is a polyphase matrix multiplication in the z-domain. That matrix contains the blocks from the Toeplitz form, just as $H(z)$ contains the coefficients from one filter. The polyphase matrix extends the familiar idea of $H(z)$ from one filter to M filters. This is the efficient order that gives the polyphase form:

- (direct time domain) Apply the analysis filters and then $(\downarrow M)$.
- (more efficient order) Put the input in blocks and filter in parallel.
- (polyphase in z-domain) Multiply by $\boldsymbol{H}_p(z) = \sum \underline{\boldsymbol{h}}_p(n)z^{-n}$.

For $M \times M$ blocks, the kth row contains coefficients from the kth filter. Column n has the M coefficients starting with $\boldsymbol{h}_k(nM)$.

In the LOT case, with filter length $2M$ and two blocks per row, the first M coefficients $\boldsymbol{h}_k(0), \ldots, \boldsymbol{h}_k(M-1)$ go into the main diagonal block $\underline{\boldsymbol{h}}(0)$. The other coefficients $\boldsymbol{h}_k(M), \ldots, \boldsymbol{h}_k(2M-1)$ go into the subdiagonal block $\underline{\boldsymbol{h}}(1)$.

The synthesis bank is also a block Toeplitz matrix. The filter lengths and the number of blocks could be different, and the blocks in $\boldsymbol{F}_b$ are transposed. The coefficients from the kth synthesis filter are in the kth *column* of the blocks (not the kth row). Thus the two $M \times M$ blocks $\underline{f}(0)$ and $\underline{f}(1)$ in the length $2M$ case would be

$$\begin{bmatrix} f_0(0) & \cdots & f_{M-1}(0) \\ \vdots & & \vdots \\ f_0(M-1) & \cdots & f_{M-1}(M-1) \end{bmatrix} \text{ and } \begin{bmatrix} f_0(M) & \cdots & f_{M-1}(M) \\ \vdots & & \vdots \\ f_0(2M-1) & \cdots & f_{M-1}(2M-1) \end{bmatrix}.$$

The simplicity in the time domain resides in the fact that the whole filter bank becomes a matrix multiplication: $\widehat{\boldsymbol{x}} = \boldsymbol{F}_b\boldsymbol{H}_b\boldsymbol{x}$. For $M > 2$ there is enough freedom to maintain orthogonality, rather than biorthogonality, while achieving other good properties. In this case we ask $\boldsymbol{H}_b$ to be orthogonal. We also ask for fast implementation, which leads us to GenLOT (with the DCT matrix). Cosine modulation is the alternative described next.

Problem Set 9.3

1. Show that $\underline{\boldsymbol{h}}(0) = \boldsymbol{PQ}$ and $\underline{\boldsymbol{h}}(1) = (\boldsymbol{I}-\boldsymbol{P})\boldsymbol{Q}$ produce an orthogonal block Toeplitz $\boldsymbol{H}_b$, for any $\boldsymbol{P}$ = symmetric projection matrix ($\boldsymbol{P}^2 = \boldsymbol{P}$) and $\boldsymbol{Q}$ = orthogonal matrix ($\boldsymbol{Q}^T\boldsymbol{Q} = \boldsymbol{I}$).

2. When the synthesis bank is anti-causal, the PR condition is $\boldsymbol{F}_b\boldsymbol{H}_b = \boldsymbol{I}$:

$$\boldsymbol{F}_b = \begin{bmatrix} \underline{f}(0) & \underline{f}(1) & 0 & \\ & \underline{f}(0) & \underline{f}(1) & \\ & & \cdot & \end{bmatrix} \text{ and } \boldsymbol{H}_b = \begin{bmatrix} \underline{\boldsymbol{h}}(0) & & \\ \underline{\boldsymbol{h}}(1) & \underline{\boldsymbol{h}}(0) & \\ 0 & \underline{\boldsymbol{h}}(1) & \cdot \end{bmatrix}.$$

(a) $\boldsymbol{F}_b\boldsymbol{H}_b = \boldsymbol{I}$ and $\boldsymbol{H}_b\boldsymbol{F}_b = \boldsymbol{I}$ give what six PR conditions on the blocks?

(b) Deduce $\underline{f}(0) = (\underline{f}(0) + \underline{f}(1))\underline{\boldsymbol{h}}(0)\underline{f}(0)$ and $\underline{\boldsymbol{h}}(0) = \underline{\boldsymbol{h}}(0)\underline{f}(0)(\underline{\boldsymbol{h}}(0) + \underline{\boldsymbol{h}}(1))$. Write the corresponding equations for $\underline{f}(1)$ and $\underline{\boldsymbol{h}}(1)$.

(c) Deduce also that $\boldsymbol{P} = \underline{\boldsymbol{h}}(0)\underline{f}(0)$ equals $\boldsymbol{P}^2$ and $\underline{\boldsymbol{h}}(1)\underline{f}(1)$ equals $\boldsymbol{I} - \boldsymbol{P}$.

(d) Show finally that $\underline{f}(0) + \underline{f}(1)$ is the inverse of $\boldsymbol{Q} = \underline{\boldsymbol{h}}(0) + \underline{\boldsymbol{h}}(1)$. Then PR for filter banks with $B = 2$ blocks has this form with $\boldsymbol{P}^2 = \boldsymbol{P}$:

$$\underline{f}(0) = \boldsymbol{Q}^{-1}\boldsymbol{P},\ \underline{f}(1) = \boldsymbol{Q}^{-1}(\boldsymbol{I}-\boldsymbol{P}),\ \underline{\boldsymbol{h}}(0) = \boldsymbol{PQ},\ \underline{\boldsymbol{h}}(1) = (\boldsymbol{I}-\boldsymbol{P})\boldsymbol{Q}.$$

3. Verify that $\boldsymbol{F}_p(z)\boldsymbol{H}_p(z) = \boldsymbol{I}$ if $\boldsymbol{P}^2 = \boldsymbol{P}$ in the formulas

$$\boldsymbol{F}_p(z) = \boldsymbol{Q}^{-1}(\boldsymbol{P} + (\boldsymbol{I}-\boldsymbol{P})z) \text{ and } \boldsymbol{H}_p(z) = (\boldsymbol{P} + (\boldsymbol{I}-\boldsymbol{P})z^{-1})\boldsymbol{Q}.$$

4. If $\boldsymbol{P}^2 = \boldsymbol{P}$ has rank r, it can be diagonalized to give r ones and $M - r$ zeros. Show that $\det \boldsymbol{H}_p(z) = z^{-(M-r)}\det \boldsymbol{Q}$ which confirms that the filter bank is FIR.

5. By choosing 2×2 matrices $\boldsymbol{P}$ and $\boldsymbol{Q}$ in Problem 2 give examples of (a) an orthogonal bank and (b) a biorthogonal bank. What are the lowpass and highpass analysis coefficients in your examples?

6. Let $H_0(z) = \sum_{l=0}^{M-1} z^{-l} H_{0,l}(z^M)$ and $H_k(z) = H_0(zW^k)$. If $H_{0,l}(z)$ are allpass, the analysis filters $H_k(z)$ in this DFT bank are power complementary:

$$\begin{bmatrix} H_0(z) \\ \vdots \\ H_{M-1}(z) \end{bmatrix} = [\text{IDFT}] \begin{bmatrix} H_{0,0}(z^M) & \cdots & 0 \\ \vdots & \ddots & \vdots \\ 0 & \cdots & H_{0,M-1}(z^M) \end{bmatrix} \begin{bmatrix} 1 \\ \vdots \\ z^{-(M-1)} \end{bmatrix}.$$

This is not PR! With $F_0(z) = H_0(z)$, the DFT bank has no distortion. It only has aliasing.

7. Let $\underline{\boldsymbol{h}}_k(z)$ and $\underline{\boldsymbol{h}}_\ell(z)$ be columns of a paraunitary matrix $\boldsymbol{H}(z)$. They are mutually orthogonal. Show that the elements $H_{k,m}(z)$ in the kth column are power complementary filters: $\sum H_{k,m}(z^{-1})H_{k,m}(z) = 1$.

8. We could design M linear phase M-th band filters $P_k(z)$ and find their spectral factors $H_k(z)$. What are the properties of this analysis bank? It is not necessarily PR. Does it have aliasing or amplitude or phase distortions?

9. Verify that the GenLOT in equation (9.40) has linear phase.

10. Find the relations between the analysis functions $H_k'(z)$ and their factors $H_{ij}(z)$ for the 5-channel system in the Glossary (Tree-structured Filter Bank). When the H_{ij} are ideal filters, sketch the ideal responses of $H_k'(z)$.

11. With filter lengths $3M$, the time domain matrix $\boldsymbol{H}_b$ will have blocks $\underline{\boldsymbol{h}}(0), \underline{\boldsymbol{h}}(1), \underline{\boldsymbol{h}}(2)$ in each row. Write down the $B = 3$ equations for orthogonality, corresponding to the two equations in (9.34).

9.4 Cosine-modulated Filter Banks

The idea of modulation keeps developing and improving. It is a way to build the whole filter bank around one filter — the prototype filter $\boldsymbol{p}(n)$. Instead of modulating by exponentials, we modulate by cosines. An exponential will shift the frequency in one direction. The cosine is a sum of two exponentials, so the frequency shift goes two ways. The frequency band is partitioned symmetrically by cosine modulation, as shown below.

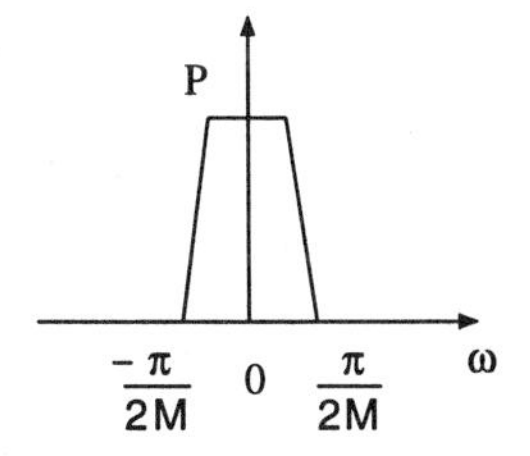

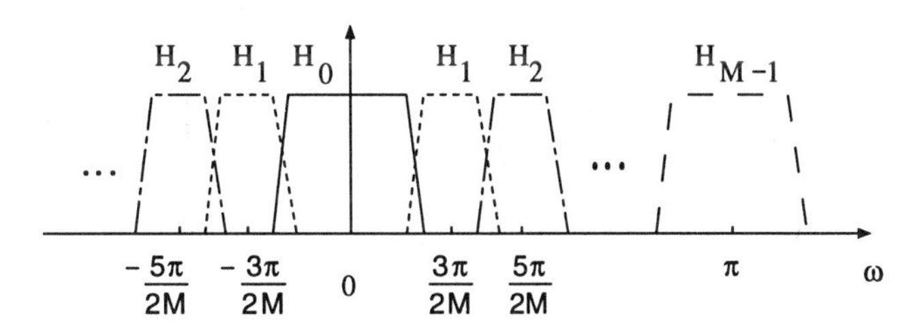

The original construction [Rothweiler] was chosen to cancel aliasing between adjacent subbands. In the z-domain, only the first aliasing error $T_1(z)$ was exactly zero. See *Cosine-modulated Filter Bank* in the Glossary. The aliasing between other channels, such as k and $k+2$, will be small when the prototype response dies quickly in the stopband (good attenuation at $|\omega| = \frac{\pi}{2M}$). But now, with better designs, all aliasing is gone. The filter bank can give perfect reconstruction.

Another step forward is in the length of the filters. The maximum length was originally $2M$, in the Modulated Lapped Transform (MLT). Now the length can be $2KM$, in the Extended Lapped Transform (ELT). Those developments by Malvar [Mr] have parallels from other authors as the efficiency of cosine modulation is appreciated. A very active area is the construction of *time-varying filter banks*, in which the filter changes length as the signal passes through. Only with a simple basic design could we maintain orthogonality while the overlapping filters change with time.

Cosine modulation can be analyzed in two domains. In the time domain, certain constant matrices are *orthogonal*. In the z-domain, certain polynomial matrices are *paraunitary*. It seems right to do both, but perhaps to emphasize the time domain. We begin there with the orthogonality conditions on these filter coefficients:

$$h_k(n) = f_k(n) = p(n)\sqrt{\frac{2}{M}}\cos\left[\left(k+\tfrac{1}{2}\right)\left(n+\frac{M+1}{2}\right)\frac{\pi}{M}\right] \tag{9.43}$$

Does $\boldsymbol{p} = (1, 1, \ldots, 1)$ give the ordinary DCT block transform? *No, because of the frequency shift.* The M in the numerator of (9.43) changes the phase of the DCT-IV. It also affects the orthogonality. In fact the usual orthogonality is gone. We see this directly for $M = 2$, by comparing the matrix $\boldsymbol{C}^{\text{IV}}$ with the new matrix $\boldsymbol{E}_0$:

$$(\boldsymbol{C}^{\text{IV}})_{kn} = \cos\left[\left(k+\tfrac{1}{2}\right)\left(n+\tfrac{1}{2}\right)\frac{\pi}{2}\right] = \begin{bmatrix} \cos(\frac{\pi}{8}) & \cos(\frac{3\pi}{8}) \\ \cos(\frac{3\pi}{8}) & \cos(\frac{9\pi}{8}) \end{bmatrix} = \begin{bmatrix} c & s \\ s & -c \end{bmatrix}$$

$$(\boldsymbol{E}_0)_{kn} = \cos\left[\left(k+\tfrac{1}{2}\right)\left(n+\tfrac{3}{2}\right)\frac{\pi}{2}\right] = \begin{bmatrix} \cos(\frac{3\pi}{8}) & \cos(\frac{5\pi}{8}) \\ \cos(\frac{9\pi}{8}) & \cos(\frac{15\pi}{8}) \end{bmatrix} = \begin{bmatrix} s & -s \\ -c & c \end{bmatrix}$$

Here $c = \cos(\frac{\pi}{8})$ and $s = \sin(\frac{\pi}{8})$. Then immediately $(\boldsymbol{C}^{\text{IV}})^T(\boldsymbol{C}^{\text{IV}}) = \boldsymbol{I}$ and $\boldsymbol{C}^{\text{IV}}$ is orthogonal. But $\boldsymbol{E}_0$ is not orthogonal:

$$\boldsymbol{E}_0^T\boldsymbol{E}_0 = \begin{bmatrix} s & -c \\ -s & c \end{bmatrix}\begin{bmatrix} s & -s \\ -c & c \end{bmatrix} = \begin{bmatrix} 1 & -1 \\ -1 & 1 \end{bmatrix} = \boldsymbol{I} - \boldsymbol{J}. \tag{9.44}$$

Here and always $\boldsymbol{J}$ is the reverse identity matrix. This convention is approaching the status of $\delta(n)$ and δ_{ij}, where definitions need not be repeated. Together with $\boldsymbol{E}_0$ in modulating a lapped transform comes the matrix $\boldsymbol{E}_1$ that saves orthogonality. With $M = 2$ this has the rest of the cosines:

$$(\boldsymbol{E}_1)_{kn} = \cos\left[\left(k+\tfrac{1}{2}\right)\left(n+2+\tfrac{3}{2}\right)\frac{\pi}{2}\right] = \begin{bmatrix} \cos(\frac{7\pi}{8}) & \cos(\frac{9\pi}{8}) \\ \cos(\frac{21\pi}{8}) & \cos(\frac{27\pi}{8}) \end{bmatrix} \tag{9.45}$$

$$\boldsymbol{E}_1^T\boldsymbol{E}_1 = \begin{bmatrix} -c & -s \\ -c & -s \end{bmatrix}\begin{bmatrix} -c & -c \\ -s & -s \end{bmatrix} = \begin{bmatrix} 1 & 1 \\ 1 & 1 \end{bmatrix} = \boldsymbol{I} + \boldsymbol{J} \tag{9.46}$$

$$\boldsymbol{E}_1^T\boldsymbol{E}_0 = \begin{bmatrix} -c & -s \\ -c & -s \end{bmatrix}\begin{bmatrix} s & -s \\ -c & c \end{bmatrix} = \begin{bmatrix} 0 & 0 \\ 0 & 0 \end{bmatrix} \tag{9.47}$$

This pattern is extended in the Lemma below to all sizes M and to a longer sequence $\boldsymbol{E}_0, \ldots, \boldsymbol{E}_{2K-1}$. Here, we stay with $M = 2$ and $K = 1$ to see orthogonality all the way. Our prototype filter will be $\boldsymbol{p} = (1, 1, 1, 1)/\sqrt{2}$. Then the $2KM = 4$ coefficients in the $M = 2$ channels will be the cosines in E_0 and E_1 divided by $\sqrt{2}$. The time-domain matrix H_b, with the analysis channels interleaved to give the block diagonal form, has E_0 along the main diagonal and E_1 beside it:

$$H_b = \frac{1}{\sqrt{2}} \begin{bmatrix} \cdot & & & \\ & E_1 & E_0 & \\ & & E_1 & E_0 \\ & & & \cdot \; \cdot \end{bmatrix} = \frac{1}{\sqrt{2}} \begin{bmatrix} \cdot & \cdot & & & & & & \\ -c & -c & s & -s & & & & \\ -s & -s & -c & c & & & & \\ & & -c & -c & s & -s & & \\ & & -s & -s & -c & c & & \\ & & & & & & \cdot & \cdot \end{bmatrix} \tag{9.48}$$

Notice how double-shift orthogonality has become *block orthogonality* (and here $M = 2$). In the block form, this orthogonality comes from $E_1^T E_0 = 0$ off the diagonal and $E_1^T E_1 + E_0^T E_0 = 2I$ on the diagonal. The J's cancel in $I + J + I - J$. So we have an orthogonal filter bank.

The prototype filter p was symmetric, as we always assume. But cosine modulation shifted the phase. The individual filters $(-c, -c, s, -s)$ and $(-s, -s, -c, c)$ are not linear phase. And they do not have the accuracy of the Daubechies filters. *We do not get high accuracy from cosine modulation.* In this present form we do not even get $p = 1$! The lowpass filter applied to the alternating vector $x(n) = (-1)^n$ does not give zero output. Nevertheless, the coding gain is adequate for compression, and the implementation is extremely fast.

M Channels with Filter Lengths 2*KM*

In general we have $2K$ matrices $E_0, \ldots, E_{2K-1}$ and each matrix is $M \times M$:

$$(E_\ell)_{kn} = \sqrt{\frac{2}{M}} \cos\left[\left(k + \tfrac{1}{2}\right)\left(n + \ell M + \frac{M+1}{2}\right)\frac{\pi}{M}\right]. \tag{9.49}$$

These are the pieces of the extended cosine matrix $E = \begin{bmatrix} E_0 \, E_1 \, \cdots \, E_{2K-1} \end{bmatrix}$, which would give the filter coefficients if the prototype filter had all $p(n) = 1$. We deal first with this case of a "box window". Then we allow other windows and discover the orthogonality conditions on $p(n)$.

The rows of E are orthogonal. Piece by piece we have the following identities.

Lemma: *The pieces of the extended cosine matrix* $\mathbf{E}$ *satisfy*

$$\begin{matrix} E_{2s} = (-1)^s E_0 \\ E_{2s+1} = (-1)^s E_1 \end{matrix} \quad \text{and} \quad \begin{matrix} E_\ell^T E_{\ell+2s} = (-1)^s \left[I + (-1)^{\ell+1} J\right] \\ E_\ell^T E_{\ell+2s+1} = \mathbf{0}. \end{matrix} \tag{9.50}$$

Proof: With $\ell = 2s$, we have added $(k + \frac{1}{2})(2sM)\frac{\pi}{M}$ to the arguments of the cosine in (9.49). This is the same as adding $s\pi$ which multiplies the cosine by $(-1)^s$. Similarly E_{2s+1} is a shift by $s\pi$ from E_1. The two identities on the left are proved. The identities on the right then quickly reduce to the first cases

$$E_0^T E_0 = I - J \;\text{ and }\; E_1^T E_1 = I + J \;\text{ and }\; E_1^T E_0 = \mathbf{0}. \tag{9.51}$$

Those were checked above for $M = 2$. For larger M, the $(n, \acute{n})$ entry of $E_0^T E_0$ is

$$\frac{2}{M} \sum_{k=0}^{M-1} \cos\left[\left(k + \tfrac{1}{2}\right)\left(n + \frac{M+1}{2}\right)\frac{\pi}{M}\right] \cos\left[\left(k + \tfrac{1}{2}\right)\left(\acute{n} + \frac{M+1}{2}\right)\frac{\pi}{M}\right] =$$

$$\sum_{k=0}^{M-1} \frac{1}{M} \cos\left[\left(k + \tfrac{1}{2}\right)\left(n - \acute{n}\right)\frac{\pi}{M}\right] + \frac{1}{M} \cos\left[\left(k + \tfrac{1}{2}\right)\left(n + \acute{n} + M + 1\right)\frac{\pi}{M}\right]. \tag{9.52}$$

On line two, we wrote $(\cos a)(\cos b)$ using $a+b$ and $a-b$. Its first sum is $\delta(n-\acute{n})$, which gives the matrix $\boldsymbol{I}$. Its second sum is zero, except when $n+\acute{n}=M-1$ and all terms are $\cos\pi=-1$. This gives the matrix $-\boldsymbol{J}$ in $\boldsymbol{E}_0^T\boldsymbol{E}_0$. Note that the reverse identity has 1's when $n+\acute{n}=M-1$, because the numbering starts at zero. The identities $\boldsymbol{E}_1^T\boldsymbol{E}_1=\boldsymbol{I}+\boldsymbol{J}$ and $\boldsymbol{E}_1^T\boldsymbol{E}_0=\boldsymbol{0}$ have similar proofs.

Orthogonality Conditions on the Prototype Filter $p(n)$

Now multiply the cosines by $\boldsymbol{p}(n)$ to get the true filter coefficients $\boldsymbol{h}_k(n)$. The cosine matrix $\boldsymbol{E}$ is M by $2KM$, with columns indexed by n. It is multiplied by the diagonal matrix $\boldsymbol{P}=\operatorname{diag}(\boldsymbol{p}(0),\boldsymbol{p}(1),\ldots,\boldsymbol{p}(2KM-1))$. Their product $\boldsymbol{EP}$ splits into $2K$ square blocks $\boldsymbol{E}_\ell\underline{\boldsymbol{P}}_\ell$ of size M:

$$\boldsymbol{EP}=\begin{bmatrix}\boldsymbol{E}_0 & \cdots & \boldsymbol{E}_{2K-1}\end{bmatrix}\begin{bmatrix}\underline{\boldsymbol{P}}(0) & & \\ & \ddots & \\ & & \underline{\boldsymbol{P}}(2K-1)\end{bmatrix}=\begin{bmatrix}\boldsymbol{E}_0\underline{\boldsymbol{P}}_0 & \cdots & \boldsymbol{E}_{2K-1}\underline{\boldsymbol{P}}_{2K-1}\end{bmatrix} \quad (9.53)$$

The entries of $\boldsymbol{EP}$ are cosines multiplied by $\boldsymbol{p}(n)$. In other words we have $\boldsymbol{h}_k(n)$.

The square block $\underline{\boldsymbol{P}}_\ell$ contains the prototype coefficients $\boldsymbol{p}(\ell M),\ldots,\boldsymbol{p}(\ell M+M-1)$. We are simply writing out a typical block row of the analysis bank matrix $\boldsymbol{H}_b$. That row is $\boldsymbol{EP}$ as in (9.53), except the columns come in reverse order because $\boldsymbol{h}_k(0)$ is at the right end of the row. This has no effect on orthogonality.

Under what condition is $\boldsymbol{H}_b$ an orthogonal matrix? The center block of $\boldsymbol{H}_b^T\boldsymbol{H}_b$ must be the identity and the off-diagonal blocks of $\boldsymbol{H}_b^T\boldsymbol{H}_b$ must be zero:

$$\sum_0^{2K-1}(\boldsymbol{E}_\ell\underline{\boldsymbol{P}}_\ell)^T(\boldsymbol{E}_{\ell+i}\underline{\boldsymbol{P}}_{\ell+i})=\delta(i)\boldsymbol{I}. \quad (9.54)$$

Inside those sums are the products $\boldsymbol{E}_\ell^T\boldsymbol{E}_{\ell+i}$ that are given by the Lemma. Substituting the identity (9.50) yields the orthogonality condition on the matrices $\underline{\boldsymbol{P}}_\ell$. That gives us the coefficients $\boldsymbol{p}(n)$ of the (symmetric) prototype filter.

Theorem 9.6 *The cosine modulated filter bank is orthogonal if and only if the diagonal blocks $\underline{\boldsymbol{P}}_l$ are double-shift orthogonal:*

$$\sum\underline{\boldsymbol{P}}_\ell^T\underline{\boldsymbol{P}}_{\ell+2s}=\delta(s)\boldsymbol{I}. \quad (9.55)$$

For the coefficients this means that we shift by double blocks:

$$\sum_{\ell=0}^{2K-2s-1}p(n+\ell M)\,p(n+\ell M+2sM)=\delta(s)\ \textit{for}\ n=0,\ldots,\tfrac{M}{2}-1. \quad (9.56)$$

The special case $K=1$ has only two blocks $\underline{\boldsymbol{P}}_0$ and $\underline{\boldsymbol{P}}_1$. So there is only $s=0$:

$$\boldsymbol{p}^2(n)+\boldsymbol{p}^2(n+M)=1. \quad (9.57)$$

That corresponds exactly to the condition $g^2(t)+g^2(t+1)=1$ on the window function in Section 8.4. That continuous-time theory only allowed neighboring windows to overlap. In discrete time this is $K=1$.

The next discrete case $K = 2$ has blocks $\underline{P}_0, \underline{P}_1, \underline{P}_2, \underline{P}_3$. Now there are conditions from $s = 0$ and $s = 1$:

$$p^2(n) + p^2(n+M) + p^2(n+2M) + p^2(n+3M) = 1$$

$$p(n)p(n+2M) + p(n+M)p(n+3M) = 0. \tag{9.58}$$

We can design $4M$ lowpass coefficients to satisfy these conditions!

Proof: Theorem 9.6 is our main result. The analysis of cosine modulation led here. The conditions (9.54) for odd i are automatic because $\boldsymbol{E}_\ell^T \boldsymbol{E}_{\ell+i} = 0$ in (9.50). The conditions for even $i = 2s$ require the substitution of $(-1)^s[\boldsymbol{I} + (-1)^{\ell+1}\boldsymbol{J}]$ for $\boldsymbol{E}_\ell^T \boldsymbol{E}_{\ell+2s}$. The $\boldsymbol{J}$'s cancel when we sum on ℓ because the prototype filter is symmetric. (You see why the filter length $2KM$ is an *even* multiple of M, to have an even number of $\boldsymbol{J}$'s with alternating sign.) Then (9.55) and (9.56) reduce to $\sum \underline{\boldsymbol{P}}_\ell^T \underline{\boldsymbol{P}}_{\ell+2s}$. For orthogonality this must give $\delta(s)$ in (9.57) and (9.58).

Two symmetric solutions to $p^2(n) + p^2(n+M) = 1$ are the sine windows

$$p(n) = \sin\left[\frac{n\pi}{2(M-1)}\right] \quad \text{and} \quad p(n) = -\sin\left[\left(n+\tfrac{1}{2}\right)\frac{\pi}{2M}\right]. \tag{9.59}$$

Malvar observed that the latter is the only prototype for which the resulting nth filter has frequency response $\delta(n)$ at $\omega = 0$. (The accuracy is $p = 1$.) The zeroth filter reproduces a DC input and the other filters null it, with no DC leakage. This normalization was expected in the rest of the book and is desirable in image coding. Malvar also gives a family of solutions when $K = 2$.

Polyphase and Lattice Structure

The polyphase coefficients are the blocks $\boldsymbol{E}_0\underline{\boldsymbol{P}}_0, \ldots, \boldsymbol{E}_{2K-1}\underline{\boldsymbol{P}}_{2K-1}$ along each row of the time domain matrix $\boldsymbol{H}_b$. The filter bank is orthogonal when this polyphase matrix $\sum \boldsymbol{E}_\ell \underline{\boldsymbol{P}}_\ell\, z^{-\ell}$ is paraunitary:

$$\boldsymbol{H}_p^T(z^{-1})\boldsymbol{H}_p(z) = \left(\sum \underline{\boldsymbol{P}}_\ell^T \boldsymbol{E}_\ell^T z^\ell\right)\left(\sum \boldsymbol{E}_\ell \underline{\boldsymbol{P}}_\ell z^{-\ell}\right) = \boldsymbol{I}. \tag{9.60}$$

The constant term in the product is $\sum \underline{\boldsymbol{P}}_\ell^T \boldsymbol{E}_\ell^T \boldsymbol{E}_\ell \underline{\boldsymbol{P}}_\ell$. Equation (9.55) makes this the identity matrix $\boldsymbol{I}$. The coefficient of z^{-i} in the product is $\sum \underline{\boldsymbol{P}}_\ell^T \boldsymbol{E}_\ell^T \boldsymbol{E}_{\ell+i} \underline{\boldsymbol{P}}_{\ell+i}$. Equation (9.55) makes this the zero matrix. Under these conditions $\boldsymbol{H}_p(z)$ is paraunitary.

For $K = 1$ the polyphase matrix simplifies to $\boldsymbol{E}_0\underline{\boldsymbol{P}}_0 + z^{-1}\boldsymbol{E}_1\underline{\boldsymbol{P}}_1$. It is paraunitary, after using $\boldsymbol{E}_1^T\boldsymbol{E}_0 = \boldsymbol{0}$ and $\boldsymbol{E}_1^T\boldsymbol{E}_1 = \boldsymbol{I} + \boldsymbol{J}$ and $\boldsymbol{E}_0^T\boldsymbol{E}_0 = \boldsymbol{I} - \boldsymbol{J}$, when

$$\underline{\boldsymbol{P}}_1^T\underline{\boldsymbol{P}}_1 + \underline{\boldsymbol{P}}_0^T\underline{\boldsymbol{P}}_0 = \boldsymbol{I}. \tag{9.61}$$

This pairs off the prototype filter coefficients into equation (9.57).

Let $G_k(z), 0 \le k \le 2M-1$ be the polyphase components of the prototype filter $P(z)$. The pairs $(G_k(z), G_{M+k}(z))$ of a paraunitary cosine-modulated filter bank satisfy

$$G_k(z^{-1})\,G_k(z) + G_{M+k}(z^{-1})\,G_{M+k}(z) = \frac{1}{2M}. \tag{9.62}$$

This agrees with the orthogonality condition in a two-channel filter bank. The conditions (9.62) extend to filters of any length [NgKoil]. Thus, the cosine-modulated bank can be implemented by two-channel paraunitary filter banks in parallel, as depicted in the Glossary (see *Tree-structured Filter Bank*). This is efficient because of the lattice structures associated with the pair $[G_k(z), G_{k+M}(z)]$ and the matrix $\widehat{C}$.

Example 9.5. Figure 9.15 shows the frequency responses of an 8-channel PR cosine-modulated filter bank. The filter length is 128. The resulting filters have stopband attenuation about 80 dB. The design procedure is based on QCLS. Chapter 10 will elaborate more on this formulation.

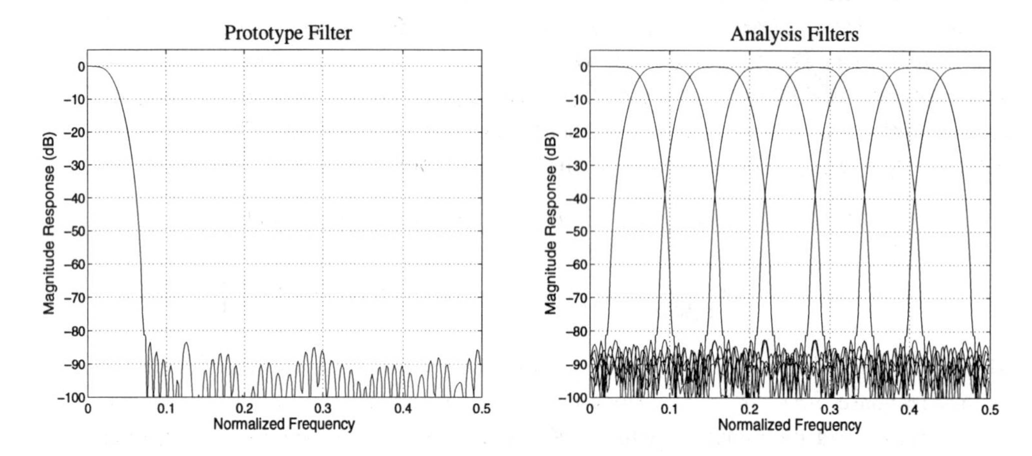

Figure 9.15: Frequency response plots for an 8-channel PR cosine-modulated filter bank.

Note 1 A pseudo-QMF bank is a cosine-modulated filter bank where the first aliasing transfer function $T_1(z)$ is cancelled. The other $T_2(z), \ldots, T_{M-1}(z)$ are minimized by constraining the stopband cutoff frequency $\omega_c \leq \frac{\pi}{M}$ of $P(z)$ such that there is no overlap between $H_k(z)$ and $H_{k\pm 2}(z)$. Consequently, the aliasing error is comparable to the stopband attenuation of the prototype filter. The distortion function $T_0(z)$ is not a delay. The design procedure is to find a prototype filter that minimizes the following weighted objective function:

$$\alpha \int_{\omega_s}^{\pi} \left|P(e^{j\omega})\right|^2 d\omega + (1-\alpha) \int_0^{\pi} \left|T_0(e^{j\omega}) - e^{-jn_0\omega}\right|^2 d\omega$$

where $0 \leq \alpha \leq 1$. One often has to trade off aliasing with distortion.

Note 2 The prototype filter $P(z)$ for an NPR Pseudo-QMF bank is a *linear phase spectral factor of a $2M$-th band filter*. There is no distortion. The aliasing errors can be minimized by high attenuation.

Example 9.6. Figure 9.16 shows the frequency responses of a 16-channel NPR cosine-modulated filter bank. The filter length is 256. By constraining $P(z)$ to be the spectral factor of a 32-band filter, the resulting filter bank has no magnitude nor phase distortion. The only distortion here is the aliasing, which is less than or equal to -72 dB.

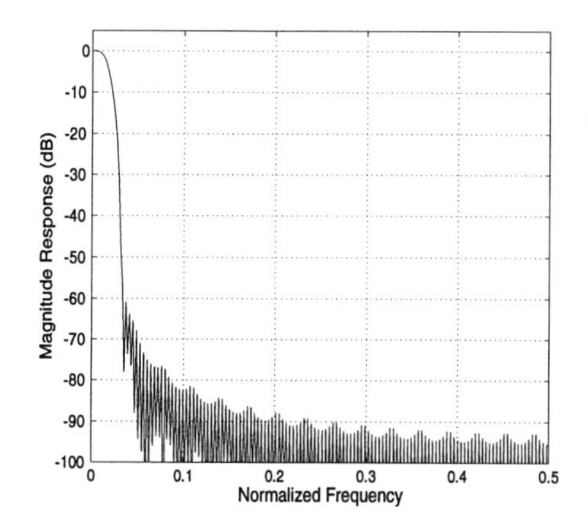

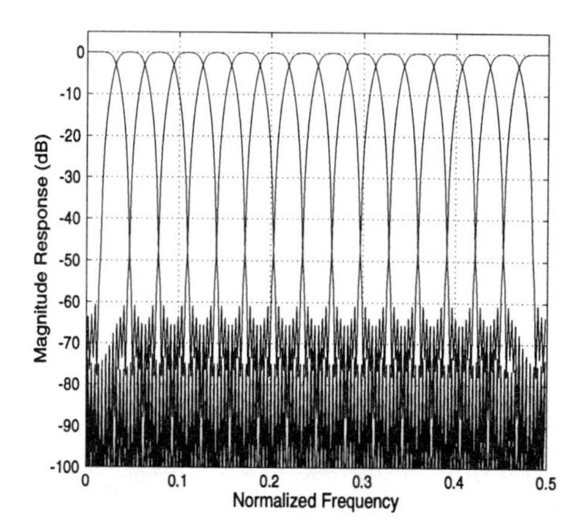

Figure 9.16: Frequency response plots for the 16-channel NPR cosine-modulated filter bank. Prototype filter $P(z)$ (left) and Analysis filters $H_k(z)$ (right).

Problem Set 9.4

1. Verify that Malvar's windows (9.59) satisfy $p^2(n) + p^2(n+M) = 1$ for orthogonality.
2. Construct explicitly (by hand or by Matlab) the 3×3 matrices $C^{\mathrm{IV}}, E_0, E_1, E_2$ for $M = 3$. Check by Matlab whether the identities (9.51) still hold for odd M.
3. For prototypes $p(n)$ in synthesis and $\widetilde{p}(n)$ in analysis (so that $H_k \neq F_k$), what will be the conditions for a perfect reconstruction cosine-modulated filter bank (filter length $2M$ and eventually $2KM$)?
4. Let $p(n)$ have linear phase. Show that $h_k(n)$ in (9.43) cannot have linear phase.
5. Let $p(n)$ be a spectral factor of a $2M$-th band filter. Show that the distortion function $T_0(z) = \sum_{k=0}^{M-1} F_k(z)H_k(z)$ is a delay.
6. Find the conditions on the prototype filters $p(n)$ for a *biorthogonal* cosine-modulated filter bank. Cheung's MIT thesis has shown that with symmetry, there are $M/2$ extra free parameters compared to the orthogonal case.

9.5 Multidimensional Filters and Wavelets

Images are two-dimensional. Processing those images is an extremely important application of subband filtering. We certainly need two-dimensional filters! Their construction is coming late in the book because it is either quite easy or quite hard—depending on the type of filter:

1. *Separable*: Products of one-dimensional filters (easy).
2. *Nonseparable*: Genuinely two-dimensional filters (hard).

Each filter bank is associated with a subsampling matrix M. In d dimensions this matrix is $d \times d$. In one dimension it contains the usual number M. In two dimensions we consider the two leading possibilities:

$$\text{(separable) } M_s = \begin{bmatrix} 2 & 0 \\ 0 & 2 \end{bmatrix} \quad \text{(nonseparable quincunx) } M_q = \begin{bmatrix} 1 & 1 \\ -1 & 1 \end{bmatrix}.$$

These matrices have eigenvalues $|\lambda| > 1$, as required. Their determinants are $M_s = 4$ and $M_q = 2$. The lightface symbol M still represents the number of channels. M is also the number of scaling functions plus wavelets.

In two dimensions, a filter is a two-dimensional convolution $\boldsymbol{y} = \boldsymbol{h} * \boldsymbol{x}$. In the $\boldsymbol{\omega} = (\omega_1, \omega_2)$ and $\boldsymbol{z} = (z_1, z_2)$ domains, we multiply by $H(\omega_1, \omega_2)$ and $H(z_1, z_2)$:

$$\boldsymbol{Hx}\,(n_1, n_2) = \sum_{k_1}\sum_{k_2} \boldsymbol{h}(k_1, k_2)\, \boldsymbol{x}(n_1 - k_1, n_2 - k_2)$$

$$H(\boldsymbol{\omega})\, X(\boldsymbol{\omega}) = \left(\sum\sum \boldsymbol{h}(k_1, k_2) e^{-i(k_1\omega_1 + k_2\omega_2)}\right)\left(\sum\sum \boldsymbol{x}(n_1, n_2) e^{-i(n_1\omega_1 + n_2\omega_2)}\right)$$

$$H(\boldsymbol{z})\, X(\boldsymbol{z}) = \left(\sum\sum \boldsymbol{h}(k_1, k_2) z_1^{-k_1} z_2^{-k_2}\right)\left(\sum\sum \boldsymbol{x}(n_1, n_2) z_1^{-n_1} z_2^{-n_2}\right).$$

The frequency response $H(\omega_1, \omega_2)$ has period 2π in both variables. A set of four brick wall filters will cover the period square with no overlap. These ideal filters are (0) low-low, (1) low-high, (2) high-low, and (3) high-high:

Support regions for four ideal 2D filters

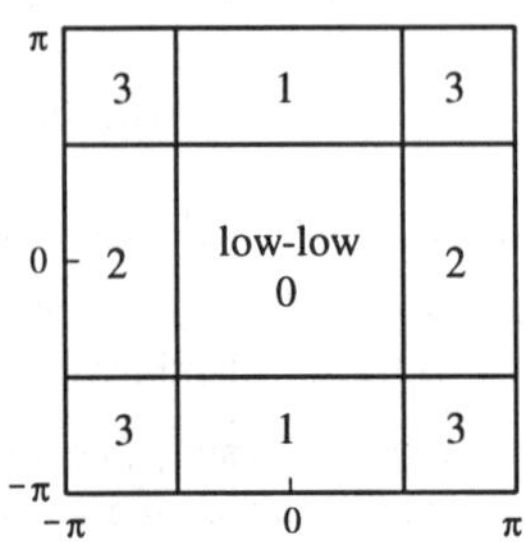

Those ideal filters are IIR. A set of four FIR separable filters is easily constructed from one-dimensional filters $\boldsymbol{h}_{low}$ and $\boldsymbol{h}_{high}$:

$$\boldsymbol{h}_0(n_1, n_2) = \boldsymbol{h}_{low}(n_1)\, \boldsymbol{h}_{low}(n_2) \qquad \boldsymbol{h}_2(n_1, n_2) = \boldsymbol{h}_{high}(n_1)\, \boldsymbol{h}_{low}(n_2)$$

$$\boldsymbol{h}_1(n_1, n_2) = \boldsymbol{h}_{low}(n_1)\, \boldsymbol{h}_{high}(n_2) \qquad \boldsymbol{h}_3(n_1, n_2) = \boldsymbol{h}_{high}(n_1)\, \boldsymbol{h}_{high}(n_2).$$

This is the easy construction. After the filters, we sample by $(\downarrow 2)$ *in each direction.* This subsampling corresponds to the separable matrix $\boldsymbol{M}_s = 2\boldsymbol{I}$:

$$(\downarrow \boldsymbol{M}_s)\, \boldsymbol{y}(n_1, n_2) = \boldsymbol{y}(\boldsymbol{Mn}) = \boldsymbol{y}(2n_1, 2n_2). \tag{9.63}$$

We are keeping one sample out of four. The four filters in four channels give critical sampling. The polyphase matrix will be 4×4 (generally $M \times M$).

Compare with the nonseparable quincunx filter bank. With $M_q = 2$ filters, the quincunx rule keeps the samples for which $n_1 + n_2$ is even:

$$(\downarrow \boldsymbol{M}_q)\, \boldsymbol{y}(n_1, n_2) = \boldsymbol{y}(\boldsymbol{M}_q\, \boldsymbol{n}) = \boldsymbol{y}(n_2 + n_1, n_2 - n_1). \tag{9.64}$$

Figure 9.17 shows the lattice of integers and the quincunx sublattice of samples:

Notice how $M_s = 4$ separable sublattices cover the whole lattice. Similarly $M_q = 2$ quincunx sublattices (*staggered grids*) cover all mesh points. The separable lattices are strongly oriented in the horizontal and vertical directions ! The quincunx lattice has an extra symmetry at 45° and −45°. It is closer to isotropic, meaning independent of direction. A diagonal edge in an

Figure 9.17: The sampling lattices for $\boldsymbol{M}_s$ (separable) and $\boldsymbol{M}_q$ (quincunx).

image will be captured far better by the quincunx. But those filters are harder to design, if we want PR and especially if we want orthogonality.

The reader will see how other sampling matrices $\boldsymbol{M}$ give the lattice points $\boldsymbol{Mn}$. Downsampling keeps all values $\boldsymbol{y}(\boldsymbol{Mn})$ on this lattice. Then the upsampling step $(\uparrow \boldsymbol{M})$ assigns a zero when (n_1, n_2) is not in the lattice. As always, *$(\uparrow \boldsymbol{M})$ is the transpose of $(\downarrow \boldsymbol{M})$*. If we apply both, we get the identity operator on the lattice and otherwise zero:

$$(\uparrow \boldsymbol{M})\,(\downarrow \boldsymbol{M})\,\boldsymbol{y}(n_1, n_2) = \begin{cases} \boldsymbol{y}(n_1, n_2) & \text{on the lattice} \\ 0 & \text{for other } (n_1, n_2). \end{cases} \tag{9.65}$$

A synthesis filter in each channel completes the filter bank: all normal.

Example 1 (Separable Haar): The Haar filter will be typical of separable filters. The four filters have coefficients $\pm\frac{1}{4}$ and the low-low response is $H_0(\omega_1, \omega_2) = \frac{1}{4}(1 + e^{-i\omega_1})(1 + e^{-i\omega_2})$. We "block" the input four samples at a time, with $\boldsymbol{x}(0,0)$, $\boldsymbol{x}(0,1)$, $\boldsymbol{x}(1,0)$, $\boldsymbol{x}(1,1)$ in the zeroth block. Or in practice, the signal goes through ordinary Haar in the x direction and then in the y direction. The two-dimensional Haar bank is a block transform with 4×4 blocks:

$$\text{Haar block} = \frac{1}{4}\boldsymbol{F}_2 \otimes \boldsymbol{F}_2 = \frac{1}{4}\begin{bmatrix} 1 & 1 & 1 & 1 \\ 1 & -1 & 1 & -1 \\ 1 & 1 & -1 & -1 \\ 1 & -1 & -1 & 1 \end{bmatrix}$$

This is just the two-dimensional DFT matrix. It is also the polyphase matrix. One constant term only, because Haar has no overlap. It is an orthogonal filter bank!

The Haar block shows a tensor product $\boldsymbol{H}_{2\times2} \otimes \boldsymbol{H}_{2\times2}$ of one-dimensional DFT's. The matrix $\boldsymbol{H}_{2\times2} = \frac{1}{2}\begin{bmatrix} 1 & 1 \\ 1 & -1 \end{bmatrix}$ is multiplied by each of its four entries to give the four 2×2 subblocks. (A tensor product has order $N_1 N_2$ when the matrices have orders N_1 and N_2. The matrix of order N_1 appears N_2 times on each block row of the tensor product, multiplied by entries from the matrix of order N_2.) For separable filters, the 4×4 polyphase matrix is always the tensor product $\boldsymbol{H}_p(z_1) \otimes \boldsymbol{H}_p(z_2)$ of 2×2 one-dimensional polyphase matrices.

Every separable filter bank inherits the properties of its one-dimensional factors. The bank is orthogonal with accuracy (p, p) if the factors are orthogonal with accuracy p. *The filters cannot also be linear phase* (except for Haar). It is possible [KarVet] to achieve linear phase and orthogonality with nonseparable filters.

Example 2 (Nonseparable): Quincunx sampling has one alias term because $M - 1 = 1$. When $(\downarrow \boldsymbol{M}_q)$ is followed by $(\uparrow \boldsymbol{M}_q)$ as in (9.65), that alias is $X(-z_1, -z_2)$. The quincunx modulation matrix is 2×2:

$$\boldsymbol{H}_m(z_1, z_2) = \begin{bmatrix} H_0(z_1, z_2) & H_0(-z_1, -z_2) \\ H_1(z_1, z_2) & H_1(-z_1, -z_2) \end{bmatrix}.$$

Orthogonality requires that $\boldsymbol{H}_m(z_1, z_2)\boldsymbol{H}_m^T(z_1^{-1}, z_2^{-1}) = 2\boldsymbol{I}$. The $(1, 1)$ entry is

$$H_0(z_1, z_2)H_0(z_1^{-1}, z_2^{-1}) + H_0(-z_1, -z_2)H_0(-z_1^{-1}, -z_2^{-1}) \equiv 2. \tag{9.66}$$

The product filter $P(z_1, z_2)$ must be halfband! The highpass H_1 comes from H_0 by an *alternating flip* in z_1 and z_2:

$$H_1(z_1, z_2) = \text{ odd 2D delay of } H_0(-z_1^{-1}, -z_2^{-1}). \tag{9.67}$$

All looks familiar. But in two dimensions there is one enormous difference. *We cannot factor most product filters*. Even if the symmetric polynomial $P(\omega_1, \omega_2)$ is nonnegative, it may be impossible to express it as the square $|H(\omega_1, \omega_2)|^2$ of a polynomial. The idea of computing zeros of $P(z)$ and separating them into two factors is strictly one-dimensional. We must design the filters directly, and not by factorization.

A cascade structure [VK, p. 182] can give orthogonality or linear phase:

$$\boldsymbol{H}_p(z_1, z_2) = \boldsymbol{R}_{2j}\begin{bmatrix} 1 & \\ & z_2^{-1} \end{bmatrix}\boldsymbol{R}_{2j-1}\begin{bmatrix} 1 & \\ & z_1^{-1} \end{bmatrix}\cdots \boldsymbol{R}_0.$$

The filter bank is orthogonal if the matrices $\boldsymbol{R}$ are orthogonal. The shortest lowpass filter $\boldsymbol{h}$ has parameters *a, b, c* in its eight nonzero coefficients:

$$\boldsymbol{h} = \begin{bmatrix} & -b & -ab & \\ -c & -ac & -a & 1 \\ & abc & -abc & \end{bmatrix}. \tag{9.68}$$

With constraints on *a, b, c* this is analogous to the Daubechies D_4 filter.

Example 3 (McClellan transformation): A convenient way to design linear phase 2D filter banks is to begin with a symmetric centered 1D filter:

$$H(\omega) = \sum_{-L}^{L} \boldsymbol{h}(n)\cos n\omega = \sum_{-L}^{L} \boldsymbol{h}(n)T_n(\cos\omega).$$

The Tchebycheff polynomial T_n produces $\cos n\omega$ from powers of $\cos\omega$. For example $\cos 2\omega = 2\cos^2\omega - 1$, so that $T_2(x) = 2x^2 - 1$. The McClellan transformation replaces $\cos\omega$ by a symmetric 2D filter response $F(\omega_1, \omega_2)$. Then $H(\omega_1, \omega_2) = \sum \boldsymbol{h}(n)T_n(F(\omega_1, \omega_2))$ is still symmetric. In the quincunx case we choose $F = \frac{1}{2}(\cos\omega_1 + \cos\omega_2)$. More general polynomials than T_n have been used effectively in [TayKing].

By designing the 1D coefficients in $H(\omega)$ we get a good linear phase 2D filter $H(\omega_1, \omega_2)$. When the former gives perfect reconstruction with FIR inverse, so does the latter. (The 2D determinant is a monomial when the 1D determinant is. We cannot have orthogonality too.) The accuracy is also preserved, because a factor $(1+\cos\omega)^p$ transforms to $(1+\frac{1}{2}(\cos\omega_1 + \cos\omega_2))^p$. There are p zeros at the alias frequency (π, π) in the (ω_1, ω_2) plane.

Note that a separable filter with response $H(\omega_1)H(\omega_2)$ has p zeros at all three of the alias frequencies $(0, \pi)$ and $(\pi, 0)$ and (π, π). The zeros are needed as in 1D for stability of the iterated lowpass filter and convergence to the scaling function.

Dilation Equation and Wavelets

The lowpass filter has coefficients $\boldsymbol{h}_0(k_1, k_2)$. When we iterate with rescaling, the cascade algorithm hopes to converge to the scaling function. The limiting equation when $\phi^{(i+1)}$ and $\phi^{(i)}$ approach $\phi(t)$ is the dilation equation:

$$\phi(t_1, t_2) = M \sum \boldsymbol{h}_0(n_1, n_2)\phi(\boldsymbol{M}\boldsymbol{t} - \boldsymbol{n}). \tag{9.69}$$

Here $\boldsymbol{t} = (t_1, t_2)$ and $\boldsymbol{n} = (n_1, n_2)$ are column vectors. The matrix $\boldsymbol{M}$ has determinant M. When we change variables from $\boldsymbol{M}\boldsymbol{t} - \boldsymbol{n}$ to $\boldsymbol{\tau}$, this determinant preserves the double integral:

$$M \iint \phi(\boldsymbol{M}\boldsymbol{t} - \boldsymbol{n}) dt_1 dt_2 = \iint \phi(\boldsymbol{\tau}) d\tau_1 d\tau_2. \tag{9.70}$$

The lowpass normalization is still $\sum \sum \boldsymbol{h}_0(n_1, n_2) = 1$. This filter leads to the scaling function. The other $M - 1$ filters lead to $M - 1$ wavelets by the usual wavelet equation:

$$w_k(t) = M \sum \sum \boldsymbol{h}_k(n_1, n_2)\phi(\boldsymbol{M}\boldsymbol{t} - \boldsymbol{n}), \quad 1 \le k < M. \tag{9.71}$$

Orthogonality will mean that the translates $\phi(\boldsymbol{t} - \boldsymbol{n})$ at scale zero are an orthonormal basis for the space V_0. The wavelet translates $w_1(\boldsymbol{t} - \boldsymbol{n}), \ldots, w_{M-1}(\boldsymbol{t} - \boldsymbol{n})$ are an orthonormal basis for W_0. The wavelet translates and dilates $M^{j/2} w_k(\boldsymbol{M}^j \boldsymbol{t} - \boldsymbol{n})$ are an orthonormal basis for the whole finite energy space $L^2(\mathbf{R}^2)$.

Notice the dilation matrix $\boldsymbol{M}$! We nearly wrote 2^j instead of $\boldsymbol{M}^j$. This would be correct only for the matrix $\boldsymbol{M} = 2\boldsymbol{I}$ that gives separable filters. The iteration then gives a separable scaling function and three separable wavelets:

$$\begin{aligned} &\text{low-low:} \quad \phi(\boldsymbol{t}) = \phi(t_1)\phi(t_2) \qquad &&\text{high-low:} \quad w_2(\boldsymbol{t}) = w(t_1)\phi(t_2) \\ &\text{low-high:} \quad w_1(\boldsymbol{t}) = \phi(t_1)w(t_2) \qquad &&\text{high-high:} \quad w_3(\boldsymbol{t}) = w(t_1)w(t_2) \end{aligned} \tag{9.72}$$

It is pleasant to verify (Problem 3) that separable filters give these separable solutions to the dilation equation and wavelet equation. With accuracy p in the one-dimensional filter and $1, t, \ldots, t^{p-1}$ in its scaling space V_0, the separable 2D filter (product filter) will inherit this accuracy. All p^2 of the polynomials $(t_1)^r (t_2)^s$ will be in V_0, for $r < p$ and $s < p$. In the special case of splines, $\phi(t)$ is a product $\phi(t_1)\phi(t_2)$ of one-dimensional B-splines. The coefficients $\boldsymbol{h}(n_1, n_2)$ are products of binomial coefficients. These are just the coefficients in $H(z_1, z_2) = (1+z_1^{-1})^p (1+z_2^{-1})^p$.

Note that a quincunx filter does not need this factor. It is $[1 + \frac{1}{2}(z_1^{-1} + z_2^{-1})]^p$ that gives accuracy p for quincunx. We would construct $\phi(t_1, t_2)$ by the cascade algorithm. Then $(t_1)^r (t_2)^s$ is locally in V_0 if $r + s < p$. For $p = 2$ we only need the linear polynomials $1, t_1, t_2$ and not the extra $t_1 t_2$ that comes with separable filters.

Remark We believe that ***box splines*** could lead to multidimensional algorithms. They are generated by filters for which $H(z_1, z_2)$ has simple factors. The factorization of H is a fundamental 2D difficulty, and box splines are direct constructions with known factors. The future will determine whether these (or other) multidimensional filters are successful.

Bounded Domains

For a bounded interval in one dimension, Section 8.5 proposed several constructions of boundary functions. Local support and the multiresolution property $V_j \subset V_{j+1}$ and even the polynomial accuracy p were preserved. The same properties are desirable in two dimensions but not so easy to achieve.

We note one immediate difficulty. The boundary scaling functions $\phi_b(t)$ were differences between monomials t^k and combinations $\sum y_k \phi(t-k)$ of interior functions. The combination reproduces t^k exactly except near the end of the interval. In one dimension, $\phi_b(t)$ will have local support. But in two dimensions the support will be a *thin ring* along the boundary—not local! A more careful construction is needed.

The new paper [CoDaDe] gives a simple local construction of boundary functions for a square domain, starting from separable wavelets. On a general domain their method is less simple. It seems better than earlier constructions, and this problem must be faced in solving partial differential equations by a wavelet method (Section 11.6).

Problem Set 9.5

1. Show that the quincunx upsampling $\boldsymbol{u}(\boldsymbol{n}) = \boldsymbol{x}(\boldsymbol{M}_q^{-1}\boldsymbol{n})$ yields
$$U(\omega_1, \omega_2) = \tfrac{1}{2}\left[X(\omega_1, \omega_2) + X(\omega_1 + \pi, \omega_2 + \pi)\right].$$
Express this also in the (z_1, z_2) domain. Which inputs give $\boldsymbol{u} = \boldsymbol{0}$?

2. For $(\downarrow\ \boldsymbol{M}_s)$ the first row of the modulation matrix contains $H_0(z_1, z_2)$, $H_0(-z_1, z_2)$, $H_0(z_1, -z_2)$, $H_0(-z_1, -z_2)$. Using products $H_0(z_1)H_0(z_2)$ of one-dimensional filters, write out the 4×4 modulation matrix $\boldsymbol{H}_m(z_1, z_2)$. How is it related to the 2×2 matrices $\boldsymbol{H}_m(z_1)$ and $\boldsymbol{H}_m(z_2)$?

3. With the separable filters in equation (9.72), show that the separable scaling function $\phi(t_1)\phi(t_2)$ solves the 2D dilation equation with $\boldsymbol{M} = 2\boldsymbol{I}$.

4. Find $M = \det \boldsymbol{M}$ and draw the lattices of vectors $\boldsymbol{Mn}$ for
$$\boldsymbol{M}_{\text{all}} = \begin{bmatrix} 1 & 1 \\ 1 & 2 \end{bmatrix} \text{ and } \boldsymbol{M} = \begin{bmatrix} 1 & 1 \\ -1 & 2 \end{bmatrix} \text{ and } \boldsymbol{M}_{\text{hex}} = \begin{bmatrix} 1 & 2 \\ -1 & 2 \end{bmatrix}$$

5. For $\boldsymbol{M}_s$ and $\boldsymbol{M}_q$ draw the *Voronoi cell* of points whose closest lattice point is the origin. What is the area of the Voronoi cell?

6. If the lowpass filter $\boldsymbol{h}_0(n_1, n_2)$ satisfies the quincunx orthogonality condition (9.66), show that $H_1(z_1, z_2) = -z_1^{-1} H_0(-z_1^{-1}, -z_2^{-1})$ gives an orthogonal highpass filter. The modulation matrix should have $\boldsymbol{H}_m \boldsymbol{H}_m^T = 2\boldsymbol{I}$.

7. A 2D analogue of the hat function $\phi(t)$ is the bilinear tent $\phi(x)\phi(y)$. Find its Fourier transform as a function of ω_x and ω_y.

8. Another 2D analogue of the hat function is the Courant finite element $C(x, y)$. It is linear in each triangle obtained from grid lines $x = k_1$, $y = k_2$, $y - x = k_3$. Draw the six triangles around the origin and show that the function that is zero outside, with $C(0, 0) = 1$, has transform
$$\widehat{C}(\omega_1, \omega_2) = \left(\frac{\sin(\omega_1/2)}{\omega_1/2}\right)\left(\frac{\sin(\omega_2/2)}{\omega_2/2}\right)\left(\frac{\sin(\omega_1 + \omega_2)/2}{(\omega_1 + \omega_2)/2}\right).$$

9. Take the Fourier transform of the dilation equation (9.69). By recursion find the infinite product formula (notice the transpose) $\widehat{\phi}(\boldsymbol{\omega}) = \prod H((\boldsymbol{M}^{-j})^T \boldsymbol{\omega})$.

10. What is the z-transform of $(\uparrow M)(\downarrow M)\boldsymbol{y}$?

Chapter 10

Design Methods

10.1 Distortions in Image Compression

One of the most successful applications of the wavelet transform is transform-based image compression (also called *image coding*). The coder in Figure 10.1 operates by transforming the data to remove redundancy, then quantizing the transform coefficients (a lossy step), and finally entropy coding the quantized output. Because of superior energy compaction and correspondence with the human visual system, wavelet compression has produced good results. Since the wavelet basis functions have short support for high frequencies and long support for low frequencies, large smooth areas of an image may be represented with very few bits. High frequency detail is added where it is needed.

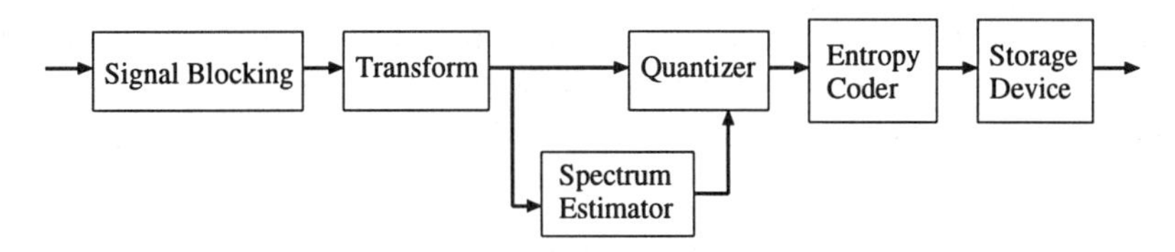

Figure 10.1: Transform-based image coder.

Figure 10.2 shows the original image and its decomposition (three levels) using the two-channel biorthogonal filter bank with $(9,7)$ coefficients. The upper left block is a smoothed approximation of the original image; it comes from three iterations of the lowpass filter with downsampling. The other subbands contain details at various scales. Since the filters have four vanishing moments, the signal energy in those subbands is small. Consequently, they have to be amplified in order to see the details.

The ten subimages are processed by a spectrum estimator for bit allocation. The statistical properties of the subimages guide this quantization step. The lowpass subband (upper left) is allocated the most bits. One strategy is based on the rate distortion curve. Given B bits, how does one allocate b_k bits per pixel to the kth subimage such that the reconstructed image has smallest distortion? Important distortion measures are mean-square-error (MSE), peak signal-to-noise (PSNR), and maximum error.

At high and medium bit rate (low and medium compression ratio), there is a strong correlation between the image quality and these objective measures. A good compression algorithm

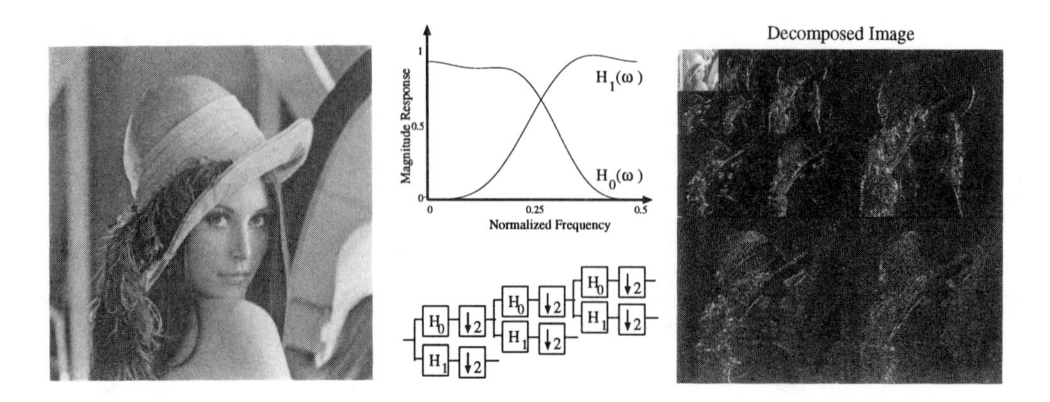

Figure 10.2: Original image "Lenna" and its multiresolution decomposition. The filter bank used is the (9, 7)-tap linear-phase filter bank.

would reconstruct the image with low MSE, high PSNR, and low Max error. *At low bit rate* (high compression), *the basis function characteristics have profound effects on the quality of the reconstructed image.* Typical distortions are ringing, blocking, and blurring artifacts.

Figure 10.3 shows the reconstructions for compression ratios of 32 :1 (0.25 bits per pixel) and 64 :1 (0.125 bpp) and 100 :1 (0.08 bpp). The objective measures for 32 :1 compression are MSE = 37.32 , PSNR = 32.41 dB, Max error = 62.66 . For 64 :1 compression the error is larger: MSE = 61.36, PSNR = 30.25 dB, Max error = 68.67; and for 100 :1 compression MSE = 104.3, PSNR = 27.95 dB, Max error = 121.3.

For medium bit rate (0.25 bpp), the objective measures are good indicators of the subjective quality of the image. At low bit rate, ***this is not always the case***! There are many artifacts in the reconstructed image at 0.08 bpp. The choice of filters is important, as well as bit allocation.

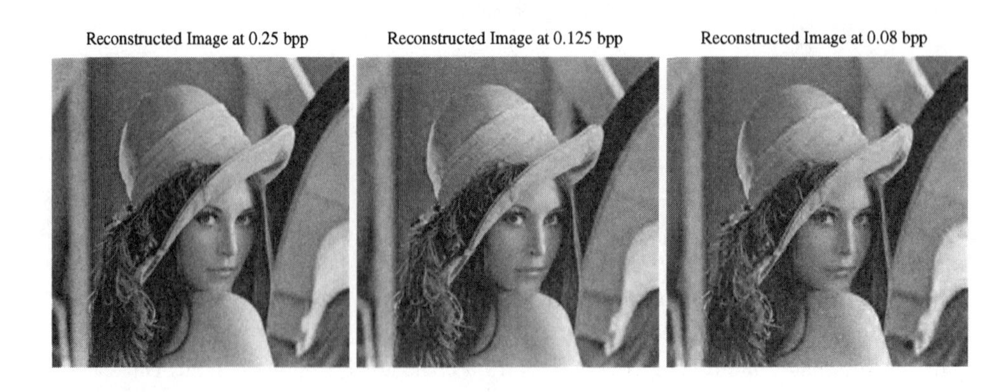

Figure 10.3: Reconstructed images for 32 :1, 64 :1, and 100 :1 compression with (9, 7) filters.

With four levels and 13 subbands, there are several artifacts in the reconstructed images: blurring, border distortions, blocking, checkerboarding, and ringing. Each artifact will be elaborated below. They are the results of the filter choices, the bit allocation algorithms, and the convolutions.

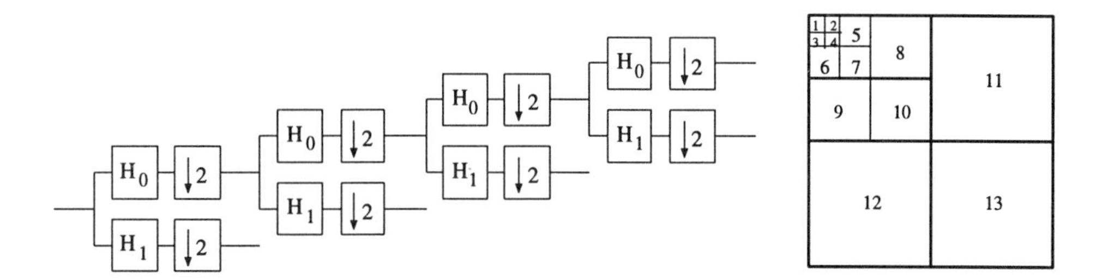

Blurring Artifacts

Blurring occurs when the algorithm does not assign enough bits to the higher subbands 5 to 13. In the extreme case, we allocate all bits to subbands 1 to 4 and none to subbands 5 to 13. The compression ratio is 48 to 1 (0.167 bpp). Figure 10.4 shows that the reconstructed image is essentially an interpolated version of subband 1. It suffers from blurring.

The blurring will decrease if one allocates more bits to the higher subbands. Figure 10.4(b) shows the reconstructed image when the bit allocations are

9.29 5.63 6.52 4.96 3.38 4.20 2.81 0.97 1.73 0.23 0.00 0.00 0.00

The resulting compression is also 48 to 1. This image is much sharper.

Figure 10.4: Blurring in image compression: Too few bits for the detailed subbands.

Border Distortions

The image and the filters have finite length. One has to be careful in computing their convolution. Border distortion is the result of choosing the wrong extension. Chapter 8 discusses the filtering operations for finite-length signals by zero-padding of the input signals, circular convolution, and symmetric extension. In image compression at 32 to 1 (0.25 bpp), with the (9, 7) symmetric filters, we can compare the effects of these three extensions.

Original

Zero - Padding

Circular Convolution

Symmetric Extension

Figure 10.5	Zero-Padding	Circular Convolution	Symmetric Extension
MSE	278.40	96.10	88.67
PSNR (dB)	23.68	28.30	28.75
Max Error	168.40	54.43	55.26

The zero-padding method assumes that images are zero outside their borders. Therefore they are discontinuous signals. The pixel intensity at the reconstructed border is very low (dark). The large error of zero-padding occurs at the boundary.

Circular convolution assumes that the image is periodic. This assumption is not valid for our original image. If the subband images are not quantized, the reconstructed image has no distortion since the filter bank is a PR system. But quantization yields errors at the border when using circular convolution. The periodic assumption results in a discontinuous intensity level at the border.

Symmetric extension extends the images in a continuous way at the border. Although the objective performance of circular convolution is close, the subjective (perceptual) quality from circular convolution is not as good as from symmetric extension.

Blocking Artifacts

One major disadvantage of the DCT in image coding is that it cannot be used for high compression because of its blocking artifacts. The basis functions typically have short length $M = 8$ for the DCT block transform with no overlap. Blocking will show up at the output. One way to improve is to use longer basis functions, which overlap neighboring blocks. Figure 10.5 also shows the reconstructed image using GenLOT (length 48). Most of the blocking artifacts are now unnoticeable.

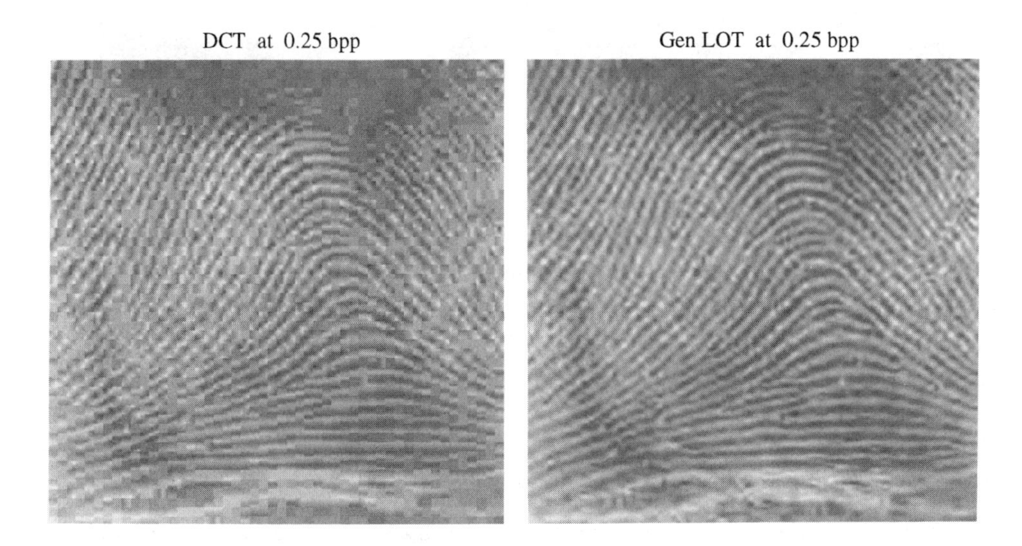

Figure 10.5: Blocking artifacts in image compression using DCT and lapped transforms.

High Frequency Distortion: Ringing

The reconstructed image is a linear combination of the scaling function and wavelets of the ***synthesis bank***. Since symmetric filters are not orthogonal, the analysis and synthesis functions are not the same. The scaling function of the synthesis filter should be smooth and the wavelet should be short. The error is larger at higher subbands, assigning fewer bits to the wavelet coefficients. When the wavelet has length N, the error at the pixel (K, L) affects its neighbors $(K \pm k, L \pm \ell)$, where $k, \ell \leq N/2$.

Figure 10.6 shows the original image and reconstructed image at 0.5 bpp. The lengths of the synthesis filters are 11 and 17. Even at medium bit rate, the ringing effect is quite severe. This is especially true at strong edges.

Nonsmooth Functions: Checkerboard Artifacts

In all these compression examples, the filters are linear-phase factors of a maximally-flat halfband filter. All scaling functions and wavelets in the synthesis bank are smooth, which is essential for image reconstruction. When the functions are not smooth, what is the effect on reconstruction?

Figure 10.6: Ringing artifacts are clearly visible at edges.

The lowpass responses $H_0(z)$ and $F_0(z)$ in Figure 10.7 are designed such that the stopband error is equiripple (a desirable property in conventional digital signal processing). *The scaling functions are not smooth.* The reconstructed image suffers from the the ***checkerboard effect***. This can be explained by observing that the scaling function from $F_0(z)$ has a large and narrow peak. The magnitudes near this narrow peak are about 10% below it. This explains the bright horizontal and vertical lines in the reconstructed image. The stronger intensity produces the checkerboard effect. This artifact can be attributed to the "roughness" of $\phi(t)$.

$$\begin{cases} \textbf{\textit{Lowpass synthesis:}} & \textbf{\textit{Long and smooth to avoid blocking and checkerboarding}} \\ \textbf{\textit{Highpass synthesis:}} & \textbf{\textit{Short to avoid ringing}} \end{cases}$$

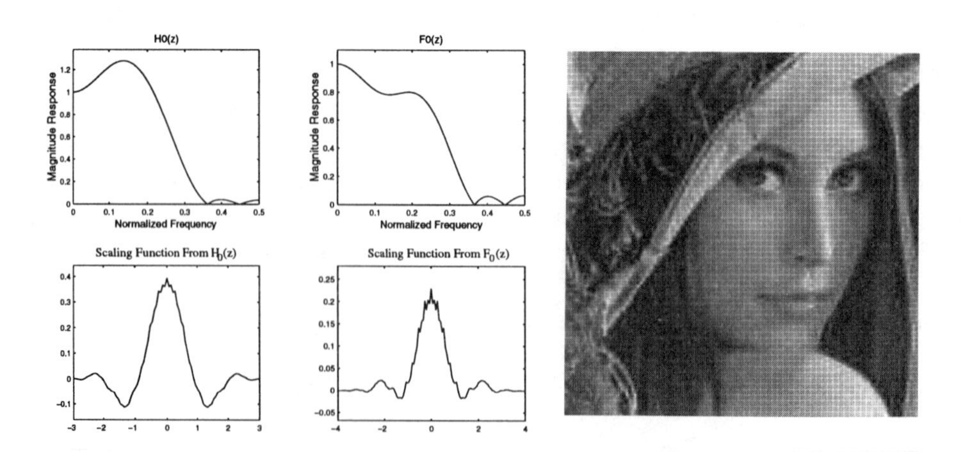

Figure 10.7: These lowpass equiripple filters are linear-phase factors of a halfband equiripple filter. The lengths are 9/7 and the rate is 0.25 bpp. The nonsmooth scaling functions create a checkerboard.

10.2 Design Methods — General Perspective

In all of our examples, one important factor for image quality is the choice of coefficients in the filter banks. For an objective function of coding gain or degree of smoothness or stopband attenuation, *one optimizes the filter coefficients.* This section will present several methods to design and optimize filter banks and wavelets. These methods are based on spectral factorizations, lattice structures and time-domain formulations. We will present several common choices for the objective function Φ, and its parameters, and the constraints to impose:

$$\textbf{\textit{Design Problem:}} \quad \text{Minimize } \Phi \text{ (parameters) subject to constraints.} \tag{10.1}$$

The design parameters could be the lattice coefficients or the filter coefficients. Recall how a lattice structure can impose the PR property and also linear phase, pairwise-mirror-image or cosine modulation. The lattice structure of a two-channel paraunitary filter bank includes $\boldsymbol{R}_k =$ rotation by θ_k. There is a one-to-one correspondence between lattice coefficients $\theta_0, \theta_1, \ldots, \theta_N$ and filter coefficients. The advantage of the lattice coefficients is that the quantized filters are still PR. The rotations are still orthogonal when the angles are rounded off.

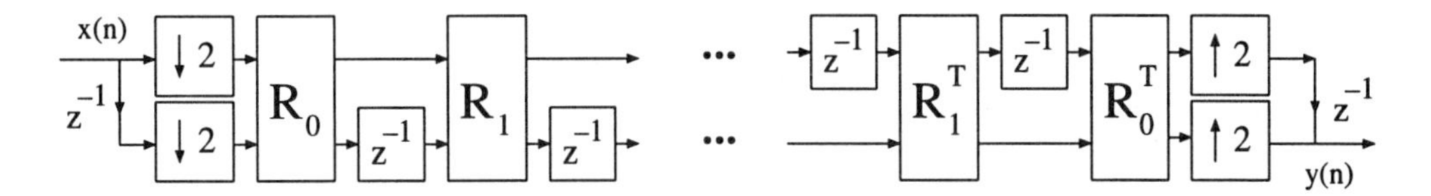

Objective Function Φ

Useful objective functions are the *stopband attenuation, coding gain,* and *degree of smoothness.* Stopband attenuation is very important in audio compression, whereas coding gain and smoothness are more important in image compression.

Stopband Attenuation: Φ measures the passband and stopband errors of $H(z)$:

$$\Phi = \begin{cases} \displaystyle\int_\Omega W(e^{j\omega})\,|H_d(e^{j\omega}) - H(e^{j\omega})|^2\,d\omega, & L_2 \text{ norm (energy)} \\[2ex] \text{Max}\,\left[W(e^{j\omega})\,|H_d(e^{j\omega}) - H(e^{j\omega})|\right], & L_\infty \text{ norm (maximum).} \end{cases} \tag{10.2}$$

The frequency region Ω consists of the passband $0 \le \omega \le \omega_p$ and stopband $\omega_s \le \omega \le \pi$. $H_d(e^{j\omega})$ is the desired response of the lowpass filter. $W(e^{j\omega})$ is a positive weighting function. Examples of typical $H_d(e^{j\omega})$ and $W(e^{j\omega})$ are:

$$H_d(e^{j\omega}) = \begin{cases} e^{-jL\omega}, & 0 \le \omega \le \omega_p \\ 0, & \omega_s \le \omega \le \pi \end{cases} \qquad W(e^{j\omega}) = \begin{cases} a, & 0 \le \omega \le \omega_p \\ b, & \omega_s \le \omega \le \pi. \end{cases} \tag{10.3}$$

If one increases the ratio b/a, the stopband error δ_s is much smaller than the passband error δ_p. In a filter bank with M channels, Φ has a term from each filter $H_k(e^{j\omega})$. The bandpass filters have two stopbands enclosing one passband.

The analysis filters in a paraunitary bank satisfy $\sum |H_k(e^{j\omega})|^2 = 1$. Consequently, it is sufficient to consider only the stopbands. With three filters, the figure shows $\frac{\pi}{3} \le \omega \le \frac{2\pi}{3}$ as a stopband of H_0 and H_2 and as the passband of H_1. In its passband, $H_1(e^{j\omega})$ must approximate unity. The passband error $\delta_{p1} = 1 - \sqrt{1 - \delta_{s0}^2 - \delta_{s2}^2}$ is very small, if both stopband errors δ_{s0} and δ_{s2} are small. Filter banks with high stopband attenuation are essential in audio compression and signal detection.

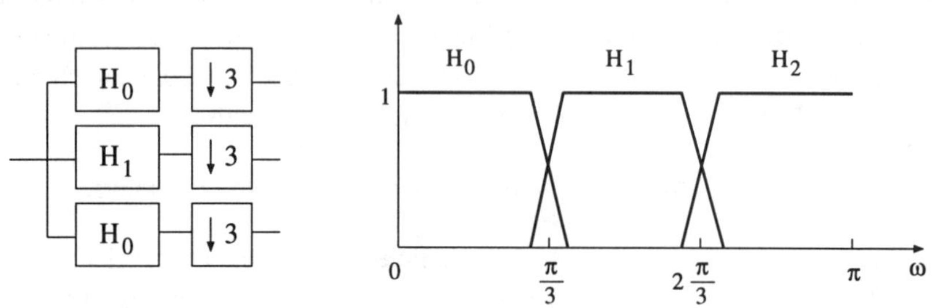

Example 10.1. Consider the design of a linear-phase filter bank with (9, 7) taps based on maximum error in the stopband. Figure 10.8(a) shows the equiripple stopband responses of $H_0(z)$ and $F_0(z)$. They are the factors of an equiripple halfband filter of length 15. There is no zero at $\omega = \pi$ and the scaling function and wavelets are *not smooth.* This pair of filters was used earlier to demonstrate checkerboard artifacts in image compression.

Figure 10.8(b) shows the frequency responses of a 9/7 maximally-flat halfband filter. *The stopband attenuation of both filters is bad.* However, each filter has four zeros at $\omega = \pi$. The scaling functions and wavelets are very smooth. They are the filters used in the FBI fingerprint image compression standard.

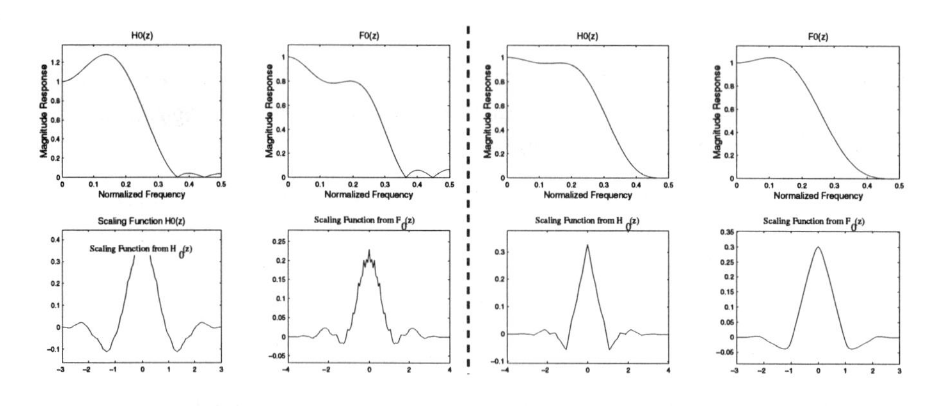

Figure 10.8: Example of equiripple design (left) and maximally-flat design (right).

Example 10.2. Figure 10.9 shows the filter responses of a cosine-modulated filter bank (8 channels, filter lengths 128). The objective function is $\int_{\omega_s}^{\pi} |H(e^{j\omega})|^2 \, d\omega$, the stopband energy of the prototype filter.

Coding Gain: A uniform quantizer has an equivalent noise model. The input signal $\boldsymbol{x}(n)$ is quantized to b bits to obtain $\boldsymbol{v}(n)$. For large b, one can assume that the quantization error $\boldsymbol{q}(n) = \boldsymbol{v}(n) - \boldsymbol{x}(n)$ is a random variable, uniformly distributed in $-2^b \le \boldsymbol{q} \le 2^b$. The error variance of $\boldsymbol{q}(n)$ is $\frac{1}{3}2^{-2b}$.

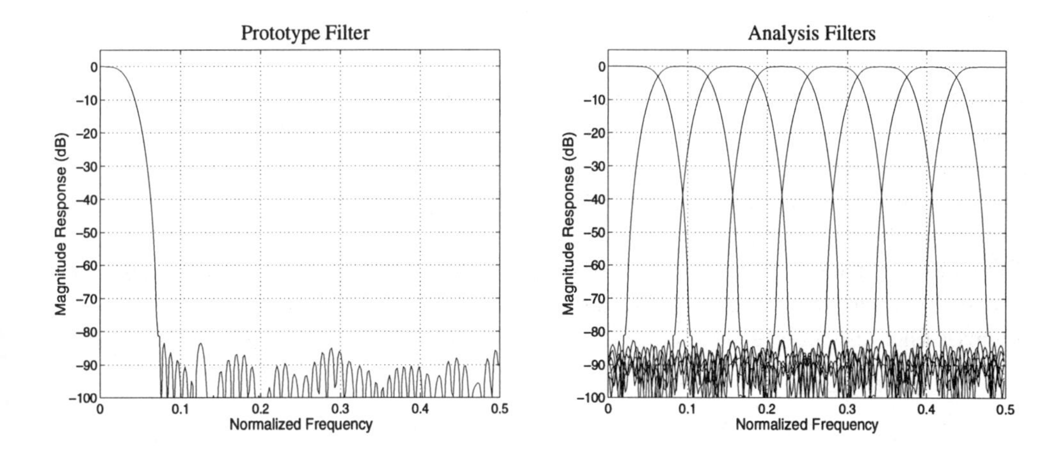

Figure 10.9: A cosine-modulated filter bank with stopband attenuation as objective Φ.

Instead of quantizing the input directly, one can transform it into another domain. The input signal is blocked into length N and transformed to length M by $\boldsymbol{P}^T$. The subband signal $\boldsymbol{y}_k(n)$ is then quantized to produce $\widehat{\boldsymbol{y}}_k(n)$. The transform should be designed so that the energy of $\boldsymbol{x}$ is concentrated in the first few coefficients of $\boldsymbol{y}_k$. The number of bits in the quantizers Q_k should be proportional to the energy of $\boldsymbol{y}_k(n)$.

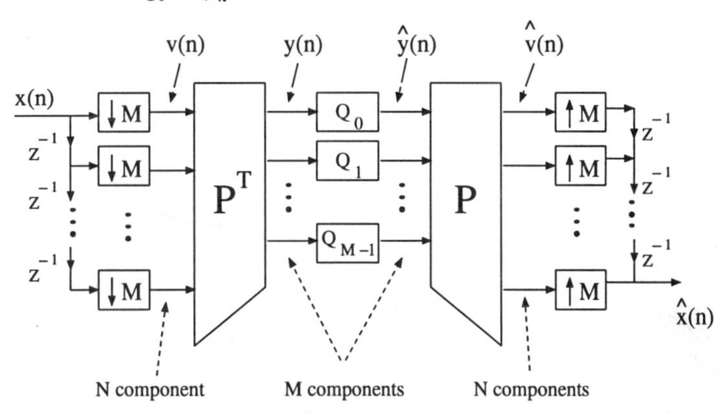

The coding gain measures the energy compaction from the transform. An orthogonal transform with high coding gain yields reconstructed images with high quality at low bit rate. Chapter 11 will discuss coding gain in more detail. For *biorthogonal* systems, [Katto-Yasuda] derived a unified formula when the input signal $\boldsymbol{x}(n)$ is modeled as a Markov process. Its inter-sample correlation factor is ρ. The coding gain is G when the sampling ratios are α_k:

$$G(\rho) = \prod_{k=0}^{M-1} (A_k B_k)^{-\alpha_k} \quad \text{where} \quad \left\{ \begin{array}{lcl} A_k & = & \sum_n \sum_m \boldsymbol{h}_k(n)\, \boldsymbol{h}_k(m)\, \rho^{|m-n|} \\ B_k & = & \sum_n (\boldsymbol{f}_k)^2(n) \end{array} \right.$$

For example, a 3-level wavelet transform has four analysis filters:

$$H_0(z)H_0(z^2)H_0(z^4) \quad H_0(z)H_0(z^2)H_1(z^4) \quad H_0(z)H_1(z^2) \quad H_1(z). \tag{10.4}$$

The coefficients in these combinations are used in the formula for coding gain. Assuming

$\rho = 0.95$, one can compute G and improve the performance. The sampling ratios are $1/8$, $1/8$, $1/4$ and $1/2$.

Example 10.3. Figure 10.10 shows the frequency responses of an 8-channel GenLOT with length 48, for maximum stopband attenuation and also for maximum coding gain. Note that the responses are different. These GenLOTs are compared at 50 : 1 compression. The objective performances with high stopband attenuation are MSE = 135, PSNR = 26.84 dB and Maximum error = 96. The corresponding numbers for the GenLOT with high coding gain are 130, 26.99 dB and 84. Even though these are comparable, maximizing the attenuation produced blocking artifact. This is not due to the block length 48, but it is due to the DC leakage. The bandpass and highpass filter responses are not zero at $\omega = 0$, so the DC term has leaked out of the lowpass.

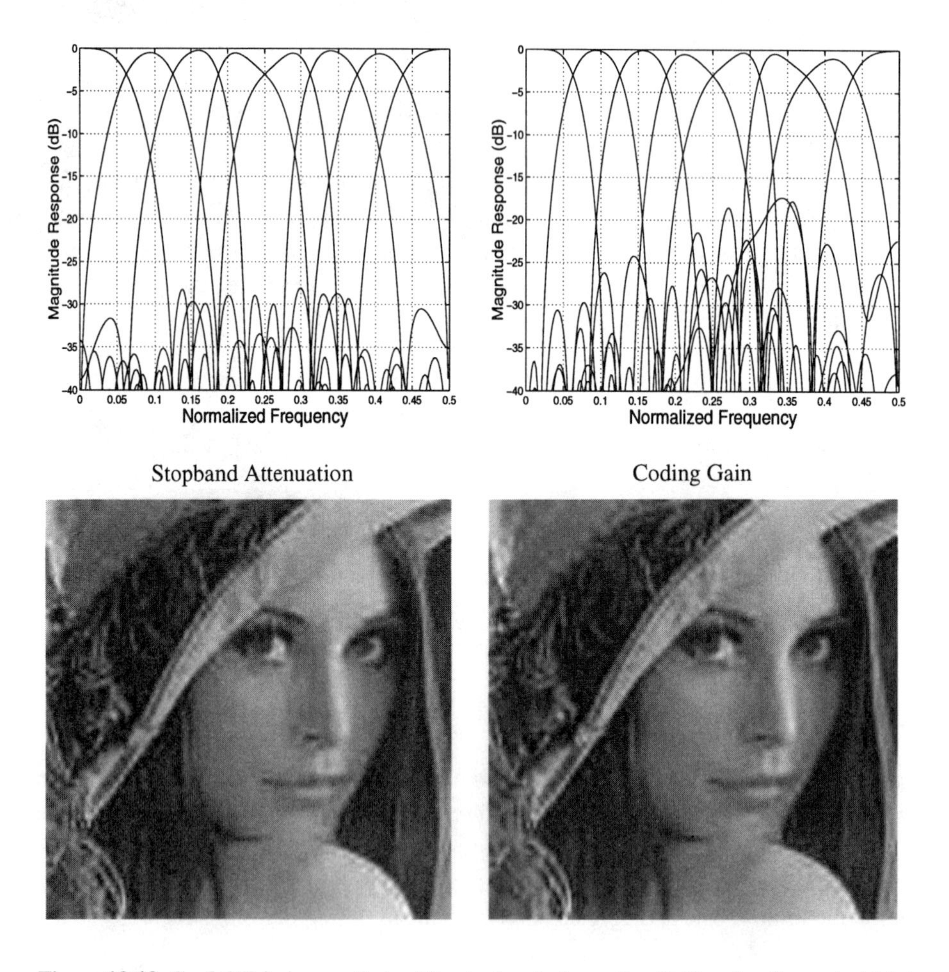

Figure 10.10: GenLOT designs optimized for stopband attenuation (left) and coding gain (right).

Constraints

PR constraints: The equivalent conditions for M-channel perfect reconstruction are

$$\begin{cases} \sum_n \boldsymbol{h}_k(n) \boldsymbol{f}_\ell(2m-n) = \delta(m)\delta(k-\ell) \\ \boldsymbol{F}_p(z)\,\boldsymbol{H}_p(z) = z^{-n_0}\boldsymbol{I}. \end{cases} \tag{10.5}$$

How does one design $H_k(z)$ and $F_k(z)$ that satisfy (10.5)? One way is to design $\boldsymbol{h}_k(n)$ first and then solve for $\boldsymbol{f}_k(n)$. For arbitrary analysis filters, synthesis filters might not exist. The time domain approach in [Nayebi1] presents an iterative method to solve for $\boldsymbol{h}_k(n)$ and $\boldsymbol{f}_k(n)$.

For FIR filters, (10.5) is true only if the determinant of $\boldsymbol{H}_p(z)$ is a delay. The special case of *paraunitary* $\boldsymbol{H}_p(z)$ can be parametrized by a set of rotation angles or Householder matrices. This is the lattice structure. The PR constraints are *automatically satisfied.*

For two channels, a simple PR condition exists for FIR systems: $H_0(z)F_0(z)$ is a halfband filter. This is the result when aliasing is cancelled. Such simplicity does not exist for systems with more channels.

Smoothness constraints: From the image compression examples, it is clear that the synthesis functions should be smooth. This relates directly to the number p of zeros at $\omega = \pi$ of $F_0(z)$. The smoothness property is also known as the DC-leakage property in traditional image processing.

Example 10.4. The lowpass Daubechies orthogonal filter $H_0(z)$ is the minimum-phase spectral factor of a maximally-flat halfband filter $P(z)$. Since $P(z)$ has $2p$ zeros at $\omega = \pi$, $H_0(z)$ and $F_0(z)$ have p zeros. Note that the functions in Figure 10.11 of D_{24} are considerably smoother than D_4, since D_{24} has 12 zeros at $\omega = \pi$.

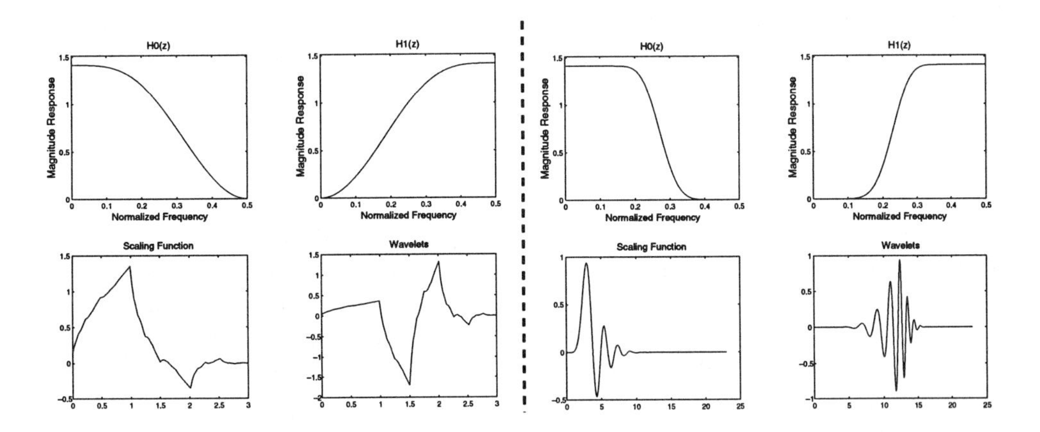

Figure 10.11: Magnitude response, scaling function and Daubechies wavelets D_4 and D_{24}.

10.3 Design of Perfect Reconstruction Filter Banks

The three main design methods extend in different ways to M channels:

Spectral factorization	factor an M-th band filter
Lattice structure method	minimize Φ (lattice coefficients)
Time-domain method	minimize Φ (filter coefficients) subject to PR.

The first approach uses spectral factorization. We find roots or factors $F_0(z)H_0(z)$ of a linear-phase polynomial $P(z)$. For applications that require large stopband attenuation, $P(z)$ has many zeros on or near the unit circle. Its factorization becomes difficult. A special class that is very useful in function approximation is the halfband maxflat Daubechies filter (Section 5.5). One can distribute its zeros in several ways to $H_0(z)$ and $F_0(z)$ to obtain:

$$\begin{cases} \text{Paraunitary filter banks} & -- & \text{Orthogonal Daubechies wavelets} \\ \text{Type A linear phase filter banks} & -- & \text{Symmetric even-length wavelets} \\ \text{Type B linear phase filter banks} & -- & \text{Symmetric odd-length wavelets.} \end{cases} \tag{10.6}$$

These wavelets have excellent approximation from the zeros at $\omega = \pi$.

The second approach is based on the lattice structure implementation of the filter bank. The unknown parameters are the lattice coefficients k_i. Since all PR filter banks are characterized by an appropriate set of k_i, the design problem reduces to minimizing the objective function $\Phi(k_i)$. Then the corresponding filters can be computed. The minimization is a challenging problem because of the *nonlinear relation* between the lattice and filter coefficients.

The time-domain approach expresses the PR conditions directly on the filter coefficients. Nonlinear quadratic constrained optimization yields good filters. One optimization method formulates the PR conditions in terms of matrices and the filter coefficients [Nayebi**1**]. The algorithm iteratively solves (10.5). An alternative is to formulate the objective function and constraints in quadratic forms:

$$\text{Minimize } \boldsymbol{h}^T \boldsymbol{P} \boldsymbol{h} \quad \text{subject to } \boldsymbol{h}^T \boldsymbol{Q}_k \boldsymbol{h} = c_k. \tag{10.7}$$

We may use Lagrange multipliers (explicitly or implicitly). The advantage of this constrained quadratic approach is that the derivatives and the Hessian of both the objective function and the PR conditions can be computed exactly.

Spectral Factorization Method

Given a polynomial of the form $P(z) = F_0(z)H_0(z)$, the objective here is to find the two unknown polynomials $F_0(z)$ and $H_0(z)$. We assume that all polynomials have real coefficients. Note that $F_0(z) = z^{-N} H_0(z^{-1})$ in a paraunitary filter bank, and the problem becomes a spectral factorization. For a low-degree polynomial $P(z)$, one can find its zeros and distribute them to $F_0(z)$ and $H_0(z)$. Since all polynomials are real-valued, the zeros have to be in complex-conjugate pairs.

For example, consider the maximally-flat halfband filter $P(z)$ of length 15. Its zeros are shown in Figure 10.12 with 8 zeros at $\omega = \pi$. Given the 14 zeros of $P(z)$, there are many ways of distributing its zeros to $F_0(z)$ and $H_0(z)$. One can assign half the zeros to $H_0(z)$ and half to $F_0(z)$ in such a way that $F_0(z) = z^{-7} H_0(z^{-1})$. This means that z_0 is a zero of $H_0(z)$ when z_0^{-1} is a zero of $F_0(z)$. The factorization assigns half of the zeros at π to $H_0(z)$, and the remaining half to $F_0(z)$. For minimum phase, $H_0(z)$ has all other zeros inside the unit circle, whereas $F_0(z)$ has all the zeros outside the unit circle. Both filters have length 8. This factorization yields a paraunitary filter bank and the Daubechies D_8 orthogonal wavelets.

Part b shows an alternate zero distribution. The zeros are chosen such that the resulting filters have odd length and linear phase. Note that $H_0(z)$ and $F_0(z)$ have the same number of zeros at $\omega = \pi$. This (9, 7) pair is used in the FBI fingerprint image compression standard.

Part c shows another zero distribution that yields linear phase with even length. Note that the zeros at π for $H_0(z)$ and $F_0(z)$ are not the same. The scaling function and wavelet from the

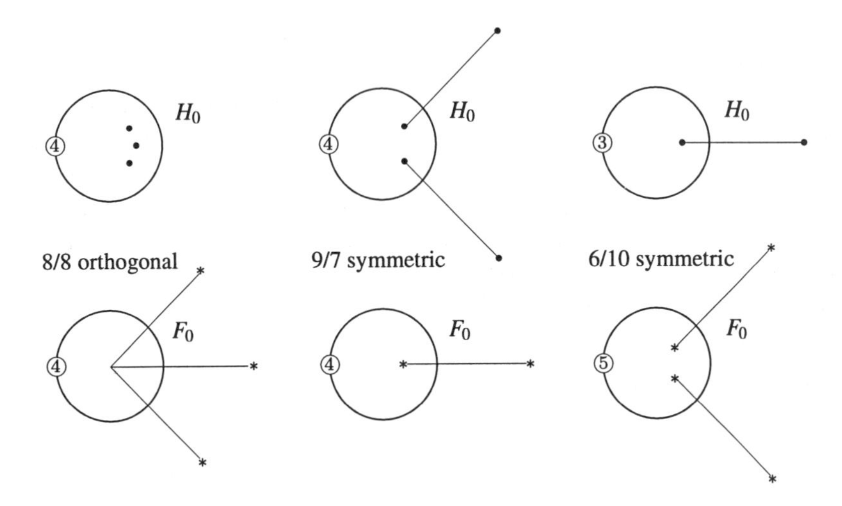

Figure 10.12: The zero distributions of $F_0(z)$ and $H_0(z)$. (a) Paraunitary filter bank (b) Linear phase filter bank (Type B, odd length) and (c) Linear phase filter bank (Type A, even length).

synthesis bank is smoother (it has more zeros at π). This is the (6, 10)-tap filter pair. Its performance is very good in image compression. The three factorizations demonstrate the essential ideas in designing two-channel paraunitary filter banks (orthogonal wavelets) and two-channel linear phase filter banks (symmetric wavelets).

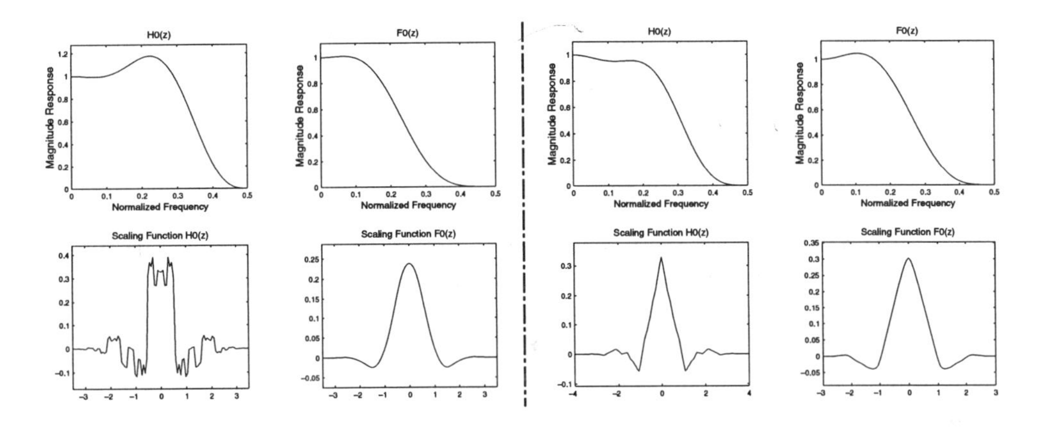

Spectral factorization applies to the design of two-channel filter banks and Near-Perfect-Reconstruction cosine-modulated filter banks. In the two-channel filter bank case, $F_0(z)$ and $H_0(z)$ are the factors of $P(z)$. In the NPR cosine-modulated case, the prototype filter $H(z)$ is a spectral factor of a $2M$-th band filter. For an arbitrary filter bank, such simple PR/NPR conditions do not exist. There is no optimization process involved in spectral factorization. In general, an optimal filter $P(z)$ does not guarantee optimal filters $H_0(z)$ and $F_0(z)$.

For orthogonal filters, the cepstral algorithm of [Mian-Nainer] is widely used. It computes the minimum phase spectral factor without finding the zeros. Section 5.4 describes this (additive) splitting of the logarithm of $P(e^{j\omega})$.

Lattice Structure

Most PR filter banks can be implemented based on lattice structure. For any set of lattice coefficients, the paraunitary, linear-phase, and cosine-modulation properties are structurally imposed. This means that those properties (and also PR) are enforced independent of the lattice coefficient values. The objective function could take many different forms depending on the applications. Examples are stopband attenuation, coding gain, and interference energy. The lattice method is used for two-channel paraunitary filter bank [VaidHoang], the two-channel biorthogonal linear-phase filter bank [V], the M-channel linear-phase paraunitary filter bank [Soman], and the paraunitary cosine-modulated filter bank [Koil**2**].

Another role for lattices begins in the time domain. Optimization yields a *Near Perfect Reconstruction (NPR)* solution:

$$\sum_n \boldsymbol{h}_k(n)\boldsymbol{f}_\ell(2m-n) < \epsilon < 10^{-10} \quad \text{for } k \neq \ell, m \neq 0. \tag{10.8}$$

From these NPR filters, one can find the lattice coefficients. *Those give PR filters* $\boldsymbol{h}_k^{PR}(n)$ *and* $\boldsymbol{f}_k^{PR}(n)$. If the reconstruction error ϵ is small, the NPR filters and PR filters are close and $\Phi^{PR} \approx \Phi$. If ϵ is not small, the filters are not close and Φ could change drastically.

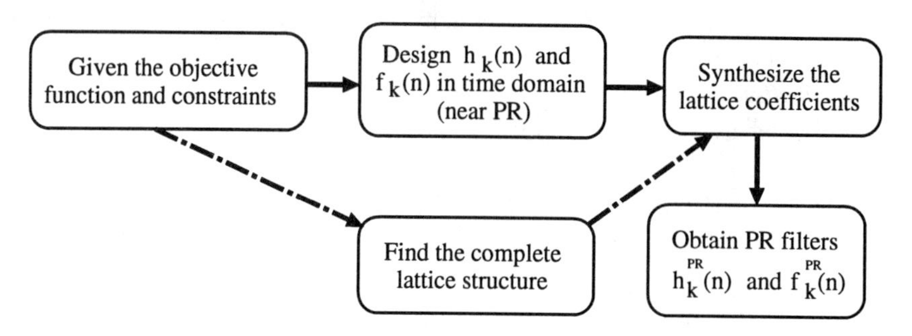

There are two key points to ensure success:

- Existence of *complete* and *minimal lattice structure*. The *completeness* is crucial here since the lattice structure must cover all filters with the given properties. *Minimal* is desirable to ensure efficient implementation.
- The reconstruction error ϵ must be sufficiently small to guarantee that $\Phi^{PR} \approx \Phi$. A typical objective function is stopband attenuation.

Example 10.5. A two-channel linear phase system of length 32 is shown in Figure 10.13(a) (solid lines). These filters are designed with $\epsilon = 10^{-4}$. They are used in the above lattice synthesis procedure to obtain a PR filter bank (broken lines). The attenuation has changed from -42 dB to -25 dB. Apparently, 10^{-4} is not small enough. Figure 10.13(b) shows the frequency response of the NPR solution with $\epsilon = 10^{-14}$ (solid line) and the PR solution (broken line). The NPR filters are very close to PR. Practically, there are no changes in the frequency responses.

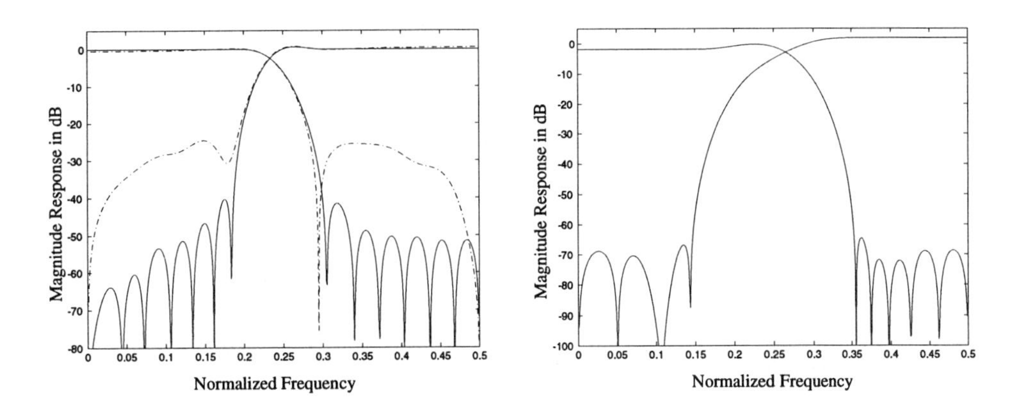

Figure 10.13: NPR and PR filters obtained by lattice synthesis for $\epsilon = 10^{-4}$ (left) and $\epsilon = 10^{-14}$ (right).

What went wrong in the lattice synthesis that changes the filter response drastically when ϵ is not small? Note the subtle difference between PR and NPR lattice structures in Figure 10.14. The delays are no longer z^{-2}, but z^{-1}. There are additional blocks $\widehat{\Gamma}_k = \begin{bmatrix} 1 & \widehat{\gamma}_k \\ \widehat{\gamma}_k & 1 \end{bmatrix}$ between Γ_k. A large ϵ will yield nonzero $\widehat{\gamma}_k$, and setting them to zero will change the filters dramatically. Since $\widehat{\gamma}_k$ are very small if $\epsilon < 10^{-14}$, setting $\widehat{\gamma}_k = 0$ does not change the frequency responses or the objective function.

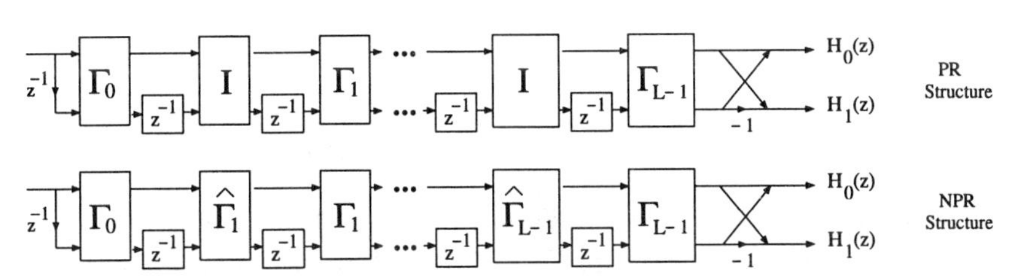

Figure 10.14: Lattice structures for PR and NPR Type-A linear phase filter banks.

Time-Domain Methods

Block matrix approach: This method is suggested by [Nayebi1] to solve the PR equations when the filter length is $N = LM$:

$$\underbrace{\begin{bmatrix} \boldsymbol{P}_0^T & \boldsymbol{0} & \cdots & \boldsymbol{0} \\ \boldsymbol{P}_1^T & \boldsymbol{P}_0^T & \cdots & \boldsymbol{0} \\ \vdots & \vdots & \ddots & \vdots \\ \boldsymbol{P}_{L-1}^T & \boldsymbol{P}_{L-2}^T & \cdots & \boldsymbol{P}_0^T \\ \boldsymbol{0} & \boldsymbol{P}_{L-1}^T & \cdots & \boldsymbol{P}_1^T \\ \vdots & \vdots & \ddots & \vdots \\ \boldsymbol{0} & \boldsymbol{0} & \cdots & \boldsymbol{P}_{L-1}^T \end{bmatrix}}_{\boldsymbol{A}} \underbrace{\begin{bmatrix} \boldsymbol{Q}_0^T \\ \boldsymbol{Q}_1^T \\ \vdots \\ \boldsymbol{Q}_{L-1}^T \end{bmatrix}}_{\boldsymbol{Q}} = \underbrace{\begin{bmatrix} \boldsymbol{0} \\ \vdots \\ \boldsymbol{0} \\ \boldsymbol{J} \\ \boldsymbol{0} \\ \vdots \\ \boldsymbol{0} \end{bmatrix}}_{\boldsymbol{B}} \tag{10.9}$$

$\boldsymbol{J}$ is the exchange matrix (reverse identity). Its location determines the delay of the system. The entries of the filter matrices are $\boldsymbol{P}_{k\ell} = \boldsymbol{h}_k(\ell)$ and $\boldsymbol{Q}_{k\ell} = \boldsymbol{f}_k(\ell)$. The $M \times M$ submatrices $\boldsymbol{P}_j$ and $\boldsymbol{Q}_j$ are defined by:

$$\boldsymbol{P} = \begin{bmatrix} \boldsymbol{P}_0 & \boldsymbol{P}_1 & \cdots & \boldsymbol{P}_{L-1} \end{bmatrix} \quad \text{and} \quad \boldsymbol{Q} = \begin{bmatrix} \boldsymbol{Q}_0 & \boldsymbol{Q}_1 & \cdots & \boldsymbol{Q}_{L-1} \end{bmatrix}. \qquad (10.10)$$

Chapter 9 used this formulation for LOT ($N = 2M$ and $L = 2$), where (10.10) reduces to $\boldsymbol{P}_1^T \boldsymbol{P}_1 + \boldsymbol{P}_0^T \boldsymbol{P}_0 = \boldsymbol{I}$ and $\boldsymbol{P}_1^T \boldsymbol{P}_0 = \boldsymbol{0}$.

In summary, the time domain PR constraints $\boldsymbol{AQ} = \boldsymbol{B}$ are general. Given the analysis filters $\boldsymbol{h}_k(n)$ (and thus $\boldsymbol{A}$), this is not always solvable. One can find "optimal" synthesis filters $\boldsymbol{Q} = (\boldsymbol{A}^T\boldsymbol{A})^{-1}\boldsymbol{A}^T\boldsymbol{B}$ by minimizing $\|\boldsymbol{A}\,\boldsymbol{Q} - \boldsymbol{B}\|$. Then fix those filters $\boldsymbol{f}_k(n)$ and reverse the equation to find optimal $\boldsymbol{h}_k(n)$. This is the iterative block matrix approach.

Quadratic Constrained Least Squares method (QCLS): For an M-channel paraunitary filter bank, the PR condition requires no distortion ($k = 0$) and no aliasing ($k > 0$):

$$T_k(z) = \frac{1}{M} \sum_{\ell=0}^{M-1} z^{-N} H_\ell(z^{-1}) H_\ell(zW^k) = \delta(k)\, z^{-\ell}.$$

Let $\boldsymbol{h}$ be a column vector consisting of all the filter coefficients $\boldsymbol{h}_k(n)$:

$$\boldsymbol{h} = (\boldsymbol{h}_0(0) \cdots \boldsymbol{h}_0(N) \cdots \boldsymbol{h}_{M-1}(0) \cdots \boldsymbol{h}_{M-1}(N)\,)^T.$$

The PR conditions can be written in the form $\boldsymbol{h}^T \boldsymbol{Q}_k \boldsymbol{h} = 0$ and $\boldsymbol{h}^T \boldsymbol{S}_k \boldsymbol{h} = 1$. Furthermore, the objective function for stopband attenuation can be formulated as a quadratic $\Phi = \boldsymbol{h}^T\boldsymbol{P}\boldsymbol{h}$, where $\boldsymbol{P}$ is symmetric and positive definite. The optimized filter is precisely $\boldsymbol{h}_{\text{opt}}$ such that

$$\boldsymbol{h}_{\text{opt}} \quad \text{minimizes} \quad \boldsymbol{h}^T\boldsymbol{P}\boldsymbol{h} \quad \text{subject to} \quad \begin{cases} \boldsymbol{h}^T \boldsymbol{Q}_k \boldsymbol{h} = 0, \\ \boldsymbol{h}^T \boldsymbol{S}_k \boldsymbol{h} = 1. \end{cases} \qquad (10.11)$$

Since $\mathbf{Q}_k$ is normally not positive definite, the solution can be difficult. However, there are effective optimization procedures that linearize the quadratic constraints [Schittkowski]. These procedures will yield an approximate solution (the constraints are not satisfied exactly). The errors are very small and can be ignored in most practical cases. The QCLS method is used to design the two-channel PR linear-phase filter bank [QCLS], the NPR Pseudo-QMF bank [Nguyen**1**], the PR cosine-modulated filter bank [QCLS], and the M-channel paraunitary linear-phase filter bank [Nguyen**2**].

10.4 Design of Two-Channel Filter Banks

This book has emphasized the construction of $H_0(z)$ and $F_0(z)$ as factors of a halfband filter $P_0(z)$. This method is conceptually simple, and good in practice. The other two approaches, based on lattice structure and quadratically constrained optimization, look awkward by comparison.

But those methods have their own advantages. The lattice structure is efficient to implement, once it is found. The optimization can yield "best" filters rather than "good" filters. This brief section will record the details for the two cases of greatest significance: *orthogonal filter banks and linear phase filter banks.*

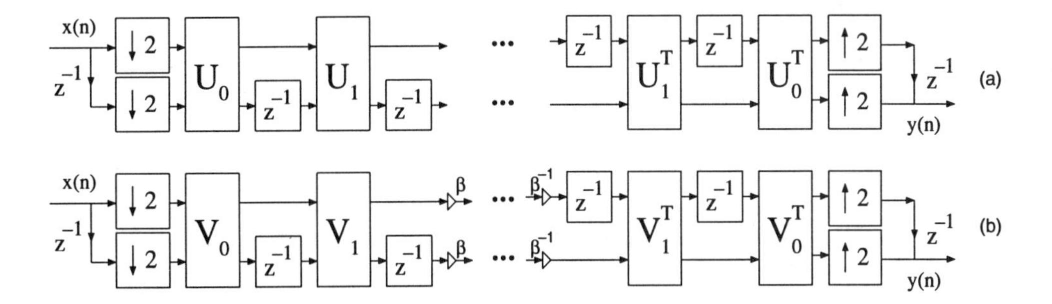

Figure 10.15: Lattice structures for the two-channel paraunitary filter bank. The first structure uses rotation angles as parameters and the second structure uses lattice coefficients.

Lattice Structure Method

Figure 10.15 shows two lattice structures for a paraunitary filter bank and orthogonal wavelets. The objective is to find the right set of angles θ_k in $\boldsymbol{U}_k$ (or lattice coefficients α_k in $\boldsymbol{V}_k$) such that an objective function is minimized. The objective function considered by [VaidHo] is the stopband error (L_2 norm). There are several tables compiled for orthogonal filters with various frequency specifications.

Figure 10.16 shows the two lattice structures for linear phase filter banks. Recall that Type A yields filters with even length (one is symmetric and the other is antisymmetric). Type B yields symmetric filters with odd length. In Type A, the parameters are the lattice coefficients in Γ_k. [NgVa1] presents design examples for both Type A and B systems, designed using the nonlinear optimization where the parameters are the lattice coefficients.

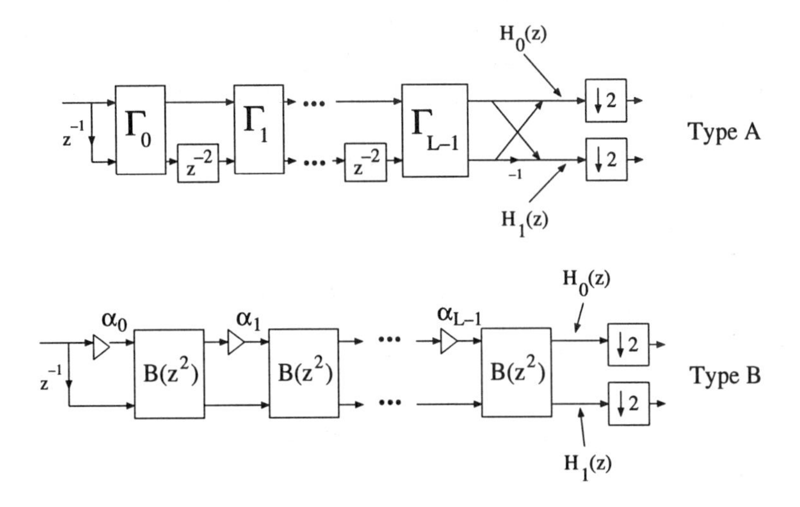

Figure 10.16: Lattice structures for symmetric filter banks, even and odd lengths.

Time-Domain Method—QCLS Formulation of Orthogonality

The orthogonality condition is that $P(z) = z^{-N} H_0(z^{-1}) H_0(z)$ is a halfband filter. We will show how to formulate this in quadratic form $\boldsymbol{h}^T \boldsymbol{Q}_k \boldsymbol{h} = c_k$ using

$$\boldsymbol{h} = [\boldsymbol{h}_0(0)\ \boldsymbol{h}_0(1)\ \cdots\ \boldsymbol{h}_0(N)]^T \quad \text{and} \quad \boldsymbol{e}(z) = \begin{bmatrix} 1 & z^{-1} & \cdots & z^{-N} \end{bmatrix}^T \quad \text{and} \quad H_0(z) = \boldsymbol{e}^T(z)\boldsymbol{h}.$$

Then $z^{-N} H_0(z^{-1}) = \boldsymbol{h}^T \boldsymbol{J} \boldsymbol{e}(z)$ where $\boldsymbol{J}$ is the reverse identity matrix (antidiagonal). Therefore $P(z) = \boldsymbol{h}^T \boldsymbol{J} \boldsymbol{e}(z) \boldsymbol{e}^T(z) \boldsymbol{h}$. Let

$$\boldsymbol{e}(z)\boldsymbol{e}^T(z) = \sum z^{-k} \boldsymbol{D}_k \quad \text{where} \quad [\boldsymbol{D}_k]_{i,j} = \begin{cases} 1; & i+j=k \\ 0; & \text{otherwise.} \end{cases} \tag{10.12}$$

Then $P(z)$ can be rewritten as

$$P(z) = \sum z^{-k} \boldsymbol{h}^T \boldsymbol{J} \boldsymbol{D}_k \boldsymbol{h} = \sum \boldsymbol{p}(k) z^{-k}. \tag{10.13}$$

In other words, the kth coefficient of $P(z)$ is $\boldsymbol{p}(k) = \boldsymbol{h}^T \boldsymbol{J} \boldsymbol{D}_k \boldsymbol{h}$. The halfband condition on $P(z)$ gives the constraints in the design problem:

$$\min_{\boldsymbol{h}}\ \boldsymbol{h}^T \boldsymbol{P} \boldsymbol{h} \quad \text{subject to} \quad \boldsymbol{h}^T \boldsymbol{J} \boldsymbol{D}_{N-1-2n} \boldsymbol{h} = \delta(n).$$

Here $\boldsymbol{h}^T \boldsymbol{P} \boldsymbol{h}$ denotes the stopband energy. See [Nguyen3] for more examples.

Lagrange Multiplier Method

With the choice of $F_0(z) = H_1(-z)$ and $F_1(z) = -H_0(z)$ to cancel aliasing, the filters $H_0(z)$ and $H_1(z)$ must satisfy the PR condition $H_0(z)H_1(-z) - H_1(z)H_0(-z) = z^{-\ell}$. This is equivalent to $\boldsymbol{C}\,\boldsymbol{h}_1 = \boldsymbol{m}$, where $\boldsymbol{C}$ is a matrix whose elements come from $\boldsymbol{h}_0(n)$. The highpass filter is $\boldsymbol{h}_1$ and the k-th element of $\boldsymbol{m}$ is $\delta(k-l)$. Given $\boldsymbol{h}_0(n)$, the minimization problem becomes:

$$\text{Minimize}\ \ \Phi\ \ \text{subject to}\ \ \boldsymbol{C}\,\boldsymbol{h}_1 = \boldsymbol{m}. \tag{10.14}$$

The minimization constructs a Lagrangian function

$$L(\boldsymbol{h}_1, \lambda) = \Phi - \lambda^T(\boldsymbol{C}\boldsymbol{h}_1 - \boldsymbol{m}). \tag{10.15}$$

The optimality conditions $\partial L/\partial \boldsymbol{h}_1 = 0$ and $\partial L/\partial \lambda = 0$ are

$$\frac{\partial \Phi}{\partial \boldsymbol{h}_1} - \boldsymbol{C}^T \lambda = \boldsymbol{0} \quad \text{and} \quad \boldsymbol{C}\boldsymbol{h}_1 - \boldsymbol{m} = \boldsymbol{0}. \tag{10.16}$$

In some cases Φ can be expressed explicitly in terms of $\boldsymbol{h}_1$. When stopband error is the objective function, Φ has the form $\frac{1}{2}\boldsymbol{h}_1^T \boldsymbol{S}_1 \boldsymbol{h}_1 + \boldsymbol{S}_2^T \boldsymbol{h}_1 + s_3$. The optimality conditions become

$$\begin{bmatrix} -\boldsymbol{S}_1 & \boldsymbol{C}^T \\ \boldsymbol{C} & \boldsymbol{0} \end{bmatrix} \begin{bmatrix} \boldsymbol{h}_1 \\ \lambda \end{bmatrix} = \begin{bmatrix} \boldsymbol{S}_2 \\ \boldsymbol{m} \end{bmatrix}. \tag{10.17}$$

QCLS formulation for Type A: The filter length is $2m$ and the PR condition is

$$T(z) = \sum_{n=0}^{4m-2} t(n)z^{-n} = -H_0(z)H_1(-z) + H_0(-z)H_1(z) = z^{-(2m-1)}. \tag{10.18}$$

Substituting $(-z)$ for z in (10.18), one obtains $t(n) = 0$ for even n. Consequently we have m conditions on the odd-numbered coefficients:

$$t(2k+1) = \begin{cases} 0, & 0 \le k \le m-2 \\ 1, & k = m-1. \end{cases} \tag{10.19}$$

Let $\boldsymbol{h} = (\, h_0(0) \ \cdots \ h_0(m-1) \ \ h_1(0) \ \cdots \ h_1(m-1)\,)^T$. Our objective is to express the m conditions in (10.19) in terms of $\boldsymbol{h}$. The polynomials $H_0(z)$, $H_1(z)$, $H_0(-z)$ and $H_1(-z)$ can be written using $\boldsymbol{e}(z) = (1 \ z^{-1} \ \cdots \ z^{-(m-1)})^T$:

$$\begin{cases} H_0(z) = \boldsymbol{h}^T \begin{pmatrix} \boldsymbol{e}(z) + z^{-m}\boldsymbol{J}\boldsymbol{e}(z) \\ \boldsymbol{0} \end{pmatrix}, & H_0(-z) = \boldsymbol{h}^T \begin{pmatrix} \boldsymbol{U}\boldsymbol{e}(z) + (-1)^{m/2} z^{-m}\boldsymbol{J}\boldsymbol{U}\boldsymbol{e}(z) \\ \boldsymbol{0} \end{pmatrix} \\ H_1(z) = \boldsymbol{h}^T \begin{pmatrix} \boldsymbol{0} \\ \boldsymbol{e}(z) - z^{-m}\boldsymbol{J}\boldsymbol{e}(z) \end{pmatrix}, & H_1(-z) = \boldsymbol{h}^T \begin{pmatrix} \boldsymbol{0} \\ \boldsymbol{U}\boldsymbol{e}(z) - (-1)^{m/2} z^{-m}\boldsymbol{J}\boldsymbol{U}\boldsymbol{e}(z) \end{pmatrix} \end{cases}$$

Here $\boldsymbol{J}$ is the exchange matrix and $\boldsymbol{U}$ is diagonal with $\boldsymbol{U}_{kk} = (-1)^k$. Substituting into (10.18), $T(z)$ is simplified to

$$T(z) = \boldsymbol{h}^T \begin{pmatrix} \boldsymbol{0} & \Gamma(z) \\ \boldsymbol{0} & \boldsymbol{0} \end{pmatrix} \boldsymbol{h} \quad \text{where} \quad \boldsymbol{E}(z) = \boldsymbol{e}(z)\boldsymbol{e}^T(z) \quad \text{and} \tag{10.20}$$

$$\Gamma(z) = [\boldsymbol{U}\boldsymbol{E}(z) - \boldsymbol{E}(z)\boldsymbol{U}] + z^{-m}\Big[(-1)^{m/2}\big(\boldsymbol{J}\boldsymbol{U}\boldsymbol{E}(z) + \boldsymbol{E}(z)\boldsymbol{U}\boldsymbol{J}\big) - \big(\boldsymbol{U}\boldsymbol{E}(z)\boldsymbol{J} + \boldsymbol{J}\boldsymbol{E}(z)\boldsymbol{U}\big)\Big] + z^{-2m}(-1)^{m/2}[\boldsymbol{J}\boldsymbol{E}(z)\boldsymbol{U}\boldsymbol{J} - \boldsymbol{J}\boldsymbol{U}\boldsymbol{E}(z)\boldsymbol{J}].$$

Substituting (10.12) for $\boldsymbol{E}(z)$ produces a polynomial $\sum_{k=0}^{4m-2} z^{-k}\boldsymbol{h}^T \begin{pmatrix} \boldsymbol{0} & \Gamma_k \\ \boldsymbol{0} & \boldsymbol{0} \end{pmatrix} \boldsymbol{h}$, where the Γ_k are constant matrices $\boldsymbol{D}_k$, $\boldsymbol{J}$ and $\boldsymbol{U}$. Comparing term by term in (10.20), (10.19) becomes

$$\begin{cases} \boldsymbol{h}^T\boldsymbol{Q}_{2k+1}\boldsymbol{h} = 0; & 0 \le k \le m-2 \\ \boldsymbol{h}^T\boldsymbol{Q}_{2m-1}\boldsymbol{h} = 1; \end{cases} \quad \text{where} \quad \boldsymbol{Q}_n = \begin{pmatrix} \boldsymbol{0} & \Gamma_n \\ \boldsymbol{0} & \boldsymbol{0} \end{pmatrix} \quad \text{and} \tag{10.21}$$

$$\Gamma_n = \begin{cases} \boldsymbol{U}\boldsymbol{D}_n - \boldsymbol{D}_n\boldsymbol{U}; & 0 \le n \le m-1 \\ \boldsymbol{U}\boldsymbol{D}_n - \boldsymbol{D}_n\boldsymbol{U} + (-1)^{m/2}(\boldsymbol{J}\boldsymbol{U}\boldsymbol{D}_{n-m} + \boldsymbol{D}_{n-m}\boldsymbol{U}\boldsymbol{J}) - (\boldsymbol{U}\boldsymbol{D}_{n-m}\boldsymbol{J} + \boldsymbol{J}\boldsymbol{D}_{n-m}\boldsymbol{U}); & m \le n \le 2m-2 \\ (-1)^{m/2}(\boldsymbol{J}\boldsymbol{U}\boldsymbol{D}_{m-1} + \boldsymbol{D}_{m-1}\boldsymbol{U}\boldsymbol{J}) - (\boldsymbol{U}\boldsymbol{D}_{m-1}\boldsymbol{J} + \boldsymbol{J}\boldsymbol{D}_{m-1}\boldsymbol{U}); & n = 2m-1. \end{cases}$$

In summary, the PR condition in (10.19) is rewritten as m quadratic constraints on $\boldsymbol{h}$.

Example 10.6. The QCLS formulation is used in the design of a Type A bank with filter lengths 30. The objective function Φ is

$$\int_0^{\omega_p} [|H_0(e^{j0}) - H_0(e^{j\omega})|^2 + |H_1(e^{j\omega})|^2]\,d\omega + \int_{\omega_s}^{\pi} [|H_0(e^{j\omega})|^2 + |H_1(e^{j\pi}) - H_1(e^{j\omega})|^2]\,d\omega$$

Note that both passband and stopband errors are included since this is not a power complementary system. This is an NPR filter bank with $\epsilon = 1.5 \times 10^{-15}$. The figure below shows the frequency responses of the filters.

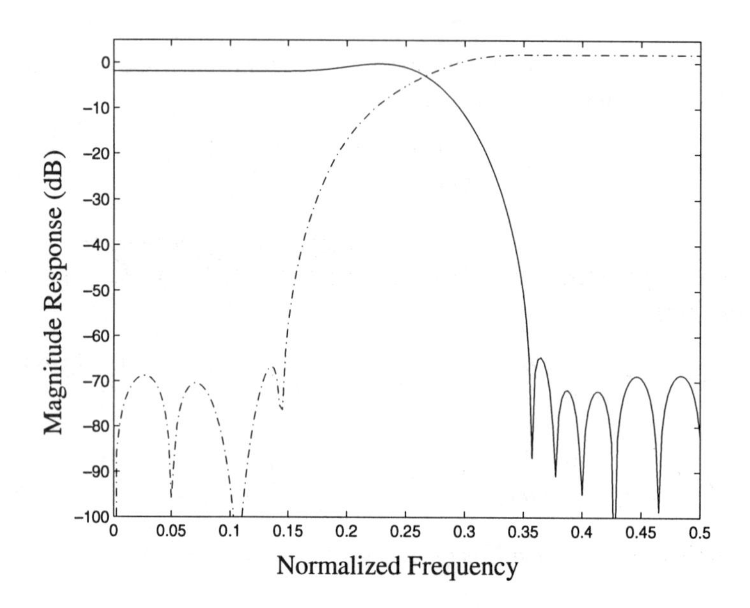

We close by noting four other design methods for two channels:

- Biorthogonal filter banks with integer coefficients divided by 2^k [AH].
- LMS-based algorithm [Roy].
- Biorthogonal linear-phase filter bank [Phoong**1**, Kiya].
- Alias-free two-channel allpass-based IIR filter bank [V, Vaid**1**, Nguyen**4**, EkPr].

10.5 Design of Cosine-modulated Filter Banks

A cosine-modulated filter bank is very efficient. It can be implemented using *one prototype filter* $H(z)$ and a cosine-modulation block. The M filters $H_k(z)$ are cosine-modulated versions of $H(z)$. As a result, a 32-channel bank with length 512 has only 256 unknowns (assuming linear phase), instead of 16, 384 unknowns in the general case. This helps the design algorithm tremendously in terms of convergence speed and results.

There are many design algorithms and they all use $H(z)$. What properties does $H(z)$ satisfy to obtain NPR solution or PR solution?

- ***Nearly PR*** If $H(z)$ is a spectral factor of a $2M$-th band filter, amplitude and phase distortions are eliminated. Aliasing is suppressed by sufficient stopband attenuation.
- ***Exactly PR*** Let $H(z)$ be linear phase. The $2M$ polyphase components $G_k(z)$ of a paraunitary cosine-modulated filter bank have to satisfy

$$G_k(z^{-1})\, G_k(z) + G_{M+k}(z^{-1})\, G_{M+k}(z) = \frac{1}{2M}. \tag{10.22}$$

We begin with NPR methods that are based on spectral factorization. Lattice structure and time-domain approaches will also be discussed, for PR designs. We will see that NPR methods tend to give better stopband attenuation.

Traditional Pseudo-QMF Bank Design

Pseudo-QMF Banks are NPR and cosine-modulated. They were proposed for communication and audio processing applications. Aliasing of the adjacent bands is cancelled by appropriate phase terms in the modulation. The nonadjacent aliasing terms are minimized by designing a prototype with good stopband attenuation δ_s. The aliasing level is comparable to δ_s, and the reconstruction error consists of distortion (Φ_d) and aliasing (Φ_a):

$$\Phi = \Phi_d + \Phi_a = (1-\alpha)\int_0^{\pi} \left|T_0(e^{j\omega}) - e^{-j\ell\omega}\right|^2 d\omega + \alpha\int_{\omega_s}^{\pi} \left|H(e^{j\omega})\right|^2 d\omega.$$

The weighting factor has $0 \leq \alpha \leq 1$, and ℓ is the system delay. There is typically a tradeoff between distortion and aliasing. Distortion is eliminated if $H(z)$ is restricted to be a spectral factor of a $2M$-th band filter $P(z)$. The objective function becomes simply Φ_a.

Spectral Factorization Evaluating $H(z)H(z^{-1})$ on the unit circle $z = e^{j\omega}$ yields a nonnegative function $P(e^{j\omega}) = |H(e^{j\omega})|^2$. How does one design $P(z)$ where $P(e^{j\omega}) \geq 0$? The design procedure is as follows:

- Design an equiripple $2M$-th band filter with $\widehat{p}(N) = \frac{1}{2M}$ and $\widehat{p}(N - 2M\ell) = 0$ for $\ell \neq 0$. N is large since $\widehat{P}(z)$ must also have good stopband attenuation δ_s.
- Form another $2M$-th band filter $P(z)$ (nonnegative!) by $P(z) = \delta_s\, z^{-N} + \widehat{P}(z)$. Factor $P(z)$.

The lowpass filter $H_0(z)$ and the highpass filter $H_{M-1}(z)$ are most affected by the choice of factor (minimum phase or near linear phase). Their passbands are combinations of two passbands and they have bumps or dips [Koil3] if the phase of $H(z)$ is not linear.

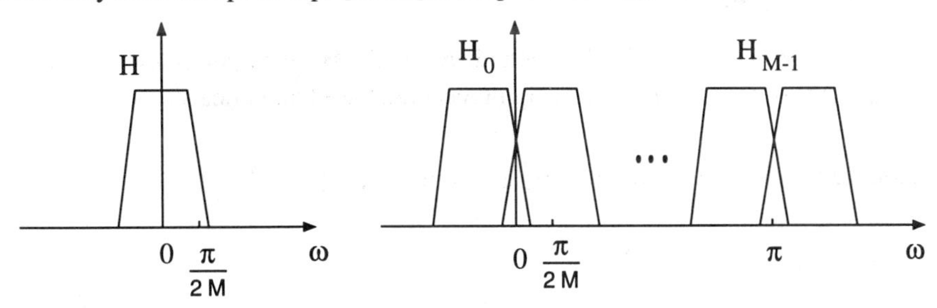

[Nguyen1] uses QCLS for linear-phase factorization. The resulting cosine-modulated filter bank has no distortion, small aliasing (comparable to stopband attenuation), and flat passband responses for H_0 and H_{M-1}. A 32-channel system of length 512 is shown, with distortion 5×10^{-11} and aliasing -83 dB.

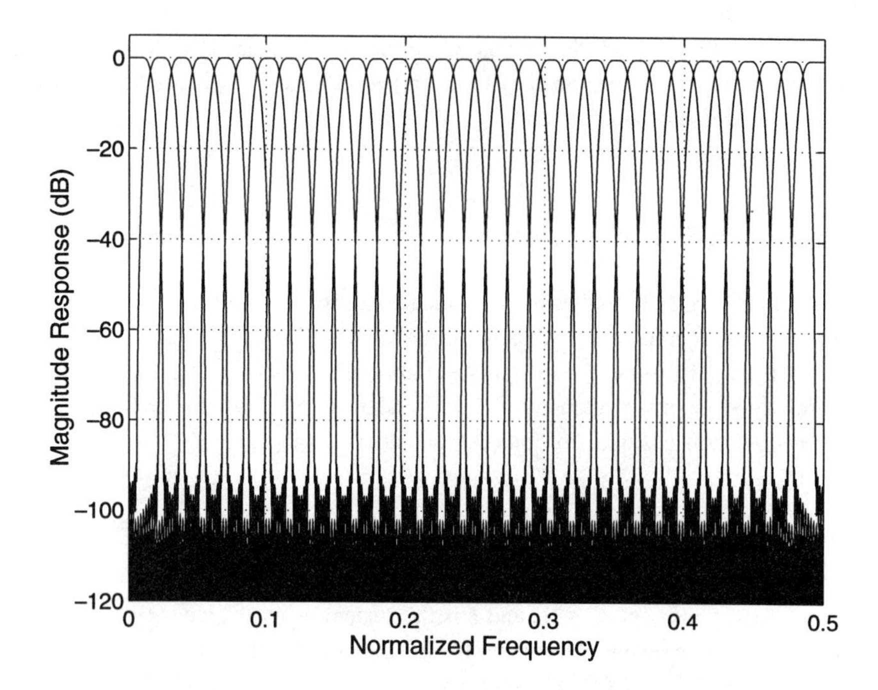

QCLS in NPR Pseudo-QMF Design

The linear phase relation $H(z) = z^{-N}H(z^{-1})$ expresses the coefficients $\boldsymbol{p}(n)$ as $\boldsymbol{h}^T\boldsymbol{D}(n)\boldsymbol{h}$. The matrix $\boldsymbol{D}(n)$ depends on the index n and the filter length [**Nguyen1**]. The design becomes a quadratic-constrained minimization:

$$\text{Minimize the stopband error } \Phi_a \text{ subject to } \quad \boldsymbol{p}(n) = \boldsymbol{h}^T\boldsymbol{D}(n)\boldsymbol{h} = 2M\text{-th band filter.}$$

The quadratic forms have simple derivatives and Hessians, which makes the optimization search very fast. A high-order system design like the example above takes about 4 minutes on a Sun Workstation 10.

Lattice Structure Method

In a PR cosine-modulated filter bank of length $N = 2mM$, each polyphase component pair $(G_k(z), G_{M+k}(z))$ is paraunitary. The design procedure based on lattice structure is:

- Initialize the angles θ_k. Compute $H_k(z)$ and $\Phi = \displaystyle\int_{\omega_s}^{\pi} \left|H(e^{j\omega})\right|^2 d\omega.$

- Iterate on θ_k to minimize Φ.

This method works well for short filters. One needs a time domain approach to design long PR systems with high attenuation. We show the frequency responses of two eight-channel PR cosine-modulated filter banks with lengths 64 and 80. The stopband attenuation improves from -46 dB to -52 dB (no distortion or aliasing). The attenuation using the QCLS is -58 dB, a significant improvement from -52 dB.

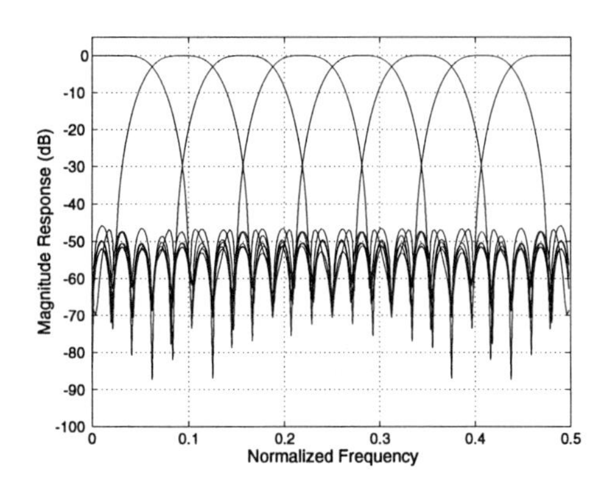

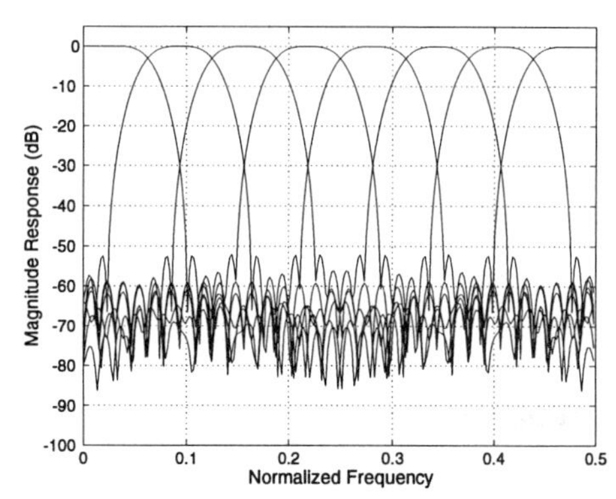

Time Domain Method in PR Design — QCLS Formulation

The objective is to design the linear phase prototype filter $H(z)$ such that its $2M$ polyphase components $G_k(z)$ satisfy the PR conditions (in pairs k and $M+k$). Let $\boldsymbol{h}$ be a vector of length mM consisting of the first half of $\boldsymbol{h}(n)$. Let $\boldsymbol{e}$ be a delay vector of length m with components z^{-k}. Then the polyphase components $G_k(z)$ can be written as $\boldsymbol{h}^T\boldsymbol{V}_k\boldsymbol{e}$, where $[\boldsymbol{V}_k]_{i,j} = 1$ only for $i = k + 2jM$. The PR condition (10.22) on the pair $G_k(z)$ and $G_{M+k}(z)$ becomes

$$\boldsymbol{h}^T\left[\boldsymbol{V}_k\boldsymbol{J}\boldsymbol{e}\boldsymbol{e}^T\boldsymbol{V}_k^T + \boldsymbol{V}_{M+k}\boldsymbol{J}\boldsymbol{e}\boldsymbol{e}^T\boldsymbol{V}_{M+k}^T\right]\boldsymbol{h} = \frac{1}{2M}z^{-(m-1)}. \tag{10.23}$$

Substituting $\boldsymbol{e}\boldsymbol{e}^T = \sum_{n=0}^{2m-2} z^{-n}\boldsymbol{D}_n$ where $[\boldsymbol{D}_n]_{i,j} = 1$ only for $i + j = n$, we obtain the

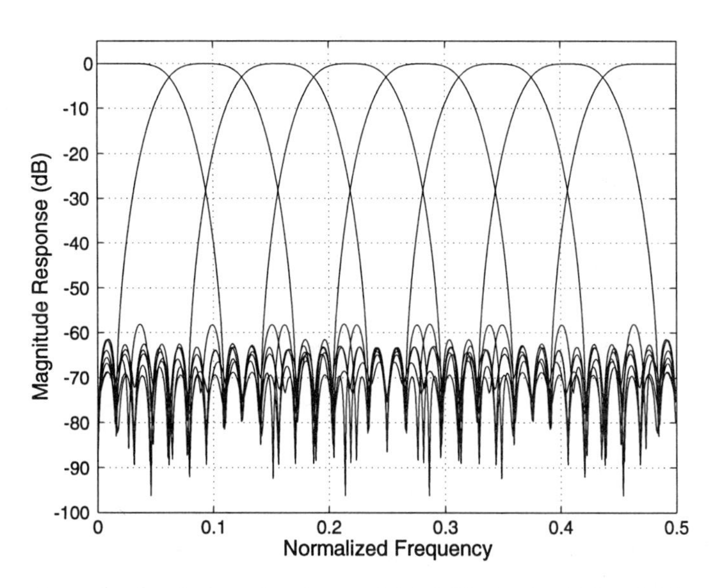

Example of an 8-channel cosine-modulated filter bank with length 80.

PR condition in terms of $\boldsymbol{h}(n)$:

$$\boldsymbol{h}^T\left[\boldsymbol{V}_k\boldsymbol{J}\boldsymbol{D}_n\boldsymbol{V}_k^T + \boldsymbol{V}_{M+k}\boldsymbol{J}\boldsymbol{D}_n\boldsymbol{V}_{M+k}^T\right]\boldsymbol{h} = \frac{1}{2M}\,\delta(n-m+1). \tag{10.24}$$

The design problem has these *quadratic constraints* [QCLS]. The objective function Φ is also quadratic in $\boldsymbol{h}$. As an example, we design an 8-channel cosine-modulated filter bank with length 80. The frequency responses of the analysis filters are shown above.

Final Design Example: *M* Channels with Linear Phase (GenLOT)

Linear phase filter banks are not cosine-modulated. For $M = 2$, they are constructed by factorization of the Daubechies polynomial or any symmetric halfband $P(z)$. They can also be designed by optimizing the lattice coefficients in the lattice structures [NgVaid**1**]. For $M > 2$, there is no simple spectral factorization method and one has to rely on lattice structure or time-domain approaches.

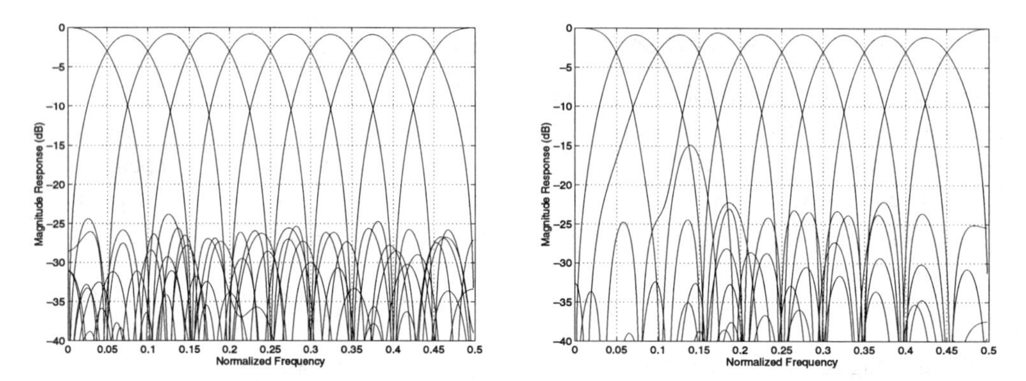

Frequency responses of a 10-channel linear-phase system with length 40.

Lattice structures exist for even M, when the length is a multiple of M (see Chapter 9 and [Soman, deQueiroz]). The figure above shows the frequency responses of a 10-channel linear-phase system with length 40. These are designed by optimizing the lattice coefficients for stopband attenuation (left) and coding gain (right).

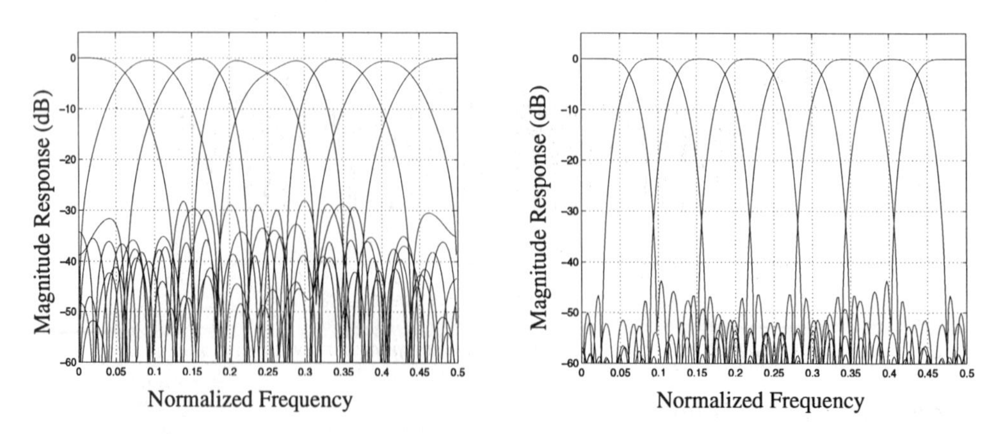

An NPR design (also linear phase) gave attenuation of 43.7dB.

One notices from those GenLOT designs that the stopband attenuation does not improve significantly as the filter length increases. One way to obtain high attenuation is to ***trade off the reconstruction error and stopband attenuation*** [Nguyen**2**]. An 8-channel PR design gave us attenuation of 28dB using the lattice structure. On the other hand, an NPR design (also linear phase) gave attenuation of 43.7dB in the last graph shown. The price was amplitude distortion of 0.0365 and aliasing of −47dB. The acceptability of a near PR design depends on the applications!

Chapter 11

Applications

11.1 Digitized Fingerprints and the FBI

The FBI has 30 million sets of fingerprints (300 million fingers) from felony arrests. I always assumed that they have my fingerprints too, because I once applied for a security clearance. (I believe it was granted. So far I have not been arrested. This is the first author, still at large.) The 40,000 sets of prints that arrive each day are divided into three groups:

5000	new prints to be saved
15000	repeat prints (recidivists) to be compared
20000	security clearances to check and return to other agencies.

Up to now, the prints are on cards in Washington. The file cabinets fill a whole floor in the FBI building. The final identification is always done by trained examiners. The "minutiae" that they watch for are ridge endings and bifurcations, which form permanently in childhood. In a British court, 12 of these indicators provide a legal match (fewer than 12 in most US courts — there are about 150 minutiae per finger). The FBI stores the cards in a special order, which begins with an automated classification using arches, loops, and whorls. Then come finer details until manual search is possible. Multiresolution is natural for fingerprints.

This data has to be digitized, and transmitted more quickly. The prints will be captured by electronic "live scans" instead of ink. At present most cards are simply mailed to the FBI. The criminal has gone from the booking station long before the fingerprints reach Washington. The new turnaround time, for digital information sent to the West Virginia office, is to be two hours (if requested).

At present, the rules do not allow the FBI to keep fingerprint files for juveniles. That group is responsible for a large fraction of breaking and entering felonies. They leave fingerprints all over the place! Some states do maintain files on juveniles, and it seems likely that changes in technology will bring changes in the law. The digital revolution is reaching criminology.

Our focus is on the specifications for the compression of gray scale images (256 levels of gray). The FBI chose a *wavelet/scalar quantization* (WSQ) algorithm. It was initially expected that the JPEG standard would win, but at compression of 15:1 and 20:1 the blocking from the DCT was severe. Ridges that are separated in the true image were found to merge during compression. This is unacceptable. It did not happen for wavelets. The linear phase 9/7 filter bank appeared just in time to be compared with the Daubechies 8/8 bank, and 9/7 was better for two main reasons:

1. Symmetry (and symmetric extension at the boundaries)
2. The image contents do not shift between the subbands.

A ridge has the same position within each subband, when filters have zero phase. The FBI confirmed that symmetric extension is better than circular convolution.

At 500 pixels per inch, ridges typically repeat every 10 to 16 pixels. The dominant frequency band is $\omega = \pi/8$ to $\omega = \pi/4$. The subband tree (Figure 11.1) goes down four levels in this range. All WSQ encoders will use the same tree. Symmetric filters up to 31 or 32 taps are permitted—and the 9/7 pair is recommended.

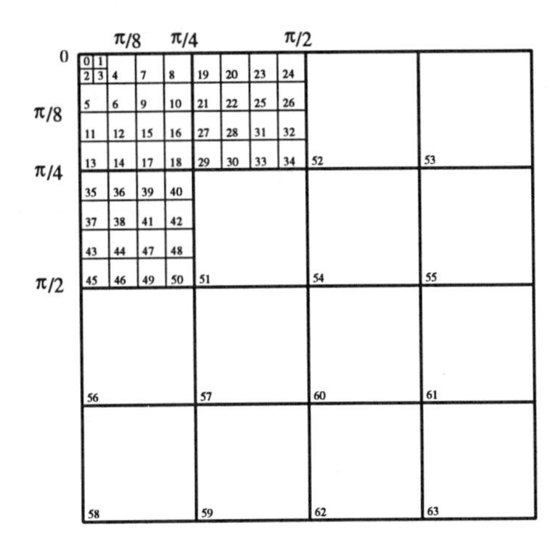

Figure 11.1: The WSQ subbands.

Each of the 64 subbands has its own uniform scalar quantization. The coefficient is set to $p = 0$ if it falls between $-Z/2$ and $Z/2$. A high percentage of the compression comes from this zeroth bin. The compressed image has strings of zeros, and the entropy coding step (Huffman coding) transmits a long string by giving its length. The other bin widths Q are smaller (about $.8Z$). Then the coefficient $a > 0$ falls into the pth bin in Figure 11.2 if $\frac{1}{2}Z + (p-1)Q \le a < \frac{1}{2}Z + pQ$.

Thus a is coded by the small integer p. The coding rule is similar for $a < 0$. The *decoder* (inverse direction) assigns to p the number a at the midpoint of the bin. The decoder also has an option to shift away from the bin center.

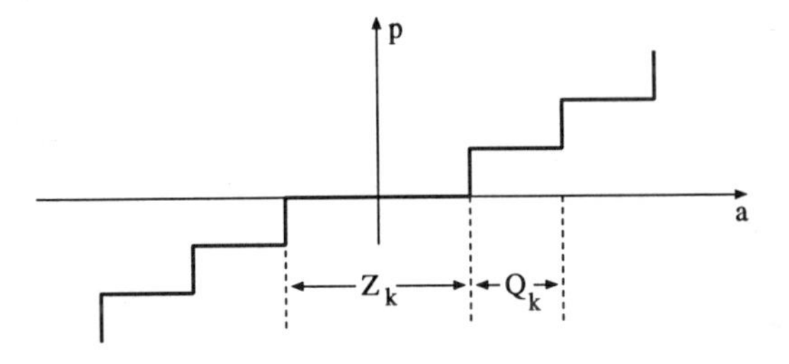

Figure 11.2: The quantization map from real numbers a to integers p.

The key question is the choice of Z and Q for each subband. This is the problem of *bit allo-*

cation in Section 11.2. The FBI uses an empirical relationship $Q_k = qC_k/\log \sigma_k$ controlled by the variance σ_k of the kth subband. Flat signals, with small variance, contain little information. The bin widths are correspondingly larger. The constant C_k can give extra emphasis to subbands that happen to be at the scale of ridges and pores. Then the constant q controls the overall compression. In the FBI system, q is partly determined by the size of the blank background that surrounds the print. The ten megabytes of data per card are compressed by 15 :1.

Theoretically, all bands should be at the same stage on their rate-distortion curves. The difficulty is to measure the distortion that our eyes actually perceive. It is not the L^2 norm! Energy is a convenient measure of error, but it does not agree well with human perception.

Experiments indicate that the improvement over the JPEG algorithm (the DCT in 8×8 blocks) ***increases with the compression ratio***. At high compression, the blocking effects in the JPEG output overwhelm the signal. It becomes impossible to follow the ridge lines, and the FBI examiners frankly said *no*. One effort of Tom Hopper, who is leading the move to digital, is to convert all the groups who deal with fingerprints — including state and local police and the registry of motor vehicles. They have to believe in the investment, and the quality of the compressed signal is what they go by. Then they build the systems.

Entropy Coding

A set of 254 symbols captures bin values p from -73 to 74 and the lengths of zero runs up to 100. A few escape symbols account for all remaining possibilities. In a highpass channel, the symbols $p = \pm 1$ might occur with frequency up to 25% each, and $p = \pm 2$ up to 5% each. Zero runs might account for 12% (length 1) and 6% (length 2) and 3% (length 3). The lowpass channels, with smaller bin widths, have more information in higher values of p. A block of subbands will use the same Huffman table to encode the quantized values. This yields the compressed image data in the figure below. The decoder reverses the entropy coding, then the quantization, and finally the wavelet transform.

Original Image

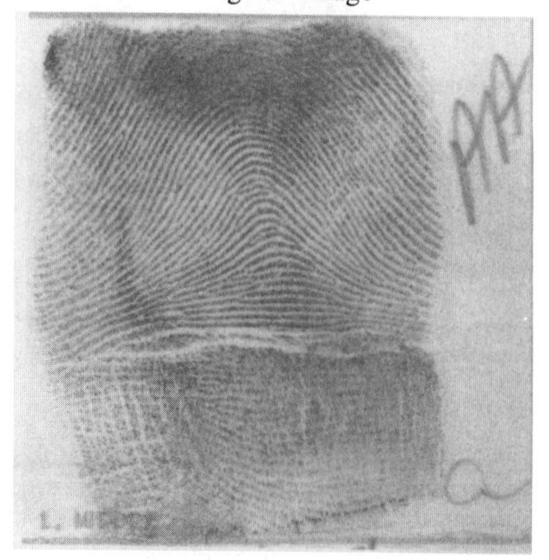

Reconstructed Image at 0.25 bpp

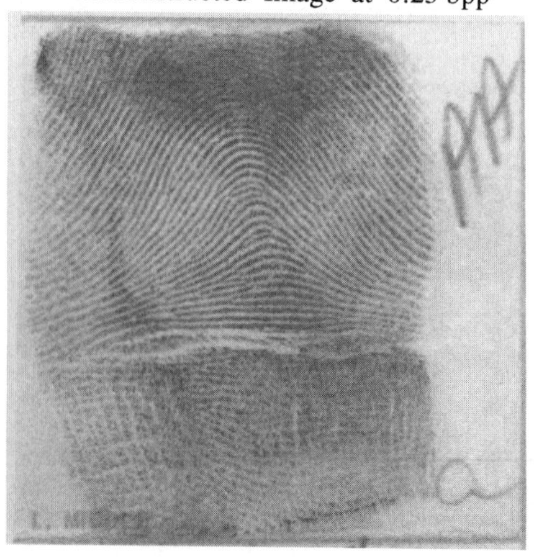

11.2 Image and Video Compression

Image Compression

"A picture is worth a thousand words." This English aphorism reminds us of the importance of images. It is especially true in the age of information highway and multimedia. Computers, fax machines, videophones, teleconferencing systems and storage devices impact our workplace. Text, data, sound, image and video clip are grouped together to send over data networks or to store. The amount of data is astronomical. Compression increases the throughput of the network and the capacity of the storage device. For satellite transmission, compression greatly reduces cost.

A 24-bit color picture with 256 by 256 pixels needs more than 0.2 MByte of storage. A high density diskette with capacity of 1.4 MB can store about 7 pictures. If the picture can be compressed by 50 to 1 without any perceptual distortion, the capacity of the diskette increases to 350 pictures. This is significant. The key point here is the notion of *perceptual* lossless compression. A good coding algorithm should study and incorporate the human visual system to exploit redundancy in the image.

There are many techniques for image coding. Subband coding is today the most successful. Pyramid coding was and is effective for high bit-rate compression. Transform coding based on the Discrete Cosine Transform became popular in the 1980s because of low complexity and effective bit allocation. This became the JPEG standard in image coding. For gray-scale images, it performs well for compression ratios up to 16 to 1. At 24 to 1 compression, JPEG's synthesized image suffers from blocking, which manifests the short basis functions used in reconstruction.

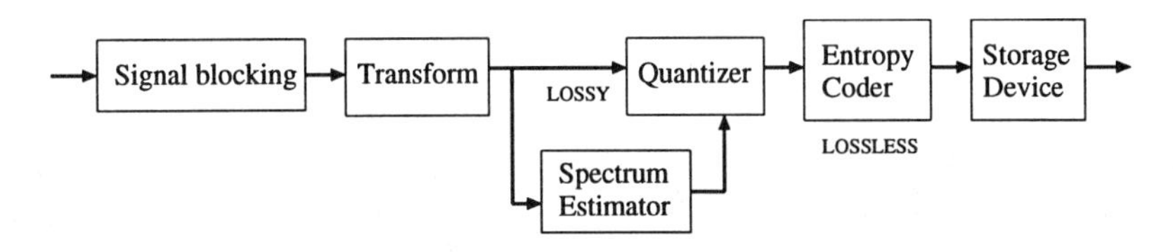

Figure 11.3: The steps of a transform-based image coder.

Subband coding using wavelets (this means tree-structured filter banks) avoids blocking at medium bit-rate, because its basis functions have variable length. Long basis functions represent flat background (low frequency). Short basis functions represent regions with texture. At a reasonable distance, one can not detect the errors easily. At low bit-rate, wavelet coding suffers from *ringing* when high frequencies (textures) are deleted. The ringing artifacts are significant around edges with high intensity. They have the shape of the basis functions that are emphasized in synthesis.

Part of the "Barbara" image and its Discrete Cosine Transform are in Figure 11.4. The image is blocked (8 by 8) and then transformed. The transformed subimage also has size 8 by 8. The intensities of the first few coefficients (the upper left corner of the transform) are the largest. The DCT preserves energy and the essential information is in those few coefficients. Quantization assigns more bits to these pixels. The objective of bit allocation is to minimize the distortion. The quantized subbands are then scanned and coded using lossless compression. This *entropy coder* watches for runs of zeros and transmits their length (roughly 3 : 1 compression for free).

Similar procedures are used in the wavelet-based transform coder. Figure 11.5 shows the

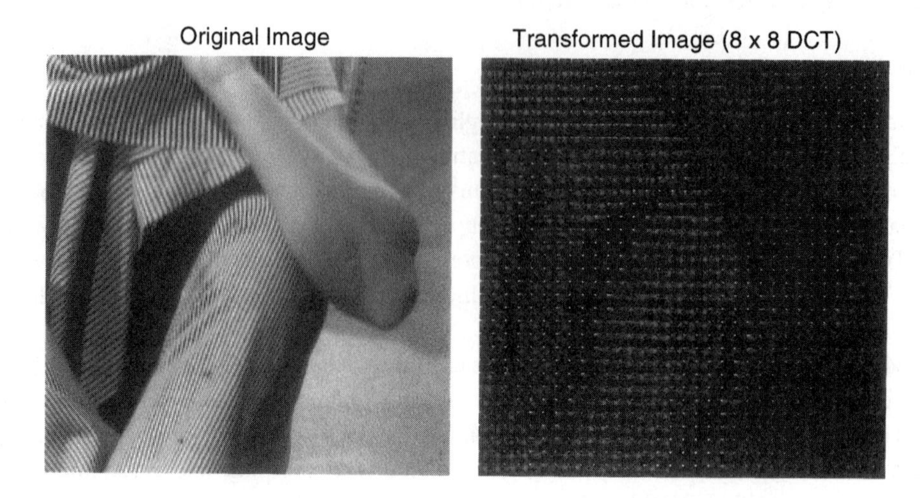

Figure 11.4: The discrete cosine transform shows most energy in the DC coefficients (bright points).

transform using a two-channel filter bank. The upper left subimage is obtained by lowpass filtering in both the horizontal and vertical directions, indicated by *LL*. The other three subimages (*much lower intensity*) have details involving high frequencies. The bit allocation algorithm will assign many bits to *LL* and few bits to *HH*. The normal number of iterations on the *LL* subimages is 4 or 5, as shown on the right side of Figure 11.5.

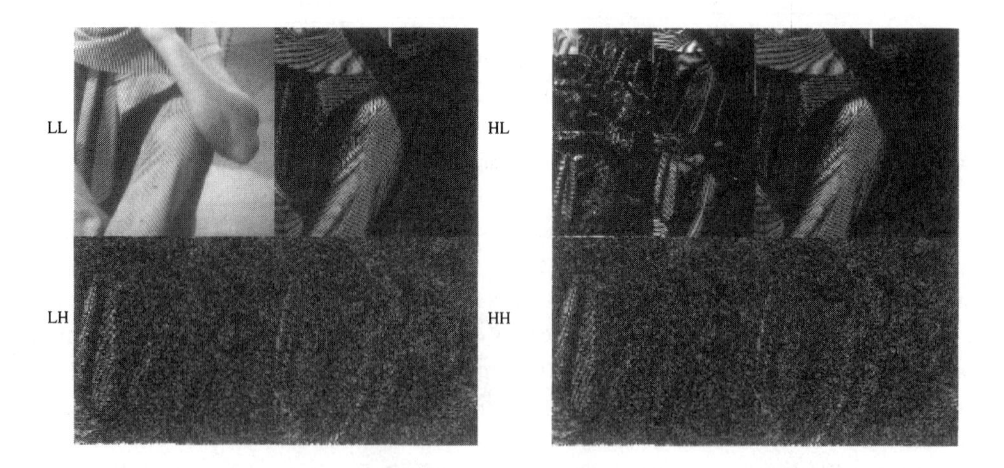

Figure 11.5: The discrete wavelet transform of "Barbara" (one level and four levels).

The subsections below discuss these building blocks in detail. The organization is:

- Choice of transformations and their effects on image coding.
- Bit allocation algorithm and quantization.
- Entropy coding algorithms.

- Error measures.
- Comparison of block transforms and wavelets.

Choice of Transformations and Their Effects on Image Coding

Consider the block transform coder in Figure 11.6. The input blocks are

$$v(n) = \left[\begin{array}{cccc} x(nM) & x(nM-1) & \cdots & x(nM-N+1) \end{array}\right]^T.$$

Each block is transformed to length M by $y(n) = P^T\, v(n)$. The figure shows a uniform filter bank with M channels. The kth row of P^T contains the filter coefficients $h_k(n)$. Examples are the DCT ($N = M$), the Lapped Orthogonal Transform ($N = 2M$), and the Generalized LOT ($N = LM$).

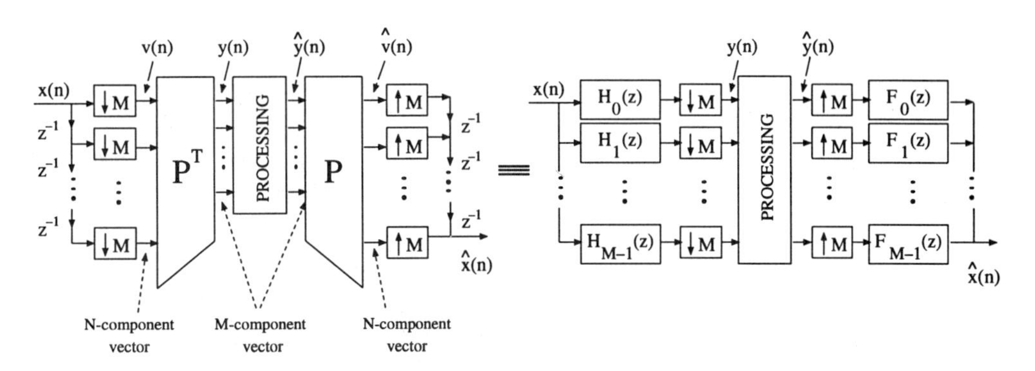

Figure 11.6: Block transform and the equivalent filter bank. The kth row of $\mathbf{P}^T$ contains $\mathbf{h}_k(n)$.

The subband signal $y(n)$ is quantized, entropy coded, and stored. In synthesis, the stored signal is decoded (to produce $\widehat{y}(n)$) and then transformed by the matrix P. In the absence of quantization, $P = \left[P_0\, P_1\, \cdots\, P_{L-1}\right]$ gives an orthogonal transform:

$$\sum_{m=0}^{L-1-\ell} P_m\, P_{m+\ell}^T = \delta(\ell)\, I; \quad 0 \le \ell \le L-1. \tag{11.1}$$

The synthesis filters are time-reversed versions of the analysis filters. The reconstructed signal $\widehat{x}(n)$ is a linear combination of the basis vectors. What properties of those vectors (columns of P) will produce a good perceptual image coder?

- *They should be smooth and symmetric (or antisymmetric).* Smoothness controls the noise in a region with constant background. Symmetry allows the use of symmetric extension to process the image's borders.

- *They should decay to zero smoothly at both ends.* Figure 11.7(a) shows the basis functions of the DCT. Note that they do not decay to zero. This creates discontinuity between blocks (subimages) when the image is compressed. This *blocking artifact* can be seen from the reconstructed image in Figure 11.7(b). The compression rate is 32 to 1 (0.25 bpp).

- *The bandpass and highpass filters should have no DC leakage.* Higher frequency bands will be quantized severely. It is desirable for the lowpass band to contain all of the DC information. Otherwise, if the bandpass and highpass responses to $\omega = 0$ are not zero, we see the *checkerboard artifact* in Figure 11.8(b).

  An equivalent form of the DC leakage condition can be derived for the lowpass filter: $H_0(2k\pi/M) = \delta(k)$ for $k < M$. These ω_k are the mirror frequencies.

- *The basis functions should be chosen to maximize coding gain.*

- *Their lengths should be reasonably short to avoid excessive ringing and reasonably long to avoid blocking.* Figure 11.9 shows transforms of length 8 (DCT) and 48 (GenLOT) used in 32 :1 compression. The first transform has blocking and the second transform has ringing. Intuitively, one would like to represent texture by short functions and background by long functions. This is offered by wavelets.

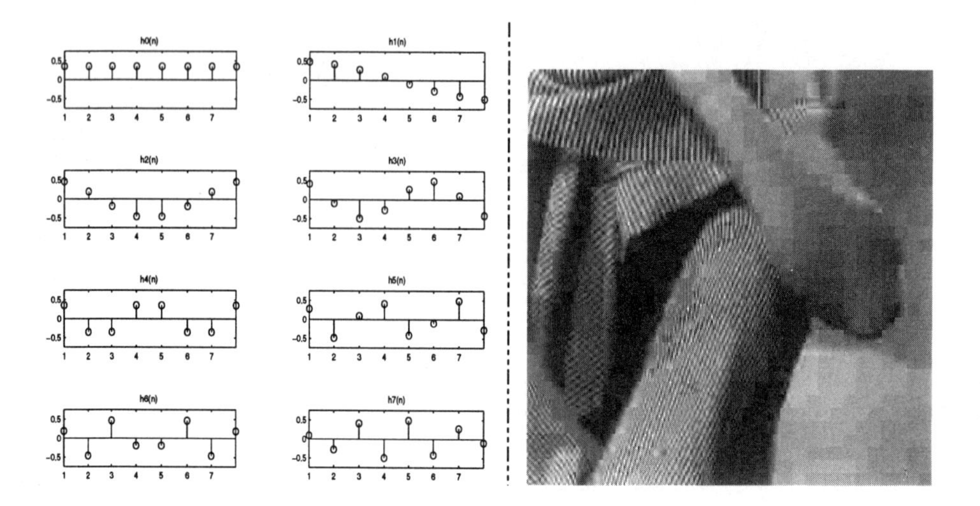

Figure 11.7: *Blocking* at 32 :1 comes from discontinuities of the 8 DCT basis functions at edges.

- *In the frequency range $|\omega| \leq \frac{\pi}{M}$, the bandpass and highpass responses should be small.* This minimizes the quantization effect on bandpass and highpass.

Wavelet-based transform One attractive property of wavelets is their ability to adjust the lengths of basis functions. A four-level wavelet decomposition and its equivalent nonuniform filter bank are in Figure 11.10. The low frequency basis function is a cascade of interpolated versions of the lowpass filter $\boldsymbol{H}_0$. Its effective length is large. Higher frequencies are iterated less; the basis functions become shorter. The signal is approximated by a few basis functions. After four levels in Figure 11.5, most of the energy is in the lowpass subband. This upper left subimage is a coarse approximation of the original. The other bands add details. *Bit allocation becomes crucial.* Clearly, subimages with low energy levels should have fewer bits.

We comment further on desirable properties of a two-channel filter bank, now emphasizing what becomes important with *iteration.*

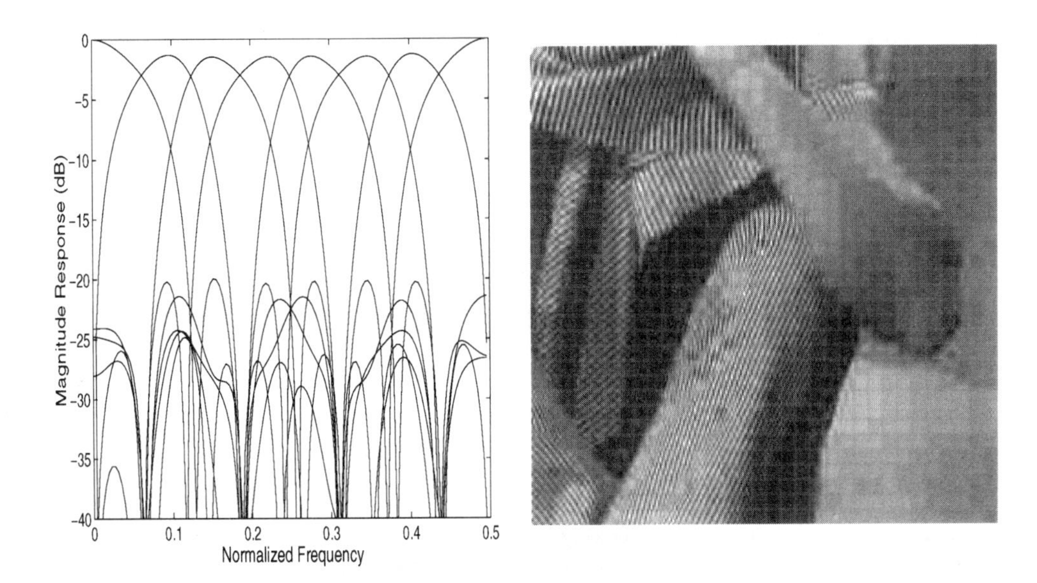

Figure 11.8: Frequency responses of GenLOT with $M = 8$ and $N = 24$. The bandpass responses at $\omega = 0$ (DC) are not zero. The DC component leaks over to produce *checkerboarding*.

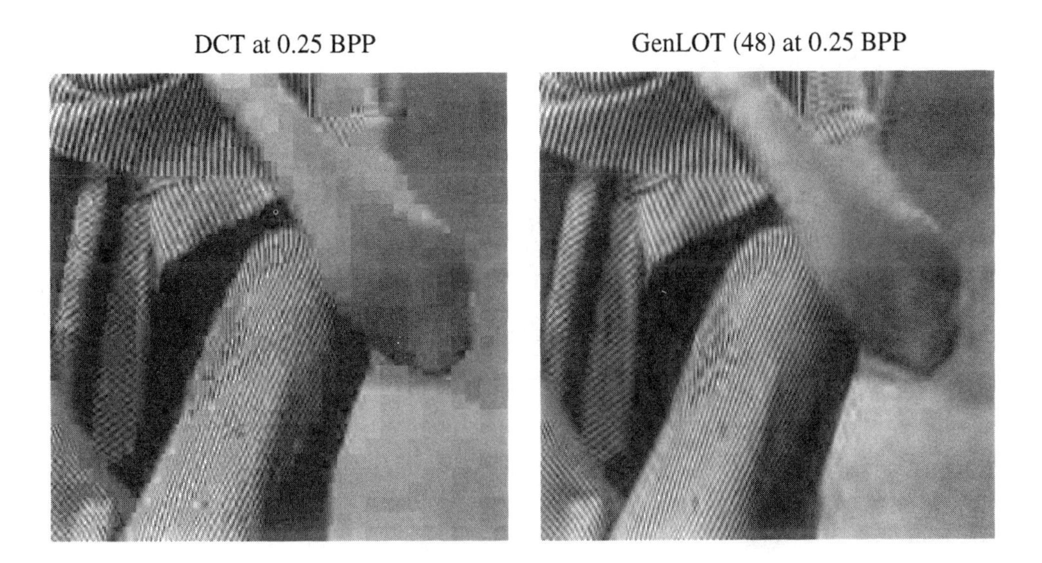

Figure 11.9: Short DCT basis vectors produce blocking. Long GenLOT vectors produce ringing.

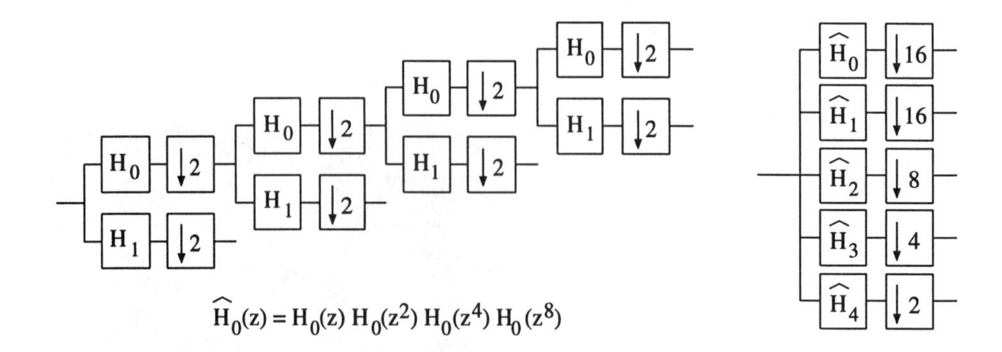

Figure 11.10: A four-level discrete wavelet transform and its equivalent nonuniform filter bank.

- *The synthesis scaling function should be smooth and symmetric.* Figure 11.11 shows several sawtooth subsections, plus two large jumps. Those jumps in the synthesis scaling function produce "blocking-like artifacts" in the reconstructed image.

- *The highpass analysis filter should have no DC leakage.* When output from this filter is quantized, we do not want to affect the DC component of the signal. It is desirable that the lowpass subband contains all the DC energy. Figure 11.12 shows the filter banks, their scaling functions, and the reconstructed image (at 1 bpp). Note the checkerboard artifact and the nonzero responses at $\omega = \pi$. Wavelet theory requires the lowpass filters to have at least one zero. This is also beneficial for image coding.

- *The analysis filters should be chosen to maximize coding gain.*

- *$F_0(z)$ should be long and $F_1(z)$ should be short.* A long $F_0(z)$ will help the coder to represent flat regions. A short $F_1(z)$ will minimize ringing due to quantization. Figure 11.13(a) shows blocking at 0.25 bpp from the short Haar filter. Figure 11.13(b) shows ringing for a longer 9-tap $F_1(z)$.

- *$H_1(z)$ should have good stopband attenuation to minimize the leakage of quantization noise into low frequencies.*

- How does one factor a given halfband $P(z)$ into $H_0(z)\ F_0(z)$? One should choose $H_0(z)$ short (and therefore $F_1(z)$ short) and $F_0(z)$ long. $H_0(z)$ should have at least one zero at $\omega = \pi$. $F_0(z)$ should have many zeros at π to obtain a smooth scaling function. An example is the filter bank with length (2, 14) (factors of the maxflat filter). The total of eight zeros at π is the same as the (9, 7) FBI filter bank. Figure 11.14 shows the reconstucted images at 0.5 bpp using the (2, 14) and FBI systems. The PSNR is 24.68 dB and 25.30 dB for (2, 14) and (9, 7). The image qualities are comparable, although (2, 14) has smaller coding gain (9.45 dB) compared to 9.787 dB of the FBI system.

Bit Allocation and Quantization

After transformation the subband signals are quantized, entropy coded and stored (or transmitted). Figure 11.15 shows typical distributions for the subband signals. For bandpass and highpass subbands, zero-mean Gaussian is a good approximation. The lowpass subband is image-dependent and its distribution is roughly uniform. *Given M subimages and a fixed bit rate R,*

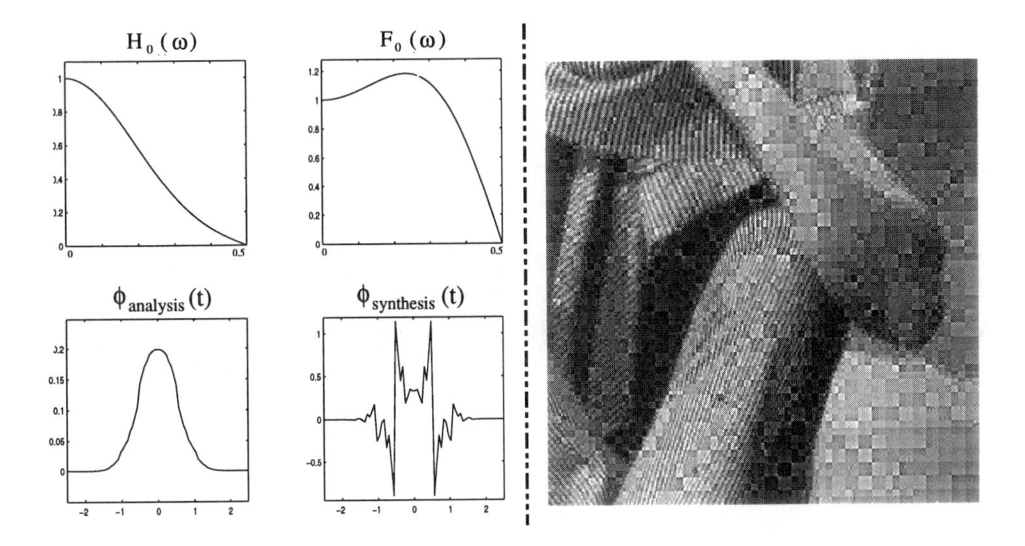

Figure 11.11: Frequency responses and scaling functions. The synthesis scaling function is rough and has large jumps. This produces blocking-like artifacts.

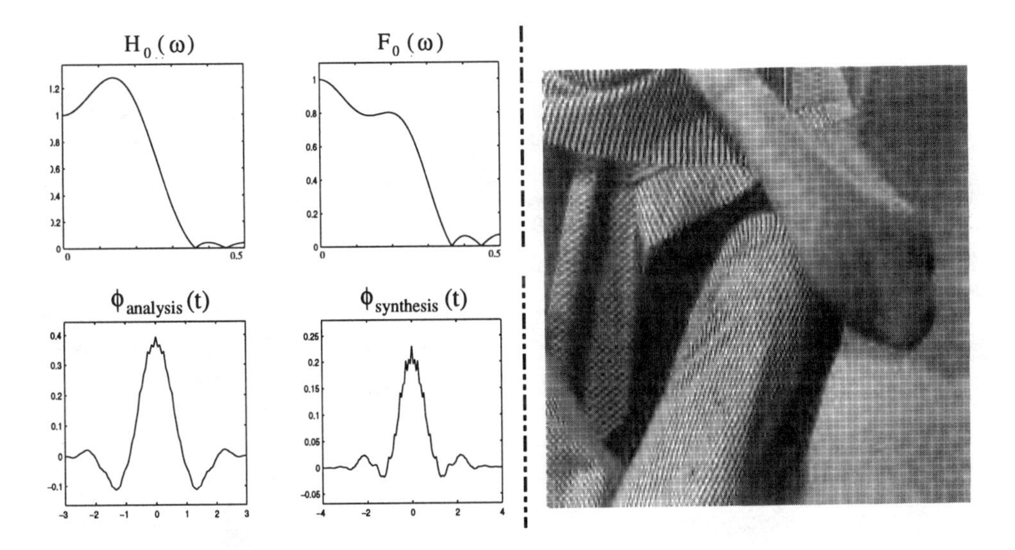

Figure 11.12: Frequency responses and scaling functions of a filter bank. The lowpass responses are not zero at $\omega = \pi$. This DC leakage yields a checkerboard in the reconstructed image.

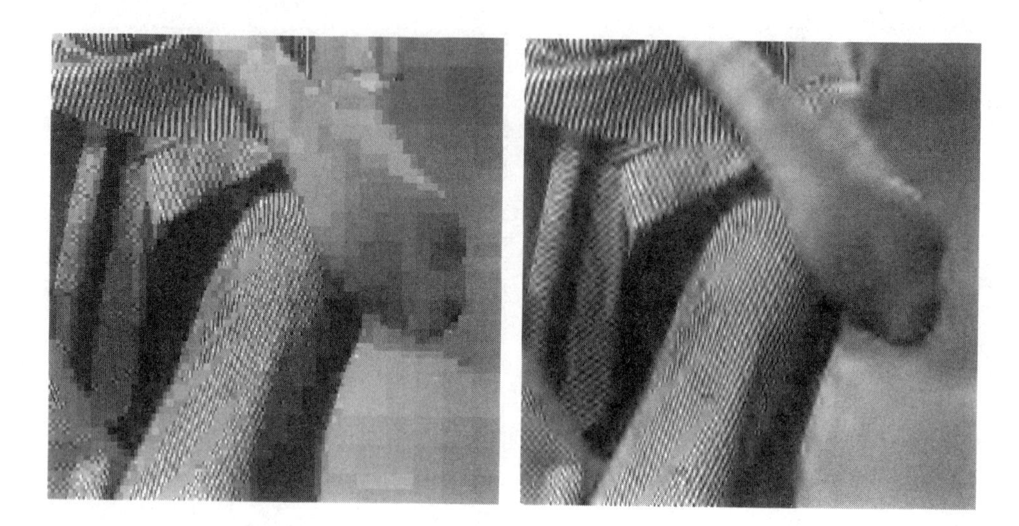

Figure 11.13: Blocking from a short synthesis lowpass filter. Ringing from a long synthesis highpass filter.

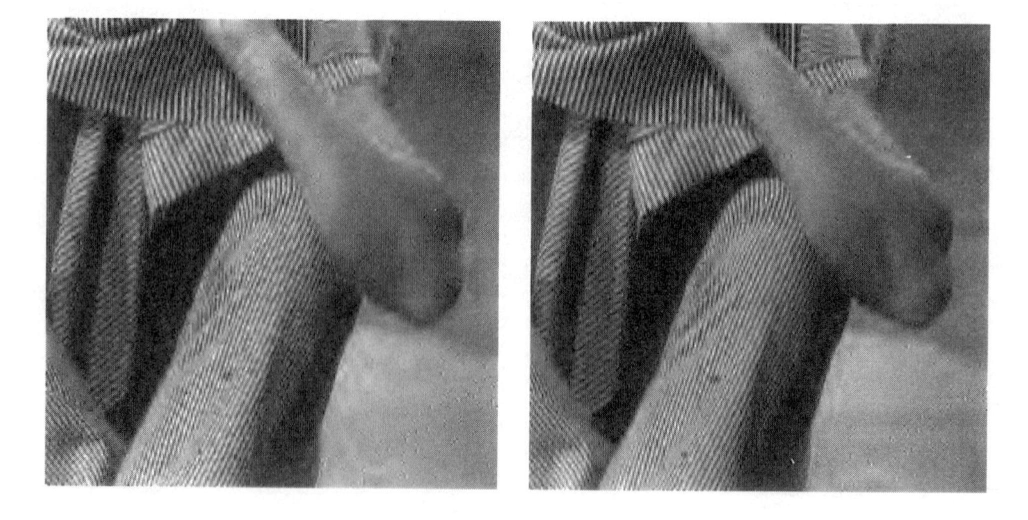

Figure 11.14: Image compression at 0.5 bpp using (2, 14) and (9, 7) filter lengths.

how does one assign bits to the subimages? The bit allocation algorithm needs a cost function D. Examples of D are distortion, energy and entropy. We discuss the minimization of distortion for a uniform quantizer with variable bit length next.

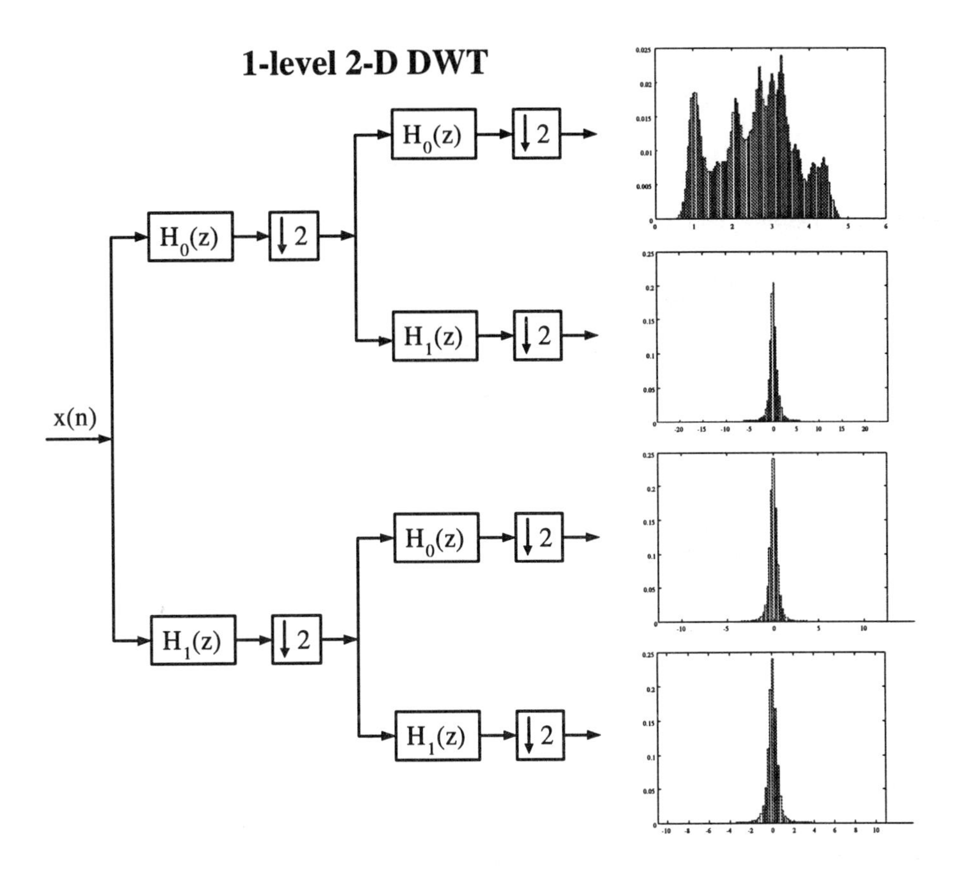

Figure 11.15: Typical distributions of subband coefficients.

Consider the 4-level wavelet decomposition in Figure 11.5. The original image size is 256 by 256 with 8 bits per pixel, for a total of 524,288 bits. How does one assign 16,384 bits to the subbands for a 32 to 1 compression? The numbers of pixels in the subbands go from 16 by 16 up to 128 by 128. Assigning one extra bit to the coefficients in the small subband does not increase the total number of bits as much as assigning the extra bit to the large subbands. Let

- N be the total number of bits in the original image
- M be the number of subbands ($M = 13$ for four levels)
- $\alpha_k = N_k/N$ be the relative subband size
- $\boldsymbol{b} = (b_1, \ldots, b_M)$ contain the bit lengths allocated to the subbands
- ϵ_k be the quantizer performance index
- σ_k^2 be the subband variance
- w_k be the weighting factor (for perceptual coding).

Define the quantization error and the total bit rate as:

$$D(\boldsymbol{b}) = \sum_{k=1}^{M} \alpha_k w_k \epsilon_k^2 \, 2^{-2b_k} \, \sigma_k^2 \quad \text{and} \quad R(\boldsymbol{b}) = \sum_{k=1}^{M} \alpha_k \, b_k.$$

The bit allocation problem is to minimize $D(\boldsymbol{b})$ subject to $R(\boldsymbol{b}) = R_c$ = fixed bit rate. This can be solved using a Lagrange multiplier λ:

$$\min \ \{D(\boldsymbol{b}) + \lambda R(\boldsymbol{b})\} = \min \sum \alpha_k \left(w_k \epsilon_k^2 \, 2^{-2b_k} \, \sigma_k^2 + \lambda \, b_k \right). \tag{11.2}$$

Assume that all $\epsilon_k = \epsilon$, and set $\widetilde{\lambda} = \lambda/\epsilon^2$. Then (11.2) reduces to

$$\min \ \left\{ w_k \, 2^{-2b_k} \, \sigma_k^2 + \widetilde{\lambda} \, b_k \right\} \quad \forall k. \tag{11.3}$$

Differentiate (11.3) with respect to b_k to obtain a closed form solution:

$$b_k = \tfrac{1}{2} \log_2 \left(\frac{(2 \log_e 2) w_k \sigma_k^2}{\widetilde{\lambda}} \right). \tag{11.4}$$

Then the bit-rate constraint $R(\boldsymbol{b}) = R_c$ becomes

$$\sum_{k=1}^{M} \alpha_k b_k = \tfrac{1}{2} \sum_{k=1}^{M} \alpha_k \log_2 \left(\frac{(2 \log_e 2) w_k \, \sigma_k^2}{\widetilde{\lambda}} \right) = R_c.$$

This yields the multiplier $\widetilde{\lambda} = \lambda/\epsilon^2$:

$$\widetilde{\lambda} = 2^{\left\{ \sum_{k=1}^{M} \alpha_k \log_2((2 \log_e 2) w_k \, \sigma_k^2) - 2R_c \right\}} \tag{11.5}$$

In summary, the average subband variances σ_k^2 lead to $\widetilde{\lambda}$, subject to the constraint. Then the allocated bit length b_k is computed from (11.4). When R_c is too small, the resulting bit length can be negative for subbands with very small signal variances. Since one can not assign negative bits, the bit allocation algorithm is restarted with these subbands removed (0 bits). The iterative algorithm is:

1. Find bit lengths b_k and set all negative b_k to zero.
2. Reduce the number of subbands appropriately, $M_{new} = M_{old} - M_{zero}$.
3. Adjust $R_c = R_{c,new}$ and repeat step 1. Stop if all $b_k \geq 0$.

For a linear-phase filter bank without orthogonality, the signal energy is not preserved. Quantization noise introduced in the subbands is amplified by the synthesis bank. Assuming that the quantization errors are uncorrelated and defining $B_k = \alpha_k \sum_n f_k^2(n)$, the reconstruction error variance is $\sigma^2 = \sum_{k=0}^{M-1} B_k \, \sigma_k^2$. B_k normally ranges from $0.9\alpha_k$ to $1.1\alpha_k$ for practical systems, and it is reasonable to assume that $B_k = 1$ (as for an orthogonal filter bank).

From b_k and σ_k, we compute the step size Δ_k for the quantizer. The maximum allowable quantization error for the kth subband is $T_k = c_k \, \sigma_k / 2^{b_k}$, where $c_k = 8$ is a reasonable choice. The step size Δ_k is normally chosen to be

$$\Delta_k = \max \, (2T_k, \Delta_{k,min}) \ \text{for} \ b_k > 0 \quad \text{and} \quad \Delta_k = 2X_{k,max} + \epsilon \ \text{for} \ b_k = 0. \tag{11.6}$$

Here $X_{k,max}$ is the largest nonzero value, $b_{k,max}$ is the maximum number of bits for coding a nonzero normalized coefficient, and $\Delta_{k,min}$ is the minimum step size to guarantee that $b_{k,max}$ is not overridden for coding $X_{k,max}$. The Huffman table of the sequential baseline coder has a limited size and should be kept small. The table is also quantized and transmitted. Therefore, Δ_k must be greater than or equal to the smallest δ_k that is not quantized to zero. Summarizing,

$$\Delta_{k,min} = \max \left\{ \frac{X_{k,max}}{2^{b_{k,max}} - 1}, \quad \delta_k \right\}; \qquad \epsilon \geq \delta_k.$$

The exception for the case where $b_k = 0$ in (11.6) assures that all coefficients are quantized to zero if no bit is assigned. $\epsilon \geq \delta_k$ is necessary to keep all normalized coefficients $\tilde{y}_k =$ round (y_k/Δ_k) smaller than 0.5.

The lowpass channel variance strongly depends on the signal. Its distribution is approximately uniform. Δ_1 computed from (11.6) would be too large since the uniform quantizer is more effective for uniform distribution (rather than Gaussian). This error is equalized appropriately by scaling the lowpass subband variance by $\tilde{\sigma}_1^2 = c_1 \, \sigma_1^2$ before bit allocation. c_1 is determined experimentally and is different for every signal. It is about 1.5 for the image Lenna.

Perceptual weighting for wavelet-based image coding It is well known that minimizing the squared error does not guarantee optimal results in the perceptual sense. At medium and low bit-rates, human eyes are less sensitive to loss of high frequencies. One needs to weight the quantization noise in the subband using the sensitivity of the human eye. But perceptual quality is complicated. We summarize here a simple and effective weighting scheme for the estimated quantization noise before bit allocation. More bits are allocated to low and medium bands, and high frequency noise is increased. The figure below shows the weight factors for three levels with $a > 1$ and the improvement for $a = 2$.

a^6 a^5 a^5 a^4 a^3 a^3 a^2 a^1 a^1 a^0

(9,7) system at 0.125 bpp, a=1

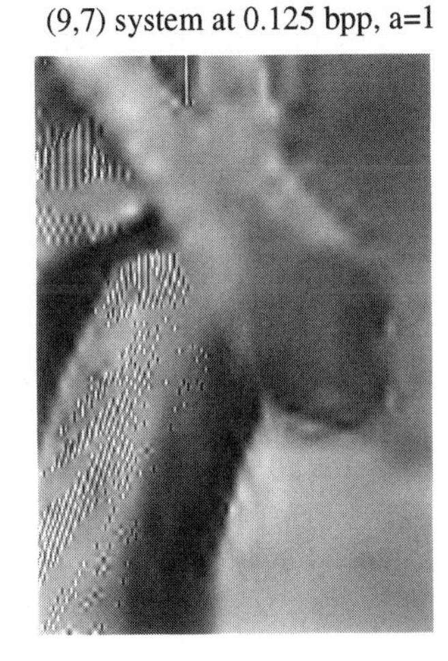

(9,7) system at 0.125 bpp, a=2

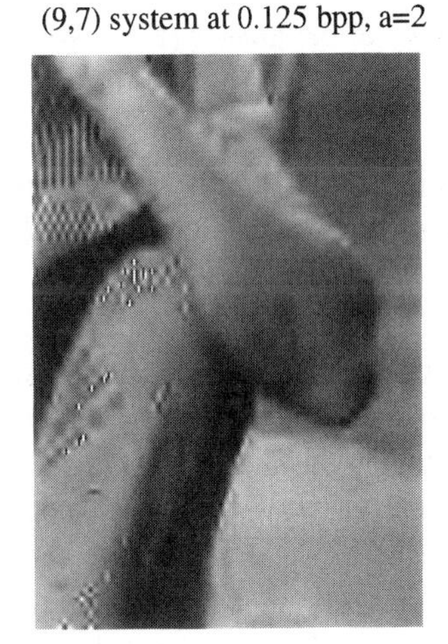

The choice of weight for an M-channel GenLOT is $\boldsymbol{w} = \begin{bmatrix} a^{M-1} & \cdots & a & 1 \end{bmatrix}^T$. Extending to 2-D yields $\boldsymbol{W} = \boldsymbol{w}\,\boldsymbol{w}^T$. Here are the reconstructed images using GenLOT of length 48, with $a = 1$ and $a = 2$. Note again the improvement with $a = 2$.

GenLOT(48) at 0.125 bpp, a=1

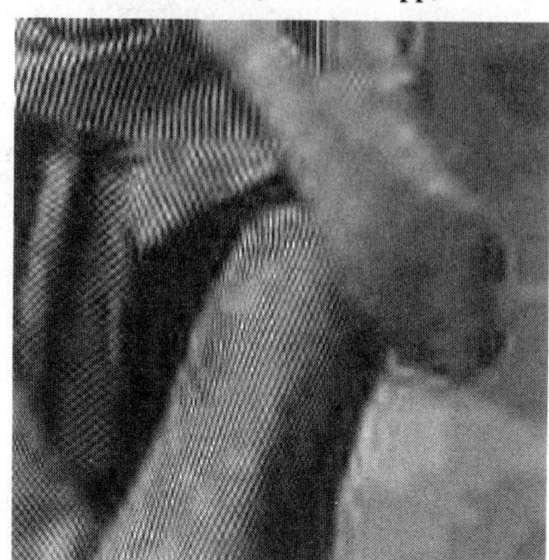

GenLOT (48) at 0.125 bpp, a=2

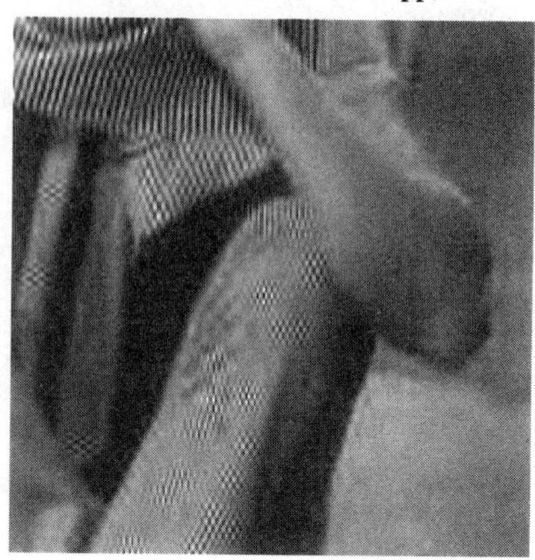

Since PSNR is maximal for $a = 1$, this weighting is a tradeoff between perceptual quality and error measure of the reconstructed image. At lower rates, perceptual quality is more important. The weighting scheme is simple, efficient, and image independent. It does not guarantee perceptually optimal results! A reliable measure for perceptual quality (and the best weight) is an ongoing research problem.

Entropy Coding

After bit allocation and quantization, we have subimages with discrete levels represented by integers. How do we store or transmit these subimages? Many highpass coefficients are zero after quantization. These coefficients should be grouped so that the entropy coder can take full advantage of long strings of zeros. This is accomplished by *scanning*.

Run-length coding or Huffman coding or a combination should be used to reduce the redundancy of the images [PM]. We will discuss the baseline entropy coder which is a combination. JPEG also uses the baseline coding method.

Scanning of the discrete wavelet coefficients To demonstrate scanning, consider the three-level wavelet transform in Figure 11.16. Subbands 2, 5 and 8 are highly correlated since 2 is the coarse approximation of 5 and 5 is the coarse approximation of 8. Suppose the pixel at the upper left corner of subband 2 is zero. *Then it is very likely that the pixels in a* 2×2 *shaded square of subband* 5 *are zero.* Similarly, the pixels in a 4×4 shaded square of subband 8 are probably zero. One can group these pixels into an "AC sequence" of length 21 ($= 1+4+16$) by vertically scanning the shaded squares. Figure 11.16 also shows the scanning patterns for subbands 3, 6 and 9 (horizontal) and for subbands 4, 7 and 10 (diagonal). When the original image has size 32, the 16 pixels each in subbands 2, 3, 4 give 48 AC sequences. The low frequency band is scanned horizontally and grouped into the DC sequence of length 16. This scanning method is similar to the *zero-tree coder* proposed by [Shapiro].

Scanning of the block transform Consider an image of size 32×32 transforms to 16 blocks of size 8×8, using an 8-channel GenLOT. The quantized coefficients are scanned and entropy-coded. The ℓ coefficients in block k are correlated with the ℓ coefficients in blocks $k \pm m$. There-

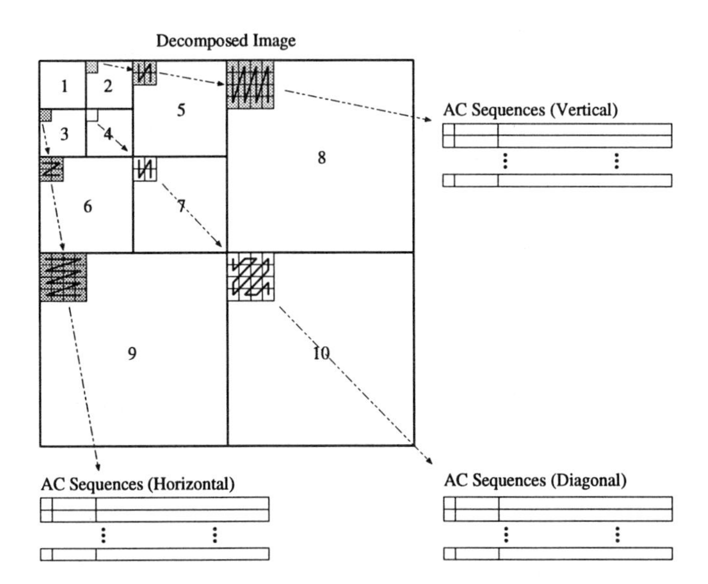

Figure 11.16: Scanning method used in the Discrete Wavelet Decomposition.

fore one should scan them in a zig-zag pattern, as shown in the figure. These scanned coefficients are grouped to sequences of length 16. There are 64 such sequences.

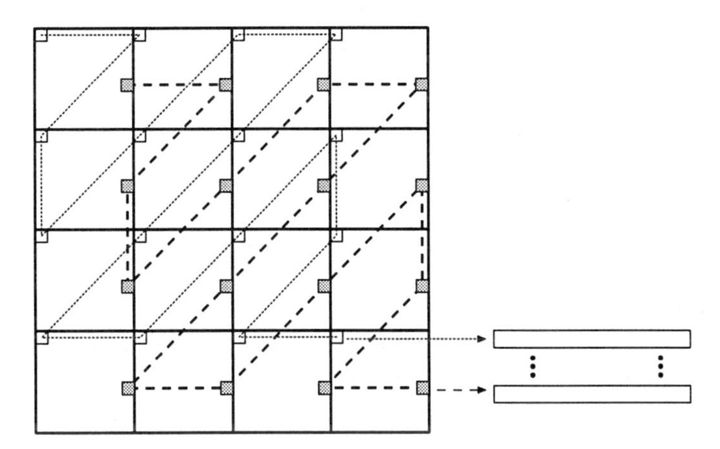

Figure 11.17: Scanning method used in the block transform.

Sequential Baseline Coding After scanning the quantized subimages, we have a set of DC and AC sequences to be stored. The baseline coder takes advantage of the correlation in the AC sequences. This algorithm combines Sequential Baseline Coding and Huffman entropy coding. The basic principle is similar to the JPEG coder, but does not restrict to the DCT.

Coding of the AC sequence: The sequence 9 0 0 2 0 1 0 0 0 -3 0 0 3 0 -1 -1 has strings of zeros interlacing with nonzeros. An efficient representation remembers the number of zeros before each nonzero. ***Symbol-1*** is the pair (*Runlength, Size*) where *Runlength* specifies the number of preceding zeros. *Size* determines the number of bits to encode the current nonzero. ***Symbol-2*** gives the (*Amplitude*) of the nonzero: *Size* = n corresponds to *Amplitude* less than 2^n (but not less than 2^{n-1}).

This representation of the example gives a string of ***symbol-1*** and ***symbol-2***:

$$(0, 1)4 \;\; (2, 2)2 \;\; (1, 1)1 \;\; (3, 2){-3} \;\; (2, 2)3 \;\; (1, 1){-1} \;\; (0, 1){-1} \;\; (0, 0).$$

Note the terminal ***symbol-1*** $(0, 0)$ at the end. Also $(1, 1)$ and $(0, 1)$ and $(2, 2)$ occur twice in the string. Huffman entropy coding (see Figure 11.18) can exploit this redundancy in ***symbol-1***. A simple way to code ***symbol-2*** "a" in binary is:

$$b = \begin{cases} a - 2^{Size-1} & \text{if } a > 0 \\ |a| & \text{otherwise.} \end{cases}$$

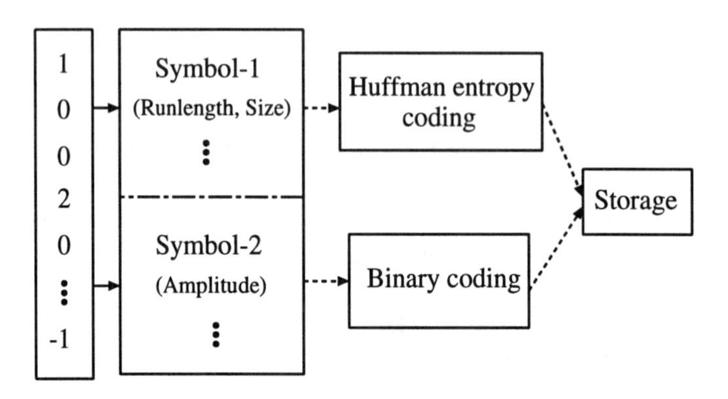

Figure 11.18: Steps in a Sequential Baseline Coder.

Coding of the DC sequence: DC coefficients measure the average energy of the input signal. There is usually a strong correlation between neighboring coefficients. For efficiency we use ***differential coding***: save the first coefficient and then the ***differences between successive coefficients***. These are coded as for AC coefficients. Since one would not expect long zero strings, *Runlength* is not used. ***Symbol-1*** only gives *Size*. An excellent source for Sequential Baseline Coder is [PM].

Three error measures are often used to compare coders and perceptual quality:

$$\begin{cases} \text{Mean Square Error} & MSE = \frac{1}{MN} \sum_{m=0}^{M-1} \sum_{n=0}^{N-1} |\boldsymbol{x}(m,n) - \widehat{\boldsymbol{x}}(m,n)|^2 \\ \text{Peak Signal Noise Ratio} & PSNR = 10 \log_{10}\left(\frac{255^2}{MSE}\right) \\ \text{Maximum Error} & MaxError = \text{Max}\, |\boldsymbol{x}(m,n) - \widehat{\boldsymbol{x}}(m,n)| \end{cases} \tag{11.7}$$

The image is $M \times N$. The MSE and PSNR are directly related, and one normally uses PSNR to measure the coder's objective performance. At high rate, images with PSNR above 32 dB are considered to be perceptually lossless. At medium and low rates, the PSNR does not agree with the quality of the image.

Comparison of image coder based on block transforms and wavelets

Block Transform Image Coder: GenLOT with M *channels* Figure 11.19(a) shows the coding gain as a function of the overlapping factor N, for various values of M. Notice the large improvement from 4 channels to 8 channels. The added improvement for $M = 10$ and 16 channels is less. For fixed M, one observes a steady increase in coding gain as N increases. GenLOT with 8 channels is a good compromise between the implementation complexity and performance.

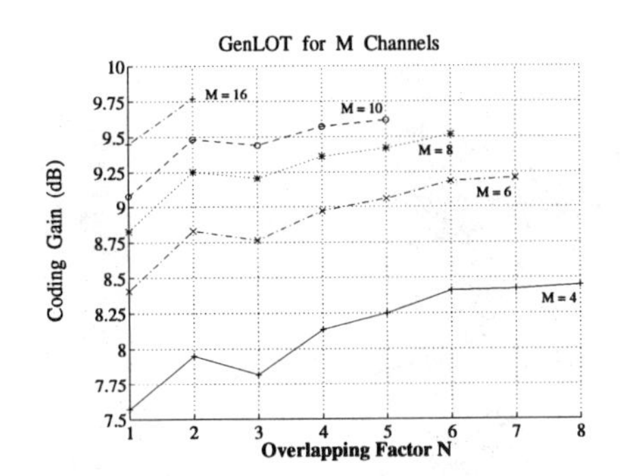

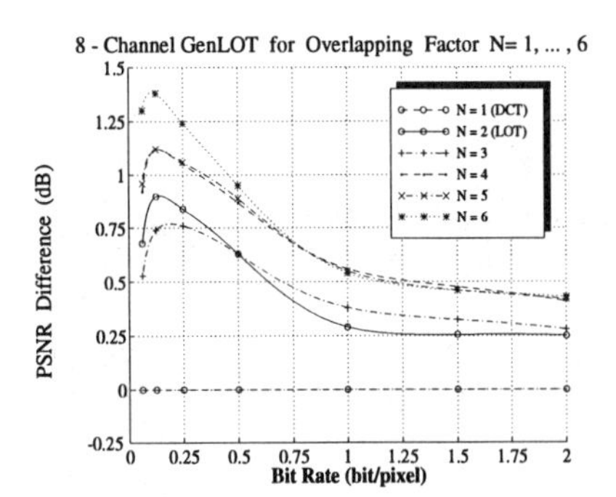

Figure 11.19: (a) Plots of coding gains as functions of overlapping factor. (b) Plots of PSNR differences for the "Lenna" image using 8-channel systems for various bit rates.

Figure 11.19(b) shows the PSNR difference between GenLOT and DCT, for several overlapping factors N. At high rate (≥ 1 bpp), the improvement is not significant. All block image coders would work well at high rate. At medium rate the improvement is larger, since DCT algorithms suffer from blocking. At low rate, blurring and ringing artifacts control the performance of the image coder.

One should also compare the *subjective* performance. Figure 11.20 shows the reconstructed images for DCT and GenLOT (length 48) at 1 bpp and 0.125 bpp. Although the PSNR improvement at 0.125 bpp is only 0.52 dB, one observes a large difference in subjective quality.

Wavelet-based Image Coder with L levels and p zeros at π All filters used in this study are factors of a halfband maximally-flat $P(z)$. One way to form symmetric $F_0(z)$ and $H_0(z)$ from $P(z)$ of length 15 is to assign four zeros at π and the four complex zeros to $F_0(z)$. Then $H_0(z)$ will have the remaining four zeros at π and the two real zeros. This yields the $(9, 7)$ system used by the FBI. To obtain filter banks with even length (Type A), move one zero at π from $H_0(z)$ to $F_0(z)$. This yields the $(10, 6)$ system.

Figure 11.21(a) shows the unified coding gain as function of the number of levels L and the p zeros at π in $P(z)$. For a given $P(z)$, we compute the unified coding gain for all linear-phase systems and plot the best value. See also [Maj**1**]. This plot assumes that the image is an AR(1) model with $\rho = 0.95$. The unified coding gain saturates around 4 or 5 levels of decomposition. ***More than* 5 *levels of decomposition help very little.*** Type A and Type B systems show a similar coding gain.

The best filter bank for a given $P(z)$ is used in a wavelet-based study with four levels of decomposition for the "Lenna" image. The PSNRs of the reconstructed images are computed for various rates and plotted in Figure 11.22 for Type A systems. Notice how Haar wavelets are worst. Odd-length B systems show a similar PSNR performance.

Figure 11.20: Reconstructed images using DCT and GenLOT (length 48) for 1 bpp and 0.125 bpp.

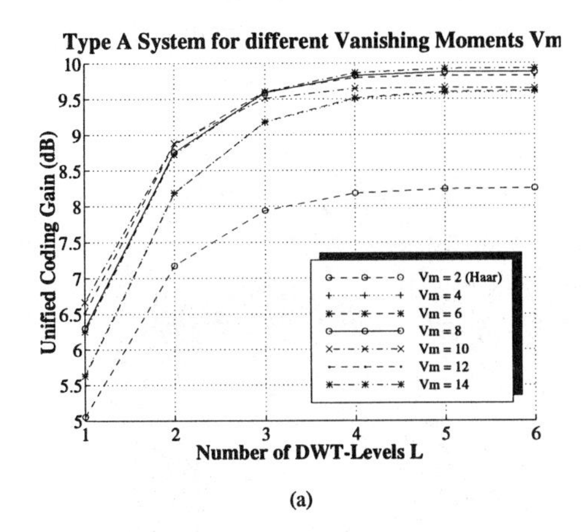

(a)

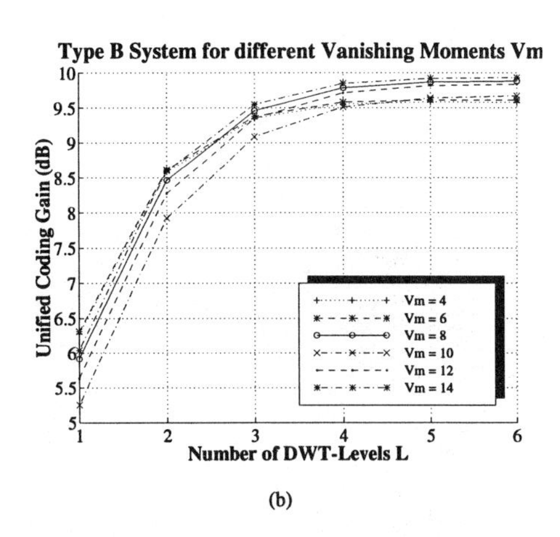

(b)

Figure 11.21: Unified coding gain for the best even-length Type-A systems and odd-length Type-B systems. Vm is the number of vanishing moments of the halfband filter $P(z)$.

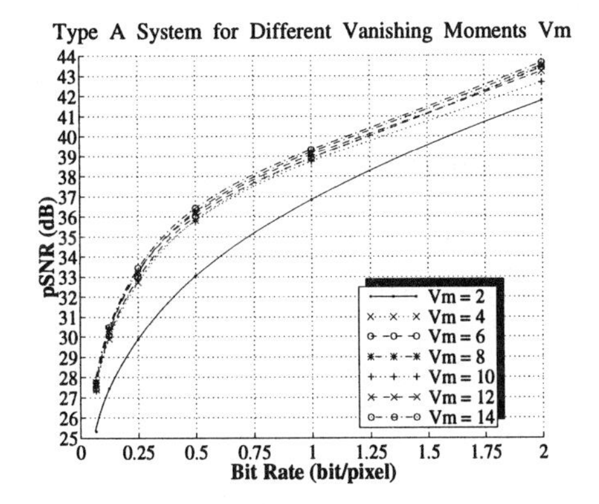

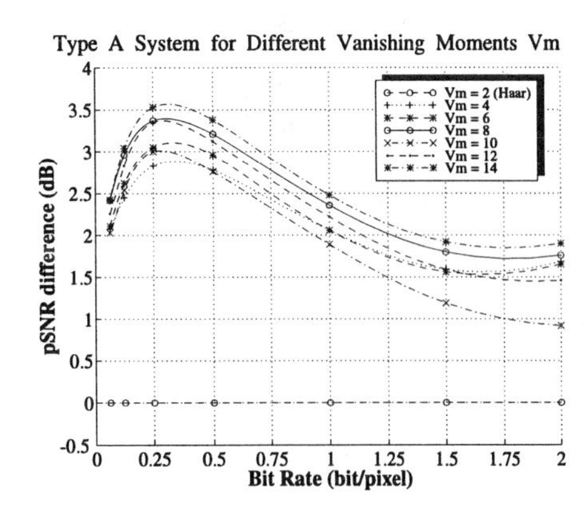

Figure 11.22: PSNR for even length filter. Difference from the PSNR of a DCT transform.

Discussion Given an image and a compression ratio, should one use a block transform (GenLOT) or a wavelet-based transform? At high rate (1 bpp), the perceptual quality is very similar. At medium rate, wavelet-based transforms perform a little better (GenLOT has ringing). At low rate, all transforms have their drawbacks and it is hard to pick one over the others. We can only display the outputs and tabulate the error measures as shown in Figure 11.23.

Type A at 0.125 bpp (a=2.5)

	(6,10)	(7,9)	GenLOT
MSE	189.1	181.7	169.3
PSNR (dB)	25.36	25.54	25.84
Max Error	88.42	93.72	90.16

Figure 11.23: Reconstructed images using Type A (6, 10) and Type B (9, 7). Error measures include GenLOT (length 48) from Figure 11.21.

Software for image compression is available at

http://saigon.ece.wisc.edu/~waveweb/QMF.html

Video Compression

Commercial systems such as video-on-demand, videophone, video conferencing and multimedia applications are being planned and manufactured for home use. The key to their success is video compression. Video on the Internet will be at a low bit-rate!

Video signals are sequences of 2D images updating at about 30 frames per second. *The new dimension is time.* One can extend separable processing from 2D to 3D. Then a video compression system would use a 3D separable filter bank in the front end. The transformed sequences are quantized and entropy coded. A bit allocation algorithm based on rate distortion theory can be used to find the optimal bit assignment.

Another approach to video compression is based on *motion estimation*. At 30 frames/second, the information in frames m and $m \pm 1$ is highly correlated. Suppose that one can estimate the motion vectors for all pixels, to indicate where each part of the image moves in the following frame. Then it is sufficient to send the first frame (compressed) and the *motion vectors*. At the synthesis bank, the first frame is reconstructed and subsequent frames are formed by using the motion vectors (plus small corrections to the image). The quality of the reconstructed image depends on the accuracy of the estimated motion vectors.

Consider the image coder based on 8×8 blocks transformed by the DCT. The first frame is quantized, entropy coded and transmitted. The second frame is transformed into blocks. For a specific block (K, L), the search algorithm considers the neighbor blocks $(K \pm 1, L \pm 1)$ to estimate the motion vectors. These are also coded and transmitted. However, an imperfect estimate reduces the quality of the reconstructed second frame. Consequently, one also needs to transmit the expected residual error. The MPEG video standard [MPEG2] employs both forward and backward predictions for motion vector estimation.

Similar algorithms based on the wavelet transform are being developed. Where MPEG deals with subblocks, the wavelet algorithm has blocks with different sizes at different resolutions. Motion estimation is more complicated since there are several scales to search. ***We estimate motion first on a coarse scale, then on finer scales.*** The support regions also depend on the filter lengths. The memory requirement for MPEG is lower, but multiresolution is more powerful. An early comparison is in [Zafar].

Problem Set 11.1

1. If A, B, C, D occur with probabilities 0.4, 0.3, 0.2, 0.1, a Huffman entropy coder will create a binary tree:

$$A, B, C, D \quad \text{have codes} \quad 0, 10, 110, 111$$

What are the Huffman codes for X, Y, Z with probabilities 0.2, 0.5, 0.3?

11.3 Speech, Audio, and ECG Compression

Psychoacoustic model of the ear To design effective algorithms for speech and audio compression, one needs to know how hearing works. The more one knows, the better the algorithms. Psychoacoustics is the study of hearing, from which quantitative models are built. The models are based on years of extensive tests on humans, which led to these conclusions:

- *Hearing is associated with **critical bands***. These nonuniform frequency bands can be approximated by tree-structured filter banks. In speech compression, the filter bank is a four-level dyadic tree. Audio compression is based on either M-band uniform or M-band tree-structure cosine-modulated filter banks (Figure 11.24). There is always a tradeoff between complexity and accuracy in the approximation of critical bands. The filter bank in Figure 11.24(c) is more complicated than the other two. However, it approximates the critical bands more accurately.
- *Around any frequency f_m there is **masking***. An adjacent frequency f with magnitude below $T(f_m, f)$ is masked by f_m and is not audible:

$$T(f_m, f) = \begin{cases} M(f_m)\left(\frac{f}{f_m}\right)^{28}; & f \le f_m \\ M(f_m)\left(\frac{f}{f_m}\right)^{-10}; & f > f_m. \end{cases} \tag{11.8}$$

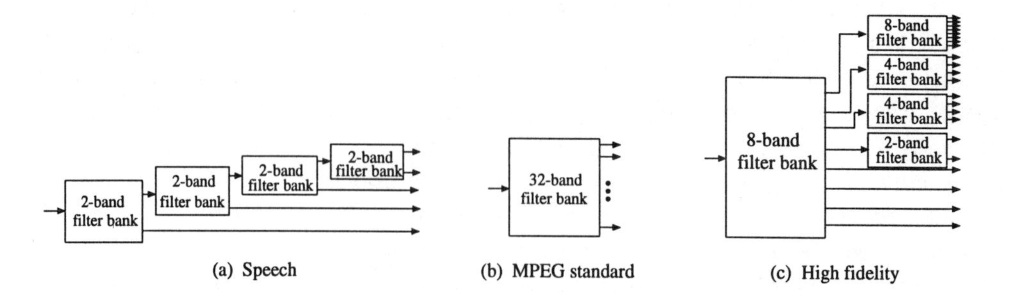

(a) Speech (b) MPEG standard (c) High fidelity

Figure 11.24: Tree-structured filter banks used to approximate the critical bands.

The quantity $M(f_m)$ is the masking threshold at frequency f_m, independent of the signal. The graph of T is linear on a log-log scale. Quantization noise can be freely added to the signal as long as it is below $T(f_m, f)$.

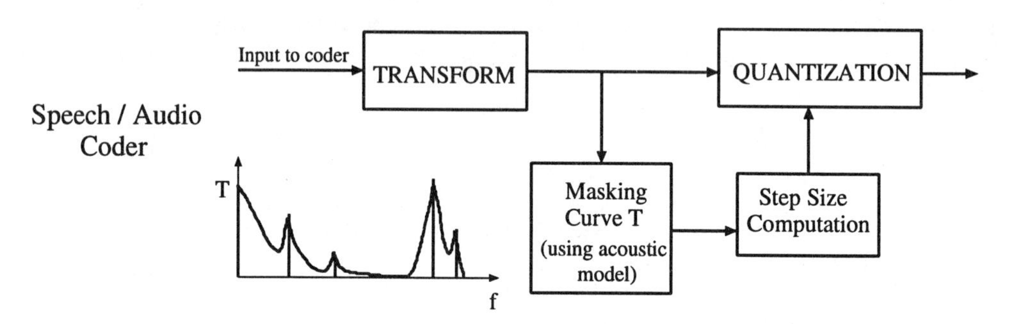

In a speech/audio compression system, the signal is transformed by a tree-structured filter bank. The frequency allocation approximates the critical bands of the human ear. The frequencies f_m with significant power are detected, and their masking envelopes $T(f_m, f)$ are computed. The combined masking envelope forms the *masking curve* of the signal. The quantization noise level is kept below the masking curve. Out-of-band masking noise is negligible as long as the frequency responses of the filter banks have high attenuation.

The allocation of B bits to the signal, assigning b_k bits to subband k, tries to keep the quantization noise inside the masking curve:

$$\text{Minimize} \quad \sum_{k=1}^{M} \frac{1}{12}\left(\frac{2\,p_k}{2^{b_k}-1}\right)^2 \frac{1}{\sigma_{m,k}^2} \quad \text{such that} \quad \sum_{k=1}^{M} b_k = B.$$

Here $\sigma_{m,k}^2$ is the masking power of the kth subband and p_k is its peak value.

Speech Compression

Speech compression is important in mobile communications, to reduce transmission time. Digital answering machines also depend on compression. The bit-rates are low, typically 2.4 kbits per second to 9.6 kbits/second. The best algorithms use either linear predictive models or sinusoidal models.

Speech is classified into *voiced* and *unvoiced* sounds. Voiced sounds are mainly low frequency. In CELP (code excitation linear predictor) the voiced sound is modeled as the output of an all-pole IIR filter with white noise as input. The filter coefficients are found by linear prediction. This filter represents the transfer function of the vocal tract. In a sinusoidal transform, the voiced sounds use a sinusoidal basis. Unvoiced sounds (like sss) have components in all frequency bands and resemble white noise. Model-based techniques achieve reasonable performance at low rates.

At more than 16 kbits/second, subband coding is effective and compatible with the models. Psychoacoustics has associated human hearing to nonuniform critical bands. These bands can be realized roughly as a four-level dyadic tree (Figure 11.24a). For sampling at 8 kHz, the frequency bands of the dyadic tree are: 0–250 Hz, 250–500 Hz, 500–1000 Hz, 1000–2000 Hz and 2000–4000 Hz. These bands can be quantized and coded depending on subband energy; the average signal to noise ratio is maximized. And the *noise masking property* is used.

High-Fidelity Audio Compression

Music signals have no models. The frequency spectrum of a harp is not similar to that of a piano. Subband coding, which makes no assumption on the signal model, is a natural choice. Consider an audio signal of CD quality, sampled at 44.1 kHz with 16 bits of resolution. The total bit-rate is 705.6 kbits/second. For multimedia applications, one would like to compress it to a range from 64 to 192 kbits/second (11:1 to 4 :1). High-fidelity audio compression implies that there is no perceptual loss in the reconstructed signal. This is crucial for digital audio broadcast and satellite TV, where sound quality is the most important feature. Applications of audio compression systems are:

- Digital audio broadcasting
- Satellite TV, High-Definition TV
- Contribution and Distribution links
- Production (tapeless studio, editing systems)
- Storage devices (studio & consumer market)
- Multimedia applications

The noise floor (stopband attenuation) of the filter banks should be below −96 dB (16 bits at 6 dB per bit), although lower attenuation is always preferred. To minimize noise from adjacent subbands, the stopband attenuation is required to have steep roll-off rate. Interested readers should consult [Dehery, Stautner, Brandenburg].

Electrocardiogram Compression

ECG waveforms are signals measured from the heart. They provide essential information to the cardiologist. Normally, a 24-hour recording is desirable to detect heart abnormalities or disorders. Storage requirements can range from 26 Mbytes, with one lead to the heart and 12 bit resolution sampled at 200 Hz, to 138 Mbytes for a system with two leads and 16 bit resolution sampled at 400 Hz. Compression is needed at a low bit-rate for both storage and telemedicine.

An ECG compression algorithm is judged by its ability to minimize the distortion while retaining all significant features of the signal. An accepted error measure is the percent root-mean-square difference (PRD). Let x_{or} and x_{re} be the original and reconstructed signals of length N:

$$\text{PRD} = \left\{ \sum [x_{or}(n) - x_{re}(n)]^2 / \sum [x_{or}(n)]^2 \right\}^{\frac{1}{2}} \times 100\%$$

Reconstruction with low PRD does not necessarily mean clinical acceptance. The crucial requirement is not to distort the diagnostic information used by the physician. This is a challenging problem since GenLOT and wavelet compression typically produce blocking and ringing. Moreover, ECG is normally used as input to classification. If compression does not change the outcome of the detection and classification algorithms, it is acceptable.

An ECG waveform (dashed line) is shown with its reconstruction (solid line). The compression ratio is 7.9 to 1, using the (13, 11)-tap filter bank. The reconstructed image preserves the significant features of the original and the PRD is 3.9%.

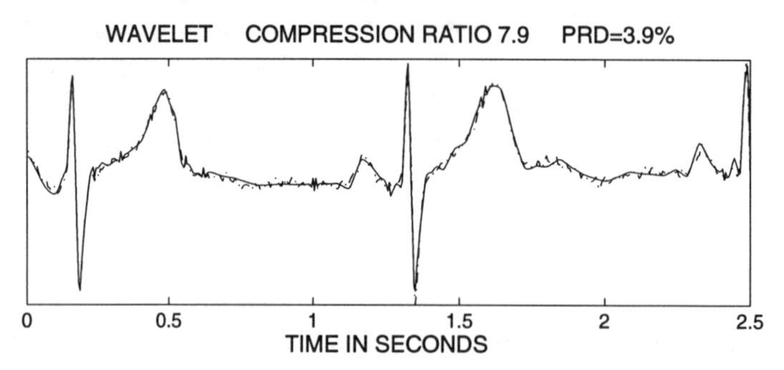

Techniques such as the turning point algorithm, amplitude zone time epoch coding (AZTEC), coordinate reduction time encoding system (CORTES), and the FAN algorithm compress the data by discarding relatively insignificant information [T]. The reconstructed signal is obtained by interpolating the stored samples. These algorithms are simple to implement, but they can produce significant distortion. Figure 11.25 shows the original and reconstructed waveforms using the AZTEC and FAN algorithms. We also show the PRD as function of average bit rate for the AZTEC, FAN and wavelet algorithms. The original signal has 12 bits per sample. In all our tests, wavelet-based compression gave the best objective measures.

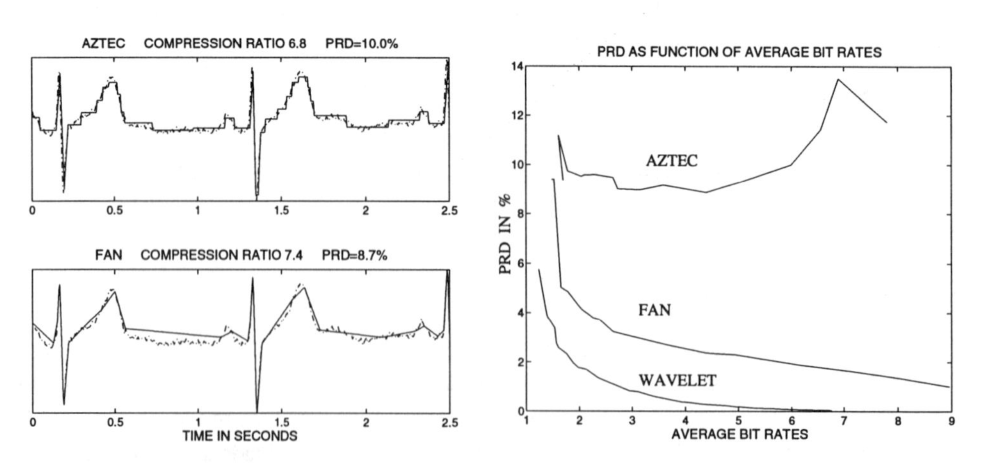

Figure 11.25: ECG Compression using the AZTEC, FAN and wavelet-based algorithms.

The blocks of the wavelet-based ECG algorithm are the same as for image coding. The input is divided into segments of N samples and these vectors are transformed (four levels, linear phase). The spectrum estimator computes a bit allocation to the subbands. They are quantized (scalar uniform) and entropy coded. The segment size N is important in determining the compression ratio and the corresponding PRD. A large N increases the variance of the subband signals and the distortion. By experimenting with many sets of ECG data, we conclude that *a block size of* 2000 *is reasonable.*

Software for ECG compression is available at

http://saigon.ece.wisc.edu/~waveweb/QMF.html

11.4 Shrinkage, Denoising, and Feature Dectection

Wavelet Shrinkage

In the L-level wavelet decomposition of a signal, the number of wavelet coefficients with significant energy is small. This is a direct consequence of the approximation property of the wavelets, assuming a sufficient number of vanishing moments (say $p > 2$). The signal can be accurately represented by a small number of coefficients. *Wavelet shrinkage*, developed by Johnstone and Donoho, selects these coefficients based on *thresholding*. The wavelet shrinkage algorithm decomposes the signal into L levels and then:

- For each level, we select a threshold and apply "hard thresholding". This will zero out many small coefficients, which results in efficient representation. Thresholding is a lossy algorithm; the original signal can not be reconstructed exactly. An alternative is soft thresholding at level δ, chosen for compression performance or relative error. The outputs $y_{hard}(t)$ and $y_{soft}(t)$ with threshold δ are

$$y_{hard}(t) = \begin{cases} x(t), & |x(t)| > \delta \\ 0, & |x(t)| \leq \delta \end{cases} \quad \textbf{\textit{Hard thresholding}}$$

$$y_{soft}(t) = \begin{cases} sign(x(t))(|x(t)| - \delta), & |x(t)| > \delta \\ 0, & |x(t)| \leq \delta \end{cases} \quad \textbf{\textit{Soft thresholding.}}$$

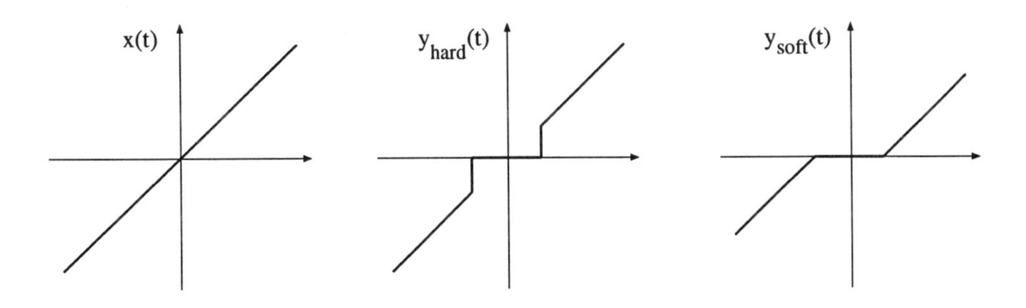

The signal in Figure 11.26 is basically lowpass with noise added. We show the reconstructed signal with threshold levels $\delta = 35$ and 100. After thresholding at these levels, 92.8% and 93.4% of the coefficients are zero. The L_2 norm recoveries are 99.99% and 99.98%, respectively. Reconstruction with $\delta = 35$ is closer to the original since the threshold is lower. But it uses more terms.

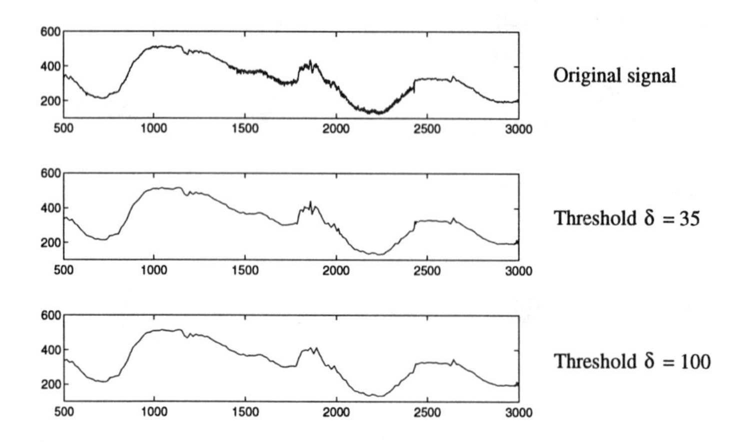

Figure 11.26: Hard thresholding generated using the *Wavelet Toolbox* from *MathWorks*.

Denoising

Thresholding generally gives a lowpass version of the original signal. By selecting δ, one can suppress the noise $w(n)$ in the signal $x(n) = u(n) + \sigma w(n)$. A simple example of $w(n)$ is Gaussian white noise $N(0, 1)$. ***For denoising we use soft thresholding.***

The noise power σ^2 is assumed to be much smaller than the signal power. Moreover, the signal is assumed to have low frequency components, whereas the noise source is white. Thresholding the detailed coefficients will also remove some of signal's power. It is generally impossible to filter out all the noise without affecting the signal. Algorithms selecting the threshold levels include Stein's Unbiased Risk Estimate, fixed threshold, Minimax criteria or a combination [Donbf1].

The piecewise constant signal below is corrupted by Gaussian white noise. The corrupted signal is decomposed using the Daubechies wavelet D_6. The coefficients at level 4 are thresholded using Stein's Unbiased Risk Estimate. Notice that the reconstruction consists of the original signal and *some* of the noise.

In both wavelet shrinkage and denoising, the output is a cleaned-up version of the input. This works only when one knows the signal characteristics in advance. The algorithm will distort the desired signals when thresholding is applied.

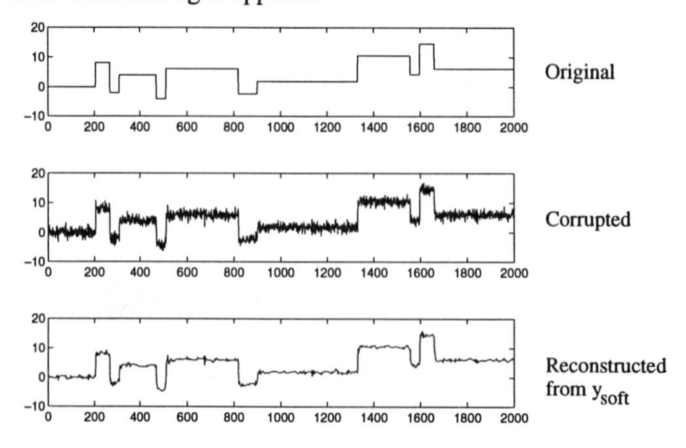

Discontinuity Detection

A discontinuity in the signal can be detected by Haar wavelets. A discontinuity in the first derivative can be detected by D_4. The signal in Figure 11.27(a) is linear, except for an interval of zero slope. The D_4 wavelet can represent straight lines exactly ($p = 2$). The two discontinuities are being captured in the wavelet coefficients. Although wavelets with longer supports could be used, D_4 is the best choice since it has good time localization.

Figure 11.27(b) shows a signal with discontinuity in the second derivative. The wavelet used is D_8, which can represent parabolas exactly (so could D_6). Even though the discontinuity can not be seen from the signal, it is clearly detected by the coefficients in the first level of the decomposition.

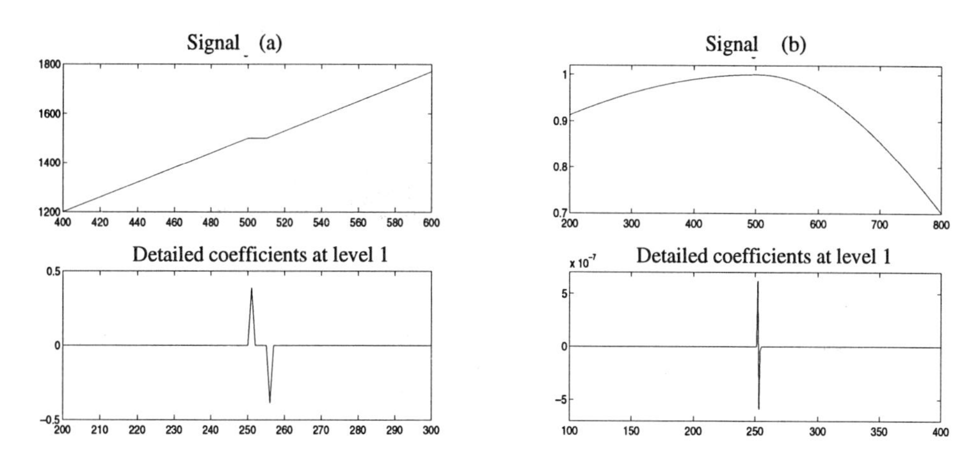

Figure 11.27: Detecting discontinuity in the first derivative and second derivative. (*These plots are generated by the Wavelet Toolbox.*)

Detecting discontinuities using a wavelet transform works well for signals with no noise. When noise dominates, one will obtain many false alarms (wrong detections). For reliable detection, one needs to follow the coefficients for several scales. Mallat pioneered this important application in continuous time for the case with no noise. The section below elaborates on the multiscale feature detection and points out its advantage comparing to the conventional feature detection.

Multiscale Feature Detection

The step edge in Figure 11.28 can be detected from the detailed coefficients using Haar wavelets. An alternative approach uses the Canny edge detector in image processing [Canny]. Both approaches work well for a signal with no noise. The output is clean and the edges can be detected. For the same input with added noise, on the right side, there are many false alarms (wrong detections).

Figure 11.29 shows the same noisy signal with its multiscale representation. Note the occurence of false alarms at each scale. ***The consistency between the peaks at all scales allows the edge locations to be detected.***

There are several issues in a multiscale feature detector. Our example uses *peak* as an indicator for the *edge* feature. However, one could also use *zero crossing* as an indicator. Given a

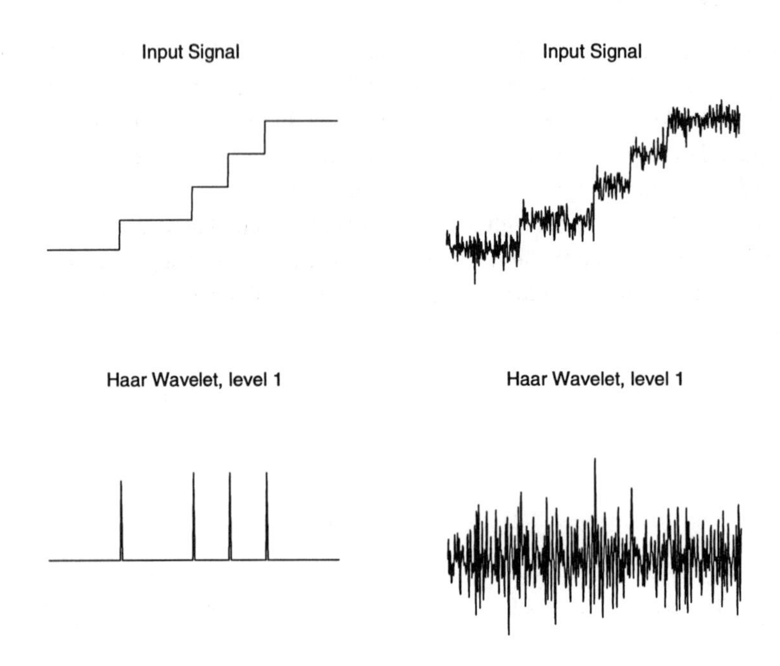

Figure 11.28: Step edge detection using Haar wavelets. The Canny edge detector is very similar. The clean step signal (left) is easy. The signal with noise (right) is difficult at one level.

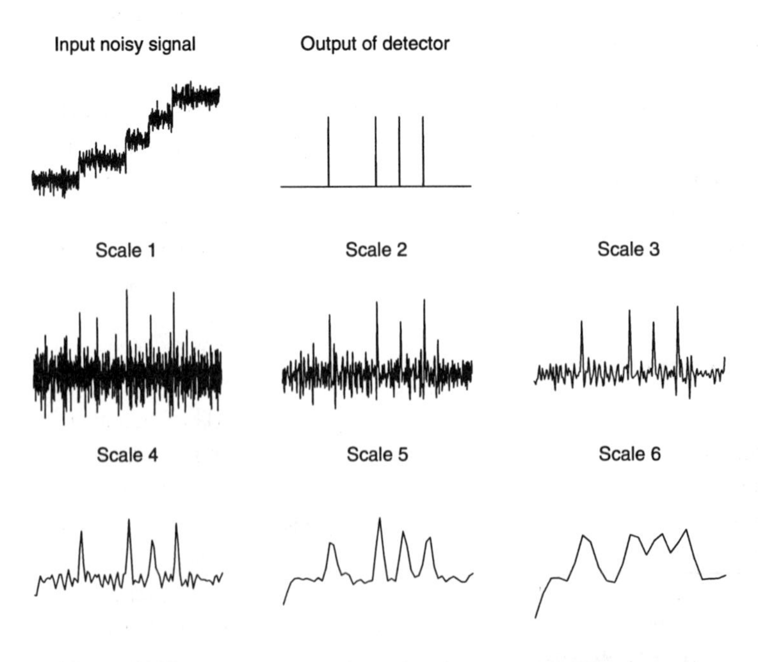

Figure 11.29: Step edge detection using the outputs at different scales.

feature, which indicator is the best one to use? The second issue is the choice of filter bank. How can one design filters that give a multiresolution representation of the indicators? The third issue is the design of an optimal detector for various noise models. Finally, many algorithms are derived in continuous time and can not be used for discrete-time implementation [Mt]. [HaNgCh] presents a general approach to discrete-time multiscale feature detection.

Problem Set 11.4

1. For the signal $x(t) = \sin t$, draw the graphs of $y_{hard}(t)$ and $y_{soft}(t)$ after thresholding with $\delta = \frac{1}{2}$.

11.5 Communication Applications and Adaptive Systems

Transmultiplexers

A transmultiplexer is a filter bank with synthesis first. This reverse order is depicted in the figure below. M signals come in, and they are combined into *one* transmitted signal. Then the receiver has to reconstruct the M separate signals.

Low-bandwidth input signals $\boldsymbol{x}_k(n)$ are upsampled and filtered. The combined signal goes through a high-bandwidth channel. The received signal is filtered and downsampled to the original rates. The outputs $\widehat{\boldsymbol{x}}_k(n)$ may suffer from ***distortion*** and ***crosstalk*** because of the decimation, interpolation, and non-ideal filtering.

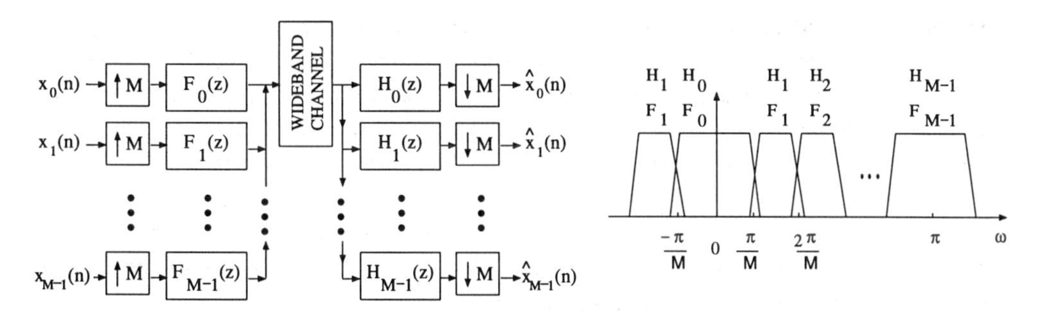

The output $\widehat{\boldsymbol{x}}_k(n)$ depends on all M inputs by $\widehat{X}_k(z) = \sum S_{k\ell}(z)X_\ell(z)$. The transfer functions $S_{kk}(z)$ give the distortion. The remaining $S_{k\ell}(z)$, $\ell \neq k$, are the crosstalk functions. The objective is to find $H_k(z)$ and $F_k(z)$ such that $\mathbf{S}(z)$ is a diagonal matrix, $S_{k\ell}(z) = \delta_{k\ell}\, z^{-n_k}$. This would be a *perfect reconstruction transmultiplexer.* The design solutions for the PR transmultiplexer are closely related to those in the PR filter bank!

Let $H_k(z)$ and $F_k(z)$ be the analysis and synthesis filters of a perfect reconstruction filter bank. [Vet3, KoNgVa] show how the filters $H_k(z)$ and $z^{-1}F_k(z)$ yield a perfect reconstruction transmultiplexer (for an ideal communication channel).

Discrete Multitone Modulation Transceivers

Transmultiplexers can be used in conjunction with discrete multitone modulation transceivers. Discrete multitone (DMT) or multicarrier modulation is a class of orthogonal frequency modulations. The transmitted data is split into several bit streams and used to modulate several carriers. The spectrum is divided into parallel orthogonal and narrowband subchannels. Each carrier occupies one subchannel. This concept was proposed thirty years ago, but did not receive much attention because of the difficulty in implementation using analog technology.

The advance of digital signal processors and related hardware has made the multicarrier modulation a preferred structure over the single-carrier QAM system. There is important interest in multicarrier modulation for high-speed data transmission over the twisted pair channel of a digital subscriber line. As a result, discrete multitone has been proposed as a standard for high-speed digital subscriber line (HDSL) and asymmetric digital subscriber line (ADSL) communication.

Consider a typical channel where the attenuation variation could be as large as 40 dB. A linear equalizer for a single-carrier QAM system would increase the noise, whereas a decision-feedback equalizer would be too complex. For many channels, multicarrier modulation approximates a constant transfer function in each subchannel. Equalization becomes very simple. Other advantages of multicarrier modulation are the reduced effect of impulse noise as a result of large symbol duration; the flexibility of not transmitting in the corrupted subchannels in case of narrowband interference; and the flexibility of transmitting important data in subchannels with high SNRs.

One disadvantage is the large peak-to-rms ratio of the transmitted signals, which might lead to nonlinear distortion. And the implementation complexity must be controlled. When the frequency separation Δf is $\frac{1}{T}$, where T is the symbol duration, the multicarrier system can use the DFT. That avoids implementing several carriers using analog technology. This is one of the main attractive features of DFT-based multicarrier modulation.

Given an ideal channel, orthogonality is ensured by the $\frac{1}{T}$ spacing between subcarriers, and they can be independently demodulated. In a dispersive channel, the carriers are no longer orthogonal. Interference depends on the spectral overlap between subchannels. DFT-based multicarrier systems, with rectangular pulse shape and -13 dB sidelobes, can have significant interchannel interference. One could employ pulse shaping to reduce spectral overlap, but the resulting transmultiplexer is not necessarily orthogonal (nor perfect reconstruction).

An alternative is to use a ***cosine-modulated*** orthogonal transmultiplexer. Its spectral containment leads to superior robustness against narrowband radio frequency interference (RFI). Interested readers should consult [RiPrNg, SaTz, TzTzPh, TzTzRe] for algorithms and comparisons between DFT and cosine-modulated multicarrier systems.

Adaptive Systems

Adaptive systems are important when the signals or environments are changing with time. Typical applications are inverse filtering, room acoustics modelling, and echo cancellation. We also mention adaptive array processing (beamforming and direction finding) and channel equalization. In these applications, an FIR filter models the signal or the environment. Its coefficients are adapted in time to minimize an error measure. Popular adaptive algorithms are LMS (Least-Mean-Square) and RLS (Recursive-Least-Square). A detailed treatment of adaptive filter theory is in [H, PRCN].

The adaptive system in Figure 11.30 is used in inverse filtering. The objective is to find $f(n)$ such that the error $e(n)$ is as small as possible. Here, the desired signal $d(n)$ is distorted by an unknown system $h(n)$ and is corrupted by noise. The z-transform of $e(n)$ is

$$E(z) = \left[F(z)\, H(z) - z^{-N}\right] D(z).$$

Assuming that $H(z)$ is an FIR filter, the ideal choice for $F(z)$ is $z^{-N}/H(z)$. But normally $F(z)$ is constrained to be FIR. Its length affects the convergence rate and the error.

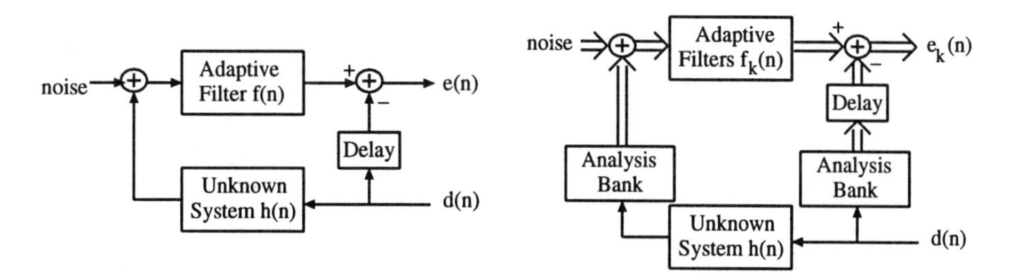

Figure 11.30: Adaptive systems in inverse filtering (left) and using a filter bank (right).

If the spectrum of $H(e^{j\omega})$ is complicated, the inverse spectrum is equally complicated and the resulting $f(n)$ is long. By separating $H(e^{j\omega})$ into many bands $H_k(e^{j\omega})$, the inverse filter in the k-th subband would not be too large. These M subbands in Figure 11.30 are adapted using M FIR adaptive filters $f_k(n)$, much shorter than $f(n)$ since the analyzing spectrum is simplified band by band. The convergence rates are faster. This is a tradeoff between hardware complexity and convergence rate.

Figure 11.31(a) shows a very different adaptive system. From the error $[H(z) - F(z)]\, D(z)$, one observes that $F(z)$ is no longer an inverse but is an approximation to $H(z)$. The length of $f(n)$ is comparable to that of $h(n)$. This system is used to cancel telephone echo coming from an impedance mismatch of the residence line and the switching center. It is most noticeable and annoying for long distance calls. Here, $s(n)$ is the speech signal, $d(n)$ is the reference signal, $y(n)$ is the echo signal received at the handset, and you hear $x(n)$.

The objective of $f(n)$ is to reproduce $s(n)$ while suppressing the echo $y(n)$. For a complicated and nonlinear echo path, an adequate $f(n)$ is too long. ***A bank of shorter filters $f_k(n)$ is better.*** The signals $s(n)$ and $d(n)$ are decomposed into M subbands, on which adaptation is performed.

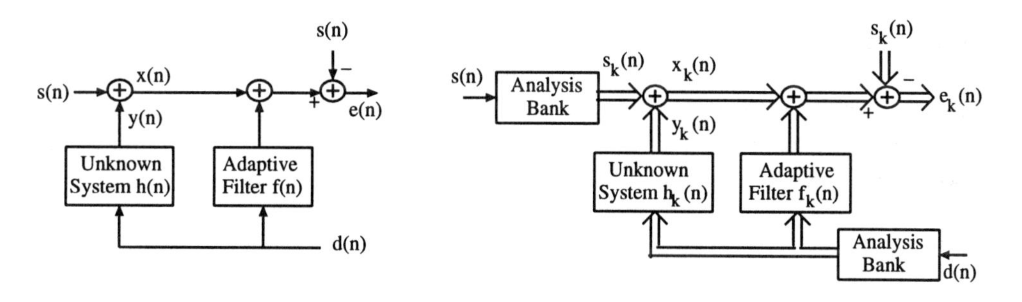

Figure 11.31: Adaptive systems for echo cancellation (left) and using a filter bank (right).

Filter bank adaptive systems give an improved convergence rate and error reduction [GilVet,

JinWoLu]. Subbands with large energy can be adapted by long filters, whereas subbands with insignificant energy can be either discarded or adjusted using short filters. The price is the additional complexity (minimized by using lattice structure or polyphase). Two-channel adaptive banks are designed in [JinWoLu] and compared with spline and Daubechies wavelets.

11.6 Wavelet Integrals for Differential Equations

The possible application of wavelets to differential equations is important. In principle, this application is available. *In practice, it is not yet a real success.* The overall approach is familiar, but the competition with other methods is severe. We do not necessarily predict that wavelets will win.

The unknown function $u(t)$ (or $u(x)$ or $u(x, y)$) is to be approximated by a finite combination $b_1\phi_1(t) + \cdots + b_N\phi_N(t)$. These "trial functions" are chosen in advance. They are polynomials or sinusoids in the spectral method. The $\phi_k(t)$ are piecewise polynomials in the finite element method. Presumably they are scaling functions and wavelets in the "wavelet method."

A key point is the option to use different scales $\Delta t = 2^{-j}$. The grid becomes a *multigrid*. The number of scale levels can vary with the position t or x, to produce an *adaptive mesh*. Flexibility comes from multiresolution. But the wavelet mesh does not achieve the total flexibility of the finite element method.

Up to this point in the book, all signals were assumed known. Now the coefficients $b_1, \ldots,$ b_N are unknowns because $u(t)$ is unknown. There are two outstanding rules for producing N equations for $b_1, \ldots, b_N$:

1. *Collocation*: Apply the differential equation at N points $t_1, \ldots, t_N$.

2. *Galerkin method*: Choose N test functions $g_1(t), \ldots, g_N(t)$. Replace an equation like $u''(t) = f(t)$ by N "weighted residual equations:"

$$\int \left(b_1\phi_1''(t) + \cdots + b_N\phi_N''(t)\right) g_j(t)\,dt = \int f(t)g_j(t)\,dt, \quad 1 \le j \le N. \tag{11.9}$$

When the test functions g_j are delta functions $\delta(t - t_j)$, this is collocation. When the g_j are the same as the ϕ_j, this is the Rayleigh-Ritz method. When the g_j are powers of t, this is the moment method. The trial functions and test functions should satisfy all *essential* boundary conditions in the problem—these are Dirichlet conditions like $u(0) = 0$. They need not satisfy *natural* boundary conditions like $u'(1) = 0$.

Collocation seems doubtful for wavelets, which are not smooth. The Galerkin method may also look doubtful, because of the integrals in (11.9). But these integrals can be computed by a quadrature formula, in which the coefficients are found once and for all. Then an integral of $f(t)\phi(t)$ does not require point values of $\phi(t)$, only of the smoother function $f(t)$:

$$\int_a^b f(t)\phi(t)\,dt \approx \sum_{k=1}^{r} a_k f(t_k). \tag{11.10}$$

This section discusses several quadrature rules. We have written $\phi(t)$ but similar rules are needed also for the wavelet $w(t)$. We will choose equally spaced evaluation points, so that overlapping integrals can share the same t_k (and $f(t)$ may be known only by those samples anyway).

Note that integrals are often computed by parts. Derivatives are moved away from rough functions and onto smooth functions. When the trial and test functions are the same ($\phi_j = g_j$), integration by parts changes $\int \phi''(t)g(t)\,dt$ into $-\int \phi'(t)g'(t)\,dt$. This needs only one derivative, not two.

We expect exact integrals for functions chosen in advance, and quadrature rules for integrals involving inputs $f(t)$. We discuss quadrature rules and then exact integrals.

Quadrature Rules

The usual test for accuracy is appropriate for smooth functions $f(t)$. The idea is to try $f(t) =$ *polynomial.* The rule (11.10) has accuracy d if it gives the correct integral for all polynomials of degree less than d. The first d terms in the Taylor series of $f(t)$ are integrated correctly, so the error is from the dth derivative:

$$\int f(t)\,\phi(t)\,dt = \sum a_k f(t_k) + O(\|f^{(d)}\|).$$

Changing from t to $2^j t$, this error will be of order $(\Delta t)^d$ when the scale is $\Delta t = 2^{-j}$:

$$\int f(t) 2^{j/2} \phi\left(2^j t\right) dt = \sum a_k 2^{-j/2} f\left(2^{-j} t_k\right) + O\left((\Delta t)^d\right). \tag{11.11}$$

We study quadrature formulas at unit scale. Then (11.11) gives the error at other scales.

Example 11.1. One-point quadrature. The rule has only one term:

$$\int f(t)w(t)\,dt \approx 0 \quad \text{and} \quad \int f(t)\phi(t)\,dt \approx f(t_1). \tag{11.12}$$

The first rule is strange but its accuracy is $d = p$, because the wavelet has p vanishing moments. The exact integral is zero up to $f(t) = t^{p-1}$.

The second rule is correct for $f(t) \equiv 1$ provided $\int \phi(t)\,dt = 1$. Thus $d = 1$ at least. The accuracy increases to $d = 2$ when the evaluation point is $t_1 = \int t\phi(t)\,dt$. (The rule is then exact for $f(t) = t$.) This is a low-level example of the Gauss idea, to double the accuracy by choosing intelligent t_k as well as a_k. But higher-order Gauss has unequally spaced t_k and is not ideal for these applications. We follow instead the excellent paper [Sweldens-Piessens], which also gives the coefficients a_k for high order quadrature rules.

How can you find this number $\int t\phi(t)\,dt$? By definition, it is the first moment M_1 of $\phi(t)$. It equals the first moment $m_1 = \sum k\boldsymbol{h}(k)$ of the filter coefficients! There is a simple recursion that gives the moments $M_n = \int t^n \phi(t)\,dt$ exactly from the moments $m_n = \sum k^n \boldsymbol{h}(k)$. As always, the answer is in those coefficients $\boldsymbol{h}(k)$, when we use the dilation equation for $\phi(t)$:

$$\begin{aligned}
2^n \int t^n \phi(t)\,dt &= \sum 2\boldsymbol{h}(k) \int (2t)^n \phi(2t-k)\,dt \\
&= \sum \boldsymbol{h}(k) \int t^n \phi(t-k)\,dt \\
&= \sum \boldsymbol{h}(k) \int (t+k)^n \phi(t)\,dt \\
&= \sum_{j=0}^{n} \binom{n}{j} \sum \boldsymbol{h}(k) k^j \int t^{n-j} \phi(t)\,dt.
\end{aligned}$$

This says that $2^n M_n = \sum_{j=0}^{n} \binom{n}{j} m_j M_{n-j}$. Bring M_n to the left side:

$$\textbf{\textit{Moment formula}} \quad M_n = \frac{1}{2^n - 1} \sum_{j=1}^{n} \binom{n}{j} m_j M_{n-j}. \tag{11.13}$$

In particular $M_1 = m_1$. This is the best evaluation point for wavelet integrals—but it is not generally an integer. The D_4 scaling function has $M_1 = m_1 = (1 + \sqrt{3})/4$.

Three comments. First, orthogonal scaling functions have $M_2 = M_1^2$ (Problem 10). The one-point quadrature $\int f(t)\phi(t)\,dt \cong f(M_1)$ is *third-order* accurate. It is correct for $f(t) = 1, t, t^2$. Second, the Daubechies polynomials are nearly "interpolating" at the points $t = M_1 + k$:

$$\phi(M_1) \approx 1.0002 \quad \text{and} \quad \phi(M_1 + 1) \approx -0.0004 \quad \text{and} \quad \phi(M_1 + 2) \approx 0.0002.$$

See [K] for graphs and discussion. Third, by using $3p$ instead of $2p$ coefficients we can construct "*coiflets*" that have *zero moments* $M_1, M_2, \ldots, M_{p-1}$. Then the one-point rule $\int f(t)\phi(t)dt \approx f(0)$ is pth order accurate.

Exact Wavelet Integrals

We have already computed the inner products $\boldsymbol{a}(k) = \int \phi(t)\phi(t+k)\,dt$ between translates of the scaling function. We will review the key idea, which gives $\boldsymbol{a}(k)$ in terms of the filter coefficients $\boldsymbol{h}(k)$. We never want to do an exact integral any other way. Generally we can't! The applications to differential equations require many more integrals, for example

$$\int w(t)w(t+k)\,dt \quad \text{and} \quad \int \phi'(t)\phi'(t+k)\,dt \quad \text{and} \quad \int t^m \phi(t)\phi(t+k)\,dt.$$

The last integrand has three factors. Freely available software now allows *four factors* [Kunoth]. These integrals may be needed on an interval I (usually of length 1 or 2^{-n}) as well as on the whole line. Then the integral has a one-zero box function $\chi_I(t)$ as an extra factor.

In all cases the key idea is to use the two-scale equation for each factor. There are two-scale equations for $\phi(t)$ and $\phi'(t)$ and $w(t)$ and the box function $\chi_I(t)$. Recall what happens when two-scale equations are substituted into typical integrals:

$$\begin{aligned} \boldsymbol{a}(k) &= \int \phi(t)\phi(t+k)\,dt = \int \left(\sum 2\boldsymbol{h}(n)\phi(2t-n)\right)\left(\sum 2\boldsymbol{h}(n)\phi(2t+2k-n)\right)dt \\ \boldsymbol{a}_w(k) &= \int \phi(t)w(t+k)\,dt = \int \left(\sum 2\boldsymbol{h}(n)\phi(2t-n)\right)\left(\sum 2\boldsymbol{h}_1(n)\phi(2t+2k-n)\right)dt \end{aligned}$$

The next step is to replace $2t$ by t on the right side. Then dt becomes $\frac{1}{2}dt$. The results are very different in our two examples:

$\boldsymbol{a}(k)$ becomes a combination of $\boldsymbol{a}(k+\ell)$: ***eigenvalue problem*** $\boldsymbol{a} = \boldsymbol{T}\boldsymbol{a}$
$\boldsymbol{a}_w(k)$ also becomes a combination of $\boldsymbol{a}(k+\ell)$: ***explicit equation*** $\boldsymbol{a}_w = \boldsymbol{T}_w\boldsymbol{a}$.

The equation $\boldsymbol{a} = \boldsymbol{T}\boldsymbol{a}$ is homogeneous. Its solution must be suitably normalized. The equation $\boldsymbol{a}_w = \boldsymbol{T}_w\boldsymbol{a}$ gives $\boldsymbol{a}_w$ in terms of the known $\boldsymbol{a}$. The matrices are doubled-shifted Toeplitz:

$$\boldsymbol{T} = (\downarrow 2)2\boldsymbol{H}\boldsymbol{H}^T \quad \text{and} \quad \boldsymbol{T}_w = (\downarrow 2)2\boldsymbol{H}_1\boldsymbol{H}^T. \tag{11.14}$$

This pattern is quite typical. Section 7.2 carried out the steps leading to $\boldsymbol{T}$. We now carry out the same steps for other important integrals (with more factors). Then we deal with integrals involving $w(t)$, which do *not* yield eigenvalue problems. For integrals involving $w(2t)$, the highpass matrix with diagonals $\boldsymbol{h}_1(0), 0, \boldsymbol{h}_1(1), 0, \boldsymbol{h}_1(2), 0, \ldots$ replaces $\boldsymbol{H}_1$.

Integrals of Products of $\phi_i(t)$ and $\phi_i'(t)$

We will describe the exact computation of a much wider class of integrals $\boldsymbol{b}(k)$. All functions to be integrated must be *refinable*. They satisfy two-scale dilation equations (also called refinement equations). The coefficients for $\phi_0(t), \phi_1(t), \phi_2(t)$ are $2\boldsymbol{h}_0(k), 2\boldsymbol{h}_1(k), 2\boldsymbol{h}_2(k)$. The dilation equation is substituted for each $\phi_i(t)$ and for any derivatives. (Each derivative of $2t$ produces an extra 2.) Changing the variable of integration brings $2t$ back to t and yields a two-scale equation for the integral:

$$\boldsymbol{b}(\boldsymbol{n}) = 2^{\alpha} \sum \boldsymbol{g}(\boldsymbol{n})\, \boldsymbol{b}(2\boldsymbol{k} - \boldsymbol{n}). \tag{11.15}$$

That is an eigenvalue problem $\boldsymbol{b} = 2^{\alpha}(\downarrow 2)\boldsymbol{G}\,\boldsymbol{b}$. The eigenvector is $\boldsymbol{b}$, and the operator $(\downarrow 2)$ changes $\boldsymbol{k}$ to $2\boldsymbol{k}$. Our first problem is to find the coefficients in $\boldsymbol{g}$ from the coefficients in $\boldsymbol{h}_0, \boldsymbol{h}_1$, and $\boldsymbol{h}_2$. *The two-scale equation for the integral* $\boldsymbol{b}$ *follows from the two-scale equations for* $\phi_0(t)$, $\phi_1(t)$, $\phi_2(t)$.

Our presentation follows [Kunoth], who has created a valuable and freely available code. She works in d dimensions with $\boldsymbol{x} = (x_1, \ldots, x_d)$ and allows a product of $\ell + 1 = 4$ functions. In principle any product of functions and their derivatives and dilations and translations is workable. We will begin in one dimension with $\ell = 2$—thus a product of three functions inside the integral. The letters $\boldsymbol{b}$ and $\boldsymbol{g}$ are different from hers, since her conventions with 2's are not the same.

The integrals to be computed are sometimes called ***connection coefficients***:

$$\boldsymbol{b}(\boldsymbol{k}) = \boldsymbol{b}(k_1, k_2) = \int_{-\infty}^{\infty} \phi_0(t)\, D^{\mu_1}\phi_1(t - k_1)\, D^{\mu_2}\phi_2(t - k_2)\, dt.$$

The function $\phi_0(t)$ might be the standard box with dilation coefficients $\frac{1}{2}, \frac{1}{2}$. The functions $\phi_1(t)$ and $\phi_2(t)$ might be scaling functions, when the integral arises from Galerkin's method. If the differential equation is nonlinear, or has variable coefficients, that gives more factors ($\ell > 2$) in the integral. We are not now integrating wavelets! It is the scaling function that has a homogeneous dilation equation and leads to an eigenvalue problem. Integrals involving $w(t)$ are computed afterward. The filters $\boldsymbol{h}_0, \boldsymbol{h}_1, \boldsymbol{h}_2$ are all *lowpass*.

Our standard example is $\boldsymbol{a}(n) = \int \phi(t)\phi(t+k)\, dt = \int \phi(t-k)\phi(t)\, dt$. (From now on we have $t - k$ instead of $t + k$; no problem.) It has $\ell = 1$ with $\mu_1 = 0$; no derivatives. Or it has $\ell = 2$ with $\phi_0(t) \equiv 1$ and $(\mu_1, \mu_2) = (0, 0)$. In this familiar example, the matrix is $\boldsymbol{G} = \boldsymbol{H}\boldsymbol{H}^T$ and the exponent is $\alpha = 1$. The double-shift matrix is $\boldsymbol{T} = (\downarrow 2)2\boldsymbol{H}\boldsymbol{H}^T$ as usual. This example will be a useful check, when the coefficients $\boldsymbol{g}(n)$ come from the autocorrelation $\boldsymbol{g} = \boldsymbol{h} * \boldsymbol{h}^T$:

$$\boldsymbol{g}(n) = \sum \boldsymbol{h}(k)\boldsymbol{h}(k - n). \tag{11.16}$$

We jump directly to the key formula with three factors ϕ_0, ϕ_1, ϕ_2:

$$\boldsymbol{g}(n_1, n_2) = \sum \boldsymbol{h}_0(k)\boldsymbol{h}_1(k - n_1)\boldsymbol{h}_2(k - n_2). \tag{11.17}$$

To establish that formula, and to allow derivatives of the functions $\phi_i(t)$, substitute the dilation equations into the integral $\boldsymbol{b}(k_1,k_2)$. The derivatives bring 2^{μ_1} and 2^{μ_2}:

$$\int_{-\infty}^{\infty}\left[\sum 2\boldsymbol{h}_0(k)\phi_0(2t-k)\right]\left[2^{\mu_1}\sum 2\boldsymbol{h}_1(m)\phi_1(2t-2k_1-m)\right]\left[2^{\mu_2}\sum 2\boldsymbol{h}_2(p)\phi_2(2t-2k_2-p)\right]dt.$$

Change $2t-k$ to T and $2\,dt$ to dT. Set $\alpha = 2+\mu_1+\mu_2$. Our integral is

$$2^{\alpha}\sum_{k,m,p}\boldsymbol{h}_0(k)\boldsymbol{h}_1(m)\boldsymbol{h}_2(p)\int_{-\infty}^{\infty}\phi_0(T)\phi_1\left(T-(2k_1+m-k)\right)\phi_2\left(T-(2k_2+p-k)\right)dT. \tag{11.18}$$

That inner integral is $\boldsymbol{b}(2k_1+m-k,\,2k_2+p-k)$. Set $n_1 = k-m$ and $n_2 = k-p$:

$$\boldsymbol{b}(k_1,k_2) = 2^{\alpha}\sum_{n_1}\sum_{n_2}\sum_{k}\boldsymbol{h}_0(k)\boldsymbol{h}_1(k-n_1)\boldsymbol{h}_2(k-n_2)\boldsymbol{b}(2k_1-n_1,2k_2-n_2). \tag{11.19}$$

The coefficients $\boldsymbol{g}(n_1,n_2)$ anticipated in (11.17) have appeared (take the sum on k). The eigenvalue equation for $\boldsymbol{b}$ is

$$\boldsymbol{b}(k_1,k_2) = 2^{\alpha}\sum_{n_1}\sum_{n_2}\boldsymbol{g}(n_1,n_2)\boldsymbol{b}(2k_1-n_1,2k_2-n_2). \tag{11.20}$$

In general $\alpha = \ell+\mu_1+\cdots+\mu_\ell$. The matrix $(\downarrow 2)\boldsymbol{G}$ must have the eigenvalue $2^{-\alpha}$. In case of multiple eigenvectors, there is a correct choice to be made for $\boldsymbol{b}$. It is given by moment conditions (11.22) discovered in [DaMi]:

Theorem 11.1 *Suppose the filters $\boldsymbol{H}_0, \boldsymbol{H}_1, \boldsymbol{H}_2$ have at least 1, $1+\mu_1$, $1+\mu_2$ zeros at π. There is a unique finite length eigenvector* **b** *satisfying* (11.21) *and* (11.22)*:*

$$\sum \boldsymbol{g}(2\boldsymbol{k}-\boldsymbol{n})\boldsymbol{b}(\boldsymbol{n}) = 2^{-\alpha}\boldsymbol{b}(\boldsymbol{k}) \quad \textit{which is} \quad (\downarrow 2)\boldsymbol{G}\boldsymbol{b} = 2^{-\alpha}\boldsymbol{b} \tag{11.21}$$

$$\sum k^{\nu_1}\,k_2^{\nu_2}\,\boldsymbol{b}(k_1,k_2) = \mu_1!\,\mu_2!\;\delta(\nu_1-\mu_1)\;\delta(\nu_2-\mu_2)\;\textit{for}\;\nu_1\le\mu_1,\;\nu_2\le\mu_2. \tag{11.22}$$

Example 11.2. Compute $\boldsymbol{b}(k) = \int\phi(t)\phi'(t-k)\,dt$ with $\ell=1$ and $\mu_1=1$ (first derivative).

The eigenvalue problem will be $2\boldsymbol{T}\boldsymbol{b} = \boldsymbol{b}$. The familiar matrix $\boldsymbol{T} = (\downarrow 2)2\boldsymbol{H}\boldsymbol{H}^T$ has an extra 2 because of the derivative ϕ'. We have $\ell=1$ and $\mu_1=1$ (there is no μ_2). Then $\boldsymbol{b}$ is the eigenvector of $\boldsymbol{T}$ with $\lambda=\frac{1}{2}$. The hat function example has eigenvector $(-0.5, 0, 0.5)$ when normalized by (11.22):

$$\boldsymbol{T}\boldsymbol{b} = \frac{1}{8}\begin{bmatrix}\cdots & 6 & 4 & 1 & & & \\ & 1 & 4 & 6 & 4 & 1 & \\ & & & 1 & 4 & 6 & \cdots\end{bmatrix}\begin{bmatrix}-0.5\\0\\0.5\end{bmatrix} = \frac{1}{2}\begin{bmatrix}-0.5\\0\\0.5\end{bmatrix} = \frac{1}{2}\boldsymbol{b}.$$

The integral of the hat function times its derivative (which is $+1$ then -1) is $\boldsymbol{b}(0)=0$. The inner products with shifts are $-.5$ and $.5$ in agreement with $\boldsymbol{b}$.

Example 11.3. Compute $\boldsymbol{b}(k) = \int\phi(t)\phi''(t-k)\,dt = -\int\phi'(t)\phi'(t-k)\,dt$.

We can take $\ell=1$ and choose $\mu_1=2$, or we can take $\ell=2$ and choose $\phi_0(t)\equiv 1$ and $(\mu_1,\mu_2)=(1,1)$. In either case $\boldsymbol{b}$ is an eigenvector of $\boldsymbol{T}$ with eigenvalue $\lambda=\frac{1}{4}$. For the hat function matrix above, that eigenvector is $\boldsymbol{b}=(-1,2,-1)$. It is pleasant to go on to third derivatives (Problem 7). The eigenvalue for $\int\phi(t)\phi'''(t-k)\,dt = -\int\phi'(t)\phi''(t-k)\,dt$ is

$\lambda = \frac{1}{8}$. The matrix T has this eigenvalue. But we must use the 5×5 central submatrix instead of 3×3.

Example 11.4. Integrate the Daubechies D_4 function $\phi(t)$ over each interval $[0, 1]$, $[1, 2]$, $[2, 3]$.

Solution. We integrate $\phi(t)$ times translates of a box function, which has dilation coefficients $\frac{1}{2}$ and $\frac{1}{2}$. The Daubechies coefficients a, b, c, d add to 1 and are convolved with $(\frac{1}{2}, \frac{1}{2})$ to find $\frac{1}{2}(a, a+b, b+c, c+d, d)$. After double shift, the eigenvector $\boldsymbol{b}$ for $\lambda = 1$ gives the integrals of $\phi(t)$ over the three intervals:

$$\begin{bmatrix} 0 & d & c+d \\ c+d & b+c & a+b \\ a+b & a & 0 \end{bmatrix} \begin{bmatrix} b_0 \\ b_1 \\ b_2 \end{bmatrix} = \begin{bmatrix} b_0 \\ b_1 \\ b_2 \end{bmatrix} \quad \text{yields} \quad \begin{bmatrix} b_0 \\ b_1 \\ b_2 \end{bmatrix} = \frac{1}{12} \begin{bmatrix} 5+3\sqrt{3} \\ 2 \\ 5-3\sqrt{3} \end{bmatrix}.$$

The theory extends directly from a single variable t to d variables $\boldsymbol{x} = (x_1, \ldots, x_d)$. Each dilation equation involves integer vectors $\boldsymbol{k} = (k_1, \ldots, k_d)$:

$$\phi(\boldsymbol{x}) = 2^d \sum \boldsymbol{h}(\boldsymbol{k}) \phi(2\boldsymbol{x} - \boldsymbol{k}). \tag{11.23}$$

The convention is still $\sum \boldsymbol{h}(\boldsymbol{k}) = 1$. The ordinary derivatives $D^\mu \phi(t)$ become mixed partial derivatives $D^{\boldsymbol{\mu}} \phi(\boldsymbol{x}) = (\partial/\partial x_1)^{\mu_1} \cdots (\partial/\partial x_d)^{\mu_d} \phi(\boldsymbol{x})$ of order $|\mu| = \mu_1 + \cdots + \mu_d$. Substitute the dilation equations into the desired integral, which is

$$\boldsymbol{b}(\boldsymbol{k}_1, \ldots, \boldsymbol{k}_\ell) = \int \cdots \int \phi_0(\boldsymbol{x})\, D^{\boldsymbol{\mu}_1} \phi_1(\boldsymbol{x} - \boldsymbol{k}_1) \cdots D^{\boldsymbol{\mu}_\ell} \phi_\ell(\boldsymbol{x} - \boldsymbol{k}_\ell)\, d\boldsymbol{x}. \tag{11.24}$$

Each integer vector $\boldsymbol{k}_1, \ldots, \boldsymbol{k}_\ell$ has d components. When $2\boldsymbol{x}$ in the dilation equation is replaced by $\boldsymbol{x}$, the multiplying factor 2^α has $\alpha = \ell d + |\mu_1| + \cdots + |\mu_\ell|$. The two-scale equation for the integrals is

$$\boldsymbol{b}(\boldsymbol{k}_1, \ldots, \boldsymbol{k}_\ell) = 2^\alpha \sum \boldsymbol{g}(\boldsymbol{n}_1, \ldots, \boldsymbol{n}_\ell)\, \boldsymbol{b}\left(2\boldsymbol{k}_1 - \boldsymbol{n}_1, \ldots, 2\boldsymbol{k}_\ell - \boldsymbol{n}_\ell\right). \tag{11.25}$$

The coefficients $\boldsymbol{g}$ follow the model (11.16). Now $\boldsymbol{k}$ is $(k_1, \ldots, k_d)$ and each vector $\boldsymbol{n}_1, \ldots, \boldsymbol{n}_\ell$ has d components:

$$\boldsymbol{g}(\boldsymbol{n}_1, \ldots, \boldsymbol{n}_\ell) = \sum_{\boldsymbol{k}} \boldsymbol{h}_0(\boldsymbol{k}) \boldsymbol{h}_1(\boldsymbol{k} - \boldsymbol{n}_1) \cdots \boldsymbol{h}_\ell\left(\boldsymbol{k} - \boldsymbol{n}_\ell\right). \tag{11.26}$$

Integrals with Wavelets

Integrals involving wavelets come directly from the integrals $\boldsymbol{b}(k)$ involving scaling functions. The user must provide the highpass coefficients for any wavelets $w_0(t)$, $w_1(t)$,..., $w_\ell(t)$ that are in the integrals (staying for now at unit scale $j = 0$). Make those highpass replacements in $\boldsymbol{g}(\boldsymbol{n})$ to get $\boldsymbol{g}_w(n)$, leaving the lowpass coefficients whenever $\phi_i(t)$ is left in the integral. Then the wavelet integrals are $\boldsymbol{b}_w = 2^\alpha (\downarrow 2) \boldsymbol{g}_w * \boldsymbol{b}$.

To emphasize: Wavelets bring no new eigenvalue problem, just a trivial matrix multiplication (trivial if you manage all the indices). Start with a basic example to see the pattern. When $w(t)$ replaces $\phi(t)$, the matrix $\boldsymbol{H}_1$ replaces $\boldsymbol{H}$:

Theorem 11.2 *The wavelet integrals $\boldsymbol{a}_w(k) = \int \phi(t)w(t+k)\,dt$ and $\boldsymbol{a}_{ww}(k) = \int w(t)w(t+k)\,dt$ are computed from $\boldsymbol{a}$ by the equations*

$$\boldsymbol{a}_w = \boldsymbol{T}_w\boldsymbol{a} = (\downarrow 2)\, 2\, \boldsymbol{H}_1\boldsymbol{H}^T\boldsymbol{a} \quad \textit{and} \quad \boldsymbol{a}_{ww} = \boldsymbol{T}_{ww}\boldsymbol{a} = (\downarrow 2)2\boldsymbol{H}_1\boldsymbol{H}_1^T\boldsymbol{a}. \tag{11.27}$$

Proof: The wavelet equation in vector form is $\boldsymbol{W}(t) = (\downarrow 2)2\boldsymbol{H}_1\boldsymbol{\Phi}(2t)$, where

$$\boldsymbol{\Phi}(t) = \begin{bmatrix} \cdot \\ \phi(t-1) \\ \phi(t) \\ \phi(t+1) \\ \cdot \end{bmatrix} \quad \text{and} \quad \boldsymbol{W}(t) = \begin{bmatrix} \cdot \\ w(t-1) \\ w(t) \\ w(t+1) \\ \cdot \end{bmatrix}. \tag{11.28}$$

This vector form leads directly to the inner product vectors:

$$\boldsymbol{a}_w = \int_{-\infty}^{\infty} \phi(t)\boldsymbol{W}(t)\,dt \quad \text{and} \quad \boldsymbol{a}_{ww} = \int_{-\infty}^{\infty} w(t)\boldsymbol{W}(t)\,dt.$$

To compute $\boldsymbol{a}_w$ substitute the dilation and wavelet equations:

$$\boldsymbol{a}_w = \int_{-\infty}^{\infty} \left(\sum 2\boldsymbol{h}(k)\phi(2t-k)\right)(\downarrow 2)2\boldsymbol{H}_1\boldsymbol{\Phi}(2t)dt. \tag{11.29}$$

For the kth term in the sum, change variables to $u = 2t - k$ (so that $du = 2dt$). That kth term becomes

$$2\boldsymbol{h}(k)(\downarrow 2)\boldsymbol{H}_1 \int_{-\infty}^{\infty} \phi(u)\boldsymbol{\Phi}(u+k)\,du = 2\boldsymbol{h}(k)(\downarrow 2)\boldsymbol{H}_1\boldsymbol{S}^{-k}\boldsymbol{a}. \tag{11.30}$$

The k-step shift gave $\boldsymbol{\Phi}(u+k) = \boldsymbol{S}^{-k}\boldsymbol{\Phi}(u)$, and then $\int \phi(t)\boldsymbol{\Phi}(t)\,dt$ is exactly $\boldsymbol{a}$. Now sum on k to complete the formula:

$$\boldsymbol{a}_w = \sum_k 2\boldsymbol{h}(k)(\downarrow 2)\boldsymbol{H}_1\boldsymbol{S}^{-k}\boldsymbol{a} = (\downarrow 2)2\boldsymbol{H}_1\boldsymbol{H}^T\boldsymbol{a}. \tag{11.31}$$

Notice that $\sum \boldsymbol{h}(k)\boldsymbol{S}^{-k}$ is the *upper triangular Toeplitz matrix* $\boldsymbol{H}^T$. The same steps for $\boldsymbol{a}_{ww}$, wavelet times wavelet, produce the double-shifted Toeplitz matrix $\boldsymbol{T}_{ww} = (\downarrow 2)2\boldsymbol{H}_1\boldsymbol{H}_1^T$.

Example 11.5. For $\boldsymbol{h}(k) = \left(\frac{1}{4}, \frac{2}{4}, \frac{1}{4}\right)$ the piecewise linear hat function on $[0, 2]$ is $\phi(t)$. Its inner products $\boldsymbol{a}(k)$ come from $\boldsymbol{T}$:

$$\boldsymbol{T} = \frac{1}{8}\begin{bmatrix} 4 & 1 & 0 \\ 4 & 6 & 4 \\ 0 & 1 & 4 \end{bmatrix} \quad \text{has eigenvector } \boldsymbol{a} = \frac{1}{6}\begin{bmatrix} 1 \\ 4 \\ 1 \end{bmatrix}.$$

For the wavelets, we choose the highpass coefficients $\boldsymbol{h}_1(k) = \frac{1}{8}(-1, -2, 6, -2, -1)$. Then $H(z)H_1(-z)$ is the 6th degree Daubechies halfband filter and the synthesis functions $\widehat{\phi}(t)$ are biorthogonal to the hats. From the wavelet equation, $w(t)$ is also piecewise linear and it is supported on $[0, 3]$. The inner products with wavelet use matrix entries from $H_1(z)H(z^{-1})$ and $H_1(z)H_1(z^{-1})$:

$$\boldsymbol{a}_w = \frac{1}{16}\begin{bmatrix} -2 & -1 & & & \\ -2 & 6 & -2 & -1 & \\ & -1 & -2 & 6 & -2 \\ & & & -1 & -2 \end{bmatrix}\begin{bmatrix} 1 & & \\ 2 & 1 & \\ 1 & 2 & 1 \\ & 1 & 2 \\ & & 1 \end{bmatrix}\begin{bmatrix} 1/6 \\ 4/6 \\ 1/6 \end{bmatrix} = \begin{bmatrix} -1/12 \\ 1/12 \\ 1/12 \\ -1/12 \end{bmatrix}$$

$$a_{ww} = \frac{1}{32}\begin{bmatrix} 4 & 1 & \\ -20 & -8 & 4 \\ -20 & 46 & -20 \\ 4 & -8 & -20 \\ & 1 & 4 \end{bmatrix}\begin{bmatrix} 1/6 \\ 4/6 \\ 1/6 \end{bmatrix} = \begin{bmatrix} 1/24 \\ -6/24 \\ 18/24 \\ -6/24 \\ 1/24 \end{bmatrix}.$$

Integrals with multiple scales are also possible. For $a_1(k) = \int \phi(t)\phi(2t-k)\,dt$ we get a different eigenvalue problem $a_1 = (\downarrow 2)2UH^T a_1$. The reason is that the dilation equation for $\phi(2t)$ involves $4t$. On the right side is $h(k)\phi(4t-k)$ which is $u(2k)\phi(4t-k)$, provided $u = (\uparrow 2)h$. The matrix U with diagonals $h(0), 0, h(1), 0, h(2), 0, \ldots$ replaces H when the integral involves $\phi(2t)$.

Problem Set 11.6

1. Integrate the hat function on $[0, 2]$ times the box function on $[0, 1]$, using their dilation coefficients $(\frac{1}{4}, \frac{2}{4}, \frac{1}{4})$ and $(\frac{1}{2}, \frac{1}{2})$.
2. For the quadratic spline that comes from $h = \frac{1}{8}(1, 3, 3, 1)$, find the integrals over $[0, 1]$ and $[1, 2]$ and $[2, 3]$.

   Compute the integrals in **3.**–**6.** by creating the coefficients $g(n)$ from h_0, h_1, h_2 and solving the eigenvalue problem $(\downarrow 2)Gb = 2^{-\alpha}b$. Normalize by (11.22) and check by direct integration.

3. $\displaystyle\int_{-\infty}^{\infty} (\text{hat function})^3\,dt$
4. $\displaystyle\int_0^1 (\text{hat function})^2\,dt$
5. $\displaystyle\int_{-\infty}^{\infty} (D_4 \text{ function } \phi(t))^2\,dt$
6. $\displaystyle\int_0^1 (D_4 \text{ function } \phi(t))^2\,dt$
7. Follow Example 3 to integrate $\phi(t)\,\phi'''(t-k)$ for $\phi(t) =$ hat function.
8. The bilinear hat function in two variables $x = (x_1, x_2)$ has what formula and what coefficients $h(n)$? Find its inner products with its translates.
9. The frequency response $\sum h(k)\,e^{-i\omega k}$ gives the moments $H(0) = m_0 = 1$ and $H'(0) = -im_1$ and $H''(0) = -m_2$. Prove that an orthonormal filter with $H(\pi) = H'(\pi) = 0$ has $m_2 = m_1^2$, by setting $\omega = 0$ in the second derivative of $|H(\omega)|^2 + |H(\omega+\pi)|^2 = 1$.
10. Put $m_2 = m_1^2$ in equation (11.13) to show that $M_2 = M_1^2$ for orthonormal scaling functions with accuracy $p > 1$. Then the one-point rule $\int f(t)\phi(t)dt \approx f(M_1)$ has second-order accuracy.

Glossary

This book is necessarily written in two languages. One is mathematics and the other is signal processing. Translation between them is not difficult, and our goal is to make it easier. A second purpose is to give the reader (in either language) a quick reference to the many terms that are constantly used in this subject.

Before the alphabetical entries, we summarize the structure and notation for filter banks. The coefficients $\boldsymbol{h}(k)$ are multiplied by $\sqrt{2}$ in the lowpass channel. They are multiplied again by $\sqrt{2}$ in the dilation equation. The operator $(\downarrow 2)$ introduces a double shift $(j \to 2j)$ in the rows of $\boldsymbol{M}$.

Lowpass filter $\boldsymbol{H}$	Lowpass channel $\boldsymbol{C}$	Matrix $\boldsymbol{M} = \sqrt{2}\boldsymbol{L} = (\downarrow 2)2\boldsymbol{H}$
$\boldsymbol{h}(0), \ldots, \boldsymbol{h}(N)$	$\boldsymbol{c}(k) = \sqrt{2}\boldsymbol{h}(k)$	$\boldsymbol{M}_{jk} = 2\boldsymbol{h}(2j-k)$

The convention is $\sum \boldsymbol{h}(k) = 1$. The frequency response at $\omega = 0$ is $H(0) = 1$. The even-odd sum rule $\sum \boldsymbol{h}(2k) = \sum \boldsymbol{h}(2k+1) = \frac{1}{2}$ is then equivalent to a zero at $\omega = \pi$: $H(\pi) = 0$. All columns of $\boldsymbol{M}$ add to 1. The highpass filter $\boldsymbol{H}_1$ has zero response at $\omega = 0$.

Analysis Bank (the input vector $\boldsymbol{x}$ yields half-length outputs $\boldsymbol{v}_0$ and $\boldsymbol{v}_1$)

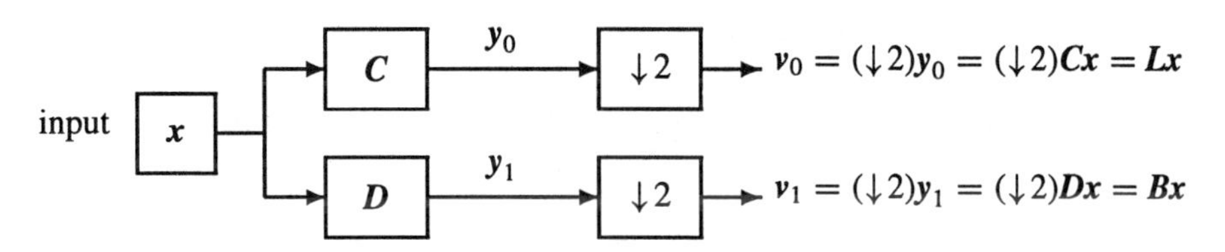

Synthesis Bank (two half-length inputs yield the output $\widehat{\boldsymbol{x}} = \boldsymbol{w}_0 + \boldsymbol{w}_1$)

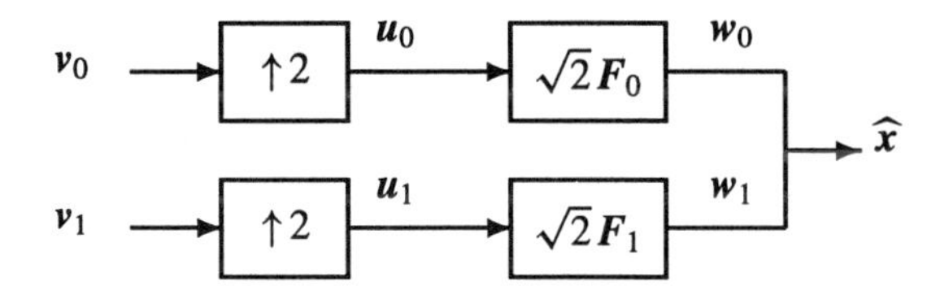

Flip Constructions (synthesis from analysis, highpass from lowpass)

$$\begin{aligned} \text{Ordinary flip:} \quad & F(z) = z^{-N}H(z^{-1}) \text{ gives } (\boldsymbol{h}(N), \boldsymbol{h}(N-1), \ldots, \boldsymbol{h}(1), \boldsymbol{h}(0)) \\ \text{Alternating flip:} \quad & H_1(z) = -z^{-N}H_0(-z^{-1}) \text{ gives } (\boldsymbol{h}(N), -\boldsymbol{h}(N-1), \ldots, -\boldsymbol{h}(0)) \\ \text{Alternating signs:} \quad & F_1(z) = -H_0(-z) \text{ gives } (-\boldsymbol{h}(0), \boldsymbol{h}(1), -\boldsymbol{h}(2), \boldsymbol{h}(3), \ldots) \end{aligned}$$

The ordinary flip transposes H, with N delays to make it causal.
The alternating flip (Smith-Barnwell, odd N) permits orthogonality.
Alternating signs in $F_1(z) = -H_0(-z)$ and $F_0(z) = H_1(-z)$ to cancel aliasing.

Perfect Reconstruction The synthesis filters cancel distortion and aliasing:

$$\begin{aligned} F_0(z)H_0(z) + F_1(z)H_1(z) &= 2z^{-\ell}, \quad \text{odd } \ell \\ F_0(z)H_0(-z) + F_1(z)H_1(-z) &= 0. \end{aligned}$$

The filters are double-shifted biorthogonal: $\sum_n \boldsymbol{f}_j(n+l)\,\boldsymbol{h}_k(n+2m) = \delta(j-k)\,\delta(m)$.

The modulation matrices satisfy $\boldsymbol{F}_m(z)\,\boldsymbol{H}_m(z) = \begin{bmatrix} z^{-\ell} & 0 \\ 0 & -z^{-\ell} \end{bmatrix}$

Orthogonal Filter Bank Synthesis is the transpose of analysis: $\boldsymbol{F}_p(z) = \boldsymbol{H}_p^T(z^{-1})$

$$\begin{bmatrix} \boldsymbol{L}^T & \boldsymbol{B}^T \end{bmatrix} \begin{bmatrix} \boldsymbol{L} \\ \boldsymbol{B} \end{bmatrix} = \boldsymbol{L}^T\boldsymbol{L} + \boldsymbol{B}^T\boldsymbol{B} = \boldsymbol{I}$$

$$\begin{bmatrix} \boldsymbol{L} \\ \boldsymbol{B} \end{bmatrix} \begin{bmatrix} \boldsymbol{L}^T & \boldsymbol{B}^T \end{bmatrix} = \begin{bmatrix} \boldsymbol{L}\boldsymbol{L}^T & \boldsymbol{L}\boldsymbol{B}^T \\ \boldsymbol{B}\boldsymbol{L}^T & \boldsymbol{B}\boldsymbol{B}^T \end{bmatrix} = \begin{bmatrix} \boldsymbol{I} & \boldsymbol{0} \\ \boldsymbol{0} & \boldsymbol{I} \end{bmatrix}$$

Dilation Equation and Wavelet Equation

$$\begin{aligned} \text{Scaling function:} \quad & \phi(t) = \sqrt{2}\sum \boldsymbol{c}(k)\phi(2t-k) = 2\sum \boldsymbol{h}_0(k)\phi(2t-k) \\ \text{Wavelet:} \quad & w(t) = \sqrt{2}\sum \boldsymbol{d}(k)\phi(2t-k) = 2\sum \boldsymbol{h}_1(k)\phi(2t-k) \end{aligned}$$

Biorthogonal Scaling Functions and Wavelets

$$\int_{-\infty}^{\infty} \phi(t)\widetilde{\phi}(t-k)\,dt = \delta(k) \quad \text{and} \quad \int_{-\infty}^{\infty} w(t)2^{j/2}\widetilde{w}(2^j t-k)\,dt = \delta(j)\delta(k)$$

$$\int_{-\infty}^{\infty} \phi(t)\widetilde{w}(t-k)\,dt = 0 \quad \text{and} \quad \int_{-\infty}^{\infty} \widetilde{\phi}(t)w(t-k)\,dt = 0$$

Orthogonal Scaling Functions and Wavelets ($\widetilde{\phi} \equiv \phi$ and $\widetilde{w} \equiv w$)

The functions $\phi(t-k)$ are an orthonormal basis for the scaling space V_0.
The wavelets $w_{jk}(t) = 2^{j/2}w(2^j t - k)$ are an orthonormal basis for $L^2(\mathbf{R})$.

Where possible, we give a description in the time domain, frequency domain, and z-domain. Note the multiple entries under **basis** (Dual basis, Orthonormal basis, Fourier basis, Haar basis, Riesz basis) and **transform** and **matrix** and **factorization**. Some definitions in the text are not repeated in this Glossary.

Aliasing (two inputs give the same output)

One input is often a modulation of the other, $X_1(\omega) = X_0(\omega+\pi)$ or $X_k(\omega) = X_0(\omega+\frac{2\pi k}{M})$. After $(\downarrow 2)$ or $(\downarrow M)$ these are aliased.

Alias Cancellation The output signal $\widehat{x}(n)$ has no aliased component of $x(n)$ if the product of the modulation matrices $\boldsymbol{F}_m(z)$ and $\boldsymbol{H}_m(z)$ is ***diagonal***. For $M = 2$ channels this means $F_0(z)H_0(-z) + F_1(z)H_1(-z) = 0$. The product of polyphase matrices is ***pseudo-circulant***:

$$\boldsymbol{F}_p(z)\boldsymbol{H}_p(z) = \begin{bmatrix} P_0(z) & P_1(z) & \cdots & P_{M-1}(z) \\ z^{-1}P_{M-1}(z) & P_0(z) & \cdots & P_{M-2}(z) \\ \vdots & \vdots & \ddots & \vdots \\ z^{-1}P_1(z) & z^{-1}P_2(z) & \cdots & P_0(z) \end{bmatrix}.$$

Then the only distortions at the output are in amplitude and phase. Examples of alias-free (non PR) filter banks are the ***two-channel Quadrature Mirror Filter (QMF) bank*** (known as the Johnston filter bank), ***the two-channel allpass-based IIR bank*** and the ***M-channel DFT bank***.

Aliasing Component (AC) Matrix Transpose of the modulation matrix $\boldsymbol{H}_m(z)$.

Allpass Filter An allpass digital filter has a rational frequency response with $|A(e^{j\omega})| = 1$. It has a sum form and a product form, so it can be implemented in a *cascaded lattice structure*.

$$A(z) = \frac{z^{-N}D(z^{-1})}{D(z)} = \frac{\sum \boldsymbol{d}(N-n)z^{-n}}{\sum \boldsymbol{d}(n)z^{-n}} = \prod_{k=1}^{N} \frac{\alpha_k^* + z^{-1}}{1+\alpha_k z^{-1}}.$$

- $\widetilde{A}(z)A(z) = 1;\ |A(e^{j\omega})| = 1.$
- If z_0 is a pole of $A(z)$, then $1/z_0^*$ is a zero of $A(z)$.
- The input and output have the same energy:

$$\sum_{n=-\infty}^{\infty} |y(n)|^2 = \sum_{n=-\infty}^{\infty} |x(n)|^2 = \frac{1}{2\pi}\int_0^{2\pi} |Y(e^{j\omega})|^2\, d\omega = \frac{1}{2\pi}\int_0^{2\pi} |X(e^{j\omega})|^2\, d\omega.$$

- The autocorrelation of its impulse response is $\boldsymbol{a}(n) * \boldsymbol{a}(-n) = \delta(n)$.
- A stable allpass filter has $|A(z)| > 1$ for $|z| < 1$ and $|A(z)| < 1$ for $|z| > 1$. The phase response $\phi(\omega)$ is decreasing. Then $\phi(0) = 0$ and $\phi(\pi) = -N\pi$.
- N first-order factors each require one multiply and two adds.

Allpass-based DFT IIR Filter Bank The figure below shows a DFT IIR filter bank where the polyphase components $A_k(z)$ are allpass functions. The analysis and synthesis filters are shifted copies $H_0(zW^k)$ and $F_0(zW^k)$ of the lowpass filters $H_0(z)$ and $F_0(z)$:

$$H_0(z) = \sum_{k=0}^{M-1} z^{-k} A_k(z^M), \quad F_0(z) = \sum_{k=0}^{M-1} z^{-(M-1-k)} B_k(z^M), \quad B(z) = \prod_{\ell \neq k} A_\ell(z)$$

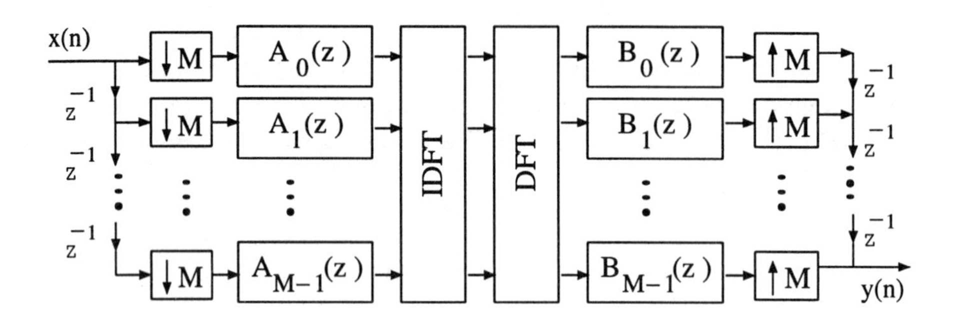

For $M > 2$, the synthesis filters are longer than the analysis filters. Aliasing is cancelled and there is no amplitude distortion:

$$T_0(z) = z^{-(M-1)} \prod_{\ell=0}^{M-1} A_\ell(z^M) \quad \text{has} \quad |T_0(e^{j\omega})| = 1.$$

The phase distortion is the sum of all the phase responses, which has to be equalized. For two channels, good lowpass filters can be designed since the stopband peak coincides with the transition band. The analysis filters $H_k(z)$ are sums and differences of $A_0(z)$ and $A_1(z)$:

$$\begin{cases} H_0(z) = \frac{1}{2}[A_0(z^2) + z^{-1}A_1(z^2)] \\ H_1(z) = \frac{1}{2}[A_0(z^2) - z^{-1}A_1(z^2)] \end{cases}$$

Butterworth, Chebyshev and elliptic filters (with appropriate specifications) can be decomposed into this form. The DFT filter bank is not PR. The distortion $T_0(z)$ is a product of $A_0(z)$ and $A_1(z)$. Let $\phi_0(\omega)$ and $\phi_1(\omega)$ be the phase responses, so that

$$|H_0(e^{j\omega})| = \tfrac{1}{2}|e^{j\phi_0(2\omega)} + e^{j(\phi_1(2\omega)-\omega)}| = \tfrac{1}{2}|1 + e^{j(\phi_1(2\omega)-\phi_0(2\omega)-\omega)}|.$$

H_0 is lowpass when ϕ_0 and ϕ_1 are in-phase in the passband and out-of-phase in the stopband. Suppose $A_0(z) = z^{-K}$ so that $\phi_0(\omega) = -K\omega$. Then $A_1(z)$ should have

$$\phi_1(2\omega) - 2K\omega - \omega = \begin{cases} 0; & \text{in the passband region} \\ \pi; & \text{in the stopband region.} \end{cases}$$

Basis (Dual basis, Haar basis, Orthonormal basis, Fourier basis, Riesz basis)

Start with ordinary three-dimensional space $\mathbf{R}^3$. A ***basis*** for $\mathbf{R}^3$ is a sequence $\boldsymbol{v}_1, \boldsymbol{v}_2, \boldsymbol{v}_3$ of vectors with a double property:

Every vector $\boldsymbol{v}$ *in the space can be written in* ***one and only one*** *way as a combination* $\boldsymbol{v} = \sum b_k \boldsymbol{v}_k$ *of the basis vectors.*

The basis vectors $\boldsymbol{v}_1, \boldsymbol{v}_2, \boldsymbol{v}_3$ are the columns of a 3 by 3 invertible matrix $\boldsymbol{A}$. Then $\boldsymbol{v} = b_1\boldsymbol{v}_1 + b_2\boldsymbol{v}_2 + b_3\boldsymbol{v}_3$ is $\boldsymbol{v} = \boldsymbol{Ab}$. Since $\boldsymbol{A}$ is invertible, there is exactly one solution $\boldsymbol{b} = \boldsymbol{A}^{-1}\boldsymbol{v}$. The three columns ***span*** the space. The only combination that produces $\boldsymbol{v} = 0$ is $b_1 = b_2 = b_3 = 0$ (the columns of $\boldsymbol{A}$ are ***independent***). The three columns are a basis for $\mathbf{R}^3$.

Similarly a sequence of functions $v_1(t), v_2(t), \ldots$ is a basis for a function space if every function can be written as $\sum b_k v_k(t)$ in one and only one way.

Biorthogonal basis (Dual basis) The *dual basis* $\boldsymbol{w}_1, \boldsymbol{w}_2, \ldots$ *is biorthogonal to the original basis*: $(\boldsymbol{v}_k, \boldsymbol{w}_\ell) = \delta_{k\ell}$. Then the numbers in $b_1\boldsymbol{v}_1 + b_2\boldsymbol{v}_2 + \ldots = \boldsymbol{v}$, when we take the inner product with each $\boldsymbol{w}_k$, must be $b_k = (\boldsymbol{v}, \boldsymbol{w}_k)$:

$$\textit{Dual bases yield } \boldsymbol{v} = \sum (\boldsymbol{v}, \boldsymbol{w}_k)\boldsymbol{v}_k \quad \text{and also} \quad \boldsymbol{v} = \sum (\boldsymbol{v}, \boldsymbol{v}_k)\boldsymbol{w}_k.$$

The ***analysis step*** finds $b_k = (\boldsymbol{v}, \boldsymbol{w}_k)$. The ***synthesis step*** combines the basis vectors into $\boldsymbol{v} = \sum b_k \boldsymbol{v}_k$. When the columns of $\boldsymbol{A}$ are the original basis $\boldsymbol{v}_1, \boldsymbol{v}_2, \boldsymbol{v}_3$, the dual basis is in *the rows of* $\boldsymbol{A}^{-1}$. The equation $\boldsymbol{A}^{-1}\boldsymbol{A} = \boldsymbol{I}$ makes the rows of $\boldsymbol{A}^{-1}$ biorthogonal to the columns of $\boldsymbol{A}$:

$$\boldsymbol{A}^{-1}\boldsymbol{A} = \begin{bmatrix} -- \boldsymbol{w}_1 -- \\ -- \boldsymbol{w}_2 -- \\ -- \boldsymbol{w}_3 -- \end{bmatrix} \begin{bmatrix} & & \\ \boldsymbol{v}_1 & \boldsymbol{v}_2 & \boldsymbol{v}_3 \\ & & \end{bmatrix} = \begin{bmatrix} 1 & 0 & 0 \\ 0 & 1 & 0 \\ 0 & 0 & 1 \end{bmatrix}$$

The numbers in $\boldsymbol{b} = \boldsymbol{A}^{-1}\boldsymbol{v}$ are the inner products of $\boldsymbol{v}$ with these rows of $\boldsymbol{A}^{-1}$: $b_k = (\boldsymbol{v}, \boldsymbol{w}_k)$. For complex vectors it is the rows of $\overline{\boldsymbol{A}}^{-1}$ that give the dual basis. The inner product takes the complex conjugate of those rows.

Haar basis This orthonormal basis starts with Haar's up-down wavelet $w(t)$. The basis function $w_{jk}(t) = 2^{j/2}w(2^j t - k)$ comes from rescaling to $2^j t$ and shifting the start to $t = 2^{-j}k$.

Orthonormal basis When the basis vectors are biorthogonal *to themselves*; the dual basis is the same as the original basis:

$$\begin{bmatrix} -- \boldsymbol{v}_1 -- \\ -- \boldsymbol{v}_2 -- \\ -- \boldsymbol{v}_3 -- \end{bmatrix} \begin{bmatrix} & & \\ \boldsymbol{v}_1 & \boldsymbol{v}_2 & \boldsymbol{v}_3 \\ & & \end{bmatrix} = \begin{bmatrix} 1 & 0 & 0 \\ 0 & 1 & 0 \\ 0 & 0 & 1 \end{bmatrix}$$

$\boldsymbol{A}^{-1}$ is the same as $\boldsymbol{A}^T$. In the complex case, a unitary matrix (orthonormal columns) has $\overline{\boldsymbol{A}}^T = \boldsymbol{A}^{-1}$. The general formula $\boldsymbol{v} = \sum (\boldsymbol{v}, \boldsymbol{w}_k)\boldsymbol{v}_k$ specializes to the orthogonal expansion $\boldsymbol{v} = \sum (\boldsymbol{v}, \boldsymbol{v}_k)\boldsymbol{v}_k$.

Example in $\mathbf{R}^N$: The N-point DFT matrix $\boldsymbol{F}_N$ (normalized by $1/\sqrt{N}$) is unitary. The 2-point DFT matrix is real and orthogonal:

$$\boldsymbol{F} = \frac{1}{\sqrt{2}}\begin{bmatrix} 1 & 1 \\ 1 & -1 \end{bmatrix} \quad \text{has} \quad \boldsymbol{F}^T\boldsymbol{F} = \boldsymbol{I}.$$

Fourier basis The complex exponentials $v_k(t) = \frac{1}{\sqrt{2\pi}}e^{ikt}$ form an orthonormal basis for all 2π-periodic functions $v(t)$ with finite energy $||v||^2$. The orthonormal basis formula $\boldsymbol{v} = \sum (\boldsymbol{v}, \boldsymbol{v}_k)\boldsymbol{v}_k$ becomes the Fourier series:

$$v(t) = \frac{1}{2\pi}\sum_{-\infty}^{\infty}(v, e^{ikt})e^{ikt} = \frac{1}{2\pi}\sum_{-\infty}^{\infty}\left(\int_0^{2\pi} v(t)e^{-ikt}dt\right)e^{ikt}.$$

Riesz basis In finite dimensions, every basis is a Riesz basis. In a Hilbert space, the requirement is that the transform between the signal (or function) $\sum b_k v_k$ and its coefficients b_k should be bounded above and below:

$$A\sum |b_k|^2 \le \|\sum b_k v_k\|^2 \le B\sum |b_k|^2 \text{ with } A > 0.$$

An *orthonormal* basis is a Riesz basis with $A = B = 1$. The translates of the hat function are a Riesz basis but not orthogonal, so $B > A$. The powers $1, t, t^2, \ldots$ are a basis but not a Riesz basis for $L^2[0, 1]$. Their inner products $(t^k, t^\ell) = 1/(k+\ell+1)$ go into the ill-conditioned Hilbert matrix. Its columns are independent but not uniformly (Riesz) independent. The infinite Hilbert matrix is not positive definite and the lower constant is $A = 0$.

$\phi(t-k)$ is a shift-invariant Riesz basis iff $A \le \sum |\widehat{\phi}(\omega + 2\pi k)|^2 \le B$.

Biorthogonal (vectors, functions, wavelets)

The vectors or functions $\boldsymbol{v}_1, \boldsymbol{v}_2, \ldots$ and $\boldsymbol{w}_1, \boldsymbol{w}_2, \ldots$ are biorthogonal if $(\boldsymbol{v}_k, \boldsymbol{w}_\ell) = \delta_{k\ell}$. Biorthogonal scaling functions are usually written as $\phi(t-k)$ and $\widetilde{\phi}(t-\ell)$. The inner product is $\int \phi(t-k)\widetilde{\phi}(t-\ell)dt = \delta_{k\ell}$. Biorthogonal wavelets $w_{jk}(t)$ and $\widetilde{w}_{JK}(t)$ have indices j, J for scaling and k, K for shift: $(w_{jk}, \widetilde{w}_{JK}) = \delta_{jJ}\delta_{kK}$. *A biorthogonal filter bank is a perfect reconstruction (PR) filter bank.*

Block Transform (no overlap when filters have length M)

The signal is blocked into vectors of length L by an S/P converter, multiplied by an $M \times L$ matrix $\boldsymbol{T}$, and coded. The processed vector is inverse-transformed by $\boldsymbol{Q}$ (of size $L \times M$) and unblocked by a P/S converter.

For $M = L$, a block transform is a simple PR filter bank when $\boldsymbol{Q} = \boldsymbol{T}^{-1}$. The filters have length M and the polyphase matrices are constant. With no processing at the transformed vector, the output is an exact copy $\widehat{\boldsymbol{x}}(n) = \boldsymbol{x}(n-M+1)$ of the input.

Block transforms have low complexity. The design objective is to find $\boldsymbol{T}$ and $\boldsymbol{Q}$ such that $\boldsymbol{QT} = \boldsymbol{I}$ and the coding gain is maximized. The DCT and LOT are orthogonal block transforms with $L = M$ and $L = 2M$.

Cascade Algorithm $\phi^{(i+1)}(t) = 2\sum \boldsymbol{h}(k)\phi^{(i)}(2t-k)$
The limit $\phi(t)$, if it exists, solves the dilation equation. Often $\phi^{(0)}(t) =$ box function. It is required that $\sum \phi^{(0)}(t-k) \equiv 1$.

Causal Filter ($\boldsymbol{h}(k) = 0$ for negative k). The response cannot precede the source.

Coding A process that converts signals, images, or video sequences to a bit stream that is stored or transmitted. It is normally referred as *compression* when essential information is extracted and stored using as few bits as possible.

Condition Number The number $c = \|A\|\,\|A^{-1}\|$ estimates the relative error in the solution to $\boldsymbol{Ax} = \boldsymbol{b}$ in terms of the relative error in $\boldsymbol{b}$:

$$\begin{array}{rcll} \boldsymbol{Ax} = \boldsymbol{b} & \text{gives} & \|\boldsymbol{b}\| & \le \|A\|\,\|\boldsymbol{x}\| \\ A\Delta\boldsymbol{x} = \Delta\boldsymbol{b} & \text{gives} & \|\Delta\boldsymbol{x}\| & \le \|A^{-1}\|\,\|\Delta\boldsymbol{b}\| \end{array}$$

Multiply the inequalities and divide by $\|\boldsymbol{b}\|\,\|\boldsymbol{x}\|$ to get relative errors:

$$\frac{\|\Delta\boldsymbol{x}\|}{\|\boldsymbol{x}\|} \le \|A\|\,\|A^{-1}\|\,\frac{\|\Delta\boldsymbol{b}\|}{\|\boldsymbol{b}\|} = c\,\frac{\|\Delta\boldsymbol{b}\|}{\|\boldsymbol{b}\|}.$$

Rule of thumb: Roundoff error loses $\log c$ decimal places in the computed solution.

Convolution The nth entry of $\boldsymbol{h} * \boldsymbol{x}$ is $\sum \boldsymbol{h}(k)\boldsymbol{x}(n-k) = \sum \boldsymbol{h}(n-k)\boldsymbol{x}(k)$.

Cosine-modulated Filter Bank (see also Paraunitary Cosine-modulated Filter Bank)

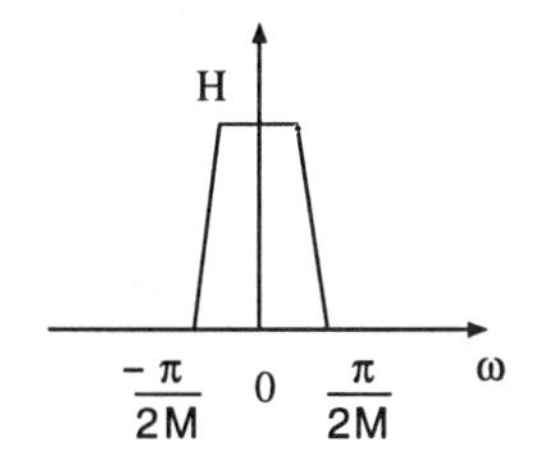

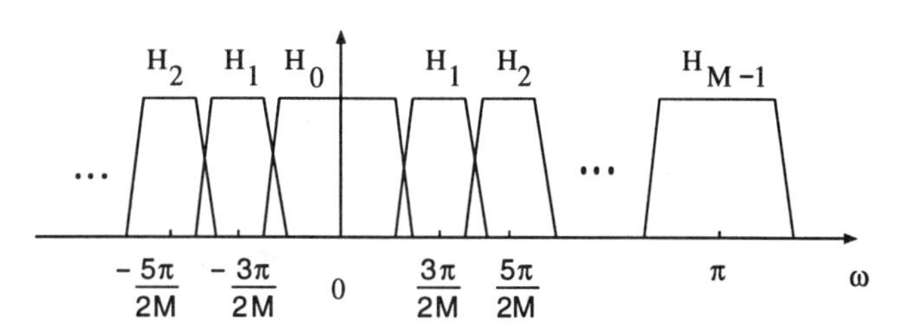

Let $H(z)$ and $F(z)$ be the prototype filters of the analysis and synthesis banks. By modulation (= frequency shifts) the M analysis filters $H_k(z)$ and synthesis filters $F_k(z)$ come from $H(z)$ and $F(z)$. The cosine-modulated filters are defined as:

$$\begin{cases} H_k(z) = a_k b_k U_k(z) + a_k^* b_k^* U_k^*(z) \\ F_k(z) = a_k^* b_k V_k(z) + a_k b_k^* V_k^*(z) \end{cases}$$

where

$$\begin{cases} U_k(z) = H(zW^{k+\frac{1}{2}}) \\ V_k(z) = F(zW^{k+\frac{1}{2}}) \end{cases} \qquad \begin{cases} a_k = e^{j(-1)^k\frac{\pi}{4}} \\ b_k = W^{-\frac{1}{2}(k+\frac{1}{2})} \end{cases}$$

The number $W = e^{-j\frac{\pi}{M}}$ shifts the frequency by π/M. Notice W^k and also $W^{1/2}$. The analysis and synthesis filter coefficients are cosine-modulated versions of the prototype coefficients:

$$\begin{cases} h_k(n) = 2h(n)\cos((2k+1)\frac{\pi}{2M}(n+\frac{1}{2}) + (-1)^k\frac{\pi}{4}) \\ f_k(n) = 2f(n)\cos((2k+1)\frac{\pi}{2M}(n+\frac{1}{2}) - (-1)^k\frac{\pi}{4}). \end{cases}$$

The only parameters to be chosen are the coefficients $h(n)$ and $f(n)$ in the (symmetric) lowpass filters $H(z)$ and $F(z)$.

Cosine Transform The four types of Discrete Cosine Transform (DCT) are

$$\begin{cases} \text{I:} & \sqrt{\frac{2}{M}}c_i c_k \cos(ik\frac{\pi}{M}), & i,k=0,\dots,M \\ \text{II:} & \sqrt{\frac{2}{M}}c_i \cos\left(i(k+\frac{1}{2})\frac{\pi}{M}\right), & i,k=0,\dots,M-1 \\ \text{III:} & \sqrt{\frac{2}{M}}c_k \cos\left(k(i+\frac{1}{2})\frac{\pi}{M}\right), & i,k=0,\dots,M-1 \\ \text{IV:} & \sqrt{\frac{2}{M}} \cos\left((i+\frac{1}{2})(k+\frac{1}{2})\frac{\pi}{M}\right), & i,k=0,\dots,M-1 \end{cases}$$

where $c_i = 1$ except for $c_0 = 1/\sqrt{2}$. The Discrete Sine Transform is defined similarly (with cos replaced by sin). All these transforms are orthogonal. The Type IV DCT and DST are symmetric:

1. $\boldsymbol{C}^T = \boldsymbol{C}$ and $\boldsymbol{S}^T = \boldsymbol{S}$ and $\boldsymbol{C}^T\boldsymbol{C} = \boldsymbol{I}$ and $\boldsymbol{S}^T\boldsymbol{S} = \boldsymbol{I}$.
2. $\boldsymbol{C} = \Gamma\boldsymbol{S}\boldsymbol{J}, \boldsymbol{S} = \Gamma\boldsymbol{C}\boldsymbol{J}$; $\boldsymbol{J}$ is the reverse identity and Γ is diagonal with $\Gamma_{kk} = (-1)^k$.

Crosstalk Crosstalk in a transmultiplexer corresponds to aliasing in a filter bank. The first stage of a transmultiplexer combines M signals into one. (The analysis stage of a filter bank separates one signal into M.) The second stage of the transmultiplexer reconstructs all M input signals from one transmitted signal, provided the system has no crosstalk (and also provided there is no amplitude or phase distortion; see the PR condition for filter banks).

Decibel (dB) The magnitude response $|H(e^{j\omega})|$ can be plotted in both normal scale and decibel (dB) scale. In dB scale, a positive number a is $20 * \log_{10}(a)$. Thus, -20 dB corresponds to $a = 0.1$ and -100 dB corresponds to $a = 10^{-5}$.

Decimation (Downsampling) $\boldsymbol{y}(n) = (\downarrow M)\boldsymbol{x}(n) = \boldsymbol{x}(nM)$

$$Y(z) = \frac{1}{M}\sum_{k=0}^{M-1} X(z^{1/M}W_M^k) \quad \text{and} \quad Y(e^{j\omega}) = \frac{1}{M}\sum_{k=0}^{M-1} X(e^{\frac{j}{M}(\omega-2\pi k)}).$$

The output of a decimator has M copies of the "stretched" input spectrum. The first term ($k = 0$) is the stretched spectrum $X(z^{1/M}) = X(e^{j\omega/M})$. The remaining $M - 1$ terms are the aliased versions. If the input spectrum $X(e^{j\omega})$ is bandlimited to $-\frac{\pi}{M} < \omega \leq \frac{\pi}{M}$, there is no contribution from the alias versions in $-\pi < \omega \leq \pi$. If $X(e^{j\omega})$ is not bandlimited to this range, the output from $(\downarrow M)$ is aliased. This is the main reason for lowpass filtering the input signal before decimation. Chapter 3 shows examples where the input is not bandlimited and aliasing occurs.

Decimation Filter (Filter then decimate by $(\downarrow M)$, or polyphase equivalent)

Using the Noble Identity, the equivalent operations are a delay chain with decimators (Serial to Parallel converters) followed by polyphase filtering at the lower rate.

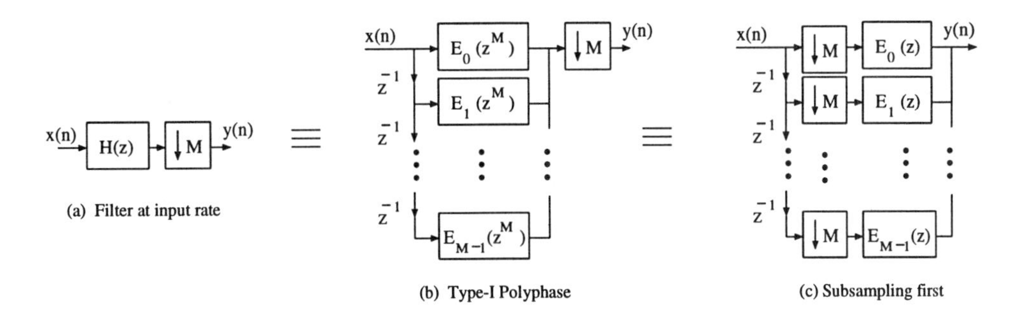

(a) Filter at input rate (b) Type-I Polyphase (c) Subsampling first

Delay and Advance (Shift S and S^{-1})

A sequence $h(n)$ is delayed to $h(n-1)$. Its z-transform is multiplied by z^{-1}. The shift matrix S that transforms $h(n)$ to $Sh(n) = h(n-1)$ has ones on the first subdiagonal.

A sequence $h(n)$ is advanced to $S^{-1}h(n) = h(n+1)$. Its transform is multiplied by $e^{j\omega} = z$. The matrix S^{-1} has ones just above the diagonal.

Delay Chain, Serial-Parallel Converter, Parallel-Serial Converter

The transfer function from the input of a delay chain to the kth output is z^{-k}.

Delay chain cascaded with decimators	=	Serial to Parallel (S/P) converter
Interpolators cascaded with delay chain	=	Parallel to Serial (P/S) converter

The output at the kth branch of the S/P converter is $x(nM-k)$, so the input sequence is selected *counter-clockwise*: $0,\ M-1, \ldots, 1,\ 0,\ M-1, \ldots$. The output rate of an S/P converter is $1/M$ of the input rate.

A P/S converter cascades expanders with a reverse-ordered delay chain. The inputs $y_k(n)$ are selected *clockwise*: $y_{M-1}(0)\cdots y_0(0)\, y_{M-1}(1)\cdots y_0(1)\cdots$.

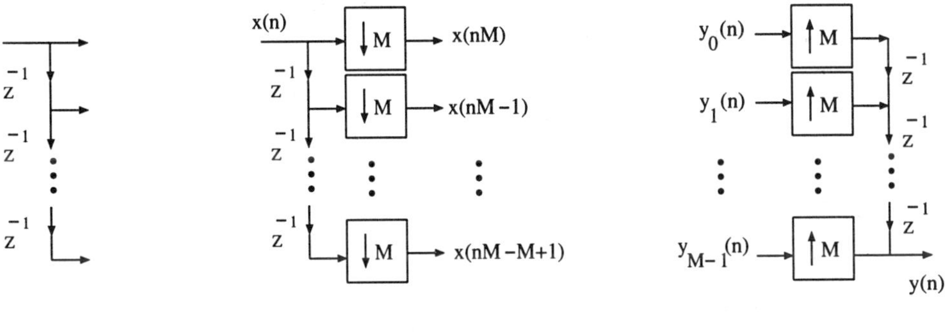

(a). Delay chain (b). Serial to parallel converter (c). Parallel to serial converter

DFT Filter Bank The analysis and synthesis filters $H_k(z)$ and $F_k(z)$ are uniformly shifted versions ($z \to zW^k$) of the lowpass filters. Perfect reconstruction can not be achieved unless the filter length is M. When the polyphase components of $F_0(z)$ and $H_0(z)$ are related by $R_k(z) = \prod_{\ell \neq k} E_\ell(z)$, the overall distortion is $T_0(z) = z^{-(M-1)} \prod E_\ell(z^M)$. The output should be equalized.

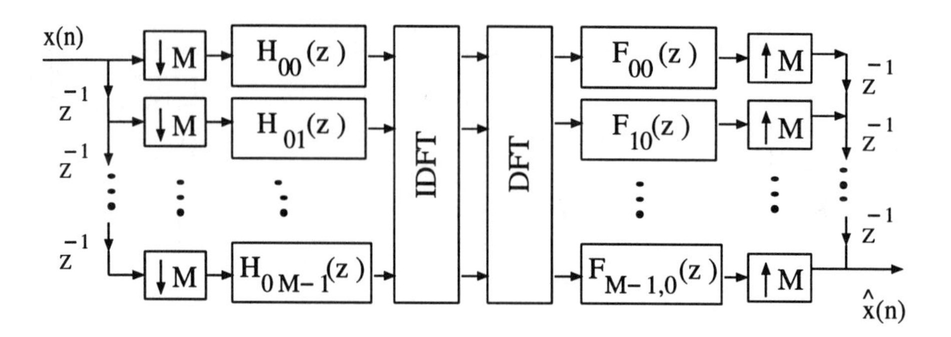

Dilation $\phi(t)$ is "dilated" to $\phi(at)$. The transform $\widehat{\phi}(\omega)$ is rescaled to $\frac{1}{a}\widehat{\phi}(\frac{\omega}{a})$.

Dilation Equation (Refinement equation; two-scale difference equation)

From a lowpass filter with $\sum \boldsymbol{h}(n) = 1$, the dilation equation for the scaling function $\phi(t)$ is

$$\phi(t) = 2\sum \boldsymbol{h}(k)\phi(2t-k) \quad \Longleftrightarrow \quad \widehat{\phi}(\omega) = H\left(\frac{\omega}{2}\right)\widehat{\phi}\left(\frac{\omega}{2}\right).$$

The infinite matrix-vector form in Section 6.2 is $\Phi_\infty(t) = \boldsymbol{M}\Phi_\infty(2t)$ with $\boldsymbol{M}_{k\ell} = 2\boldsymbol{h}(2k-\ell)$. The two-coefficient vector form is $\Phi(t) = \boldsymbol{m}(0)\Phi(2t) + \boldsymbol{m}(1)\Phi(2t-1)$. This applies for $0 \le t < 1$ and gives the unit slices of $\phi(t)$.

Eigenfilter Let $D(\omega)$, $H(e^{j\omega})$ and $W(e^{j\omega})$ denote the complex-valued desired function, the designed function and a weighting function. The eigenfilter approach minimizes

$$E_{LS} = \int_R \left|W(e^{j\omega})\right| \left|\gamma D(\omega) - H(e^{j\omega})\right|^2 d\omega$$

where R excludes the transition band. $\gamma = H(e^{j0})$ is chosen such that E_{LS} can be expressed in the quadratic form $E_{LS} = \boldsymbol{h}^t\boldsymbol{P}\boldsymbol{h}$, where $\boldsymbol{h}$ is the vector of filter coefficients and $\boldsymbol{P}$ is symmetric positive definite. The vector $\boldsymbol{h}$ that minimizes the least-square error E_{LS} is the eigenvector corresponding to the minimum eigenvalue of $\boldsymbol{P}$. Applications of eigenfilter design:

1. Both time- and frequency-domain specifications
2. Arbitrary magnitude and phase response
3. Hilbert Transformer, Differentiator, Beamformer
4. Allpass filter and Nyquist filter

Eigenvalues and Eigenvectors (Diagonalization of a matrix $\boldsymbol{A}$)

For eigenvectors, $\boldsymbol{Ax}$ is parallel to $\boldsymbol{x}$. Thus, $\boldsymbol{Ax} = \lambda\boldsymbol{x}$. The eigenvalues satisfy the polynomial equation $\det(\boldsymbol{A} - \lambda\boldsymbol{I}) = 0$. There are n eigenvalues $\lambda_1, \ldots, \lambda_n$ (including possible repetitions), and the matrix $\boldsymbol{S}$ of eigenvectors has $\boldsymbol{AS} = \boldsymbol{S}\Lambda$:

$$\boldsymbol{A}\begin{bmatrix} \boldsymbol{x}_1 & \cdots & \boldsymbol{x}_n \end{bmatrix} = \begin{bmatrix} \lambda_1\boldsymbol{x}_1 & \cdots & \lambda_n\boldsymbol{x}_n \end{bmatrix} = \begin{bmatrix} \boldsymbol{x}_1 & \cdots & \boldsymbol{x}_n \end{bmatrix}\begin{bmatrix} \lambda_1 & & \\ & \ddots & \\ & & \lambda_n \end{bmatrix}.$$

If the eigenvectors are independent, this equation $AS = S\Lambda$ gives $A = S\Lambda S^{-1}$. The differential equation $u' = Au$ is uncoupled to $(S^{-1}u)' = \Lambda(S^{-1}u)$.

Equiripple Filter The specifications on $H(\omega)$ are ω_p (the end of the passband region), ω_s (the start of the stopband region), δ_p (the passband error), δ_s (the stopband error), and $N+1$ (the filter length). The same maximum error is reached in every ripple of the frequency response $H(\omega)$. This equiripple condition (with $N+2$ ripples) is necessary and sufficient to minimize the maximum error in the bands. It is possible to specify different errors δ_p and δ_s and then to estimate the filter length by

$$N = \frac{-20\log_{10}\sqrt{\delta_1\delta_2} - 13}{14.6\Delta f}$$

where $\Delta f = (\omega_s - \omega_p)/(2\pi)$. The design algorithm uses the Remez exchange method, iteratively reducing the highest ripple until all ripples are equal. This is the *Parks-McClellan algorithm.* Of all problems in Chebyshev (minimax) approximation, this is the most frequently solved.

Here are equiripple responses in normal and dB scales.

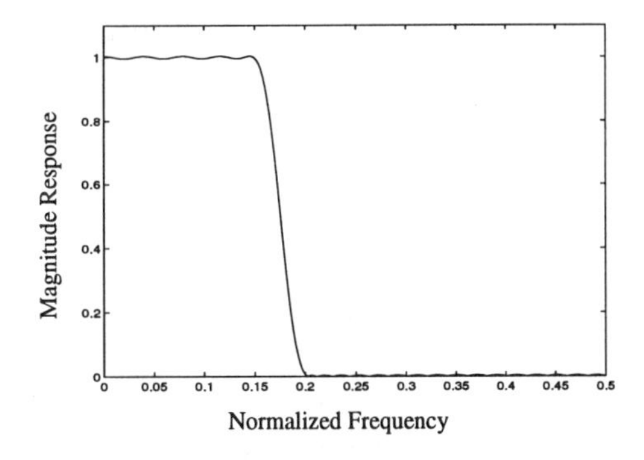

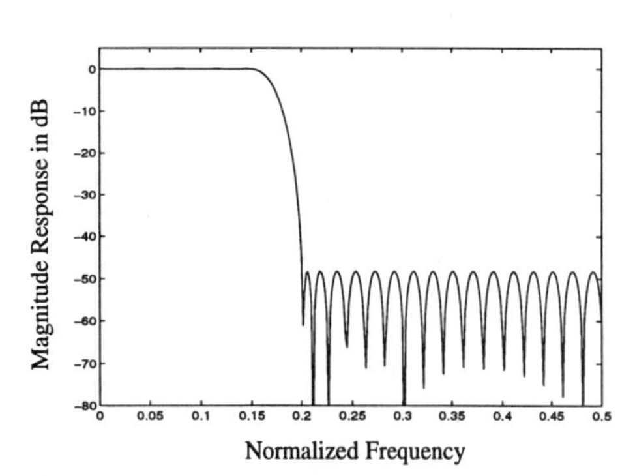

Expander (Upsampling Operator)

$$u(n) = \begin{cases} x(n/M), & n \text{ is a multiple of } M \\ 0, & \text{otherwise,} \end{cases} \qquad \begin{cases} U(z) &= X(z^M), \\ U(e^{j\omega}) &= X(e^{jM\omega}) \end{cases}$$

$u(n)$ is a "compressed" copy of $x(n)$. There are images repeated at $\omega_k = \frac{2\pi k}{M}$. It is necessary to lowpass the output to suppress the images.

Factorization of Matrices (see also Filter Factorizations)

• $A = LU = \begin{bmatrix} \text{lower triangular } L \\ \text{1's on the diagonal} \end{bmatrix} \begin{bmatrix} \text{upper triangular } U \\ \text{pivots on the diagonal} \end{bmatrix}$

There must be no row exchanges as Gaussian elimination reduces A to U. The upper left matrices $\begin{bmatrix} a_{11} \end{bmatrix}$, $\begin{bmatrix} a_{11} & a_{12} \\ a_{21} & a_{22} \end{bmatrix}$, $\ldots, A$ must all be invertible. Pivots are always nonzero!

• $A = C^TC = \begin{bmatrix} \text{lower triangular} \\ \text{positive diagonal} \end{bmatrix} \begin{bmatrix} \text{upper triangular} \\ \text{positive diagonal} \end{bmatrix}$ = Cholesky factors A is symmetric and positive definite (all n pivots are positive). The square roots of the pivots are on the diagonal of C and C^T.

• $A = QR = \begin{bmatrix} \text{orthogonal} \\ \text{columns in } Q \end{bmatrix} \begin{bmatrix} \text{upper triangular } R \\ \text{with positive diagonal} \end{bmatrix}$

A must have independent columns (rank n). Those are *orthogonalized* in Q by the Gram-Schmidt process. If A is square then Q is square and $Q^{-1} = Q^T$.

• $A = S\Lambda S^{-1} = \begin{bmatrix} \text{eigenvectors in } S \end{bmatrix} \begin{bmatrix} \text{eigenvalues in } \Lambda \end{bmatrix} \begin{bmatrix} \text{left eigenvectors in } S^{-1} \end{bmatrix}$

For this reduction to diagonal Λ, A must have n linearly independent eigenvectors (in the columns of S). *Diagonalization* is always possible when A has n different eigenvalues.

• $A = Q\Lambda Q^T = \begin{bmatrix} \text{orthogonal matrix } Q \end{bmatrix} \begin{bmatrix} \text{real eigenvalue matrix } \Lambda \end{bmatrix} \begin{bmatrix} Q^T \text{ is } Q^{-1} \end{bmatrix}$

A is real and symmetric. Eigenvectors are orthogonal and eigenvalues are real.

• $A = MJM^{-1} = \begin{bmatrix} \text{generalized eigenvectors} \end{bmatrix} \begin{bmatrix} \text{Jordan blocks} \end{bmatrix} \begin{bmatrix} \text{inverse of } M \end{bmatrix}$

A is any square matrix. Number of independent eigenvectors of A = number of blocks in the *Jordan form* J. Each block is $\lambda I +$ (superdiagonal of ones).

• $A = U\Sigma V^T = \begin{bmatrix} \text{orthogonal} \\ m \times m \text{ matrix} \end{bmatrix} \begin{bmatrix} m \times n \text{ singular value matrix} \\ \sigma_1, \ldots, \sigma_r \text{ on its diagonal} \end{bmatrix} \begin{bmatrix} \text{orthogonal} \\ n \times n \text{ matrix} \end{bmatrix}$

This *singular value decomposition* (SVD) has the eigenvectors of A^TA in U and of AA^T in V. These columns give orthonormal bases for $\mathbf{R}^m$ and $\mathbf{R}^n$. These bases diagonalize A to Σ, containing the singular values $\sigma_i = \sqrt{\lambda_i(A^TA)} = \sqrt{\lambda_i(AA^T)}$.

• $F_n = \begin{bmatrix} I & D \\ I & -D \end{bmatrix} \begin{bmatrix} F_{n/2} & \\ & F_{n/2} \end{bmatrix} \begin{bmatrix} \text{even-odd} \\ \text{permutation} \end{bmatrix}$ = Fast Fourier Transform

F_n = Fourier matrix = DFT matrix with entries $W^{k\ell}$, where $W = e^{2\pi i/n}$ and $W^n = 1$. D has $1, W, W^2, \ldots$ on its diagonal. For $n = 2^m$, the FFT has $\frac{1}{2}nm = \frac{1}{2}n \log_2 n$ multiplications from m stages of D's. The flow graph has m stages of butterflies = 2-point DFT's.

Fast Wavelet Transform (FWT) = Pyramid Algorithm = Mallat Algorithm

Iterate with $L = \sqrt{2}(\downarrow 2)H_0$ and $B = \sqrt{2}(\downarrow 2)H_1$ on the lowpass output only. Continue to any level in $2LN$ operations because $1 + \frac{1}{2} + \cdots = 2$. The signal and filter lengths are L and N.

$$\begin{array}{ccccccccc} x & \to & Lx & \to & LLx & \to & \cdots & \to & LL\cdots Lx \\ & \searrow & & \searrow & & \searrow & & \searrow & \\ & & Bx & & BLx & & \cdots & & BL\cdots Lx \end{array}$$

Inverse Fast Wavelet Transform (IFWT) (Reverse Pyramid Algorithm)

$$\begin{array}{ccccccccc} LL\cdots Lx & \to & \cdots & \to & LLx & \to & Lx & \to & x \\ & \nearrow & & \nearrow & & \nearrow & & \nearrow & \\ BL\cdots Lx & & \cdots & & BLx & & Bx & & \end{array}$$

Filter An LTI system (Linear Time Invariant) is characterized by its impulse response $\boldsymbol{h}(n)$, and its frequency response $H(e^{j\omega}) = \sum \boldsymbol{h}(n)e^{-j\omega n}$, and its transfer function $H(z) = \sum \boldsymbol{h}(n)z^{-n}$. For an exponential input a^n, the output $y(n) = H(a)a^n$ is also an exponential. An input $\boldsymbol{x}(n)$ is filtered (convolved) by $\boldsymbol{y}(n) = \sum_m \boldsymbol{h}(m)\boldsymbol{x}(n-m)$. The frequency response and the z-transform of the output are $Y(e^{j\omega}) = H(e^{j\omega})X(e^{j\omega})$ and $Y(z) = H(z)X(z)$. In the time domain $\boldsymbol{H}$ is a Toeplitz matrix. The figure shows idealized magnitude responses.

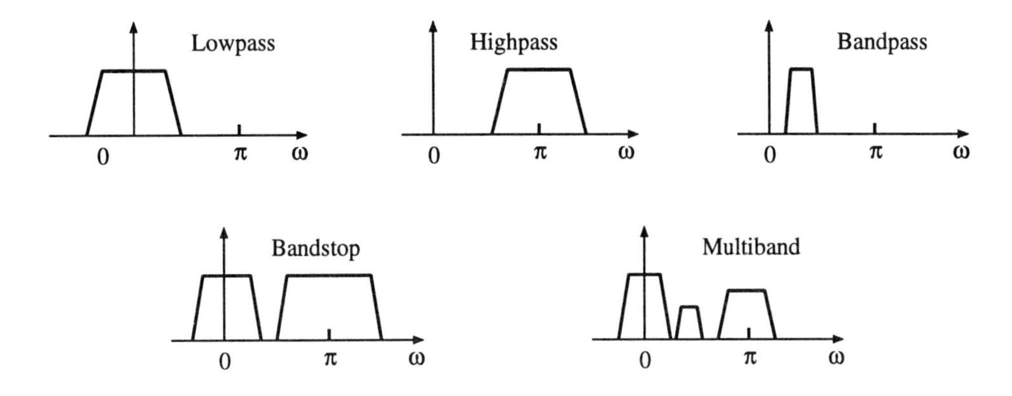

Causality & Stability ($\boldsymbol{h}(n) = 0$ for $n < 0$ and $\sum |\boldsymbol{h}(n)| < \infty$)

The system is *causal* if the output $\boldsymbol{y}(n)$ depends only on the input at time n and earlier. Then $\boldsymbol{h}(n) = 0$ for $n < 0$. For a stable LTI system, the output is bounded. The impulse response must be *absolutely summable*: $\sum |\boldsymbol{h}(n)| < \infty$. A *causal* and *stable* rational transfer function $H(z)$ has all of its poles inside the unit circle.

Filter Bank A *maximally decimated uniform* filter bank is shown below. The analysis bank channels the input signal into M subbands using a parallel set of bandpass filters. The synthesis bank reconstructs $\boldsymbol{x}(n)$ from the subband signals. The essential information is extracted from the subband signals in the processing block between banks. In audio/image compression, the spectral contents of the subband signals are coded depending on their energies. In a radar system, the subband signals might be used to null out a narrow-band interference adaptively. Filter banks are applied in audio/image compression, image analysis and enhancement, robotics, computer vision, echo cancellation, radar, voice privacy.

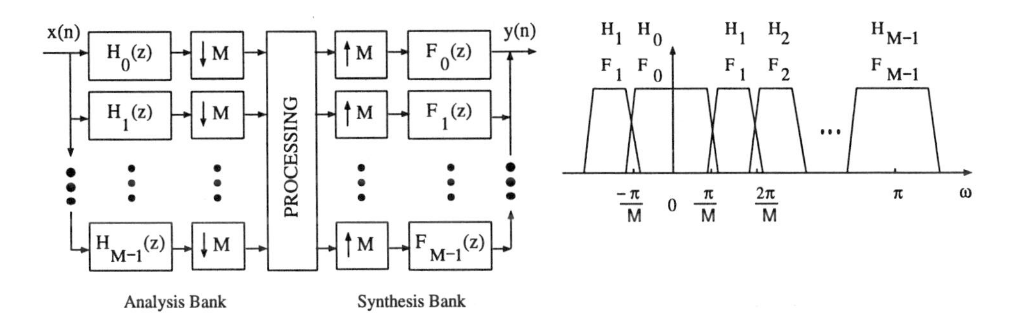

Since the analysis filters $H_k(z)$ are not ideal bandpass filters, the outputs are not strictly bandlimited to π/M. After $(\downarrow M)$, the aliased level depends on the stopband attenuation of $H_k(e^{j\omega})$. The interpolated signals from $(\uparrow M)$ have M "images" of the compressed spectrum $V_k(e^{j\omega})$. These images are filtered by the synthesis filters $F_k(z)$.

Filter Factorizations (leading to lattice structures)

• Givens Factorization: Orthogonal matrix $\boldsymbol{Q}$ into $\binom{n}{2}$ plane rotations $\boldsymbol{R}$.

$$\boldsymbol{Q} = \left[\ \boldsymbol{R}_{n-2,n-1}\ \right] \cdots \left[\ \boldsymbol{R}_{1,n-1}\ \ \cdots\ \ \boldsymbol{R}_{12}\ \right]\left[\ \boldsymbol{R}_{0,n-1}\ \ \cdots\ \ \boldsymbol{R}_{01}\ \right].$$

A 3×3 orthonormal matrix has three (Euler angles) factors:

$$\begin{bmatrix} 1 & 0 & 0 \\ 0 & \cos\theta_{12} & \sin\theta_{12} \\ 0 & -\sin\theta_{12} & \cos\theta_{12} \end{bmatrix} \begin{bmatrix} \cos\theta_{02} & 0 & \sin\theta_{02} \\ 0 & 1 & 0 \\ -\sin\theta_{02} & 0 & \cos\theta_{02} \end{bmatrix} \begin{bmatrix} \cos\theta_{01} & \sin\theta_{01} & 0 \\ -\sin\theta_{01} & \cos\theta_{01} & 0 \\ 0 & 0 & 1 \end{bmatrix}.$$

• Householder Factorization: (Orthogonal matrix $\boldsymbol{Q}$ into plane reflections).

$$\boldsymbol{Q} = \boldsymbol{H}_{n-1} \cdots \boldsymbol{H}_2 \boldsymbol{H}_1 \quad \text{with} \quad \boldsymbol{H}_j = \boldsymbol{I} - 2\frac{\boldsymbol{v}_j \boldsymbol{v}_j^T}{\boldsymbol{v}_j^T \boldsymbol{v}_j}$$

• Smith-McMillan Factorization: $\boldsymbol{P}(z) = \boldsymbol{W}(z)\Gamma(z)\boldsymbol{V}(z)$. $\boldsymbol{W}(z)$ and $\boldsymbol{V}(z)$ are square polynomial matrices of determinant 1 and $\Gamma(z) = \operatorname{diag}(\gamma_0(z), \gamma_1(z), \ldots)$ with γ_k dividing γ_{k-1}.

Flatness Constraints $H'(e^{j\omega})$ has a zero of order p at ω_0

Fourier Transform (between the time domain and frequency domain)

Continuous: $\widehat{f}(\omega) = \int_{-\infty}^{\infty} f(t)e^{-i\omega t}\,dt$ and $f(t) = \frac{1}{2\pi}\int_{-\infty}^{\infty} \widehat{f}(\omega)e^{i\omega t}\,d\omega$.

Discrete: $\widehat{\boldsymbol{x}}(k) = \sum_{\ell=0}^{N-1} \overline{W}^{k\ell}\boldsymbol{x}(\ell)$ and $\boldsymbol{x}(\ell) = \frac{1}{N}\sum_{k=0}^{N-1} W^{k\ell}\widehat{\boldsymbol{x}}(k)$ with $W = e^{2\pi i/N}$

Series: $\boldsymbol{c}(n) = \frac{1}{2\pi}\int_0^{2\pi} f(t)e^{-int}dt$ and 2π-periodic $f(t) = \sum_{-\infty}^{\infty} \boldsymbol{c}(n)e^{int}$.

Orthogonality gives Parseval's energy equality $\int |f(t)|^2\,dt = \frac{1}{2\pi}\int |\widehat{f}(\omega)|^2\,d\omega$ and $\sum |\boldsymbol{x}(\ell)|^2 = \frac{1}{N}\sum |\widehat{\boldsymbol{x}}(k)|^2$ and $\sum |\boldsymbol{c}(n)|^2 = \frac{1}{2\pi}\int |f(t)|^2\,dt$.

Discrete Time Fourier Transform and z Transform

$$X(e^{j\omega}) = \sum_{-\infty}^{\infty} \boldsymbol{x}(n)e^{-j\omega n} \quad X(z) = \sum_{-\infty}^{\infty} \boldsymbol{x}(n)z^{-n} \quad \boldsymbol{x}(n) = \frac{1}{2\pi}\int_0^{2\pi} X(e^{j\omega})e^{j\omega n}d\omega$$

Frequency Response $H(e^{j\omega}) = \sum_n \boldsymbol{h}(n)e^{-j\omega n}$

The frequency response is a complex quantity $H(e^{j\omega}) = |H(e^{j\omega})|e^{j\phi(\omega)}$, where $|H(e^{j\omega})|$ and $\phi(\omega)$ are the (real) *magnitude* and *phase responses* of $\boldsymbol{h}(n)$. For real coefficients $\boldsymbol{h}(n)$, the magnitude response $|H(e^{j\omega})|$ is an even function of ω and the phase response $\phi(\omega)$ is an odd function. Consequently, one only needs to show the responses for the frequency region $0 \le \omega \le \pi$.

$H(z) = 1 + az^{-1}$ has frequency response $H(e^{j\omega}) = 1 + ae^{-j\omega}$. If a is real then

$$\left|H(e^{j\omega})\right| = \sqrt{1 + 2a\cos\omega + a^2}, \quad \phi(e^{j\omega}) = \tan^{-1}\left(\frac{a \sin\omega}{1 + a\cos\omega}\right).$$

Fundamental Theorem of Linear Algebra

If $\boldsymbol{Ax} = 0$, the rows of $\boldsymbol{A}$ are perpendicular to $\boldsymbol{x}$. In general: The row space of an m by n matrix $\boldsymbol{A}$ is the orthogonal complement of the null space.

These are subspaces of $\mathbf{R}^n$. The row space has dimension r = rank of $\boldsymbol{A}$ = dimension of column space. The null space, containing all solutions to $\boldsymbol{Ax} = 0$, has dimension $n - r$.

Gabor Transform (Windowed Fourier transform with $g(t) = g(0)e^{-at^2}$)

GenLOT (Generalized Lapped Orthogonal Transform) (See LOT)

The basis functions (analysis filters) of the GenLOT are linear phase and orthonormal. They yield a linear-phase paraunitary filter bank. For any $N > 1$, the polyphase transfer matrix of a GenLOT can be expressed as

$$\boldsymbol{E}(z) = \boldsymbol{K}_{N-1}(z)\boldsymbol{K}_{N-2}(z)\cdots\boldsymbol{K}_1(z)\boldsymbol{D} \quad \text{where} \quad \boldsymbol{K}_\ell(z) = \Phi_\ell \boldsymbol{W}\Lambda(z)\boldsymbol{W}.$$

$\boldsymbol{D}$ is the DCT transform matrix. The first $M/2$ rows of $\boldsymbol{D}$ are symmetric and the other $M/2$ rows are antisymmetric. The flow-graphs for the analysis and synthesis sections are shown below. Each branch carries $M/2$ samples.

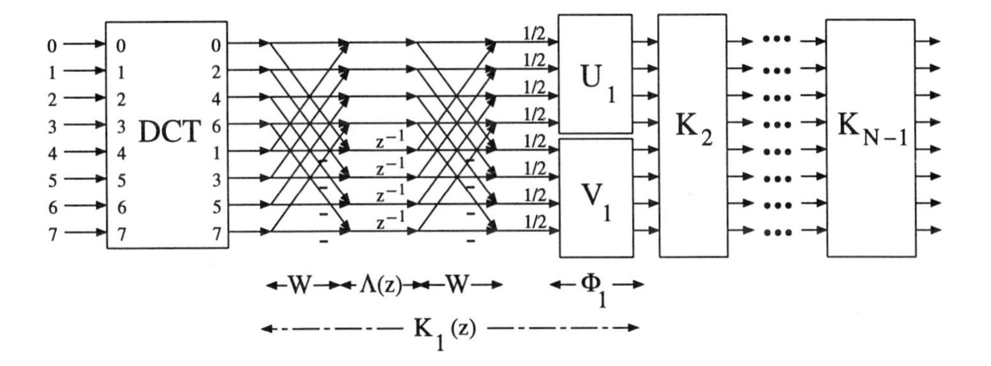

The class of GenLOTs includes the DCT and LOT as special cases for $N = 1$ and $N = 2$, respectively. The degrees of freedom reside in $\boldsymbol{U}_\ell$ and $\boldsymbol{V}_\ell$ which are only restricted to be real $M/2 \times M/2$ orthogonal matrices. Thus, each one can be described by $M(M-2)/8$ plane rotations (or less, for constrained solutions).

Group Delay $\tau(\omega) = -\frac{d\phi(\omega)}{d\omega}$, where $\phi(\omega)$ is the *phase* of $H(e^{j\omega}) = |H(e^{j\omega})|e^{j\phi(\omega)}$.

At a particular frequency ω_0, $\tau(\omega_0)$ measures the delay. Suppose $\boldsymbol{x}(n) = e^{j\omega_1 n} + e^{j\omega_2 n}$ is a sum of exponentials and the group delays at ω_1 and ω_2 are $\phi(\omega_1) = 1$ and $\phi(\omega_2) = 2$.

With unity magnitude responses the output is

$$\begin{aligned} y(n) &= H(e^{j\omega_1})e^{j\omega_1 n} + H(e^{j\omega_2})e^{j\omega_1 n} \\ &= e^{-j\omega_1}e^{j\omega_1 n} + e^{-j2\omega_2}e^{j\omega_2 n} \\ &= e^{j\omega_1(n-1)} + e^{j\omega_2(n-2)}. \end{aligned}$$

The exponentials will be delayed by a different number of samples. It is important in some applications to have constant group delay. Then $\frac{d\phi(\omega)}{d\omega}$ = constant and $\phi(\omega)$ is linear. *Constant group delay means linear phase*—a symmetric or antisymmetric filter.

Halfband Filter A centered halfband filter has $\boldsymbol{h}(0) = 0.5$ and other even components $\boldsymbol{h}(2k) = 0$. An uncentered halfband filter has $\boldsymbol{h}(2k+\ell) = 0.5\,\delta(\ell)$. Its polyphase component $H_{odd}(z)$ has one term. The frequency response $H(\omega)$ is odd around $H(\pi/2)$ and $H(z) + H(-z) = z^{-\ell}$. *See Mth Band Filter.*

Hartley Transform (an alternative to the DFT that uses real basis functions)

The HT matrix has entries $\boldsymbol{H}_{kn} = \cos\frac{2\pi kn}{L} + \sin\frac{2\pi kn}{L}$. This is a cosine with the origin translated by one eighth of a period. The Hartley matrix has a recursive factorization like the Fourier DFT matrix. Thus the FHT corresponds to the FFT, but with real multiplications.

Hilbert Space (also Function space C^k, Hölder space C^α, Sobolev space H^s)

The Hilbert space $L^2(\mathbf{R})$ contains all $f(t)$ with finite energy:

$$\|f\|^2 = \int_{-\infty}^{\infty} |f(t)|^2 dt < \infty.$$

The discrete Hilbert space $\ell^2 = L^2(\mathbf{Z})$ contains all sequences with $\sum |f(n)|^2 < \infty$.

Main point: Hilbert spaces have an *inner product* (f, g) for which $(f, f) = \|f\|^2$.

The space C^k contains all functions with k continuous derivatives: $f^{(k)}$ is in C^0.

The Hölder space $C^\alpha (0 < \alpha \le 1)$ contains $f(t)$ if $|f(t) - f(s)| \le \text{const}\,|t-s|^\alpha$.

The Sobolev space $H^s (s > 0)$ contains functions in L^2 whose derivative $f^{(s)}(t)$ is in L^2. In the Fourier domain $\int |\omega|^{2s} |\widehat{f}(\omega)|^2 d\omega < \infty$ (s need not be an integer).

Impulse Response $\boldsymbol{h} = (\boldsymbol{h}(0), \boldsymbol{h}(1), \ldots)$ is the response to the impulse $\boldsymbol{x} = \boldsymbol{\delta}$.

Interpolation Filter (Upsampling followed by filtering)

The figure shows an interpolation filter before and after application of the second Noble Identity. The equivalent operation is polyphase filtering at the lower rate, then an S/P converter, and combining at the output.

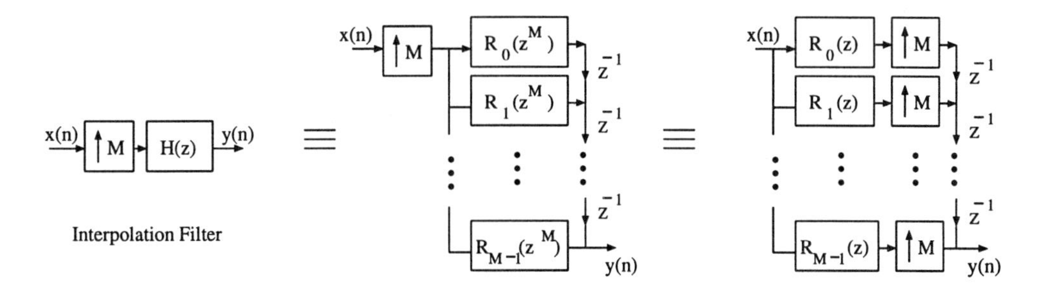

Karhunen-Loève Transform (KLT) The basis functions are eigenvectors of the covariance matrix of the input signal; the basis is fully decorrelated.

Lazy Filter Bank (PR system with no filtering)

The system blocks the input signal into even and odd sequences using an S/P converter and combines them using a P/S converter. The overall delay is one sample. If the orders are switched (after delaying $y_0(n)$ by a sample), the overall delay is two samples. These unfiltered PR systems do not process the subband signals $y_k(n)$, and thus are not useful in practice.

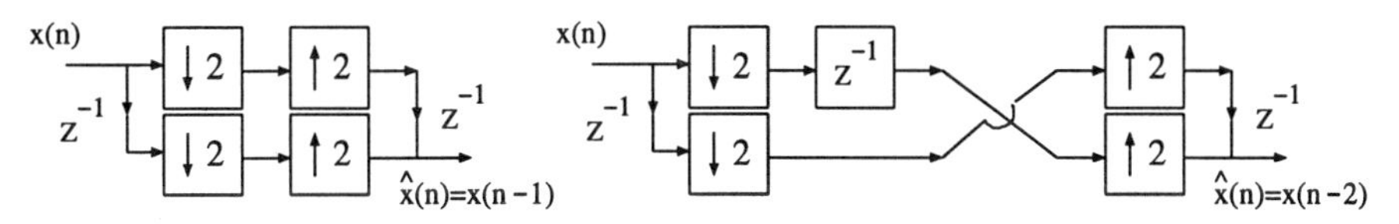

Linear Phase Filter The coefficients $h(n)$ are either symmetric or antisymmetric about $n = N/2$. Thus $h(n) = h(N-n)$ or $h(n) = -h(N-n)$. The phase of $H(\omega)$ is $\phi(\omega) = N\omega/2$ or $\phi(\omega) = N\omega/2 + \frac{\pi}{2}$. The group delay $\tau(\omega)$ is a constant.

LOT (Lapped Orthogonal Transform)

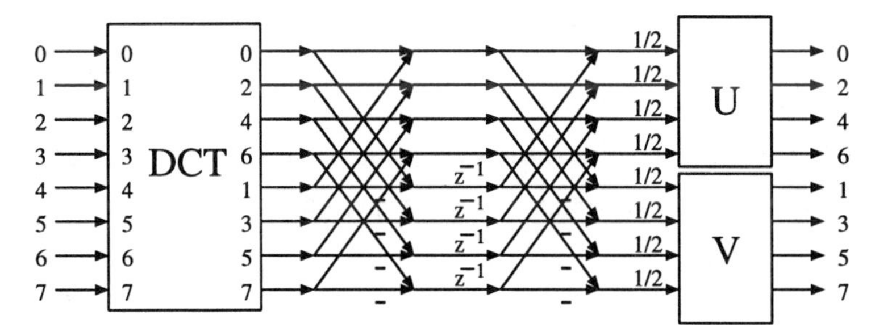

The basis functions of the LOT are linear phase and orthonormal. The polyphase transfer matrix allows arbitrary orthogonal matrices U and V of size $M/2$, parametrized by a set of rotation angles. Another interpretation of LOT is overlapping block transform. The filters have lengths $2M$. The output at block B depends on the input at blocks B and $B-1$. LOT eliminates part of the blocking effect in the DCT.

Magnitude Response $|H(\omega)|$ or $|H(f)|$ or $|H(e^{j\omega})|$

Use the normal scale to see passband and the dB scale to see stopband.

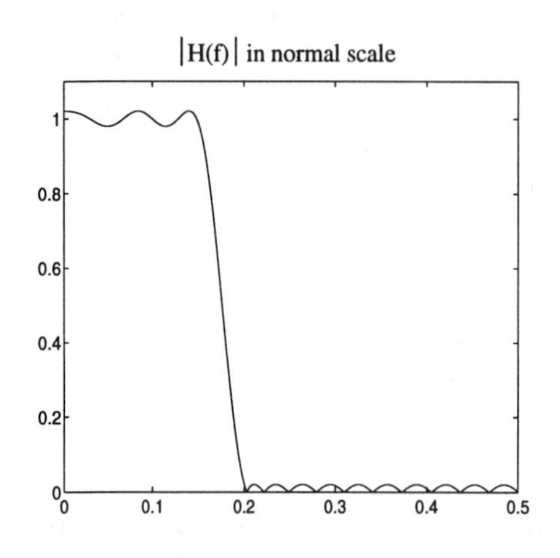

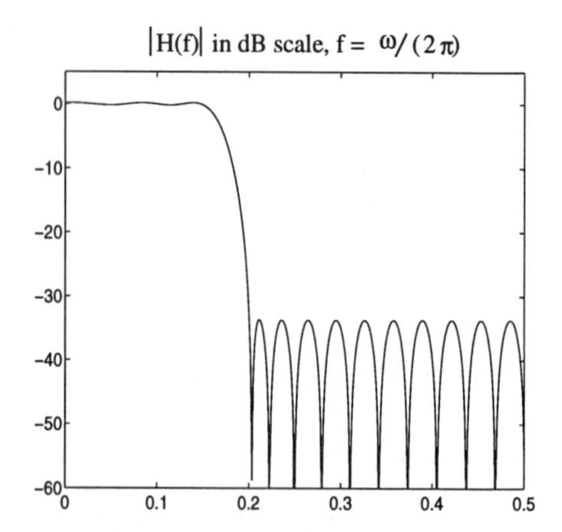

Mask Vector of coefficients $\boldsymbol{h}(n)$ in the literature on approximation theory

Matrix (Special matrices, each with a 2×2 or 3×3 or doubly infinite example)

Block matrix $\begin{bmatrix} \boldsymbol{A} & \boldsymbol{B} \\ \boldsymbol{C} & \boldsymbol{D} \end{bmatrix}$ Multiplication by blocks is correct when their sizes permit. 2D problems often separate into 1D blocks. The tensor product (Kronecker product) of matrices $\boldsymbol{P}$ and $\boldsymbol{Q}$ is the big matrix $\boldsymbol{P} \otimes \boldsymbol{Q}$ whose blocks are $\boldsymbol{P}_{ij}\boldsymbol{Q}$.

Circulant matrix $\begin{bmatrix} a & c & b \\ b & a & c \\ c & b & a \end{bmatrix}$. The first column is permuted cyclically to produce the other columns. The entries $C_{k\ell}$ depend only on $(k-\ell) \pmod n$. The eigenvectors of every circulant are the columns of the Fourier matrix $\boldsymbol{F}$ (the DFT matrix). All circulants have the form $\boldsymbol{F}\Lambda\boldsymbol{F}^*$ with Λ = diagonal eigenvalue matrix.

Fourier matrix $= N$-point DFT matrix with entries $W^{k\ell}$, where $W = e^{2\pi i/N} = N$th root of 1. The indices k and ℓ go from 0 to $N-1$. The inverse DFT matrix $\boldsymbol{F}^{-1} = \overline{\boldsymbol{F}}/N$ contains the powers of $\overline{W} = e^{-2\pi i/N}$.

Hermitian matrix (complex version of symmetric matrix: $\overline{a}_{\ell k} = a_{k\ell}$ and $\overline{\boldsymbol{A}}^T = \boldsymbol{A}$)

Householder matrix Reflection in the plane perpendicular to $\boldsymbol{v}$, so $\boldsymbol{H}\boldsymbol{v} = -\boldsymbol{v}$.

$$\boldsymbol{H} = \boldsymbol{I} - 2\frac{\boldsymbol{v}\boldsymbol{v}^T}{\boldsymbol{v}^T\boldsymbol{v}} = \boldsymbol{H}^T = \boldsymbol{H}^{-1}. \text{ If } \boldsymbol{v} = \begin{bmatrix} \cos\theta \\ \sin\theta \end{bmatrix} \text{ then } \boldsymbol{H} = \begin{bmatrix} \cos 2\theta & \sin 2\theta \\ \sin 2\theta & -\cos 2\theta \end{bmatrix}.$$

Paraunitary matrix $\boldsymbol{H}(z)$ is (normalized) paraunitary if $\boldsymbol{H}^T(z^{-1})\boldsymbol{H}(z) = \boldsymbol{I}$. This assumes real coefficients. On the unit circle, $\overline{\boldsymbol{H}}^T\boldsymbol{H} = \boldsymbol{I}$. This *unitary* matrix is a function of the *para*meter z.

Positive definite (symmetric) matrix $\boldsymbol{x}^T\boldsymbol{A}\boldsymbol{x} > 0$ except if $\boldsymbol{x} = \boldsymbol{0}$

$\boldsymbol{A} = \begin{bmatrix} 1 & 1 \\ 1 & 5 \end{bmatrix}$ has $\boldsymbol{x}^T\boldsymbol{A}\boldsymbol{x} = x_1^2 + 2x_1x_2 + 5x_2^2 > 0$. Equivalent tests:

All eigenvalues are positive. Here $\lambda = 3 \pm \sqrt{5}$.

All upper left subdeterminants are positive. Here $D_1 = 1$ and $D_2 = 4$.

All pivots are positive: Here $d_1 = 1$ and $d_2 = 4$.

Elimination gives a Cholesky factorization $A = LL^T$.

Projection matrix: $P^2 = P$. For orthogonal projection also $P^T = P$. If A has independent columns, the projection onto its column space is $P = A(A^TA)^{-1}A^T$.

Reversal matrix = Exchange matrix. J has the rows of I in reverse order. Then JA reverses the rows of A. For diagonal Λ, $J\Lambda J$ has the diagonal entries reversed.

Rotation matrix (= orthogonal matrix with determinant $+1$)

$$\text{1-3 Plane rotation} = \text{1-3 Givens rotation} = \begin{bmatrix} \cos\theta & 0 & -\sin\theta \\ 0 & 1 & 0 \\ \sin\theta & 0 & \cos\theta \end{bmatrix}$$

Similar matrix If M is invertible, then $M^{-1}AM$ is similar to A.

Similar matrices have the same eigenvalues. They represent the same transformation but with respect to different bases. The eigenvectors x_i change to $M^{-1}x_i$.

Symmetric matrix $a_{k\ell} = a_{\ell k}$ and $A^T = A$. (Antisymmetric means $A^T = -A$.)

Toeplitz matrix = Convolution matrix = Filter matrix: The entries $H_{k\ell} = h(k-\ell)$ are constant along each diagonal. They depend only on $k - \ell$.

Unimodular matrix A square polynomial matrix $U(z)$ with determinant ± 1.

Inverses and products $U(z)V(z)$ are also unimodular. A unimodular matrix can be factorized as cascade of elementary matrices. The same holds for unimodular matrices with integer elements:

$$\begin{bmatrix} 0 & 1 & 0 \\ 0 & 0 & 1 \\ 1 & 0 & 0 \end{bmatrix}, \begin{bmatrix} 1 & 0 & 0 \\ a(z) & 1 & 0 \\ b(z) & c(z) & 1 \end{bmatrix}, \begin{bmatrix} 1 - z^{-2} & z^{-1} \\ 2 - z^{-1} - 2z^{-2} & 1 + 2z^{-1} \end{bmatrix}.$$

Unitary matrix (= complex version of orthogonal matrix).

The columns are orthonormal. Then $\overline{U}^T U = I$ or $U^{-1} = \overline{U}^T = U^+$.

Vandermonde matrix $\begin{bmatrix} 1 & a_0 & a_0^2 \\ 1 & a_1 & a_1^2 \\ 1 & a_2 & a_2^2 \end{bmatrix}$. The entries are $V_{jk} = (a_j)^k$.

Maximally Flat (Maxflat) Filter

There is a unique polynomial $P(y)$ with $p + r$ coefficients that has a pth order zero at $y = 1$ and rth order flatness $1 + O(y^r)$ at $y = 0$. That polynomial is

$$P(y) = (1-y)^p\, B_{p,r}(y) = (1-y)^p \sum_{k=0}^{r-1} \binom{p+k-1}{k} y^k.$$

$B_{p,r}(y)$ consists of the first r terms of the binomial series for $(1-y)^{-p}$. Thus

$$P(y) = (1-y)^p\left[(1-y)^{-p} + O(y^r)\right] = 1 + O(y^r) \text{ as required.}$$

$P(y)$ has the correct degree $p+r-1$ and a pth order zero from $(1-y)^p$. $P(y)$ must decrease monotonically, because all $p+r-2$ zeros of P' are at the end points.

For orthogonal filter banks and Daubechies filters, the important case is $p = r$. The polynomial has pth order flatness at $y = 0$ and a pth order zero at $y = 1$. It is "odd" around the middle value $P(\frac{1}{2}) = \frac{1}{2}$. Replacing y by $1-y$, the endpoints reverse. Then $P(y) + P(1-y)$ is flat at both ends, and those $2p$ conditions on its $2p$ coefficients leave only one possible sum:

$$P(y) + P(1-y) = 1 \quad \text{and} \quad P\left(\sin^2\frac{\omega}{2}\right) + P\left(\cos^2\frac{\omega}{2}\right) = 1.$$

For the frequency response, set $y = \sin^2\frac{\omega}{2} = \frac{1-\cos\omega}{2}$. Then $1-y = \cos^2\frac{\omega}{2} = \frac{1+\cos\omega}{2}$. The response of the maxflat filter of length $2p$ is:

$$P(\omega) = \left(\frac{1+\cos\omega}{2}\right)^p \sum_{0}^{p-1} \binom{p+k-1}{k}\left(\frac{1-\cos\omega}{2}\right)^k.$$

All terms are nonnegative. $P(\omega)$ decreases ***monotonically*** from $P(0) = 1$ to $P(\pi) = 0$. Section 5.5 describes $P(z)$ when $z = e^{j\omega}$. The filter has $4p-1$ coefficients. It is halfband with $2p$ zeros at $z = -1$. The other $2p-2$ zeros are close to the moon-shaped curve $|z - z^{-1}| = 2$, half inside and half outside the unit circle. (This curve combines $|z+1| = \sqrt{2}$ and $|z-1| = \sqrt{2}$.) The spectral factor in $P(z) = C(z)C(z^{-1})$ collects half the roots (none outside the circle) for the minimum phase $C(z)$. The MATLAB exercises associated with Section 5.5 are recommended as leading to new ideas and experiments.

Minimum Phase / Maximum Phase All zeros of a minimum phase $H(z)$ are in the unit circle. The energy is concentrated at the beginning samples of $\boldsymbol{h}(n)$:

$$\sum_{n=0}^{M} |\boldsymbol{h}(n)|^2 > \sum_{n=0}^{M} |\boldsymbol{g}(n)|^2 \text{ for } M < N, \text{ if } \boldsymbol{g}(n) \text{ has the same energy as } \boldsymbol{h}(n).$$

The zeros of a *maximum-phase* $H(z)$ are all *outside* the unit circle and the energy of the sequence is concentrated in the *last* few coefficients.

Modulation Matrix The (k, ℓ) entry of $\boldsymbol{H}_m(z)$ is $H_k(zW^\ell)$, $0 \le k, \ell < M$. $\boldsymbol{F}_m(z)$ is transposed, in synthesis. The entries of $\boldsymbol{F}_m(z)\,\boldsymbol{H}_m(z)$ give the functions $T_k(z)$.

M-th Band Filter (Nyquist Filter)

M-th band filters are used in signal processing, communication, and multirate systems such as oversampling delta sigma A/D converters and decimation and interpolation filters. In many digital systems, sampling rate change is essential to obtain an efficient implementation in real time. The design of suitable filters for the rate changing operations is critical. A centered M-th band filter (noncausal) has $\boldsymbol{h}(nM) = \delta(n)$. Then $\sum H(zW^k) = 1$. A symmetric lowpass M-th band filter must have odd length, with

$$\omega_p + \omega_s = 2\frac{\pi}{M} \quad \text{and} \quad \delta_p \le (M-1)\delta_s.$$

Here ω_p, ω_s are the passband and stopband cutoff frequencies and δ_p, δ_s are the maximum errors in the bands. These properties do not hold for M-th band filter with arbitrary phase response.

Several design methods for linear phase M-th band FIR filters have been reported. One is based on the Remez exchange algorithm, to design equiripple filters. The algorithm requires exact specifications in the frequency domain. The resulting equiripple filter does not exactly satisfy the M-th band properties. An exact M-th band FIR filter can be designed as an eigenfilter with constraint $\boldsymbol{h}(nM) = \delta(n)$. A set of MATLAB m.files for eigenfilter and M-th band eigenfilter is available at

`http://saigon.ece.wisc.edu/~waveweb/QMF.html`

See *Allpass-based DFT IIR Filter Bank* for the condition on $\phi_k(\omega)$ such that the resulting M-th band filter has good frequency characteristics. One drawback is the peak in the stopband.

An orthogonal filter bank uses spectral factors of an M-th band filter.

Multiresolution Subspaces $V_j \subset V_{j+1}$ with four required properties:

1. Completeness: The projections $f_j(t)$ onto V_j approach $f(t)$ as $j \to \infty$.
2. Scale-invariance: $f(t)$ is in $V_j \Leftrightarrow f(2t)$ is in V_{j+1}.
3. Shift-invariance: $f(t)$ is in $V_0 \Leftrightarrow f(t-k)$ is in V_0.
4. Basis of translates: V_0 has a Riesz basis of the form $\{\phi(t-k)\}$.

When $\{\phi(t-k)\}$ are orthonormal, the wavelet subspace W_j is the orthogonal complement of V_j within V_{j+1}: $V_j \oplus W_j = V_{j+1}$. All other W_J are perpendicular to W_j. Backward recursion gives $V_0 \oplus W_0 \oplus \cdots \oplus W_j = V_{j+1}$ for $j \geq 0$. The translates $\{w(t-k)\}$ are an orthonormal basis for W_0. Rescaling and shifting (dilation and translation) will yield two orthonormal bases for the whole space L^2:

1. Full wavelet basis $w_{jk}(t) = 2^{j/2} w(2^j t - k)$, $-\infty < j, k < \infty$.
2. Practical wavelet basis $\phi(t-k)$ and $w_{jk}(t)$ for $j \geq 0$ and $-\infty < k < \infty$.

Biorthogonal multiresolution has two sequences V_j and $\widetilde{V}_j$ with $V_j + W_j = V_{j+1}$ and $\widetilde{V}_j + \widetilde{W}_j = \widetilde{V}_{j+1}$. The biorthogonality is $V_j \perp \widetilde{W}_j$ and $\widetilde{V}_j \perp W_j$. The Riesz bases are biorthogonal.

- The coefficients a_{jk} in the projection $f_j(t) = \sum a_{jk}\, \phi(t-k)$ satisfy the same wavelet transform as the subband signals in a tree-structured filter bank. This is the Mallat pyramid algorithm.
- Many processes including hearing and vision can be modelled by multiresolution.

Noble Identities

$x(n) \to \boxed{H(z^M)} \to \boxed{\downarrow M} \to y_1(n) \equiv x(n) \to \boxed{\downarrow M} \to \boxed{H(z)} \to y_1(n)$

$x(n) \to \boxed{\uparrow L} \to \boxed{H(z^L)} \to y_2(n) \equiv x(n) \to \boxed{H(z)} \to \boxed{\uparrow L} \to y_2(n)$

The components of a multirate system are filters, decimators and expanders. With delay chains the last two become S/P and P/S converters. The two *Noble Identities* allow commutation of filters with decimators and expanders. Filtering with $H(z^M)$ and then decimating the result is the same as decimating the input first and filtering with $H(z)$. The second Noble Identity allows expanders ($\uparrow L$) to be moved "outside" filters of the form $H(z^L)$.

Nonuniform Filter Bank (The decimation factors are n_k)

The input signal is channelized by a set of filters with unequal bandwidths, and downsampled. If the reciprocals add to $\sum 1/n_k = 1$, the filter bank is maximally-decimated. An example is a ***tree-structured filter bank***.

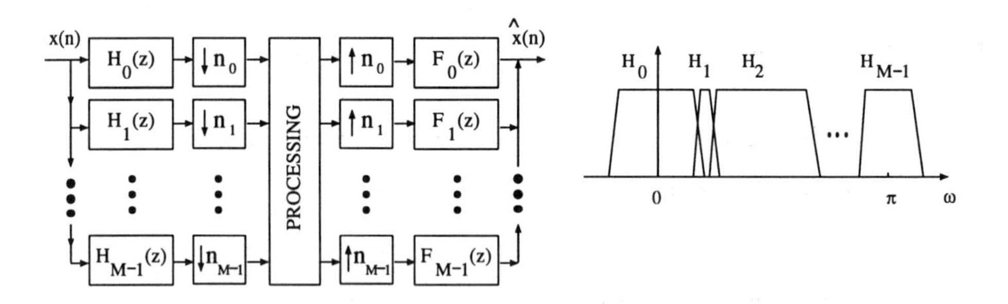

Norm and Spectral Radius of a Matrix The *spectral radius* is $\rho(A) = \max |\lambda_i|$.

$\rho(A)$ is the magnitude of the largest eigenvalue. The spectral radius is not a matrix norm because it violates the requirements (i) and (iii):

$$(i)\ \|A\| > 0 \text{ if } A \neq \mathbf{0} \quad (ii)\ \|cA\| = |c|\ \|A\| \quad (iii)\ \|A+B\| \le \|A\| + \|B\|.$$

Among the important matrix norms are these four:

$$\|A\|_1 = \text{max column sum } \textstyle\sum_i |a_{ij}| \qquad \|A\|_\infty = \text{max row sum } \textstyle\sum_j |a_{ij}|$$

$$\|A\|_2 = \text{max singular value } = \sqrt{\rho(A^T A)} \qquad \|A\|_F = \text{Frobenius } = \sqrt{\textstyle\sum_{i,j} |a_{ij}|^2}$$

Orthogonal Complement / Direct Sum Suppose V_0 is a subspace of V_1. The ***orthogonal complement*** W_0 of V_0 contains all vectors in V_1 that are orthogonal to every vector in V_0. Every vector in V_1 can be separated into $v =$ (projection onto V_0) + (projection onto W_0).

A ***direct sum*** does not require W_0 to be orthogonal to V_0. It does require that a basis for V_0 together with a basis for W_0 is a basis for V_1. Then every $v =$ (part in V_0) + (part in W_0). A plane and a perpendicular line are orthogonal complements in $\mathbf{R}^3$. A plane and a non-parallel line have direct sum $V_0 \oplus W_0 = \mathbf{R}^3$.

Pairwise-Mirror-Image Filter Bank The filter bank is PMI if

$$H_{M-1-k}(z) = H_k(-z), \quad \text{or} \quad H_{M-1-k}(z) = z^{-N} H_k(-z^{-1}).$$

The frequency responses are mirror images with respect to $\frac{\pi}{2}$.

Paraunitary Cosine-modulated Filter Bank Let $H(z) = \sum z^{-\ell} G_\ell(z^{2M})$ be the prototype filter. Its $2M$ polyphase components are $G_\ell(z)$. Then the polyphase matrix for the complete analysis bank can be expressed in terms of $G_\ell(z)$ and the M by $2M$ matrix $\widehat{\boldsymbol{C}}$:

$$\boldsymbol{H}_p(z) = \widehat{\boldsymbol{C}} \begin{bmatrix} \boldsymbol{g}_0(-z^2) \\ z^{-1}\boldsymbol{g}_1(-z^2) \end{bmatrix} \text{ with } \begin{cases} \boldsymbol{g}_0(z) &= \text{diag}\left[G_0(z)\, G_1(z) \cdots G_{M-1}(z)\right] \\ \boldsymbol{g}_1(z) &= \text{diag}\left[G_M(z)\, G_{M+1}(z) \cdots G_{2M-1}(z)\right] \\ \widehat{\boldsymbol{C}}_{k,\ell} &= 2\cos\left((2k+1)\frac{\pi}{2M}(\ell - \frac{1}{2}) + (-1)^k \frac{\pi}{4}\right). \end{cases}$$

Substitute into $\boldsymbol{P}(z) = \boldsymbol{H}_p^T(z^{-1})\,\boldsymbol{H}_p(z)$ to find

$$\frac{1}{2M}\boldsymbol{P}(z) = \widetilde{\boldsymbol{g}}_0(-z^2)\boldsymbol{g}_0(-z^2) + \widetilde{\boldsymbol{g}}_1(-z^2)\boldsymbol{g}_1(-z^2).$$

The equation $\boldsymbol{P}(z) = \frac{1}{2M} z^{-2s}$ is necessary and sufficient for a paraunitary cosine-modulated filter bank. The equivalent condition on $G_\ell(z)$ is *pairwise orthogonality*:

$$\widetilde{G}_\ell(z)G_\ell(z) + \widetilde{G}_{M+\ell}(z)G_{M+\ell}(z) = \frac{1}{2M}(-z)^{-s}.$$

We must design a linear phase filter $H(z)$ of length $2mM$ with high stopband attenuation whose polyphase components satisfy this condition. Thus the cosine-modulated filter bank can be implemented using M lattice structures of two-channel (ℓ and $\ell + M$) orthogonal filters.

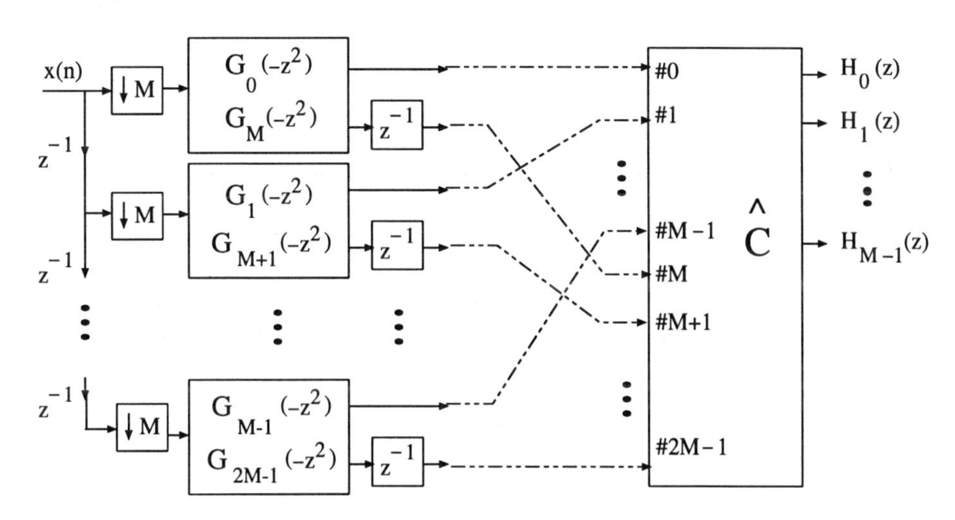

Paraunitary Filter Bank (Orthonormal Filter Bank)

A paraunitary filter bank is a special case of a biorthogonal (perfect reconstruction) filter bank, where $\boldsymbol{H}_p(z)$ and $\boldsymbol{F}_p(z)$ are related by transposition:

$$\boldsymbol{F}_p(z) = \boldsymbol{H}_p^{-1}(z) = \widetilde{\boldsymbol{H}}_p(z) = \boldsymbol{H}_p^T(z^{-1}).$$

The analysis and synthesis filters are related by $\boldsymbol{f}_k(n) = \boldsymbol{h}_k(N-n)$.

- *Factorization*: Any causal FIR paraunitary $\boldsymbol{E}(z)$ of degree J factors into

$$\boldsymbol{E}(z) = \boldsymbol{V}_J(z)\cdots\boldsymbol{V}_1(z)\,\boldsymbol{U} \qquad \text{where} \qquad \begin{cases} \boldsymbol{V}_k(z) &= \boldsymbol{I} - \boldsymbol{v}_k\boldsymbol{v}_k^T + z^{-1}\boldsymbol{v}_k\boldsymbol{v}_k^T \\ \boldsymbol{U}^T\boldsymbol{U} &= \boldsymbol{I} \text{ and } \boldsymbol{v}_k^T\boldsymbol{v}_k = 1. \end{cases}$$

Paraunitary Linear Phase Filter Bank $(M > 2)$

Linear phase and paraunitary properties can coexist for filter banks with $M > 2$. The DCT and LOT are examples. The number of symmetric and antisymmetric filters depends on the filter length $KM + \beta$:

- Even M and even β: $M/2$ symmetric and $M/2$ antisymmetric filters.
- Even M and odd β: $M/2+1$ symmetric and $M/2-1$ antisymmetric filters.
- Odd M: $(M+1)/2$ symmetric and $(M-1)/2$ antisymmetric filters.

For even M and $\beta = 0$, $\boldsymbol{H}_p(z) = \boldsymbol{SQT}_{K-1}\Lambda(z)\boldsymbol{T}_{K-2}\Lambda(z)\cdots\Lambda(z)\boldsymbol{T}_0\boldsymbol{Q}$, where

$$\boldsymbol{Q} = \begin{bmatrix} \boldsymbol{I} & \boldsymbol{0} \\ \boldsymbol{0} & \boldsymbol{J} \end{bmatrix}, \quad \boldsymbol{T}_i = \begin{bmatrix} \boldsymbol{I} & \boldsymbol{I} \\ \boldsymbol{I} & -\boldsymbol{I} \end{bmatrix}\begin{bmatrix} \boldsymbol{U}_i & \boldsymbol{0} \\ \boldsymbol{0} & \boldsymbol{V}_i \end{bmatrix}\begin{bmatrix} \boldsymbol{I} & \boldsymbol{I} \\ \boldsymbol{I} & -\boldsymbol{I} \end{bmatrix}$$

$$\Lambda(z) = \begin{bmatrix} \boldsymbol{I} & \boldsymbol{0} \\ \boldsymbol{0} & z^{-1}\boldsymbol{I} \end{bmatrix}, \quad \boldsymbol{S} = \frac{1}{\sqrt{2}}\begin{bmatrix} \boldsymbol{U}_0 & \boldsymbol{0} \\ \boldsymbol{0} & \boldsymbol{U}_1 \end{bmatrix}\begin{bmatrix} \boldsymbol{I} & \boldsymbol{J} \\ \boldsymbol{I} & -\boldsymbol{J} \end{bmatrix}$$

$\boldsymbol{J}$ is the exchange matrix and $\boldsymbol{U}_i$ and $\boldsymbol{V}_i$ are arbitrary orthogonal matrices. The complete factorization for odd M is not available. The above form is not modular and one can not easily verify that DCT and LOT are special cases. In the regular factorization $\Phi_{K-1}\,\Phi_{K-2}\,\ldots\,\Phi_1\boldsymbol{E}_0$ of Chapter 9, DCT is a special case of $\boldsymbol{E}_0$ and LOT is the product $\Phi_1\boldsymbol{E}_0$.

Perfect Reconstruction (PR) Filter Bank (Biorthogonal Filter Bank)

The product of the polyphase transfer matrices of a PR filter bank must be of the form

$$\boldsymbol{P}(z) = \boldsymbol{F}_p(z)\boldsymbol{H}_p(z) = z^{-m_0}\begin{pmatrix} \boldsymbol{0} & \boldsymbol{I}_{M-r} \\ z^{-1}\boldsymbol{I}_r & \boldsymbol{0} \end{pmatrix}$$

A diagonal $\boldsymbol{P}(z)$ corresponds to $r = 0$. The overall delay is $d = r + M - 1 + m_0 M$. Thus, $T_0(z) = cz^{-d}$. When $M = 2$, the two choices are $\boldsymbol{P}(z) = z^{-m_0}\boldsymbol{I}$ or $\boldsymbol{P}(z) = z^{-m_0}\begin{pmatrix} 0 & 1 \\ z^{-1} & 0 \end{pmatrix}$.

Polyphase Representation (Type I and Type II)

A sequence $\boldsymbol{h}(n)$ can be represented by M sequences $\boldsymbol{e}_k(n)$ or $\boldsymbol{r}_k(n)$:

$$\begin{cases} \boldsymbol{e}_k(n) = \boldsymbol{h}(nM+k), & \text{Type-I polyphase} \\ \boldsymbol{r}_k(n) = \boldsymbol{h}(nM+M-1-k) & \text{Type-II polyphase} \end{cases} \qquad \begin{cases} 0 \le k \le M-1, \\ 0 \le n \le \lceil \frac{N}{M} \rceil. \end{cases}$$

$$H(z) = \sum_{k=0}^{M-1} z^{-k}E_k(z^M) = \sum_{k=0}^{M-1} z^{-(M-1-k)}R_k(z^M), \qquad R_k(z) = E_{M-1-k}(z).$$

Example: Suppose $\boldsymbol{h}(n) = n,\ 0 \le n \le 13$. The polyphase form for $M = 3$ is

$$\text{Type I}\ \begin{cases} e_0 = 0,\ 3,\ 6,\ 9,\ 12 \\ e_1 = 1,\ 4,\ 7,\ 10,\ 13 \\ e_2 = 2,\ 5,\ 8,\ 11 \end{cases} \qquad \text{Type II}\ \begin{cases} r_0 = 2,\ 5,\ 8,\ 11 \\ r_1 = 1,\ 4,\ 7,\ 10,\ 13 \\ r_2 = 0,\ 3,\ 6,\ 9,\ 12 \end{cases}$$

Type-I polyphase for the analysis bank yields $\boldsymbol{H}_p(z^M)$ times a delay chain:

$$\begin{bmatrix} H_0(z) \\ H_1(z) \\ \vdots \\ H_{M-1}(z) \end{bmatrix} = \begin{bmatrix} E_{00}(z^M) & E_{01}(z^M) & \cdots & E_{0,M-1}(z^M) \\ E_{10}(z^M) & E_{11}(z^M) & \cdots & E_{1,M-1}(z^M) \\ \vdots & \vdots & \ddots & \vdots \\ E_{M-1,0}(z^M) & E_{M-1,1}(z^M) & \cdots & E_{M-1,M-1}(z^M) \end{bmatrix} \begin{bmatrix} 1 \\ z^{-1} \\ \vdots \\ z^{-(M-1)} \end{bmatrix}$$

The Type-II polyphase representation for the synthesis bank yields

$$\begin{bmatrix} F_0(z) & F_1(z) & \cdots & F_{M-1}(z) \end{bmatrix} = \begin{bmatrix} z^{-(M-1)} & \cdots & z^{-1} & 1 \end{bmatrix} \boldsymbol{F}_p(z^M).$$

The Noble Identities move the decimators to the left and the interpolators to the right. The filtering operations in the original bank are done at the input rate where polyphase filtering increases the flow rate by M.

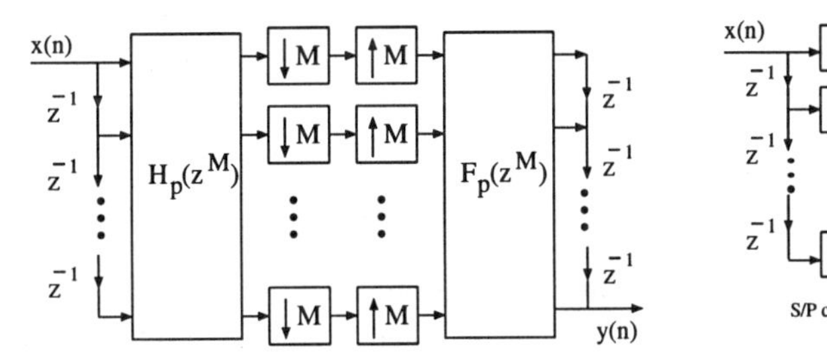

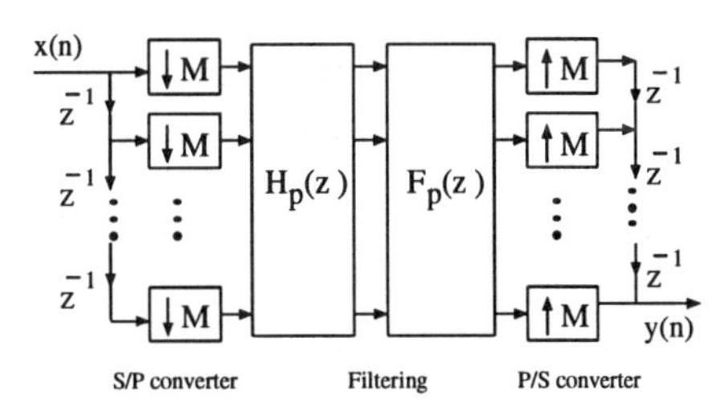

Power Complementary Filters $\sum_k \left|H_k(e^{j\omega})\right|^2 = \text{constant}$

A paraunitary filter bank has power complementary filters H_k (and F_k).

Pseudo-Circulant Matrix A circulant with extra factor z^{-1} below the diagonal.

Pseudo-QMF Bank (Pseudo-Quadrature Mirror Filter Bank)

A pseudo-QMF bank is a cosine-modulated filter bank where the prototype filters $H(z) = F(z)$ are even-length symmetric spectral factors of an M-th band filter. The aliasing from ***adjacent*** bands is cancelled (i.e., $T_1(z) \approx T_{M-1}(z) \approx 0$). The aliasing levels from $T_2(z)$ to $T_{M-2}(z)$ are comparable to the stopband attenuation of $H(e^{j\omega})$. The distortion function $T_0(z)$ is a delay. The only reconstruction error is in the (small) aliasing components.

Quantization Mapping of real numbers to values in a prescribed set

Quincunx Staggered plane grid of points (m, n) with $m + n$ even

Regularity (used for ***smoothness*** of $\phi(t)$ and also to count ***zeros at*** π)

The two numbers s and p are different. Always $s \leq p - \frac{1}{2}$, but s can be much smaller than p. We do not use the word "regularity" for this reason.

Scaling Function Solution of a multiscale equation with $\int \phi(t)dt \neq 0$

Scalogram (squared magnitude of the wavelet transform)

The scalogram can be interpreted as a spectrogram with *constant Q*.

Singular Values and Singular Vectors (of an m by n matrix $\boldsymbol{A}$)

The singular values σ_k are the square roots of the nonzero eigenvalues $\lambda_k(\boldsymbol{A}^T\boldsymbol{A}) = \lambda_k(\boldsymbol{A}\boldsymbol{A}^T)$. The singular vectors $\boldsymbol{v}_k$ are the orthogonal eigenvectors of $\boldsymbol{A}^T\boldsymbol{A}$. The left singular vectors $\boldsymbol{u}_k$ are the orthogonal eigenvectors $\boldsymbol{A}\boldsymbol{v}_k/\sigma_k$ of $\boldsymbol{A}\boldsymbol{A}^T$. In matrix form this is $\boldsymbol{A}\boldsymbol{V} = \boldsymbol{U}\Sigma$ or $\boldsymbol{A} = \boldsymbol{U}\Sigma\boldsymbol{V}^T$ = Singular Value Decomposition of $\boldsymbol{A}$ (the SVD).

Smoothness ($\phi(t)$ in L^2 has s derivatives in L^2 for all $s < s_{max}$)

A box function has $s_{max} = \frac{1}{2}$. The hat function has $s_{max} = \frac{3}{2}$. Each convolution with the box function increases s_{max} by 1. Always $s_{max} \leq p - \frac{1}{2}$ when $\phi(t)$ comes from $H(\omega)$ with p zeros at π. The number s_{max} is $-\log|\lambda_{max}|/\log 4$, where λ_{max} is the largest eigenvalue of $\boldsymbol{T}$ excluding the special eigenvalues $1, \frac{1}{2}, \ldots, (\frac{1}{2})^{2p-1}$.

Sobolev Space (also Besov Space)

The Sobolev space $W^{p,s}(\mathbf{R}^d)$ contains all functions $f(x_1, \ldots, x_d)$ with s derivatives in the space $L^p = W^{p,0}$: the pth power of $f^{(s)}(t)$ has a finite integral. The important case $p = 2$ has the special notation $W^{2,s} = H^s$. A Besov space refines the smoothness condition by a third parameter that enters especially for $p = 1$ and $p = \infty$.

Spectrogram Squared magnitude of the *Short Time Fourier Transform*

Spectral Factorization The figure below shows three factorizations $P_0(z) = F_0(z)H_0(z)$ of a polynomial of degree 14. It is the Daubechies maxflat halfband filter. $P_0(z)$ has eight zeros at $z = -1$, two real zeros z and $1/z$, and four complex zeros $z, \bar{z}, 1/z, 1/\bar{z}$. These 14 zeros are distributed into $F_0(z)$ and $H_0(z)$ in different ways to give different filter banks.

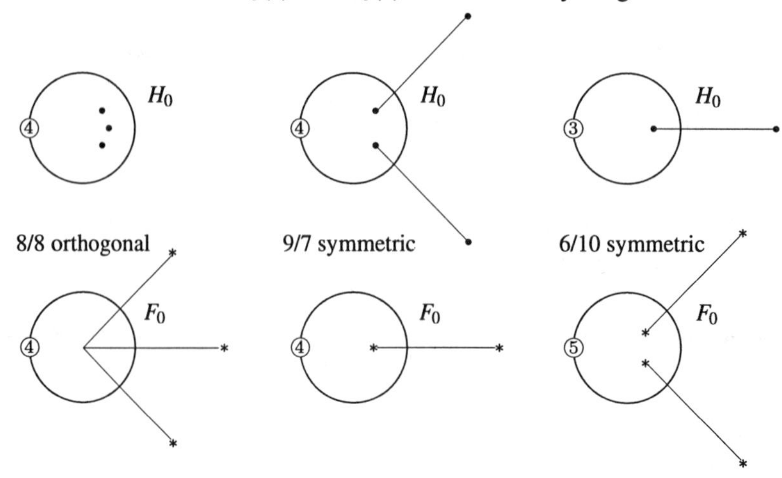

At the left are the orthogonal filters. If z is a root of $H_0(z)$, then z^{-1} is a root of $F_0(z) = z^{-7}H_0(z^{-1})$. Each polynomial has 7 roots and 8 coefficients.

In the center are the "FBI" linear phase 9/7 filters. By symmetry $H_0(z) = z^{-8}H_0(z^{-1})$ and $F_0(z) = z^{-6}F_0(z^{-1})$. If z is a root of $H(z)$, so is z^{-1}.

At the right is the 6/10 pair, also symmetric. One root at $z = -1$ was moved between the 9/7 factors; then F_0 and H_0 were reversed. The reversal gives more zeros in $F_0(z)$ and smoother functions for synthesis.

• *Nonuniqueness of the spectral factor*: There are many ways of choosing $H(z)$ with one zero from each pair z, z^{-1}. The spectral factor where all zeros are inside (outside) the unit circle is called minimum-phase (maximum-phase). The words ***spectral factorization*** are often applied to any factorization $P(z) = H(z)F(z)$, not always requiring $F(z) = H(z^{-1})$.

Subband Coding (transforms created by filtering and subsampling)

The signal is separated (approximately) into frequency bands for efficient coding. The subband signals are quantized such that the objective measure is maximized.

Symmetric Extension (of an input signal $\boldsymbol{x}(n)$ of length L)

Convolution with a filter of length $N+1$ results in a sequence of length $N+L$. In image processing, it is desirable to preserve the size L of the picture after processing. There are several techniques such as symmetric extension, circular filtering, time-varying filtering and nonlinear-phase filtering. Symmetric extension is the most popular. The input signal $\boldsymbol{x}(n)$ is symmetrically extended and convolved with symmetric filters $H_k(z)$. The signal is downsampled and windowed (it is sufficient to save half of the samples since the signals are symmetric). The total length of the windowed subband signals is still L.

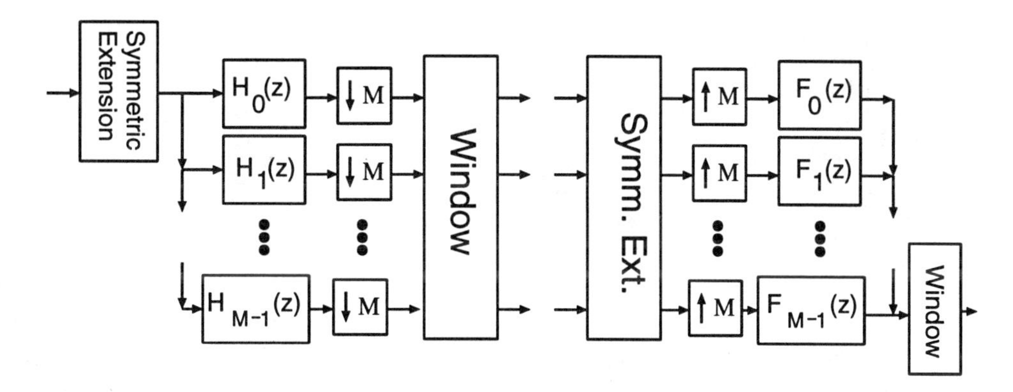

The analysis filters $H_k(z)$ must have equal lengths modulo M. For a two-channel PR filter bank, this condition always holds. The whole-sample **W** extension and half-sample **H** extension are $\boldsymbol{x}(-n) = \boldsymbol{x}(n)$ and $\boldsymbol{x}(-n) = \boldsymbol{x}(n-1),\ n > 0$.

Taps (filter coefficients $\boldsymbol{h}(n)$)

Time Invariance The discrete-time operator $\boldsymbol{H}$ is *time-invariant* if a delay of the input $\boldsymbol{x}$ produces a delay of the output $\boldsymbol{y} = \boldsymbol{H}\boldsymbol{x}$. Then $\boldsymbol{H}$ is represented by a constant-diagonal matrix (Toeplitz matrix or convolution matrix). $\boldsymbol{H}$ is a ***filter***.

Transform (See Cosine Transform, Fourier Transform, Windowed Transform, Gabor Transform, Hartley Transform, Wavelet Transform)

The transform $F = \boldsymbol{T}f$ is an expression of the original f with respect to a new basis. The inverse transform (= synthesis of f) is $f = \boldsymbol{T}^{-1}F$. The basis is in the columns of $\boldsymbol{T}^{-1}$.

If $\boldsymbol{T}$ is an orthogonal or unitary matrix, the new basis is still orthonormal. The columns of $\boldsymbol{T}^{-1}$ are the rows of $\overline{\boldsymbol{T}}$ (complex conjugate). The energy in f equals the energy in F, because $(F, F) = (\boldsymbol{T}f, \boldsymbol{T}f) = (f, f)$ when $\boldsymbol{T}$ is unitary. This is the case for all Fourier transforms and the orthonormal wavelet transform.

Translation (also Shift, Delay, or Advance)

$\phi(t)$ is "translated" to $\phi(t-k)$. Its transform $\widehat{\phi}(\omega)$ is "modulated" to $e^{-i\omega k}\widehat{\phi}(\omega)$.

Transmultiplexer (a filter bank in reverse order)

Transmultiplexers combine lower-bandwidth signals $\boldsymbol{x}_k(n)$ (by interpolation and filtering by $z^{-1}F_k(z)$) in order to send them through a high-bandwidth channel. The received signal is filtered and downsampled to the original rates. The output signals $\widehat{\boldsymbol{x}}_k(n)$ suffer from distortion and crosstalk because of the decimation, interpolation and nonideal filtering. A PR transmultiplexer comes from a PR filter bank by reversing H_k and F_k and delaying $F_k(z)$ by one sample.

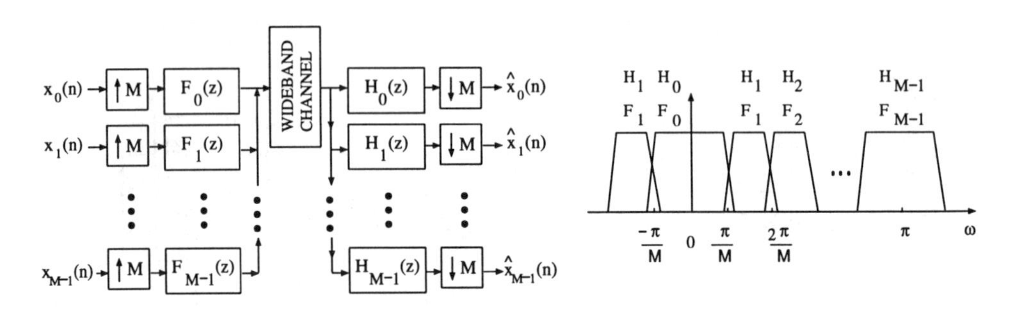

Tree-structured Filter Bank The figure shows a first branch with filters $\{H_{00}(z), H_{01}(z)\}$. The subband signals are further channeled by a three-channel analysis bank with filters $\{H_{10}(z), H_{11}(z), H_{12}(z)\}$ and a two-channel analysis bank with filters $\{H_{13}(z), H_{14}(z)\}$. This a tree-structured analysis bank where additional analysis banks are added on at the subband signals. By moving the decimators $(\downarrow 2)$ of the first analysis bank to the right and combining decimations, the factors become $(6, 6, 6, 4, 4)$.

The synthesis bank has a similar tree (in reverse order). The entire bank is PR if and only if every bank in the tree structure is PR. Tree structures are useful in creating a filter bank with nonuniform decimation/interpolation factors. The main disadvantage is the long overall delay.

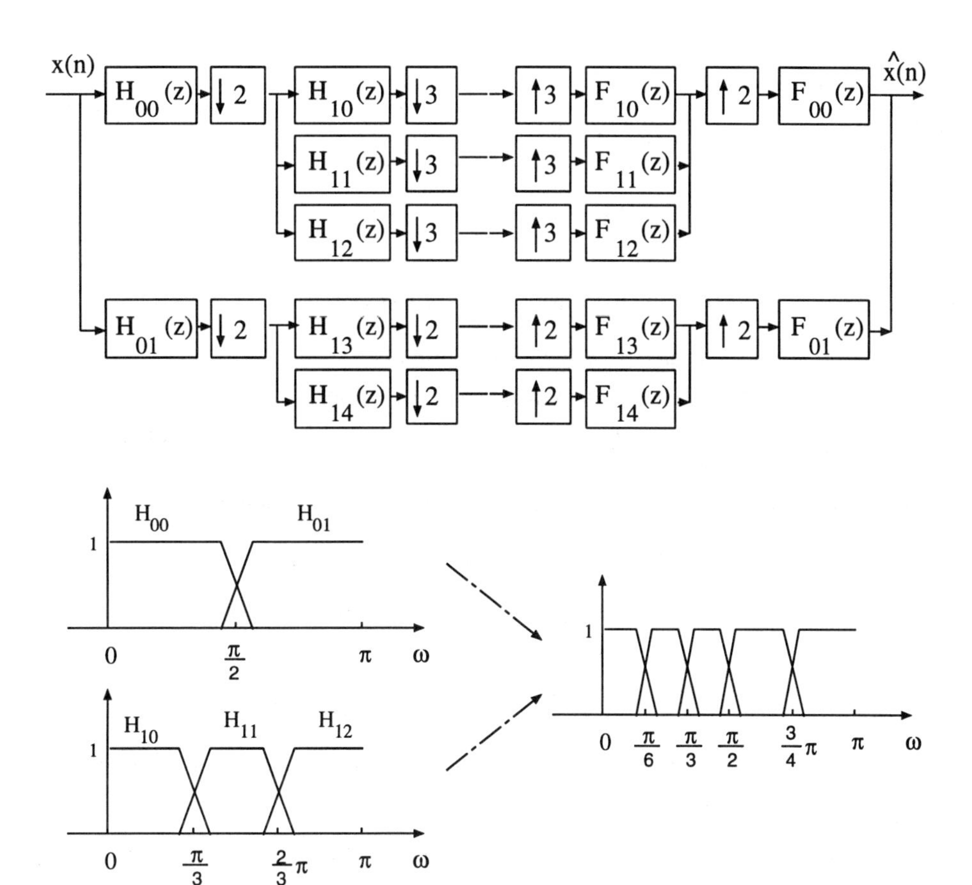

Tree-structured Filter Bank

Two-Channel Biorthogonal Linear-Phase Filter Bank

- Type A (SAOO): Both filters have odd orders (even length).
  $H_0(z)$ is symmetric and $H_1(z)$ is antisymmetric.
- Type B (SSEE): Both filters have even orders (odd length).
  Both filters have symmetric impulse responses.

Two-Channel QMF (Johnston) Filter Bank (Symmetric with $H_1(z) = H_0(-z)$)

$$\begin{bmatrix} H_0(z) \\ H_1(z) \end{bmatrix} = \begin{bmatrix} E_0(z^2) & E_1(z^2) \\ E_0(z^2) & -E_1(z^2) \end{bmatrix} \begin{bmatrix} 1 \\ z^{-1} \end{bmatrix} = \begin{bmatrix} 1 & 1 \\ 1 & -1 \end{bmatrix} \begin{bmatrix} E_0(z^2) & 0 \\ 0 & E_1(z^2) \end{bmatrix} \begin{bmatrix} 1 \\ z^{-1} \end{bmatrix}$$

$E_k(z)$ are the polyphase components of $H_0(z)$. The synthesis filters are

$$\begin{bmatrix} F_0(z) & F_1(z) \end{bmatrix} = \begin{bmatrix} H_1(-z) & -H_0(-z) \end{bmatrix} = \begin{bmatrix} z^{-1} & 1 \end{bmatrix} \begin{bmatrix} E_1(z^2) & 0 \\ 0 & E_0(z^2) \end{bmatrix} \begin{bmatrix} 1 & 1 \\ 1 & -1 \end{bmatrix}$$

The overall distortion function is $T_0(z) = z^{-1}E_0(z^2)E_1(z^2)$, which can be equalized using either IIR or FIR filters.

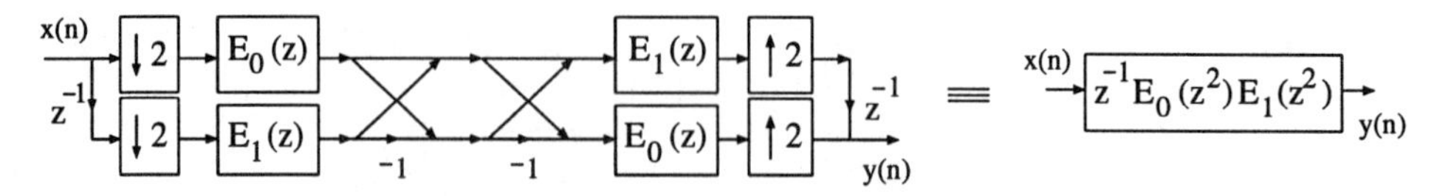

Two-Channel QMF (Johnston) Filter Bank

Two-Channel Paraunitary Linear-Phase Filter Bank (FIR)

$$\begin{cases} H_0(z) & \text{is an even-length symmetric filter} \\ H_1(z) & \text{is an even-length antisymmetric filter} \\ F_0(z) & = z^{-N} H_0(z^{-1}) = H_0(z),\ \ F_1(z) = z^{-N} H_1(z^{-1}) = -H_1(z) \end{cases}$$

With these filters, aliasing is cancelled and the distortion function becomes

$$T_0(z) = H_0^2(z) - H_1^2(z) = (H_0(z) + H_1(z))(H_0(z) - H_1(z)).$$

This should be a delay (z^{-n_0}) for a PR filter bank. Thus, both $(H_0(z)+H_1(z))$ and $(H_0(z)-H_1(z))$ should be delays, which implies that

$$H_0(z) = \tfrac{1}{2}(z^{-n_1} + z^{-n_2}) \qquad \text{and} \qquad H_1(z) = \tfrac{1}{2}(z^{-n_1} - z^{-n_2}).$$

This shows that all two-channel linear phase paraunitary filter banks are “trivial”.

Uncertainty Principle (Heisenberg)

In quantum mechanics, position $\boldsymbol{P}$ and momentum $\boldsymbol{Q}$ are operators with $\boldsymbol{PQ} - \boldsymbol{QP} = \boldsymbol{I}$. Then the Schwarz inequality gives

$$\|\boldsymbol{x}\|^2 = \boldsymbol{x}^*(\boldsymbol{PQ} - \boldsymbol{QP})\boldsymbol{x} \le 2\|\boldsymbol{Qx}\|\,\|\boldsymbol{Px}\|.$$

Similarly in Fourier analysis $\|f(t)\|^2 \le 2\,\|tf(t)\|\;\|\omega\widehat{f}(\omega)\|$. The spread in f and spread in $\widehat{f}$ cannot both be small. The Gaussian function $f(t) = e^{-t^2/2}$ gives equality in the Heisenberg inequality.

Wavelet A localized function with mean zero.

Wavelet Packet A family of scaling functions and wavelets constructed by following a binary tree of dilations/translations.

Wavelet Transform

Continuous: $w_{a,b}(t) = \frac{1}{\sqrt{a}} w(\frac{t-b}{a})$ for $a > 0$, and $C = \int |\widehat{w}(\omega)|^2 \frac{d\omega}{|\omega|}$.

$$f(a,b) = \int \overline{w_{a,b}(t)} f(t)dt \quad \text{and} \quad f(t) = C^{-1} \iint \overline{f(a,b)} w_{a,b}(t) \frac{da\,db}{a^2}$$

Example: Morlet wavelet $w(t) = \frac{1}{\sqrt{2\pi}} e^{-i\alpha t} e^{-t^2/2}$ with $\alpha = \pi\sqrt{\frac{2}{ln2}}$.

Discrete: Series with $w_{jk}(t) = 2^{j/2} w(2^j t - k)$. Transform to $b_{jk} = \int f(t) w_{jk}(t)dt$.

Appendix 1

Wavelets

Gilbert Strang *American Scientist* **82** (April 1994) 250–255

I was listening to Mozart, but my mind was wandering. Normally one doesn't admit such a thing. Possibly this was not Mozart at his best. Symphony Hall was full, we were late, and parking had been very illegal. But the real reason for inattention was an idle thought, and it led me toward this article. The thought was really a question: *How did he write it all down?*

For Mozart that must have been a serious problem. Musical notation is fairly efficient for its purpose, but a full score for all instruments equals a small book. The *notation* is crucial. To a musician it may seem the only possible notation — but it's not. When that signal is recorded (say digitally), the same symphony is expressed in a totally different way. The score that Mozart wrote is not transmitted.

The complete signal is not only the sequence of notes, but the nuances that the performer adds. The touch on the keys or the pedal or the bow — these distinguish a master. There may be an extra word saying *forte* or *pianissimo*. This is not too precise! Mathematically, the notes are an indication but not a complete code.

Our world is full of sounds and images and bit strings that have to be compressed and transmitted and recovered:

- Audio — music and speech
- Video — still images and television
- Data — numbers, letters, books, even zip codes.

Zip codes seem trivial, but this short signal illustrates the choices to be made. I think that England and Canada overdid the compression. Strings like M5S 1A4 and V6T 1W5 give extra precision, but they are too hard to list and remember. (Does any reader knows how those countries chose alphanumeric?) Compression is delicate, because we have to *process* the signal, not just code it. It is the same for a student taking notes in class: compress the lecture but stay readable.

Television has been avoiding compression, but that is about to change. Each "pixel" on the TV screen needs 24 zeros or ones to mix three colors. The shading levels go from 0 to 255, making 2^8 possibilities — eight bits for each color. A very sharp image needs more pixels with more bits than we can afford to send or receive. Compression is required; information has to be thrown away. Otherwise the TV can't keep up with real time!

The problem is really severe for *high definition television*. It has an enormous bit rate. We will soon see an amazingly clear picture with many more pixels, provided the compression is well done. The processing of HDTV signals is a billion-dollar decision that the biggest companies are competing for. At the end I will mention a personal experience with this mathematical and engineering and increasingly political competition.

Fourier Transform and Wavelet Transform

This article is about several methods — none perfect, all being improved — to analyze and synthesize a signal. I mention three ways to transform a symphony:

1. Into cosine waves (by Fourier transform)
2. Into pieces of cosines (short time Fourier transform)
3. Into wavelets (the new way).

The classical method is to separate the whole symphony into pure harmonics. Each instrument plays one steady note. The signal becomes a sum $a_0 + a_1 \cos t + \ldots$ of cosine waves. This is the *Fourier transform*, published 172 years ago by Joseph Fourier in Paris. He assumed an infinite orchestra: Conductor not needed, musicians totally bored.

All music and all signals can be analyzed into pure harmonic waves by the Fourier transform. Engineers do that almost instinctively — the signal strength at each moment in time is replaced by the amplitude of each wave. The big question is how many frequencies are needed for a high-fidelity signal. Probably too many for good compression.

A second possibility is the *Short Time Fourier Transform.* Short segments of the symphony are transformed separately. In each segment, the signal is separated into cosine waves as before. The musicians still play one note each, but they change amplitude in each segment. This is the way most long signals are carved up.

One disadvantage: There are sudden breaks between segments. You might or might not hear that. In a visual image you could probably see it as an edge. This "blocking effect" is the nemesis of the Short Time Fourier Transform.

A new idea in signal processing is closer to the score that Mozart originally wrote. Instead of cosine waves that go on forever or get chopped off, the new building blocks are "*wavelets*". These are little waves that start and stop. They all come from one basic wavelet $w(t)$, which gives the sound level of the standard tune at time t. The basses spread that tune over the longest time. The cellos play the same tune but in half the time, at doubled frequency. Mathematically, this speed-up replaces the time variable t by $2t$. The first bass plays $b_1 w(t)$ and the first cello plays $c_1 w(2t)$, both starting at time $t = 0$.

The next bass and cello play $b_2 w(t-1)$ and $c_2 w(2t-1)$, with amplitudes b_2 and c_2. The bass starts when $t - 1 = 0$, at time $t = 1$. The cello starts earlier, when $2t - 1 = 0$ and $t = \frac{1}{2}$. There are twice as many cellos as basses, to fill the total length of the symphony. Violas and violins play the same passage but faster and faster and all overlapping. At every level the frequency is doubled (up one octave) and there is new richness of detail. "Hyperviolins" are playing at 16 and 32 and 64 times the bass frequency, probably with very small amplitudes. Somehow these wavelets add up to a complete symphony. This article will attempt to show how.

Summary: A long signal is broken into a *basis of possible signals*. Those can be notes or waves or wavelets. The wavelets come from a single function $w(t)$ by speed-ups and time delays. Amplitudes are sent to the receiver, which reconstructs the symphony. (Very small amplitudes are discarded; more later about this crucial compression step.) We now explain the idea of a basis.

Choosing the Best Basis

Some readers will know about vectors and matrices. This section is for you. You remember that every vector (x, y) is a combination of the basis vectors $(1, 0)$ and $(0, 1)$. The vector $(4, 2)$ is 4 times the first basis vector plus 2 times the second.

Another pair of basis vectors is $(1, 1)$ and $(1, -1)$. Again they produce all plane vectors, including $(4, 2)$. The requirement met by both bases is this: *Each vector in the space can be expressed in one way only as a combination of the basis vectors.* The vectors $(1, 0)$ and $(2, 0)$ would not be a basis. They lie on the same line; no combination can give $(4, 2)$. The vectors $(1, 0)$ and $(0, 1)$ and $(3, 1)$ are not a basis — three vectors is too many. The signal $(4, 2)$ can be produced from any two of them (and it is also the sum of all three). Any two vectors in different directions, like $(5, 1)$ and $(3, 1)$, form a basis for the whole plane.

The best bases have a valuable extra property: The vectors are perpendicular. For the standard basis $(1, 0)$ and $(0, 1)$, across is perpendicular to up. The vectors $(1, 1)$ and $(1, -1)$ also pass the test for a $90°$ angle: *their dot product is zero*. The dot product of (x, y) with (X, Y) is $xX + yY$. The dot product of $(1, 1)$ with $(1, -1)$ is $1 - 1 = 0$. The dot product of $(5, 1)$ and $(3, 1)$ is 16 — not perpendicular. More to the point, a vector with 1000 components separates quickly into 1000 perpendicular pieces — speed is crucial. These steps are at the heart of linear algebra.

For engineers and social and physical scientists, *linear algebra* now fills a place that is often more important than calculus. My generation of students, and certainly my teachers, did not see this change coming. It is partly the move from analog to digital; functions are replaced by vectors. Linear algebra combines the insight of n-dimensional space with the applications of matrices.

Higher Dimensions

Move now to four dimensions. Think of the components (x_1, x_2, x_3, x_4) as strengths of a signal. If the signal is $x = (1, 1, 1, 1)$ you are producing a steady note. With $x = (1, 2, 4, 8)$ the volume is increasing.

The easy way to send the signal is to give its four components. Implicitly, you have chosen the standard basis in four-dimensional space. The basis vectors are $(1, 0, 0, 0)$ and $(0, 1, 0, 0)$ and $(0, 0, 1, 0)$ and $(0, 0, 0, 1)$. You are sending the numbers x_1, x_2, x_3, x_4 that multiply your basis vectors. How could this be less than optimal? The answer depends on the signal.

Suppose you hold the same note. This vector $(1, 1, 1, 1)$ is very possible and very common, especially if the time interval is short. Good sound reproduction will certainly use short intervals. With the standard basis, you have to send 1 and 1 and 1 and 1. It would be better if the useful vector $(1, 1, 1, 1)$ was actually in the basis, so one signal would do. *A good basis allows accurate reproduction with only a few vectors.*

Compression discards basis vectors that are absent (or barely present) in the signal. When the vectors represent different frequencies — the *Fourier basis* — high frequencies can often be discarded. A "lowpass filter" removes them. With the standard basis, we can discard $(0, 1, 0, 0)$ only when the second note is not played. Otherwise there is a very audible (or inaudible) gap in the broadcast.

Musical notation has met this problem. It represents $(1, 1, 1, 1)$ by a full note and also by two half-notes. The same signal has several forms. This isn't a basis! Full notes and half-notes and quarter-notes are fine for the musician, but the mathematician only wants four vectors in the basis.

It is quite pleasant to devise a basis that includes the constant vector $(1, 1, 1, 1)$. The second basis vector can be $(1, 1, -1, -1)$. This is a square wavelet, once up and once down. The

German mathematician Alfred Haar completed a perpendicular basis in 1910, by *dilation* (or squeezing) and *translation* (or shifting). He squeezed $(1, 1, -1, -1)$ to produce the third vector $(1, -1, 0, 0)$. Then he shifted by two time intervals to produce the fourth vector $(0, 0, 1, -1)$. The basis vectors go into the columns of the four-by-four "Haar matrix":

$$H_4 = \begin{bmatrix} 1 & 1 & 1 & 0 \\ 1 & 1 & -1 & 0 \\ 1 & -1 & 0 & 1 \\ 1 & -1 & 0 & -1 \end{bmatrix}.$$

Can you guess the matrix H_8? Its first column is the constant vector $(1, 1, 1, 1, 1, 1, 1, 1)$. Electrical engineers would call this the zeroth column; they start from zero, like English lifts. The next column is $(1, 1, 1, 1, -1, -1, -1, -1)$, the wavelet that is played by the bass. It is squeezed to $(1, 1, -1, -1, 0, 0, 0, 0)$ for the first cello, and shifted for the second cello. The last four columns of H_8 are viola notes like $(1, -1, 0, 0, 0, 0, 0, 0)$, with higher frequency over a shorter time. All these vectors are perpendicular!

The violins enter when the dimension reaches 16. I don't recommend writing down the matrix H_{16}, but you would know how. This is typical of linear algebra: When four dimensions are understood, higher dimensions are no problem. There are four violas and eight violins. The count $1+2+4+8$ equals 15, one short. The constant vector $(1, 1, \ldots, 1, 1)$ makes the 16th or zeroth basis vector, which is the scaling function.

Fast Wavelet Transform

The very smallest wavelet transform we saw already. $(4, 2)$ is 3 times $(1, 1)$ plus 1 times $(1, -1)$. From the amplitudes 3 and 1, we recover the vector $(4, 2)$. To find those amplitudes for longer signals, we use *averages* and *differences*. The averages move up in a pyramid, to be averaged again. The differences give the fine details at each level. When a signal changes quickly, those details are important. When the signal is almost steady, we might not transmit them. That is how compression works, by ignoring what the eye can't see and the ear can't hear.

The test case is in four dimensions, for a vector like (4,2,5,5). We know how 4,2 in the first half yields 3 and 1. The second half 5,5 yields 5 (the average) and 0 (the difference). The two averages 3 and 5 move up the pyramid. The two differences 1 and 0 are the details that go with the basis vectors $(1, -1, 0, 0)$ and $(0, 0, 1, -1)$.

The averages give a coarse signal $(3, 5)$. Treat that the usual way. Compute $\frac{1}{2}(3+5) = 4$ and the difference $\frac{1}{2}(3-5) = -1$. Consistency is the key to a good algorithm! The wavelet transform of $(4, 2, 5, 5)$ is the set of averages and differences $(4, -1, 1, 0)$. This is transmited to the receiver, which does an inverse transform back to $(4, 2, 5, 5)$. The decoder goes down the pyramid, first using the coarse parts 4 and -1 to recover 3 and 5. Those are combined with the fine details 1 and 0 to produce 4, 2 and 5, 5. We have reconstructed the signal.

This gives hope for a signal with 256 numbers. The pyramid has 8 levels because $256 = 2^8$. the lowest level starts as always with $A = \frac{1}{2}(x_1 + x_2)$ and $D = \frac{1}{2}(x_1 - x_2)$. The average A moves up to the next level. The difference D is the detail at this finest level.

The key word behind this pyramid is ***multiresolution***. We are seeing the signal at multiple scales—fine, medium, and coarse. We resolve it at each scale by details and averages. Possibly our eyes do the same, to see the target and then the bull's-eye. Certainly our ears are designed to pick off higher frequencies as the sound travels into the cochlea. The art of the engineer is imitating nature.

Note: We called this wavelet transform "*fast*". A vector of N numbers is to be expressed as a combination of N basis vectors. In matrix language, we are multiplying by the Haar matrix H or its inverse. Normally such a multiplication takes N^2 steps. The pyramid does it with only N averages and N differences.

Question: Is there also a fast Fourier transform? Can we quickly find the amplitudes of N cosine waves? Instead of H we use the Fourier matrix F, to transform between "time domain" and "frequency domain". This fast transform is the most important numerical algorithm in our lifetime, a very tough competitor for wavelets. It is wired into computers and explained in linear algebra textbooks. The key is to produce F out of matrices whose entries are almost all zeros. The heart of numerical linear algebra has become matrix factorizations — H and F are outstanding examples.

The Daubechies Wavelets

Wavelets are a hot topic, but the Haar wavelets are cold. They were invented in 1910. Their graphs are made from flat pieces; anything that simple has already been thought of. Approximation to most signals is very poor. We need many many flat pieces to represent even a sloping line to decent accuracy. The basis will not give compression ratios of 20 :1 or 100 :1, as desired. So we turn to a better basis.

The new wavelets are more intricate, and their only formula is an infinite product, but eventually mathematics had to find them. I will touch on this advance by describing the discovery made by Ingrid Daubechies. In 1988, at AT&T Bell Laboratories, she found a pulse that starts and stops and is perpendicular to all its dilations and shifts. It is based on four "magic" numbers h_0, h_1, h_2, h_3. Her scaling vector S uses them in that order. Her wavelet uses them in the order $W = (h_3, -h_2, h_1, -h_0)$. Do you see why W is perpendicular to S? Multiply and add: the dot product $S \cdot W$ cancels itself to give zero. She also wanted $(1, 1, 1, 1)$ and $(1, 2, 3, 4)$ to have zero component along W, so constant and linear signals can be greatly compressed. Their dot products with W must be zero:

$$h_3 - h_2 + h_1 - h_0 = 0 \quad \text{and} \quad h_3 - 2h_2 + 3h_1 - 4h_0 = 0.$$

Those are two equations for the h's; we need two more. The third equation makes the first bass tune $(h_3, -h_2, h_1, -h_0, 0, 0)$ perpendicular to the second bass tune $(0, 0, h_3, -h_2, h_1, -h_0)$. Their dot product is required to be $h_1h_3+h_0h_2 = 0$. Then a fourth equation $h_0+h_1+h_2+h_3 = 2$ sets the size of the h's. Her numbers from solving the four equations give a much better filter than Haar: $4h_0 = 1+\sqrt{3}, 4h_1 = 3+\sqrt{3}, 4h_2 = 3-\sqrt{3}, 4h_3 = 1-\sqrt{3}$. I must mention that this is not the ultimate. Six numbers or eight numbers can be even better, but also more work. Video filters tend to be short. Audio filters are amazingly long, because music is generally smoother than a picture.

The key step from discrete vectors to continuous functions is the *dilation equation.* This uses the magic numbers h_0, h_1, h_2, h_3 in a curious but natural way. The equation for the scaling function $\phi(t)$ involves t and $2t$ and the magic h's:

$$\phi(t) = h_0\phi(2t) + h_1\phi(2t - 1) + h_2\phi(2t - 2) + h_3\phi(2t - 3).$$

Substitute $t = 1$ and $t = 2$ to find $\phi(1)$ and $\phi(2)$. Then the equation gives ϕ at $t = \frac{1}{2}, \frac{3}{2}, \frac{5}{2}$ because $2t$ on the right side is a whole number. From these half-integer times we go to quarter-integers. Eventually we have enough values to draw the graph from $t = 0$ to $t = 3$. This has become famous (to some people) as a function with entirely new properties, all built into the dilation equation.

Where Fourier used a cosine wave and Haar used a square wave, Daubechies begins with this scaling function. Her wavelet $w(t)$ has the same right side as the equation above, but with coefficients $h_3, -h_2, h_1, -h_0$. It has a highly irregular graph, a type of fractal. Squeezing and shifting it gives the complete wavelet basis. But all computations go back to the four numbers.

High Definition Television

"The shot heard round the world" was a metaphor in 1775. The signal seen around the world is now a reality. It is television. You may think that the signal could be improved (by better programs). Engineers also want to improve it (by more pixels and better compression). Probably this second improvement is the one we will get, with HDTV.

The battle to set the standard for high definition television has been played out in a surprisingly public way. Europe invested a billion dollars, then stopped. It was decided to follow the North American standard. A competition was organized by the Federal Communications Commission, and there were four finalists. One big issue is *motion estimation*, to predict where the picture will move. Then you only have to transmit a small correction, and you can keep up with all the pixels in real time.

The New York Times expected that the Japanese entry would win, because HDTV is already available in Japan. But it is based on analog signals, not 0's and 1's. Digital filters are now well developed, to separate the high frequencies and compress that information. But sharp edges have to stay sharp. Experts know exactly what to look for, like tasting wine or tea.

To the great credit of the New York Times, the competition was explained to the public. Every month it took a new turn. The FCC finished its tests, but one group of companies wanted a makeup test. They had stayed up the night before and their system wasn't feeling too well. (All professors recognize these symptoms, but the government is softer. They allowed a retest.) We were expecting one winner, but the FCC urged cooperation in the end. We all hope it works.

I must explain how I learned about this. As usual, it was from a student. He came into my office and said "Do you remember me?" Well, I tried to think fast. Too many students, too weak a memory, and better to be absolutely honest. "Not completely," I replied — never expecting to hear what he told me next. "You are my thesis advisor."

What do you say to a thesis student you don't remember? In that position I suggest something very short: "Tell me more." The most amazing part was his thesis topic. "I am designing the filter bank for MIT's entry in the HDTV competition." Some days you can't lose, even if you deserve to.

It is true that I am his advisor. He is a mathematics student who looked for a thesis in electrical engineering. He found Jae Lim, who masterminded MIT's joint entry with General Instruments. A formal committee was required by the mathematics department. I must have agreed, since my signature was there on the paper. But my very own student would not tell me the coefficients $h_0, \ldots, h_6$ in his filters. He expressed confidence in his advisor, but not enough for a billion-dollar secret.

Summary

A family of wavelets is an orthogonal basis. By combining the wavelets we represent any signal. Music has one independent variable: time. A photograph has two variables, across and up. Then wavelets in x multiply wavelets in y. A video image has three variables, x, y, and t — a flow of pictures with very important predictability. The competition between bases is fierce. Where computers race for faster calculations, mathematics races for quicker algorithms. An idea that cuts in half the *number* of steps is as good as a chip that doubles the speed. I don't know if the idea of wavelets is in that league, but it might be.

The HDTV standard is based on Fourier transforms. Wavelets came too late to have a real chance in that video race, but they are highly competitive for audio. Four stereo channels need only a small fraction of a TV band. In a completely different competition, run by the FBI and worth describing here, wavelets have won.

The FBI has 30 million sets of fingerprints, more every day. They need to be digitized. When a driver is stopped, or a thief is apprehended, fingerprints will be recorded electronically. A central computer will look for a match. As it is now, with thumb prints in enormous files in Washington, your chance to escape has been pretty good. Soon the "minutiae" of ridge endings and bifurcations will catch you.

It was expected that the older Fourier methods would succeed here too. But with 20:1 compression, the fingerprint ridges couldn't be followed. Lines were broken between one 8 by 8 square and the next. The short time Fourier transform introduced too much blocking. Wavelets gave a better picture. Tom Hopper at the FBI has chosen that new basis.

So when you get caught, you can blame it on wavelets.

Appendix 2

Wavelets and Dilation Equations

Gilbert Strang

Siam Review **31** (1989) 613–627

Abstract

Wavelets are new families of basis functions that yield the representation $f(x) = \sum b_{ij} W(2^j x - k)$. Their construction begins with the solution $\phi(x)$ to a dilation equation with coefficients c_k. Then W comes from ϕ, and the basis comes by translation and dilation of W. It is shown in Part 1 how conditions on the c_k lead to approximation properties and orthogonality properties of the wavelets. Part 2 describes the recursive algorithms (also based on the c_k) that decompose and reconstruct f. The object of wavelets is to localize as far as possible in both time and frequency, with efficient algorithms.

Wavelets are based on translation ($W(x) \to W(x+1)$) and above all on dilation ($W(x) \to W(2x)$). It is remarkable how long it has taken for "*dilation equations*" to be mentioned beside differential equations and difference equations. True, they are hardly in the same league. But ideas about wavelets are coming fast. The mathematics is attractive and several important applications seem to fit—I hope this survey will be helpful. You should know that its author is neither an expert nor an evangelist.

The goal is a new way to represent functions—especially functions that are local in time and frequency (or space and wave number). Compare with Fourier series. Sines and cosines are perfectly local in frequency, but global in x or t. A short pulse has slowly decaying coefficients that are hard to measure. To reconstruct the pulse, a Fourier series depends heavily on cancellation. The whole of Fourier analysis, relating properties of functions to properties of coefficients, is made difficult (some say interesting) by the nonlocal support of $\sin x$.

In achieving local support we lose the greatest property of the basis $\{e^{inx}\}$. With respect to a wavelet basis the differentiation operator is not diagonal. Wavelets are not eigenfunctions of $\partial/\partial x$, and frequencies are mixed up. The uncertainty principle imposes limits on what is possible in x and ξ together. The commutator $(\partial/\partial x)(\partial/\partial \xi) - (\partial/\partial \xi)(\partial/\partial x)$ is a multiple of the identity (since $(\partial/\partial x)(xu) - x(\partial u/\partial x) = u$), so we cannot diagonalize both operators. But a good "microlocalization" leaves $\partial/\partial x$ nearly diagonal, and at the same time nearly diagonalizes $\partial/\partial \xi$ (which is multiplication by x). To connect dilation with multiplication by x, differentiate $f(cx)$ with respect to c at $c = 1$.

The second important property of $\{e^{inx}\}$ is orthogonality. That can be saved. Wavelets can be made orthogonal to their own dilations and translations. Then $\int W(x)W(2^j x - k)\, dx = 0$ for all integers j and k. The wavelet basis has two indices, in which k is translation and j is dilation or compression. It suggests multigrid. A wavelet expansion $\sum b_{jk} W_{jk}(x)$ is a *multiresolution* of $f(x)$, in which b_{jk} carries information about f near $\xi = 2^j$ and $x = 2^{-j}k$. The sum on k is the detail at the scaling level $h = 2^{-j}$.

Orthogonality is not easy to achieve with local support. Truncated at zero and 2π, a sine wave $\phi(x)$ is orthogonal to $\phi(2x)$ but not to $\phi(4x)$. The "windowed Fourier transform" combines smoothness with local support by bringing $e^{i\xi x}$ gradually to zero, but it is not fully satisfactory. The price of orthogonality with compact support is irregular basis functions. We live with these wavelets by doing all computations recursively (this subject is recursion heaven). And it is important to recognize that orthogonality and even linear independence (!) are not essential in the representation of functions. Wavelets need not be orthogonal.

This brief introduction cannot do justice to the applications. Nor can we attempt a proper history — it would be mostly in French. The idea of wavelets grew out of seismic analysis. Their development has been led by Yves Meyer, whose book will describe a new chapter in harmonic analysis (connecting to work of Calderòn, Grossmann, Morlet, Coifman, Weiss, and many others). The interest in wavelets is both pure and applied — like the interest in splines.

Part 1 of this paper establishes the properties of wavelets — approximation through Condition A and orthogonality through Condition O. Since we never see wavelets as functions (only recursively), their properties have to be discovered indirectly. We absolutely need these properties in order to have any idea what the algorithms are producing. Then Part 2 begins with a piecewise constant example (ϕ is a box function, the wavelet is Haar's). The example reveals a lot with no deep analysis. You could go directly to Part 2, about algorithms, and then return to dilation equations.

1. Dilation Equations: Construction of ϕ

The basic dilation equation is a two-scale difference equation:

$$\phi(x) = \sum c_k \phi(2x - k). \tag{A.1}$$

We look for a solution normalized by $\sum \phi\, dx = 1$. The first requirement on the coefficients c_k comes from multiplying by 2 and integrating:

$$2 \int \phi\, dx = \sum c_k \int \phi(2x-k)\, d(2x-k) \quad \text{yields} \quad \sum c_k = 2.$$

Uniqueness of ϕ is ensured by $\sum c_k = 2$. A smooth solution is not ensured. For a striking example, set $c_0 = 2$:

$$\text{The delta function} \quad \phi = \delta \quad \text{satisfies} \quad \delta(x) = 2\delta(2x).$$

That dilation of δ is unfamiliar (but somehow very pleasing). For other c's, spline functions appear:

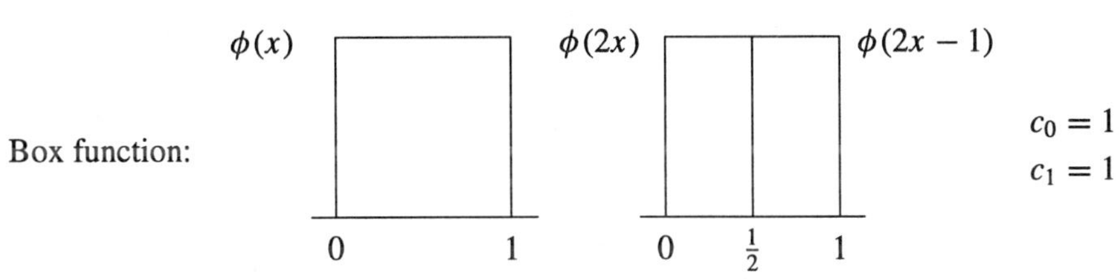

Hat function: $\phi(x)$, $\frac{1}{2}\phi(2x)$, 0, 1, 2 $\quad c_0 = \frac{1}{2}$, $c_1 = 1$, $c_2 = \frac{1}{2}$

We now outline three constructions of the "scaling function" ϕ. Those constructions display very clearly the mathematics of dilation. Then we turn to wavelets, their properties and their purpose. A wavelet $W(x)$ is a second combination (involving the same recursion coefficients c_k) of the translates $\phi(2x-k)$.

Construction 1. Iterate $\phi_j(x) = \sum c_k \phi_{j-1}(2x-k)$ with the box function as $\phi_0(x)$. When $c_0 = 2$ the boxes get taller and thinner, approximating the delta function. For $c_0 = c_1 = 1$ the box is invariant: $\phi_j = \phi_0$. For $\frac{1}{2}, 1, \frac{1}{2}$ the hat function appears as $j \to \infty$, and $\frac{1}{8}, \frac{4}{8}, \frac{6}{8}, \frac{4}{8}, \frac{1}{8}$ yields the cubic B-spline. An example that will be important (an inspiration of Daubechies—we propose the notation D_4) has coefficients $\frac{1}{4}\left(1+\sqrt{3}\right), \frac{1}{4}\left(3+\sqrt{3}\right), \frac{1}{4}\left(3-\sqrt{3}\right)$, and $\frac{1}{4}\left(1-\sqrt{3}\right)$:

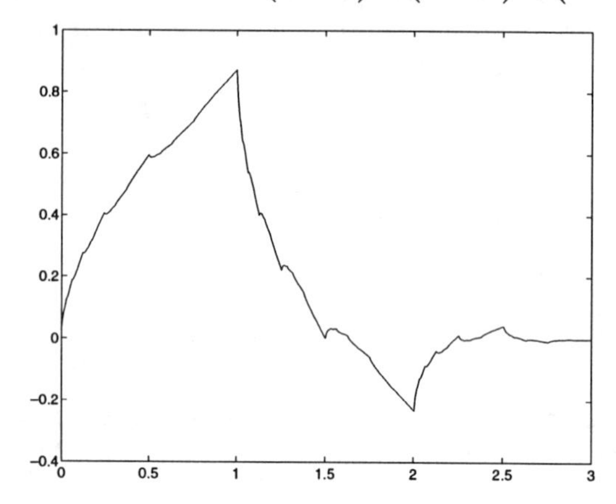

This scaling function D_4 leads to orthogonal wavelets. *It is not as smooth as it looks.* Note that the Weierstrass nowhere differentiable function, which is $\sum b^n \cos(3^n x)$, involves dilation by 3. So does de Rham's function, which has $c_k = \frac{2}{3}, \frac{1}{3}, 1, \frac{1}{3}, \frac{2}{3}$ adding to 3. Resnikoff has found a connection between Weierstrass functions and wavelets.

Construction 2. The second construction takes the Fourier transform of (1):

$$\begin{aligned} \widehat{\phi}(\xi) &= \sum c_k \int \phi(2x-k) e^{i\xi x}\, dx \\ &= \tfrac{1}{2}\left(\sum c_k e^{ik\xi/2}\right) \int \phi(y) e^{iy\xi/2}\, dy = P\left(\frac{\xi}{2}\right)\widehat{\phi}\left(\frac{\xi}{2}\right). \end{aligned} \tag{A.2}$$

The symbol $P(\xi) = \frac{1}{2}\sum c_k e^{ik\xi}$ is the crucial function in this theory. Note that $P(0) = 1$. Now repeat (2) at $\xi/2, \xi/4, \ldots$ and recall $\widehat{\phi}(0) = \int \phi\, dx = 1$:

$$\widehat{\phi}(\xi) = \left[\prod_1^n P\left(\frac{\xi}{2^j}\right)\right]\widehat{\phi}\left(\frac{\xi}{2^N}\right) \quad \text{approaches} \quad \prod_1^\infty P\left(\frac{\xi}{2^j}\right). \tag{A.3}$$

For $c_0 = 2$ we find $P \equiv 1$ and $\widehat{\phi} \equiv 1$, the transform of the delta function. For $c_0 = c_1 = 1$ the products of the P's are geometric series:

$$P\left(\frac{\xi}{2}\right)P\left(\frac{\xi}{4}\right) = \tfrac{1}{4}\left(1+e^{i\xi/2}\right)\left(1+e^{i\xi/4}\right) = \frac{1-e^{i\xi}}{4\left(1-e^{i\xi/4}\right)}.$$

As $N \to \infty$ this approaches the infinite product $\left(1-e^{i\xi}\right)/(-i\xi)$. This is $\int_0^1 e^{i\xi x}\,dx$, the transform of the box function. The hat function comes from squaring $P(\xi)$ which by (3) also squares $\widehat{\phi}(\xi)$. (Multiplication of P's is $\frac{1}{2}$ times convolution of c's). The cubic B-spline comes from squaring again.

Construction 3. This construction of ϕ works directly with the recursion. Suppose ϕ is known at the integers $x = j$. The recursion (1) gives ϕ at the half-integers. Then it gives ϕ at the quarter-integers, and ultimately at all dyadic points $x = k/2^j$. This is fast to program. *All good wavelet calculations use recursion.*

The values of ϕ at the integers come from an eigenvector. With the four Daubechies coefficients, set $x = 1$ and $x = 2$ in the dilation equation (1) and use the fact that $\phi = 0$ unless $0 < x < 3$:

$$\begin{aligned} \phi(1) &= \tfrac{1}{4}(3+\sqrt{3})\phi(1) + \tfrac{1}{4}(1+\sqrt{3})\phi(2) \\ \phi(2) &= \tfrac{1}{4}(1-\sqrt{3})\phi(1) + \tfrac{1}{4}(3-\sqrt{3})\phi(2). \end{aligned} \tag{A.4}$$

This is $\phi = L\phi$, with matrix entries $L_{ij} = c_{2i-j}$. Compare with c_{i-j} for an ordinary difference equation. The eigenvalues are 1 and $\frac{1}{2}$. The eigenvector for $\lambda = 1$ has components $\phi(1) = \frac{1}{2}\left(1+\sqrt{3}\right)$ and $\phi(2) = \frac{1}{2}\left(1-\sqrt{3}\right)$, which are the heights on our graph of D_4. The other eigenvalue $\lambda = \frac{1}{2}$ means that the recursion can be differentiated: $\phi'(x) = \sum c_k 2\phi'(2x-k)$ leads similarly to $\phi'(1)$ and $\phi'(2)$. In some weak sense, $\phi = D_4$ has a "dilational derivative." For the hat function, the recursion matrix (see below) again has $\lambda = 1, \frac{1}{2}$. For the cubic spline the eigenvalues are $1, \frac{1}{2}, \frac{1}{4}, \frac{1}{8}$.

To repeat for emphasis: From $\phi(1)$ and $\phi(2)$ the recursion gives everything.

In these constructions the properties of $P(\xi) = \frac{1}{2}\sum c_k e^{ik\xi}$ are decisive. The precise hypotheses are in flux, and infinitely many c_k can be allowed. One basic property will bring together the theory of dilation equations, before we go on to wavelets.

1.1. Dilation equations: Fundamental theorem. The accuracy of piecewise polynomial approximation, by splines or finite elements, depends on the answer to this question: To what degree $p-1$ can the polynomials $1, x, x^2, \ldots, x^{p-1}$ be reproduced exactly by the approximating functions? When the polynomials are "in the space," the approximation error is of order h^p. In our case, the approximating functions are $\phi(x)$ and its translates. Splines are the best at approximation, and finite elements have the narrowest support—but both are weeded out when we require orthogonality. There is already a theory of approximation by translates. It connects p with the properties of $\widehat{\phi}$. The link is the Poisson summation formula. When ϕ solves a dilation equation, that throws new questions into the theory—it is extremely satisfying that these new questions have the same answers.

For approximation with accuracy h^p, the Fourier transform $\widehat{\phi}$ must have zeros of order p at all points $\xi = 2\pi n$ (except at $\xi = 0$ where $\widehat{\phi} = 1$). Notice how easily that converts to a condition on the symbol P. According to (3), the transform $\widehat{\phi}$ is the infinite product of $P\left(\xi/2^j\right)$. At $\xi = 2\pi$ the first factor is $P(\pi)$. At $\xi = 4\pi$ the second factor becomes $P(\pi)$. At $\xi = 6\pi$ the first factor is $P(3\pi)$, which by periodicity is the same as $P(\pi)$. The zeros of P produce zeros of $\widehat{\phi}$:

Condition A. The symbol $P = \frac{1}{2}\sum c_k e^{ik\xi}$ has a zero of order p at $\xi = \pi$. Equivalently, the coefficients c_k satisfy the sum rules that yield $P^{(m)}(\pi) = 0$:

$$\sum(-1)^k k^m c_k = 0, \quad m = 0, 1, \ldots, p-1. \tag{A.5}$$

The box function has $P = \frac{1}{2}\left(1 + e^{i\xi}\right)$ and $p = 1$. The hat function has $p = 2$ and so does D_4. The cubic spline has $p = 4$.

A zero at $\xi = \pi/2$ (instead of π) would also produce the desired zeros in the product $\widehat{\phi}$. Thus Condition A is not strictly necessary in what follows. Choosing $c_0 = 1$ and $c_2 = 1$ and $P = \frac{1}{2}\left(1 + e^{2i\xi}\right)$ stretches out the box function—it becomes $\phi = \frac{1}{2}$ on the double interval $0 < x \le 2$. But $P(\pi/2) = 0$ produces instability and linear dependence—the alternating sum of stretched boxes is $\sum(-1)^k \phi(x-k) = 0$. With the added requirement of stability, the condition is exactly right.

The fundamental theorem states the consequences of Condition A:

1. The polynomials $1, x, \ldots, x^{p-1}$ are linear combinations of the translates $\phi(x-k)$.

2. Smooth functions can be approximated with error $O(h^p)$ by combinations at every scale $h = 2^{-j}$:
$$\left\| f - \sum_k a_k \phi(2^j x - k)\right\| \le C 2^{-jp} \left\| f^{(p)}\right\| \quad \text{for suitable } a_k.$$

3. The first p moments of the wavelet $W(x)$ (see below) are zero:
$$\int x^m W(x)\, dx = 0 \quad \text{for } m = 0, \ldots, p-1.$$

4. The wavelet coefficients of a smooth function decay like $|\int f(x) W\left(2^j x\right) dx| \le C 2^{-jp}$.

5. The recursion matrix M_N determining ϕ at the integers has eigenvalues $1, \frac{1}{2}, \ldots, \left(\frac{1}{2}\right)^{p-1}$.

1 and **2** come from approximation theory. The combination of ϕ's at scale j is also a combination $\sum b_{jk} W\left(2^j x - k\right)$ down to scale j. **3** and **4** are easy once wavelets are defined. Mallat gives a sharp result, with properly stated requirements on the smoothness and decay of ϕ: The H^p norm of f is equivalent to the corresponding norm of its coefficients b_{jk}. Wavelets lead to unconditional bases, suitable for a wide range of function spaces.

It is **5** that makes $\phi(x)$ smoother as p increases and also makes the constructions successful. The smoothness is weaker than $\phi \in C^{p-1}$, but it is striking that "dilational derivatives" come at the same time as higher degrees of approximation. What remains to be studied is orthogonality—which imposes an entirely different condition on the c_k.

Remark 1. Suppose the basic recursion has coefficients $c_0, \ldots, c_N$. Then ϕ is zero outside the interval $[0, N]$. With continuity it follows that $\phi(0) = 0$ and $\phi(N) = 0$. Those were assumed in (4) when we determined $\phi = D_4$ at the integers. For the box function with $N = 1$, $\phi(0)$ and $\phi(N)$ cannot both be dropped. Our recursion matrix will be $(M_N)_{ij} = c_{2i-j}$ with $i, j = 0, \ldots, N-1$. For the box function $M_1 = [1]$ has eigenvalue $\lambda = 1$, as expected in **5** above.

The spectrum of the infinite matrix M (allowing all i, j) is an attractive problem in operator theory. Notice that M is *convolution followed by decimation*—multiplication by the matrix c_{i-j} followed by projection onto even-numbered coordinates. By contrast with the usual Toeplitz case, eigenfunctions can have compact support! Homogeneous difference equations with zero boundary conditions lead to $\phi = 0$, but not so for dilation equations.

Remark 2. The minimum requirement is $p = 1$. Then $P(\pi) = 0$, which means that $\sum c_{2k} = \sum c_{2k+1}$. Since $\sum c_k = 2$, the columns of M add to 1:

$$M_N = \begin{bmatrix} c_0 & & & \\ c_2 & c_1 & c_0 & \\ & c_3 & c_2 & c_1 \\ & & & c_3 \end{bmatrix} \quad \begin{array}{l} \text{steps of 2 down columns} \\ \text{steps of 1 across rows} \\ \text{here } N = 4 \end{array}$$

$(1, 1, 1, 1)$ is a left eigenvector with $\lambda = 1$. The right eigenvector yields the values $\phi(0), \ldots, \phi(N-1)$ at the integers. The recursion determines ϕ at all dyadic points. Values at other points are never used.

1.2. Wavelets and orthogonality. Finally we define a wavelet. It comes from the scaling function ϕ by taking "differences":

$$W(x) = \sum (-1)^k c_{1-k} \phi(2x - k). \tag{A.6}$$

We write W in place of the usual ψ, to distinguish more clearly from ϕ. Notice $2x$ on the right, and especially $(-1)^k$. Examples show the effect of alternating signs:

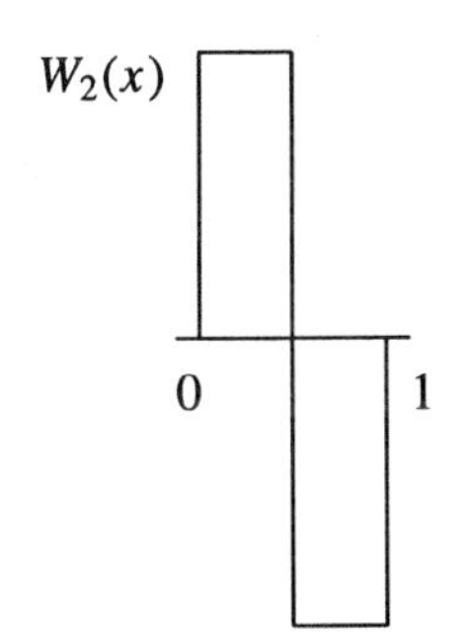

Haar wavelet from box function
$W_2(x) = \phi(2x) - \phi(2x - 1)$

"Wavelet" from hat function
$W = \phi(2x) - \frac{1}{2}\phi(2x - 1) - \frac{1}{2}\phi(2x + 1)$

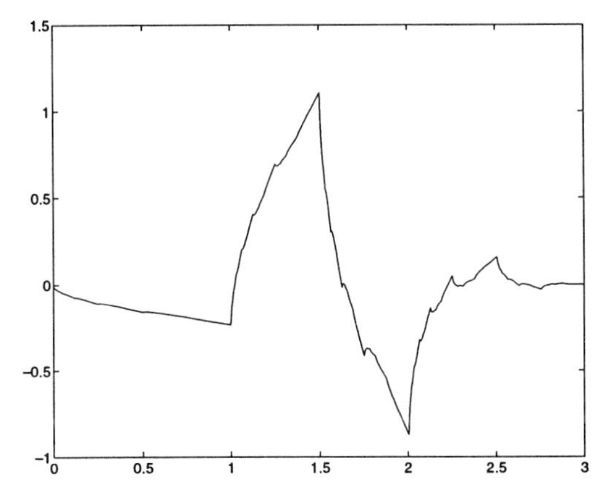

$W_4(x)$ from $\phi = D_4$ Orthogonal wavelet

The wavelet from the hat function does not belong here. It is not orthogonal to $W(x+1)$. The point is that the other two do belong. The Haar function is orthogonal to its own translations and dilations. Historically it was the original wavelet (but with $p=1$ and poor approximation). The orthogonal wavelet W_4 has $p=2$ and second-order approximation.

Without formulas for D_4 and W_4, how is the orthogonality of their translates known? We need a test that applies to the recursion coefficients c_k, or to the symbol $P(\xi)=\frac{1}{2}\sum c_k e^{ik\xi}$.

Condition O.

$$|P(\xi)|^2+|P(\xi+\pi)|^2\equiv 1 \quad \text{or} \quad \sum c_k c_{k-2m}=2\delta_{0m}.$$

With this condition, the infinite matrix L^*L in Part 2 is an orthogonal projection. To see now the role of Condition O, suppose the functions $\phi_0(2x-k)$ are orthogonal. Then so are the translates of $\phi_1(x)=\sum c_k\phi_0(2x-k)$:

$$\begin{aligned}\int \phi_1(x)\phi_1(x-m)\,dx &= \int \Big(\sum c_k\phi_0(2x-k)\Big)\Big(\sum c_l\phi_0(2x-2m-l)\Big)\,dx\\ &= \sum c_k c_{k-2m}\int \phi_0^2(2x)\,dx=0 \quad \text{for } m\neq 0. \qquad \text{(A.7)}\end{aligned}$$

Construction 1 creates ϕ by iteration from the box function, which is orthogonal to its translates. Therefore (as Daubechies observed) so is ϕ.

The wavelet $W(x)$ in (6) is also orthogonal to $\phi(x-m)$. This is simple but neat, not involving Condition O. The sum in (7) changes to

$$\sum(-1)^k c_{1-k}c_{k-2m}, \quad \text{which is identically zero!} \qquad \text{(A.8)}$$

Just replace k by $1-n+2m$. This identity is $HL^*=0$ in Part 2. Then (6) makes $W(x)$ orthogonal to $W(2x-m)$. The orthogonality of $W(x)$ and $W(x-m)$ comes back to Condition O.

The goal in constructing wavelets is to satisfy Conditions A and O. The basic family W_2, W_4, $W_6,\ldots$ was discovered by Daubechies, following Haar's W_2. The accuracies are $p=1,2,3,\ldots$ and there are $2,4,6,\ldots$ nonzero coefficients c_k. The smoothness also increases with p—but only by about $\frac{1}{2}$ derivative each time. D_4 and W_4 are Hölder continuous with exponent $.550\cdots$. In Galerkin's method for solving differential equations, it is natural for these wavelets to be the trial functions—broader support than splines, nonsymmetric but orthogonal, multigrid built in, all computations based on recursion, difficulty to be expected at boundaries. The first experiments by Glowinski, Lawton, and Ravachol are particularly interesting for Burgers' equation.

2. Algorithms for wavelet expansions

Now comes a change of direction. Instead of discussing the properties of wavelets, we describe algorithms. The main question is how to *decompose* a signal into its wavelet coefficients, and how to *reconstruct* the signal from the coefficients. There is a "tree algorithm" or "pyramid algorithm" that makes these steps simple and fast. It does for the discrete wavelet transform what the Fast Fourier Transform (FFT) does for the discrete Fourier transform. The algorithm is fully recursive.

The user chooses a specific wavelet. We begin with the simplest choice, based on the box function. It satisfies the orthogonality property (Condition O), so all pieces of the decomposition are orthogonal. The approximation property (Condition A which preserves polynomials) determines how quickly the coefficients decay—for efficiency we want to stop the decomposition early. In that respect the box function is poor. Efficiency is the reason for working with higher wavelets $W_4, W_6, W_8, \ldots$, and simplicity is the reason for starting with W_2. This is Haar's wavelet $[1\ -1]$.

The discussion will be discrete—for vectors not functions. We are given $n = 2^j$ values $f_1, \ldots, f_n$. They may be equally spaced values of a function $f(x)$ on a unit interval. The goal is to split this vector f into its components at different scales, indexed by j. At each new level the meshwidth h is cut in half and the number of wavelet coefficients is doubled. The decomposition is

$$f = f^{\phi} + f^{(0)} + \cdots + f^{(J-1)}.$$

The "detail" $f^{(j)}$ is a combination of 2^j wavelets at scale 2^{-j}, and f^{ϕ} is a multiple of the scaling function ϕ. For a numerical example take $J = 2$. Then the finest detail $f^{(1)}$ is the sum of two terms, here with coefficients $b_{11} = 4$ and $b_{12} = 1$:

$$f = \begin{bmatrix} 9 \\ 1 \\ 2 \\ 0 \end{bmatrix} = 3 \begin{bmatrix} 1 \\ 1 \\ 1 \\ 1 \end{bmatrix} + 2 \begin{bmatrix} 1 \\ 1 \\ -1 \\ -1 \end{bmatrix} + 4 \begin{bmatrix} 1 \\ -1 \\ 0 \\ 0 \end{bmatrix} + 1 \begin{bmatrix} 0 \\ 0 \\ 1 \\ -1 \end{bmatrix}. \tag{A.9}$$

Notice that the four components are mutually orthogonal. There are $1 + 2 + \cdots + 2^{J-1}$ wavelet coefficients, and the one from f^{ϕ} makes 2^J.

How are the coefficients 3, 2, 4, 1 computed from f? *On the finest scale first.* As in the FFT, the decomposition begins with a "butterfly":

$$\begin{bmatrix} \frac{1}{2} & \frac{1}{2} & & \\ \frac{1}{2} & -\frac{1}{2} & & \\ & & \frac{1}{2} & \frac{1}{2} \\ & & \frac{1}{2} & -\frac{1}{2} \end{bmatrix} \begin{bmatrix} 9 \\ 1 \\ 2 \\ 0 \end{bmatrix} = \begin{bmatrix} 5 \\ 4 \\ 1 \\ 1 \end{bmatrix}. \tag{A.10}$$

This is followed by a permutation, in which high frequencies go to the bottom:

$$\begin{bmatrix} 1 & & & \\ & & 1 & \\ & 1 & & \\ & & & 1 \end{bmatrix} \begin{bmatrix} 5 \\ 1 \\ 1 \\ 4 \end{bmatrix} = \begin{bmatrix} 5 \\ 1 \\ 4 \\ 1 \end{bmatrix}. \tag{A.11}$$

The next step is another butterfly, on low frequencies only:

$$\begin{bmatrix} \frac{1}{2} & \frac{1}{2} & & \\ \frac{1}{2} & -\frac{1}{2} & & \\ & & 1 & \\ & & & 1 \end{bmatrix} \begin{bmatrix} 5 \\ 1 \\ 4 \\ 1 \end{bmatrix} = \begin{bmatrix} 3 \\ 2 \\ 4 \\ 1 \end{bmatrix}. \tag{A.12}$$

The result is the set of wavelet coefficients 3, 2, 4, 1. The product of the three matrices in (10–12) is the decomposition matrix D. Its inverse is the reconstruction matrix R:

$$D = \begin{bmatrix} \frac{1}{4} & \frac{1}{4} & \frac{1}{4} & \frac{1}{4} \\ \frac{1}{4} & \frac{1}{4} & -\frac{1}{4} & -\frac{1}{4} \\ \frac{1}{2} & -\frac{1}{2} & 0 & 0 \\ 0 & 0 & \frac{1}{2} & -\frac{1}{2} \end{bmatrix} \quad \text{has} \quad D^{-1} = R = \begin{bmatrix} 1 & 1 & 1 & 0 \\ 1 & 1 & -1 & 0 \\ 1 & -1 & 0 & 1 \\ 1 & -1 & 0 & -1 \end{bmatrix}.$$

The coefficients 3, 2, 4, 1 enter the vector $b = (b_{\phi}, b_{01}, b_{11}, b_{12})$. The wavelet expansion in (9) is $f = R\,b$. The coefficients are $b = R^{-1}f = Df$. This product Df was computed recursively, from two butterfly matrices with a permutation between. In general there will be J matrices with permutations between.

The reconstruction is also recursive. It inverts (12) then (11) then (10). The global matrix R is the product of these local inverse matrices.

Notice that the operation count is proportional to n. It is best possible (the FFT count is $n \log_2 n$). There are only $n-1$ individual 2-by-2 matrix multiplications, since high frequency coefficients (here 4 and 1) are settled and not reused. The Walsh functions give a different piecewise constant representation, in which the last two basis vectors are $(1, -1, 1, -1)$ and $(1, -1, -1, 1)$. In that case 4 and 1 enter another butterfly to produce the Walsh coefficients $\frac{5}{2}$ and $\frac{3}{2}$. The Walsh basis is global. The wavelet basis is local, but scaled — its support has width $O(2^{-J})$ at the finest scale and $O(1)$ at the coarsest scale.

Notice also the normalizing factors $\frac{1}{2}$ in decomposition (and 1's in reconstruction). The alternative is to introduce $1/\sqrt{2}$ for both. This has the advantage of normalizing the wavelets $W_{jk} = 2^{j/2} W(2^j x - k)$ at every scale. The whole basis is orthonormal (when $\|W\| = 1$). In the discrete case R and D become orthogonal matrices:

$$\widehat{D} = \begin{bmatrix} 1/2 & 1/2 & 1/2 & 1/2 \\ 1/2 & 1/2 & -1/2 & -1/2 \\ 1/\sqrt{2} & -1/\sqrt{2} & 0 & 0 \\ 0 & 0 & 1/\sqrt{2} & -1/\sqrt{2} \end{bmatrix} \quad \text{has } \widehat{R} = \widehat{D}^{-1} = \text{transpose of } \widehat{D}.$$

Based on the Haar example, we now start on Mallat's beautiful *tree algorithm* for wavelets. The simple average from $\left[\frac{1}{2}\ \frac{1}{2}\right]$ is replaced by a discrete filter based on ϕ. The difference $\left[\frac{1}{2}\ -\frac{1}{2}\right]$ is replaced by a filter based on W. The filters use the same recursion coefficients c_k that led to ϕ and W in the first place.

Decomposition. The given n-vector f is on the finest scale $h = 2^{-J}$. The *fine-to-coarse filter* (the "restriction operator" in multigrid language, the lowpass filter in signal processing language) is L. It produces a vector with half as many entries:

$$(Lf)_i = \frac{1}{2} \sum c_{2i-j} f_j, \quad i = 1, \ldots, \frac{n}{2}. \tag{A.13}$$

In the Haar example with $c_0 = c_1 = 1$, the entries of Lf are $\frac{1}{2}(f_1 + f_2)$ and $\frac{1}{2}(f_3 + f_4)$. The recursion continues to coarser scales, and after J steps it reaches a single number — the coefficient b_ϕ in f^ϕ at the coarsest scale $h = 1$. Here $b_\phi = \frac{1}{4}(f_1 + f_2 + f_3 + f_4)$.

The dual to L is the *coarse-to-fine map* L^* (the "interpolation operator" in multigrid language). Notice the change of index and the disappearance of $\frac{1}{2}$:

$$(L^* g)_j = \sum c_{2i-j} g_i, \quad j = 1, \ldots, n. \tag{A.14}$$

In the Haar example L^*Lf has entries $\frac{1}{2}(f_1 + f_2), \frac{1}{2}(f_1 - f_2), \frac{1}{2}(f_3 + f_4), \frac{1}{2}(f_3 - f_4)$. It is the projection of f onto the subspace that is piecewise constant at scale $2h$. It gives a blurred picture, with details lost.

The decomposition picks out these details, orthogonal to the average. The projection onto the wavelet subspace is the high frequency component:

$$f^{(J-1)} = f - L^* L f. \tag{A.15}$$

This repeats at every stage. There is an "average" or "blurred picture" $a^{(j-1)} = La^{(j)}$, starting from $a^{(J)} = f$. The detail lost in that average is the component of f at that stage:

$$f^{(j-1)} = (I - L^* L) a^{(j)} = a^{(j)} - L^* a^{(j-1)}. \tag{A.16}$$

This is a first statement of the decomposition algorithm. We will see how Condition O simplifies the formula.

Reconstruction. To produce f from its details $f^{(j)}$, run the recursion (16) in reverse:

$$a^{(j)} = f^{(j-1)} + L^* a^{(j-1)}. \tag{A.17}$$

This starts from the coarsest detail $f^{(0)}$ and the totally blurred picture $a^{(0)} = f^{\phi}$. It returns to $f = a^{(J)}$.

Apply orthogonality. The most elegant part of the algorithm is still to come. It is not necessary to compute the detail vector $f^{(j)}$ from (16), and then to compute its wavelet coefficients b_{jk}. Those are the numbers we want (4 and 1 in the example at level $j = 1$). *These numbers can be found directly from* $a^{(j)}$.

Review the Haar example first. The lowpass filter gave $a^{(1)}$ from $f = a^{(2)}$:

$$Lf = \frac{1}{2}\begin{bmatrix} c_1 & c_0 & & \\ & & c_1 & c_0 \end{bmatrix}\begin{bmatrix} f_1 \\ f_2 \\ f_3 \\ f_4 \end{bmatrix} = \begin{bmatrix} \frac{1}{2} & \frac{1}{2} & & \\ & & \frac{1}{2} & \frac{1}{2} \end{bmatrix}\begin{bmatrix} 9 \\ 1 \\ 2 \\ 0 \end{bmatrix} = \begin{bmatrix} 5 \\ 1 \end{bmatrix}.$$

The blurred picture is $a^{(1)} = (5, 5, 1, 1)$. At the next level the low-pass filter leaves 3, the coefficient of $(1, 1, 1, 1)$. *We now want the orthogonal filter—the highpass filter* H. In the Haar example it produces

$$Hf = \frac{1}{2}\begin{bmatrix} c_0 & -c_1 & & \\ & & c_0 & -c_1 \end{bmatrix}\begin{bmatrix} f_1 \\ f_2 \\ f_3 \\ f_4 \end{bmatrix} = \begin{bmatrix} \frac{1}{2} & -\frac{1}{2} & & \\ & & \frac{1}{2} & -\frac{1}{2} \end{bmatrix}\begin{bmatrix} 9 \\ 1 \\ 2 \\ 0 \end{bmatrix} = \begin{bmatrix} 4 \\ 1 \end{bmatrix}.$$

Those coefficients 4 and 1 represent the detail $f^{(1)} = (4, -4, 1, -1)$, which is lost when $a^{(2)}$ is blurred to $a^{(1)}$. At the next level H is applied to $a^{(1)}$. That produces $\frac{1}{2}(5) - \frac{1}{2}(1) = 2$. This is the coefficient b_{01}, representing the detail $(2, 2, -2, -2)$ lost when $a^{(1)}$ is blurred to $a^{(0)}$. We now put these pieces together into *Mallat's pyramid algorithm*:

Decomposition. Initialize $a^J = f$. For $j = J, \ldots, 1$ compute

$$a^{j-1} = La^j \quad \text{and} \quad b^{j-1} = Ha^j. \tag{A.18}$$

Reconstruction. Start with a^0 and $b^0, \ldots, b^{J-1}$. For $j = 1, \ldots, J$ compute

$$a^j = L^* a^{j-1} + H^* b^{j-1}. \tag{A.19}$$

The full decomposition is represented by a tree of filters:

$$\begin{array}{ccccccccc} a^J & \xrightarrow{L} & a^{J-1} & \xrightarrow{L} & a^{J-2} & \cdots & \xrightarrow{L} & a^0 \\ & \searrow^{H} & & \searrow^{H} & & & \searrow^{H} & \\ & & b^{J-1} & & b^{J-2} & & & b^0 \end{array}$$

The reconstruction goes from the branches of the tree back to the root:

$$\begin{array}{cccccccc} a^0 & \xrightarrow{L^*} & a^1 & \xrightarrow{L^*} & a^2 & \cdots & \xrightarrow{L^*} & a^J = f \\ & \nearrow_{H^*} & & \nearrow_{H^*} & & & \nearrow_{H^*} & \\ b^0 & & b^1 & & & & & \end{array}$$

The next step is to identify these filter matrices L and H for examples other than "box and Haar."

Note. The filter matrices L and H have half as many rows as columns. By dropping the parentheses around j, we distinguish the vector a^j with only 2^j components from the vector $a^{(j)}$ with the full $w^J = n$ components. The vector a^j contains the expansion coefficients of $a^{(j)}$ with respect to the translates $\phi(2^j x - k)$. See the example above and the multiresolution below!

2.1. The filter matrices *L* and *H*. The matrix L is known from the first part of the paper. Its entries $L_{ij} = c_{2i-j}$ are the recursion coefficients for the scaling function. Rows 1, 2 and columns $-1, 0, 1, 2$ are displayed with $N = 3$:

$$L = \frac{1}{2}\begin{bmatrix} c_3 & c_2 & c_1 & c_0 & & \\ & & c_3 & c_2 & c_1 & c_0 \end{bmatrix}.$$

The beautiful thing is that the highpass filter (strictly speaking it is band-pass) uses the same coefficients. H is associated with the wavelet W just as L is associated with the scaling function ϕ. Equation (6) for W uses the same c_k, but with alternating signs and reversed order. The wavelet filter has

$$H_{ij} = (-1)^{j+1} c_{j+1-2i}. \tag{A.20}$$

Rows 1, 2 and columns 1, 2, 3, 4 are displayed:

$$H = \frac{1}{2}\begin{bmatrix} c_0 & -c_1 & c_2 & -c_3 & & \\ & & c_0 & -c_1 & c_2 & -c_3 \end{bmatrix}.$$

The indices were chosen to match the Haar example (variants are possible). The transposed matrices, without the factor $\frac{1}{2}$, represent the dual filters L^* and H^*. The important points now come quickly, and matrix multiplication is the best proof.

Theorem 1. *By their construction the filters are orthogonal*:

$$HL^* = 0. \tag{A.21}$$

This multiplication is the reason behind the construction of H—alternating signs, reversed order, index shifted by one. See equation (8).

We finally come to the reward for Condition O: $\sum c_k c_{k+2m} = 2\delta_{0m}$. The reason for that condition is in the reward. Remember that the box function and D_4 satisfied this requirement but not the hat function or the cubic spline. Condition O can be stated and understood in transform space, but I believe that the matrix interpretation is again the clearest.

Theorem 2. *If condition* O *holds then*

1. $LL^* = I$ *and* $HH^* = I$. (A.22)
2. L^*L *and* H^*H *are mutually orthogonal projections with*

$$L^*L + H^*H = I. \tag{A.23}$$

Remember that L and H map into subspaces half as large as the original. L^* and H^* map back. The identity operators in (22) are on the half-sized subspaces.

The proof of (22) is by direct matrix manipulation. Condition O gives the result. Then it follows that $L^*LL^*L = L^*L$, so L^*L is a projection—and similarly for H^*H. The property $HL^* = 0$ in (21) yields $H(L^*L + H^*H) = H$. The transpose $LH^* = 0$ yields $L(L^*L + H^*H) = L$. The operator in (23) is the identity on both orthogonal components—the ranges of L and H—so it is the identity. We have an orthogonal decomposition by "*quadrature mirror filters*" L and H at every step.

2.2. Multiresolution of L^2. The last paragraphs changed quietly from functions to vectors. That was for the sake of algorithms, which use values of ϕ and W at dyadic points $k/2^J$. The Haar example began with f at equally spaced points on $(0, 1]$. But the filter matrices really apply to discrete values along the whole line—they are infinite matrices. More than that, the decomposition $f = \sum f^{(j)}$ is just as valuable for functions in L^2 as for vectors in l^2.

This multiresolution yields the details of f at all scalings 2^{-j}. On the whole line we take $j = 0, \pm 1, \pm 2, \ldots$. The decomposition develops an idea that was already present in approximation theory—to put frequencies together in "octaves." (Besov spaces combine frequencies $2^j \leq \xi < 2^{j+1}$. It seems that the ear also receives frequencies on a logarithmic scale.) For functional analysis the starting point is the subspace S_j spanned by the translates $\phi(2^j x - k)$. If a function $g(x)$ is in S_j, then $g(2x)$ is in S_{j+1}. The dilation equation writes $\phi(x)$ as a combination of $\phi(2x - k)$, which assures that $S_0 \subset S_1$. At all scales we have

$$\cdots S_{-1} \subset S_0 \subset S_1 \subset S_2 \cdots \text{ with } \cup S_j \text{ dense in } L^2 \text{ and } \cap S_j = \{0\}.$$

Now turn to the *wavelet subspace* W_j. It is spanned by the translates $W(2^j x - k)$. It is invariant under translation by multiples of 2^{-j}. If $g(x)$ is in W_j then $g(2x)$ is in W_{j+1}. The construction $W(x) = \sum (-1)^k x_{1-k} \phi(2x - k)$ puts W and its translates into S_1, and makes them orthogonal to S_0. *In fact, W_0 and S_0 are orthogonal complements in S_1.* At every scale $W_j \bigoplus S_j = S_{j+1}$. *The spaces S_j give the "partial sums" of the differences W_j:*

$$\cdots \bigoplus W_{-1} \bigoplus W_0 \bigoplus \cdots \bigoplus W_j = S_{j+1} \quad \text{and} \quad \bigoplus_{-\infty}^{\infty} W_j = L^2.$$

The multiresolution of f is a splitting into components $f^{(j)} \in W_j$:

$$f = \sum_{-\infty}^{\infty} f^{(j)} \quad \text{or} \quad f = f^{\phi} + \sum_{0}^{\infty} f^{(j)}, \quad f^{\phi} \in S_0. \tag{A.24}$$

This is a very satisfying decomposition of L^2 functions, classical but with new subspaces. The coefficients b^j in Mallat's pyramid algorithm corresponded to $f^{(j)} \in W_j$, and a^j corresponded to $a^{(j)} \in S_j$.

The analogue of the discrete Fourier transform was in the algorithm. The analogue of ordinary Fourier series is (24). The analogue of the Fourier integral formula is the *integral wavelet transform.* Representations of different groups give rise to different transforms.

2.3. Applications. Image processing works with $F(x, y)$, so it is natural to look for *two-dimensional wavelets.* The simplest construction uses the products $\phi(x)\phi(y)$, $\phi(x)W(y)$, $W(x)\phi(y)$, $W(x)W(y)$. Orthogonality is clear. New constructions have been invented that are genuinely two-dimensional, but it is useful to start with the tensor products of "box and Haar." The given two-dimensional array F yields a two-dimensional array B of wavelet coefficients.

For pattern recognition, a major difficulty with the wavelet transform B is the lack of translation invariance. If the pattern is shifted by a fraction of h, its wavelet model is changed. A higher sampling rate is possible but expensive. Mallat studies instead the *zero-crossings* of the wavelet transform, which locate the signal edges. Now the difficulty is to make the reconstruction stable. In edge detection the first wavelets were Laplacians of shifted Gaussians, introduced by Gabor. The orthogonal wavelets of Meyer are C^{∞} with polynomial decay, the Battle-Lemarié wavelets based on splines are C^n with exponential decay, and the Daubechies wavelets are C^n (smaller n) with compact support.

In closing we recall the original problem — to localize in time and frequency. Geophysics needs to represent short high-frequency pulses. Physics needs to divide up phase space. The coherent states $g_{pq} = e^{ipx} g(x-q)$ give a "Weyl-Heisenberg" frame, with some redundancy — but still f can be reconstructed from $\iint (f, g_{pq}) g_{pq} \, dp \, dq$. Mathematics needs (or wants) an orthogonal decomposition, better than g_{pq} at high frequencies and with no redundancy. The answer for now is wavelets.

It is a pleasure to thank Ingrid Daubechies and Howard Resnikoff for introducing me to wavelets.

MATLAB and the Wavelet Toolbox

This section consists of 11 sets of MATLAB exercises and experiments, one set for each chapter. They provide hands-on experience that is essential to the understanding of this subject. We use the MATLAB Wavelet Toolbox to explore scaling functions, wavelets, and multiresolution analysis. The exercises also include GenLOT design and image compression (and many other topics) with codes from our homepage.

The Wavelet Toolbox provides functions for both continuous and discrete wavelet analysis. Analysis and synthesis of one-dimensional signals and two-dimensional images are supported, as well as denoising and compression. Wavelet packets are similarly supported for one and two dimensions, with associated denoising and compression techniques. The compression routines extract the minimum number of wavelet coefficients that represent the signal accurately, which is the first stage of a complete compression system.

The Wavelet Toolbox was created by Michel Misiti, Yves Misiti, Georges Oppenheim, and Jean-Michel Poggi. It includes a comprehensive user's guide. The toolbox also provides an extensive mouse-driven graphical user interface (GUI) for interactive exploration and display, with the flexibility to add new wavelets. The MATLAB Wavelet Toolbox is available from

The MathWorks, Inc.
24 Prime Park Way
Natick MA 01760 USA
Phone: (508) 647-7000

WWW: `http://www.mathworks.com`
E-mail: `info@mathworks.com`
FTP: `ftp.mathworks.com`
Fax: (508) 647-7001

Our homepage at

`http://saigon.ece.wisc.edu/~waveweb/Tutorials/book.html`

has additional MATLAB exercises and codes. Many wavelet and GenLOT coefficients and programs, including the ones used in Chapters 10 and 11, are also available at

`http://saigon.ece.wisc.edu/~waveweb/QMF/software.html`

We would like to thank Steffen Trautmann for writing the programs that are used extensively in this book. These include m-files to generate and plot scaling functions and wavelets; to design and display GenLOT; to design wavelets; to display multiresolution representation of images; and to compress images using dyadic wavelet transforms and GenLOTs. The one-dimensional compression program was written by Adriannus Djohan.

MATLAB Exercises – Chapter 1

The first assignment is designed to familiarize the reader with the MATLAB Wavelet Toolbox and its GUI. The assignment includes Wavelet Display, Continuous Wavelet Transform, 1-D and 2-D Discrete Wavelet Transform. It is assumed that the Wavelet Toolbox is installed. At the MATLAB prompt, type `wavetool`. A window should pop up with choices ranging from `Wavelet 1-D` to `Wavelet Packet 2-D` to `Continuous Wavelet 1-D`.

1. **Wavelet Display**

   Select the third option `Wavelet Display` and wait for the figure to be created.

   - Press `Display` — This will calculate and then display the Haar wavelet. Press `Information on Haar wavelet` to learn about its properties and the conventions used in the Toolbox. Press `Close Info`.
   - Press `Information on Wavelet`. Use the scroll bar on the right to see the complete documents, or use the page selection on the right. Press `Close` when you are done.
   - Select another wavelet family such as `db` (Daubechies). Select `db 2` and press `Display`. This will calculate and display the scaling function, wavelet, and filters for the D_4 orthonormal wavelets.
   - To zoom in on a region
     * Move the mouse onto an axis of interest, for example, the red scaling function.
     * Select a bounding box by pressing down the left mouse button and dragging. When you release the button, the box remains.
     * Move down to the visualization tool at the bottom of the figure. Click on `XY+` to zoom in on both axes. To zoom in on one axis only click on `X+` or `Y+`.
   - To zoom out, either
     * Select a box again and press `XY-`, `X-` or `Y-` as appropriate.
     * Press the <-button next to the word `History`. This allows one to take a step back from the zoom operations.
   - The keys to the visualization tool (press `close` to leave) are

     `<--` go back in zoom history buffer.

     `-->` go forward in zoom history buffer.

     `<<-` go back to the beginning of the zoom history buffer.

2. **Continuous Wavelet Transform**

   Select the seventh option `Continuous Wavelet 1-D` and wait for the figure.

   - Under `File` menu option, select `Load Signal`. Choose the file `freqbrk.mat`.
   - Select the `db 1` Haar wavelet and press `Analyze` to perform the Continuous Wavelet Transform (CWT) of the signal. The coefficients are presented at the bottom figure with absolute values color-coded. Select a colormap using the `colormap` button.
   - For normal coloration, select the `Coloration Mode` pop-up and `Normal Mode`.
   - Use the `File` menu option to load another signal ***quachirp.mat***. Press `Analyze` to perform the CWT. To see the different features of the signal, change the colormap to `"hot"`.

- Press Close to leave the tool.

3. **One-Dimensional Discrete Wavelet Transform**
   Select the first option Wavelet 1-D and wait for the figure.

   - Under File menu option, select Load Signal. Choose noisdopp.mat.
   - Select the db 1 wavelet and Level 6. Press Analyze to perform the Discrete Wavelet Transform (DWT) of the signal. The complete decompositions of the signal (coarse approximation at level 6 and the detail coefficients at levels 1 to 6) are shown with the input signal.
   - Display Modes
     * Change the Display Mode to Superimpose Mode. This will show the signal and all approximations on one plot. The second plot includes all the details.
     * Change the Display Mode to Show and Scroll Mode. You can look at the approximation and details at every level. Use the control under Show and Scroll Mode popup to select the appropriate level.
     * Change to Separate Mode to see all approximations and details.
     * Change to Tree Mode to get a graphical representation of the decomposition tree. Click on 'a3' to see the approximation at level 3. For the detail at level 5, click on 'd5'. Try other nodes.
     * Change the Display Mode back to Full Decomposition.
   - Change the Full Decomposition from level 6 to level 2, in the middle of the GUI on the right. Take care not to change the main decomposition level closer to the top of the GUI. This will only show the decomposition to level 2 with the detail at level 1 and 2, as well as the approximation at level 2.
   - Restore to Full Decomposition at level 6. Explore the additional capabilities Statistics, Histograms, Compress and Denoise.

4. **Two-Dimensional Discrete Wavelet Transform**
   Select the second option Wavelet 2-D and wait for the figure.

   - Under File menu option, select Load Signal. Choose woman2.mat.
   - Select the symm 4 wavelet and Level 2. Press Analyze to perform the Discrete Wavelet Transform (DWT) of the signal. The original image is displayed in the top left. The decomposed image is displayed in the bottom right, with the green Image Selection box around it. The reconstructed image is on the bottom left.
   - Press the Full Size 4 button. This will enlarge the bottom right portion of the screen. To see the approximation at level 1, change Decomposition at level 2 to level 1. The approximation at level 1 is 1/4 the size of the original image. Changing back to level 2 will reduce the approximation and details at level 2 to 1/16 the size of the original image. Press the 'end 4' button to restore the display.
   - To see the enlarged version of the approximation at level 2, select the small picture in the top left of the Image Selection box. The selected image should have a green border. Now press Visualize for an enlarged version. Note that the image is subsampled.

- Changing the display mode to `View Mode: Tree` shows all approximation and detail images at the same size. Any image can be selected and displayed with the `Visualize` button.
- Change back to `View Mode: Square`. Explore the additional capabilities `Statistics`, `Histograms`, `Compress` and `Denoise`.

MATLAB Exercises – Chapter 2

1. **Aliasing of Sinusoidal Signals**
   The signal $x(t) = \cos(2\pi\ f_0\, t)$ can be sampled at the rate $f_s = \frac{1}{T_s}$ to yield $x(n) = \cos(2\pi \frac{f_0}{f_s} n)$. Aliasing will be observed for various f_0.
   a. Let $f_s = 10$ KHz and $f_0 = 1$ KHz. Compute and plot $x(n)$ using *stem.m*. One can see the sinusoidal envelope.
   b. Use `subplot` to plot $x(n)$ for $f_0 = 300$ Hz, 700 Hz, 1100 Hz and 1500 Hz. The frequencies increase as expected.
   c. Use `subplot` to plot $x(n)$ for $f_0 = 8500$ Hz, 8900 Hz, 9300 Hz and 9700 Hz. Do the frequencies increase or decrease? Explain.
2. **IIR Filter and Gibbs Phenomenon**
   a. The lowpass IIR filter $h(n) = (0.9)^n u(n) = (\ldots, 0, 0, 1, 0.9, 0.81, \ldots)$ has transform $H(e^{j\omega}) = 1/(1 - 0.9e^{-j\omega})$. Plot $|H(e^{j\omega})|$. Normalize all plots so that the gain is $H = 1$ at DC.
   b. Truncate $h(n)$ to length 11 to obtain $h_{11}(n)$. Compute and plot $|H_{11}(e^{j\omega})|$ on the same graph with $|H(e^{j\omega})|$. Describe the Gibbs phenomenon resulting from the discontinuity of the rectangular window.
   c. Use a rectangular window of length 21. Compute and plot $|H_{21}(e^{j\omega})|$ on the same graph. Compare the ripples in $|H_{11}(e^{j\omega})|$ and $|H_{21}(e^{j\omega})|$.
   d. To reduce the ripple, repeat using the Hanning window. Compare the ripples to those in b and c.
3. **Window Design and Filter Design**
   Plot and compare the magnitude responses of the rectangular, Hanning, and Kaiser windows for $N = 20$. Use those windows to truncate the ideal filter $h_I(n) = \frac{1}{\pi n} \sin(\frac{\pi}{3} n)$. Plot the magnitude responses of the resulting filters. Find the error (difference from h_I) in the passbands and stopbands.
4. **IIR Lowpass and Highpass Relation**
   $h(n) = (0.9)^n u(n)$ is IIR lowpass. Compute $H(e^{j\omega})$ using *freqz.m* and plot its magnitude and phase responses. This problem compares alternating flip and alternating sign:
   a. $g_1(n) = (-1)^n h(n)$ is a highpass filter. Find the relation of $G_1(z)$ to $G(z)$. Plot the magnitude and phase responses of $G_1(e^{j\omega})$. Where is the pole of $G_1(z)$?
   b. An alternating flip also yields a highpass filter $G_2(z) = H(-z^{-1})$. Find $g_2(n)$. Compute $G_2(e^{j\omega})$ and plot its magnitude and phase responses. Where is the pole of $G_2(z)$?
   c. Plot the phase responses of $G_1(z)$ and $G_2(z)$ on one graph.

5. FIR Lowpass and Highpass Relation

Use *remez.m* to design an equiripple lowpass filter $\boldsymbol{h}(n)$ of length 15 with $\omega_p = 0.2\pi$, $\omega_s = 0.4\pi$ and $\delta_p = \delta_s$. Then use alternating flip and alternating sign:

a. Plot the magnitude and phase responses of $H(e^{j\omega})$. Plot the zeros of $H(z)$ on the z-plane using *zplane.m*.

b. Let $G_1(z) = H(-z)$. Plot its magnitude and phase responses. Plot the zeros of $G_1(z)$ on the z-plane. What is the relation between the zeros of $H(z)$ and $G_1(z)$?

c. Repeat Part b for $G_2(z) = z^{-14}H(-z^{-1})$.

d. *Effect of Weighting*: Use *remez.m* to design an equiripple lowpass filter $\boldsymbol{h}(n)$ of length 15 with $\omega_p = 0.2\pi$, $\omega_s = 0.35\pi$ for $\delta_p = \delta_s$ and $\delta_p = 0.1\,\delta_s$ and $\delta_p = 10\,\delta_s$. Plot the responses $H(e^{j\omega})$ of both designs on one graph.

6. Equiripple Halfband Filter

Design an equiripple halfband filter of length 23 with $\omega_p = 0.4\pi$.

a. What is ω_s and what weighting do you use?

b. Use *stem.m* to plot the impulse response. Verify the halfband property $\boldsymbol{h}(11 \pm 2k) = 0.5\delta(k)$.

c. The halfband condition in the z-domain is $H(z) + H(-z) = z^{-11}$. What is the corresponding condition on $\boldsymbol{h}(n)$? Verify it.

d. $G(e^{j\omega}) = H(e^{j\omega}) + H(e^{j(\omega+\pi)}) = e^{-j11\omega}$. Compute and plot $|G(e^{j\omega})|$.

7. Equiripple M-th band Filter

An M-th band filter of length $2N + 1$ satisfies $\boldsymbol{h}(N \pm Mk) = \frac{1}{M}\,\delta(k)$, or equivalently, $\sum_{k=0}^{M-1} H(ze^{j2\pi k/M}) = z^{-N}$. Consequently, $\omega_p + \omega_s = \frac{2\pi}{M}$ and $\delta_p < (M-1)\delta_s$.

a. Use *remez.m* to design a 4-th band filter $\boldsymbol{h}(n)$ with length 23 and cutoff frequencies 0.2π and 0.3π. Choose $\delta_p = \delta_s$. Verify that $\boldsymbol{h}(11 \pm 4k) \approx 0$.

b. Do any other weighting functions make $\boldsymbol{h}(11 \pm 4k)$ smaller?

c. Verify that $\sum_{k=0}^{3} H(e^{j\omega}\, e^{j2\pi k/4}) = e^{-j\,11\omega}$.

8. Eigenfilter from
`http://saigon.ece.wisc.edu/~waveweb/software.html`

a. Use *eigenfilt.m* to design a lowpass filter with the specifications in Problem 5. Plot its magnitude response and compare the stopband attenuation with 5a.

b. Design a halfband filter using *nyqfilt.m* with the specifications in Problem 6. Plot the magnitude response and compare to the result in 6a. Repeat for 6b and 6c.

c. Use *nyqfilt.m* to design a fourth-band filter with the same specifications as in Problem 7a. Use *stem.m* to plot the impulse response and verify that $\boldsymbol{h}(11 \pm 4k) = 0$, exactly for $k \neq 0$! Plot $|H(e^{j\omega})|$ and compare with 7b.

9. Maxflat Filter Design

Write a program to compute and plot $|H(e^{j\omega})|$ for the maxflat filters with $p = 3, 7, 11$ (on the same plot). Describe the effect of increasing p on the transition band. Verify that the filters are halfband.

MATLAB Exercises – Chapter 3

1. Downsampling of Bandlimited Signal

a. Create an approximate bandlimited signal $\boldsymbol{x}(n)$ by designing a lowpass filter of length 256, with cutoff frequencies at $\omega_p = 0.25\pi$ and $\omega_s = 0.4\pi$. Choose equal weights for passband and stopband. Transform to the frequency domain by a 1024-pt FFT and measure the passband and stopband errors δ_p and δ_s. Plot the frequency response.

b. Downsample $\boldsymbol{x}(n)$ by $\boldsymbol{x}_1(n) = \boldsymbol{x}(2n)$. Compute a 1024-pt FFT of $\boldsymbol{x}_1(n)$ and plot the frequency response. Measure the passband and stopband errors δ_{p1} and δ_{s1}. Are they approximately the same as δ_p and δ_s? Explain.

c. Shift the signal by one sample and downsample to $\boldsymbol{x}_2(n) = \boldsymbol{x}(2n-1)$. Plot $|X_2(e^{j\omega})|$ and find δ_{p2} and δ_{s2}. Compare them to δ_{p1} and δ_{s1}.

d. Is $\boldsymbol{x}_1(n)$ the same as $\boldsymbol{x}_2(n)$? Explain by plotting the magnitude and phase responses of $X_1(e^{j\omega})$ and $X_2(e^{j\omega})$.

e. Shift the signal by two samples and downsample, $\boldsymbol{x}_3(n) = \boldsymbol{x}(2n-2)$. Compare $\boldsymbol{x}_3(n)$ to $\boldsymbol{x}_1(n)$ and $\boldsymbol{x}_2(n)$. Explain using the Noble Identity.

2. Downsampling and Aliasing

a. Downsample the same signal $\boldsymbol{x}(n)$ by a factor of 3 to $\boldsymbol{y}(n) = \boldsymbol{x}(3n)$. Compute a 1024-pt FFT and plot the frequency response. Measure the passband and stopband errors δ_{p3} and δ_{s3}. Are they approximately the same as δ_p and δ_s? Explain.

b. Is there aliasing in $\boldsymbol{y}(n)$? Can we reconstruct $\boldsymbol{x}(n)$? How?

c. Downsample $\boldsymbol{x}(n)$ by 4 to $\boldsymbol{z}(n) = \boldsymbol{x}(4n)$ and repeat part a.

d. Is there any aliasing in $\boldsymbol{z}(n)$? Can we reconstruct $\boldsymbol{x}(n)$ from $\boldsymbol{z}(n)$?

3. Upsampling and Interpolation Filter

Upsampling increases the sampling rate of $\boldsymbol{x}(n)$ by interlacing $M-1$ zero samples between the nonzeros (M is the interpolation factor). These zero samples should then be interpolated by a lowpass filter $\boldsymbol{f}(n)$, keeping the nonzero samples unchanged. The cutoff frequency ω_c of $F(e^{j\omega})$ is $\frac{\pi}{M}$. The effect of ω_c in interpolated samples is discussed in this problem.

a. Create an approximate bandlimited signal $\boldsymbol{x}(n)$ by designing a lowpass filter of length 256, cutoff frequencies at $\omega_p = 0.9\pi$ and $\omega_s = 0.95\pi$. Choose equal weights for passband and stopband. Compute a 1024-pt FFT and measure the passband and stopband errors δ_p and δ_s. Plot the frequency response.

b. Upsample $\boldsymbol{x}(n)$ by a factor of 2 to obtain $\boldsymbol{u}(n)$. Compute a 1024-pt FFT and plot the frequency response. Describe $U(e^{j\omega})$ and compare it with $X(e^{j\omega})$.

c. Design an interpolation filter $\boldsymbol{f}(n)$ for $\boldsymbol{u}(n)$. (Note that $\boldsymbol{u}(2n+1) = 0$ and $\boldsymbol{u}(2n) \neq 0$.) What is the cutoff frequency ω_{c1} of $F(e^{j\omega})$? Compute $\boldsymbol{y}(n) = \boldsymbol{u}(n) * \boldsymbol{f}(n)$. Graph $\boldsymbol{y}(n)$ and $\boldsymbol{u}(n)$ on the same plot. Verify that $\boldsymbol{y}(2n) = \boldsymbol{u}(2n)$. Plot $|Y(e^{j\omega})|$ and $|U(e^{j\omega})|$. Describe their relation.

d. Design another interpolation filter $\boldsymbol{h}_2(n)$ with cutoff frequency $\omega_{c2} = \frac{3\pi}{4}$. Compute $\boldsymbol{y}_2(n) = \boldsymbol{x}_1(n) * \boldsymbol{h}_2(n)$. Plot $\boldsymbol{y}_2(n)$ and $\boldsymbol{x}_1(n)$ on the same plot. Is $\boldsymbol{y}_1(2n) = \boldsymbol{x}_1(2n)$? Why?

e. Upsample $\boldsymbol{x}(n)$ by a factor of 3 to obtain $\boldsymbol{x}_2(n)$. Repeat Parts b and c above for $\boldsymbol{x}_2(n)$.

4. **Downsampling and Decimation Filter**

   The object of a decimation filter $\boldsymbol{h}(n)$ is to bandlimit the signal $\boldsymbol{x}(n)$ before downsampling. The cutoff frequency ω_c of $H(e^{j\omega})$ is $\frac{\pi}{M}$ where M is the decimation factor. This problem studies the relation between ω_c and aliasing.

   a. Downsample by 2 the signal $\boldsymbol{x}(n)$ from Problem 3 to obtain $\boldsymbol{x}_1(n)$. Compute a 1024-pt FFT and plot the frequency response. Describe $X_1(e^{j\omega})$. Is there severe aliasing?
   b. Design a decimation filter $\boldsymbol{h}_1(n)$ to be used in Part a. What is the cutoff frequency ω_{c1}? Compute $\boldsymbol{y}_1(n) = \boldsymbol{x}(n) * \boldsymbol{h}_1(n)$. Compute and plot $|Y_1(e^{j\omega})|$. Is $\boldsymbol{y}_1(n)$ lowpass?
   c. Downsample $\boldsymbol{y}_1(n)$ by 2 to obtain $\boldsymbol{v}_1(n)$. Compute a 1024-pt FFT and plot the frequency response. Describe $V_1(e^{j\omega})$. Is there severe aliasing? Why?
   d. Repeat a, b, c for downsampling by a factor of 3.

5. **Commutation of ($\downarrow M$) and ($\uparrow L$)**

   Let $L = 3$ and $M = 2$. Use the same signal $\boldsymbol{x}(n)$ as in Problem 3. Find $y_1(n) = (\downarrow 2)(\uparrow 3)\boldsymbol{x}(n)$ and $\boldsymbol{y}_2(n) = (\uparrow 3)(\downarrow 2)\boldsymbol{x}(n)$. Verify that $\boldsymbol{y}_1(n) = y_2(n)$.

6. **Noble Identities**

   The Noble Identities allow one to commute filtering with downsampling (or upsampling). The two Noble Identities are shown:

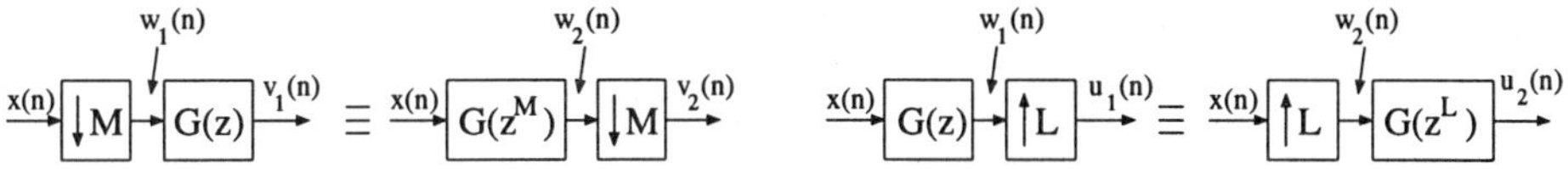

   x(n) → [↓M] → $w_1(n)$ → [G(z)] → $v_1(n)$ ≡ x(n) → [G(z^M)] → $w_2(n)$ → [↓M] → $v_2(n)$

   x(n) → [G(z)] → $w_1(n)$ → [↑L] → $u_1(n)$ ≡ x(n) → [↑L] → $w_2(n)$ → [G(z^L)] → $u_2(n)$

   a. Design a bandstop filter $\boldsymbol{g}(n)$ of length 81 and cutoff frequencies at 0.4π and 0.5π; 0.7π and 0.8π. Plot $|G(e^{j\omega})|$ and $\arg[H(e^{j\omega})]$. Let $\boldsymbol{x}(n) = (\ \cdots\ \ 0\ \ 1\ \ -1\ \ 1\ \ 1\ \ 1\ \ 1\ \ -1\ \ 0\ \ \cdots\ )$ be the input to the system. Compute and plot $\boldsymbol{w}_1(n)$ and $\boldsymbol{v}_1(n)$ for $M = 2$.
   b. Using the same filter, compute and plot $\boldsymbol{w}_2(n)$ and $\boldsymbol{v}_2(n)$. Verify that $\boldsymbol{v}_1(n) = \boldsymbol{v}_2(n)$.
   c. Use the same bandstop filter $\boldsymbol{g}(n)$ and the same input $\boldsymbol{x}(n)$. Compute and plot $\boldsymbol{w}_1(n)$ and $\boldsymbol{u}_1(n)$ for $L = 3$. Compute and plot $\boldsymbol{w}_2(n)$ and $\boldsymbol{u}_2(n)$. Verify that $\boldsymbol{u}_1(n) = \boldsymbol{u}_2(n)$.

MATLAB Exercises – Chapter 4

1. **Polyphase Representation of Filter**

   a. Design a linear-phase halfband filter $\boldsymbol{h}(n)$ of length 51 and cut-off frequencies at 0.4π and 0.6π. Plot the magnitude and phase responses of $H(e^{j\omega})$. What is the group delay of $H(e^{j\omega})$?
   b. Find the two polyphase components $\boldsymbol{h}_{\text{even}}(n)$ and $\boldsymbol{h}_{\text{odd}}(n)$. Plot their magnitude and phase responses. Verify that the magnitude and phase responses of $H_{\text{even}}(z^2) + z^{-1}H_{\text{odd}}(z^2)$ agree with $H(z)$.

c. Find the three polyphase components of $\boldsymbol{h}(n)$ and plot their magnitude and phase responses. Describe them. What equation relates them to $H(z)$?

2. Efficient Implementation of Decimation Filter by Polyphase

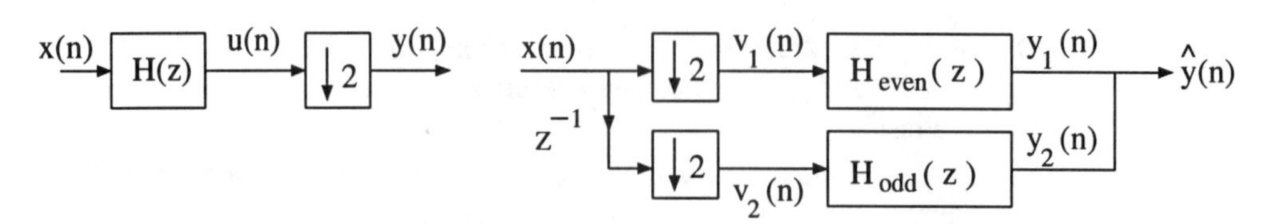

a. Design a lowpass filter $\boldsymbol{h}(n)$ of length 51 with cutoff frequencies at 0.7π and 0.8π. Plot $|H(e^{j\omega})|$ and $\arg[H(e^{j\omega})]$. Compute and plot $\boldsymbol{u}(n)$ and $\boldsymbol{y}(n)$ from the input $\boldsymbol{x}(n) = (\; \cdots \quad 0 \quad 1 \quad 1 \quad 1 \quad 1 \quad 1 \quad 1 \quad 1 \quad 0 \quad \cdots \;)$.

b. Find the polyphase components $\boldsymbol{h}_{\text{even}}(n)$ and $\boldsymbol{h}_{\text{odd}}(n)$. Find and plot $\boldsymbol{v}_1(n)$ and $\boldsymbol{v}_2(n)$. Compute and plot $\boldsymbol{y}_1(n)$ and $\boldsymbol{y}_2(n)$. Find $\widehat{\boldsymbol{y}}(n)$ and compare it with $\boldsymbol{y}(n)$.

c. Compare the number of multiplications and additions per unit time for parts a and b.

3. Efficient Implementation of Interpolation Filter by Polyphase

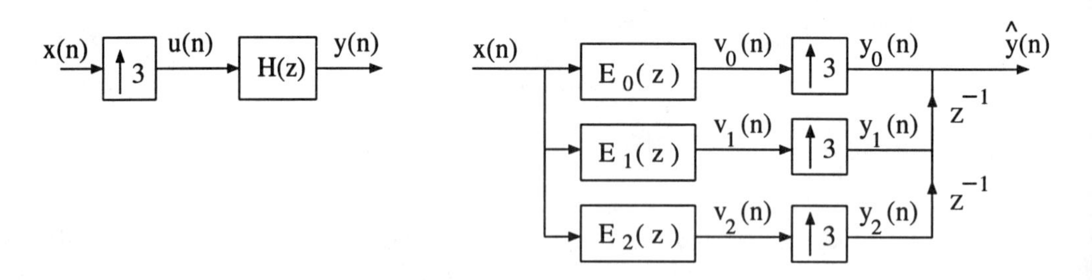

a. Design a bandpass filter $\boldsymbol{h}(n)$ of length 81 and cutoff frequencies at 0.2π and 0.3π; 0.7π and 0.8π. Plot $|H(e^{j\omega})|$ and $\arg[H(e^{j\omega})]$. Compute and plot $\boldsymbol{u}(n)$ and $\boldsymbol{y}(n)$ from
$\boldsymbol{x}(n) = (\; \cdots \quad 0 \quad 1 \quad 1 \quad 1 \quad 1 \quad 1 \quad 1 \quad 1 \quad 0 \quad \cdots \;)$.

b. Find the polyphase components $\boldsymbol{e}_0(n)$, $\boldsymbol{e}_1(n)$, and $\boldsymbol{e}_2(n)$. Find and plot $\boldsymbol{v}_k(n)$. Compute and plot $\boldsymbol{y}_k(n)$. Find $\widehat{\boldsymbol{y}}(n)$ and compare it with $\boldsymbol{y}(n)$.

c. Compare the number of multiplications and additions per unit time for parts a and b.

4. Polyphase Matrix and Analysis Bank

Given the analysis filters $H_0(z)$ and $H_1(z)$ of a two-channel analysis bank, one can find the polyphase matrix $\boldsymbol{H}_p(z)$ such that $\begin{bmatrix} H_0(z) \\ H_1(z) \end{bmatrix} = \boldsymbol{H}_p(z^2) \begin{bmatrix} 1 \\ z^{-1} \end{bmatrix}$.

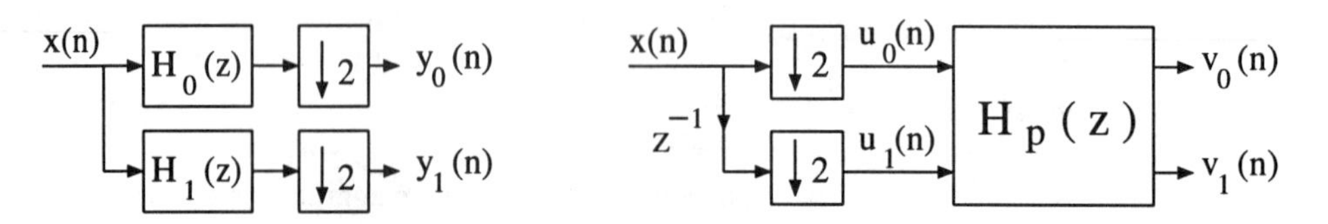

a. Design lowpass and highpass filters $H_0(z)$ and $H_1(z)$ of length 50 with cutoff frequencies at 0.4π and 0.6π. Let $\boldsymbol{x}(n) = \left\{ \begin{array}{ll} n; & 0 \le n \le 256 \\ 512 - n; & 257 \le n \le 512 \end{array} \right\}$.
Compute $\boldsymbol{y}_1(n)$ and $\boldsymbol{y}_2(n)$.

b. Find the polyphase components of $H_k(z)$ and the polyphase matrix $\boldsymbol{H}_p(z)$. Find and plot $\boldsymbol{u}_k(n)$ and $\boldsymbol{v}_k(n)$. Verify that $\boldsymbol{y}_k(n) = \boldsymbol{v}_k(n)$.

5. Polyphase Matrix and Synthesis Bank

Given the synthesis filters $G_0(z)$ and $G_1(z)$, one can find the polyphase matrix $\boldsymbol{G}_p(z)$ such that $\left[\begin{array}{cc} G_0(z) & G_1(z) \end{array}\right] = \left[\begin{array}{cc} z^{-1} & 1 \end{array}\right] \boldsymbol{G}_p(z^2)$.

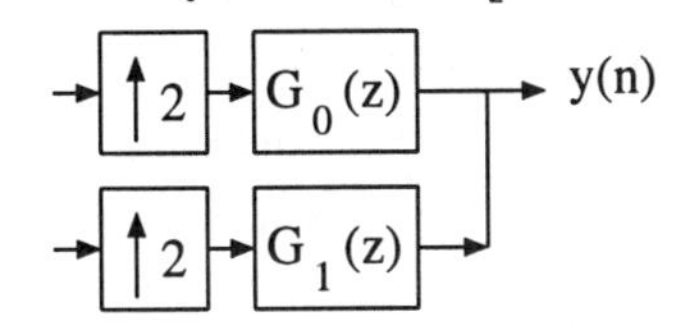

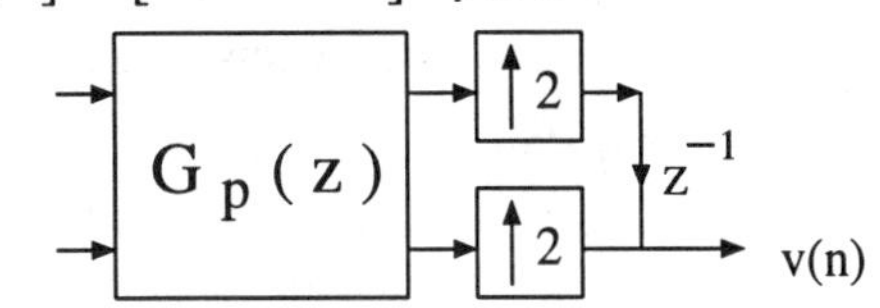

a. Use the lowpass and highpass filters from Problem 5. Create two arbitrary inputs to the synthesis bank. Compute and plot $\boldsymbol{y}(n)$.

b. Find the polyphase components $G_k(z)$ and the polyphase transfer matrix $\boldsymbol{G}_p(z)$. Compute and plot $\boldsymbol{v}(n)$. Verify that $\boldsymbol{y}(n) = \boldsymbol{v}(n)$.

MATLAB Exercises – Chapter 5

These MATLAB assignments study orthonormal filter banks and their lattice structure. We use the m-files provided by the MATLAB Wavelet Toolbox:

dbwavf.m:	Design minimum-phase Daubechies orthonormal filter.
qmf.m:	Find the highpass filter given the lowpass filter.
symwavf.m:	Design near-symmetric Daubechies orthonormal filter.

1. Daubechies Filters

a. Create the minimum-phase Daubechies filters $\boldsymbol{c}(n)$ and $\boldsymbol{d}(n)$ of length 8. Plot $\boldsymbol{c}(n)$ and $\boldsymbol{d}(n)$. Plot the magnitude and phase responses of $C(e^{j\omega})$ and $D(e^{j\omega})$. Find the zeros of $C(z)$ and $D(z)$ and verify that they are inside/outside the unit circle. What is the relation between the zeros of $C(z)$ and $D(z)$?

b. $C(z)$ is the spectral factor of a halfband filter $P(z)$. Compute $P(z) = z^{-7}C(z^{-1})C(z)$ and its impulse response $\boldsymbol{p}(n)$. Verify that $P(\omega) + P(\omega + \pi) = 1$. Find and plot the zeros of $P(z)$.

c. Create the near-symmetric Daubechies filter $\boldsymbol{c}_1(n)$ and $\boldsymbol{d}_1(n)$ of length 8. Plot $\boldsymbol{c}_1(n)$ and $\boldsymbol{d}_1(n)$. Plot the magnitude and phase responses of $C_1(e^{j\omega})$ and $D_1(e^{j\omega})$. Compare them to those of $C(z)$ and $D(z)$. Find the zeros of $C_1(z)$. Are they all inside $|z| = 1$? Explain.

d. Compute $P_1(z) = z^{-7}C_1(z^{-1})C_1(z)$ and its impulse response $\boldsymbol{p}_1(n)$. Verify that $\boldsymbol{p}_1(n) = \boldsymbol{p}(n)$.

2. Daubechies Filters and Lattice Coefficients

a. Create the minimum-phase Daubechies filters $\boldsymbol{c}(n)$ and $\boldsymbol{d}(n)$ of length 16. Plot $\boldsymbol{c}(n)$ and $\boldsymbol{d}(n)$. Find the lattice coefficients using *filtolat.m* (from the book homepage).

b. Create the near-symmetric Daubechies filter $\boldsymbol{c}_1(n)$ and $\boldsymbol{d}_1(n)$ of length 16. Plot $\boldsymbol{c}_1(n)$ and $\boldsymbol{d}_1(n)$. Find the lattice coefficients. Are these lattice coefficients the same as those in Part a?

3. Lattice Coefficients and Quantization

The filter bank remains PR even after the lattice coefficients are quantized. This problem studies the effects of quantization on the frequency responses of the filters.

a. Write an m-file to quantize x to y where x is real (∞ bits) and y has b bits.

b. Create the minimum-phase Daubechies filters $\boldsymbol{c}(n)$ and $\boldsymbol{d}(n)$ of length 6. Plot $\boldsymbol{c}(n)$, $\boldsymbol{d}(n)$ and their frequency responses. Find the lattice coefficients γ_k.

c. Quantize γ_k to 3 bits to obtain $\gamma_{q,k}$. Find the corresponding filters $\boldsymbol{c}_q(n)$ and $\boldsymbol{d}_q(n)$. Plot $\boldsymbol{c}_q(n)$, $\boldsymbol{d}_q(n)$ and their frequency responses. Are they still maximally flat filters? Are the zeros of $C_q(z)$ still inside $|z| = 1$? Verify that the resulting filter banks are still PR.

d. Repeat a and b for near-symmetric Daubechies filters with length 6. Compare the quantization effects on the filters' responses with those in Parts a and b.

e. Repeat a and b for a minimum-phase Daubechies filters of length 18. *Quantize the filter coefficients, then the lattice coefficients, and plot the corresponding frequency responses.* Are the quantization effects more severe for long filters?

f. Repeat a and b for near-symmetric Daubechies filters with length 18. Compare the coefficient quantization effects with those in Part e.

4. Lattice Coefficients and Quantization

Problem 3 quantized all lattice coefficients to the same number of bits. In reality the first lattice section affects the frequency response more than the last lattice section. The responses are nonlinear functions of γ_k. How does one distribute B bits to $\gamma_{q,k}$ to minimize the quantization distortions?

a. Create the minimum-phase Daubechies filters $\boldsymbol{c}(n)$ and $\boldsymbol{d}(n)$ of length 8. Find the lattice coefficients γ_k. Quantize γ_k to 3 bits to obtain $\gamma_{q,k}$. Find the corresponding filters $\boldsymbol{c}_q(n)$ and $\boldsymbol{d}_q(n)$. Plot $\boldsymbol{c}_q(n)$, $\boldsymbol{d}_q(n)$ and their frequency responses.

b. Compute the total number of available bits B (3 bits $\times$ number of sections). Find a different allocation to reduce distortion.

5. Maximally-Flat Halfband Filter

Write an m-file to compute the impulse response $\boldsymbol{p}(n)$ of a maxflat halfband filter. The input is N (number of zeros at $\omega = \pi$). The output is $\boldsymbol{p}(n)$.

6. Spectral Factorization of Maximally-Flat Halfband Filter

This problem compares factorization algorithms for $P(z) = C(z^{-1})C(z)$.

a. Use the m-file in Problem 5 to design a maximally-flat halfband filter with $N = 8$. Plot $\boldsymbol{p}(n)$ and its frequency response.

b. ***Method A (zeros of a polynomial)*** Use the MATLAB function *roots.m* to find the zeros of $P(z)$. How many possible choices for $C(z)$ and $z^{-N}C(z^{-1})$? Find the minimum-phase spectral factor and plot its frequency response.

c. ***Method B (solve quadratic equations)*** Use the same halfband filter $P(z)$ and the provided m-file for quadratic-constrained spectral factorization to compute the spectral factor $C(z)$. Plot its frequency response.

d. ***Method C (matrix factorization)*** Section 5.4 describes an algorithm in terms of matrix (Cholesky) factorization. Write an m-file to implement this algorithm. Use the same halfband filter $P(z)$ to compute the spectral factor $C(z)$. Plot its frequency response.

e. ***Method D (cepstral method: Take logarithms)*** Use the same halfband filter $P(z)$ and the provided m-file for cepstral-based spectral factorization to compute the spectral factor $C(z)$. Plot its frequency response.

f. Compare the results in Parts b–e. Then repeat for $N = 40$. At which N do the factorization algorithms break down?

7. Spectral Factorization of Equiripple Halfband Filter

a. Design an equiripple symmetric halfband (lowpass) filter $\widehat{P}(z)$ of length 15, using *remez.m*. The cutoff frequencies are 0.4π and 0.6π, and the passband and stopband weights are equal.

b. Measure the stopband attenuation δ_S of $|\widehat{P}(e^{j\omega})|$ and construct a new nonnegative $P(z) = \widehat{P}(z) + \delta_S z^{-7}$. Find its zeros and assign them into $P(z) = z^{-7}H_0(z^{-1})H_0(z)$ such that no zeros of $H_0(z)$ are outside the unit circle. Find the highpass filter $H_1(z) = z^{-7}H_0(-z^{-1})$ by alternating flip. Plot $\boldsymbol{h}_0(n)$, $\boldsymbol{h}_1(n)$, $|H_0(e^{j\omega})|$ and $|H_1(e^{j\omega})|$. Compare them to the plots of $|C(e^{j\omega})|$ and $|D(e^{j\omega})|$ in Problem 6b. What are the differences?

c. Verify the PR condition. What is the overall delay from the input to the output?

d. Let $C'(z) = \sqrt{2}H_0(z)$ and $D'(z) = \sqrt{2}H_1(z)$. Compute the lattice coefficients of the filter bank, associated with the filters $\boldsymbol{c}'(n)$ and $\boldsymbol{d}'(n)$. Any significant changes from the lattice coefficients for $\boldsymbol{c}(n)$ and $\boldsymbol{d}(n)$ in Problem 6b?

8. Daubechies Wavelets

Select the third option `Wavelet Display` in the Toolbox.

a. Display and plot the Daubechies scaling function `db3` and wavelet for 5 levels. Increase to 10 levels and plot again. Compare.

b. Change to `db6` with 12 taps and repeat Part a. What differences?

c. Change to `sym6`. Plot its scaling function and "symlet" and verify that they are nearly symmetric. How to measure the distance from symmetry?

MATLAB Exercises – Chapter 6

MATLAB *Wavelet Toolbox:*

idwt.m:	Dyadic discrete wavelet transform (1D).
qmf.m:	Dyadic inverse discrete wavelet transform (1D).
dbwavf.m:	Design minimum-phase Daubechies orthonormal filter.
biorfilt.m:	Design biorthogonal wavelets.
qmf.m:	Find the highpass filter given the lowpass filter.
orthfilt.m:	Design analysis and synthesis filters.
wavefun.m:	Compute the scaling function and wavelets.

1. Daubechies Filters, Scaling Function and Wavelets

a. Design the minimum-phase Daubechies filter of length 8. Compute the corresponding scaling function and wavelets for 3 and 6 levels. Plot and compare $\phi(t)$ and $w(t)$ for these two levels.

b. Repeat Part a for Daubechies filters with length 16. Compare their scaling function and wavelets.

c. Design the near-symmetric Daubechies filter of length 8. Compute the corresponding scaling function and wavelets for 3 and 6 levels. Plot and compare $\phi(t)$ and $w(t)$ for these two levels. Compare the near-symmetric scaling functions with the minimum-phase scaling function.

d. Repeat Part a for near-symmetric Daubechies filters with length 16. Compare their scaling functions and wavelets.

2. Equiripple Paraunitary Filter Banks, Scaling Functions and Wavelets

MATLAB Problem 7 in Chapter 5 outlines an orthonormal design based on equiripple filters. Use the same design (length 8) and compute the scaling function and wavelets for 3 and 6 levels. Describe the scaling function and wavelets. (Note that there are no zeros at $\omega = \pi$.) Compare the results with those in Problem 1a above.

3. Daubechies Filters and Signal Representation

a. The input signal with length 544 to the Haar filters is

$$x(n) = \left\{ \begin{array}{ll} n - 16; & 16 \leq n \leq 272 \\ 528 - n; & 272 \leq n \leq 528 \\ 0; & \text{otherwise} \end{array} \right\}.$$

Compute the discrete wavelet transform (4 levels). Plot all coarse approximations and details. Describe the results (from Daubechies of length 2).

b. Compute the inverse discrete wavelet transform to obtain $y(n)$. What is the relation between $y(n)$ and $x(n)$?

c. Repeat Parts a and b for Daubechies filters with length 8.

4. Equiripple Paraunitary Filter Banks and Signal Representation

a. Use the equiripple filter bank in Problem 2 and compute the discrete wavelet transform for 4 levels. Plot and describe all coarse approximations and details. Compare them to the results in Problem 3c.

b. Compute the inverse discrete wavelet transform to obtain $y(n)$. What is the relation between $y(n)$ and $x(n)$? What can you conclude about the use of equiripple filters in signal representation?

5. **Daubechies Filters and Multiresolution**

   a. Design the minimum-phase Daubechies filters of length 6. Let the input be a Gaussian noise sequence with length 1024. Compute the discrete wavelet transform (4 levels). Plot all coarse approximations and details.

   b. Compute the inverse discrete wavelet transform to obtain $y(n)$. What is the relation between $y(n)$ and $x(n)$?

   c. Increase the filter length to 16 and repeat Parts a and b. Are the detail coefficients different from those in Part a? Why?

   d. Repeat a, b, c using the equiripple filter bank from Problem 2.

6. **Even-length Biorthogonal Symmetric Wavelets**

   Design a maxflat halfband filter $p(n)$ of length 15. Find its zeros.

   a. Distribute the zeros of $P(z)$ such that $H_0(z)$ and $F_0(z)$ are even-length. How many choices of grouping the zeros? Find all possible choices that yield (8, 8) filters. Plot their impulse responses.

   b. For each choice above, compute and plot its scaling function and wavelet. Which solution gives the smoothest scaling function and wavelets? Plot the impulse and frequency response of this choice. Check the result using the program biorfilt.m.

   c. Let $x(n)$ be the hat function in Problem 3. Compute the discrete wavelet transform (4 levels) using the smoothest biorthogonal wavelet. Plot and describe all coarse approximations and detail coefficients. Compare them to those in Part a of Problem 4.

   d. Repeat Part c on the Gaussian noise in Problem 5.

7. **Odd-length Biorthogonal Symmetric Wavelets**

   a. Distribute the zeros of $P(z)$ in Problem 6 such that $H_0(z)$ and $F_0(z)$ are odd-length. How many choices of grouping the zeros? Find all possible choices that yield (9, 7) taps. Plot their impulse responses.

   b. For each choice above, compute and plot its scaling function and wavelet. Which solution gives the smoothest scaling function and wavelets? Plot the impulse and frequency response of this choice. Check the result using the program biorfilt.m.

   c. Let $x(n)$ be the hat function in Problem 3. Compute the discrete wavelet transform (4 levels) using the smoothest biorthogonal wavelet. Plot and describe all coarse approximations and detail coefficients. Compare with Problem 3a.

   d. Repeat Part c on the Gaussian noise in Problem 5.

8. **Binlets**

   Section 6.5 constructs *binlets* with integer coefficients; see also the book homepage.

   a. Consider the (9, 7)-tap and (8, 8)-tap *binlets*. Compute and plot their scaling functions and wavelet. Describe and compare the results.

   b. Compute the discrete wavelet transforms (4 levels) for the hat function in Problem 3. Plot and describe all coarse approximations and detail coefficients.

9. Biorthogonal Wavelets

a. In the Toolbox `Wavelet Display`, choose the biorthogonal wavelets `bior(2,2)` with (5, 3) taps. Display and plot the two scaling functions and wavelets for 5 levels. Increase to 10 levels and compare.

b. Change the wavelet choice to `bior(3,3)` with (8, 4) taps and repeat Part a. What are the differences between a and b?

10. Multiresolution Analysis in 1D

a. Under `File` option in the Toolbox `Wavelet 1-D`, select `Load Signal`. Choose `heavysin.mat` and the `db 1` wavelet and `Level 4`. Press `Analyze` for the Discrete Wavelet Transform (DWT) of the signal. The decomposition (coarse at level 4 and details at levels 1 to 4) are shown with the input signal. Display using `Full Decomposition`.

b. Repeat with `db 8`. Compare the coarse approximations. Do the detail signals get smaller? Why?

c. Change to `symlets` and compute the full decomposition with 4 levels. `File` and `Demo Analysis` offers many examples of multiresolution. Try them out.

11. Wavelet Packet in 1D

a. Under `File` menu option in `Wavelet Packets 1-D`, select `Load Signal`. Choose the file `heavysin.mat` and the `db 1` wavelet and `Level 4`. Press `Analyze` to perform the Full Transform (at every node on the tree). Prune the tree using Shannon entropy by selecting `Best Tree`. Is the best tree a dyadic wavelet tree?

b. Select the `db 8` wavelet and `Level 4`. Press `Analyze` to perform the Full Transform, and prune the tree by selecting `Best Tree`. Is the best tree different for `db 8` and `db 1`?

c. Select the `sym8` wavelet and `Level 4`. Compare the best tree with that in Part b.

12. Multiresolution Analysis in 2D

a. Under `File` menu option in `Wavelet 2-D`, select `Load Image` and choose `woman.mat`. Select the `db 1` wavelet and `Level 2`. Press `Analyze` for the DWT. Display the coarse approximation at level 2 and the details at levels 1 to 2 using `Square View` mode.

b. Repeat Part a with `db 8`. Do the detail images have fewer features? Why? Compare also using the symlets.

MATLAB Exercises – Chapter 7

1. Construction and Eigenvalues of $M = (\downarrow 2)2H$

a. Use the MATLAB code at the beginning of Chapter 7 or the homepage to construct M (more precisely its $N \times N$ submatrix $m(0)$) for the D_4, D_6, and D_8 filters. Compute the eigenvalues by `eig(M)` and the eigenvector for $\lambda = 1$.

b. Repeat Part a for the filters $\boldsymbol{h} = (-1, 3, 3, -1)/4$ and $\boldsymbol{f} = (1, 3, 3, 1)/8$. Which one fails to yield a bounded $\phi(t)$?

2. **Cascade Algorithm at Dyadic Points**

a. Write a code or use *wavefun.m* (or our homepage) to execute the cascade algorithm starting from $\phi^{(0)}(n) = \delta(n)$. The input is $\boldsymbol{h}$ and the number I of iterations.

b. Apply the code to the five filters in Problem 1. Choose I to give a reasonable graph of $\phi(t)$.

c. Start the cascade from the eigenvector Φ of $\boldsymbol{m}(0)$ for $\lambda = 1$. Show $I = 2$ iterations for the D_4 filter and the spline filter $(1, 3, 3, 1)/8$.

d. Use the code to graph the derivative $\phi'(t)$ for D_4. Why does $\boldsymbol{M}$ become $2\boldsymbol{M}$?

3. **Construction and Eigenvalues of $\boldsymbol{T} = (\downarrow 2)2\boldsymbol{H}\boldsymbol{H}^T$**

a. Find the autocorrelation $\boldsymbol{p} = \boldsymbol{h} * \boldsymbol{h}^T$ of the five filters in Problem 1. Construct the matrix $\boldsymbol{T}$ of order $2N - 1$ (not $2N + 1$) for each filter and compute its eigenvalues. How many powers of $\frac{1}{2}$ for each $\boldsymbol{T}$?

b. What is the smoothness $s_{\max}$ (in L^2) of the corresponding $\phi(t)$?

c. What are the inner products $\boldsymbol{a}(k) = \langle\phi(t), \phi(t + k)\rangle$ in each case?

4. **Improvement from zeros at π**

a. Add an extra zero at π to the filter $\boldsymbol{h}$ in Problem 1b, and remove a zero from $\boldsymbol{f}$. Decide whether both filters now satisfy Condition E.

b. Construct $\phi(t)$ and compute $s_{\max}$ for the new $\boldsymbol{h}$ and $\boldsymbol{f}$. How are they related to $\phi(t)$ and $s_{\max}$ for the original $\boldsymbol{h}$ and $\boldsymbol{f}$?

c. Among all orthogonal 4-tap filters with at least *one* zero $H(\pi) = 0$, does the D_4 filter give the smallest $|\lambda_{\max}(T)|$ excluding $\lambda = 1$?

MATLAB Exercises – Chapter 8

1. **Periodic Extension of the Signal**

Create a symmetric signal $\boldsymbol{x}(n) = \left\{ \begin{array}{ll} n; & 0 \leq n < 256 \\ 511 - n; & 256 \leq n < 512 \end{array} \right\}$. Design a lowpass filter $\boldsymbol{h}(n)$ with $\omega_p = 0.4\pi$, $\omega_s = 0.6\pi$ and length 21.

a. Use `conv(h,x)` to compute the convolution $\boldsymbol{y}(n)$. What is its length? Truncate $\boldsymbol{y}(n)$ to the same size as $\boldsymbol{x}(n)$ to obtain $\boldsymbol{y}_T(n)$.

b. Process $\boldsymbol{x}(n)$ in two blocks of length 256 using periodic extension. Let $\boldsymbol{x}_1(n)$ be the first 256 values of $\boldsymbol{x}(n)$. Periodically extend $\boldsymbol{x}_1(n)$ to obtain $\widehat{\boldsymbol{x}}_1(n)$ and filter it with $\boldsymbol{h}(n)$ to obtain $\widehat{\boldsymbol{y}}_1(n)$. Truncate $\widehat{\boldsymbol{y}}_1(n)$ to obtain $\boldsymbol{y}_1(n)$. Repeat for the second block $\boldsymbol{x}_2(n)$. Concatenate $\boldsymbol{y}_1(n)$ and $\boldsymbol{y}_2(n)$ into a vector $\boldsymbol{y}_C(n)$. Plot $\boldsymbol{y}_T(n)$ and $\boldsymbol{y}_C(n)$ on the same graph. Plot the frequency responses for $\boldsymbol{y}_T(n)$ and $\boldsymbol{y}_C(n)$ and compare them.

c. Repeat Parts a and b for the Daubechies filter D_8.

d. Create the circulant matrix associated with D_8. Find its eigenvalues and eigenvectors. Verify that the eigenvectors are columns of the DFT matrix.

2. Zero Padding of the Signal

Use the same signal $\boldsymbol{x}(n)$ and same filter $\boldsymbol{h}(n)$ as in Problem 1. The signal $\boldsymbol{y}_T(n)$ is the same as that in Problem 1, part a.

a. Process $\boldsymbol{x}(n)$ in blocks of length 256 using zero padding. Extend $\boldsymbol{x}_1(n)$ to obtain $\widehat{\boldsymbol{x}}_1(n)$ by zero padding and filter it with $\boldsymbol{h}(n)$ to obtain $\widehat{\boldsymbol{y}}_1(n)$. Truncate $\widehat{\boldsymbol{y}}_1(n)$ to obtain $\boldsymbol{y}_1(n)$. Repeat for the second block $\boldsymbol{x}_2(n)$. Concatenate $\boldsymbol{y}_1(n)$ and $\boldsymbol{y}_2(n)$ into a vector $\boldsymbol{y}_C(n)$. Plot $\boldsymbol{y}_T(n)$ and $\boldsymbol{y}_C(n)$. Plot the frequency responses for $\boldsymbol{y}_T(n)$ and $\boldsymbol{y}_C(n)$.

b. Compare with the results in Problem 1b. Which is better? Repeat for the filter D_8.

3. Symmetric Extension of the Signal

Use the same signal $\boldsymbol{x}(n)$ and filter $\boldsymbol{h}(n)$ and output $\boldsymbol{y}_T(n)$ as in Problem 1.

a. Process $\boldsymbol{x}(n)$ in blocks of length 256 using symmetric extension. How many choices of symmetric extension can one use? Extend $\boldsymbol{x}_1(n)$ to obtain $\widehat{\boldsymbol{x}}_1(n)$ by symmetric extension. Filter with $\boldsymbol{h}(n)$ to $\widehat{\boldsymbol{y}}_1(n)$ and truncate to obtain $\boldsymbol{y}_1(n)$. Repeat for $\boldsymbol{x}_2(n)$. Concatenate $\boldsymbol{y}_1(n)$ and $\boldsymbol{y}_2(n)$ into $\boldsymbol{y}_C(n)$. Plot $\boldsymbol{y}_T(n)$ and $\boldsymbol{y}_C(n)$ and describe the differences. Compare the frequency responses.

b. Compare with Problems 1b and 2a. *Which method is best*?

c. Repeat Part a for D_8. The error is much larger now, which confirms that one can not use symmetric extension if the filters are not symmetric.

4. Comparison of Extension Methods on Image Compression

This problem uses the image coder `uicoder.m` available at our webpage.

a. Choose wavelets as the transform and load in the FBI (9, 7) system. and Barbara image. Choose *zero padding* and compress Barbara at 1 bpp and unity weighting. Display the reconstructed and error images using `Extract to Separate Window`. Describe the border effect.

b. Decrease the rate to 0.25 bpp. Does the error at the border increase?

c. Choose *circular extension.* Compress the image at both 1 bpp and 0.25 bpp. Compare the error images to those using zero-padding. Which is better (objectively and subjectively)?

d. Choose *symmetric extension* and compress at 1 bpp and 0.25 bpp. Compare the error images to zero-padding and circular extension. Which method yields smallest error? Compile a table of MSE, PSNR and Max Error for the three methods.

e. Repeat for the Baboon image. Does the difference in PSNR decrease? Why?

MATLAB Exercises – Chapter 9

1. Polyphase Representation of FIR filter

Use *remez.m* to design a third band lowpass filter $H(z)$ of length 61, with $\omega_p = \frac{\pi}{4}$ and $\omega_s = \frac{5\pi}{12}$.

a. Find the three polyphase components $E_\ell(z)$. Plot their impulse, magnitude, and phase responses (similar to Figure 9.6). Verify that $|E_\ell(e^{j\omega})| \approx \frac{1}{3}$ and the phase offset $\Delta_\ell(\omega) \approx -\frac{\ell\omega}{3}$.

b. Create a highpass filter $G(z)$ by alternating the sign to $\boldsymbol{g}(n) = (-1)^n \boldsymbol{h}(n)$. Find the three polyphase components $E'_\ell(z)$ of $G(z)$. Plot their impulse, magnitude, and phase responses. What are the properties of $|E'_\ell(e^{j\omega})|$ and $\Delta'_\ell(\omega)$?

2. **Multirate System and Polyphase Components**

Write a MATLAB m-file to simulate the system $(\downarrow M)\boldsymbol{H}(\uparrow M)$ in Problem 9.2.7.

a. Let $M = 4$ and $H(z)$ be an equiripple lowpass filter of length 81. Verify your solution in Problem 9.2.7 with $\boldsymbol{x}(n) = \delta(n)$.

b. Change the filter length to 83 and repeat Part a. What are the effects of the filter length on the results?

3. **Filtering in DFT Filter Bank**

Design an equiripple $H_0(z)$ with length 21, $\omega_p = 0.15\pi$ and $\omega_s = 0.35\pi$.

a. Let $M = 4$. Create the other three analysis filters by $\boldsymbol{h}_k(n) = \boldsymbol{h}_0(n)\, e^{j2\pi kn/4}$. Compute and plot $|H_k(e^{j\omega})|$, $0 \le \omega < 2\pi$ on the same graph. Verify that this is a DFT filter bank.

b. Compute and plot the phase responses of $H_k(e^{j\omega})$. Are they linear?

4. **Polyphase Form of DFT Filter Bank**

Design an equiripple lowpass $H_0(z)$ with length 31, $\omega_p = \frac{\pi}{4}$ and $\omega_s = \frac{5\pi}{12}$.

a. Let $M = 3$. Plot the magnitude responses of $H_k(z) = H_0(ze^{-j2\pi k/3})$.

b. The $H_k(z)$ can be implemented with the polyphase components of $H_0(z)$:

$$\begin{bmatrix} H_0(z) \\ H_1(z) \\ H_2(z) \end{bmatrix} = (IDFT) \begin{bmatrix} E_0(z^3) \\ z^{-1}E_1(z^3) \\ z^{-2}E_2(z^3) \end{bmatrix}$$

Find $E_\ell(z)$ and compute $H_k(z)$. Plot $|H_k(e^{j\omega})|$ and compare with Part a.

5. **Simulation of M-Channel Filter Bank**

Write *filterbank.m* to simulate the M-channel filter bank in Figure 9.5. The inputs are $\boldsymbol{x}(n)$, the filter coefficients $\boldsymbol{h}(n)$ and $\boldsymbol{f}(n)$, and the decimation factor M. The output is $\widehat{\boldsymbol{x}}(n)$. All filters have the same length (zero-pad the short filters) and the subband signals are not processed. Test your program with $\boldsymbol{x}(n) = \delta(n)$. Plot each input below on the same graph with the output.

a. $M = 2$ and the Haar filters, expecting $\widehat{\boldsymbol{x}}(n) = \boldsymbol{x}(n-1)$.

b. $M = 2$ and the Daubechies filters of length 8 (use *dbwavf.m* in the MATLAB Toolbox, expecting $\widehat{\boldsymbol{x}}(n) = \boldsymbol{x}(n-7)$).

c. $M = 2$ and the linear-phase biorthogonal FBI (9,7) filters (coefficients from our homepage or *biorfilt.m* in the Toolbox). $\widehat{\boldsymbol{x}}(n)$ should be $\boldsymbol{x}(n-7)$.

d. $M = 8$ and filters from a DFT filter bank with length 8: $\boldsymbol{h}_0(n) = 1, 0 \le n \le 7$, and $\boldsymbol{h}_k(n) = \boldsymbol{h}_0(n)\, e^{j2\pi kn/8}$. Then $\widehat{\boldsymbol{x}}(n) = \boldsymbol{x}(n-7)$.

6. **Reconstruction Error in the DFT Filter Bank (Johnston: $M = 2$)**

$H_0(z)$ is an equiripple filter of length 21 with $\omega_p = 0.4\pi$ and $\omega_s = 0.6\pi$ and $H_1(z) = H_0(-z)$.

a. Which $F_k(z)$ cancel aliasing ? Plot the magnitude response of $T_0(z)$. Compute and verify that $T_1(z) = 0$. For $\boldsymbol{x}(n) = \delta(n)$, find the output using *filterbank.m* in Problem 5. Verify that this is the impulse response $\boldsymbol{t}_0(n)$. Compute and plot the error $\widehat{\boldsymbol{x}}(n) - \boldsymbol{x}(n)$.

b. Find the polyphase components of $H_0(z)$ and verify linear phase. Equation (9.29) does not give stable IIR synthesis filters.
c. Repeat a and b for length 22 (the system delay should be 21).

7. Reconstruction Error in 4-channel Filter Bank

$H_0(z)$ is equiripple with length 41 and $\omega_p = 0.2\pi$, $\omega_s = 0.3\pi$. Let $H_k(z) = H_0(ze^{-j2\pi k/4})$.

a. Which $F_k(z)$ cancel aliasing? Use $E_\ell(z)$ (the polyphase components of $H_0(z)$) and (9.29) to find the polyphase components $R_\ell(z)$ of $F_0(z)$. What is the length of $F_0(z)$? Plot $|F_k(e^{j\omega})|$.
b. Verify that $T_1(z) = T_2(z) = T_3(z) = 0$. Find and plot the magnitude and phase responses of $T_0(e^{j\omega})$. For $x(n) = \delta(n)$, find the output using *filterbank.m* from Problem 5. Verify that $\widehat{x}(n) = t_0(n)$.
c. For PR synthesis filters (IIR), all zeros of $E_\ell(z)$ must be inside the unit circle. Use *subplot* and *zplane.m* for those zeros. Find the polyphase components that have zeros outside the unit circle.

8. Paraunitary Filter Bank

a. Interchanging the analysis and synthesis does not change the PR property. Verify this for the Daubechies filter D8.
b. Shifting the analysis filters $H_k(z)$ to get $H_k(ze^{j\theta})$ does not change the PR property. Verify this for several choices of θ.

9. Transmultiplexer

Write *transmultiplexer.m* to simulate the two-channel transmultiplexer in Problem 9.2.11. The inputs are $(x_0(n), x_1(n))$ and the filter coefficients are $(h_0(n), h_1(n))$, $(f_0(n), f_1(n))$. The output is $(\widehat{x}_0(n), \widehat{x}_1(n))$.

a. Design the Daubechies filters $H_k'(z)$ and $F_k'(z)$ with length 10.
b. For the transmultiplexer use $H_k(z) = H_k'(z)$ and $F_k(z) = z^{-1}F_k'(z)$. Verify that the transmultiplexer is PR. (Use any choices for $x_k(n)$).

10. Cosine-Modulated Filter Bank

Obtain the prototype filter $p(n)$ of a 4-channel cosine-modulated filter bank from

`http://saigon.ece.wisc.edu/~waveweb/Tutorials/books/matlab/chap9.html`

a. Plot $|P(e^{j\omega})|$. Find and plot the magnitude and phase responses of $h_k(n)$. Verify that $h_0(n)$ is not symmetric.
b. Verify that the system is PR by computing the transfer functions $T_k(z)$.
c. Verify that the system is PR using *filterbank.m* in Problem 5 with $x(n) = \delta(n)$.

11. GenLOT

The program *uigenlot.m* to design GenLOT is available at

`http://saigon.ece.wisc.edu/~waveweb/QMF/software.html`

a. Use it to design two 8-channel GenLOTs with length 32 (overlapping factor 4), for maximizing coding gain and for maximizing stopband attenuation. Plot and compare the frequency responses.
b. Include the PMI property in the design for stopband attenuation. Compare the result with the one in Part a.

12. Tree-structured Filter Bank

Let $H_k(z)$ be the Daubechies filters of length 10. Let $G_k(z)$ be a three-channel DFT filter bank with equiripple $H_0(z)$ of length 31, $\omega_p = \frac{\pi}{4}$ and $\omega_s = \frac{5\pi}{12}$.

a. Compute and plot the magnitude responses of the filters $H_k^{'}(z)$ of the 6-channel uniform analysis bank at the end of Section 9.3.

b. Compute and plot the magnitude responses of the filters $H_k^{'}(z)$ of the 5-channel *non*uniform bank in the Glossary (See Tree-structured Filter Bank).

MATLAB Exercises – Chapter 10

The image coder (*uicoder*) and GenLOT design program (*uigenlot*) are available at our homepage, together with README files.

1. DCT and Blocking Artifact

Use *uigenlot.m* to design and plot the impulse responses and frequency responses of an 8-channel DCT filter bank. Verify that the basis function have length 8. What is their stopband attenuation?

a. Use *uicoder.m* to compress the *Barbara* image using DCT at 1 bpp (unit weighting). Display the original and reconstructed images. At this bit rate, there should be no visible artifact. Describe the error image.

b. Decrease the bit rate to 0.125 bpp at intervals of 0.125 bpp. Display each signal and reconstructed image on the same plot. At which bit rate does blocking start to show? Describe the error image.

c. Repeat part b for the image *Baboon*. Do *Barbara* and *Baboon* start to show blocking at the same bit rates? Why or why not?

2. GenLOT and Ringing Artifact

GenLOT has longer basis functions than DCT and blocking is eliminated. *Ringing* is introduced by long bandpass basis functions.

a. Use *uigenlot.m* to plot the impulse and frequency responses of an 8-channel GenLOT with length 48 (use *lot86-cgdc*). Verify that the filters have linear phase.

b. For bit rates from 1 bpp to .125 bpp, compress the image *Barbara*. Describe the quality of the images compared to the DCT. Does blocking occur? At what rate does ringing start to show? Describe the error image at this bit rate.

3. Border Distortion

a. Use the GenLOT in Problem 2 to compress *Barbara* at 0.25 bpp with zero-padding, circular convolution and symmetric extension. Compare the results both objectively and subjectively.

b. Is the border distortion for *Baboon* less visible? Why? Repeat using a wavelet transform.

4. **Ringing Artifact in Image Compression**
   a. Use the wavelet system (13, 11) to compress *Yogi* at 0.25 bpp. Is there ringing in the reconstructed image?
   b. Repeat with the FBI (9, 7) system. Is ringing reduced?
5. **Checkerboard & Canvas Artifacts**
   a. Use the wavelet *checker.mat* to compress the image *boats* at 0.25 bpp. Describe the reconstructed image. Plot $\phi(t)$ and $w(t)$ for this system. Explain the artifact in the reconstructed image using the shape of $\phi(t)$.
   b. Use the wavelet *canvas.mat* and repeat Part a. Is the artifact the same?
6. **GenLOT Design**
   a. Use *uigenlot.m* to design a 10-channel GenLOT with length 40 using stopband attenuation as the objective function. Plot $\boldsymbol{h}_k(n)$ and $H_k(e^{i\omega})$.
   b. Repeat for coding gain as the objective function. Compare the stopband attenuation here with the results in Part a.

MATLAB Exercises – Chapter 11

The image coder (*uicoder*) and ECG compression program (*wvgui*) are available at our homepage. Use the README files to become familiar with these programs.

1. **Weighting in DCT-based Image Compression**
   At medium and low bit rates, the reconstructed images from DCT suffer from blocking. This can be reduced using the weighting scheme that we discuss. This problem analyzes the tradeoff between reduced blocking and loss of texture.
   a. Verify the blocking artifact by compressing *Barbara* at 0.25 bpp with unity weighting in *uicoder.m.* Display the original and reconstructed images on the same plot. Describe the blocking artifact.
   b. Display four reconstructed images with weights 1, 1.5, 2.0 and 2.5 on the same plot. Does the blocking artifact decrease as the weight increases? What happens to the texture details in the scarf? Which weight gives the best subjective performance (most pleasing to look at)?
   c. Compile the objective measures (MSE, PSNR, and Max Error) of the four designs. Is there a correlation between low MSE and high subjective performance? Does the MSE decrease or increase as the weight increases?
2. **Weighting in GenLOT-based Image Compression**
   The basis functions of GenLOT are longer than in DCT, and blocking is almost eliminated. But ringing is introduced by long bandpass basis functions.
   a. Compress *Barbara* at 0.25 bpp using GenLOT (*lot86-cgdc)* with weights 1, 1.5, 2 and 2.5. Compare the effects of the weights on the ringing artifacts.
   b. Compile and compare the objective measures (MSE, PSNR, Max Error) of the four compressions. Which weight is best perceptually?

3. **Choice of Transforms**
   Compare the DCT and LOT (Type II-Malvar) with GenLOT (*lot86-cgdc*): weight = 1.
   a. Compress *Barbara* using the three transforms at 0.25 bpp. Compile a table of objective measures. Compare and explain the artifacts in the three reconstructed images.
   b. Change the weight to 1.5 and repeat. Then repeat at 0.5 bpp.
4. **Weighting in Wavelet-based Image Compression**
   a. Compress *Barbara* at 0.25 bpp using the FBI $(9, 7)$ system with weights $1, 1.25, 1.5, 1.75$. Compile the objective measures. Which image is best perceptually?
   b. Repeat for the image *Baboon*. Repeat using the $(5, 3)$ biorthogonal pair.
5. **Choice of Wavelets and Artifacts**
   a. Plot the scaling functions and wavelets of the $(5, 3)$, $(9, 7)$, and $(17, 11)$ filters.
   b. Compress *Barbara* at 0.25 bpp using these systems. Compare the objective measures. Which system give the best reconstruction? Which has ringing? Compare the texture details on the scarf.
6. **Comparison of Type A and B in Image Compression**
   a. Compress *Barbara* using the $(6, 10)$ and $(9, 7)$ systems at 0.5 bpp. Compare the reconstructed images objectively and subjectively. Which one has lower MSE? Which one looks best?
   b. Repeat for bit rates of 0.25 bpp and 0.125 bpp. Do the same conclusions hold? Repeat also for the image *Boats*. Which system is the best overall?
7. **Artifacts in Wavelet Compression**
   Choose *Barbara*, unity weighting, and bit rate 0.5 bpp. Plot the scaling functions and wavelets of the $(2, 10)$ and $(10, 2)$ systems. Compress *Barbara* and compare the reconstructed images objectively and subjectively. Which has more blocking? Which has more ringing? Explain.
8. **Comparison of Wavelets and GenLOT in Image Compression**
   a. Plot the $(9, 7)$ and $(6, 10)$ scaling functions and wavelets. Plot the impulse and frequency responses of the GenLOT *lot86-cgdc*.
   b. Compress *Barbara* at 1 and 0.5 and 0.25 bpp using these systems. Summarize the objective measures by plotting the PSNR as a function of bit rate.
9. **ECG Compression**
   a. Invoke the program *wvgui* from our homepage on ECG compression. Open a file by clicking on *File*. Use the default setting to compress the ECG signal at $2, 4, 6$ and 8 compression ratios. Compile the MSE and PRD of these four compressions. Plot MSE and also PRD as a function of compression ratio. Is this what you expect?
   b. Part a uses the 4-level wavelet system with the FBI $(9, 7)$ wavelet. Choose the $(11, 13)$ system by clicking on *Method-Change* and compress at $4 : 1$. Do the errors decrease?
   c. The $(9, 7)$ and $(11, 13)$ systems each allow four choices (4 and 5 levels of log-tree decomposition and 4 and 5 levels of best-tree decomposition). At $4 : 1$ compression, which of the eight choices gives the best objective measures?

10. Denoising in 1D

Denoising keeps the largest wavelet coefficients (*thresholding*). The basis functions play an important role since the denoised (reconstructed) signal is a combination of those functions.

a. Under `File` menu select `Wavelet 1-D` and `Load Signal`. Choose `heavysin.mat`. Select the `db 1` wavelet and `Level 4`. Press `Analyze` to perform the DWT. Display using `Full Decomposition`.

b. Press `Denoise` for Automatic Soft Thresholding. Describe the denoised signal. Is this a good approximation of the original? Does the Haar wavelet affect the reconstructed signal?

c. Repeat with `db 8`. Is the denoised signal much better?

d. Repeat using `Automatic Hard Thresholding`. Compare the denoised signals!

11. Discontinuity Detection

One can detect discontinuities in a signal using multiresolution analysis.

a. Under `File` menu option in `Wavelet 1-D`, select `Load Signal`. Choose the file `scddvbrk.mat`. This signal has a change in the second derivative. Select the `db 1` wavelet and `Level 2`. Press `Analyze` for the DWT. Display the `Full Decomposition`. Can the discontinuity be detected?

b. Select the `db 2` wavelet and `Level 2`. Press `Analyze` and use `Full Decomposition`. Can the discontinuity (in second derivative) be detected? Why?

References

Books on Wavelets and Filter Banks

[AH] A. N. Akansu and R. A. Haddad, *Multiresolution Signal Decomposition*, Academic Press (1992).

[AS] A. N. Akansu and M. J. T. Smith, eds., *Subband and Wavelet Transforms*, Kluwer (1995).

[AO] A. Antoniadis and G. Oppenheim, eds., *Wavelets and Statistics*, Springer (1995).

[BF] J. Benedetto and M. Frazier, eds., *Wavelets: Mathematics and Applications*, CRC Press (1993).

[Ba] M. Barlaud, ed., *Wavelets in Image Communication*, Elsevier (1994).

[By] J. Byrnes, ed., *Proceedings 1992 NATO ASI on Wavelets*, Kluwer (1994).

[CDM] A. S. Cavaretta, W. Dahmen, and C. A. Micchelli, *Stationary Subdivision*, Amer. Math. Soc. Memoirs 453 (1991).

[C1] C. K. Chui, *An Introduction to Wavelets*, Academic Press (1992).

[C2] C. K. Chui, ed., *Wavelets: A Tutorial in Theory and Applications*, Academic Press (1992).

[CR] A. Cohen and R. D. Ryan, *Wavelets and Multiscale Signal Processing*, Chapman and Hall (1995).

[CGT] J. M. Combes, A. Grossmann, and Ph. Tchamitchian, eds., *Wavelets, Time-Frequency Methods, and Phase Space*, Springer (1990).

[CRa] R. E. Crochiere and L. R. Rabiner, *Multirate Digital Signal Processing*, Prentice-Hall (1983).

[D] I. Daubechies, *Ten Lectures on Wavelets*, SIAM (1992).

[F] N. J. Fliege, *Multirate Digital Signal Processing*, John Wiley (1994).

[G] R. M. Gray, *Source Coding Theory*, Kluwer (1990).

[GG] A. Gersho and R. M. Gray, *Vector Quantization and Signal Compression*, Kluwer (1992).

[H] M. Holschneider, *Wavelets: An Analysis Tool*, Oxford (1995).

[IEEE1] Special Issue on Wavelets, *IEEE Transactions on Signal Processing* **41** (12/93).

[IEEE2] Special Issue on Wavelet Transforms, *IEEE Transactions on Information Theory* **38** (3/92).

[IEEE3] W. K. Chen, ed., *The Circuits and Filters Handbook*, chapters by T. Nguyen, I. Djokovic, and P.P. Vaidyanathan, IEEE Press (1995).

[MSSP] Special Issue on Wavelets and Multiresolution Signal Processing, *Multidimensional Systems and Signal Processing* (1996).

[JN] N. J. Jayant and P. Noll, *Digital Coding of Waveforms*, Prentice-Hall (1984).

[K] G. Kaiser, *A Friendly Guide to Wavelets*, Birkhäuser-Boston (1994).

[LMR] A. Louis, P. Maas, and A. Rieder, *Wavelets: Theorie und Anwendungen*, Teubner (1994).

[Mt] S. Mallat, *Wavelet Signal Processing*, Academic Press (1996).

[Mr] H. S. Malvar, *Signal Processing with Lapped Transforms*, Artech House (1992).

[M1] Y. Meyer, *Wavelets and Operators*, Translation of *Ondelettes et Opérateurs* (Hermann, 1990), Cambridge University Press (1993).

[M2] Y. Meyer, *Wavelets: Algorithms and Applications*, SIAM (1993).

[M3] Y. Meyer, ed., *Wavelets and Applications: Proceedings of the Marseille Conference on Wavelets*, Springer and Masson (1993).

[N] D. E. Newland, *Random Vibrations, Spectral and Wavelet Analysis*, Longmans and John Wiley (1993).

[OS] A. V. Oppenheim and R. W. Schafer, *Discrete-Time Signal Processing*, Prentice-Hall (1989).

[PM] W. B. Pennebaker and J. L. Mitchell, *JPEG: Still Image Compression Standard*, Van Nostrand Reinhold (1993).

[RBCDM] M. B. Ruskai et al., eds., *Wavelets and Their Applications*, Jones and Bartlett (1992).

[SW] L. L. Schumaker and G. Webb, eds., *Recent Advances in Wavelet Analysis*, Academic Press (1994).

[SPIE1] A. Laine, ed., *Mathematical Imaging: Wavelet Applications*, vol. **2334**, SPIE (1993).

[SPIE2] A. Laine and M. Unser, eds., *Wavelet Applications II*, vol. **2303**, SPIE (1994).

[SF] G. Strang and G. Fix, *An Analysis of the Finite Element Method*, Wellesley-Cambridge Press (1973).

[S] G. Strang, *Introduction to Applied Mathematics*, Wellesley-Cambridge Press (1986); *Linear Algebra and Its Applications*, Saunders (1988); *Introduction to Linear Algebra*, Wellesley-Cambridge Press (1993).

[SN] G. Strang and T. Nguyen, *Wavelets and Filter Banks*, Wellesley-Cambridge Press (1996).

[V] P. P. Vaidyanathan, *Multirate Systems and Filter Banks*, Prentice-Hall (1992).

[VK] M. Vetterli and J. Kovacevic, *Wavelets and Subband Coding*, Prentice-Hall (1995).

[Wa] G. G. Walter, *Wavelets and Other Orthogonal Systems with Applications*, CRC Press (1994).

[Wi] M. V. Wickerhauser, *Adapted Wavelet Analysis from Theory to Software*, AK Peters (1994).

[Wo] J. W. Woods, ed., *Subband Image Coding*, Kluwer (1991).

[Wl] G. W. Wornell, *Signal Processing with Fractals: A Wavelet-Based Approach*, Prentice-Hall (1995).

The next pages contain articles that we have found useful — many are mentioned in the text. The reader will understand that this list is very far from complete. With such rapid growth, the only hope is to establish an electronic bibliography (including ftp and web sites). Please look on the homepage

`http://saigon.ece.wisc.edu/~waveweb/QMF.html`

for an announcement. We refer also to the homepage

`http://www.scarolina.edu/~wavelet/`

of the *Wavelet Digest*. The *Digest* serves more than 5000 readers from

`wavelet@math.scarolina.edu`

(subject *subscribe*). Ultimately there could be a catalogue of available software for wavelet calculations.

Articles on Wavelets and Filter Banks

[BaEdNu] R. H. Bamberger, S. L. Eddins, and V. Nuri, Generalized symmetrical extensions for size-limited multirate filter banks, *IEEE Trans. Image Proc.* **3** (1994) 82–87.

[Bat] G. Battle, A block spin construction of ondelettes. Part I: Lemarié functions, *Commun. Math. Phys.* **110** (1987) 601–615; Part II: the QFT connection, *Commun. Math. Phys.* **114** (1988) 93–102.

[BeHeWa] J. Benedetto, C. Heil, and D. Walnut, Differentiation and the Balian-Low Theorem, *J. Fourier Anal. Appl.* **4** (1995).

[BeCoRo] G. Beylkin, R. Coifman, and V. Rokhlin, Fast wavelet transforms and numerical algorithms, *Comm. Pure Appl. Math.* **44** (1991) 141–183.

[BrJo] K. Brandenburg and J. Johnston, Second generation perceptual audio coding: the hybrid coder, *88th Audio Eng. Soc. Conv.* (1990).

[Brislawn] C. Brislawn, Classification of nonexpansive symmetric extension transforms for multirate filter banks, *Los Alamos Report LA-UR-94-1747* (1994); Preservation of subband symmetry in multirate signal processing, *IEEE Trans. SP* (1996).

[Brower] B. W. Brower, Low-bit-rate image compression evaluations, *Proc. SPIE* (1994).

[BuAd] P. Burt and E. Adelson, The Laplacian pyramid as a compact image code, *IEEE Trans. Comm.* **31** (1983) 482–540.

[Canny] J. Canny, A computational approach to edge detection, *IEEE Trans. Pattern Anal. Machine Intell.* **6** (1986).

[CaDaPe] J. M. Carnicer, W. Dahmen, and J. M. Peña, Local decompositions of refinable spaces, *RWTH Aachen* (1994).

[ChSt] R. Chan and G. Strang, Toeplitz equations by conjugate gradients with circulant preconditioner, *SIAM J. Sci. Comp.* **10** (1989) 104–119.

[ChenDon] S. Chen and D. Donoho, Atomic decomposition by basis pursuit, *Stanford preprint* (1995).

[ChuiLi] C. K. Chui and C. Li, A general framework of multivariate wavelets with duals, *Appl. Comp. Harm. Anal.* **1** (1994) 368–390.

[ChuiWang1] C. K. Chui and J. Z. Wang, A cardinal spline approach to wavelets, *Proc. AMS* **113** (1991) 785–793.

[ChuiWang2] C. K. Chui and J. Z. Wang, On compactly supported spline wavelets and a duality principle, *Trans. Amer. Math. Soc.* **330** (1992) 903–915.

[Cohen] A. Cohen, Ondelettes, analyses multiresolutions et filtres miroirs en quadrature, *Ann. Inst. H. Poincaré* **7** (1990) 439–459.

[CoDaDe] A. Cohen, W. Dahmen, and R. DeVore, Multiscale decompositions on bounded domains, preprint (1995).

[CoDa] A. Cohen and I. Daubechies, A stability criterion for biorthogonal wavelet bases and their related subband coding schemes, *Duke Math. J.* **68** (1992) 313–335.

[CoDaFe] A. Cohen, I. Daubechies, and J.C. Feauveau, Biorthogonal bases of compactly supported wavelets, *Comm. Pure Appl. Math.* **45** (1992) 485–560.

[CoDaPl] A. Cohen, I. Daubechies, and G. Plonka, Regularity of refinable function vectors, preprint (1995).

[CoDaVi] A. Cohen, I. Daubechies, and P. Vial, Wavelets and fast wavelet transforms on an interval, *Appl. Comp. Harm. Anal.* **1** (1993) 54–81.

[Coh] L. Cohen, Time-frequency distributions: A review, *Proc. IEEE* **77** (1989) 941–981.

[CoMey] R. Coifman and Y. Meyer, Remarques sur l'analyse de Fourier à fenêtre, *C. R. Acad. Sci. Paris* **312** (1991) 259–261.

[ConRa] J. P. Conze and A. Raugi, Fonctions harmoniques pour un opérateur de transition et applications, *Bull. Soc. Math. Fr.* **118** (1990) 273–310.

[CoWi] R. Coifman and V. Wickerhauser, Entropy-based algorithms for best-basis selection, *IEEE Trans. Inf. Th.* **38** (1992) 713–718.

[CoHe] D. Colella and C. Heil, Characterizations of scaling functions. Continuous solutions, *SIAM J. Math. Anal.* **15** (1994) 496–518.

[DaMi] W. Dahmen and C. Micchelli, Using the refinement equation for evaluating the integrals of wavelets, *SIAM J. Num. Anal.* **30** (1993) 507–537.

[Dau1] I. Daubechies, Time-frequency localization operators: a geometric phase space approach, *IEEE Trans. Inf. Th.* **34** (1988) 605–612.

[Dau2] I. Daubechies, Orthonormal bases of compactly supported wavelets, *Comm. Pure Appl. Math.* **41** (1988) 909–996.

[Dau3] I. Daubechies, The wavelet transform, time-frequency localization and signal analysis, *IEEE Trans. Inf. Theory* **36** (1990) 961–1005.

[Dau4] I. Daubechies, Orthonormal bases of compactly supported wavelets II: Variations on a theme, *SIAM J. Math. Anal.* **24** (1993) 499–519.

[DauHu] I. Daubechies and Y. Huang, How does truncation of the mask affect a refinable function?, *Constr. Approx.* **11** (1995) 365–380.

[DauLag] I. Daubechies and J. Lagarias, Two-scale difference equations I. Existence and global regularity of solutions, *SIAM J. Math. Anal.* **22** (1991) 1388–1410; II. Local regularity, infinite products of matrices and fractals, *SIAM J. Math. Anal.* **24** (1992) 1031–1079.

[DaMaZh] G. Davis, S. Mallat, and Z. Zhang, Adaptive time-frequency approximations with matching pursuits, *NYU preprint* (1995).

[DeLeUr] Y. Dehery, M. Lever, and P. Urcun, A MUSICAM source coder for digital audio broadcasting and storage, *Proc. IEEE ICASSP* (1991).

[DelJud] B. Delyon and A. Juditsky, On the computation of wavelet coefficients, *IRISA Rennes preprint* (1994); published in [AO].

[DesDu1] G. Deslauriers and S. Dubuc, Interpolation dyadique, *Fractals, dimensions non entières et applications*, Masson (1987) 44–55.

[DesDu2] G. Deslauriers and S. Dubuc, Symmetric iterative interpolation processes, *Constr. Approx.* **5** (1989) 49–68.

[DeJaPo] R. DeVore, B. Jawerth, and V. Popov, Compression of wavelet decompositions, *Amer. J. Math.* **114** (1992) 737–785.

[DeJaLu] R. DeVore, B. Jawerth, and B. Lucier, Image compression through transform coding, *IEEE Trans. Inf. Th.* **38** (1992) 719–746.

[DeLu] R. DeVore and B. Lucier, Wavelets, *Acta Numerica* **1** (1991) 1–56.

[DjNgTo] A. Djohan, T. Q. Nguyen, and W. Tompkins, ECG compression using discrete symmetric wavelet transform, *International Conf. EMBS* (1995).

[DjVaid] I. Djokovic and P. P. Vaidyanathan, On optimal analysis/synthesis filters for coding gain maximization, *IEEE Trans. SP*, to appear.

[Don1] D. Donoho, De-noising by soft thresholding, *IEEE Trans. Inf. Th.* **41** (1995) 613–627.

[Don2] D. L. Donoho, Interpolating wavelet transforms, *Stanford preprint* (1992).

[Don3] D. L. Donoho, Unconditional bases are optimal bases for data compression and for statistical estimation, *Appl. Comp. Harmonic Analysis* **1** (1993) 100–115.

[Donoho] D. Donoho and I. Johnstone, Empirical atomic decomposition, *Stanford preprint* (1995).

[DoJoHS] D. Donoho, I. Johnstone, J. Hoch, and A. Stern, Maximum entropy and the nearly black object, *J. Roy. Stat. Soc.* **54** (1992) 41–81.

[DoJoKP] D. Donoho, I. Johnstone, G. Kerkyacharian, and D. Picard, Wavelet shrinkage: Asymptopia?, *J. Roy. Stat. Soc.* **57** (1995) 301–369.

[Eirola] T. Eirola, Sobolev characterization of solutions of dilation equations, *SIAM J. Math. Anal.* **23** (1992) 1015–1030.

[EkPr] M. Ekanayake and K. Premaratne, Two-channel IIR QMF banks with approximately linear-phase analysis and synthesis filters, *Proc. 28th Asilomar Conf.* (1994).

[EstGal] D. Estaban and C. Galand, Application of quadrature mirror filters to split-band voice coding schemes, *Proc. IEEE Int. Conf. ASSP* (1977) 191–195.

[FeiGr] H. Feichtinger and K. Gröchenig, Theory and practice of irregular sampling, in [BF].

[FlaAbr] P. Flandrin and P. Abry, On the initialization of the discrete wavelet transform, *IEEE Sig. Proc. Letters* **1** (1994) 32–34.

[Gabor] D. Gabor, Theory of communication, *Journ. IEE* **93** (1946) 429–457.

[GHM1] J. Geronimo, D. Hardin, and P. R. Massopust, Fractal functions and wavelet expansions based on several functions, *J. Approx. Theory* **78** (1994) 373–401.

[GHM2] J. Geronimo, D. Hardin, P. R. Massopust, and G. Donovan, Construction of orthogonal wavelets using fractal interpolation functions, *SIAM J. Math. Anal.* (1996).

[GilVet] A. Gilloire and M. Vetterli, Adaptive filtering in subbands with critical sampling, *IEEE Trans. SP* **40** (1992) 1862–1874.

[GoBu1] R. Gopinath and S. Burrus, On upsampling, downsampling, and rational sampling rate filter banks, *IEEE Trans. SP* **42** (1994) 812–824.

[GoBu2] R. Gopinath and S. Burrus, On cosine-modulated wavelet orthonormal bases, *IEEE Trans. Image Proc.* **2** (1995) 162–176.

[Gro1] K. Gröchenig, Acceleration of the frame algorithm, *IEEE Trans. SP* **41** (1993) 3331–3340.

[Gro2] K. Gröchenig, Analyse multi-echelle et bases d'ondelettes, *C. R. Acad. Sci.* **305** (1987) 13–17.

[GrMor] A. Grossmann and J. Morlet, Decomposition of Hardy functions into square integrable wavelets of constant shape, *SIAM J. Math. Anal.* **15** (1984) 723–736.

[Haar] A. Haar, Zur Theorie der orthogonalen Funktionen-Systeme, *Math. Ann.* **69** (1910) 331–371.

[HaNgCh] H. Hajj, T. Nguyen, and R. Chin, On multiscale feature detection using filter banks, *Proc. Asilomar Conf.* (1995).

[HeStSt] C. Heil, G. Strang, and V. Strela, Approximation by translates of refinable functions, *Numerische Mathematik* (1996).

[HeWa] C. Heil and D. Walnut, Continuous and discrete wavelet transforms, *SIAM Rev.* **31** (1989) 628–666.

[Heller1] P. Heller, The construction of rank M wavelet matrices with N vanishing moments, *SIAM J. Matrix Anal.* **16** (1995).

[Heller2] P. Heller, V. Strela, et al., Multiwavelet filter banks for data compression, *Proc. IEEE ISCAS* (1995); see also [SHSTH].

[Her] C. Herley, Exact interpolation and iterative subdivision schemes, *IEEE Trans. SP* **43** (1995) 1348–1359.

[HeKRV] C. Herley, J. Kovacevic, K. Ramchandran, and M. Vetterli, Tilings of the time-frequency plane, *IEEE Trans. SP* **41** (1993) 3341–3359.

[HerVet] C. Herley and M. Vetterli, Wavelets and recursive filter banks, *IEEE Trans. SP* **41** (1993) 2536–2556.

[Herr] O. Herrmann, On the approximation problem in nonrecursive digital filter design, *IEEE Trans. Circ. Theory* **18** (1971) 411–413.

[HoVai] P. H. Hoang and P. P. Vaidyanathan, Nonuniform multirate filter banks: theory and design, *Proc. ISCAS* (1989) 371–374.

[Jaf] S. Jaffard, Exposants de Hölder et coefficients d'ondelettes, *C. R. Acad. Sci.* **308** (1989) 79–81.

[Jan] A. J. E. M. Janssen, The Zak transform and sampling theorems for wavelet subspaces, *Philips Res. Lab. preprint.*

[JaLiuSw] B. Jawerth, J. Liu, and W. Sweldens, Signal compression with smooth local trigonometric bases, *J. Optical Eng.* **33** (1994) 2125–2135.

[JaSw] B. Jawerth and W. Sweldens, An overview of wavelet based multiresolution analyses, *SIAM Rev.* **36** (1994) 377–412.

[JiaWang] R. Q. Jia and J. Z. Wang, Stability and linear independence associated with wavelet decompositions, *Proc. Amer. Math. Soc.* **117** (1993) 1115–1124.

[JinWoLu] Q. Jin, K. M. Wong, and Z. Q. Luo, Design of an optimum wavelet for cancellation of long echoes in telephony, *Proc. IEEE SP Int. Symp. on Time-Freq.* (1994) 488–491.

[JPEG] G. Wallace, The JPEG still picture compression standard, *IEEE TCE* **38** (1992).

[KatYas] J. Katto and Y. Yasuda, Performance evaluation of subband coding and optimization of its filter coefficients, *SPIE Proc. Visual Comm. and Image Proc.* (1991) 95–106.

[Kei] F. Keinert, Biorthogonal wavelets for fast matrix computations, *Appl. Comp. Harm. Anal.* **1** (1994) 147–156.

[KimAn] C. Kim and R. Ansari, FIR/IIR exact reconstruction filter banks with applications to subband coding of images, *Proc. Midwest Circuits and Syst. Symp.* (1991).

[KiYaIw] H. Kiya, M. Yae, and M. Iwahashi, A linear-phase two-channel filter bank allowing perfect reconstruction, *Proc. IEEE ISCAS* (1992) 951–954.

[KoNgVa] R. D. Koilpillai, T. Q. Nguyen, and P. P. Vaidyanathan, New results in the theory of crosstalk-free transmultiplexer, *IEEE Trans. SP* **39** (1991) 2174–2183.

[KoVa1] R. D. Koilpillai and P. P. Vaidyanathan, Cosine-modulated FIR filter banks satisfying perfect reconstruction, *IEEE Trans. SP* **40** (1992).

[KoVa2] R. D. Koilpillai and P. P. Vaidyanathan, A spectral factorization approach to pseudo-QMF design, *IEEE Trans. SP* **41** (1993) 82–92.

[KovVet] J. Kovacevic and M. Vetterli, Nonseparable multidimensional PR filter banks and wavelets for R^n, *IEEE Trans. Inf. Th.* **38** (1992) 533–555.

[Kunoth] A. Kunoth, Computing refinable integrals, *Texas A&M preprint* (1995).

[KwTa] M. Kwong and P. Tang, W-matrices, nonorthogonal multiresolution analysis, and finite signals of arbitrary length, *Argonne preprint MCS-P449-0794* (1994).

[LaFr] M. Lang and B.-C. Frenzel, Polynomial root finding, *IEEE SP Letters* (1994); *Rice Univ. ECE Report 9308.*

[Law] W. Lawton, Necessary and sufficient conditions for constructing orthonormal wavelet bases, *J. Math. Phys.* **32** (1991) 57–61.

[LaLeSh] W. Lawton, S. L. Lee, and Z. Shen, Convergence of multidimensional cascade algorithm, *Univ. Singapore preprint* (1995).

[LeKa] P.G. Lemarié and D. Kateb, The phase of the Daubechies filters, *Orsay preprint* (1994), *Rev. Mat. Iberoamer.* (1996); Asymptotic behavior of the Daubechies filters, *Appl. Comp. Harm. Anal.* **4** (1995) 398–399.

[Lem] P. G. Lemarié, Fonctions d'échelle pour les ondelettes de dimension n, *C. R. Acad. Sci.* **316** (1993) 145–148.

[LiMaMi] M. Lightstone, E. Majani, and S. Mitra, Low bit-rate considerations for wavelet-based image coding, *Multidim. Sys. and Sig. Proc.*, to appear.

[LinPhVa] Y. P. Lin, S-M. Phoong, and P. P. Vaidyanathan, New results on the multidimensional remainder theorem, *IEEE Sig. Proc. Letters* (1994) 176–178.

[LinVa1] Y. P. Lin and P. P. Vaidyanathan, Linear phase cosine modulated maximally decimated filter banks, *IEEE Trans. SP* (1996).

[LinVa2] Y. P. Lin and P. P. Vaidyanathan, Theory and design of two-dimensional cosine modulated filter banks, *IEEE Trans. SP* (1996).

[LinVa3] Y. P. Lin and P. P. Vaidyanathan, A review of two-dimensional nonseparable filter bank design techniques, *Multidim. Sys. and Sig. Proc.* (1995).

[LiMa] J.-M. Lina and M. Mayrand, Complex Daubechies wavelets, *Appl. Comp. Harm. Anal.* **2** (1995) 219–229.

[LoDeWa] M. Lounsbery, T. DeRose, and J. Warren, Multiresolution surfaces of arbitrary topological type, *Comp. Sci. preprint, Univ. of Wash.* (1993).

[Maj1] E. Majani, Low-complexity wavelet filter design for image compression, *NASA report 42–119* (1994) 181–200.

[Maj2] E. Majani, Biorthogonal wavelets for image compression, *Proc. SPIE, VCIP'94* (1994).

[Mt] S. Mallat, Multifrequency channel decomposition of images and wavelet models, *IEEE Trans. ASSP* **37** (1989) 2091–2110.

[Mal2] S. Mallat, Multiresolution approximations and wavelet orthonormal bases of $L^2(\mathbf{R})$, *Trans. Amer. Math. Soc.* **315** (1989) 69–87.

[Mal3] S. Mallat, A theory of multiresolution signal decomposition: the wavelet representation, *IEEE Trans. PAMI* **11** (1989) 674–693.

[MalHw] S. Mallat and W. Hwang, Singularity detection and processing with wavelets, *IEEE Trans. Inf. Th.* **38** (1992) 617–643.

[Matv] G. Matviyenko, Optimized local trigonometric bases, *Yale Comp. Sci. Report 1041* (1994).

[MiNa] G. A. Mian and A. P. Nainer, A fast procedure to design equiripple minimum-phase FIR filters, *IEEE Trans. Circuits Syst.* **39** (1992).

[MPEG2] Recommendation H.262, ISO/IEC 13818, Generic coding of moving picture and associated audio, *Draft international standard.*

[NaSmBa1] K. Nayebi, T. Barnwell, and M. Smith, Time-domain filter bank analysis: A new design theory, *IEEE Trans. SP* **40** (1992).

[NaBaSm2] K. Nayebi, T. Barnwell, and M. Smith, On the design of FIR analysis-synthesis banks with high computational efficiency, *IEEE Trans. SP* **42** (1994) 825–834.

[New] D. Newland, Harmonic wavelet analysis, *Proc. R. Soc. London* **443** (1993) 203–225.

[QCLS] T. Q. Nguyen, Digital filter banks design — quadratic-constrained formulation, *IEEE Trans. SP* **43** (1995) 2103–2108.

[Ng1] T. Q. Nguyen, Near-perfect-reconstruction pseudo-QMF banks, *IEEE Trans. SP* **42** (1994) 65–76.

[Ng2] T. Q. Nguyen, A class of generalized cosine-modulated filter banks, *Proc. ISCAS* (1992) 1344–1347.

[Ng3] T. Q. Nguyen, A novel spectral factorization method and its application in the design of filter banks and wavelets, *Proc. IEEE-SP Int. Symp. on Time-Frequency* (1992) 303–306.

[NgKo] T. Q. Nguyen and R. D. Koilpillai, The design of arbitrary-length cosine-modulated filter banks and wavelets, *Proc. IEEE-SP Int. Symp. on Time-Frequency* (1992) 299–302.

[NgLaKo] T. Q. Nguyen, T. I. Laakso, and R. D. Koilpillai, Eigenfilter approach for the design of allpass filters approximating a given phase response, *IEEE Trans. SP* **42** (1994).

[NgSoVa] T. Q. Nguyen, A. K. Soman, and P. P. Vaidyanathan, A quadratic-constrained least-squares approach to linear-phase orthonormal filter bank design, *Proc. ISCAS* (1993) 383–386.

[NgVa1] T. Q. Nguyen and P. P. Vaidyanathan, Two channel perfect reconstruction FIR QMF structures which yield linear phase analysis and synthesis filters, *IEEE Trans. ASSP* **37** (1989) 676–690.

[NgVa2] T. Q. Nguyen and P. P. Vaidyanathan, Structures for M-channel perfect-reconstruction FIR QMF banks which yield linear-phase analysis filters, *IEEE Trans. ASSP* **38** (1990) 433–446.

[NgVa3] T. Q. Nguyen and P. P. Vaidyanathan, Maximally decimated perfect reconstruction FIR filter banks with pairwise mirror-image responses, *IEEE Trans. ASSP* **36** (1988) 693–706.

[PhKiVA] S.-M. Phoong, C. W. Kim, P. P. Vaidyanathan, and R. Ansari, A new class of two-channel biorthogonal filter banks and wavelet bases, *IEEE Trans. SP* **43** (1995) 649–665.

[PhVa] S.-M. Phoong and P. P. Vaidyanathan, Time-varying filters and filter banks: Some basic principles; Factorizability of lossless time-varying filters and filter banks; *IEEE Trans. SP*, to appear.

[PrBr] J. P. Princen and A. B. Bradley, Analysis/synthesis filter bank design based on time domain aliasing cancellation, *IEEE Trans. ASSP* **34** (1986) 1153–1161.

[deQNgRa] R. deQueiroz, T. Nguyen, and K. Rao, Generalized lapped orthogonal transform, *Proc. ISCAS 94* (1994).

[RamVet] K. Ramchandran and M. Vetterli, Best wavelet packet bases in a rate-distortion sense, *IEEE Trans. Image Proc.* **2** (1994) 160–175.

[Rioul] O. Rioul, Simple regularity criteria for subdivision schemes, *SIAM J. Math. Anal.* **23** (1992) 1544–1576.

[RiDu] O. Rioul and P. Duhamel, Fast algorithms for discrete and continuous wavelet transforms, *IEEE Trans. Inf. Th.* **38** (1992) 569–586.

[RiVet] O. Rioul and M. Vetterli, Wavelets and signal processing, *IEEE Signal Proc. Mag.* (1991) 14–38.

[RiPrNg] A. D. Rizos, J. G. Proakis, and T. Q. Nguyen, Comparison of DFT and cosine-modulated filter banks in multicarrier modulation, *Globecom 94* (1994).

[Rot] J. H. Rothweiler, Polyphase quadrature filters — a new subband coding technique, *IEEE Int. Conf. ASSP* (1983) 1280–1283.

[Roy] S. Roy, Design of linear phase FIR filters using the LMS algorithm, *IEEE Trans. SP* **41** (1993) 1685–1689.

[SaiBey] N. Saito and G. Beylkin, Multiresolution representations using the auto-correlation functions of compactly supported wavelets, *IEEE Trans. SP* **41** (1993) 3584–3590.

[SaTz] S. Sandberg and M. Tzannes, Overlapped discrete multitone modulation for high speed copper wire communications, *IEEE JSAC* (1995).

[Schitt] K. Schittkowski, NLPQL: A FORTRAN subroutine solving constrained nonlinear programming problems, *Annals Op. Res.* **5** (1986).

[SchSw] P. Schröder and W. Sweldens, Spherical wavelets: Efficiently representing functions on the sphere, *Comp. Graphics, SIGGRAPH '95* (1995).

[Shapiro] J. M. Shapiro, Embedded image coding using zerotrees of wavelet coefficients, *IEEE Trans. SP* **41** (1993) 3445–3462.

[ShenSt] J. Shen and G. Strang, The zeros of the Daubechies polynomials, *Proc. Amer. Math. Soc.* (1996).

[Shensa] M. Shensa, Wedding the à trous and Mallat algorithms, *IEEE Trans. SP* **40** (1992) 2464–2482.

[SiFrAH] E. Simoncelli, W. Freeman, E. Adelson, and D. Heeger, Shiftable multiscale transforms, *IEEE Trans. Inf. Th.* **38** (1992) 587–607.

[SmBa] M. J. T. Smith and T. P. Barnwell, Exact reconstruction techniques for tree-structured subband coders, *IEEE Trans. ASSP* **34** (1986) 434–441.

[SmEd] M. J. T. Smith and S. L. Eddins, Analysis-synthesis techniques for subband image coding, *IEEE Trans. ASSP* **38** (1990) 1446–1456.

[SoVa] A. K. Soman and P. P. Vaidyanathan, A complete factorization of paraunitary matrices with pairwise mirror-image symmetry, *IEEE Trans. SP* **43** (1995).

[SoVaNg] A. K. Soman, P. P. Vaidyanathan, and T. Q. Nguyen, Linear-phase paraunitary filter banks: Theory, factorizations and applications, *IEEE Trans. SP* **41** (1993).

[Stau] J. Stautner, High quality audio compression for broadcast and computer applications, *SMPTE Adv. TV Conf.* (1992) (*Aware Report AD920207*).

[StHeGB] P. Steffen, P. N. Heller, R. A. Gopinath, and C. S. Burrus, Theory of m-band wavelet bases, *IEEE Trans. SP* **41** (1993) 3497–3511.

[St1] G. Strang, Wavelets and dilation equations: A brief introduction, *SIAM Review* **31** (1989) 614–627.

[St2] G. Strang, Wavelet transforms vs. Fourier transforms, *Bull. Amer. Math. Soc.* **28** (1993) 288–305.

[St3] G. Strang, Eigenvalues of $(\downarrow 2)H$ and convergence of the cascade algorithm, *IEEE Trans. SP* **44** (1996).

[StFi] G. Strang and G. Fix, A Fourier analysis of the finite element variational method, in *Constructive Aspects of Functional Analysis* (1971) 796–830.

[StSt1] G. Strang and V. Strela, Short wavelets and matrix dilation equations, *IEEE Trans. SP* **43** (1995) 108–115.

[StSt2] G. Strang and V. Strela, Orthogonal multiwavelets with vanishing moments, *J. Optical Eng.* **33** (1994) 2104–2107.

[SHSTH] V. Strela, P. Heller, et al., The application of multiwavelet filter banks to image processing, *IEEE Trans. Image. Proc.* (1996).

[Stri] R. Strichartz, How to make wavelets, *Amer. Math. Monthly* **100** (1993) 539–556.

[Swel] W. Sweldens, The lifting scheme: A construction of second generation wavelets, *Univ. So. Carolina preprint* (1994).

[SwelPi] W. Sweldens and R. Piessens, Quadrature formulae and asymptotic error expansions for wavelet approximations of smooth functions I, II, *SIAM J. Numer. Anal.* **31** (1994) 1240–1264, *Numer. Math.* **68** (1994) 377–401.

[TayKing] D. Tay and N. Kingsbury, Flexible design of multi-dimensional filters using transformations of variables, *IEEE Trans. Image Proc.* **2** (1993) 466–480.

[Tch] P. Tchamitchian, Biorthogonalité et théorie des opérateurs, *Rev. Mat. Iberoamer.* **3** (1987) 163–189.

[TranNg] T. Tran and T. Q. Nguyen, On arbitrary-length M-channel linear phase FIR filter banks, *Asilomar Conf.* (1995).

[TrauNg] S. Trautmann and T. Nguyen, Comparison of linear-phase perfect-reconstruction filter banks in wavelet-transform-based image compression, *Proc. Conf. Inf. Sci. and Sys.* (1995).

[TzTzPH] M. A. Tzannes, M. C. Tzannes, J. Proakis, and P. N. Heller, DMT systems, DWMT systems, and digital filter banks, *Proc. IEEE ICC* (1994).

[TzTzRe] M. A. Tzannes, M. C. Tzannes, and H. L. Resnikoff, The DWMT: A multicarrier transceiver for ADSL using M-band wavelet transforms, *ANSI T1E1.4 93-067* (1993).

[Uns1] M. Unser, Efficient dyadic wavelet transformation of images using interpolating filters, *Proc. Int. Conf. ASSP* **5** (1993) 149–152.

[Uns2] M. Unser, Approximation power of biorthogonal wavelet expansions, *NIH preprint* (1995); *IEEE Trans. SP*, to appear.

[UnAlEd] M. Unser, A. Aldroubi, and M. Eden, A family of polynomial spline wavelet functions, *Signal Proc.* **30** (1993) 141–162.

[Va] P. P. Vaidyanathan, Theory and design of M-channel maximally decimated quadrature mirror filters, *IEEE Trans. ASSP* **35** (1987) 476–492.

[VaChen] P. P. Vaidyanathan and T. Chen, Role of anticausal inverses in multirate filter banks I, II, *IEEE Trans. SP* **43** (1995) 1090–1115.

[VaHo] P. P. Vaidyanathan and P. Q. Hoang, Lattice structures for optimal design of two-channel perfect-reconstruction QMF banks, *IEEE Trans. ASSP* **36** (1988).

[VaNgDS] P. P. Vaidyanathan, T. Q. Nguyen, Z. Doğanata, and T. Saramäki, Improved technique for design with lossless polyphase matrices, *IEEE Trans. ASSP* **37** (1989) 1042–1055.

[VaReMi] P. P. Vaidyanathan, P. Regalia, and S. K. Mitra, Design of doubly complementary IIR digital filters, *IEEE Trans. Circ. Syst.* **34** (1987) 378–389.

[Vet1] M. Vetterli, Multidimensional subband coding: some theory and algorithms, *Signal Proc.* **6** (1984) 97–112.

[Vet2] M. Vetterli, Filter banks allowing perfect reconstruction, *Signal Proc.* **10** (1986) 219–244.

[Vet3] M. Vetterli, Perfect transmultiplexers, *Proc. ICASSP* (1986) 2567–2570.

[VetHer] M. Vetterli and C. Herley, Wavelets and filter banks: theory and design, *IEEE Trans. ASSP* **40** (1992) 2207–2232.

[Villemoes] L. Villemoes, Energy moments in time and frequency for two-scale difference equation solutions and wavelets, *SIAM J. Math. Anal.* **23** (1992) 1519–1543; Wavelet analysis of refinement equations, *SIAM J. Math. Anal.* **25** (1994) 1433–1466.

[WiAm] J. Williams and K. Amaratunga, High order wavelet extrapolation schemes for initial value problems and boundary value problems, *MIT IESL Report 94-07* (1994); also *SPIE Proc.* **2491** (1995) 894–902.

[WoOn] J. W. Woods and S. D. O'Neil, Subband coding of images, *IEEE Trans. ASSP* **34** (1986) 1278–1288.

[XiaG] X.-G. Xia, J. Geronimo, D. Hardin, and B. Suter, Computations of multiwavelet transforms, *Proc. SPIE* **2569** (1995).

[XuSh] J.-C. Xu and W.-C. Shann, Galerkin-wavelet methods for two-point boundary value problems, *Numer. Math.* **63** (1992) 123–142.

[YamOhk] M. Yamada and K. Ohkitani, Orthonormal wavelet expansion and its application to turbulence, *Progr. Theor. Physics* **83** (1990) 819–823.

[Zafar] S. Zafar, Multiscale video representation using multiresolution motion compensation and wavelet decomposition, *Argonne Wavelets Workshop* (1994).

Additional References in the Second Edition

Two important new journals publish papers on wavelet theory:

Archive for Computational Harmonic Analysis

Journal of Fourier Analysis and Applications

We add next a few short notes on recent developments.

Since the first edition of this book (1996), the "lifting scheme" has been intensively developed. The new Appendix 4 explains this approach, with examples. Two software libraries are mentioned below. The smoothness of multidimensional filters (and multifilters) has been established in major papers by Cohen-Grochenig-Villemoes, Jia, Micchelli-Sauer, and Shen. They have largely achieved the goal of extending formula (7.58). Multiwavelets continue to be an active research area.

In compression, the Modulated Lapped Transform (Section 9.4) is very successful for audio. It is used in the Sony mini disc, MPEG, and Dolby AC-3. The multilevel wavelet transform will enter the new JPEG and MPEG standards for images and video. Appendix 5 proposes a combination of uniform M-band filter banks with wavelet iteration on the DC band that gives even better quality reconstructions.

Wavelet applications like *feature detection* using multiple scales are promising and already working; this uses an analysis filter bank but not synthesis (so perfect reconstruction is not required). The IEEE Journals and SPIE proceedings contain some truly important papers. It is just as impossible as ever for our bibliography to be complete, and we hope these recent books and papers will supplement this textbook.

[Hu] B.B. Hubbard, *The World According to Wavelets*, AK Peters (1996).

[C3] C.K. Chui, *Wavelets: A Mathematical Tool for Signal Analysis*, SIAM (1997).

[BGG] C.S. Burrus, R.A. Gopinath, and H. Guo, *Introduction to Wavelets and Wavelet Transforms*, Prentice-Hall (1998).

[T] C. Taswell, *Handbook of Wavelet Transform Algorithms*, Birkhäuser (1997).

[MS] J.-M. Morel and S. Solimini, *Variational Methods in Image Segmentation*, Birkhäuser (1995).

[PI] L. Prasad and S.S. Iyengar, *Wavelet Analysis with Applications to Image Processing*, CRC Press (1997).

[RAH] T. Ramstad, S.O. Aase, and J.H. Husoy, *Subband Compression of Images*, Elsevier (1995).

[Su] B. Suter, Multirate and Wavelet Signal Processing, Academic Press (1997).

[CDSY] R. Calderbank, I. Daubechies, W. Sweldens, and B.-L. Yeo, Wavelet transforms that map integers to integers, *preprint* (1996).

[DaSw] I. Daubechies and W. Sweldens, Factoring wavelet transforms into lifting steps, *J. Fourier Anal. Appl.*, to appear.

[HNC] H. Hajj, T. Nguyen, and R. Chin, Theory, design and analysis of filter banks for feature detection, submitted to *IEEE Trans. SP.*

[IHWL] N. Intrator, Q. Huynh, Y. Wong, and B.H.K. Lee, Wavelet feature extraction for discrimination tasks, *Proc. 1997 Canadian Workshop on Inf. Theory*, 83–86, Toronto.

[KaTu**1**] J. Kautsky and R. Turcajova, A matrix approach to discrete wavelets, in *Wavelets: Theory, Algorithms and Applications* 5, C.K. Chui et al., Academic Press (1994) 117–136.

[KaTu**2**] J. Kautsky and R. Turcajova, Discrete biorthogonal wavelet transforms as block circulant matrices, *Linear Algebra and Its Applications*, **223** (1995) 393–413.

[Mal] H. Malvar, Lapped biorthogonal transforms for transform coding with reduced blocking and ringing artifacts, *Proc. Int. Conf. on Acoustics, Speech, and Signal Processing*, Munich (1997) 2421–2424.

[Mar] S. Martucci, Symmetric convolution and the discrete sine and cosine transforms, *IEEE Trans. on Signal Proc.* **42** (1994) 1038–1051.

[SaPe] A. Said and W.A. Pearlman, A new fast and efficient image codec on set partitioning in hierarchical trees, *IEEE Trans. Circuits Syst. Video Tech.* **6** (1996) 243–250.

[St4] G. Strang, The search for a good basis, *Numerical Analysis 1997*, D. Griffiths, D. Higham & G. Watson, eds., Pitman Lecture Notes, Addison Wesley Longman.

[TrNg] T. Tran and T.Q. Nguyen, A progressive transmission image coder using linear phase uniform filter banks as block transforms, *preprint* (1997).

[TuKa] R. Turcajova and J. Kautsky, Shift products and factorizations of wavelet matrices, *Numerical Algorithms* **8** (1994) 27–45.

[XRO] Z. Xiong, K. Ramchandran and M.T. Orchard, Space frequency quantization for wavelet image coding, *IEEE Trans. Image Proc.* (1997) 677–693.

We mention two software libraries for the lifting scheme in Appendix 4:

LIFTPACK, *Wavelet Applications in Signal and Image Processing IV*, SPIE Proc. **2825**, 396–408.

WAILI — Wavelets with Integer Lifting, G. Uytterhoeven et al.,
`http://www.cs.kuleuven.ac.be/~wavelets`

Appendix 3

The Discrete Cosine Transform

Gilbert Strang

Submitted to *SIAM Review*

Abstract

Each Discrete Cosine Transform uses N real basis vectors whose components are cosines. In the DCT-4, for example, the components of v_k are $\cos(j+\frac{1}{2})(k+\frac{1}{2})\frac{\pi}{N}$. These basis vectors are orthogonal and the transform is extremely useful in image processing. It can be implemented by an FFT. But a direct proof of orthogonality, by calculating inner products, does not reveal how natural these cosine vectors are.

We suggest a proof that exhibits the DCT basis as the eigenvectors of symmetric "second difference" matrices. By varying the boundary conditions we get the established transforms DCT-1 through DCT-4. Other combinations lead to four additional cosine transforms that we call DCT-5 to DCT-8. The type of boundary condition (Dirichlet or Neumann, centered at a meshpoint or a midpoint) determines the applications that are appropriate for the transform. The centering also determines the period: $N-1$ or N in the established transforms, $N-\frac{1}{2}$ or $N+\frac{1}{2}$ in the second group. The key point is that all these "eigenvectors of cosines" come from simple and familiar matrices.

Just as the Fourier series is the starting point in transforming and analyzing functions, the basic step for vectors is the Discrete Fourier Transform (DFT). It maps the "time domain" to the "frequency domain". A vector with N components is written as a combination of N special basis vectors v_k. Those are constructed from powers of the complex number $w = e^{2\pi i/N}$:

$$v_k = \left(1, w^k, w^{2k}, \ldots, w^{(N-1)k}\right), \quad k = 0, 1, \ldots, N-1.$$

The vectors v_k are the columns of the Fourier matrix $F = F_N$. *They are orthogonal.* So the inverse of F is its conjugate transpose, divided by $\| v_k \|^2 = N$.

Two points to mention, about orthogonality and speed, before we come to the purpose of this note. For these DFT basis vectors, a direct proof of orthogonality is very efficient:

$$(v_k, v_\ell) = \sum_{j=0}^{N-1} (w^k)^j (\overline{w}^\ell)^j = \frac{(w^k \overline{w}^\ell)^N - 1}{w^k \overline{w}^\ell - 1}.$$

The numerator is zero because $w^N = 1$. The denominator is nonzero because $k \neq \ell$. This proof of $(v_k, v_\ell) = 0$ is short but not very revealing. I want to recommend a different proof, which recognizes the v_k as *eigenvectors*. We could work with any circulant matrix, and we will choose below a symmetric A_0. Then linear algebra guarantees that its eigenvectors v_k are orthogonal.

Actually this second proof, verifying that $A_0 v_k = \lambda_k v_k$, brings out a central point of Fourier analysis. The Fourier basis diagonalizes every periodic constant coefficient operator. Each frequency k (or $2\pi k/N$) has its own frequency response λ_k. The complex exponential vectors v_k are important in applied mathematics because they are eigenvectors!

The second key point is speed of calculation. The matrices F and F^{-1} are full, which normally means N^2 multiplications for the transform and the inverse transform: $y = Fx$ and $x = F^{-1}y$. But the special form $F_{jk} = w^{jk}$ of the Fourier matrix allows a factorization into very sparse and simple matrices. This is the Fast Fourier Transform (FFT). It is easiest when N is a power 2^L. The operation count drops from N^2 to $\frac{1}{2}NL$, which is an enormous saving. But the matrix entries (powers of w) are complex.

The purpose of this note is to consider *real transforms that involve cosines.* Each matrix of cosines yields a Discrete Cosine Transform (DCT). There are four established types, DCT-1 through DCT-4, which differ in the boundary conditions at the ends of the interval. (This difference is crucial. The DCT-2 and DCT-4 are constantly applied in image processing; they have an FFT implementation and they are truly useful.) All four types of DCT are orthogonal transforms. The usual proof is a direct calculation of inner products of the N basis vectors, using trigonometric identities.

We want to prove this orthogonality in the second (indirect) *way*. The basis vectors of cosines are actually eigenvectors of symmetric second-difference matrices. This proof seems more attractive, and ultimately more useful. It also leads us, by selecting different boundary conditions, to new cosine transforms. Four additional types (DCT-5 to DCT-8) are introduced and discussed; their history is mentioned at the end of this note, and they will be studied more fully in a separate paper. We begin now with the DFT.

1. The Periodic Case and the DFT

The second difference matrix corresponding to periodic boundary conditions is a circulant matrix:

$$A_0 = \begin{bmatrix} 2 & -1 & & & -1 \\ -1 & 2 & -1 & & \\ & & \ddots & & \\ & & -1 & 2 & -1 \\ -1 & & & -1 & 2 \end{bmatrix}.$$

The jth entry of $A_0 x$ is $-x_{j-1}+2x_j-x_{j+1}$. At the first and last rows ($j = 0$ and $j = N-1$), this second difference extends beyond the boundary. Then the periodicity $x_N = x_0$ and $x_{N-1} = x_{-1}$ produces the -1 entries that appear alone in the corners. (*The numbering throughout this paper goes from* 0 *to* $N-1$, since SIAM is on very friendly terms with the IEEE. But we still believe the evidence of our eyes, that the word imaginary begins with $i = \sqrt{-1}$ and not j. Anyway, the DCT is real.)

The basis vector $v_k = (1, w^k, w^{2k}, \ldots)$ is periodic because $w = e^{2\pi i/N}$ and $w^N = 1$. We now verify that v_k is an eigenvector of A_0. The jth component of $A_0 v_k = \lambda_k v_k$ is the second difference

$$\begin{aligned} -w^{(j-1)k} + 2w^{jk} - w^{(j+1)k} &= \left(-w^{-k} + 2 - w^k\right) w^{jk} \\ &= \left(-e^{-2\pi ik/N} + 2 - e^{2\pi ik/N}\right) w^{jk} \\ &= \left(2 - 2\cos\frac{2k\pi}{N}\right) w^{jk}. \end{aligned}$$

A_0 is symmetric and those eigenvalues $\lambda_k = 2 - 2\cos\frac{2k\pi}{N}$ are real. The smallest is $\lambda_0 = 0$, corresponding to the eigenvector $v_0 = (1, 1, \ldots, 1)$. It is very useful to have this flat DC vector (direct current in circuit theory, constant gray level in image processing) as one of the basis vectors.

Since A_0 is a real symmetric matrix, its orthogonal eigenvectors can be chosen real. In fact the real and imaginary parts of the v_k must be eigenvectors:

$$c_k = \operatorname{Re} v_k = \left(1, \cos\frac{2k\pi}{N}, \cos\frac{4k\pi}{N}, \ldots, \cos\frac{2(N-1)k\pi}{N}\right)$$

$$s_k = \operatorname{Im} v_k = \left(0, \sin\frac{2k\pi}{N}, \sin\frac{4k\pi}{N}, \ldots, \sin\frac{2(N-1)k\pi}{N}\right).$$

The equal pair of eigenvalues $\lambda_k = \lambda_{N-k}$ gives the two eigenvectors c_k and s_k. The exceptions are $\lambda_0 = 0$ with one eigenvector $c_0 = (1, 1, \ldots, 1)$, and for even N also $\lambda_{N/2} = 4$ with $c_{N/2} = (1, -1, \ldots, 1, -1)$. These eigenvectors have length $\sqrt{N}$, while the other c_k and s_k have length $\sqrt{N/2}$. It is these exceptions that make the real DFT (sines together with cosines) less attractive than the complex form. That factor $\sqrt{2}$ is familiar from ordinary Fourier series. It will appear in the $k = 0$ term for the DCT-1 and DCT-2, always with the flat basis vector $(1, 1, \ldots, 1)$.

We expect the cosines alone, without sines, to be complete over a half-period. In Fourier series this changes the interval from $[-\pi, \pi]$ to $[0, \pi]$. Periodicity is gone because $\cos 0 \neq \cos \pi$. In the discrete case, the 2 will disappear from $\cos\frac{2jk\pi}{N}$. The images we meet in practice are not periodic.

2. The Discrete Cosine Transform

The next step is so natural, and almost inevitable, that it is really astonishing that the DCT was not discovered until 1974. Perhaps this time delay illustrates an underlying principle. Each continuous problem (differential equation) has many discrete approximations (difference equations). The discrete case has a new level of variety and complexity, often appearing in the boundary conditions.

In fact the original paper by Ahmed, Natarajan, and Rao derived the DCT-2 basis as approximations to the eigenvectors of an important matrix, with entries $\rho^{|j-k|}$. This is the covariance matrix for a useful class of signals; the number ρ (near 1) measures the correlation between nearest neighbors. The true eigenvectors would give an optimal "Karhunen-Loève basis" for compressing those signals, and the simpler DCT vectors are close (and independent of ρ). The four types of DCT are now studied directly from their basis vectors (recall that j and k go from 0 to $N-1$):

$$c_k^1 = \left(\frac{1}{\sqrt{2}}, \cos\frac{k\pi}{N-1}, \ldots, \cos\frac{jk\pi}{N-1}, \ldots, \frac{1}{\sqrt{2}}\cos k\pi\right)$$

$$c_k^2 = b_k\left(\cos\frac{1}{2}k\frac{\pi}{N}, \cos\frac{3}{2}k\frac{\pi}{N}, \ldots, \cos(j+\frac{1}{2})k\frac{\pi}{N}, \ldots\right)$$

$$c_k^3 = \left(\frac{1}{\sqrt{2}}, \cos(k+\frac{1}{2})\frac{\pi}{N}, \ldots, \cos j(k+\frac{1}{2})\frac{\pi}{N}, \ldots\right)$$

$$c_k^4 = \left(\cos\frac{1}{2}(k+\frac{1}{2})\frac{\pi}{N}, \ldots, \cos(j+\frac{1}{2})(k+\frac{1}{2})\frac{\pi}{N}, \ldots\right)$$

Those are the orthogonal columns of the DCT matrices C_1, C_2, C_3, C_4. The number b_k is 1 except that $b_0 = 1/\sqrt{2}$; this is the correction at frequency $k = 0$, when the cosine stays at 1.

The matrix C_3 with top row $\frac{1}{\sqrt{2}}(1, 1, \ldots, 1)$ is the transpose of C_2. All columns of C_2, C_3, C_4 have length $\sqrt{N/2}$. The immediate goal is to prove orthogonality.

Proof! These four bases are eigenvectors of symmetric second difference matrices A_1, A_2, A_3, A_4. The first three matrices were studied in an unpublished manuscript by David Zachmann, who wrote down the explicit eigenvectors. His paper is very useful. He noted earlier references for the eigenvalues; a complete history would be virtually impossible. We have seen that the periodic matrix A_0, with $-1, 2, -1$ in every row, shares the same eigenvectors $\cos kt$ and $\sin kt$ as the second derivative. The cosines are picked out by the zero-slope boundary condition at $t = 0$ in the following matrices.

The exact eigenvectors (and the discrete frequencies involving k or $k + \frac{1}{2}$) are determined by the first and last rows. Here are the four symmetric second-difference matrices (the second one is the purest):

$$A_1 = \begin{bmatrix} 2 & -\sqrt{2} & & & \\ -\sqrt{2} & 2 & -1 & & \\ & & \ddots & & \\ & & -1 & 2 & -\sqrt{2} \\ & & & -\sqrt{2} & 2 \end{bmatrix}$$

$$A_2 = \begin{bmatrix} 1 & -1 & & & \\ -1 & 2 & -1 & & \\ & & \ddots & & \\ & & -1 & 2 & -1 \\ & & & -1 & 1 \end{bmatrix}$$

$$A_3 = \begin{bmatrix} 2 & -\sqrt{2} & & & \\ -\sqrt{2} & 2 & -1 & & \\ & & \ddots & & \\ & & -1 & 2 & -1 \\ & & & -1 & 2 \end{bmatrix}$$

$$A_4 = \begin{bmatrix} 1 & -1 & & & \\ -1 & 2 & -1 & & \\ & & \ddots & & \\ & & -1 & 2 & -1 \\ & & & -1 & 3 \end{bmatrix}$$

Perhaps each matrix deserves a quick comment, starting with the unexpected $\sqrt{2}$ in A_3. The top row of Zachmann's A_3^Z contains 2 and -2, and the next row is the normal $-1, 2, -1$. This looks better than A_3 but symmetry is lost; the eigenvectors $\cos j(k + \frac{1}{2})\frac{\pi}{N}$ are not quite orthogonal. We symmetrized $D^{-1} A_3^Z D = A_3$ by the diagonal matrix $D = \text{diag}\,(\sqrt{2}, 1, \ldots, 1)$. Then $\sqrt{2}$ appears in A_3 and the eigenvectors are multiplied by D^{-1}. The first entry of the orthogonal eigenvectors c_k^3 is $\frac{1}{\sqrt{2}}$.

Similarly Zachmann's A_1^Z has 2 and -2 (not $-\sqrt{2}$) in its first and last rows. This matrix is symmetrized using $D = \text{diag}\,(\sqrt{2}, 1, \ldots, 1, \sqrt{2})$. Then $\sqrt{2}$ divides the first and last components of the cosine eigenvectors. Notice how the eigenvectors become orthogonal, whether N

is even or odd:

$$N = 3 \qquad (\tfrac{1}{\sqrt{2}}, 1, \tfrac{1}{\sqrt{2}}) \qquad (\tfrac{1}{\sqrt{2}}, 0, -\tfrac{1}{\sqrt{2}}) \qquad (\tfrac{1}{\sqrt{2}}, -1, \tfrac{1}{\sqrt{2}}) \quad \text{for} \quad k = 0, 1, 2$$

$$N = 4 \qquad (\tfrac{1}{\sqrt{2}}, 1, 1, \tfrac{1}{\sqrt{2}}) \qquad \cdots \qquad (\tfrac{1}{\sqrt{2}}, -1, 1, -\tfrac{1}{\sqrt{2}}) \quad \text{for} \quad k = 0, 1, 2, 3\,.$$

The first and last eigenvectors have length $\sqrt{N-1}$, the others have length $\sqrt{(N-1)/2}$.

The DCT-2 eigenvectors are the most popular of all. Their first and last components are not exceptional. The boundary condition at the left end of the interval is $x_{-1} = x_0$, which is a zero derivative centered on a *midpoint*. Similarly the right end has $x_{N-1} = x_N$. When the outside values x_{-1} and x_N are eliminated from the $-1, 2, -1$ second difference operator, this leaves the neat 1 and -1 in the boundary rows of A_2.

I believe that this DCT-2 (often just called DCT) should be in our mathematics courses along with the DFT. It is the basis for the JPEG-2 algorithm in image compression. The image is broken into 8×8 blocks of pixels, and each block is transformed by a two-dimensional DCT. We comment below on the blocking artifacts that appear when the transform coefficients are compressed.

The final entry "3" in the matrix A_4 is more surprising. We had not seen it before, but MATLAB insisted it was right. Now we realize that it comes from a Dirichlet boundary condition (function = zero), *imposed at the midpoint between nodes* $N-1$ *and* N. Therefore $x_{N-1} \approx -x_N$. This folds the final -1 in the second difference at the right hand boundary back as $+1$. Then -1, $2, -1$ becomes $-1, 3$. The eigenvectors c_k^4 are *even* at the left end and *odd* at the right end. This attractive property leads to $j + \frac{1}{2}$ and $k + \frac{1}{2}$ and a symmetric eigenvector matrix C_4.

Our proof of orthogonality is a verification that the basis vectors are the eigenvectors of A_1, A_2, A_3, A_4. For all the $-1, 2, -1$ rows this needs to be done only once (and it reveals the eigenvalues $\lambda = 2 - 2\cos\theta$). There is an irreducible minimum of trigonometry when the jth component of c_k is $\cos j\theta$ in Types 1 and 3, and $\cos(j + \frac{1}{2})\theta$ in Types 2 and 4:

$$-\cos(j-1)\theta + 2\cos j\theta - \cos(j+1)\theta = (2 - 2\cos\theta)\cos j\theta$$

$$-\cos(j-\tfrac{1}{2})\theta + 2\cos(j+\tfrac{1}{2})\theta - \cos(j+\tfrac{3}{2})\theta = (2 - 2\cos\theta)\cos(j+\tfrac{1}{2})\theta\,.$$

This is $Ax = \lambda x$ on all interior rows. The angle is $\theta = k\frac{\pi}{N-1}$ for the eigenvectors of Type 1, and $\theta = k\frac{\pi}{N}$ for Type 2. It is $\theta = \left(k + \frac{1}{2}\right)\frac{\pi}{N}$ for A_3 (or actually A_3^Z) and A_4. This leaves only the first and last components of $Ac_k = \lambda_k c_k$ to be verified in each case.

Let us do only the fourth case, for the last row $-1, 3$ of the matrix A_4. A last row of $-1, 1$ would subtract the $j = N-2$ component of c_k^4 from the $j = N-1$ component. Trigonometry gives those components as

$$j = N-1: \quad \cos\left(N - \tfrac{1}{2}\right)\left(k + \tfrac{1}{2}\right)\tfrac{\pi}{N} = \sin\tfrac{1}{2}\left(k + \tfrac{1}{2}\right)\tfrac{\pi}{N}$$

$$j = N-2: \quad \cos\left(N - \tfrac{3}{2}\right)\left(k + \tfrac{1}{2}\right)\tfrac{\pi}{N} = \sin\tfrac{3}{2}\left(k + \tfrac{1}{2}\right)\tfrac{\pi}{N}\,.$$

We subtract using $\sin a - \sin b = -2\cos\left(\frac{b+a}{2}\right)\sin\left(\frac{b-a}{2}\right)$. The difference is

$$-2\cos\left(k + \tfrac{1}{2}\right)\frac{\pi}{N}\ \sin\tfrac{1}{2}\left(k + \tfrac{1}{2}\right)\frac{\pi}{N}\,. \tag{A3.1}$$

The last row of A_4 actually ends with 3, so we still have 2 times the last component ($j = N-1$) to include with (1):

$$\left(2 - 2\cos\left(k + \tfrac{1}{2}\right)\frac{\pi}{N}\right)\ \sin\tfrac{1}{2}\left(k + \tfrac{1}{2}\right)\frac{\pi}{N}\,. \tag{A3.2}$$

This is just λ_k times the last component of c_k^4. The final row of $A_4 c_k^4 = \lambda_k c_k^4$ is verified.

There are also Discrete Sine Transforms DST-1 through DST-4. The entries of the basis vectors s_k are sines instead of cosines. These s_k are orthogonal because they are eigenvectors of symmetric second difference matrices, with a Dirichlet (instead of Neumann) condition at the left boundary. On page 279 we presented a third proof of orthogonality—which simultaneously covers the DCT and the DST, and shows their fast connection to the DFT matrix of order $2N$.

This is achieved by a neat matrix factorization given by Wickerhauser:

$$e^{-\pi i/4N} R^T F_{2N} R = \begin{bmatrix} C_4 & 0 \\ 0 & -iS_4 \end{bmatrix}.$$

The entries of S_4 are $\sin(j+\frac{1}{2})(k+\frac{1}{2})\frac{\pi}{N}$. The connection matrix R is very sparse, with $w = e^{\pi i/2N}$:

$$R = \frac{1}{\sqrt{2}} \begin{bmatrix} D & D \\ E & -E \end{bmatrix} \quad \text{with} \quad \begin{array}{l} D = \operatorname{diag}(1, \overline{w}, \ldots, \overline{w}^{N-1}) \\ E = \operatorname{antidiag}(w, w^2, \ldots, w^N). \end{array}$$

Since R^T and F_{2N} and R have orthogonal columns, so do C_4 and S_4.

3. Cosine Transforms with *N* – *1/2* and *N* + *1/2*

There are other combinations of discrete boundary conditions, and we describe four possibilities here. Every combination that produces a symmetric matrix will also produce (from the eigenvectors of that matrix) an orthogonal transform. The matrix A_5 imposes a Neumann condition (derivative = zero) at both boundaries.

First we write it non-symmetrically:

$$A_5 = \begin{bmatrix} 1 & -1 & & & \\ -1 & 2 & -1 & & \\ & \cdot & \cdot & \cdot & \\ & & -1 & 2 & -1 \\ & & & -2 & 2 \end{bmatrix}.$$

This has the desirable flat eigenvector $(1, 1, \ldots, 1)$ corresponding to $\lambda = 0$. The other eigenvectors also look familiar, but in their denominators you will see something new:

$$c_k^5 = \left(\cos \tfrac{1}{2} \frac{k\pi}{N-\frac{1}{2}}, \ldots, \cos\left(j+\tfrac{1}{2}\right) \frac{k\pi}{N-\frac{1}{2}}, \ldots\right).$$

The period is $N - \frac{1}{2}$. The boundary condition $x_1 = x_0$ is centered at a midpoint, and $x_{N-2} = x_N$ is centered at the last meshpoint. (It changes $-1, 2, -1$ into the last row $-2, 2$.) The number of meshwidths between those centers is $N - \frac{1}{2}$.

A_5 is symmetrized by the diagonal matrix $D = (1, \ldots, 1, \sqrt{2})$. Then $D^{-1} A_5 D$ has $-\sqrt{2}$ in the two lower right entries, where A_5 has -1 and -2. The last components of the eigenvectors c_k^5 are divided by $\sqrt{2}$; they are orthogonal but less beautiful. We expect to implement the DCT-5 by keeping the matrix C_5 with pure cosine entries as above, and accounting for the correction factors by diagonal matrices:

$$\frac{4}{2N-1} \quad C_5 \quad \operatorname{diag}\left(\tfrac{1}{2}, 1, \ldots, 1\right) \quad C_5^T \quad \operatorname{diag}\left(1, \ldots, 1, \tfrac{1}{2}\right) = I. \tag{A3.3}$$

The cosine vectors c_k^5 have squared length $\frac{2N-1}{4}$. The all-ones vector ($k = 0$) is adjusted by the first diagonal matrix. The last diagonal matrix corrects the last components. The inverse of C_5 is not quite C_5^T (analysis is not quite the transpose of synthesis, as in an orthogonal transform) but the corrections have trivial cost. For $N = 2$, the matrix identity (A3.3) involves $\cos \frac{1}{2} \frac{\pi}{3/2} = \frac{1}{2}$ and $\cos \frac{3}{2} \frac{\pi}{3/2} = -1$ in C_5:

$$\frac{4}{3} \begin{bmatrix} 1 & \frac{1}{2} \\ 1 & -1 \end{bmatrix} \begin{bmatrix} \frac{1}{2} & \\ & 1 \end{bmatrix} \begin{bmatrix} 1 & 1 \\ \frac{1}{2} & -1 \end{bmatrix} \begin{bmatrix} 1 & \\ & \frac{1}{2} \end{bmatrix} = \begin{bmatrix} 1 & \\ & 1 \end{bmatrix}.$$

Malvar has added a further good suggestion: Orthogonalize the last $N - 1$ basis vectors against the all-ones vector. Otherwise the DC component (which is usually largest) leaks into the other components. Thus we subtract from each c_k^5 (with $k > 0$) its projection onto the flat c_0^5:

$$\widetilde{c}_k^5 = c_k^5 - \frac{(c_k^5, c_0^5)}{(c_0^5, c_0^5)} c_0^5 = c_k^5 - \frac{(-1)^k}{2N} (1, 1, \ldots, 1). \tag{A3.4}$$

The adjusted basis vectors are now the columns of $\widetilde{C}_5$, and (A3.4) becomes

$$C_5 = \widetilde{C}_5 \begin{bmatrix} 1 & \frac{-1}{2N} & \frac{+1}{2N} & \cdots \\ & 1 & & \\ & & \ddots & \\ & & & 1 \end{bmatrix}.$$

This replacement in equation (A3.3) also has trivial cost, and that identity becomes $\widetilde{C}_5 \widetilde{C}_5^{-1} = I$. The coefficients of a vector x are $y = \widetilde{C}_5^{-1} x$, and then $\widetilde{C}_5 y$ reconstructs x (possibly after compressing y). You see how we search for a good basis...

Exchanging the left and right boundary conditions will produce $2, -2$ in the top row of A and $-1, 1$ in the bottom row. We denote this new matrix by A_6 and the transform by DCT-6. The basis vectors c_k^6 are just the c_k^5 with components in reverse order.

The matrices A_7 and A_8 combine a Neumann condition at the left boundary with a Dirichlet condition (function = zero) at the right boundary. The first one is classical:

$$A_7 = \begin{bmatrix} 1 & -1 & & & \\ -1 & 2 & -1 & & \\ & & \ddots & & \\ & & -1 & 2 & -1 \\ & & & -1 & 2 \end{bmatrix}.$$

The eigenvectors were given by Zachmann, and this matrix must have a very long history. We could study A_7 by reflection across the left boundary, to produce the pure Toeplitz $-1, 2, -1$ matrix (which is my favorite example in teaching). The eigenvectors become discrete sines on a double interval—*almost*. The length of the double interval is not $2N$, because the matrix from reflection has *odd order*. This leads to the new "period length" $N + \frac{1}{2}$ in the eigenvectors of A_7:

$$c_k^7 = \left(\cos \tfrac{1}{2} \left(k + \tfrac{1}{2}\right) \frac{\pi}{N + \frac{1}{2}}, \ \ldots, \cos \left(j + \tfrac{1}{2}\right) \left(k + \tfrac{1}{2}\right) \frac{\pi}{N + \frac{1}{2}}, \ \ldots \right), \ j = 0, \ldots, N - 1.$$

The component $\cos \left(k + \frac{1}{2}\right) \pi = 0$ that satisfies the Dirichlet condition would come from the next value $j = N$ beyond the right boundary. So A_7 has a midpoint condition at the left boundary and a meshpoint condition at the right. This is reversed in A_8, to meshpoint at the left and midpoint at the right:

$$A_8 = \begin{bmatrix} 2 & -2 & & & \\ -1 & 2 & -1 & & \\ & \cdot & \cdot & \cdot & \\ & & -1 & 2 & -1 \\ & & & -1 & 3 \end{bmatrix}.$$

Now the eigenvectors have components $\cos j(k+\frac{1}{2})\frac{\pi}{N-\frac{1}{2}}$.

The half-integer periods would seem to be a disadvantage, but reflection offers a possible way out. Martucci has shown how this matches the *circular convolution of symmetric extensions*. The reflected vectors have an integer "double period" and *they overlap*.

A major improvement for compression and image coding was Malvar's extension [Mr] of the ordinary DCT to a *lapped transform*. Instead of dividing the image into completely separate blocks for compression, his basis vectors overlap two or more blocks. The JPEG standard uses the DCT-2 and has no overlap, which produces blocking artifacts (discontinuities) in the compressed image. Nobody is very happy with JPEG, although it is extremely very fast. The overlapping has been easiest to develop for the DCT-4, using its even-odd boundary conditions— which the DCT-7 shares! Those conditions maintain orthogonality between the tail of one vector and the head of another. Coifman-Meyer found the analogous construction [CoMey] for continuous wavelets.

We only note one property of the DCT-7 compared to the DCT-4. The theoretical *coding gain* is significantly better. (Coding includes the quantization step that puts real numbers into discrete bins; only the bin number is coded and transmitted. This more subtle and statistical form of roundoff should have applications elsewhere in numerical analysis.) The DCT-2 and the DCT-5 have much greater (and almost equal) coding gains, because their bases include the constant vector $(1, 1, \ldots, 1)$ that represents flat backgrounds.

The success of any transform in image coding depends on a combination of properties— mathematical, computational, and *visual*. The relation to the human visual system is decided above all by experience. This article was devoted to the mathematical property of orthogonality (which helps the computations). We hope that the eigenvector approach will suggest more new transforms, and that at least one of them will be fast and visually attractive.

4. Summary

For each transform we give the first and last rows of A, which determine the center points for the boundary conditions. Zachmann makes the important observation that *all those boundary conditions give second-order accuracy around their center points*. Finite differences are one-sided and less accurate only with respect to the wrong center! The distance between center points is the period. We use the unsymmetric forms that have $-1, 2, -1$ in all other rows of A, so the eigenvectors are pure cosines:

DCT-1	$2, -2$ (center 0)	and	$-2, 2$ (center $N-1$):	period $N-1$
DCT-2	$1, -1$ (center $-\frac{1}{2}$)	and	$-1, 1$ (center $N-\frac{1}{2}$):	period N
DCT-3	$2, -2$ (center 0)	and	$-1, 2$ (center N):	period N
DCT-4	$1, -1$ (center $-\frac{1}{2}$)	and	$-1, 3$ (center $N-\frac{1}{2}$):	period N
DCT-5	$1, -1$ (center $-\frac{1}{2}$)	and	$-2, 2$ (center $N-1$):	period $N-\frac{1}{2}$
DCT-6	$2, -2$ (center 0)	and	$-1, 1$ (center $N-\frac{1}{2}$):	period $N-\frac{1}{2}$
DCT-7	$1, -1$ (center $-\frac{1}{2}$)	and	$-1, 2$ (center N):	period $N+\frac{1}{2}$
DCT-8	$2, -2$ (center 0)	and	$-1, 3$ (center $N-\frac{1}{2}$):	period $N-\frac{1}{2}$

The eight Discrete Sine Transforms would have the same four last rows, with 2, -1 or 3, -1 in the top row. Mixed boundary conditions corresponding to $x'(0) = cx(0)$ may produce the (cosine + sine) eigenvectors that give the Hartley Transform. And there is no absolute restriction to second difference matrices. They do seem to have the best eigenvectors.

Added in Proof: We have just located an excellent paper by Martucci [IEEE Trans. SP **42** (1994) 1038–1051] that studies in detail these eight DCT's and DST's. The complete set was first recorded in 1985 by Wang. Martucci develops convolution-multiplication rules that generalize the convolution rule for the DFT. (Signal processing is based on that convolution rule; see the first page of Chapter 1!) His generalized convolution applies to the symmetric extensions of Chapter 8. Each transform is identified by its whole-sample or half-sample (**W** or **H**) symmetry at the endpoints.

The early DCT's discovered by Ahmed et al., Jain, and Kitajima were identified as the eigenvectors of particular matrices. Recently Sanchez et al. [IEEE Trans. SP **43** (1995) 2631–2641] successfully described *all* symmetric matrices with these eight sets of eigenvectors: they have the form "Toeplitz + near-Hankel". Our observation that all eight DCT's are eigenvectors of simple second-difference matrices, with odd or even and ***W*** or ***H*** symmetry at the ends, seems to complete this circle of ideas — and open up new applications in numerical analysis.

Appendix 4

The Lifting Scheme

A central theme in mathematics is to reduce every construction to a sequence of simple steps. We factor numbers into primes, we factor polynomials into $(x - x_1)\ldots(x - x_n)$, and we factor matrices into LU or QR or the SVD. Since a filter bank is represented by a *matrix polynomial*—the polyphase matrix $\boldsymbol{H}_p(z)$—the next step is to factor that matrix into simpler pieces. One approach was the lattice structure in Section 4.5, where each step is a "butterfly" operation: multiples of the lowpass and highpass signals are added into the highpass and lowpass channels *at the same time*. Now we study *lifting*, which employs butterflies with only one wing. The signal from one channel is filtered and added to the other channel. Each step is either lifting or dual lifting:

Lifting Filter the highpass signal and add to the lowpass channel.

Dual lifting Filter the lowpass signal and add to the highpass channel.

The filter bank begins with Haar or even the Lazy Filter (which just separates $\boldsymbol{x}$ into $\boldsymbol{x}_{\text{even}}$ and $\boldsymbol{x}_{\text{odd}}$). After lifting we end with a *perfect reconstruction filter bank* of higher order. This representation is complete and this lifting factorization is always achievable: *Every PR filter bank (for one-dimensional signals) can be constructed from lifting steps.*

We illustrate one dual lifting step to create $\boldsymbol{x}'_{\text{odd}}$ followed by a lifting step to create $\boldsymbol{x}'_{\text{even}}$. The filters $\boldsymbol{S}_0$ and $\boldsymbol{S}_1$ are usually low order:

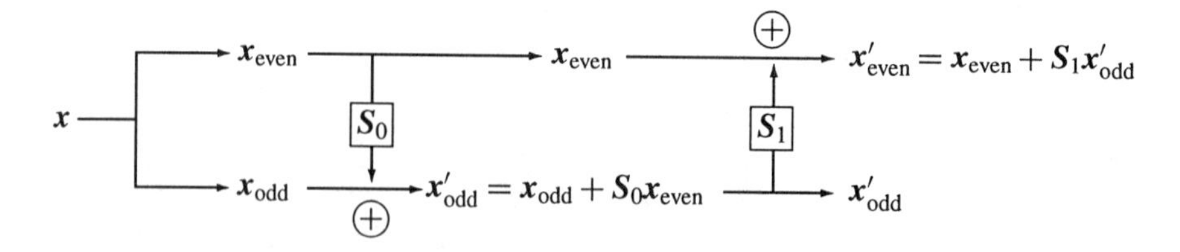

Each inverse step is absolutely simple. We just subtract what we added. Those inverses come in the opposite order, so lifting is first in the synthesis bank:

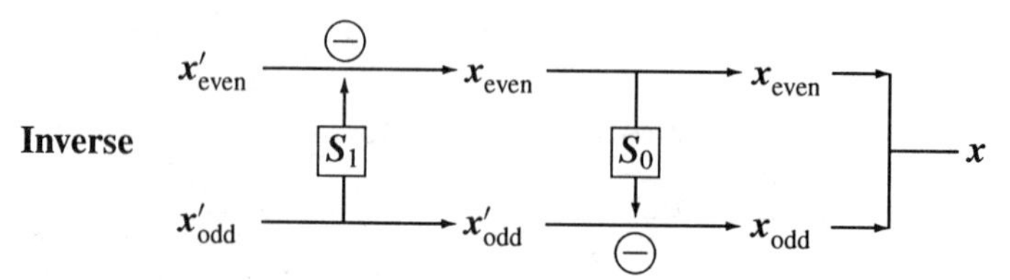

The inverse lifting step is $\boldsymbol{x}_{\text{even}} = \boldsymbol{x}'_{\text{even}} - \boldsymbol{S}_1\boldsymbol{x}'_{\text{odd}}$. Because of the minus sign, this inverts the last step $\boldsymbol{x}'_{\text{even}} = \boldsymbol{x}_{\text{even}} + \boldsymbol{S}_1\boldsymbol{x}'_{\text{odd}}$ in the analysis bank.

Every analysis step is inverted just by changing a plus sign to a minus sign (and moving that term to the other side of the equation). The lifting scheme is also called a *ladder structure* in the engineering literature, because these block diagrams look like ladders. You will correctly expect that each step multiplies the polyphase matrix by an elementary factor (described below).

First we show that the Haar filter bank can be lifted from the Lazy Filter. The filters S_0 and S_1 are just multiplications by -1 and $+\frac{1}{2}$:

Split $\boldsymbol{x}$ into $\boldsymbol{x}_{\text{even}}$ and $\boldsymbol{x}_{\text{odd}}$
Dual lifting: $\boldsymbol{x}'_{\text{odd}} = \boldsymbol{x}_{\text{odd}} - \boldsymbol{x}_{\text{even}}$
Lifting: $\boldsymbol{x}'_{\text{even}} = \boldsymbol{x}_{\text{even}} + \frac{1}{2}\boldsymbol{x}'_{\text{odd}}$.

The combination of those steps produces one form of Haar:

$$\boldsymbol{x}'_{\text{even}} = \text{average} = \frac{\boldsymbol{x}_{\text{even}} + \boldsymbol{x}_{\text{odd}}}{2} \qquad \boldsymbol{x}'_{\text{odd}} = \text{difference} = \boldsymbol{x}_{\text{odd}} - \boldsymbol{x}_{\text{even}}$$

Here we insert an important comment about *integer* lifting. We can create an *Integer Wavelet Transform* by rounding off the lifted parts before they enter their new channels. The point is that this transform is reversible, where ordinary roundoff within a channel would be irreversible. The inverse always subtracts what was added!

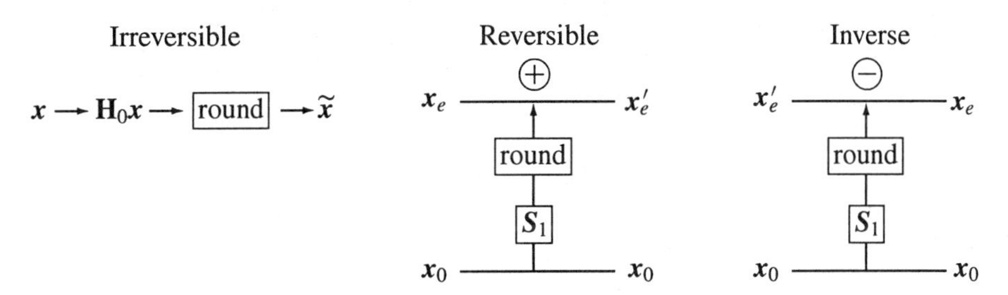

Adding roundoff to the Haar example, the final lifting step becomes $\boldsymbol{x}'_{\text{even}} = \boldsymbol{x}_{\text{even}} + \{\frac{1}{2}\boldsymbol{x}'_{\text{odd}}\}$. This produces an integer. The inverse step puts $\{\frac{1}{2}\boldsymbol{x}'_{\text{odd}}\}$ on the other side of the equation, with a minus sign. Thus the integer Haar Transform has an integer inverse. Notice that these transforms are nonlinear.

It may be useful to list some of the main advantages of lifting:

1. The wavelet transform (the filter bank) can be executed *in place*. No auxiliary memory is required.

2. Integer transforms are easily constructed, for *lossless coding*.

3. The inverse transform has exactly the same complexity as the forward transform.

Those properties are clear from the lifting construction. Other advantages that come from lifting will be listed here but not developed in detail:

4. Lifting can speed up the implementation of a high order filter bank by roughly a factor of 2.

5. The flexibility can simplify constructions in the time or space domain that would be difficult in the frequency domain:

   (a) Boundary filters (for finite length signals)

(b) Adaptive transforms (finer levels can be localized)

(c) Irregularly spaced samples

(d) Wavelets on a sphere (or with weight functions).

First-generation wavelets come from translations and dilations to $w(2^j t - k)$. The *second generation* allows more general forms of multiresolution.

One disadvantage is connected with orthogonal transforms. *The lifting factors are never orthogonal.* This was true already for the Haar transform: the separate lifting steps are *bi*orthogonal transforms. They are easy to invert, but the inverses are not the transposes. A typical step is expressed by an elementary but *non-orthogonal* matrix:

$$\textbf{Elementary matrix} \quad \begin{bmatrix} 1 & 0 \\ S_0 & 1 \end{bmatrix} \quad \textbf{Inverse} \quad \begin{bmatrix} 1 & 0 \\ -S_0 & 1 \end{bmatrix}.$$

Several new references on lifting, including software, have been appended to our bibliography in this second edition. The history and the connections to other research are in the basic paper [DaSw] by Daubechies and Sweldens.

The Mathematics of Lifting

We began the mathematical analysis on pages 214–215. Here we start again, to see the simple changes in the polyphase matrices after a lifting step. Remember that there is an *analysis* polyphase matrix, denoted by $\boldsymbol{H}_p(z)$, and a *synthesis* polyphase matrix $\boldsymbol{F}_p(z)$. Both will change after a lifting step. One is multiplied by an elementary matrix and the other is multiplied (of course) by the inverse elementary matrix.

The main step in the theory is that every polyphase matrix from a PR filter bank can be factored into these elementary matrices. The factorization is constructed by the Euclidean algorithm. To find the greatest common divisor of two polynomials $a_0(z)$ and $b_0(z)$, Euclid iterates a division of polynomials:

$$a_{i+1}(z) = b_i(z) \quad \text{and} \quad b_{i+1}(z) = \text{remainder from } \frac{a_i(z)}{b_i(z)}.$$

Since the degrees are decreasing, eventually the remainder is $b_n(z) = 0$. At that point $a_n(z)$ is the greatest common divisor of $a_0(z)$ and $b_0(z)$. For an extreme example, when $b_0(z)$ divides $a_0(z)$ exactly, the remainder is $b_1(z) = 0$ and then $a_1(z) = b_0(z)$ is the *gcd*. This Euclidean algorithm is extended in [DaSw] to polynomials in z and z^{-1}.

It is useful to see the flow diagram for the Euclidean algorithm. Suppose the division a_i/b_i produces quotient q_i and remainder b_{i+1}. Then $a_i = q_i b_i + b_{i+1}$. Watch how this appears (and the channels cross to exchange a's and b's):

Two Euclidean Steps

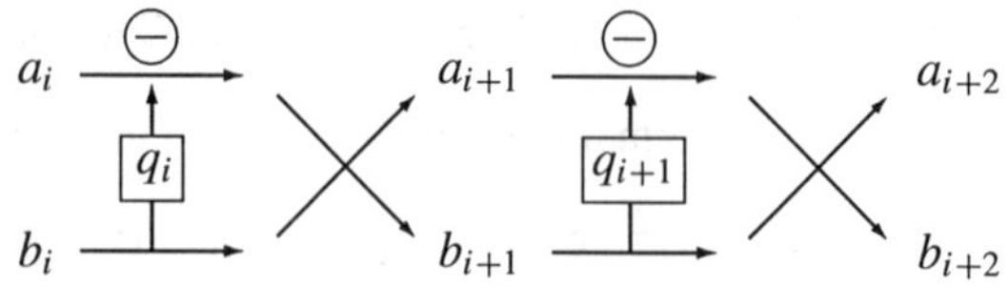

a_i ⊖ a_{i+1} ⊖ a_{i+2}

q_i q_{i+1}

b_i b_{i+1} b_{i+2}

We can combine those two steps by omitting the cross-over and cross-back and the $a - b$ exchange. Then the second step of the Euclidean algorithm becomes a dual lifting!

= Two Lifting Steps

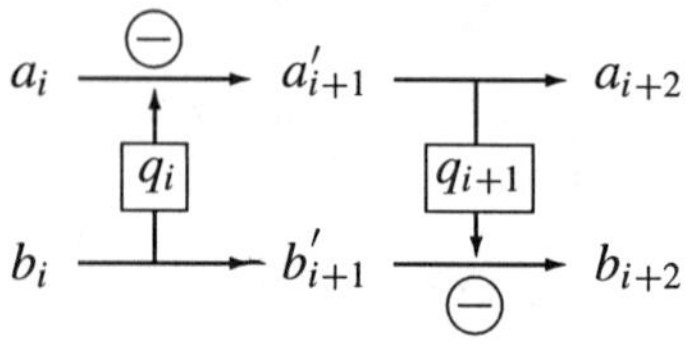

a_i ⊖ a'_{i+1} a_{i+2}

q_i q_{i+1}

b_i b'_{i+1} ⊖ b_{i+2}

These diagrams bring out what we might have missed in the algebra. *The lifting scheme is a form of the Euclidean algorithm*, adapted to the polynomials $H_0(z)$ and $H_1(z)$ that represent the filter bank.

We briefly explain the proof that every PR filter bank (for one-dimensional signals) can be expressed as a product of lifting steps — possibly with a final rescaling and delay to match the filter bank exactly.

Apply the Euclidean algorithm to the even phases $H_{0,\text{even}}$ and $H_{1,\text{even}}$ of the lowpass and highpass filters. These polynomials have no common factor (they are coprime), because the determinant of the polyphase matrix $\boldsymbol{H}_p(z)$ is a monomial:

$$\det\begin{bmatrix} H_{0,\text{even}}(z) & H_{0,\text{odd}}(z) \\ H_{1,\text{even}}(z) & H_{1,\text{odd}}(z) \end{bmatrix} = Cz^{-L}$$

A common factor of the first column would be a factor in the determinant, not allowed. So the Euclidean algorithm will produce a nonzero constant C_1 as the greatest common divisor of $H_{0,\text{even}}$ and $H_{1,\text{even}}$. The steps of that algorithm are lifting and dual lifting (in the flow diagram above). Those steps are expressed by elementary matrices with determinant one:

$$E_i^{\text{lift}} = \begin{bmatrix} 1 & -q_i(z) \\ 0 & 1 \end{bmatrix} \quad \text{and} \quad E_{i+1}^{d\,\text{lift}} = \begin{bmatrix} 1 & 0 \\ -q_{i+1}(z) & 1 \end{bmatrix}.$$

The lifting matrices multiply the first column of $\boldsymbol{H}_p(z)$ to produce two constants C_1 and 0 out of $H_{0,\text{even}}$ and $H_{1,\text{even}}$. Since the determinant is unchanged, the *second column* must have a special form too:

$$\text{lifting matrices} \times (\boldsymbol{H}_p(z)) = \begin{bmatrix} C_1 & s(z) \\ 0 & C_2 z^{-L} \end{bmatrix}.$$

The $(2,2)$ entry cannot be a polynomial because the determinant cannot be a polynomial. (The inverse bank is assumed to be FIR.) Now one final lifting step will account for the $s(z)$ in the $(1,2)$ position, leaving only the scaling factors C_1, C_2 and the delays.

The matrix $\boldsymbol{F}_p(z) = (\boldsymbol{H}_p(z))^{-1}$ is now a product of lifting and dual lifting steps. Since the inverses all have the same form, the whole filter bank is factored as required. We recall that these factors are not orthogonal matrices; an orthogonal factorization comes from the lattice structure and not the ladder structure.

In mathematical terms, an $n \times n$ invertible matrix belongs to the general linear group $GL(n)$. It factors into elementary matrices (this is Gauss-Jordan elimination). Here we are dealing with invertible 2×2 matrix *polynomials*, and the factorization extends. It also extends to $M \times M$ matrix polynomials, so that any invertible M-band FIR filter bank can be factored into lifting steps. Those steps filter and add a multiple of one channel into another channel. The number of steps increases with M, and lifting is most efficient in the 2-channel case.

Example We will factor the 5/3 analysis bank with

$$\boldsymbol{h}_0 = \tfrac{1}{8}(-1, 2, 6, 2, -1) \quad \text{and} \quad \boldsymbol{h}_1 = \tfrac{1}{4}(1, -2, 1).$$

The lazy filter separates the input signal into even and odd:

$$\boldsymbol{x}_{\text{even}}(n) = \boldsymbol{x}(2n) \quad \text{and} \quad \boldsymbol{x}_{\text{odd}}(n) = \boldsymbol{x}(2n+1).$$

Now execute dual lifting and lifting. (For an integer transform, round each increment before including it.) The steps are

$$\boldsymbol{x}'_{\text{odd}}(n) = \boldsymbol{x}_{\text{odd}}(n) - \tfrac{1}{2}(\boldsymbol{x}_{\text{even}}(n) + \boldsymbol{x}_{\text{even}}(n+1))$$

$$\boldsymbol{x}'_{\text{even}}(n) = \boldsymbol{x}_{\text{even}}(n) + \tfrac{1}{4}(\boldsymbol{x}'_{\text{odd}}(n-1) + \boldsymbol{x}'_{\text{odd}}(n)).$$

The first step implements the highpass filter $(1, -2, 1)$ with a scaling factor. The corresponding elementary matrices are

$$\begin{bmatrix} 1 & 0 \\ -\frac{1}{2}z - \frac{1}{2} & 1 \end{bmatrix} \quad \text{and} \quad \begin{bmatrix} 1 & \frac{1}{4} + \frac{1}{4}z^{-1} \\ 0 & 1 \end{bmatrix}.$$

We check the off-diagonal entry in their product:

$$1 + \left(\tfrac{1}{4} + \tfrac{1}{4}z^{-1}\right)\left(-\tfrac{1}{2}z - \tfrac{1}{2}\right) = 1 - \tfrac{1}{8}\left(1 + z^{-1}\right)(1 + z) \approx \left(-\tfrac{1}{8}, \tfrac{6}{8}, -\tfrac{1}{8}\right)$$

which is the even part of the lowpass filter $h_0 = \frac{1}{8}(-1, 2, 6, 2, -1)$.

The Daubechies filters D_4 and D_6 are factored in [DaSw]. That paper also includes the example of a cubic spline, where $h_0 = \frac{1}{16}(1, 4, 6, 4, 1)$. It is the simplicity of each step that makes lifting so flexible. Here we discussed the standard filter banks (and we refer to Sweldens for the scaling function and wavelet formulas). The beauty of the idea appears when we have a nonstandard problem, like the construction of wavelets on spheres.

Appendix 5

Block Transforms in Progressive Image Coding

Trac Tran and Truong Nguyen

Block transform coding and subband coding are two dominant techniques in image compression. Both methods rely on a transform to convert the input image to a more decorrelated representation. Then they use the same building blocks such as bit allocator, quantizer, and entropy coder to achieve compression.

Block transform coders act separately on blocks of the image. Their low complexity and reasonable performance brought early success. The current compression standard JPEG utilizes the 8×8 Discrete Cosine Transform (DCT). At high bit rates (1 bpp and up), JPEG offers almost visually lossless reconstruction. However, when more compression is needed (at lower bit rates), annoying blocking artifacts show up because the DCT bases are short, non-overlapped, and have discontinuities at the ends. Inter-block correlation has been completely lost.

The lapped orthogonal transform and its extension to GenLOT reduce this blocking by borrowing pixels from the adjacent blocks. The lapped transform outperforms DCT on two counts: (i) from the analysis viewpoint, inter-block correlation provides better energy compaction that leads to more efficient entropy coding of the coefficients; (ii) from the synthesis viewpoint, its basis functions decay to zero at the ends, reducing discontinuities drastically. However, earlier lapped-transform coders have not used global information to full advantage. Quantization and entropy coding are still independent from block to block.

Subband coding has recently emerged as the leading standardization candidate thanks to the development of the discrete wavelet transform. The implicit overlapping and variable-length basis functions from wavelets produce smoother and more pleasant reconstructions. Multiresolution has created an intuitive foundation for simple, yet sophisticated, methods of encoding the transform coefficients — starting with *zero tree coding*. By relating the parent to its offspring coefficients in a wavelet tree, progressive wavelet coders [Shapiro,SaPe,XRO] can order the coefficients by bit planes. (The first bit plane has the most significant bit from each pixel, to be transmitted first. There are 8 bit planes if each pixel uses 8 bits.) This ordering results in an embedded bit stream, optimal wherever it is cut off. It gives exact bit rate control (and perfect idempotency when the transform maps integers to integers). Global information is taken into account fully.

In the frequency domain, the wavelet transform provides an *octave-band* representation. The wavelet transform is a non-uniform-band lapped transform. It can decorrelate smooth images. But it has problems with localized high frequency components. This appendix shows how the embedded framework can be extended to uniform-band lapped transforms. *An optimized lapped transform coupled with several wavelet levels for the DC band* can provide much finer frequency spectrum partitioning. This gives a significant improvement over current wavelet coders.

The wavelet transform and progressive image transmission

Progressive image transmission is perfect for the recent explosion of the internet. The wavelet-based progressive coding approach introduced by Shapiro relies on transmitting more important information first. The embedded bit stream can be truncated at any point by the decoder to yield an optimal reconstruction. The algorithm combines a scalar quantizer with power-of-two step-sizes and an entropy coder for wavelet coefficients.

This embedded algorithm relies on the *wavelet tree*, which connects coefficients from different scales that belong in the same spatial locality. The *DC band* or *reference signal* contains the lowest frequency coefficients (upper left corner). Each later *parent node* has four higher-frequency *offspring nodes*. All later descendants stay in the same spatial locality. Image statistics have shown that if a coefficient is insignificant (it falls below a threshold T) then it is very likely that its offspring and descendents are insignificant. Exploiting this fact, the sophisticated embedded wavelet coder SPIHT [SaPe] can output a single binary marker to represent very efficiently a large, smooth image area (an insignificant branch of the wavelet tree).

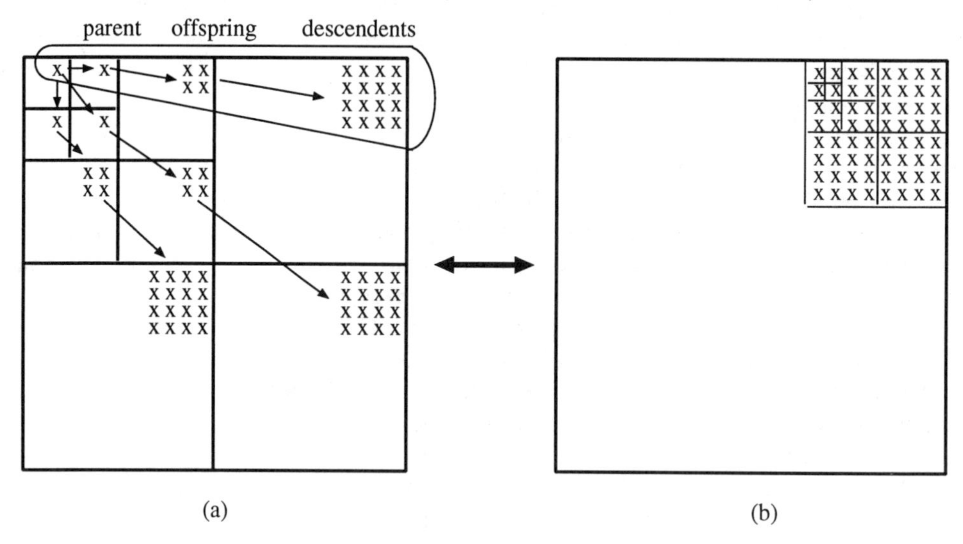

Figure A5.1: Wavelet and block transform analogy.

The wavelet tree provides an elegant hierachical data structure, but the efficiency depends heavily on the transform's ability to produce insignificant trees. For non-smooth images with a lot of texture, the wavelet transform does not decorrelate as well as transforms with finer frequency selectivity and superior energy compaction. Uniform-band lapped transforms hold the edge in this area.

Wavelet and block transform analogy

Instead of octave bands, one can have finer uniform bands. This compacts more signal energy into fewer coefficients. The performance of the zero tree algorithm is enhanced. But if a uniform bank has uniform downsampling (all subbands have the same size), then a parent node does not

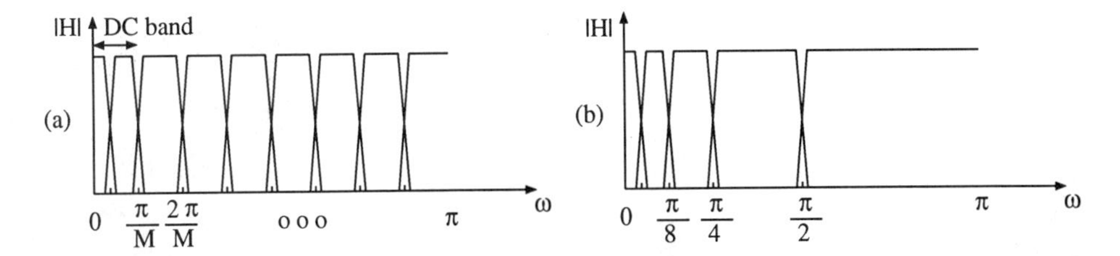

Figure A5.2: Frequency spectrum partitioning of (a) M-channel uniform-band transform (b) dyadic wavelet transform.

have four offspring nodes. How would one create a new tree structure that takes full advantage of the inter-scale correlation between block-transform coefficients?

We investigate an analogy between the wavelet transform and block transform. The parent, offspring, and descendents cover the same spatial locality, so we can view them as the coefficients of a transform block. In an L-level decomposition, the block has size $M = 2^L$. The difference lies in the 2^L filters that generate these coefficients. A wavelet decomposition, if implemented as a lapped transform, has the following coefficient matrix:

$$\mathbf{P}_L = \begin{bmatrix} h_0[n] * h_0[\frac{n}{2}] * \cdots * h_0[\frac{n}{2^{L-2}}] * h_0[\frac{n}{2^{L-1}}] \\ h_0[n] * h_0[\frac{n}{2}] * \cdots * h_0[\frac{n}{2^{L-2}}] * h_1[\frac{n}{2^{L-1}}] \\ h_0[n] * h_0[\frac{n}{2}] * \cdots * h_1[\frac{n}{2^{L-2}}] \\ h_0[n] * h_0[\frac{n}{2}] * \cdots * h_1[\frac{n}{2^{L-2}}] \\ \vdots \\ h_1[n] \\ h_1[n] \\ h_1[n] \\ h_1[n] \end{bmatrix}.$$

This coefficient matrix $\mathbf{P}_L$ shows important characteristics of the wavelet transform through the block transform's prism:

- The wavelet transform can be viewed as a lapped transform with filters of variable lengths.
- Each basis function has linear phase; however, they do not share the same center of symmetry.
- The block size is defined by the length of the longest filter. For the biorthogonal 9/7 wavelet pair and $L = 3$, the eight basis functions have lengths $57, 49, 21, 21, 7, 7, 7$, and 7. The number 57 is $(2^L - 1)(N) + 1 = (7 \times 8) + 1$. This is longest since h_0 with length $N + 1 = 9$ is longer than h_1.
- For a 6-level decomposition using the same 9/7 pair, the longest basis function grows to 505 taps! The huge overlapping of pixels explains the smoothness and the complete elimination of blocking artifacts in wavelet-based coders.

Each block of lapped transform coefficients represents a spatial locality, as in a wavelet tree. The offspring of the node (i, j) in an M-channel block transform have coordinates $(2i, 2j)$, $(2i, 2j+1)$, $(2i+1, 2j)$ and $(2i+1, 2j+1)$. The only requirement here is that M has to be a power of two.

All $(0, 0)$ coefficients from all transform blocks form the DC band, which is similar to the wavelet transform's reference signal. Each of these nodes has only three offspring: $(0, 1)$, $(1, 0)$,

Figure A5.3: Demonstration of the analogy between uniform-band transform and wavelet representation.

and $(1,1)$. Figures A5.1 and A5.3 demonstrate through a simple rearrangement of the block transform coefficients that the redefined tree structure above does possess a wavelet-like multiscale representation. The quadtree grouping of the coefficients is far from optimal in the rate-distortion sense; however, other parent-offspring relationships for uniform-band transforms do not help entropy coders to increase the coding efficiency.

Transform Design

A mere replacement of the wavelet transform by low-complexity block transforms is not enough to compete with SPIHT [SaPe] as testified in [XRO, Mal]. We propose several novel criteria in designing high-performance lapped transforms. We optimize a combination of coding gain, DC attenuation, attenuation around the mirror frequencies, weighted stopband attenuation, and unequal-length constraint on filter responses:

$$C_{\text{overall}} = \alpha_1 C_{\text{coding gain}} + \alpha_2 C_{\text{DC}} + \alpha_3 C_{\text{mirror}} + \alpha_4 C_{\text{weighted-stopband}} + \alpha_5 C_{\text{unequal-length}}.$$

The first three cost functions are well-known criteria for image compression. Higher coding gain correlates most consistently with objective performance (PSNR). Transforms with higher coding gain compact more energy into fewer coefficients, and the more significant bits get transmitted first. All our designs use a version of the coding gain formula in [KatYas]. Low DC leakage and high attenuation near the mirror frequencies are less essential, but they do improve the visual quality of the reconstructed image [RAH].

The ramp-weighted stopband attenuation cost is defined as

$$C_{\text{weighted-stopband}} = \sum_{k=1}^{M-1} \int_{\omega\in\Omega_S} |W_k(e^{j\omega}) H_k(e^{j\omega})|^2 d\omega,$$

where $W_k(e^{j\omega})$ decays linearly from one at the peak frequency response to zero at DC. This frequency weighting forces the highband filters to pick up as little energy as possible, producing insignificant trees (and also improving the coding gains).

The unequal-length constraint forces the tail ends of the high-frequency filter responses to be very small (not necessarily zero). The higher the frequency band, the shorter the effective length of the filter. This constraint minimizes ringing around strong image edges at low bit rates, a typical characteristic of long filters.

High-performance lapped transforms designed specifically for progressive image coding are presented in Figures A5.5(c)-(d). Figures A5.4(a)-(b) show the popular DCT and LOT for comparison purposes. The frequency response and the basis functions of the 8-channel 40-tap GenLOT in A5.5(c) exemplify a well-optimized filter bank: high coding gain and low attenuation near DC for best energy compaction, smoothly decaying impulse responses to eliminate blocking artifacts, and unequal-length filters to suppress ringing.

Figure A5.3 shows that there still exists correlation between DC coefficients. To decorrelate the DC band even more, several levels of wavelet decomposition can be used depending on the image size. Besides the obvious increase in coding efficiency thanks to a deeper tree, wavelets provide variably longer bases for the signal's DC component. Blocking artifacts are further reduced. The complete coder's diagram is depicted in Figure A5.6.

Coding Results

The objective coding results (PSNR in dB) for standard 512×512 Lena and Barbara test images are tabulated using several different transforms:

- DCT, 8-channel 8-tap filters, shown in Figure A5.4(a).
- LOT 8-channel 16-tap filters, shown in Figure A5.4(b).
- GenLOT, 8-channel 40-tap filters, shown in Figure A5.5(c).
- LOT, 16-channel 32-tap filters, shown in Figure A5.5(d).

Lena	Progressive Transmission Coders					
Comp. Ratio	SPIHT (9-7WL)	Xiong et al (DCT)	8 x 8 DCT	8 x 16 LOT	8 x 40 GenLOT	16 x 32 LOT
1:8	40.41	39.62	39.91	40.09	40.43	40.16
1:16	37.21	36.00	36.38	36.75	37.32	36.96
1:32	34.11	32.25	32.90	33.57	34.23	33.87
1:64	31.10	--	29.67	30.48	31.16	30.85
1:100	29.35	--	27.80	28.61	29.31	28.98
1:128	28.38	--	26.91	27.61	28.35	27.99

(a)

Barbara	Progressive Transmission Coders					
Comp. Ratio	SPIHT (9-7WL)	Xiong et al (DCT)	8 x 8 DCT	8 x 16 LOT	8 x 40 GenLOT	16 x 32 GLBT
1:8	36.41	36.10	36.31	37.43	38.08	38.02
1:16	31.40	30.82	31.11	32.70	33.47	33.47
1:32	27.58	26.83	27.28	28.80	29.53	29.70
1:64	24.86	--	24.58	25.70	26.37	26.63
1:100	23.76	--	23.42	24.34	24.95	25.14
1:128	23.35	--	22.68	23.37	24.01	24.09

(b)

Table A5.1: Coding results of various progressive coders (a) for Lena (b) for Barbara.

The block transform coders are compared to the best progressive wavelet coder [SaPe] and an earlier DCT-based embedded coder [XGO]. Each PSNR is obtained from a real compressed bit stream with all overheads included. The rate-distortion curves in Figure A5.7 and the tabulated coding results in Table A5.1 clearly demonstrate the superiority of well-optimized lapped transforms over pure wavelets. For a smooth image like Lena where the wavelet transform can sufficiently decorrelate, SPIHT offers a comparable performance. For a highly-textured image like Barbara, the 8×40 GenLOT and the 16×32 LOT coder can provide a PSNR gain of around 2 dB over a wide range of bit rates. Unlike other block transform coders whose performance dramatically drops at very high compression ratios, the new progressive coders are consistent throughout as illustrated in Figure A5.7. Lastly, better decorrelation of the DC band provides around $0.3 - 0.5$ dB improvement over the earlier DCT embedded coder in [XGO].

The actual reconstructions confirm the superiority of lapped transforms in image quality. Figure A5.8 shows reconstructed Barbara images at 1:32 by various block transforms. Comparing to JPEG, blocking artifacts are already remarkably reduced in the DCT-based coder in Figure A5.8(a). Blocking is completely eliminated when DCT is replaced by better lapped transforms as shown in Figure A5.8(c)-(d), and Figure A5.9.

A closer look in Figure A5.10(a)-(c) (where 256×256 image portions are shown so that artifacts can be more easily seen) reveals that a good lapped transform can preserve texture (in the table cloth and the clothes) while keeping the edges relatively clean. The absence of excessive ringing despite long filters should not come as a surprise: the GenLOT time responses in Figure A5.5(c) reveal that the high-frequency filters are carefully designed, with lengths essentially under 16-tap. Compared to [SaPe], the reconstructed images have a sharper and more natural look with more defining edges and evenly reconstructed texture.

The PSNR difference is not as striking in the Goldhill image, but the perceptual quality is improved. Even at 1:100, the reconstructed Goldhill image in Figure A5.9(d) is visually pleasant: no blocking and not much ringing. More objective and subjective evaluation of transform-based progressive coding can be found at *http://saigon.ece.wisc.edu/~waveweb/Coder/index.html* .

The improvement over wavelets depends on the lapped transform's ability to capture and separate localized signal components in the frequency domain. In the spatial domain, this corresponds to images with directional repetitive texture patterns. To illustrate this point, the lapped-transform-based coder is compared against the FBI Wavelet Scalar Quantization (WSQ) standard. When the original 768×768 gray-scale fingerprint image is shown in Figure A5.11(a) is compressed at 1 : 13.6, WSQ and [XRO] reported a PSNR of 36.05 dB and 37.30 dB. Using the 16×32 LOT, a PSNR of 37.87 dB can be achieved at the same compression ratio. For the same PSNR, the LOT coder can compress the image down to 1 : 19 where the reconstructed image is shown in Figure A5.11(b). At 1 : 18.036 reconstruction by WSQ in Figure A5.11(c) has a PSNR of 34.42 dB while the LOT coder produces 36.32 dB. At the same distortion, we can compress the image down to 1:26 as shown in Figure A5.11(d). Notice the high perceptual image quality in Figure A5.11(b) and (d): no visually disturbing blocking and ringing artifacts.

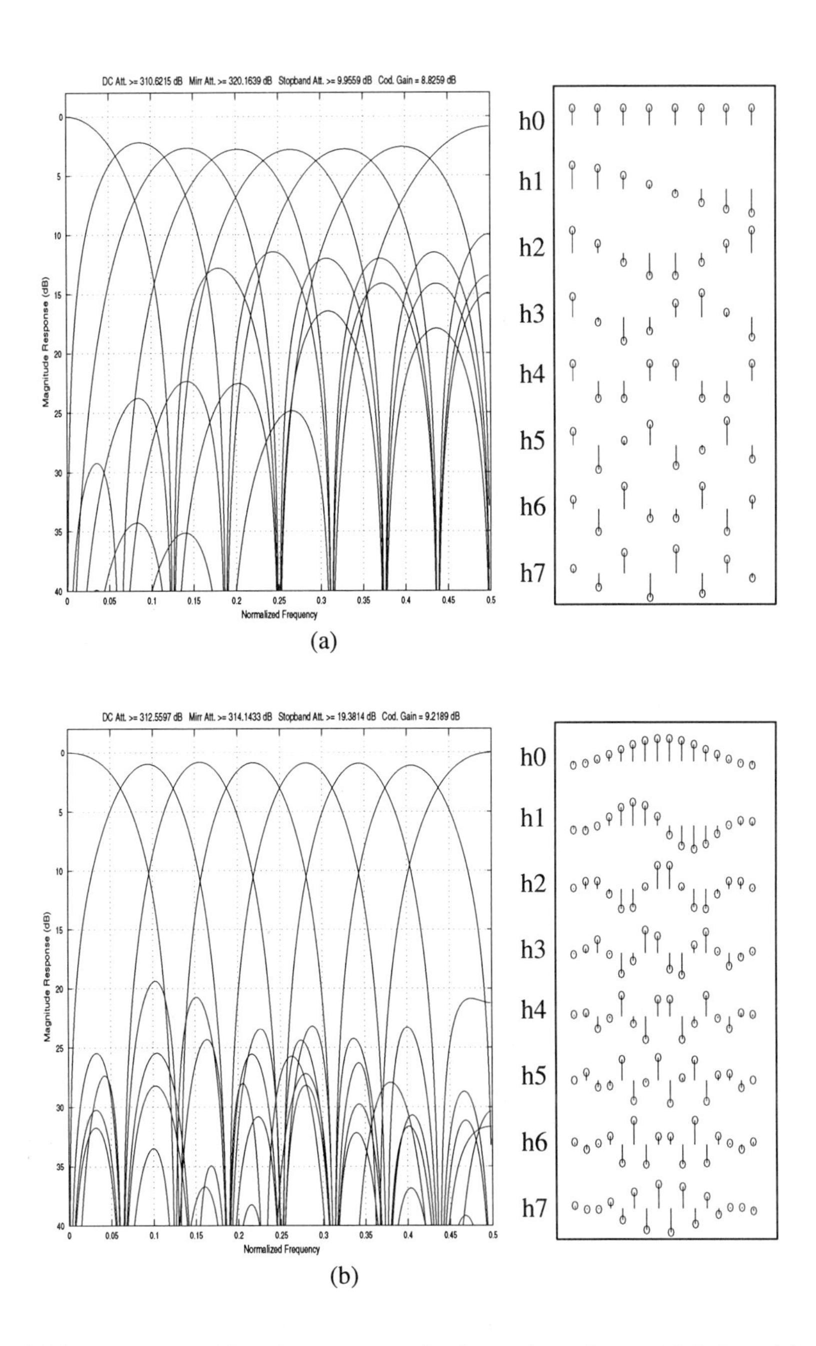

Figure A5.4: Frequency and impulse responses of orthogonal transforms: (a) 8-channel 8-tap DCT (b) 8-channel 16-tap LOT.

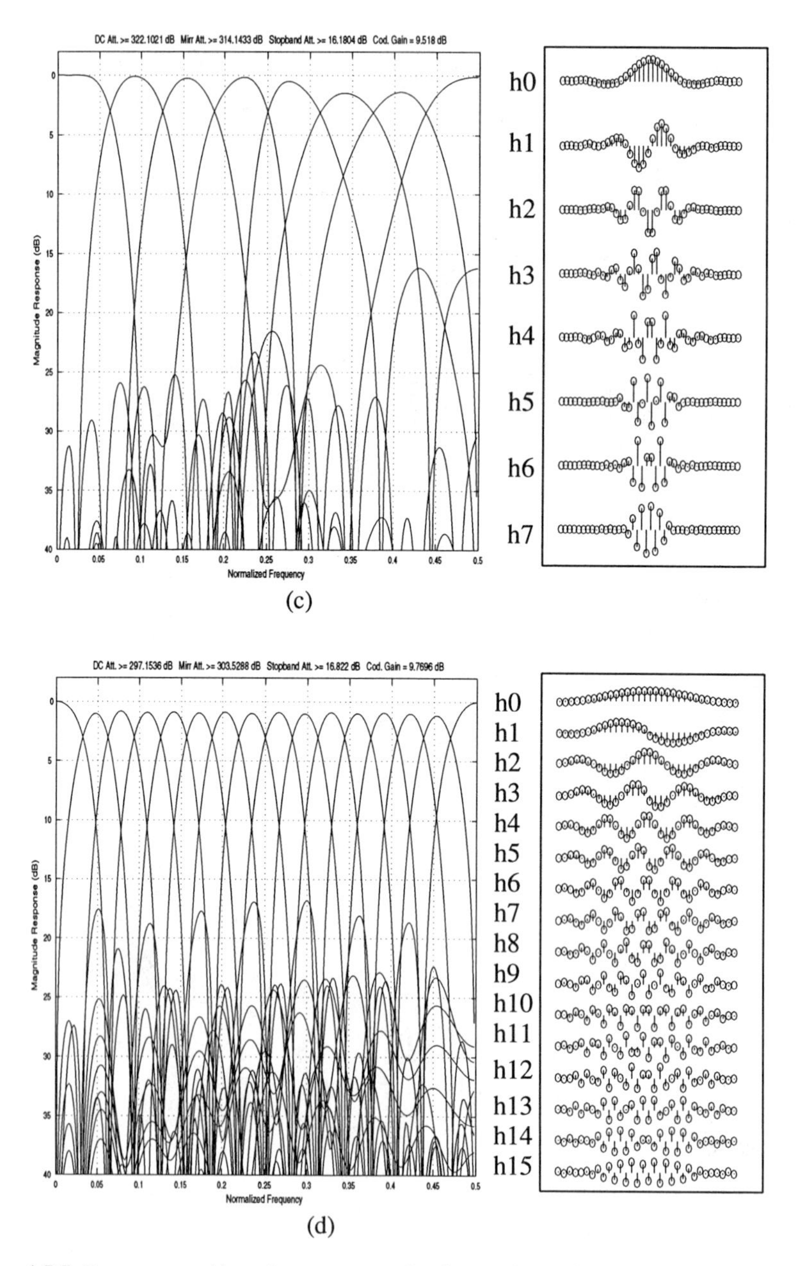

Figure A5.5: Frequency and impulse responses of orthogonal transforms: (c) 8-channel 40-tap GenLOT (d) 16-channel 32-tap LOT.

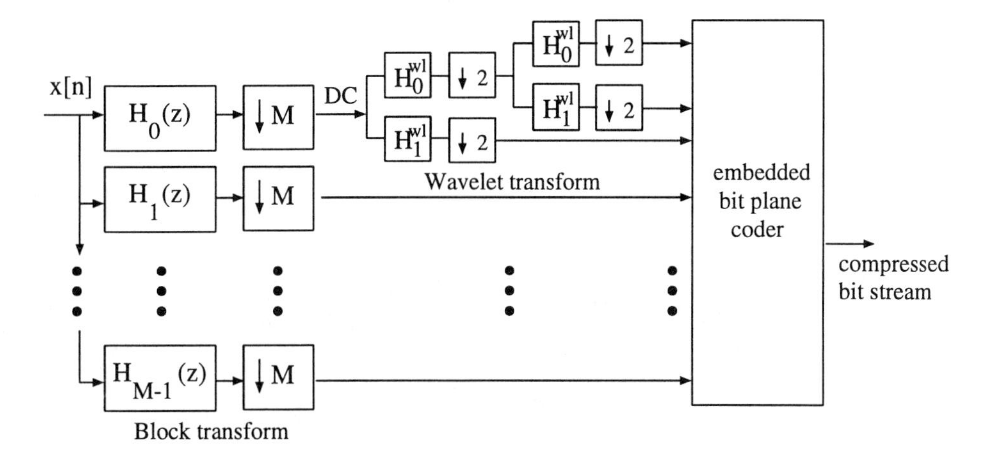

Figure A5.6: Complete coder's diagram.

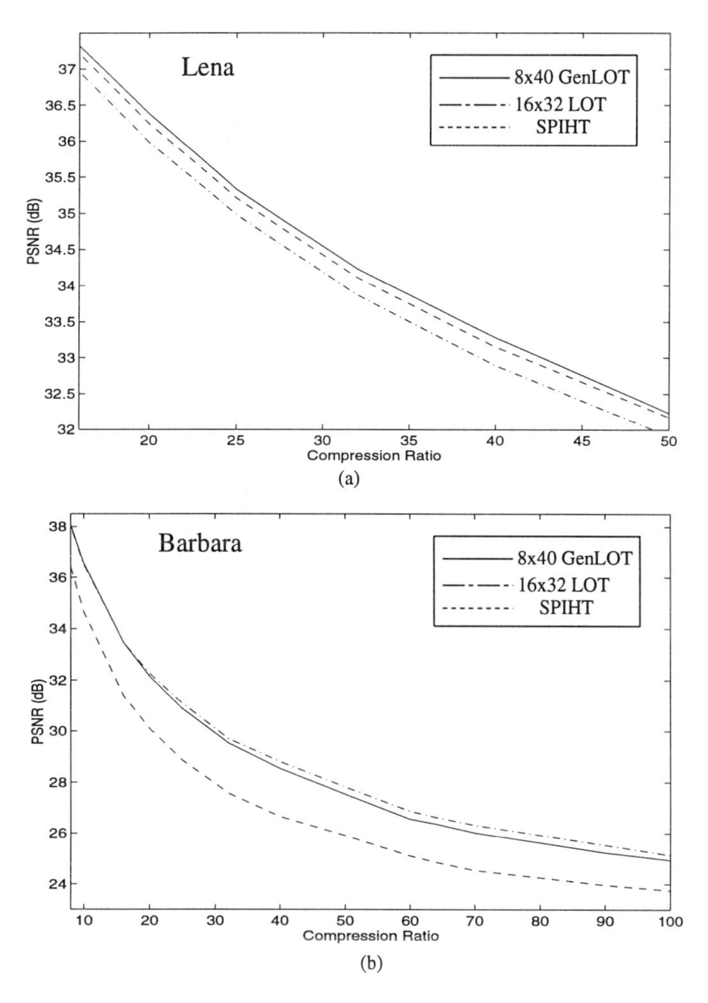

Figure A5.7: Rate-distortion curves of various progressive coders (a) for Lena (b) for Barbara.

Figure A5.8: Barbara coded at 1:32 by (a) 8×8 DCT (b) 8×16 LOT (c) 8×40 GenLOT (d) 16×32 LOT.

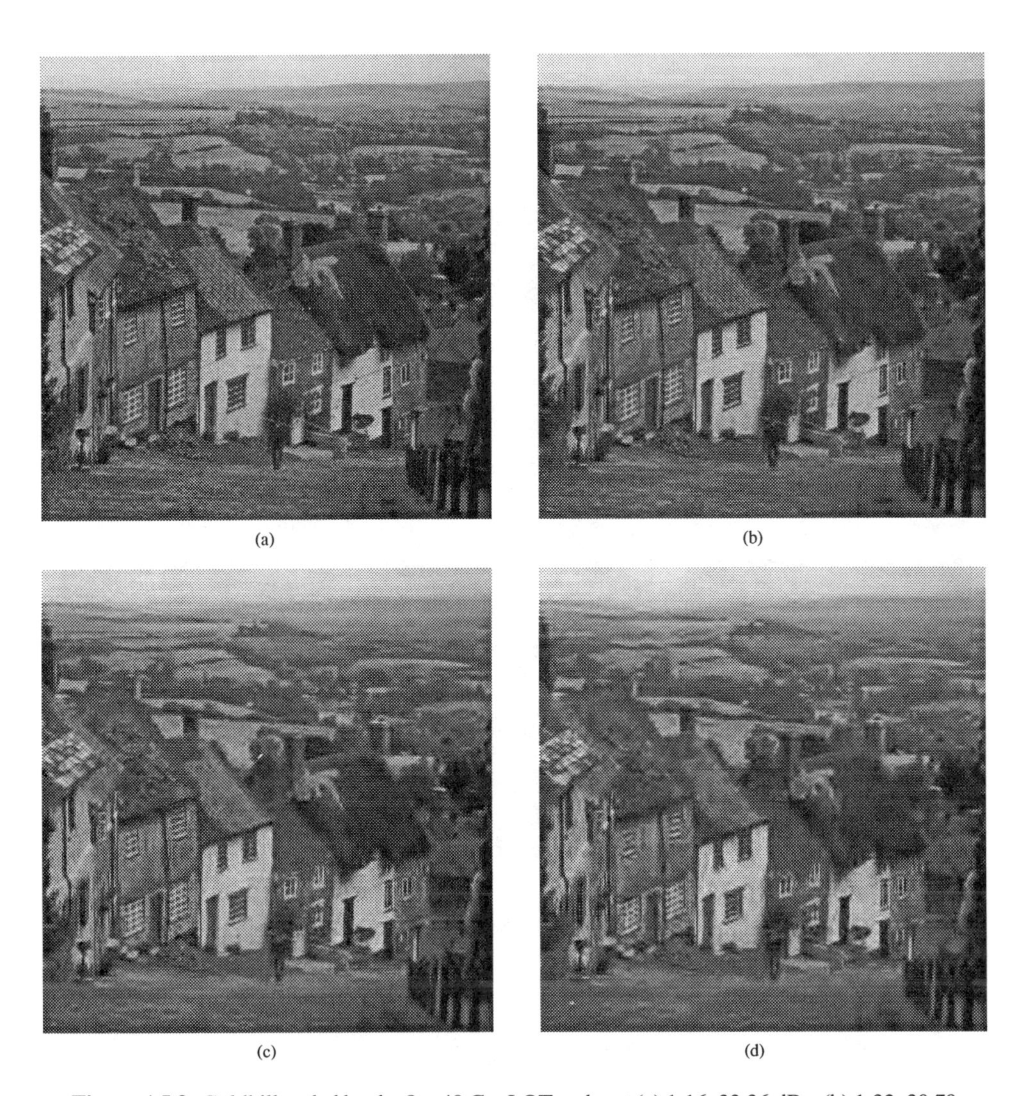

Figure A5.9: Goldhill coded by the 8×40 GenLOT coder at (a) 1:16, 33.36 dB (b) 1:32, 30.79 dB (c) 1:64, 28.60 dB (d) 1:100, 27.40 dB.

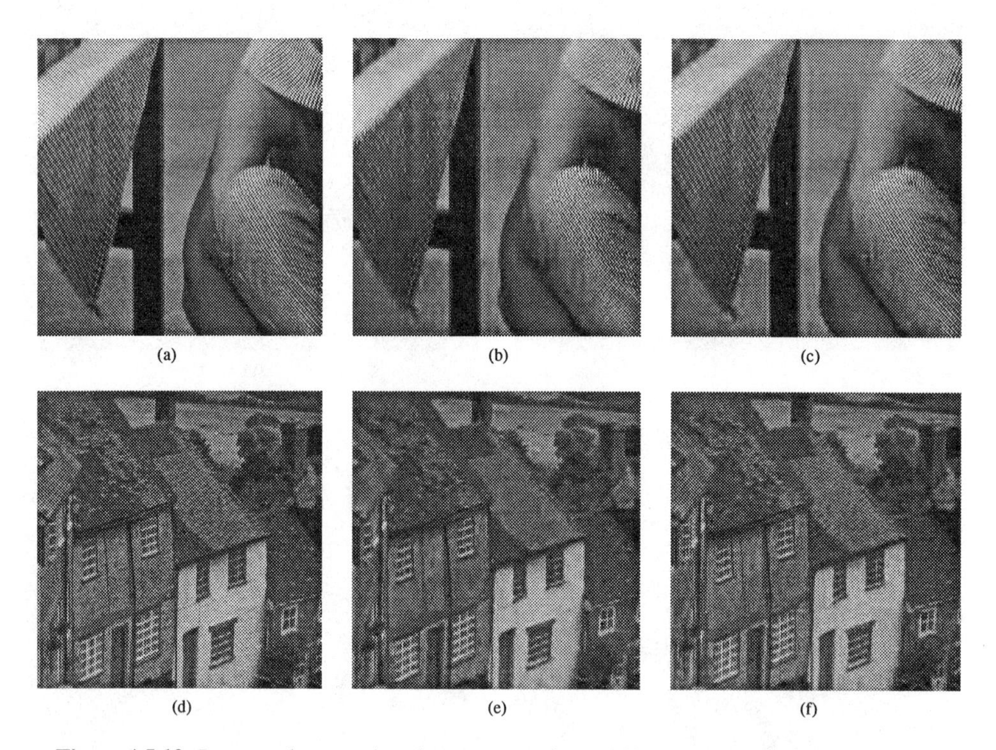

Figure A5.10: Perceptual comparison between wavelet and block transform embedded coder. Zoom-in portion of (a) original Barbara (b) SPIHT at 1:32 (c) 8×40 GenLOT embedded coder at 1:32 (d) original Goldhill (e) SPIHT at 1:32 (c) 8×40 GenLOT embedded coder at 1:32.

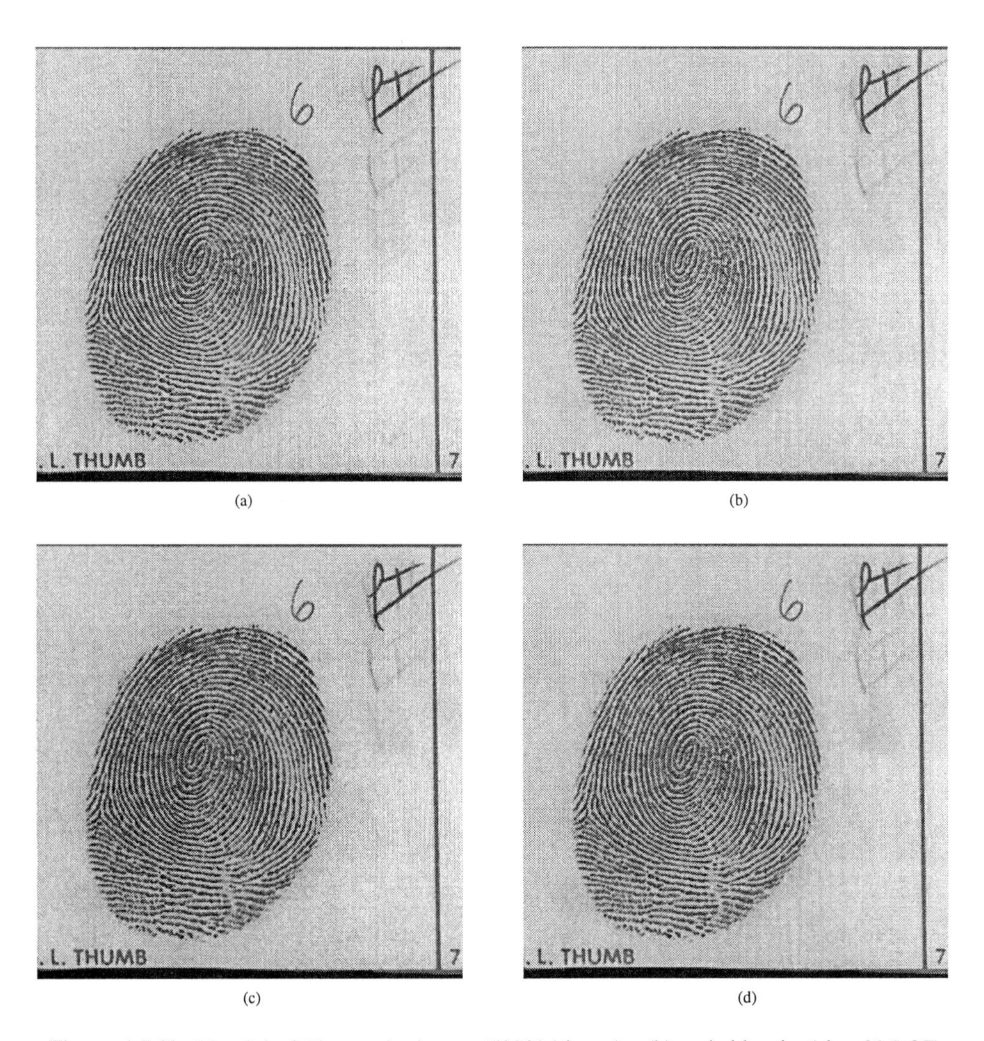

Figure A5.11: (a) original Fingerprint image (589824 bytes) (b) coded by the 16 × 32 LOT coder at 1:19 (31043 bytes), 36.05 dB (c) coded by WSQ coder at 1:18 (32702 bytes), 34.43 dB (d) coded by the 16 × 32 LOT coder at 1:26 (22685 bytes), 34.42 dB.

Index

Multiresolution and Wavelets

- $\phi(t)$ from the dilation equation (refinement equation) $\phi(t) = \sum 2\boldsymbol{h}(k)\phi(2t-k)$
- $w(t)$ from the wavelet equation $w(t) = \sum 2\boldsymbol{h}_1(k)\phi(2t-k)$
- $w_{jk}(t) = 2^{j/2}w(2^j t - k)$: Normalized wavelet on $[k\Delta t, (k+N)\Delta t]$
- Scale parameter j for stepsize $\Delta t = 2^{-j}$ ($\Delta t = 2^j$ in [D] and MATLAB Toolbox)
- Shift-invariant subspaces $V_j \subset V_{j+1}$ with $V_j \oplus W_j = V_{j+1} : f(t) \in V_0 \leftrightarrow f(2^j t) \in V_j$
- $\{\phi(t-k)\}$ An orthonormal basis (or only a Riesz basis) for V_0
- $\{w(t-k)\}$ An orthonormal basis (or only a Riesz basis) for W_0
- $\left\{2^{j/2}\phi(2^j t - k)\right\}$ and $\left\{2^{j/2}w(2^j t - k)\right\}$ Bases for V_j and W_j. Joint basis for V_{j+1}
- Orthogonal multiresolution Orthonormal bases with V_j perpendicular to W_j
- Semiorthogonal multiresolution Riesz bases with V_j perpendicular to W_j
- Biorthogonal multiresolution Biorthogonal bases with $V_j \perp \widetilde{W}_j$ and $W_j \perp \widetilde{V}_j$
- $\widetilde{\phi}(t)$ and $\widetilde{w}(t)$ from the analysis filters $\boldsymbol{h}_0$ and $\boldsymbol{h}_1$ generate $\widetilde{V}_0 \oplus \widetilde{W}_0 = \widetilde{V}_1$
- $\phi(t)$ and $w(t)$ from the synthesis filters $\boldsymbol{f}_0$ and $\boldsymbol{f}_1$ generate $V_0 \oplus W_0 = V_1$
- $\langle \phi(t-k), \widetilde{\phi}(t-\ell)\rangle = \delta(k-l)$ Biorthogonal (dual) bases for V_0 and $\widetilde{V}_0$
- $\langle w(t-k), \widetilde{w}(t-\ell)\rangle = \delta(k-l)$ Biorthogonal (dual) bases for W_0 and $\widetilde{W}_0$
- $a_{jk} = \langle f(t), \widetilde{\phi}_{jk}(t)\rangle$ Coefficients in $f_j(t) = \sum_k a_{jk}\phi_{jk}(t) =$ projection of $f(t)$ onto V_j
- $b_{jk} = \langle f(t), \widetilde{w}_{jk}(t)\rangle$ Wavelet coefficients in $f(t) = \sum\sum b_{jk}w_{jk}(t)$
- $a_{jk} = \sum \boldsymbol{h}_0(\ell - 2k)\, a_{j+1,\ell}$ and $b_{jk} = \sum \boldsymbol{h}_1(\ell - 2k)\, a_{j+1,\ell}$ Mallat Fast Wavelet Transform
- $a_{j+1,\ell} = \sum \boldsymbol{f}_0(\ell - 2k)\, a_{jk} + \sum \boldsymbol{f}_1(\ell - 2k)\, b_{jk}$ Fast Inverse Wavelet Transform